Integrating Disaster Science and Management

Global Case Studies in Mitigation and Recovery

Integrating Disaster Science and Management

Global Case Studies in Mitigation and Recovery

Edited by

Pijush Samui
Dookie Kim
Chandan Ghosh

Elsevier
Radarweg 29, PO Box 211, 1000 AE Amsterdam, Netherlands
The Boulevard, Langford Lane, Kidlington, Oxford OX5 1GB, United Kingdom
50 Hampshire Street, 5th Floor, Cambridge, MA 02139, United States

Library of Congress Cataloging-in-Publication Data
A catalog record for this book is available from the Library of Congress

British Library Cataloguing-in-Publication Data
A catalogue record for this book is available from the British Library

ISBN: 978-0-12-812056-9

For information on all Elsevier publications visit our website at
https://www.elsevier.com/books-and-journals

Publisher: Candice Janco
Acquisition Editor: Tom Stover
Editorial Project Manager: Michael Lutz
Production Project Manager: Prem Kumar Kaliamoorthi
Designer: Victoria Pearson

Typeset by Thomson Digital

Dedicated to Dr. Tarapada Mandal

Contents

List of Contributors

Ranja Bandyopadhyaya
National Institute of Technology, Patna, Bihar, India

Behrouz Behnam
Amirkabir University of Technology, Tehran, Iran

Jitendra Bothara
Miyamoto International NZ Ltd, Christchurch, New Zealand

Ali Bozorgi-Amiri
School of Industrial Engineering, College of Engineering, University of Tehran, Tehran, Iran

Lauren Brockie
School of Design, Queensland University of Technology, Brisbane, QLD, Australia

Leopoldo Carro-Calvo
University of Alcalá, Alcala de Henares, Spain

Cüneyt Çalişkan
Emergency Aid and Disaster Management School of Health, Çanakkale Onsekiz Mart University, Çanakkale, Turkey

Nurcihan Ceryan
Balikesir University, Balikesir, Turkey

Sener Ceryan
Balikesir University, Balikesir, Turkey

Prashant Kumar Champatiray
Indian Institute of Remote Sensing, Dehradun, India

Vinay Kumar Dadhwal
Indian Institute of Space Science and Technology, Thiruvananthapuram, India

Ravinesh C. Deo
University of Southern Queensland, Springfield, QLD, Australia

Karunakaran Akhil Dev
Mahatma Gandhi University; CHAERT (Centre for Humanitarian Assistance and Emergency Response Training), Kottayam, Kerala, India

Dmytro Dizhur
University of Auckland, Auckland, New Zealand

Jayakumar Drisya
National Institute of Technology, Calicut, India

T.I. Eldho
Indian Institute of Technology, Mumbai, India

Gianfranco Galli
National Institute of Geophysics and Volcanology, Rome, Italy

Anthony T.C. Goh
Nanyang Technological University, Singapore

William N. Holden
University of Calgary, Calgary, Alberta, Canada

Jason Ingham
University of Auckland, Auckland, New Zealand

Naveen James
Indian Institute of Technology Ropar, Punjab, India

Sujoy Kumar Jana
Department of Surveying & Land Studies, The Papua New Guinea University of Technology, Papua New Guinea

Ramakrishna Jayangondaperumal
Wadia Institute of Himalayan Geology, Dehradun, India

Biju John
National Institute of Rock Mechanics, Kolar Gold Fields, India

Fariborz Jolai
School of Industrial Engineering, College of Engineering, University of Tehran, Tehran, Iran

Joice K. Joseph
Mahatma Gandhi University; CHAERT (Centre for Humanitarian Assistance and Emergency Response Training), Kottayam, Kerala, India

Ayhan Kesimal
Karadeniz Technical University, Trabzon, Turkey

A.T. Kulkarni
Risk Management Solution India Pvt. Ltd., India

L. Lourenço
University of Coimbra, Coimbra, Portugal

Shawn J. Marshall
University of Calgary, Calgary, Alberta, Canada

Evonne Miller
QUT Design Lab, Brisbane, QLD, Australia

Mahesh Mohan
Mahatma Gandhi University, Kottayam, Kerala, India

Silvio Mollo
Sapienza University of Rome, Rome, Italy

Tatsuya Nogami
Japan Fire and Crisis Management Association, Tokyo, Japan

A.N. Nunes
University of Coimbra, Coimbra, Portugal

Takeyuki Okubo
Ritsumeikan University, Kyoto, Japan

Indrajit Pal
Disaster Preparedness, Mitigation and Management (DPMM),
Asian Institute of Technology, Thailand

Irshad Parvaiz
Yanbu Industrial College (YIC), Saudi Arabia

Dilip Kumar Pal
Department of Surveying & Land Studies, The Papua New Guinea University of Technology, Papua New Guinea

A.P. Pradeepkumar
University of Kerala, Trivandrum; CHAERT (Centre for Humanitarian Assistance and Emergency Response Training), Kottayam, Kerala, India

Zohreh Raziei
School of Industrial Engineering, College of Engineering, University of Tehran, Tehran, Iran

Mohammad Rezaei-Malek
School of Industrial Engineering, College of Engineering, University of Tehran, Tehran, Iran;
LCFC, Arts et Métiers ParisTech, Metz, France

Thendiyath Roshni
National Institute of Technology, Patna, India

Ruhizal Roosli
Universiti Sains Malaysia, School of Housing, Building and Planning, Malaysia;
Northumbria University of Newcastle, United Kingdom

Beatriz Saavedra-Moreno
University of Southern Queensland, Springfield, QLD, Australia;
University of Alcalá, Alcala de Henares, Spain

Hakim Saibi
United Arab Emirates University, Al-Ain, UAE

Sancho Salcedo-Sanz
University of Alcalá, Alcala de Henares, Spain

Shraban Sarkar
Department of Geography, Cooch Behar Panchanan Barma University, Cooch Behar, West Bengal, India

Sathish Kumar D
National Institute of Technology, Calicut, India

Hüseyin KoÇak
Çanakkale Onsekiz Mart University, School of Health, Department of Emergency and Disaster Management;
Bezmialem Vakif University, Instıtute of Health Science, Disaster Medicine Doctorate Program, Turkey

Piergiorgio Scarlato
National Institute of Geophysics and Volcanology, Rome, Italy

Go Shimada
Meiji University, Tokyo, Japan; Columbia University, NY, USA; JICA Research Institute, Tokyo, Japan

Fahimeh Shojaei
Structural Engineer, Independent Researcher

T.G. Sitharam
Indian Institute of Science, Bangalore, India

Michele Soligo
Università "Roma Tre", Rome, Italy

Lailan Syaufina
Bogor Agricultural University, Bogor, Indonesia

Reza Tavakkoli-Moghaddam
School of Industrial Engineering, College of Engineering, University of Tehran, Tehran, Iran; LCFC, Arts et Métiers ParisTech, Metz, France

Vikram Chandra Thakur
Wadia Institute of Himalayan Geology, Dehradun, India

Paola Tuccimei
Università "Roma Tre", Rome, Italy

Pongpaiboon Tularug
Disaster Preparedness, Mitigation and Management (DPMM), Asian Institute of Technology, Thailand

Wengang Zhang
Chongqing University, Chongqing, China

P.E. Zope
Indian Institute of Technology, Mumbai, India

Introduction

Throughout our history, humans have had to deal with different types of disaster (earthquake, landslide, flood, tsunami, cyclone, etc). The rapid growth of the world's population has increased both the frequency and severity of disasters. Disasters have exacted a high toll in terms of lives and property. Therefore, development of different techniques for disaster mitigation is an imperative task in human civilization. Any book on disaster has great relevance for human mankind. This book will try to give the advanced techniques for forecasting the occurrence of the disaster. It will be also very helpful for risk analysis. The book will cover the different topics of disaster such as earthquake, landslide, flood, fire, cyclone, etc. It is also expected that the proposed book will open new area of research in disaster mitigation and management. The proposed book will give the new computational techniques for better understanding of mechanism of different disasters. It will also give robust model for prediction of effects of disaster. Practionar engineers always want new techniques for disaster mitigation. The proposed book will serve this purpose. Eminent scientists will give innovation techniques for disaster mitigation. This collection of chapters from several authors will be an excellent analysis of different mitigation strategies. An attempt will be made in each chapter for approaching the problems of disaster more holistically. The proposed book will also cover the effect of social and economic conditions on different disasters. The effect of climate change on disaster will be also discussed.

The main contribution of the proposed book will be that it will not only deal with the forecasting and description of the various disasters, but also will stress the management aspect that is, mitigation, preparedness, response and recovery. The nature of disaster management depends on local economic and social conditions. The effect of local economic and social conditions on disaster management will be described. The different techniques for coastal disaster management will also be discussed. Editors will stress the importance of social processes and human–environmental interactions on disaster management. The book will present the contributions of the authors and other persons and covers a wide spectrum of disaster management problems that extend over the last four decades or so. An important focus of the book is on damage reduction through prevention, preparedness, mitigation, response, recovery, rehabilitation and reconstruction.

The proposed book will discuss the advantages of past data analysis for disaster mitigation and management. The knowledge from past data analysis will be an important parameter for disaster mitigation and management. Data analysis will be also useful for forecasting of disaster.

Part I

Assessment and Mitigation

Chapter 1

A Risk Index for Mitigating Earthquake Damage in Urban Structures

Behrouz Behnam*, Fahimeh Shojaei**
**Amirkabir University of Technology, Tehran, Iran;*
***Structural Engineer, Independent Researcher*

1.1 INTRODUCTION

Large earthquakes are very destructive and can lead to considerable human and financial loss. In order to reduce the risk posed by earthquakes, pre- and postdisaster strategies are often adopted. While postdisaster strategies respond to an earthquake in order to alleviate the consequences after the fact, predisaster strategies provide resources to support authorities as they work to reduce the associated risks to people, structures, and infrastructure from future earthquake hazards (Baas et al., 2008). These strategies include identifying the points of weakness in elements exposed to a possible earthquake. In other words, preparedness and prevention plans are made to increase the efficacy of operational capabilities. Among the preparedness and prevention plans, a vulnerability assessment of buildings under seismic loads is of paramount importance. From such an assessment, it can be understood whether a building does or does not need to be retrofitted. If the required retrofitting plans impose considerable budget, reconstruction plans might be substituted. Vulnerability assessments of buildings can provide information on the possible weaknesses of different structural systems. Furthermore, it can clarify whether a dictated architectural aspect of a building can make it more vulnerable to damage. If so, a preparedness plan would specifically be provided to address that aspect. From a different point of view, a building that has collapsed following an earthquake can be analyzed in order to discover the reasons for the failure. The results of such an investigation can be employed as a lesson learned when designing structures that could face future disasters.

Whether a building is analyzed in order to predict the level of damage in future earthquakes or to understand the reasons for damage sustained, the damage itself is often defined based on an index, where it can be expressed qualitatively or quantitatively. Qualitative-based damage indices (DIs) are classified using qualitative terms, such as "minor damage" or "extensive damage". When quantitative-based DIs are used, the damage level is given a numerical value, often over a range from 0.0 to 1.0, representing no damage to collapse, respectively. It is possible to combine qualitative and quantitative DIs.

A quantitative-based DI is expressed locally or globally, such that a local DI refers to the damage of a single element, whereas a global DI refers to a structure. Whether discussing a local or global DI, it can be accounted for in different ways. Overall, there are three types of DIs: energy-based, displacement-based, and cumulative, the last including both the energy dissipation and the displacement experienced. Hence, it provides more information on the seismic structural response than either the first or second type of DI. A detailed review has been provided by Blong (2003), where the differences between the three types of DIs are highlighted.

A cumulative DI is based on three parameters, which are stiffness deterioration (α), strength degradation (β), and the pinching of response (γ) resulting from slippage. One of the best employed cumulative DIs is Park and Ang's equation (1985), which is a linear combination of normalized values computed for the maximum deformation and the hysteretic energy. This index is used to account for local and global damage by combining the DI computed for different elements, based on the ratio of total energy absorbed (EA) in each story. Although the equation was first developed for reinforced concrete (RC) structures, it can also be usefully employed for other structural types. A modified

Integrating Disaster Science and Management. http://dx.doi.org/10.1016/B978-0-12-812056-9.00001-4

TABLE 1.1 Damage Categorization Based on Park and Ang's Modified DI Equation (Stone and Taylor, 1993)

Group	Range	Description
1	DI < 0.11	No damage or very minor damage
2	0.11 < DI < 0.44	Considerable damage but the element is repairable
3	0.44 < DI < 0.77	Extensive damage but the element is irreparable
4	DI > 0.77	Collapse

version of Park and Ang's equation for computing the DI of an element is shown in the following equation (Stone and Taylor, 1993):

$$D = \frac{\phi_{\mathrm{m}} - \phi_{\mathrm{y}}}{\phi_{\mathrm{u}} - \phi_{\mathrm{y}}} + \beta \frac{\int \mathrm{d}E}{M_{\mathrm{y}} \phi_{\mathrm{u}}} \tag{1.1}$$

In this equation, ϕ_{m} and ϕ_{y} represent the maximum curvature and yield curvature, respectively, experienced by the element during the cyclic loading; ϕ_{u} is the ultimate curvature experienced by the element in a monotonic loading; M_{y} is the yield moment of the element, $\int \mathrm{d}E$ is the energy dissipated by the member throughout an excitation, and β is a calibration coefficient between 0.1 and 0.15. The damage determined based on Eq. (1.1) is then categorized into four groups, as shown in Table 1.1.

The damage computed at element level is then combined for a story and a structure, using Eqs. (1.2) and (1.3), respectively:

$$\mathrm{DI}_j^s = \sum_{k=1}^{m_j} \lambda_{kj} \cdot \mathrm{DI}_{kj} \tag{1.2}$$

$$\lambda_{kj} = \frac{E_{kj}}{E_j} \tag{1.3}$$

where DI_j^s is the damage calculated for the jth story, DI_{kj} means the damage calculated for the kth element of the jth story, $E_j = \sum_{i=1}^{m_j} E_{ij}$ indicates the sum of energy dissipated at the jth story, and m_j is the number of elements in the jth story. The DI at the structure level, DI_G, is accounted for using Eqs. (1.4) and (1.5) where $E_T = \sum_{s=1}^{N} E_s$ represents the sum of the EA in an N-story structure.

$$\mathrm{DI}_G = \sum_{k=1}^{N} \lambda_i \cdot \mathrm{DI}_i^s \tag{1.5}$$

$$\lambda_i = \frac{E_i}{E_T} \tag{1.4}$$

It is worth mentioning that in RC structures even hairline cracks are considered as damage, although they have no considerable effect on the structural performance (Banon et al., 1981). Fig. 1.1 shows a typical force–deformation curve for an assumed hinge. As seen, although after point A, a member might experience some cracking, there would not be any meaningful damage until it passes point B, corresponding to the yield response where the DI is equal to DI_{y}. The damage then increases between points B and C, where its severity can be categorized into three known performance levels: immediate occupancy (IO), life safety (LS), and collapse prevention (CP). In terms of a qualitative explanation, these performance levels can, respectively, be named as light, moderate, and severe damage states.

From a different point of view, architectural specifications for buildings, such as regularity or irregularity, can significantly change the level of damage sustained over an earthquake excitation. Statistics from previous earthquakes and studies performed have shown that, on average, irregular structures are more susceptible to sustaining earthquake damage than are

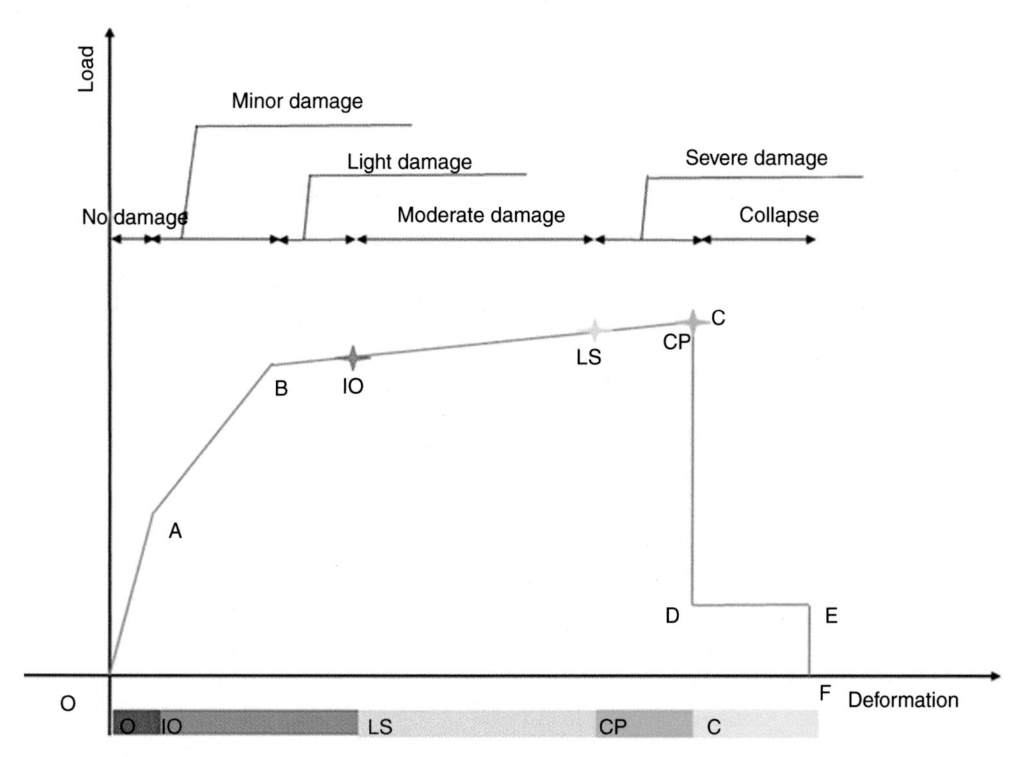

FIGURE 1.1 **Damage states and performance level.**

regular structures (Chintanapakdee and Chopra, 2004; D'Ambrisi et al., 2013). Although there are many kinds of irregularities, the main irregularities are vertical geometric irregularity, diaphragm discontinuity, soft-story, short-column, and mass irregularity. Fig. 1.2 shows some conceptual examples of these irregularities.

The term "soft story" is used when the lateral stiffness of a story is lower than 70% of that of the above story or lower than 80% of that of the average stiffness of the above three stories. A column is considered "short" if the ratio of shear span to column depth becomes lower than 2.50. This ratio is important, because it directly affects the brittle behavior of a column in such a way that for ratios lower than 1.50, a very fragile manner should be expected. This is because, when the ratio decreases, the column's moment (*M*) decreases while the shear force (*V*) remains the same. Hence, the column would fail in a shear fashion. It is understood that *M* is minimized when a load is applied to the midspan of the column where the point of inflation is also there. The synchronization of these two happenings may result in the displacement being doubled (a double curvature, as shown in Fig. 1.3A) with a very harsh consequence. Fig. 1.3B shows the seismic response of a short column.

In addition, a story is considered as being vertically irregular, if the horizontal dimension of the lateral force-resisting systems in a story is more than 130% of that of the adjacent story (FEMA310, 1998).

While the response of regular and irregular structures under gravity loads is greatly similar to one another, numerous investigations have proved that the seismic response of irregular structures is not as predictable as that of regular structures (Hosseinpour and Abdelnaby, 2017; Lavan and De Stefano, 2013). The difference is correlated with the torsional component—while in regular structures it can be ignored, in irregular structures, it is often considerable. It has also been shown in some studies that irregular structures are more susceptible to exceeding a defined performance level than regular structures (Behnam, 2015). In addition, some studies have shown that building irregularities cause a considerable increase in the probability of exceedance (Bhosale et al., 2017). For these reasons, most codes and provisions provide very rigorous procedures for the design of irregular structures, such that the more irregular a structure is, the more is that structure from the restrictions of the applicability of analysis and design methods. It is also worth noting that previous earthquakes have proved that, in general, irregular structures with concave plans are more susceptible to sustaining earthquake damage than those with convex plans (Elnashai and Di Sarno, 2015). A plan is convex if any arbitrary line drawn between two arbitrary points in the plan remains within the plan; otherwise, the plan is concave. Some examples of convex and concave plans are shown in Fig. 1.4.

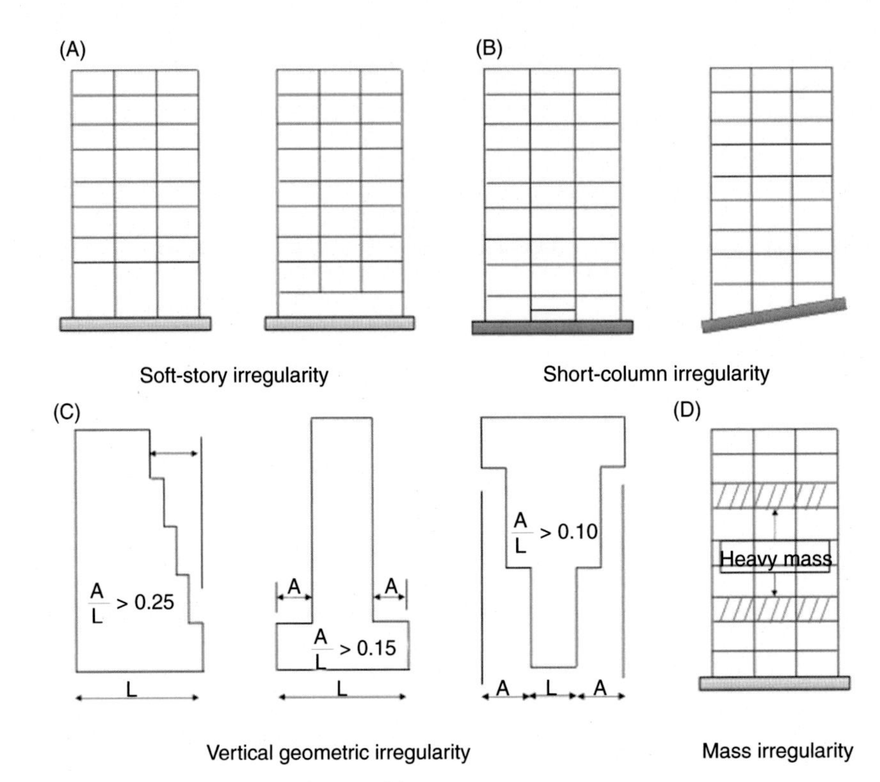

FIGURE 1.2 **Some conceptual examples of structural irregularities.**

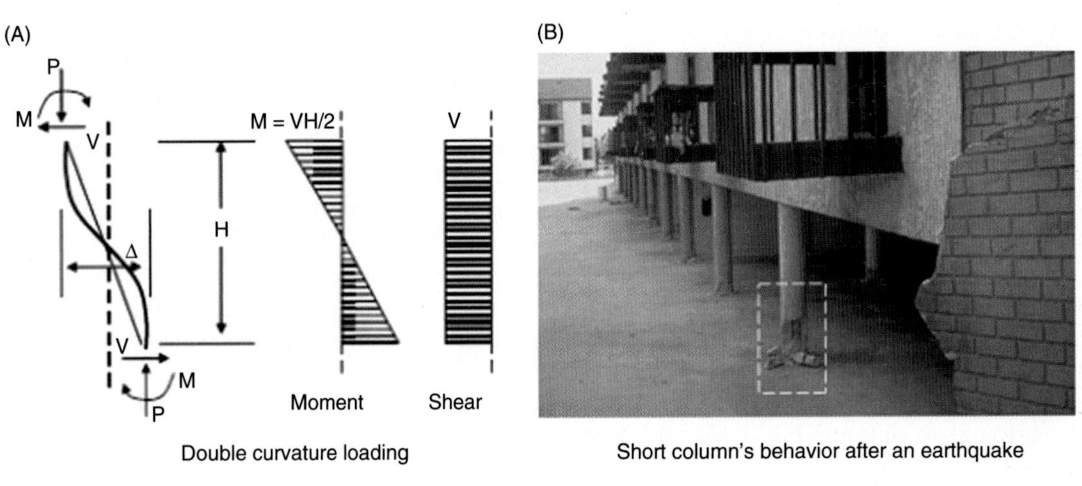

FIGURE 1.3 **Short column.**

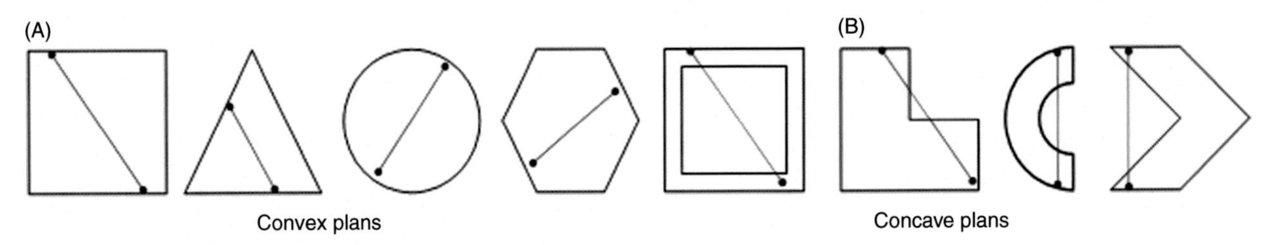

FIGURE 1.4 **Examples of convex and concave plans (Behnam, 2006).**

1.2 FAULT-TREE ANALYSIS

Quantitative DIs can provide detailed information about the seismic response of a single structure. However, no adequate information is given as to the probability of failure. It is also difficult to provide a graphical illustration of the damage sustained by a building. Also, it is not possible to combine different DIs for a building in order to propose a global DI, since every DI equation follows its own procedure. To address these limitations, fault-tree analysis (FTA) can be an appropriate tool—on the one hand, it is possible to provide a graphical illustration of the extent of damage, while on the other hand, the probability of failure can be quantified (Ericson, 1999). In addition, FTA allows for the determination of reliable parameters to improve the safety of a system.

FTA is often used in complex systems in order to determine the probability of unacceptable damage, which may result in a critical situation (LeBeau and Wadia-Fascetti, 2007). In other words, FTA is a failure study technique. FTA is a top-to-bottom event subjective-based technique that allows qualification of the reasons for failure versus the consequences of failure over different levels. The top event is often termed as failure. When the tree is created, various combinations that may lead to a failure, as well as the probability of the failure, are determined first, and then the most critical failure routes are identified. The analysis is performed qualitatively and quantitatively. The qualitative analysis is based on a Boolean depiction of events, in order to illustrate what is happening in the system. Over the tree, the basic events are linked to the upper events via symbolic logical gates, such as OR, AND, EXCLUSIVE OR, and PRIORITY AND (some shown in Table 1.2 for example). The quantitative analysis is to provide numerical information on the possible failure process. FTA has dual applications: it can be used to predict a possible failure in a system before it occurs or it can be used to realize the failure reasons in a system when that system has already failed.

The qualitative analysis starts with the hazard identification process, where the possible causes of failure are developed. The minimal cut sets (MCSs), which are the shortest combinations of basic events reaching to the top event, are then identified. This means that, using MCSs, the most critical routes leading to a failure are identified (Hadipriono and Toh, 1989). The MCSs are solved using Boolean algebra. In the quantitative analysis, on the other hand, in addition to the identification of MCSs, the probability of the top event is also determined. Yet, referring to Table 1.2, it can be understood that other than for very simple events, it is hard to identify the combinations of all the possible events that may result in a failure. Hence, it is very important to first use a technique for creating a combination of possible deficiencies that can develop MCSs. To clarify this, in a structural system, for example, the structural elements and their potential earthquake damage, when taken all together, can be considered as an MCS. Here, in order to determine the basic events, which are the damage of structural elements, and their correlations, Boolean algebra is employed. Also, in order to simulate the failure process, the quantitative DI of Park and Ang is used. The process is further explained in the case study section. It is evident that when using an AND or an OR gate, the calculations of the failure probability at the top event are different. In an FTA, although a gate may have more than one input, it always has only one output. An AND gate means the output event would occur only if all the input events occur simultaneously. An OR gate, on the other hand, means the output event would occur if just one of the input events occurs. The following equations show how an AND and OR gate work in an FTA.

TABLE 1.2 Symbolic Logical Gates Used in FTA

Symbol	Name	Usage
	Event	Top and intermediate positions
	Basic event	Bottom positions
	Condition event	Used with INHIBIT gate
	Undeveloped event	Fault expanded no further
	OR gate	Union of two or more events
	AND gate	Intersection of two or more events
	INHIBIT gate	Conditional AND gate

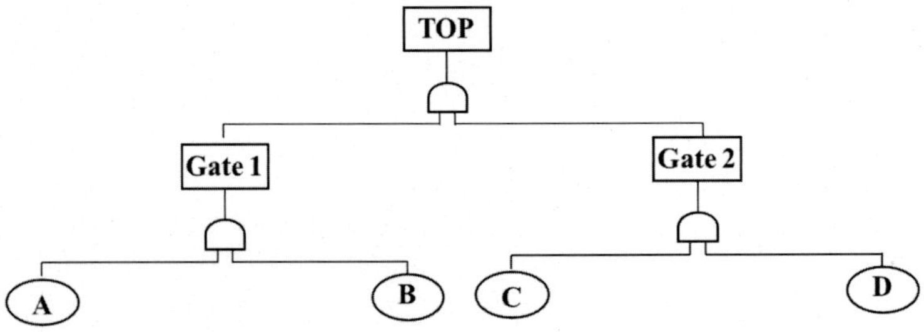

FIGURE 1.5 **An example of application of FTA in a system.**

$$p = \prod_{i=1}^{n} p_i \tag{1.6}$$

$$p = 1 - \prod_{i=1}^{n} (1 - p_i) \tag{1.7}$$

In these equations, n is the number of input events to the gate and p_i is the probability of failure of input event i, assuming that it is independent. Hence, when the probability of failure and the graphical model are available, the quantification process can be performed. For example, in Fig. 1.5, assuming that all of the basic events are independent and their probability is 0.1, the following equations can be written. The probability of failure in the example system is 0.199.

P (Gate 1) = P (A).P (B) = 0.1 × 0.1 = 0.01

P (Gate 2) = P (C).P (D) = 0.1 × 0.1 = 0.01

AND P (TOP) = P (Gate 1) OR P (Gate 2) = P (Gate 1) + P (Gate 2) − P (Gate 1).P (Gate 2) = 0.01 + 0.01 − 0.0001 = 0.199

1.3 METHODOLOGY

The application of FTA combined with quantitative DIs is shown schematically in Fig. 1.6. As seen, the study starts with the generation of the structural model, where a regular structure (as a reference model) is first designed for gravity and seismic loads. The structure is designed to meet the LS level of performance, based on FEMA356 code (FEMA356, 2000), where the maximum inter-story drifts have to be limited to 2.0% and the rotations of beam-to-column-connections shall be limited to $6\theta_y$ (θ_y is the yield rotation).

A number of deliberate changes are then made to the regular structure, so that it becomes irregular. Three types of irregularities are considered: short-column, soft-story, and setback irregularity. Over the next step, all of the structures are analyzed through nonlinear dynamic analysis (NDA). For performing this, a set of different accelerations from previous earthquakes are considered, in order to cover a wide range of different frequencies, time intervals, and amplitudes. The accelerations are then scaled for performing the required NDA. Based on the results of NDAs, the quantity of damage sustained is accounted for, using the widely used DI of Park and Ang. The damage is calculated first locally, that is, the damage of single elements, and then globally, where the damage sustained by the structure is determined. In the third step, the fault tree is created in order to perform the qualitative analysis. Identifying MCSs and employing Boolean algebra, the quantitative FTA is performed. Then, in order to validate the two different methods used, a comparison is made between the results obtained from Park and Ang's DI and the FTA results. This would allow us to determine the possibility of failure, based on the FTA results.

1.4 CASE STUDY

The procedure explained above is used here to determine the probability of seismic failure of regular and irregular RC structures. Fig. 1.7 shows the plan view, side view, and the geometric properties of beams and columns of a three-story moment resisting regular RC structure designed for a peak ground acceleration (PGA) of 0.30 g and for LS level of performance. The structure is designed based on ACI 318-08 code (ACI318, 2008), where the dead and live loads are assumed to be 5.5 and 1.5 kPa, respectively. A load combination of 100% dead load and 20% live load is used to find the required mass for determining the earthquake loads. The compressive strengths of concrete and the yield rebar's longitudinal and shear are assumed to be 25 and 400 MPa, respectively. The flooring system is a 100 mm one-way concrete slab. Poisson's ratios of the steel and concrete are, respectively, 0.2 and 0.23. The columns are introduced with an initial sinusoidal imperfection

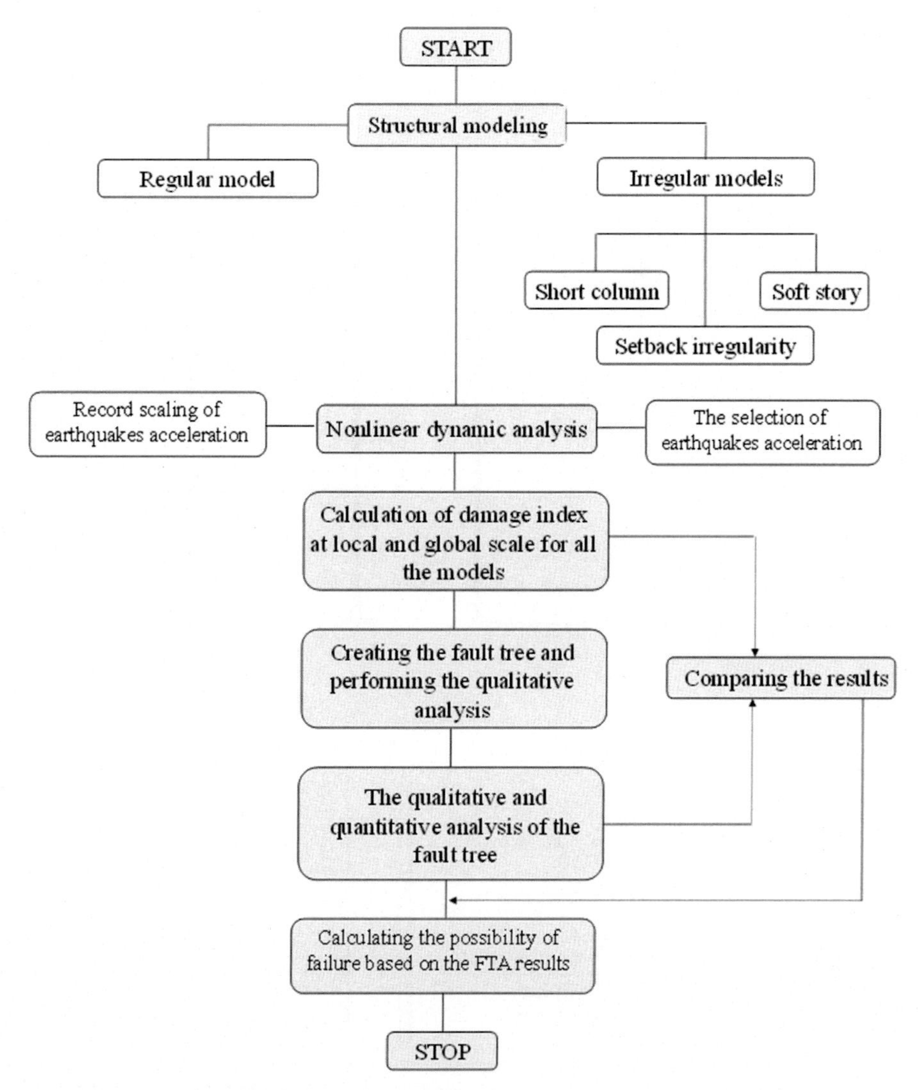

FIGURE 1.6 **Application of FTA analysis in determining the structural failure.**

amplitude of *H*/300 (*H* is the story height) so that the probable buckling and postbuckling strength can be captured during the analysis. This imperfection is applied over the unfavorable direction of the columns. Co-rotational transformation of the geometric stiffness matrix is also considered; hence, it is possible to involve large displacement, $P-\Delta$ effects, and the residual deformation effects into the analysis. Plastification of beams and columns over both the member's length and the cross-sections is considered, using fiber elements. In addition, to ensure a smooth transaction between the elastic and plastic parts, a transaction curve is introduced at the intersection of the first and second tangents. The regular structure is now changed to become irregular. Three irregular structures are created, displaying setback, short-column, and soft-story irregularity, as shown in Fig. 1.8.

It is evident that a DI at the structural level is accounted for based on the damage at the story level and, in a similar vein, the DI at the story level is determined based on the damage sustained by every element. Based on this, in order to perform the qualitative FTA, the structural failure can be introduced as the top event. The middle events are considered to be the damage at the story level and the basic events are considered to be the damage to the structural elements, that is, beams and columns. In addition, since the possible degrees of damage sustained by beams and columns are not necessarily identical to one another, the damage at the story level is divided into the damage sustained by beams and columns. Fig. 1.9 shows the fault tree of the regular structure. In this figure, the gate G0 represents the top event, that is, the structural failure, and G1, G2, and G3 show the story damage introduced as the middle event. Similarly, G4–G9 represent the damage to beams and columns. The basic events, which are the damage to single elements of beams and columns, are shown based on B_{ij} and C_{ij} (i is the story number and j is the element number). It is also worth mentioning that using an OR gate means the effect of all of the elements on the possibility of structural failure (top event) is considered.

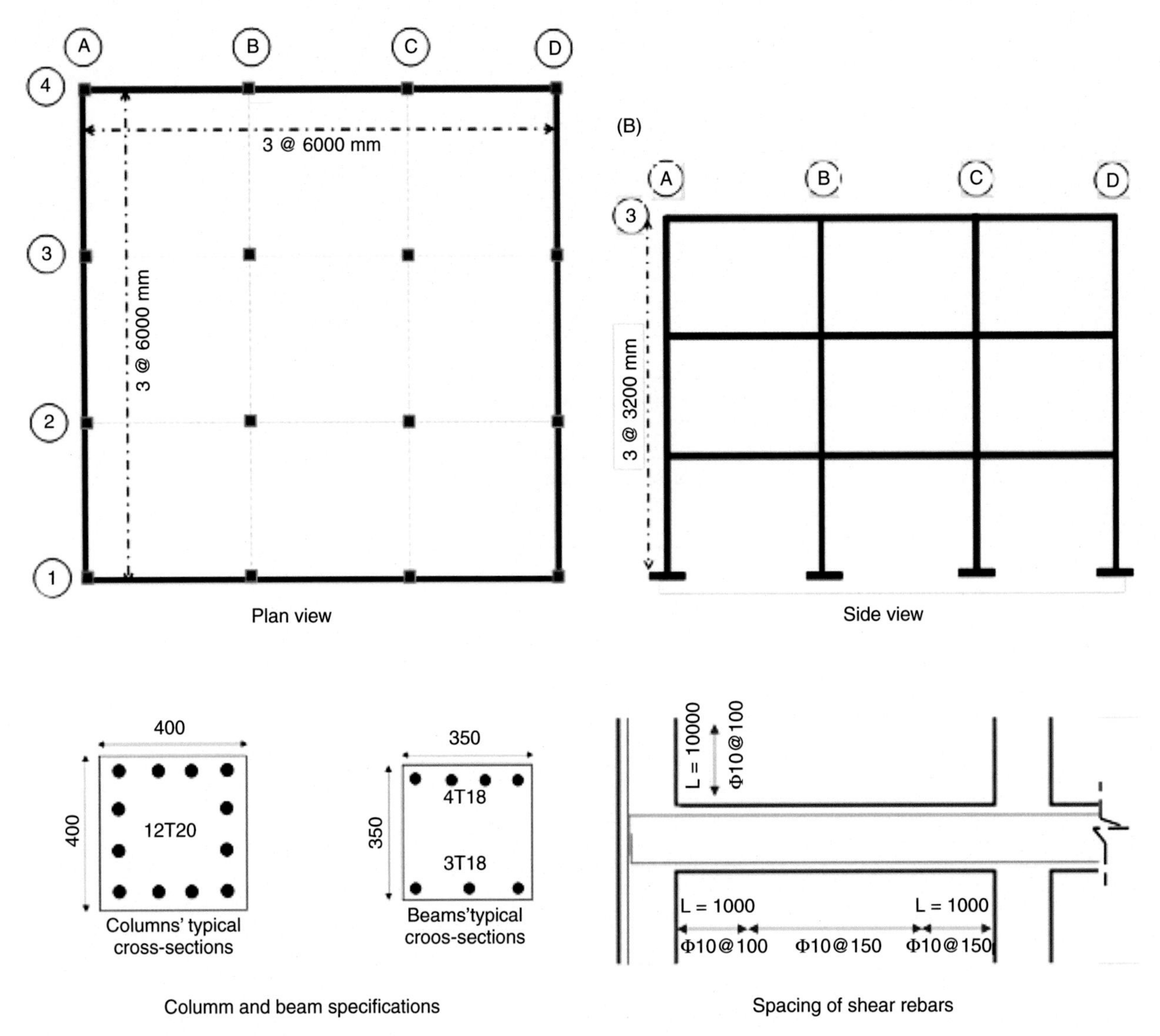

FIGURE 1.7 **The regular structure.**

Based on the fault tree shown in Fig. 1.9, the MCSs are found, as shown in Table 1.3. In the table, the unfavorable event (G0) is first listed in the first column, which is then extended to the middle events (G1, G2, and G3), shown in the second column. G1 is completed, adding G4 OR G5 in such a way that they show, respectively, the damage to the beams and columns of the first story, as shown in column 3. The events G6–G9, which represent the damage to the beams and columns of the second and third stories, are also shown in column 3. In column 4, the basic events show the damage to beams and columns separately. Thus, the final MCSs are listed in column 4.

In a similar vein, fault trees of the three introduced irregular structures are created. Figs. 1.10 and 1.11 show the fault trees of the setback irregular structure and the irregular structure with short column, respectively. As the fault tree of the irregular structure with soft story is similar to that of the regular structure, it is not plotted here. The clouds in the figures are to highlight the locations of the irregularities. The MCSs of the irregular structures are shown in Table 1.4.

In order to perform the quantitative analysis of the fault trees and then to determine the failure possibility of the structures, the amounts of damage sustained by every single element and their effect on the global damage should be accounted for. This is achieved here by using the modified DI of Park and Ang. In this respect, a wide range of earthquake accelerations are selected for performing NDAs. The information for the earthquake accelerations is shown in Table 1.5. The information includes only the far-field records, since the near-field records often cause considerable fluctuations in the analysis results (Mavroeidis and Papageorgiou, 2003). The earthquake should first be scaled to a spectral acceleration by a period defined in the seismic code employed. This process requires some programming, which is done here using the MATLAB package.

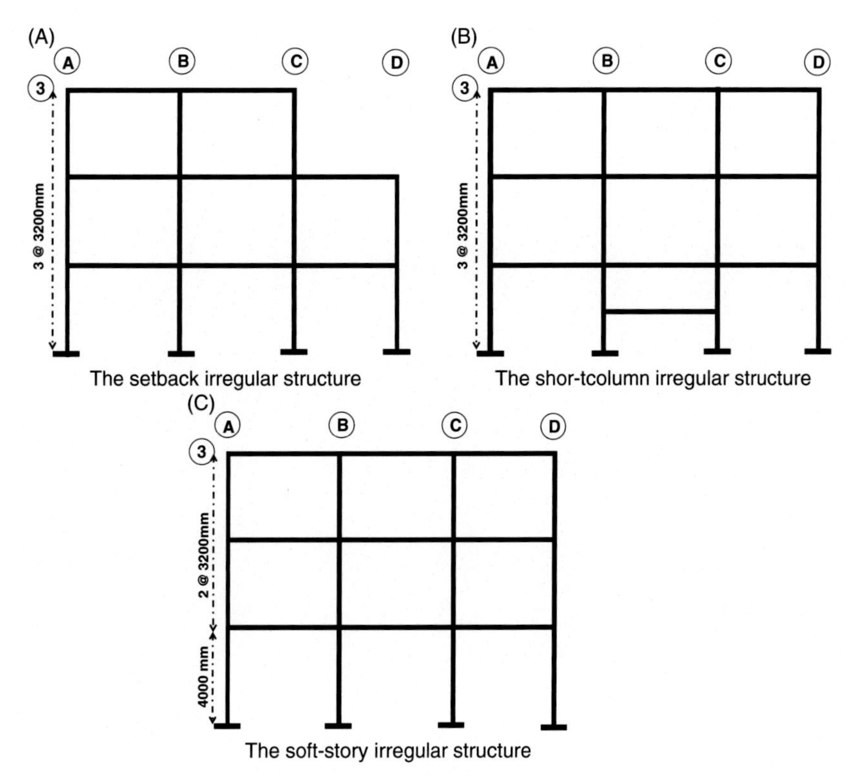

FIGURE 1.8 The irregular structures.

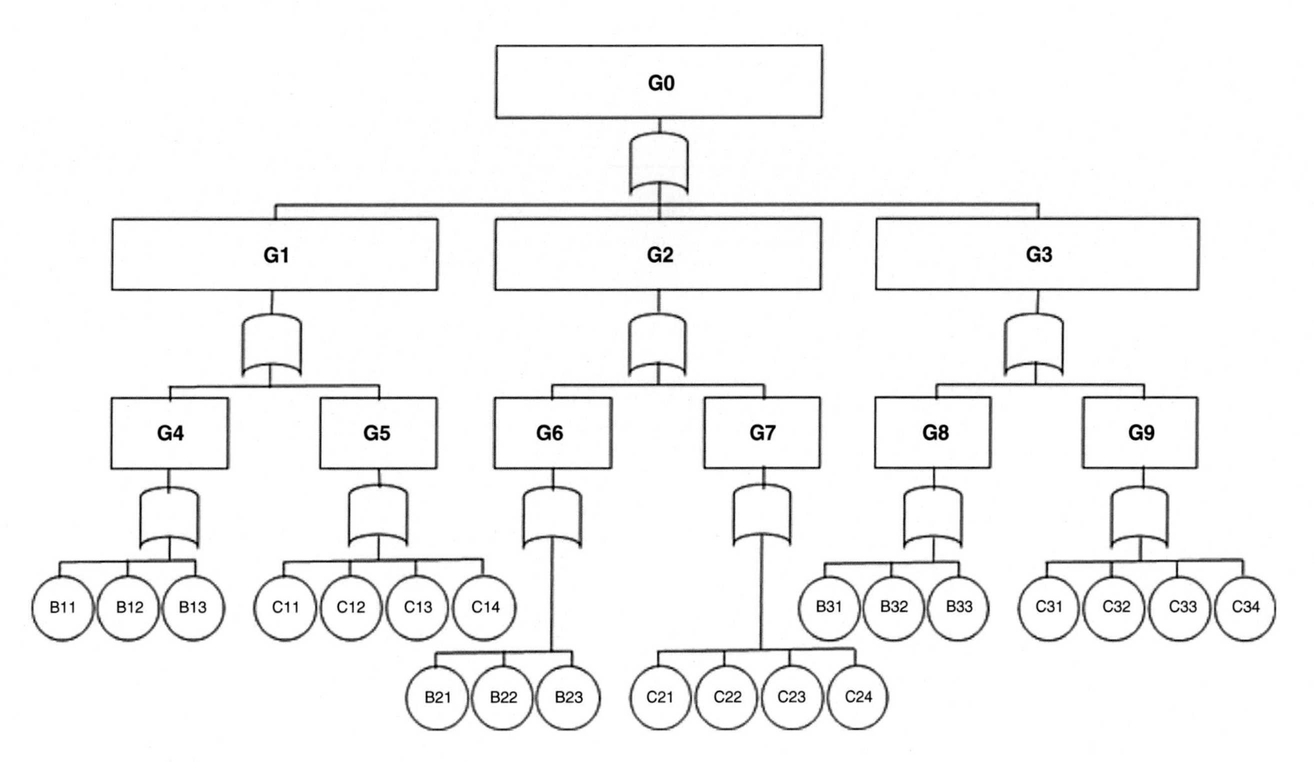

FIGURE 1.9 The fault tree of the regular structure.

TABLE 1.3 MCS Algorithm for the Fault Tree of the Regular Structure

No.			
1	**2**	**3**	**4**
G0	G1	G4	B11
			B12
			B13
		G5	C11
			C12
			C13
			C14
	G2	G6	B21
			B22
			B23
		G7	C21
			C22
			C23
			C24
	G3	G8	B31
			B32
			B33
		G9	C31
			C32
			C33
			C34

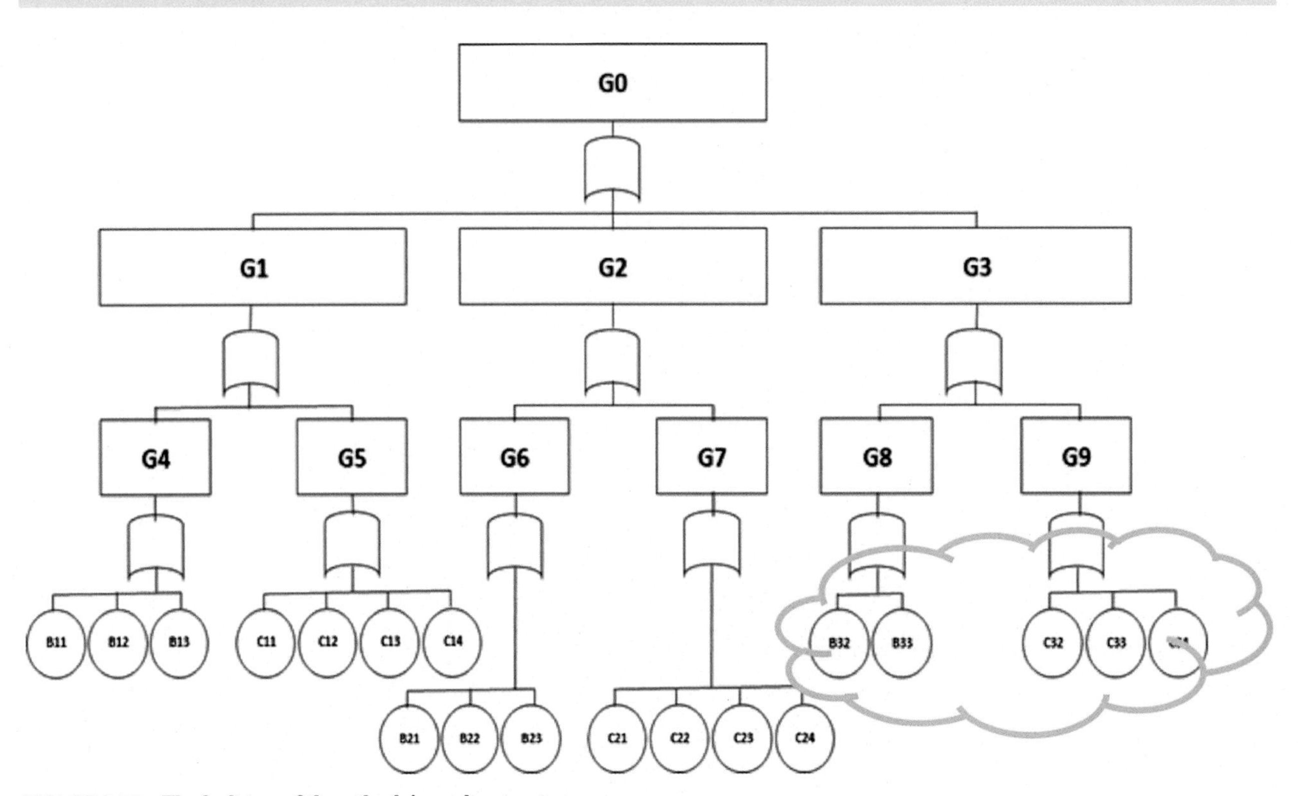

FIGURE 1.10 **The fault tree of the setback irregular structure.**

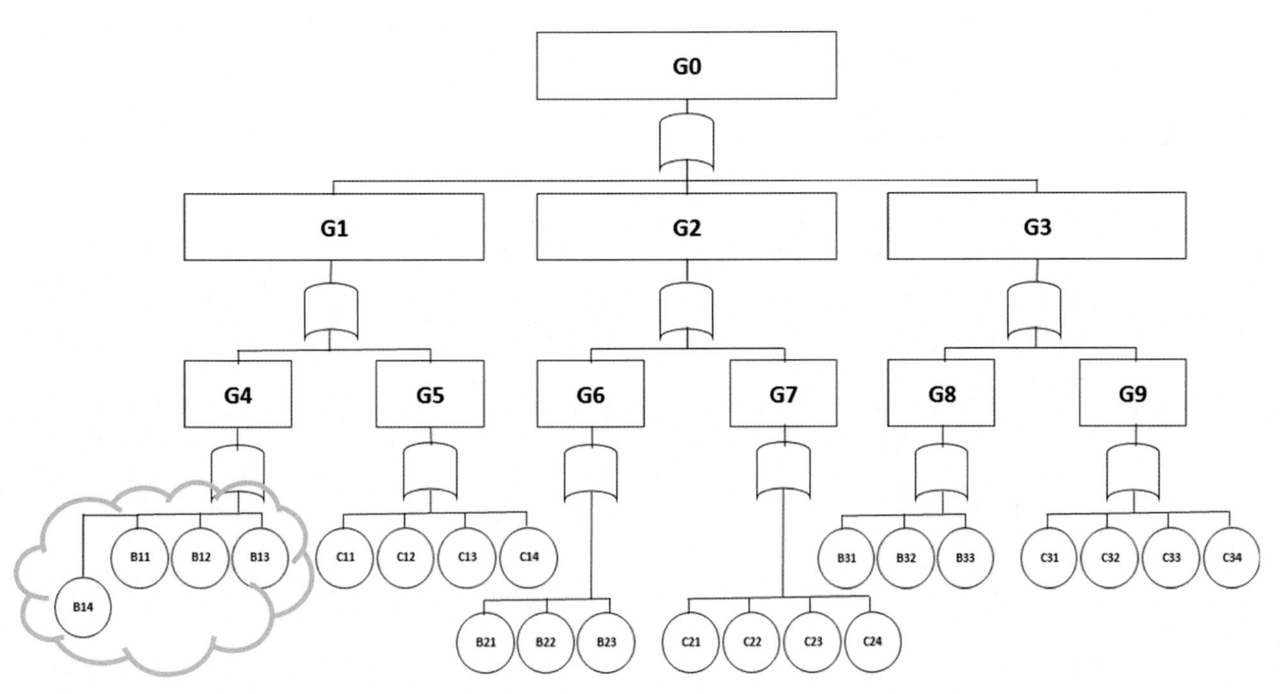

FIGURE 1.11 **The fault tree of the irregular structure with short column.**

TABLE 1.4 MCS Algorithm for the Fault Trees of the Irregular Structures

1	2	3	4	1	2	3	4
G0	G1	G4	B11	G0	G1	G4	B11
			B12				B12
			B13				B13
		G5	C11				B14
			C12			G5	C11
			C13				C12
			C14				C13
	G2	G6	B21				C14
			B22		G2	G6	B21
			B23				B22
		G7	C21				B23
			C22			G7	C21
			C23				C22
			C24				C23
	G3	G8	B32				C24
			B33		G3	G8	B31
			C32				B32
		G9	C33				B33
			C34			G9	C31
							C32
							C33
							C34

(Continued)

TABLE 1.4 MCS Algorithm for the Fault Trees of the Irregular Structures (*cont.*)

1	2	3	4	1	2	3	4
(a) Setback irregular structure				(b) The irregular structure with short column			
1	2	3	4				
G0	G1	G4	B11				
			B12				
			B13				
		G5	C11				
			C12				
			C13				
			C14				
	G2	G6	B21				
			B22				
			B23				
		G7	C21				
			C22				
			C23				
			C24				
	G3	G8	B31				
			B32				
			B33				
		G9	C31				
			C32				
			C33				
			C34				
(c) The irregular structure with soft story							

TABLE 1.5 Earthquake Data

No	Earthquake	Year	Station	Mw	PGA	PGV	Component
1	Cape Mendocino	1992	Rio Dell overpass	7.0	0.55	44	RIO270
2	Hector mine	1999	Hector	7.1	0.34	42	HEC000
3	Imperial valley	1979	El Centro array#11	6.5	0.38	42	H-E11140
4	Kocaeli Turkey	1999	Duzce	7.5	0.36	59	DZC180
5	Landers	1992	Cool water	7.3	0.42	42	CLW-LN
6	Loma Prieta	1989	Capitola	6.9	0.53	35	CAP000
7	Northridge	1994	Canyon country-WLC	6.7	0.48	45	LOS000

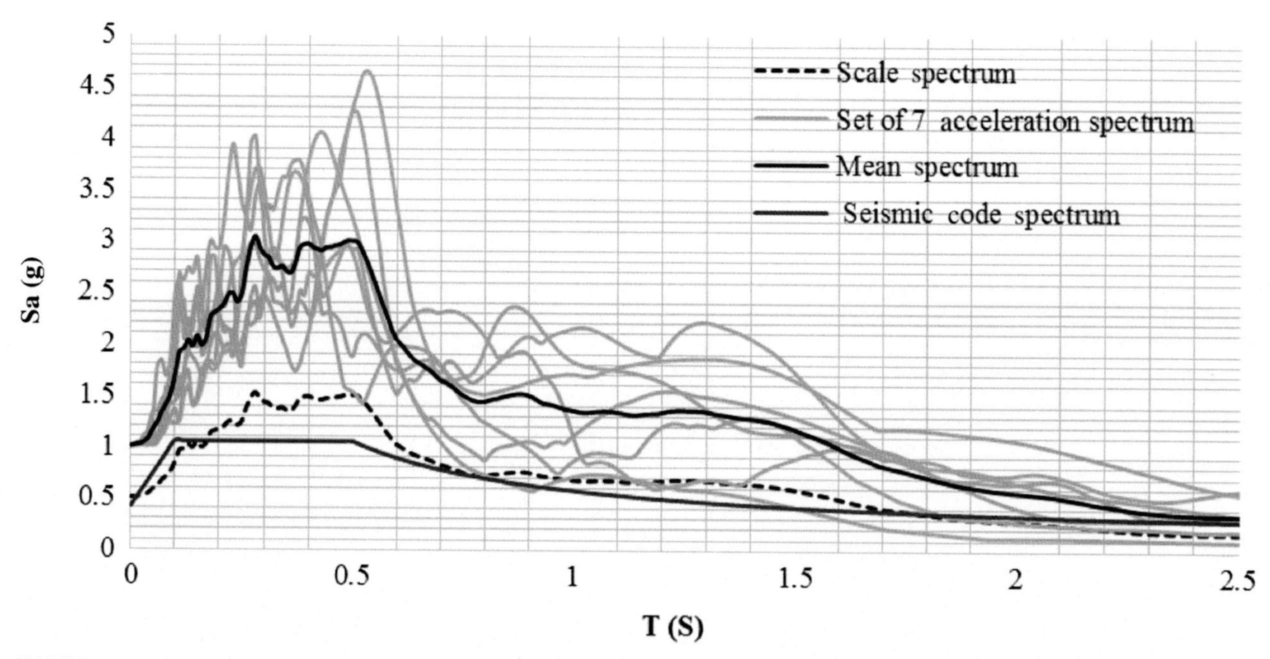

FIGURE 1.12 **The acceleration spectrum for PGA = 0.3 g and soil type B (spectra for individual accelerograms and mean spectrum).**

Maximum acceleration occurs in the 5%-damped spectra of two horizontal components, which are used to normalize both spectra. The normalization process is required for applying the perpendicular components when a two-dimensional analysis is performed. The normalized spectra is then averaged and scaled in such a way that the values exceed the maximum considered earthquake (MCE) spectrum, in a range between 0.2T and 1.5T, where T stands for the fundamental period of the structure. The records are finally multiplied by the scale factor where applicable, as shown in Fig. 1.12.

Employing SAP2000 software (SAP2000-V14, 2002), NDAs are performed where the hysteresis behavior of each element under separate seismic loading is stored. The DI of each element is then accounted for based on the modified version of Park and Ang's equation. This information is used to perform the quantitative FTAs. It is evident that, as there are seven earthquake accelerations, the quantitative FTA is hence repeated seven times for each structure. This means that, overall, there will be seven failure criteria for every structure.

1.5 RESULTS

The structural models for the regular and irregular structures were explained in the previous section. The irregularity types were also explained over the previous sections. Also, the FTA at the qualitative and quantitative levels was introduced. The analysis initiates with the application of gravity loads to the models, followed by earthquake loads simulated using the data provided in Table 1.5 and Fig. 1.12. Performing a series of NDAs for each structure and under all seven earthquake accelerations, the DI of every single element (i.e. local DI) is determined first, via which the global DI (at the building scale) is accounted for. Table 1.6 provides DIs of the structural elements of the regular structure under the seven earthquake accelerations. The participation ratio (PR) of each single element to the global DI is also shown in the table. In a similar way, the DIs of the three irregular structures are also determined, as shown in Tables 1.7–1.9.

Based on the information provided in Tables 1.6–1.9, the DIs of the cases at the story level first, and accordingly at the building level, are determined as shown in Table 1.10. Referring to the information provided in Table 1.1, where the damage boundaries were introduced, Table 1.10 shows that the regular structure meets the criteria at the LS performance level under all seven of the earthquakes. Similar results can also be seen in the setback irregular structure, although the DIs are higher than those of the regular structure. This is particularly the case where the setback irregularity occurs at the third story. The results for the irregular structure with short column show that, although the beam located within the short columns experiences local damage of more than 1.0 under all of the earthquakes, it does not have a considerable effect on the global damage, such that the global DI of this structure remains under the LS performance level. When the irregular structure with soft story is taken into consideration, the results show that the structure sustains considerable damage beyond the limit defined at the LS level of performance.

TABLE 1.6 The Damage Indices of All of the Elements of the Regular Structure

Earthquakes															
		Cape		Hector		Imperial		Kocaeli		Lander		Loma prieta		Northridge	
St.	Element	DI	PR	DI	PR	DI	PR	DI	PR	DI	PR	DI	PR	DI	PR
3rd	B31	0.26	0.1	0.22	0.05	0.31	0.05	0.28	0.11	0.38	0.08	0.2	0	0.36	0.07
	B32	0.23	0.1	0.22	0.04	0.32	0.05	0.26	0.09	0.36	0.06	0.2	0	0.34	0.06
	B33	0.23	0.1	0.23	0.07	0.34	0.07	0.27	0.09	0.36	0.09	0.3	0.1	0.34	0.06
	C31	0	0	0	0	0	0	0	0	0	0	0	0	0	0
	C32	0	0	0	0	0	0	0	0	0	0	0	0	0	0
	C33	0	0	0	0	0	0	0	0	0	0	0	0	0	0
	C34	0	0	0	0	0	0	0	0	0	0	0	0	0	0
2nd	B21	0.23	0.1	0.27	0.05	0.31	0.05	0.3	0.1	0.37	0.08	0.3	0.1	0.4	0.07
	B22	0.25	0.1	0.28	0.05	0.31	0.05	0.3	0.09	0.37	0.07	0.3	0.1	0.4	0.06
	B23	0.25	0.1	0.28	0.06	0.31	0.06	0.3	0.09	0.37	0.07	0.3	0.1	0.4	0.06
	C21	0	0	0	0	0	0	0	0	0	0	0	0	0	0
	C22	0	0	0	0	0	0	0	0	0	0	0	0	0	0
	C23	0	0	0	0	0	0	0	0	0	0	0	0	0	0
	C24	0	0	0	0	0	0	0	0	0	0	0	0	0	0
1st	B11	0.23	0.1	0.26	0.05	0.26	0.06	0.26	0.1	0.3	0.08	0.3	0	0.36	0.07
	B12	0.22	0.1	0.26	0.05	0.26	0.05	0.25	0.09	0.3	0.07	0.3	0	0.34	0.06
	B13	0.22	0.1	0.26	0.06	0.27	0.06	0.25	0.09	0.29	0.07	0.3	0.1	0.34	0.06
	C11	0	0	0	0	0	0	0	0	0	0	0	0	0	0
	C12	0	0	0	0	0	0	0	0	0	0	0	0	0	0
	C13	0	0	0	0	0	0	0	0	0	0	0	0	0	0
	C14	0	0	0	0	0	0	0	0	0	0	0	0	0	0

TABLE 1.7 The Damage Indices of All of the Elements of the Irregular Setback Structure

Earthquakes														
	Cape		Hector		Imperial		Kocaeli		Lander		Loma prieta		Northridge	
Element	DI	PR	DI	PR	DI	PR	DI	PR	DI	PR	DI	PR	DI	PR
B31	0	0	0	0	0	0	0	0	0	0	0	0	0	0
B32	0.31	0.1	0.25	0.09	0.39	0.05	0.27	0.12	0.42	0.1	0.2	0.1	0.34	0.11
B33	0.29	0.3	0.25	0.08	0.4	0.05	0.25	0.1	0.4	0.08	0.2	0.1	0.32	0.08
C31	0	0	0	0	0	0	0	0	0	0	0	0	0	0
C32	0	0	0	0	0	0	0	0	0	0	0	0	0	0
C33	0	0	0	0	0	0	0	0	0	0	0	0	0	0
C34	0	0	0	0	0	0	0	0	0	0	0	0	0	0
B21	0.28	0.1	0.27	0.05	0.36	0.03	0.28	0.08	0.35	0.06	0.3	0	0.33	0.06
B22	0.29	0.1	0.27	0.06	0.36	0.04	0.29	0.07	0.39	0.06	0.3	0	0.36	0.06
B23	0.29	0.1	0.27	0.06	0.37	0.04	0.29	0.08	0.4	0.06	0.3	0.1	0.35	0.06
C21	0	0	0	0	0	0	0	0	0	0	0	0	0	0
C22	0	0	0	0	0	0	0	0	0	0	0	0	0	0
C23	0	0	0	0	0	0	0	0	0	0	0	0	0	0
C24	0	0	0	0	0	0	0	0	0	0	0	0	0	0
B11	0.25	0.1	0.25	0.05	0.28	0.03	0.26	0.08	0.3	0.07	0.2	0	0.3	0.07
B12	0.24	0.1	0.25	0.06	0.28	0.03	0.26	0.07	0.3	0.06	0.2	0	0.3	0.06
B13	0.25	0.1	0.25	0.06	0.29	0.04	0.26	0.07	0.3	0.06	0.2	0.1	0.29	0.06
C11	0	0	0	0	0	0	0	0	0	0	0	0	0	0
C12	0	0	0	0	0	0	0	0	0	0	0	0	0	0
C13	0	0	0	0	0	0	0	0	0	0	0	0	0	0

The data shown in Tables 1.6–1.10 are now used to complete the fault trees of the cases studied, in order to determine the probability of failure. Fig. 1.13A shows the fault tree of the regular structure, where the MCSs are determined. Similarly, the fault trees of the irregular structures are also completed and then the MCSs are identified as shown in Fig. 1.13B–D. A summary of these calculations is shown in Table 1.11. In the table, PG0 represents the vulnerability level of the structures under the earthquake accelerations.

The information provided in Table 1.11 shows that there is a close association between the structural vulnerability determined via the FTA and the damage determined via the modified version of Park and Ang's equation. This comparison is also graphically shown in Fig. 1.14. In the figure, the horizontal axis represents the earthquakes' names as shown in Table 1.11. Based on the results of Table 1.11, a new damage level, which is based on the FTA, can be defined, as given in Table 1.12. For example, in the range between $0.11 < DI < 0.44$, the value of PG0 is lower than 0.33. As well, when $PG0 > 0.77$, the structure has to be considered as collapsed, because in that case, $DI > 0.77$. It is also worth mentioning that as both graphs in Fig. 1.14 follow a similar pattern, a calibration factor might be defined in such a way that the graphs are overlapped.

Based on the above results, the probability of failure in the structures can be calculated using Fig. 1.15. In the figure, the probability of failure (PoF) is dependent on the values of $PG0i$ ($i = 1$–7) for every earthquake acceleration. Accordingly, the ratio of EA in every earthquake to the total EA is determined (shown as EA in Table 1.13). EA plays the role of PR in Tables 1.6–1.9 and is required to perform FTA.

Repeating FTA for all the structures, the PoF is determined, as shown in Table 1.14. For example, it shows that the PoF of the irregular structure with soft story is 0.42. This number becomes meaningful if compared with the PG0 values in Table 1.12, showing that the soft-story irregularity can result in severe damage under earthquake excitations.

TABLE 1.8 The Damage Indices of All of the Elements of the Irregular Structure With Short Column

Earthquakes															
		Cape		Hector		Imperial		Kocaeli		Lander		Loma prieta		Northridge	
St.	Element	DI	PR	DI	PR	DI	PR	DI	PR	DI	PR	DI	PR	DI	PR
3rd	B31	0.26	0.1	0.27	0.06	0.35	0.05	0.28	0.09	0.4	0.08	0.3	0	0.42	0.06
	B32	0.25	0.1	0.27	0.04	0.33	0.05	0.26	0.07	0.38	0.06	0.3	0	0.42	0.05
	B33	0.24	0.1	0.25	0.04	0.36	0.06	0.26	0.07	0.38	0.06	0.3	0.1	0.41	0.05
	C31	0	0	0	0	0	0	0	0	0	0	0	0	0	0
	C32	0	0	0	0	0	0	0	0	0	0	0	0	0	0
	C33	0	0	0	0	0	0	0	0	0	0	0	0	0	0
	C34	0	0	0	0	0	0	0	0	0	0	0	0	0	0
2nd	B21	0.26	0.1	0.27	0.05	0.3	0.05	0.29	0.08	0.37	0.07	0.3	0	0.48	0.05
	B22	0.27	0.1	0.25	0.05	0.35	0.05	0.29	0.08	0.36	0.06	0.3	0	0.45	0.05
	B23	0.27	0.1	0.3	0.05	0.34	0.05	0.29	0.08	0.39	0.06	0.3	0	0.49	0.05
	C21	0	0	0	0	0	0	0	0	0	0	0	0	0	0
	C22	0	0	0	0	0	0	0	0	0	0	0	0	0.06	0
	C23	0	0	0	0	0	0	0	0	0	0	0	0	0.11	0.01
	C24	0	0	0	0	0	0	0	0	0	0	0	0	0	0
1st	B11	0.22	0.1	0.22	0.03	0.35	0.04	0.25	0.05	0.29	0.04	0.3	0	0.44	0.04
	B12	0.23	0.1	0.35	0.03	0.28	0.04	0.25	0.05	0.36	0.04	0.2	0	0.51	0.04
	B13	0.22	0.1	0.25	0.03	0.3	0.04	0.24	0.05	0.32	0.04	0.3	0	0.3	0.03
	C11	0	0	0	0	0	0	0	0	0	0	0	0	0	0
	C12	0	0	0.1	0.01	0.24	0.02	0	0	0.28	0.02	0	0	0.33	0.01
	C13	0	0	0.22	0.02	0.06	0	0	0	0.27	0.02	0	0	0.52	0.01
	C14	0	0	0	0	0	0	0	0	0	0	0	0	0	0
	B14	0.20	0.03	1.00	0.02	1.00	0.02	0.30	0.08	1.00	0.03	1.00	0.05	1.00	0.03

TABLE 1.9 The Damage Indices of All of the Elements of the Irregular Structure With Soft Story

Earthquakes																
		Cape		Hector		Imperial		Kocaeli		Lander		Loma prieta		Northridge	
St.	Element	DI	PR	DI	PR	DI	PR	DI	PR	DI	PR	DI	PR	DI	PR
3rd	B31	0.12	0	1	0.03	1	0.03	1	0.06	1	0.02	1	0	1	0.03
	B32	0.02	0	1	0.02	1	0.03	0.21	0.03	1	0.03	0.4	0	1	0.03
	B33	0.15	0	0.77	0.02	1	0.03	0.36	0.04	1	0.02	0.8	0	0.77	0.02
	C31	0	0	1	0.12	1	0.14	0	0	0	0	0.5	0.1	0.07	0
	C32	0	0	0	0.04	0	0	0	0	0	0	0	0	0.47	0.06
	C33	0	0	0.25	0.04	0.22	0.04	0.13	0.03	0.45	0.06	0	0	1	0.11
	C34	0.56	0.2	0.26	0.04	0.55	0.08	0.03	0	0	0	0	0	1	0.11
2nd	B21	0.61	0	1	0.02	1	0.03	1	0.02	1	0.01	1	0	1	0.06
	B22	0.31	0	1	0.02	1	0.03	1	0.03	1	0.01	1	0	1	0.06
	B23	1	0	1	0.03	1	0.03	1	0.02	1	0.01	1	0	1	0.06
	C21	0	0	1	0.11	1	0.15	0.37	0.03	0.09	0	1	0.1	0.2	0.03
	C22	0.12	0	0	0	0.02	0.01	0.04	0	0.8	0.05	0	0	0.09	0.02
	C23	0.07	0	0.28	0.05	0	0	0.26	0.02	1	0.04	0	0	0	0
	C24	0.97	0.1	0.47	0.08	0.52	0.08	0.4	0.04	0.04	0	0.2	0	1	0.13
1st	B11	0.72	0	1	0.02	1	0.04	1	0.02	1	0.01	1	0	1	0.02
	B12	0.59	0	1	0.02	1	0.04	1	0.03	1	0.01	1	0	1	0.03
	B13	1	0	1	0.04	1	0.04	1	0.02	1	0.01	1	0	1	0.04
	C11	0	0	1	0.14	0.48	0.08	0.03	0	0.24	0.02	0.8	0.1	0.14	0.04
	C12	0.09	0	0	0	0	0	0	0	0.22	0.03	0	0	0.23	0.06
	C13	0.05	0	0.15	0.04	0	0	0.42	0.05	0.52	0.04	0	0	0.15	0.03
	C14	0.87	0.1	0.32	0.04	1	0.14	0.27	0.03	0.09	0.02	0.2	0	0.68	0.14

TABLE 1.10 The Global Damage Indices of the Regular and Irregular Structures

		Regular structure		Setback irregular		Short column		Soft story	
Earthquake	**Story**	**DI**	**Global DI**	**DI**	**Global DI**	**DI**	**Global DI**	**DI**	**Global DI**
Cape	3rd	0.24	0.23	0.29	0.27	0.25	0.25	0.51	0.72
	2nd	0.24		0.28		0.27		0.78	
	1st	0.22		0.24		0.22		0.76	
Hector	3rd	0.22	0.26	0.24	0.26	0.26	0.33	0.67	0.74
	2nd	0.27		0.25		0.27		0.76	
	1st	0.26		0.27		0.38		0.8	
Imperial	3rd	0.32	0.29	0.39	0.33	0.34	0.35	0.8	0.85
	2nd	0.31		0.36		0.33		0.85	
	1st	0.26		0.28		0.37		0.88	
Kocaeli	3rd	0.27	0.27	0.26	0.27	0.26	0.27	0.51	0.6
	2nd	0.3		0.28		0.29		0.6	
	1st	0.25		0.26		0.26		0.64	
Landers	3rd	0.36	0.34	0.4	0.35	0.38	0.39	0.75	0.68
	2nd	0.37		0.38		0.37		0.87	
	1st	0.3		0.3		0.4		0.68	
Loma Prieta	3rd	0.25	0.27	0.22	0.24	0.26	0.41	0.57	0.77
	2nd	0.3		0.26		0.31		0.86	
	1st	0.26		0.24		0.52		0.8	
Northridge	3rd	0.35	0.37	0.33	0.32	0.41	0.48	0.88	0.78
	2nd	0.39		0.34		0.45		0.88	
	1st	0.35		0.3		0.51		0.77	

(A)

$$P_{G4}=(P_{B11}+P_{B12}+P_{B13})\ \ (P_{B11}\ \ P_{B12}+P_{B11}\ \ P_{B13}+P_{B12}\ \ P_{B13})+(P_{B11}\ \ P_{B12}\ \ P_{B13})$$

$$P_{G5}=(P_{C11}+P_{C12}+P_{C13}+P_{C14})\ \ (P_{C11}\ \ P_{C12}+P_{C11}\ \ P_{C13}+P_{C11}\ \ P_{C14}+P_{C12}\ \ P_{C13}+P_{C12}\ \ P_{C14}+P_{C13}\ \ P_{C14})$$
$$+(P_{C11}\ \ P_{C12}\ \ P_{C13}+P_{C11}\ \ P_{C12}\ \ P_{C14}+P_{C12}\ \ P_{C13}\ \ P_{C14})\ \ (P_{C11}\ \ P_{C12}\ \ P_{C13}\ \ P_{C14})$$

$$P_{G6}=(P_{B21}+P_{B22}+P_{B23})\ \ (P_{B21}\ \ P_{B22}+P_{B21}\ \ P_{B23}+P_{B22}\ \ P_{B23})+(P_{B21}\ \ P_{B22}\ \ P_{B23})$$

$$P_{G7}=(P_{C21}+P_{C22}+P_{C23}+P_{C24})\ \ (P_{C21}\ \ P_{C22}+P_{C21}\ \ P_{C23}+P_{C21}\ \ P_{C24}+P_{C22}\ \ P_{C23}+P_{C22}\ \ P_{C24}+P_{C23}\ \ P_{C24})$$
$$+(P_{C21}\ \ P_{C22}\ \ P_{C23}+P_{C21}\ \ P_{C22}\ \ P_{C24}+P_{C22}\ \ P_{C23}\ \ P_{C24})\ \ (P_{C21}\ \ P_{C22}\ \ P_{C23}\ \ P_{C24})$$

$$P_{G8}=(P_{B31}+P_{B32}+P_{B33})\ \ (P_{B31}\ \ P_{B32}+P_{B31}\ \ P_{B33}+P_{B32}\ \ P_{B33})+(P_{B31}\ \ P_{B32}\ \ P_{B33})$$

$$P_{G9}=(P_{C31}+P_{C32}+P_{C33}+P_{C34})\ \ (P_{C31}\ \ P_{C32}+P_{C31}\ \ P_{C33}+P_{C31}\ \ P_{C34}+P_{C32}\ \ P_{C33}+P_{C32}\ \ P_{C34}+P_{C33}\ \ P_{C34})$$
$$+(P_{C31}\ \ P_{C32}\ \ P_{C33}+P_{C31}\ \ P_{C32}\ \ P_{C34}+P_{C32}\ \ P_{C33}\ \ P_{C34})\ \ (P_{C31}\ \ P_{C32}\ \ P_{C33}\ \ P_{C34})$$

$$P_{G1}=(P_{G4}+P_{G5})\ \ (P_{G4}\ \ P_{G5})$$

$$P_{G2}=(P_{G6}+P_{G7})\ \ (P_{G6}\ \ P_{G7})$$

$$P_{G3}=(P_{G8}+P_{G9})\ \ (P_{G8}\ \ P_{G9})$$

$$P_{G0}=(P_{G1}+P_{G2}+P_{G3})\ \ (P_{G1}\ \ P_{G2}+P_{G1}\ \ P_{G3}+P_{G2}\ \ P_{G3})+(P_{G1}\ \ P_{G2}\ \ P_{G3})$$

(B)

$$P_{G4}=(P_{B11}+P_{B12}+P_{B13})\ \ (P_{B11}\ \ P_{B12}+P_{B11}\ \ P_{B13}+P_{B12}\ \ P_{B13})+(P_{B11}\ \ P_{B12}\ \ P_{B13})$$

$$P_{G5}=(P_{C11}+P_{C12}+P_{C13}+P_{C14})\ \ (P_{C11}\ \ P_{C12}+P_{C11}\ \ P_{C13}+P_{C11}\ \ P_{C14}+P_{C12}\ \ P_{C13}+P_{C12}\ \ P_{C14}+P_{C13}\ \ P_{C14})$$
$$+(P_{C11}\ \ P_{C12}\ \ P_{C13}+P_{C11}\ \ P_{C12}\ \ P_{C14}+P_{C12}\ \ P_{C13}\ \ P_{C14})\ \ (P_{C11}\ \ P_{C12}\ \ P_{C13}\ \ P_{C14})$$

$$P_{G6}=(P_{B21}+P_{B22}+P_{B23})\ \ (P_{B21}\ \ P_{B22}+P_{B21}\ \ P_{B23}+P_{B22}\ \ P_{B23})+(P_{B21}\ \ P_{B22}\ \ P_{B23})$$

$$P_{G7}=(P_{C21}+P_{C22}+P_{C23}+P_{C24})\ \ (P_{C21}\ \ P_{C22}+P_{C21}\ \ P_{C23}+P_{C21}\ \ P_{C24}+P_{C22}\ \ P_{C23}+P_{C22}\ \ P_{C24}+P_{C23}\ \ P_{C24})$$
$$+(P_{C21}\ \ P_{C22}\ \ P_{C23}+P_{C21}\ \ P_{C22}\ \ P_{C24}+P_{C22}\ \ P_{C23}\ \ P_{C24})\ \ (P_{C21}\ \ P_{C22}\ \ P_{C23}\ \ P_{C24})$$

$$P_{G8}=(P_{B32}+P_{B33})\ \ (P_{B32}\ \ P_{B33})$$

$$P_{G9}=(P_{C31}+P_{C32}+P_{C33}+P_{C34})\ \ (P_{C31}\ \ P_{C32}+P_{C31}\ \ P_{C33}+P_{C31}\ \ P_{C34}+P_{C32}\ \ P_{C33}+P_{C32}\ \ P_{C34}+P_{C33}\ \ P_{C34})$$
$$+(P_{C31}\ \ P_{C32}\ \ P_{C33}+P_{C31}\ \ P_{C32}\ \ P_{C34}+P_{C32}\ \ P_{C33}\ \ P_{C34})\ \ (P_{C31}\ \ P_{C32}\ \ P_{C33}\ \ P_{C34})$$

$$P_{G1}=(P_{G4}+P_{G5})\ \ (P_{G4}\ \ P_{G5})$$

$$P_{G2}=(P_{G6}+P_{G7})\ \ (P_{G6}\ \ P_{G7})$$

$$P_{G3}=(P_{G8}+P_{G9})\ \ (P_{G8}\ \ P_{G9})$$

$$P_{G0}=(P_{G1}+P_{G2}+P_{G3})\ \ (P_{G1}\ \ P_{G2}+P_{G1}\ \ P_{G3}+P_{G2}\ \ P_{G3})+(P_{G1}\ \ P_{G2}\ \ P_{G3})$$

FIGURE 1.13 (A) The fault tree of the regular structure with the identified MCS. (B) The fault tree of the setback irregular structure with the identified MCS. (C) The fault tree of the irregular structure with short columns with the identified MCS. (D) The fault tree of the irregular structure with soft story with the identified MCS.

Similar results can also be achieved for other irregularities. This shows that the PoF of the setback irregular structure and of the irregular structure with short column is not much higher than that of the regular structure, although these irregularities cause some damage (mostly local) to be sustained by the structures. As pointed out earlier, FTA is a functional tool for quantifying the probability of failure in structural systems, while this cannot be achieved through quantitative-based DIs.

(C)

$$P_{G4}=(P_{B11}+P_{B12}+P_{B13}+P_{B14})\quad(P_{B11}\ P_{B12}+P_{B11}\ P_{B13}+P_{B11}\ P_{B14}+P_{B12}\ P_{B13}+P_{B12}\ P_{B14}+P_{B13}\ P_{B14})$$
$$+(P_{B11}\ P_{B12}\ P_{B13}+P_{B11}\ P_{B12}\ P_{B14}+P_{B12}\ P_{B13}\ P_{B14})\quad(P_{B11}\ P_{B12}\ P_{B13}\ P_{B14})$$

$$P_{G5}=(P_{C11}+P_{C12}+P_{C13}+P_{C14})\quad(P_{C11}\ P_{C12}+P_{C11}\ P_{C13}+P_{C11}\ P_{C14}+P_{C12}\ P_{C13}+P_{C12}\ P_{C14}+P_{C13}\ P_{C14})$$
$$+(P_{C11}\ P_{C12}\ P_{C13}+P_{C11}\ P_{C12}\ P_{C14}+P_{C12}\ P_{C13}\ P_{C14})\quad(P_{C11}\ P_{C12}\ P_{C13}\ P_{C14})$$

$$P_{G6}=(P_{B21}+P_{B22}+P_{B23})\quad(P_{B21}\ P_{B22}+P_{B21}\ P_{B23}+P_{B22}\ P_{B23})+(P_{B21}\ P_{B22}\ P_{B23})$$

$$P_{G7}=(P_{C21}+P_{C22}+P_{C23}+P_{C24})\quad(P_{C21}\ P_{C22}+P_{C21}\ P_{C23}+P_{C21}\ P_{C24}+P_{C22}\ P_{C23}+P_{C22}\ P_{C24}+P_{C23}\ P_{C24})$$
$$+(P_{C21}\ P_{C22}\ P_{C23}+P_{C21}\ P_{C22}\ P_{C24}+P_{C22}\ P_{C23}\ P_{C24})\quad(P_{C21}\ P_{C22}\ P_{C23}\ P_{C24})$$

$$P_{G8}=(P_{B31}+P_{B32}+P_{B33})\quad(P_{B31}\ P_{B32}+P_{B31}\ P_{B33}+P_{B32}\ P_{B33})+(P_{B31}\ P_{B32}\ P_{B33})$$

$$P_{G9}=(P_{C31}+P_{C32}+P_{C33}+P_{C34})\quad(P_{C31}\ P_{C32}+P_{C31}\ P_{C33}+P_{C31}\ P_{C34}+P_{C32}\ P_{C33}+P_{C32}\ P_{C34}+P_{C33}\ P_{C34})$$
$$+(P_{C31}\ P_{C32}\ P_{C33}+P_{C31}\ P_{C32}\ P_{C34}+P_{C32}\ P_{C33}\ P_{C34})\quad(P_{C31}\ P_{C32}\ P_{C33}\ P_{C34})$$

$$P_{G1}=(P_{G4}+P_{G5})\quad(P_{G4}\ P_{G5})$$

$$P_{G2}=(P_{G6}+P_{G7})\quad(P_{G6}\ P_{G7})$$

$$P_{G3}=(P_{G8}+P_{G9})\quad(P_{G8}\ P_{G9})$$

$$P_{G0}=(P_{G1}+P_{G2}+P_{G3})\quad(P_{G1}\ P_{G2}+P_{G1}\ P_{G3}+P_{G2}\ P_{G3})+(P_{G1}\ P_{G2}\ P_{G3})$$

(D)

$$P_{G4}=(P_{B11}+P_{B12}+P_{B13})\quad(P_{B11}\ P_{B12}+P_{B11}\ P_{B13}+P_{B12}\ P_{B13})+(P_{B11}\ P_{B12}\ P_{B13})$$

$$P_{G5}=(P_{C11}+P_{C12}+P_{C13}+P_{C14})\quad(P_{C11}\ P_{C12}+P_{C11}\ P_{C13}+P_{C11}\ P_{C14}+P_{C12}\ P_{C13}+P_{C12}\ P_{C14}+P_{C13}\ P_{C14})$$
$$+(P_{C11}\ P_{C12}\ P_{C13}+P_{C11}\ P_{C12}\ P_{C14}+P_{C12}\ P_{C13}\ P_{C14})\quad(P_{C11}\ P_{C12}\ P_{C13}\ P_{C14})$$

$$P_{G6}=(P_{B21}+P_{B22}+P_{B23})\quad(P_{B21}\ P_{B22}+P_{B21}\ P_{B23}+P_{B22}\ P_{B23})+(P_{B21}\ P_{B22}\ P_{B23})$$

$$P_{G7}=(P_{C21}+P_{C22}+P_{C23}+P_{C24})\quad(P_{C21}\ P_{C22}+P_{C21}\ P_{C23}+P_{C21}\ P_{C24}+P_{C22}\ P_{C23}+P_{C22}\ P_{C24}+P_{C23}\ P_{C24})$$
$$+(P_{C21}\ P_{C22}\ P_{C23}+P_{C21}\ P_{C22}\ P_{C24}+P_{C22}\ P_{C23}\ P_{C24})\quad(P_{C21}\ P_{C22}\ P_{C23}\ P_{C24})$$

$$P_{G8}=(P_{B31}+P_{B32}+P_{B33})\quad(P_{B31}\ P_{B32}+P_{B31}\ P_{B33}+P_{B32}\ P_{B33})+(P_{B31}\ P_{B32}\ P_{B33})$$

$$P_{G9}=(P_{C31}+P_{C32}+P_{C33}+P_{C34})\quad(P_{C31}\ P_{C32}+P_{C31}\ P_{C33}+P_{C31}\ P_{C34}+P_{C32}\ P_{C33}+P_{C32}\ P_{C34}+P_{C33}\ P_{C34})$$
$$+(P_{C31}\ P_{C32}\ P_{C33}+P_{C31}\ P_{C32}\ P_{C34}+P_{C32}\ P_{C33}\ P_{C34})\quad(P_{C31}\ P_{C32}\ P_{C33}\ P_{C34})$$

$$P_{G1}=(P_{G4}+P_{G5})\quad(P_{G4}\ P_{G5})$$

$$P_{G2}=(P_{G6}+P_{G7})\quad(P_{G6}\ P_{G7})$$

$$P_{G3}=(P_{G8}+P_{G9})\quad(P_{G8}\ P_{G9})$$

$$P_{G0}=(P_{G1}+P_{G2}+P_{G3})\quad(P_{G1}\ P_{G2}+P_{G1}\ P_{G3}+P_{G2}\ P_{G3})+(P_{G1}\ P_{G2}\ P_{G3})$$

FIGURE 1.13 (*Cont.*)

TABLE 1.11 The Results of the FTA Performed for the Cases Studied

		Regular structure		Setback irregular		Short column		Soft story	
No.	Earthquake	PG0	Global DI	PG0	Global DI	PG0	Global DI	PG0	Global DI
1	Cape	0.20	0.23	0.24	0.28	0.22	0.25	0.53	0.72
2	Hector	0.23	0.26	0.23	0.26	0.28	0.33	0.53	0.75
3	Imperial	0.25	0.29	0.28	0.33	0.29	0.35	0.58	0.85
4	Kocaeli	0.20	0.28	0.23	0.27	0.33	0.27	0.45	0.60
5	Landers	0.29	0.34	0.29	0.35	0.32	0.39	0.50	0.68
6	Loma Prieta	0.19	0.27	0.22	0.24	0.35	0.41	0.52	0.77
7	Northridge	0.32	0.37	0.28	0.32	0.38	0.48	0.56	0.77

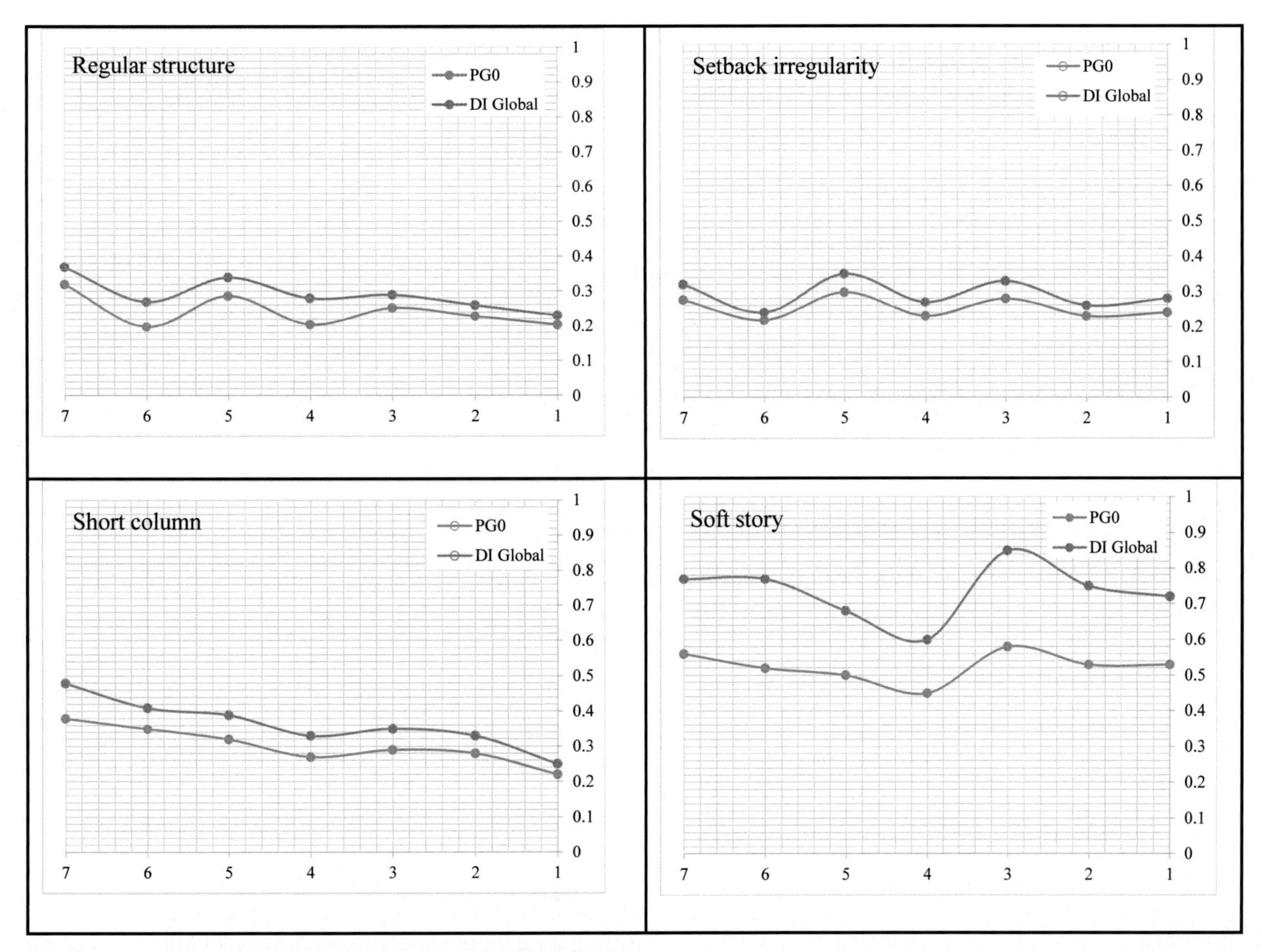

FIGURE 1.14 **The vulnerablity of the structures versus the global DIs.**

TABLE 1.12 Damage Levels Based on the Probablity of Failure

Description	Damage index	PG0
No damage or very minor damage	0	0
Light (serviceable)	DI < 0.11	0
Moderate (repairable)	0.11 < DI < 0.44	$P < 0.33$
Severe (irreparable)	0.44 < DI < 0.77	$0.33 < P < 0.55$
Collapse	DI > 0.77	$P > 0.55$

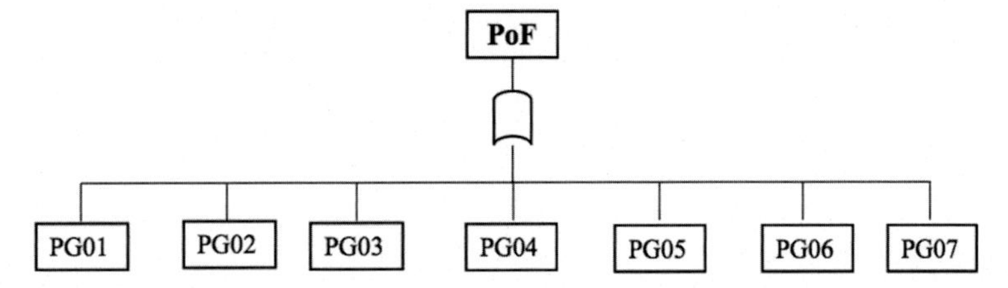

FIGURE 1.15 **The probability of failure.**

TABLE 1.13 Ratios of the Energy Absorbed by the Structures Under the Seven Earthquakes

	Regular structure		Setback irregular		Short column		Soft story	
Earthquake	PG0	EA	PG0	EA	PG0	EA	PG0	EA
Cape	0.20	0.038	0.24	0.0486	0.22	0.028	0.53	0.105
Hector	0.23	0.145	0.23	0.163	0.28	0.140	0.53	0.175
Imperial	0.25	0.200	0.28	0.252	0.29	0.253	0.58	0.155
Kacoeli	0.20	0.112	0.23	0.091	0.33	0.088	0.45	0.110
Landers	0.29	0.190	0.29	0.150	0.32	0.199	0.50	0.114
Loma Prieta	0.19	0.092	0.22	0.084	0.35	0.090	0.52	0.118
Northridge	0.32	0.220	0.28	0.211	0.38	0.202	0.56	0.223

TABLE 1.14 The Probability of Failure (PoF) of the Cases Studied Under the Earthquake Accelerations

Structure	PoF
Regular	0.24
Setback irregular	0.23
Short column	0.28
Soft story	0.42

1.6 SUMMARY

In this chapter, the failure probability of regular and irregular structures was determined using FTA. Three types of irregularities were considered: setback irregularity, short-column irregularity, and soft-story irregularity. The structures were first subjected to seven earthquake accelerations, covering a range of different intensities, amplitudes, and time intervals. Using the quantitative-based DI based on Park and Ang's equation, the damage sustained by the structures under the seven earthquakes was then determined. Based on Boolean algebra, fault trees of the structures were created, in which the damage accounted for by Park and Ang's equation was used to perform the quantitative analysis. The results showed that a risk index based on the probability of failure can be proposed which can be substituted for quantitative-based DI. The risk index proposed is particularity useful to illustrate how a regular structure is susceptible to collapse if it becomes irregular. The results of the FTA performed in this study showed that among the three irregularities considered, soft-story irregularity endangers the safety of structures to a higher degree under different earthquake accelerations.

Finally, the results of the study performed here can be used by authorities as a predisaster mitigation strategy, so that the probability of failure of different urban structures can be known in advance.

REFERENCES

ACI318, 2008. Building Code Requirements for Structural Concrete (ACI 318-08) and Commentary. American Concrete Institute, USA.

Baas, S., Ramasamy, S., DePryck, J.D., Battista, F., 2008. Disaster Risk Management Systems Analysis—A Guide Book. Food and Agriculture Organization of the United Nations, Rome, Italy, pp. 1–90.

Banon, H., Biggs, J.M., Irvine, H.M., 1981. Seismic damage in reinforced concrete members. ASCE, J. Struct. Eng. Struct. 9, 1713–1729.

Behnam, B., 2006. Retrofitting Management for Residential Buildings [Research]. Tehran Polytechnic, Tehran, Iran, pp. 32–46.

Behnam, B., 2015. Structural response of vertically irregular tall moment-resisting steel frames under pre- and post-earthquake fire. The Structural Design of Tall and Special Buildings 25 (12), 543–557.

Bhosale, A., Davis, R., Sarkar, P., 2017. Vertical irregularity of buildings: regularity index versus seismic risk. ASCE-ASME Journal of Risk and Uncertainty in Engineering Systems, Part A: Civil Engineering 3 (3), 04017001.

Blong, R., 2003. A review of damage intensity scales. Nat. Hazards 29, 57–76.

Chintanapakdee, C., Chopra, A., 2004. Seismic response of vertically irregular frames: response history and modal pushover analyses. J. Struct. Eng. 130, 1177–1185.

D'Ambrisi, A., De Stefano, M., Tanganelli, M., Viti, S., 2013. The effect of common irregularities on the seismic performance of existing RC framed buildings. In: Lavan, O., De Stefano, M. (Eds.), Seismic Behaviour and Design of Irregular and Complex Civil Structures. Springer, Dordrecht, The Netherlands, pp. 47–58.

Elnashai, A.S., Di Sarno, L., 2015. Fundamentals of Earthquake Engineering: From Source to Fragility. Wiley, West Sussex, UK, pp. 72–88.

Ericson, C.A., Ll, C., 1999. Fault tree analysis. System Safety Conference. Orlando, FL, pp. 1–9.

FEMA310, 1998. Handbook for the Seismic Evaluation of Buildings—A Pre-standard. American Society of Civil Engineering, Washington, DC, pp. 4–11.

FEMA356, 2000. Prestandard and Commentary for the Seismic Rehabilitation of Buildings Rehabilitation Requirements. American Society of Civil Engineers, Washington, DC, pp. 42–46.

Hadipriono, F.C., Toh, H.S., 1989. Modified fault tree analysis for structural safety. Civil Eng. Syst. 6, 190–199.

Hosseinpour, F., Abdelnaby, A.E., 2017. Effect of different aspects of multiple earthquakes on the nonlinear behavior of RC structures. Soil Dynamics and Earthquake Engineering 92, 706–725.

Lavan, O., De Stefano, M., 2013. Seismic Behaviour and Design of Irregular and Complex Civil Structures. Springer, Istanbul, Turkey, pp. 48–58.

LeBeau, K.H., Wadia-Fascetti, S.J., 2007. Fault tree analysis of Schoharie Creek bridge collapse. Journal of Performance of Constructed Facilities 21, 320–326.

Mavroeidis, G.P., Papageorgiou, A.S., 2003. A mathematical representation of near-fault ground motions. Bulletin of the Seismological Society of America 93, 1099–1131.

Park, Y., Ang, A., 1985. Seismic damage analysis of reinforced concrete buildings. J. Struct. Eng. 111, 740–757.

SAP2000-V14, 2002. Integrated Finite Element Analysis and Design of Structures Basic Analysis Reference Manual Berkeley, CA, USA.

Stone, W.C., Taylor, A.W., 2013. Seismic Performance of Circular Bridge Columns Designed in Accordance With AASHTO/CALTRANS Standards. U.S. Dept. of Commerce, National Institute of Standards and Technology, Washingtonp, p. 136.

Chapter 2

Importance of Geological Studies in Earthquake Hazard Assessment

Biju John
National Institute of Rock Mechanics, Kolar Gold Fields, India

2.1 INTRODUCTION

Earthquakes lead the list of natural disasters in terms of damage and human loss and they affect very large areas, causing death and destruction on a massive scale. Seismicity of any region depends on the state of tectonic stress across structural discontinuities/faults, which are remarkably different when comparing interplate and intraplate settings. The return period of damaging earthquakes are much shorter in interpolate regions compared to intraplate settings. The Indian landmass, having plate boundaries and intraplate settings, has generated a large number of destructive earthquakes in the recent past (Rajendran and Rajendran, 2004). This chapter, however, is not going into the details of tectonic setup for describing the geological data required for seismic hazard assessment.

The history of earthquake recording and the historical documentation itself are too short to represent the long-term activity of the faults. Considering the variable recurrence intervals on different faults, earthquakes are generally occurring at unexpected locations. The locations of the upcoming developmental projects like nuclear power plants and their repositories, major dams with underground power house components, underground crude oil storages, etc. are greatly influenced by the seismic shaking as well as the displacements and deformation along preexisting faults, fractures, and shears (John and Rao, 2014). The assessment of seismic hazard depends upon our understanding of how earthquakes are generated and distributed and how they recur in space and time. To solve this problem, earthquake geologists focuses to trace and evaluate past earthquakes. The scope of such an investigation is dependent not only on the complexity and economics of a project, but also on the level of risk acceptable for the proposed structures like nuclear establishment and dams or development like hospitals, high-rise buildings, etc.

The existing seismic zoning strategy is primarily constrained by incomplete database, especially in regions where seismic intervals are much longer than the historic records. Besides, significant gaps in the data exist with regard to effects of earthquakes (ground shaking, structural responses, etc.), site-specific data on ground acceleration, soil conditions, and liquefaction potential, which are also important in hazard estimation. Thus, an integration of geological information such as identification and locations of active faults or potentially active faults, information on maximum credible earthquake (maximum earthquake potential) in a region, recurrence periods, the proximity to fault zones, and local site lithology is essential for earthquake hazard assessment of an area (Fig. 2.1).

The purpose of this chapter is to highlight the importance of geological studies for a meaningful earthquake hazard assessment of an area. The scope of this chapter is limited to discussion based on Indian examples of the various basic geological information required for a long-term earthquake hazard assessment.

2.2 IDENTIFICATION OF SUSPECTED ACTIVE FAULTS

The first step toward quantitative seismic hazard analysis for the area is the recognition and identification of active faults in the region of interest. The definition of active faults varies based on the requirement and tectonic setup. For example, for nuclear power site evaluation active fault is defined as that fault moved repeatedly in the present stress regime (AERB, 1990). Many active faults are complex, consisting of multiple breaks and rupture length. In addition to that, the evidence for identifying active fault traces is generally subtle or obscure and the distinction between recently active and long inactive faults may be difficult to make. In many cases, the majority of the faults are hidden or erosional agents erase its surface evidence. Tectonic

Integrating Disaster Science and Management. http://dx.doi.org/10.1016/B978-0-12-812056-9.00002-6

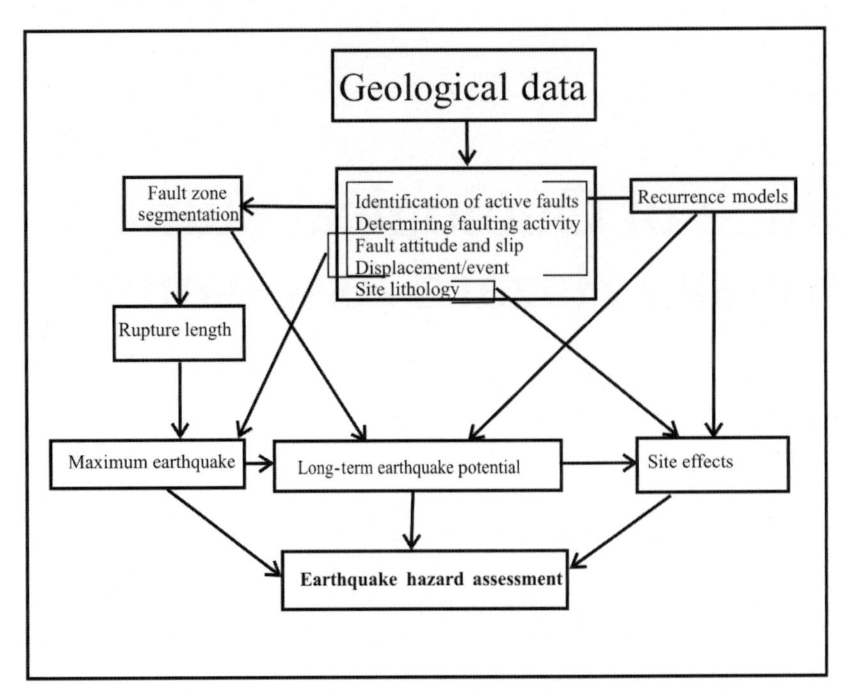

FIGURE 2.1 **Diagram showing input from the geological studies to seismic hazard evaluation (basic concept from Schwartz and Coppersmith, 1986).**

geomorphology, an interdisciplinary field at the boundary between structural geology/tectonics and surface processes, has proved to be a useful tool for identifying and quantifying geologically recent tectonic deformation. The most common goal of tectonic geomorphology research is to use quaternary landforms and stratigraphy to infer the nature, patterns, rates, and history of near-surface tectonic processes. In many geological settings around the world, the field of tectonic geomorphology has provided tools for evaluating the nature, pattern, and rates of active deformation. The following are the steps used by geologists to identify active faults.

2.2.1 Delineation of Lineaments

The application of remote-sensing technology, over the years, has shown its immense utility for large-scale geological mapping related to steep slopes, straight valley segments, abrupt changes in vegetation coverage and sudden bends along river courses. Remote sensing has the advantage of providing synoptic overviews of the region and thus it can provide information on structural geological features extending over large areas. The availability of multispectral and high-resolution data as well as the various digital image-processing techniques in generating enhanced images further widens the scope of remote-sensing studies in delineating the lithological contacts and geological structure with more detail and better accuracy.

Faults which are observed as either linear or curvilinear features on the earth surface can be easily demarcated as linears through remote-sensing studies. These linear zones of different contrast are commonly referred to as lineaments and this may range from a few meters to tens of kilometers in length. The lithological boundaries, boundaries between different land uses, drainage lines, etc. are also detected as lineaments. The interpretation of lineaments is based on the spatial correlation of remotely sensed images of geological objects as well as on the density of the available geological–geophysical data. In comparison to fieldwork investigations, remote-sensing techniques are quicker and more cost effective for fault detection. However, remote-sensing studies are neither sufficient nor a substitute for field investigations, but both studies can complement each other (Praseeda et al., 2015; Singh et al., 2016).

2.2.2 Regional Geomorphic Analysis

Tectonic geomorphology investigates the relationship between tectonics, climate, erosion-deposition process, and landscape (Caputo and Pavlides, 2008). The evolution of a landform is the result of interactions involving climatic, tectonic, and surface processes. Tectonic deformation of the Earth's surface, in general, takes place over thousands of years or longer. In the last couple of decades, tectonic geomorphology increasingly being used as one of the principal tools in identifying active tectonic features (Cox, 1994; Keller and Pinter, 2002). Presence of active tectonic deformation will normally be reflected on the disposition of drainage pattern.

Rivers are largely responsible for shaping the Earth's continental landscapes and river patterns, the spatial arrangements of channels in the landscape, are determined by slope and structure (Twidale, 2004). Rivers have a characteristic pattern and life histories with an youthful stage in the hills, a mature stage in the plains, and the old stage in the coastal zones, in general, and are controlled by the base level of erosion or mean sea level. However, rivers are sensitive to active tectonic deformation of the Earth's surface that induced subtle changes induced by vertical and lateral tectonic movements (Amos and Burbank, 2007; Holbrook and Schumm, 1999; Jain and Sinha, 2005; Ouchi, 1985; Seeber and Gornitz, 1983). Thus, the analysis of the drainage pattern is an important tool in the study of active fault systems (Audemard, 1999; Keller and Pinter, 2002).

Several studies have documented the influence of vertical and strike-slip crustal movements on the channel pattern (Holbrook and Schumm, 1999; Jorgensen, 1990; Ouchi, 1985). Some features that could be identified from these kinds of studies are river diversion, beheaded and captured streams (Grapes and Wellman, 1993), pinched valley (Marple, 1994), local meandering (Howard, 1967), compressed meandering, and knick points in longitudinal profiles (Jain and Sinha, 2005). The other geomorphologic characteristics such as wind gaps, sag ponds, tectonic gutters, broom-shaped river patterns, deflected drainages, and shutter ridges (Audemard, 1999) are also associated with active faulting. Such features generally occur when there is a change in the slope of riverbed induced by structures across it and may act as strong evidence for the presence of active faults/neotectonic lineaments. Following the global trends, there were also attempts to delineate the signatures of active faulting from drainage anomalies in Peninsular India (John and Rajendran, 2008; Praseeda et al., 2015; Ramasamy et al., 2011; Subrahmanya, 1996). Recent studies in Peninsular India shows that even the drainage segments of small rivers are showing change in directions and a drastic change in sinuosity compared to subsequent segments where they are controlled by segments of active faults (Singh et al., 2016).

The basic principle behind morphometry is based on the points that landforms are created by erosional and depositional processes and the geometry of the resulting landform is controlled by the processes that shape them. Over the years, various workers have used quantitative analysis to objectively compare different landforms to identify particular characteristics of the area. Geomorphic variables like bifurcation ratio (Schumm, 1956), drainage density (Horton, 1945), and elongation ratio (Schumm, 1956) have also been found useful for categorizing the area in terms of relative active tectonics (Singh et al., 2016).

2.2.3 Geomorphic Indices

Morphometric analysis of drainage network is found to be an appropriate tool for identifying active faults even from low-relief, densely populated areas (Cox, 1994; John and Rajendran, 2008). Geomorphic indices have been developed as basic reconnaissance tool to identify areas experiencing rapid tectonic deformation. These indices are useful in tectonic studies for quick evaluation of larger areas from topographic maps and/or aerial photographs. Some of the geomorphic indices that are widely used for studies of active tectonics are:

Hypsometric integral—Strahler, 1952
Drainage basin asymmetry—Hare and Gardner, 1985
Stream length gradient index—Hack, 1973
Valley floor valley height ratio—Bull and McFadden, 1977
Mountain front sinuosity—Bull and McFadden, 1977
Sinuosity of river—Leopold et al., 1964
Transverse topography symmetry factor—Cox, 1994

The results of several indices need to be combined with the other geological information to classify the areas as being very active, moderately active, or inactive. Such classification is useful in delineating areas where more detailed field studies could unravel active structures and calculate rates of active tectonic process.

The technique is found useful in both interplate and intraplate settings. Considering the low strain rate of the intraplate settings, geomorphic indices play a very vital role in detecting active faults from continental interiors.

2.2.4 Active Fault Identification in Peninsular India

Southern Peninsular India, part of stable continental regions, is devoid of any dramatic expressions of active fault, though a number of shear zones of Precambrian origin exist. But in 1990, a $M = 6.0$ earthquake occurred near Coimbatore, within one of the major shear zones called Palghat Cauvery shear zone. Several studies were initiated subsequently to $M = 4.3$ Wadakkancheri event to identify causative structure in the wake of heavy damage and loss of property caused by Killari earthquake. The geological study initially identified the sharp turn of the major river in the epicentral area (Fig. 2.2). It is also observed that the East to West flowing Bharathapuzha River changes its direction to Northwest after the turn and is devoid of any meandering where the river is flowing below 20 m msl. A northwest–southeast (NW–SE) trending hillock observed parallel to the trend of the downstream segment. On the southern side of the river, the drainage network shows a subtle adjustment while comparing topography mapped 15 years apart. The study further employed geomorphic indices by dividing the area into 62 subbasins having third- and fourth-order streams to objectively locate the causative structure. Analysis of the satellite image combined with field verification identified a south dipping fault. The geomorphic analysis indicates that the drainage channels in the hanging wall were

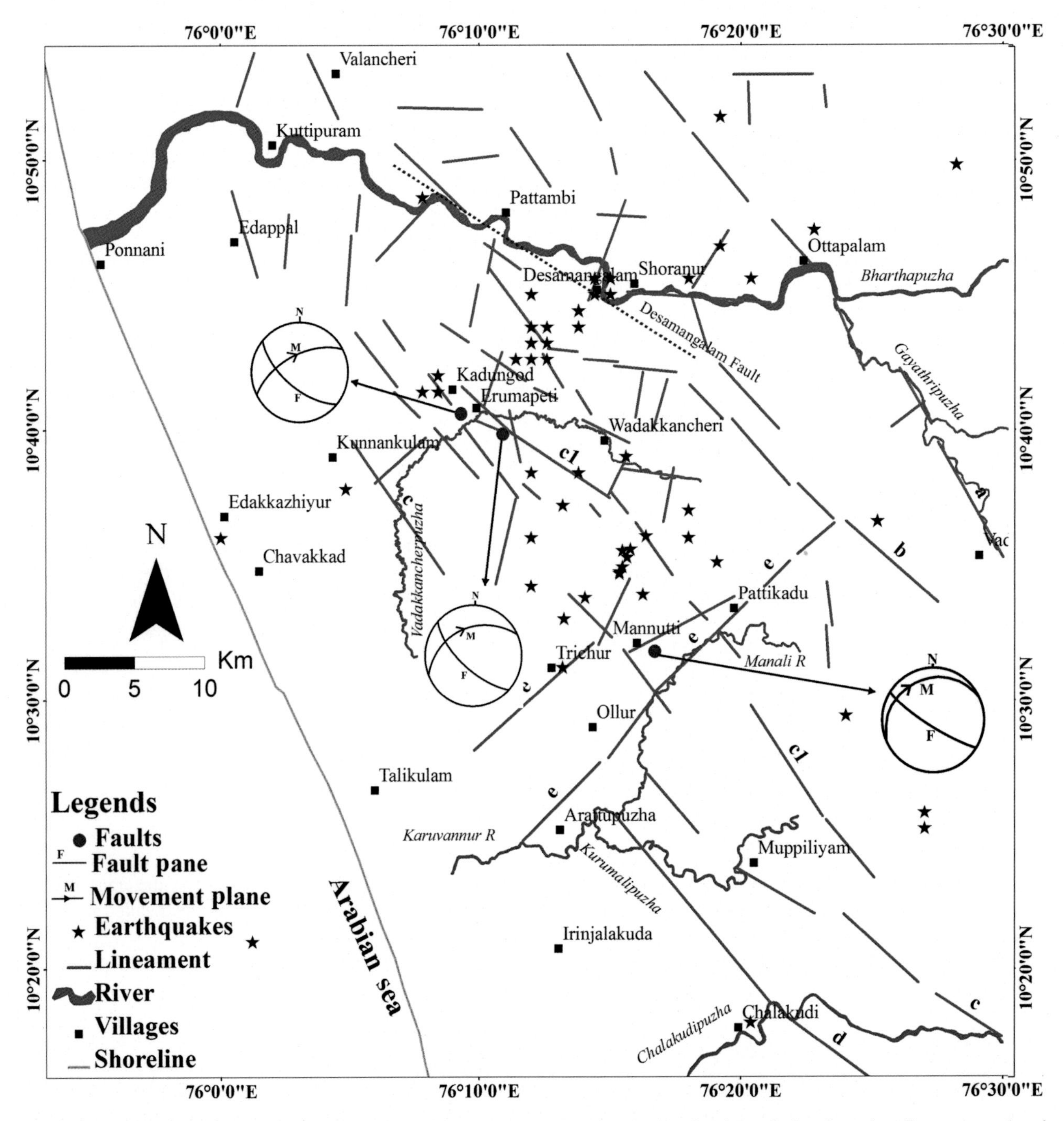

FIGURE 2.2 **Map showing major rivers of the area and lineaments.** Note the change in river direction at the locations where lineaments c, c1, and d cross the rivers and associated change in meandering (see Singh et al., 2016 for further details).

shifted from their original course, forming paleochannels, to adjust with the change in slope induced by the reverse movement of the fault. The valley floor width to valley height ratio calculated for different reaches indicate that the NW–SE segment of the Bharathapuzha shows relatively low values in comparison to the upper reaches of the river, pointing to rejuvenation of the river caused by the neotectonic adjustments of the structure (Fig. 2.2).

The studies continued further south as frequent occurrence of low-magnitude earthquakes noticed around Trissur (Singh et al., 2016). Vadakkancheripuzha, Karuvannur, and Chalakudipuzha are the rivers draining the area. The area is mostly low relief, and all rivers flow with gentle gradient. The study identified segments of NW–SE trending en echelon nature of Periyar fault entering in the study area that intersect the small rivers. If this behavior is influenced by tectonic movements, it is expected to be reflected in river channel parameters. Within a given range of channel gradients, the meandering pattern changes as vertical tectonic movements influence the valley slope. In stable tectonic conditions, alluvial rivers tend to show meandering unless it is affected with active tectonism. The study calculated sinuosity along 22 stretches on different rivers which are intersected by the NW–SE trending lineaments (Fig. 2.2). Geomorphic analysis categorizes the area into six zones based on various geomorphic parameters. The study identified that the bifurcation ratio is relatively high even in low-lying areas where the segments of Periyar lineaments influenced the drainage basins. Elongation ratio too indicates anomalies in the basins close to these lineaments.

2.3 READING PAST SEISMIC EVENTS FROM GEOLOGICAL RECORDS

The determination of the fault activity and the fault classification in active or possibly active faults is based on fault geometry, geological age, morphotectonic signature, and relation with ongoing seismicity. Understanding a fault behavior is important for constraining the magnitude range and frequency of earthquakes that a particular fault is likely to produce. The following sections will highlight the geological information for characterizing fault.

Historical data provide information on past earthquakes only for a few centuries back in time, as the human memory is too short. However, the more recent technique of paleoseismology can trace back to several thousands of years, to obtain the timing and frequency of past events (McCalpin, 2009). Paleoseismological interpretation of trench exposures provide information about recurrence rates of faults, the magnitudes of past earthquakes, time since last earthquake, and the slip rate (Crone and Wheeler, 2000). Most importantly, this geological information can provide a better insight into the source of the earthquake.

Geological investigations applied to the study of past earthquakes include the study of active faulting, seismites (sedimentary structures produced by shaking), liquefaction features, and other effects of earthquake shaking (e.g. rockfalls and landslides) that can be recorded in superficial geological units (McCalpin, 2009). The distinguishable geological indicators of past seismicity can be on-fault and/or off-fault features. On-fault features include fault scarps, sag, or fault ponds, pressure/shutter ridges, landslides, sand blows, etc. The off-fault features include tilted surfaces, uplifted, or subsided shorelines, tsunami deposits sand blows, landslides, broken speleothems, etc.

2.3.1 Timing of the Events

Once the evidences of paleoearthquakes are identified, the detailed geological investigations are to be carried out to determine its timing. Accurate age determination is essential information to evaluate fault activity and recurrence interval. The commonly used isotopic dating methods for stratigraphically controlled sample related with faulting are as follows.

The radiocarbon dating using the accelerator mass spectrometry (AMS)	Gianoa et al. (2000)
The uranium series for samples with carbonate calcium ($CaCo_3$) content	Richards and Dorale (2003)
The electron spin resonance (ESR) for fault gouge	Schwarcz and Lee (2000)
The thermo-luminescence and optically stimulated luminescence for sandy units (quartz and feldspar content)	Forman et al. (1991)
The dating of geomorphic surfaces by cosmogenic Al–Be and chlorine	Cockburn and Summerfield (2004)
The lead-210 isotopes for very young sediments (<100 years)	Noller (2000)

Based on the time of faulting, the faults shall be characterized into active, potentially active, or inactive (Table 2.1). An active fault is the one that has been activated in the Quaternary, or in the Late Pleistocene (0.1–1.2 Ma), and is usually expressed as structural or geomorphological markers on the surface (Yeats, 2012). Although the geological definition of

TABLE 2.1 Terminology Related with Fault Activity. Ages Mentioned as per the International Commission on Stratigraphy

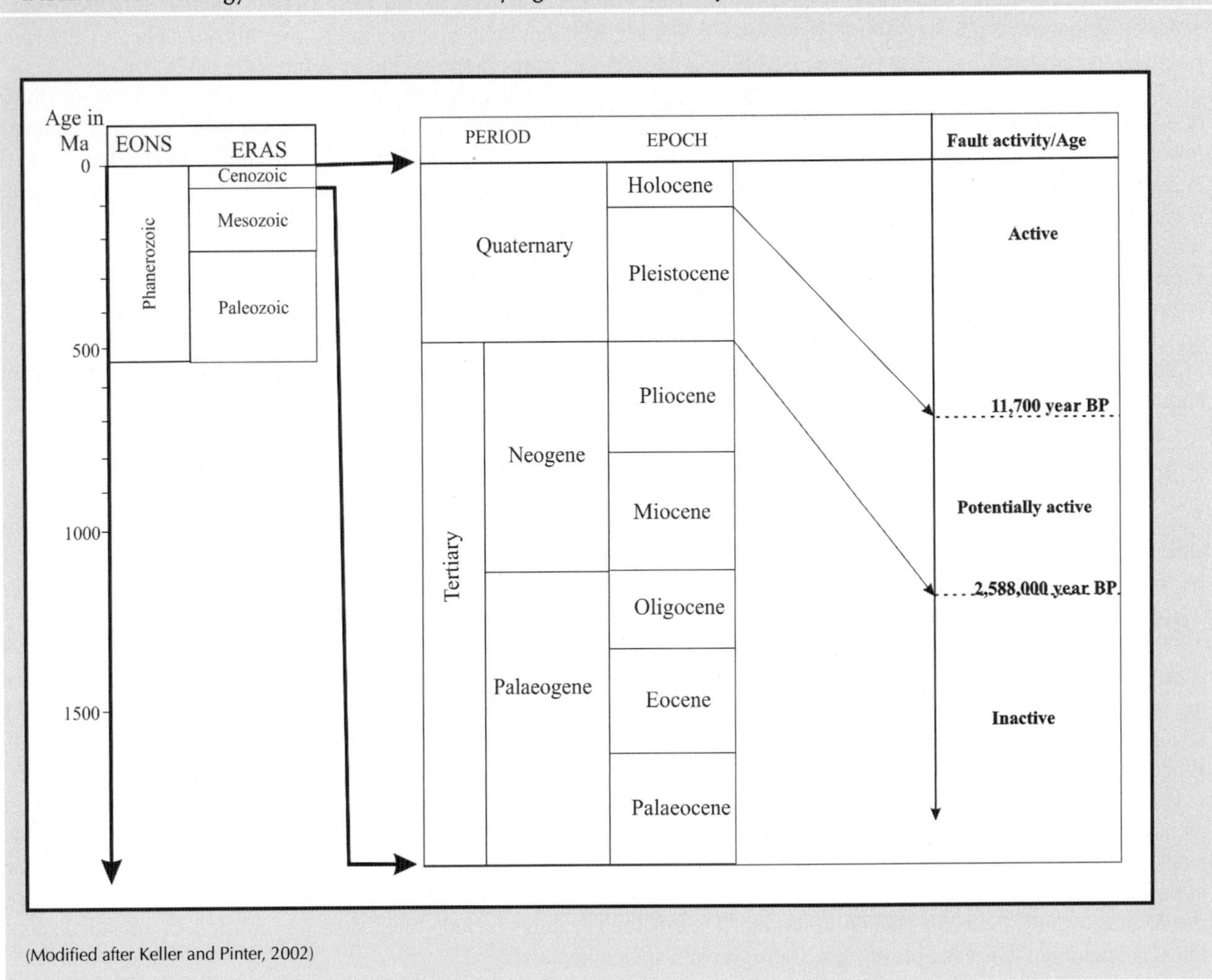

(Modified after Keller and Pinter, 2002)

active fault may span a broader time window, assessing a fault from its potential for reactivation and generation of significant earthquakes entail a more practical approach.

2.4. ON FAULT STUDIES

Damaging earthquakes often produce surface rupture along a fault up to the ground surface, by offsetting layered sediments. Subsequently, new sediments will be deposited across the rupture, creating recognizable horizon that is younger than the earthquake. Fault scarp is defined as a geomorphical expression of fault that dislocated the ground surface. The fault scarp ranges from small, ephemeral slopes created by a single increment of displacement to high bed rock escarpment formed during repeated slip (Stewart and Hancock, 1990). They usually identify initially through remote-sensing studies (Hunter et al., 2011; Oskin et al., 2012) followed by field surveying (Bucknam and Anderson, 1979; Burbank and Anderson, 2011). Since the fault scarps are the most common direct evidence of the fault's capacity to produce damaging earthquakes, they are widely assessed as part of seismic hazard assessment. Fault scarp profiles are the main source of data for estimating vertical displacement and age of faulting in dip-slip faults. However, fault scarp formed during earthquakes are soon exposed to weathering and erosional processes.

Morphological dating is a method that is derived from the simple observation that the scarp is function of its age (Wallace, 1977). The geomorphical relations between fault displacement, original scarp morphology, and present (degraded) scarp morphology need to be understood to derive constistent fault displacement data (Avouac, 1993; McCalpin, 2009).

2.4.1 Studies from Sedimentary Terrains

The techniques or basic method of geological investigations applied to primary evidence along continental fault traces in unconsolidated sediments are well described by McCalpin (2009). The study begins with the mapping of geomorphical features related to understanding the nature of faulting. Once the direct evidence of faulting is identified, geologist's will use trenching technique to evaluate the fault. Over the years, such studies have successfully updated the history of earthquakes over the past several hundred to a few thousand years on many active faults around the world.

2.4.2 Studies in the Central Seismic Gap of the Himalayas

The Himalayas, formed due to the collision between Indian and Eurasian plates, is one of the most active interplate regions of the Earth which has witnessed many great and large earthquakes in the recent past. Space–time patterns of seismicity in the Himalayan fault system identified regions of seismic quiescence in the historic period (Khattri, 1987). The 500–800 km segment of fault system between the great earthquakes of 1934 Bihar–Nepal and 1095 Kangara earthquake is called central seismic gap and received much attention for understanding the seismic hazard, due to the absence of great earthquakes in the historic period, to compensate the strain release that was calculated across the collision zone (Bilham et al., 2001; Rajendran and Rajendran, 2005).

As part of understanding the status of this segment, a 34-m long 3-m wide and 2.5–4.5-m deep trench was dug across a 15-m scarp in the frontal belt of Himalaya (Fig. 2.3). The trench exposed both older and younger deposits. The bottommost layer is thick, well-sorted, rounded to subrounded boulders and gravel representing high energy channel bed deposit. Beds of sandy slit and silty sand occurred over the boulder bed. The upper units are well developed only in the southern part of the trench. The laterally traceable sedimentary units in the trench section are offset by distinct fault strands. The fault strands exposed were buried by thin sedimentary layers. Since the 15-m fault scarp is a question of magnitude (implying to $M = 9.0$), the nature of the strata studied carefully and made a retrodeformation analysis to understand the amount of displacement across the faults (Fig. 2.4). The analysis found that even after leveling top layers, bottom layers show offset, indicating the presence of a previous displacement along the faults. The study found two events of movements separated by colluvium and sedimentation (Rajendran et al., 2015). The study proved the ability of geological data in contribution to the solving of the issue of seismic hazard in central seismic gap.

2.4.3 Fault Studies from Crystalline Rocks

Faulting in consolidated sedimentary rocks and crystalline rocks produces a variety of fault rocks ranging from mylonite to fault gouge (Scholz, 2002). A synoptic view of fault rocks with depth and temperature is shown in Fig. 2.4. In crystalline

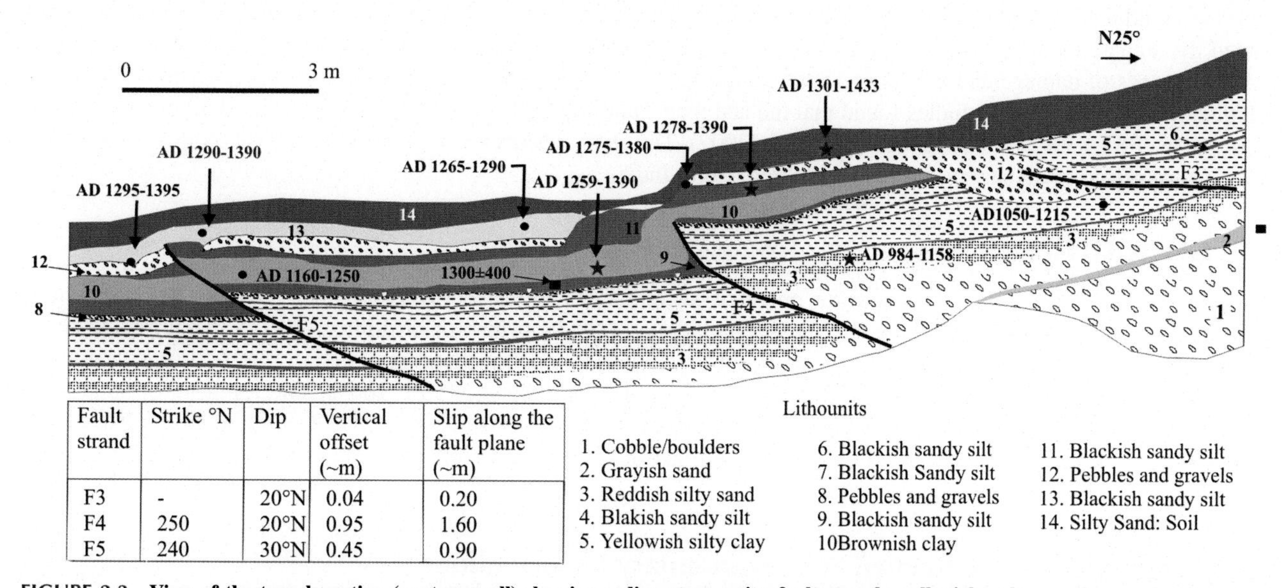

Fault strand	Strike °N	Dip	Vertical offset (~m)	Slip along the fault plane (~m)
F3	-	20°N	0.04	0.20
F4	250	20°N	0.95	1.60
F5	240	30°N	0.45	0.90

FIGURE 2.3 **View of the trench section (western wall) showing sedimentary units, fault strands, colluvial wedges, and ages of sedimentary strata.** Offset measurements on the fault strands are shown in a table (see Rajendran et al., 2015 for further details).

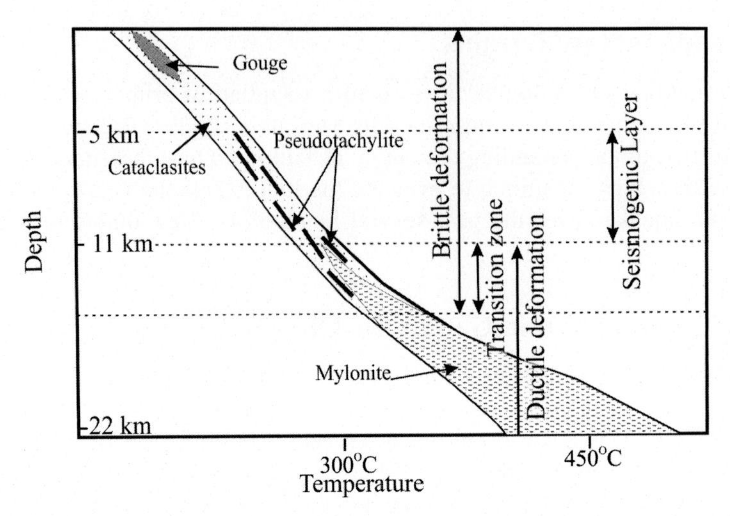

FIGURE 2.4 **Model of continental fault zone representing the fault rocks with respect to depth, concept after Scholz (2002).** Note that the tip of the fault zone is represented by clay gouge.

terrains, the study of fault rocks (breccia, cataclasite, and gouge) in crystalline terrains could reveal the deformation mechanisms and pressure temperature conditions. At upper levels of the crust, the deformation is dominated by cataclasis, which involves the brittle fragmentation of mineral grains with rotation of grain fragments accompanied by frictional grain boundary sliding and dilatancy (Fig. 2.4). The fine-grained output of faulting near the surface is the gouge, which is essentially the rock powder formed during faulting. The brittle fault zones in crystalline bedrocks are, in general, consists of a major slip planes (fault core) bounded by a zone of fractured rocks (damage zone) (Caine et al., 1996).

2.4.4 Studies in Desamangalam Fault

As mentioned earlier, geological studies subsequent to the 1994 Wadakkancheri earthquake identified geomorphic anomalies associated with the drainage network of one of the major rivers (Bharathapuzha) in Peninsular India (Fig. 2.2) and identified the 30-m long causative fault. The faulting created small water fall and rock ridge along the strike direction. A small second-order drainage running across the fault also created wide valley in the upstream and a narrow valley in downstream (John and Rajendran, 2009). The brittle fault exposed at Desamangalam (Fig. 2.5) shows distinct fault core with gouge zone of consolidated and unconsolidated nature. The damage zone shows numerous fractures with many of them sealed with secondary minerals. It identified the evidence of earlier faulting that was disrupted by subsequent events. However, at this site, we could recognize the evidence of earlier episodes through close examination of the fault rocks. The study discriminated the signatures both macroscopically and through mineralogical and microscopic studies. The study also evaluated microfractures of damage and the consolidated gouge zones to unravel the sequence of events that shaped the fault zone (Fig. 2.6). Interestingly, the studies found that the last major faulting event along this fault is dated as around 430 ± 43 ka through the ESR dating of loose gouge (Rao et al., 2002). The above results are of great consequences considering the very long return period of damaging earthquakes in Peninsular India.

2.5 OFF FAULT FEATURES

Earthquake shaking would create a wide range of secondary features in the geological record. Soft sedimentary deformations, including liquefaction, are the main evidence of paleoevents (McCalpin, 2009). The nature and size of these features also depend on the size of the event as well as epicentral distance. Though rare, earthquake-induced landslides are useful in identifying the time of paleoearthquakes (Jibson, 1996, 2009). The toppling of stalagmites is also used as an evidence for paleoseismic interpretation (Forti and Postpischl, 1984; Rajendran et al., 2016a). The following section will concentrate on the use of liquefaction and related soft sedimentary deformation for identification of paleoseismic events.

2.5.1 Liquefaction and Related Soft Sedimentary Deformations

Soft-sediment deformations are created in unconsolidated sediment and are relatively common in sandy sediments (Lowe, 1975, 1976; Owen and Moretti, 2011; Van Loon, 2009). These are essentially governed by the size of earthquake,

FIGURE 2.5 **Fault exposed near the river bend shown in Fig. 2.2.** The fault zone comprises the main slip plane "F2" (fault core) and two subparallel fractures "F1 and F3" which bound the damage zone (John and Rajendran, 2009).

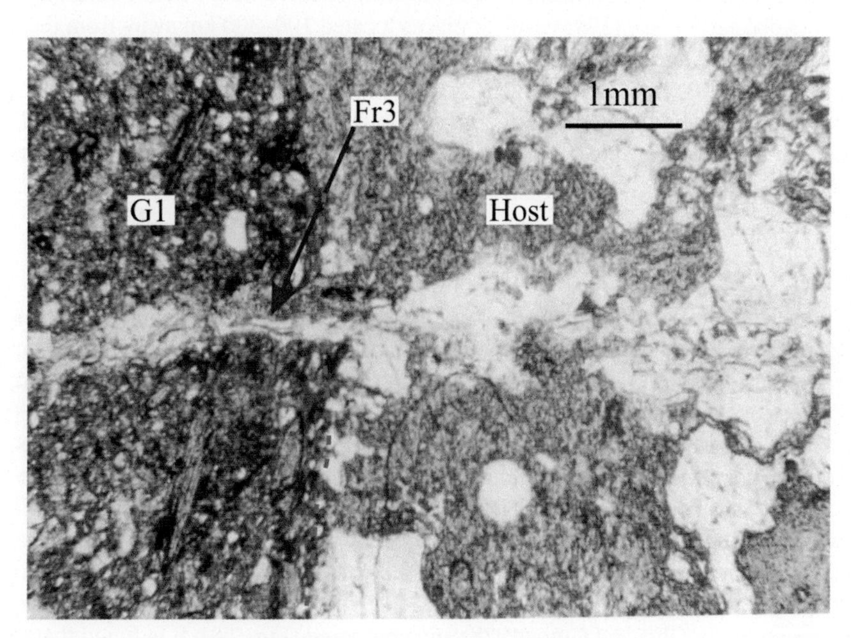

FIGURE 2.6 **Photomicrograph showing clinoptilolite bearing fracture cutting both host rock and gouge zone (G1).** The contact zone between consolidated gouge (G1) and host rock is shown by red dashed line (John and Rajendran, 2009).

epicentral distance, and geology of the area. For liquefaction-based paleoseismic analyses, proper interpretation of paleoliquefaction evidence is critical, with the difficulty of interpretation increasing for sites of recurrent liquefaction induced by earthquakes spaced closely in time. The combination of well-documented liquefaction response, detailed site specifically recorded ground motions, and detailed understanding of subsurface lithology can overcome the uncertainties related to paleoliquefaction studies. This information can provide vital input for a quantitative back-analysis to estimate the causative earthquake's magnitude. Identifying different generations of multiple features requires deciphering the crosscutting relationships between the sedimentary structures or features and their order of superposition (Rajendran et al., 2016b).

2.5.2 Studies in Gangetic Plains

The study sites in northern Bihar were selected based on the historical accounts and reports of the liquefaction during the 1934 earthquake (Rajendran et al., 2016a,b) considering the fact that there are examples of multiple earthquakes occurring at different time intervals by recurring liquefaction in a given site (Obermeier et al., 1995; Sims and Garvin, 1995).

Therefore, the studies were focused at different sites in northern Bihar alluvial plains, where the 1934 Bihar–Nepal earthquake generated liquefaction, to understand the return period of Himalayan earthquakes. Different generations of seismically induced paleoliquefaction features or sand blows (dikes, sills, sand vents, and cones) were identified based on their cross-cutting relationships or from the beds representing the periods of interseismic deposition that separate the event layers (Fig. 2.7). The study suggests that the combined recurrence interval for moderate-to-large earthquakes in the Bihar segment could be 124 ± 63 years (Rajendran et al., 2016b).

2.6 LOCAL SITE CONDITIONS

Seismic hazard analyses for major civil engineering structures are usually conducted for particular sites of interest. Thus, the hazard assessments should be site-specific and as fault-specific as possible. This requires that each fault be characterized by its own earthquake recurrence behavior. To mitigate the seismic hazard, it is necessary to define a correct response in terms of both peak ground acceleration and spectral amplification. These factors are highly dependent on the local soil conditions and on the source characteristics of the expected earthquakes (Pande and Parvez, 2008).

The ground motion amplification and soil liquefaction are the two main factors responsible for severe damage to the built environment. The damages due to large earthquakes are controlled by geological, subsurface soil. Wood (1908) was the first one to notice the role of local geology in enhancement of earthquake ground motions based on the damage pattern of 1906 San Francisco earthquake. The thickness and geotechnical properties of the soil, and the nature of the underlying rock and frequency content of seismic waves, are some of the factors that govern the site amplification (Nakagawa et al., 1996). The enhancement of ground motions due to geological conditions were found during Mexico earthquake (1985), Loma Prieta earthquake (1989), and Kobe earthquake (1995) even at places located 100–300 km away from the epicenter. In India too, some recent moderate earthquakes, occurred during last three decades, viz. 1988 Bihar–Nepal (Mw 6.6), 1991 Uttarakashi (Mw 6.8), 1999 Chamoli (Mw 6.8), 1993 Latur (Mw 6.2), and 1997 Jabalpur (Mw. 6.0), etc. produced heavy damages due to seismically induced ground motion amplification and soil liquefaction. During 2001, in Bhuj Earthquake too, higher damages are reported 250 km from the epicenter around Ahmedabad due to enhanced ground motions in soft geological formations (Pande and Kayal, 2003).

2.6.1 Chandigarh Microzonation Studies

As part of site-specific seismic hazard analysis of Union territory of Chandigarh, geotechnical studies were undertaken. Chandigarh falls in Zone IV of the Seismic Zoning Map of India (IS, 1893–2002), which implies that the maximum earthquake shaking of the area corresponds to a seismic intensity of VIII on MSK-64 scale. Corresponding to the above intensities, the expected ground motions in terms of ground acceleration may vary between 0.05 and 0.2 g (Kandpal et al., 2009).

The Himalayan frontal thrust (HFT) is the nearest tectonic structure to the area and it passes through the northern boundary of the Union Territory which separates Indo-Gangetic Alluvium from Siwalik rocks. The main boundary thrust passes at about 30 km east and the Main Central Thrust lies about 130 km NE of the area are the other major tectonic structures. The area witnessed high intensity (VI–VII on MM Scale) during 1905 Kangra earthquake (Pande et al., 1999).

Twenty-four shallow boreholes up to 30 m depth were conducted in the area of study to determine the geotechnical properties of subsurface strata (Kandpal et al., 2009). Shear wave velocities of the subsurface layers have been computed based on the formula proposed by Ohta and Goto (1978) and contoured (Fig. 2.8). It is found that the average shear wave velocity varies from less than 216–305 m/s and predominant frequencies vary from 1.8 to 2.54 Hz. The subsurface lithological units encountered along the bore holes, the geotechnical properties of the sediments, and the ground table conditions in the area reveal that there are chances of liquefaction in some of the sectors in the Central and Western parts of Chandigarh. The grain size analysis of the subsurface sediments of these locations indicates that many sand layers fall within the limits of the most liquefiable sands as suggested by Tsuchida and Hayashi (1971).

2.7 SUMMARY

This chapter focused on the geological information required for seismic hazard assessment. The hazards associated with earthquakes include ground motion, ground breaks, land slide, and liquefaction. An assessment of the earthquake hazard for a location of interest is a prerequisite for any major civil engineering structure. But in most cases, historic and instrumental data are insufficient for assessing long-term earthquake potential. In order to extend the earthquake catalog beyond historic period, geological information is widely used.

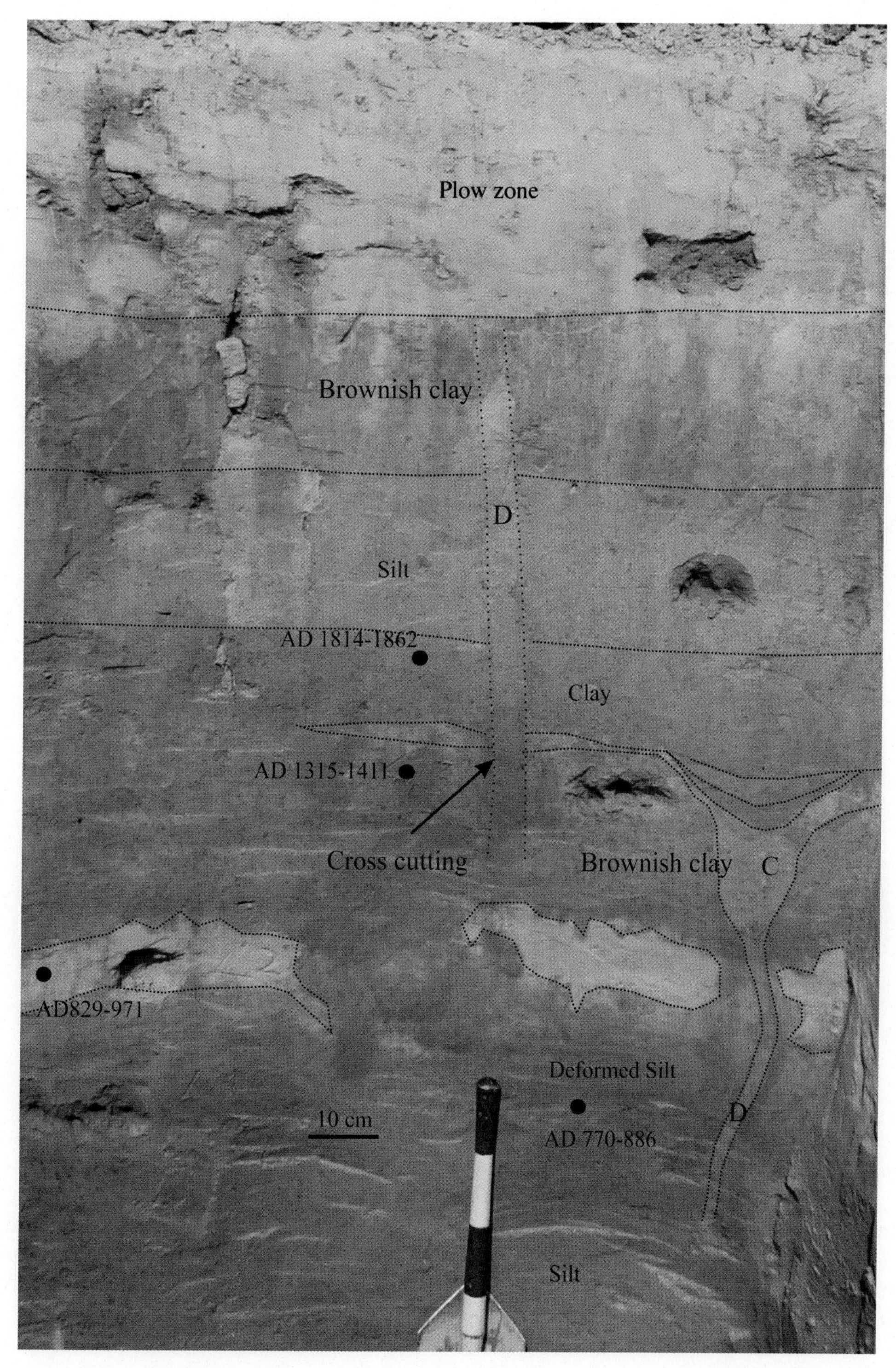

FIGURE 2.7 **Section (southern wall) at Simra Village; sketch highlighting the features (dikes (D) and craters (C)) (more details in Rajendran et al., 2016b).**

To read the past seismic information, geologists use an integrated approach comprising geomorphology, structural geology, geochronology, sedimentology, pedology, etc. Once the on-fault feature is identified through geomorphological studies, geologists usually adopt trenching investigations to unravel the stratigraphic relations and chronology of the events. Similarly, off-fault features are also studied and located through detailed evaluation of geomorphology.

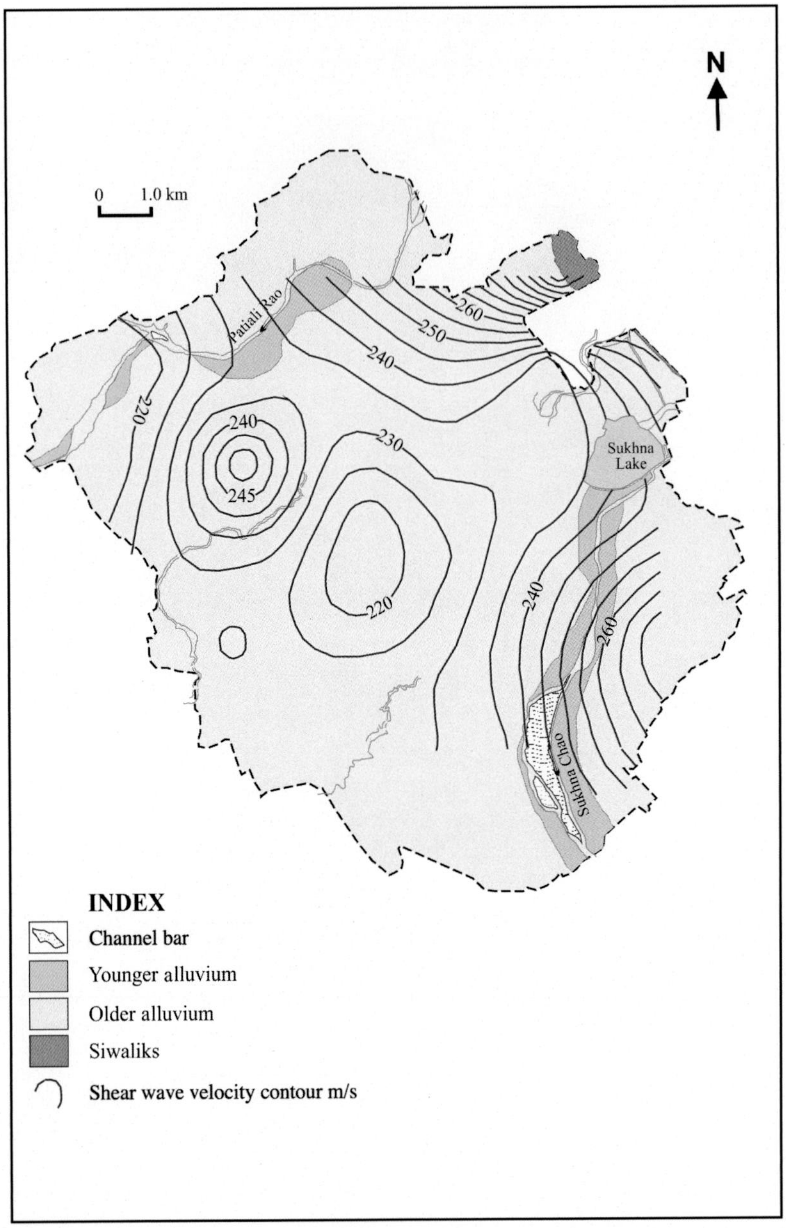

FIGURE 2.8 **Shear wave velocity map up to 30 m depth, Chandigarh area (Kandpal et al., 2009).**

It is also known that variations of ground shaking in space, amplitude, and duration during an earthquake, at a given site, are influenced by the geological formations and their geotechnical properties. The strong ground shaking is also modified by the basin effects and lateral geological discontinuities. Thus, regional and local geological information forms vital input for any earthquake hazard assessment program.

REFERENCES

AERB, 1990. Seismic studies and design basis ground motion for nuclear power plant studies. Guide No. AERB/SG/S-11.

Amos, C.B., Burbank, D.W., 2007. Channel width response to differential uplift. J. Geophys. Res. 112, F02010. doi: 10.1029/2006JF000672.

Audemard, F.A., 1999. Morpho-structural expression of active thrust fault systems in the humid tropical foothills of Colombia and Venezuela. Z. Geomorphol. 118, 1–18.

Avouac, J.-P., 1993. Analysis of scarp profiles: evaluation of errors in morphologic dating. J. Geophys. Res. 98, 6745–6754.

Bilham, R., Gaur, V.K., Molnar, P., 2001. Himalayan seismic hazard. Science 293, 1442–1444.

Bucknam, R.C., Anderson, R.E., 1979. Estimation of fault scarp ages from a scarp height–slope angle relationship. Geology 7, 11–14.

Bull, W.B., McFadden, L., 1977. Tectonic geomorphology north and south of the Garlock fault, California. In: Dohering, D.O. (Ed.), Geomorphology in Arid Regions. Geomorphology State University, Binghamton, NY, pp. 115–138.

Burbank, D.W., Anderson, R.S., 2011. Tectonic Geomorphology, second ed. Wiley-Blackwell. doi: 10.1002/9781444345063, p. 274.

Caine, J.S., Evans, J.P., Forster, C.B., 1996. Fault zone architecture and permeability structure. Geology 24, 1025–1028.

Caputo, R., Pavlides, S.B., 2008. Earthquake geology: methods and applications. Tectonophysics 453, 1–6.

Cockburn, H.A.P., Summerfield, M.A., 2004. Geomorphological applications of cosmogenic isotope analysis. Prog. Phys. Geogr. 28, 1–42.

Cox, R.T., 1994. Analysis of drainage basin asymmetry as a rapid technique to identify areas of possible Quaternary tilt-block tectonics: an example from the Mississippi embayment. Geol. Soc. Am. Bull. 106, 571–581.

Crone, A.J., Wheeler, R.L., 2000. Data for Quaternary faults, liquefaction features, and possible tectonic features in the Central and Eastern United States, east of the Rocky Mountain front. US Geological Survey Open-File Report. 00-0260, 341 pp. (May be downloaded free as a 9 MB PDF file from URL http://greenwood.cr.usgs.gov/pub/open-file-reports/ofr-OO-0260/).

Forman, S.L., Nelson, A.R., McCalpin, J.P., 1991. Thermoluminescence dating of fault-scarp-derived colluvium: deciphering the timing of paleoearthquakes on the Weber Segment of the Wasatch Fault Zone, North Central Utah. J. Geophys. Res. 96 (B1), 595–605.

Forti, P., Postpischl, D., 1984. Seismotectonic and paleoseismic analysis using Karst sediments. Mar. Geol. 55, 145–161.

Gianoa, S.I., Maschioa, L., Alessiob, M., Ferranti, L., Improta, S., Schiattarella, M., 2000. Radiocarbon dating of active faulting in the Agri high valley, Southern Italy. J. Geodyn. 29, 371–386.

Grapes, R.H., Wellman, H.W., 2000. Field guide to the Wharekauhau Thrust (Palliser Bay) and Wairarapa Fault (Pigeon Bush). Geol. Soc. N. Z. Misc. Publ. 79B, 27–44.

Hack, J.T., 1973. Stream-profile analysis and stream-gradient indices. J. Res. U. S. Geol. Surv. 1, 421–429.

Hare, P.W., Gardner, T.W., 1985. Geomorphic indicators of vertical neotectonism along converging plate margins, Nicoya Peninsula, Costa Rica. In: Morisawa, M., Hack, J.T. (Eds.), Tectonic Geomorphology. Allen and Unwin, Boston, MA, pp. 75–104.

Holbrook, J., Schumm, S.A., 1999. Geomorphic and sedimentary response of the river to tectonic deformation: a brief review and critique of a tool for recognizing subtle epierogenic deformation in modern and ancient setting. Tectonophysics 305, 287–306.

Horton, R.E., 1945. Erosional development of streams and their drainage basins: hydrophysical approach to quantitative morphology. Geol. Soc. Am. Bull. 56, 275–370.

Howard, A.D., 1967. Drainage analysis in geologic interpretations: a summation. Am. Assoc. Pet. Geol. Bull. 51, 2246–2259.

Hunter, L.E., Howle, J.F., Rose, R.S., Bawden, G.W., 2011. LiDAR-assisted identification of an active fault near Truckee, California. Bull. Seismol. Soc. Am. 101, 1162–1181.

IS, 1893–2002. Criteria for earthquake resistance design of structures. Bureau of Indian Standards (BIS), New Delhi.

Jain, V., Sinha, R., 2005. Response of active tectonics on the alluvial Baghmati River, Himalayan foreland basin, Eastern India. Geomorphology 70, 339–356.

Jibson, R.W., 1996. Use of landslides for paleoseismic analysis. Eng. Geol. 43, 291–323.

Jibson, R.W., 2009. Using landslides for paleoseismic analysis. McCalpin, J.P. (Ed.), Paleoeismology, International Geophysics Series, 95, Elsevier, pp. 565–601.

John, B., Rajendran, C.P., 2008. Geomorphic indicators of Neotectonism from the Precambrian Terrain of Peninsular India: a study from the Bharathapuzha Basin, Kerala. Journal of Geological Society India 71, 827–840.

John, B., Rajendran, C.P., 2009. Evidence of episodic brittle faulting in the cratonic part of the Peninsular India and its implications for seismic hazard in slow deforming regions. Tectonophysics 471, 240–252.

John, B., Rao, D.T., 2014. Seismotectonic evaluation of critical civil engineering facilities: the standard practice. In: National Seminar on Innovative Practices in Rock Mechanics. pp. 133–138.

Jorgensen, D.W., 1990. Adjustment of alluvial river morphology and process to localized active tectonics. PhD thesis, Colorado State University, Fort Collins, CO, USA.

Kandpal, G.C., John, B., Joshi, K.C., 2009. Geotechnical studies in relation to seismic microzonation of Union Territory of Chandigarh. J. Indian Geophys. Union 13, 75–83.

Keller, E.A., Pinter, N., 2002. Active Tectonics Earthquakes—Uplift and Landscape. Prentice-Hall, New Jersey, p. 362.

Khattri, K.N., 1987. Great earthquakes, seismicity gaps and potential for earthquake disaster along the Himalaya plate boundary. Tectonophysics 138, 79–92.

Leopold, L.B., Wolman, M.G., Miller, J.P., 1964. Fluvial Processes in Geomorphology. Freeman, San Francisco, 511.

Lowe, D.R., 1975. Water escape structures in coarse-grained sediments. Sedimentology 22, 157–204.

Lowe, D.R., 1976. Subaqueous liquefied and fluidized sediment flows and their deposits. Sedimentology 23, 285–308.

Marple, R.T., 1994. Discovery of a possible seismogenic fault system beneath the Coastal Plain of South and North Carolina from integration of river morphology and geological and geophysical data. PhD dissertion, University of South Carolina, Columbia, 354 pp.

McCalpin, J. (Ed.), 2009. Paleoseismology. Academic Press, London, UK, p. 588.

Nakagawa, K., Shiono, K., Inoue, N., Sano, M., Jan 1996. Geological characteristics and problems in and around Osaka basin as a basis for assessment of seismic hazards. Soils Found. 36, 15–28.

Noller, J.S., 2000. Lead-210 geochronology. In: Noller, J.S., Sowers, J.M., Lettis, W.R. (Eds.), Quaternary Geochronology—Methods and Applications. AGU Reference Shelf, pp. 115–120, 4.

Obermeier, S.F., Jacobson, R.B., Smoot, J.P., Weems, R.E., Gohn, G.S., Monroe, J.E., Powars, D.S., 1995. Earthquake-induced liquefaction features in the coastal setting of South Carolina and in the fluvial setting of the New Madrid seismic zone. US Geol Surv Prof Paper 1504, p. 44.

Ohta, Y., Goto, N., 1978. Emperical shear wave velocity equation in terms of characteristics soil indexes. J. Earthquake Eng. Struct. Dyn. 6, 167–187.

Oskin, M.E., Arrowsmith, J.R., Hinojosa Corona, A., Elliott, A.J., Fletcher, J.M., Fielding, E.J., Gold, P.O., Gonzalez Garcia, J.J., Hudnut, K.W., Liu-Zeng, J., Teran, O.J., 2012. Near-field deformation from the El Mayor–Cucapah earthquake revealed by differential LIDAR. Science 335, 702–705.

Ouchi, S., 1985. Response of alluvial rivers to slow active tectonic movement. Geol. Soc. Am. Bull. 96, 504–515.

Owen, G., Moretti, M., 2011. Identifying triggers for liquefaction-induced soft-sediment deformation in sands. Sedim. Geol. 235, 141–147.

Pande, P., Kayal, J.R. (Eds.), 2003. Kutch (Bhuj) earthquake, 26th January 2001. Geol. Surv. India Spec. Publ. 76, p. 272.

Pande, P., Parvez, I.A., 2011. Seismic microzonation: the Indian scene. Glimpses Geosci. Res. India 235, 141–147.

Pande, P., Gupta, S.K., Singh, B.K., Joshi, K.C., Sharda, Y.P., Singh, J., 1999. A report on seismotectonic evaluation of Kangra Block, Himachal Pradesh. Unpublished GSI Report.

Praseeda, E., John, B., Srinivasan, C., Singh, Y., Divyalakshmi, K.S., Samui, P., 2015. Thenmala fault system, Southern India: implication to Neotectonics. J. Geol. Soc. India 86, 391–398.

Rajendran, C.P., Rajendran, K., 2004. Towards better seismic hazard assessment: need for an integrated approach. In: Valdiya, K.S. (Ed.), Special Volume "Coping with Natural Hazards: Indian Context". The National Academy of Sciences, Allahabad, pp. 59–70.

Rajendran, C.P., Rajendran, K., 2005. The status of central seismic gap: a perspective based on the spatial and temporal aspects of the large Himalayan earthquakes. Tectonophysics 395, 19–39.

Rajendran, C.P., John, B., Rajendran, K., 2015. Medieval pulse of great earthquakes in the central Himalaya: viewing past activities on the frontal Belt. J. Geophys. Res.—Solid Earth 120 (3), 1623–1641. doi: 10.1002/2014JB011015.

Rajendran, C.P., Sanwal, J., Morell, K.D., Sandiford, M., Kotlia, B.S., Hellstrom, J., Rajendran, K., 2016a. Stalagmite growth perturbations from the Kumaun Himalaya as potential earthquake recorders. J. Seismol. doi: 10.1007/s10950-015-9545-5.

Rajendran, C.P., John, B., Rajendran, K., Sanwal, J., 2016b. Liquefaction record of the great 1934 earthquake predecessors from the North Bihar alluvial plains of India. J. Seismol. doi: 10.1007/s10950-016-9554-z.

Ramasamy, S.M., Kumaran, C.J., Selvakumar, R., Saravanavel, J., 2011. Remote sensing revealed drainage anomalies and related tectonics of South India. Tectonophysics 501, 4151.

Rao, T.K.G., Rajendran, C.P., Mathew, G., John, B., 2002. Electron spin resonance dating of fault gouge from Desamangalam, Kerala: evidence for Quaternary movement in Palghat gap shear zone. Proc. Indian Acad. Sci. (Earth Planet. Sci.) 111, 103–113.

Richards, D.A., Dorale, J.A., 2003. Uranium-series chronology and environmental applications of speleothems. Rev. Mineral. Geochem. 52, 407–460.

Scholz, C.H., 2002. The Mechanics of Earthquakes and Faulting. Cambridge University Press, Cambridge. doi: 10.1017/CBO9780511818516, 471.

Schumm, S.A., 1956. Evolution of drainage systems and slopes in badlands at Perth Amboy, New Jersey. Geol. Soc. Am. Bull. 67, 597–646.

Schwarcz, H.P., Lee, H.-K., 2000. Electron Spin resonance for fault rocks. In: Noller, J.S., Sowers, J.M., Lettis, W.R. (Eds.), Quaternary Geochronology—Methods and Applications. AGU Reference Shelf 4, pp. 177–186.

Schwartz, D.P., Coppersmith, K.J., 1986. Seismic hazards: New trends in analysis using geologic data. In: Wallace Chairman, R.E. (Ed.), Active tectonics: Studies in geophysics, Natl. Acad. Press, Washington, DC, pp. 215–230.

Seeber, L., Gornitz, V., 1983. River profiles along the Himalayan arc as indicators of active tectonics. Tectonophysics 92, 335–367.

Sims, J.D., Garvin, C.D., 1995. Recurrent liquefaction induced by the 1989 Loma Prieta earthquake and 1990 and 1991 aftershocks: implications for paleoseismicity studies. Bull. Seismol. Soc. Am. 85, 51–65.

Singh, Y., John, B., Ganapathy, G.P., George, A., Harisanth, S., Divyalakshmi, K.S., Sreekumari, K., 2016. Geomorphic observations from southwestern terminus of Palghat Gap, South India and their tectonic implications. J. Earth Syst. Sci. doi: 10.1007/s12040-016-0695-9.

Stewart, I.S., Hancock, P.L., 1990. What is fault scarp? Episodes 13, 25–253.

Strahler, A.N., 1952. Hypsometric (area-altitude) analysis of erosional topography. Geol. Soc. Am. Bull 63, 1117–1142.

Subrahmanya, K.R., 1996. Active intraplate deformation in south India. Tectonophysics 262, 231–241.

Tsuchida, H., Hayashi, S., 1971. Estimation of liquefaction potential of sandy soils. In: Proceedings of the Third Joint Meeting, US–Japan Panel on Wind and Seismic Effects. UJNR, Tokyo. pp. 91–101.

Twidale, C.R., 2004. River patterns and their meaning. Earth Sci. Rev. 67, 159–218.

Van Loon, A.J., 2009. Soft-sediment deformation structures in siliciclastic sediments: an overview. Geologos 15, 3–55.

Wallace, R.E., 1977. Profiles and ages of young fault scarps, north-central Nevada. Geol. Soc. Am. Bull. 88, 1267–1281.

Wood, H.O., 1908. Distribution of apparent intensity in San Francisco, in the California Earthquake of April 18, 1906. Report of the state earthquake investigation commission, vol. 87, Carnegie Institute of Washington Publishing, Washington, DC, pp. 220–245.

Yeats, R., 2012. Active Faults of the World. Cambridge University Press doi: 10.1017/CBO9781139035644.

Chapter 3

Assessment of Soil Liquefaction Based on Capacity Energy Concept and Back-Propagation Neural Networks

Wengang Zhang*, Anthony T.C. Goh**
**Chongqing University, Chongqing, China; **Nanyang Technological University, Singapore*

3.1 INTRODUCTION

One of the major causes of damage to civil engineering structures during earthquakes is due to liquefaction of loose saturated sand and silty sand deposits. Several procedures have been developed to evaluate the liquefaction potential in the field. The available evaluation procedures can be categorized into three main groups: (1) stress-based procedures, (2) strain-based procedures, and (3) energy-based procedures.

The stress-based procedure (Seed and Idriss, 1971; Whitman, 1971) is the most widely adopted method for liquefaction assessment. The method is mainly empirical and is based on laboratory and field observations. The shear stress level and the number of cycles are the major criteria in this approach. In order to correlate the earthquake actual motion to laboratory harmonic loading conditions, the equivalent stress intensity and the number of cycles have to be defined (Seed and Idriss, 1971). Seed et al. (1975) selected the equivalent stress as 65% of the maximum shear stress induced in the earth structure while Ishihara and Yasuda (1975) proposed 57% rather than 65% for 20 cycles of loading. Some probabilistic frameworks for assessing liquefaction potential of soils based on in situ tests such as the cone penetration test or standard penetration test have also been proposed (Juang et al., 1999, 2001, 2012; Moss et al., 2006; Boulanger and Idriss, 2012). Despite the fact that the stress-based procedure has been continuously revised and extended in subsequent studies and the database of liquefaction case histories expanded, the uncertainty concerning random loading still persists (Green, 2001; Baziar and Jafarian, 2007).

Dobry et al. (1982) proposed the strain-based procedure as an alternative to the empirical stress-based procedure. This method was derived from the mechanics of two interacting idealized sand grains and then generalized for natural soil deposits (Green, 2001; Baziar and Jafarian, 2007). It is based on the hypothesis that pore water pressure initiates to develop when the shear strain surpasses a threshold shear strain, which is shown to be approximately 0.01%, irrespective of sand type, relative density, initial effective confining pressure and sample preparation method. Although this strain-based approach is theoretically reasonable, it is less popular than the stress-based procedure due to the fact that the strain approach only estimates the initiation of pore pressure buildup which is essential for liquefaction to occur, but does not necessarily imply that liquefaction will occur. The main deficiency of this method is the greater difficulty of estimating the cyclic strain compared with the cyclic shear stress (Seed, 1980).

Davis and Berrill (1982) introduced an energy-based approach for liquefaction potential assessment in which the energy content of an earthquake is compared with the amount of dissipated energy required for soil liquefaction, known as "capacity energy". The basic elements of both the stress and strain methods are incorporated in the formulation of the energy-based method. The amount of total strain energy at the onset of liquefaction can be obtained from laboratory testing or field records. In a typical cyclic laboratory test, the stress, strain, and pore pressure time histories are recorded. Hysteresis loops can be generated from these stress and strain time histories. Fig. 3.1 illustrates a typical hysteresis loop (Green, 2001).

Integrating Disaster Science and Management. http://dx.doi.org/10.1016/B978-0-12-812056-9.00003-8

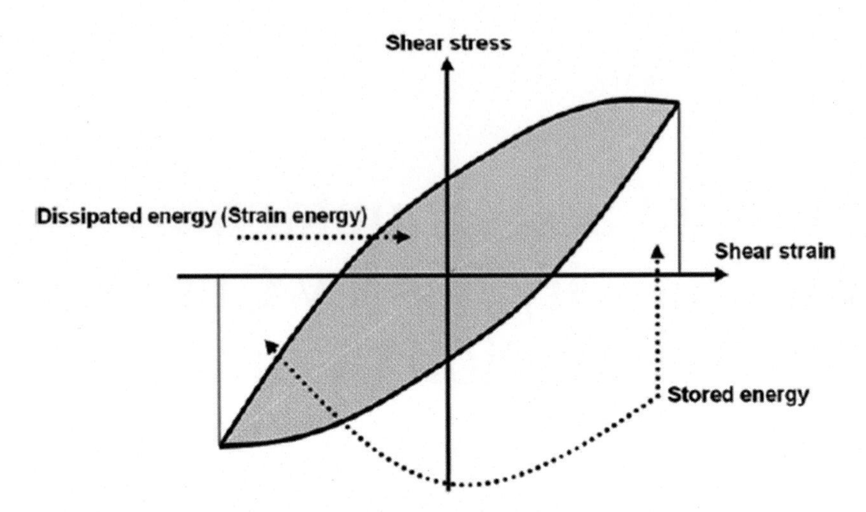

FIGURE 3.1 **A typical hysteresis shear stress–strain loop (modified from Green, 2001).**

In other words, this area represents the dissipated energy per unit volume of the soil mass (Ostadan et al., 1996). This is based on the idea that during deformation of cohesionless soils under dynamic loads part of the energy is dissipated into the soil (Nemat-Nasser and Shokooh, 1979). The instantaneous energy and its summation over time intervals are computed until the onset of liquefaction. The summation of the energy at this time is used as the measure of the capacity of the soil sample against initial liquefaction occurrence in terms of the strain energy (capacity energy).

The energy-based approach has the following advantages in comparison with the other existing methods to evaluate the liquefaction potential of soils (Baziar and Jafarian, 2007; Baziar et al., 2011):

1. Energy is associated with the quality of both shear stress and shear strain.
2. Energy is a scalar quantity which can be associated with the main characterizing earthquake parameters such as source to site distance and magnitude of the earthquake while it considers the entire spectrum of ground motions as opposed to the stress-based approach, which uses only the peak value of ground acceleration.
3. It is capable of accounting for the effects of a complicated stress–strain history on pore water pressure buildup.

The energy-based liquefaction evaluation procedures are mainly grouped into approaches developed using earthquake case histories and those developed from laboratory data (Green, 2001). Several models were developed relating the soil capacity energy to initial effective mean confining pressure and initial relative density on the basis of a series of laboratory cyclic shear and centrifuge tests (Figueroa et al., 1994; Liang, 1995; Dief and Figueroa, 2001) and most of these relationships were derived by performing multiple linear regression (MLR) analysis. Although high correlation coefficient R values can be obtained from these models, these relationships were only based on a limited number of tests and failed to take into account the important role of the fines content in the evaluation of the liquefaction behavior. Furthermore, Baziar and Jafarian (2007) demonstrated that such relationships developed based on a limited number of data, could not reasonably work in a large data set of various types of sand. Using their compiled database, Baziar and Jafarian (2007) developed a new MLR-based relationship to emphasize the necessity of developing an artificial neural network (ANN)-based model. In addition, Chen et al. (2005) presented a seismic wave energy-based method with back-propagation neural networks (BPNNs) to assess the liquefaction potential. Despite the good performance of the ANN-based models, the nature of a black-box framework restricts the practical applications of ANN. Expanding the database collected by Baziar and Jafarian (2007), Baziar et al. (2011) utilized an evolutionary approach based on genetic programming (GP) for estimation of capacity energy of liquefiable soils. Using the same database as Baziar and Jafarian (2007), Alavi and Gandomi (2012) presented promising variants of GP, namely the linear genetic programming (LGP) and multiexpression programming (MEP) to evaluate the liquefaction resistance of sandy soils. Cabalar et al. (2012) presented an alternative rule-based simulation for the prediction of liquefaction triggering through a novel neuro-fuzzy (NF) approach which possesses the natural language description of fuzzy systems and the learning capability of neural networks. Zhang et al. (2015) presented a nonparametric regression procedure known as multivariate adaptive regression splines (MARS) to assess the capacity energy required to trigger liquefaction in sand and silty sands, using a total of 302 previously published tests by Baziar and Jafarian (2007).

The current study aims to propose a relationship between the capacity energy dissipated during liquefaction and the soil initial parameters for assessment of soil liquefaction of sand deposits based on BPNNs and a wide-ranging database of laboratory tests. First the BPNN methodology and its associated procedures are explained in detail. Analyses of the database indicated that the developed BPNN model is reasonably accurate in predicting the capacity energy. The interpretability of the built BPNN model was also discussed. The prediction performance of the derived BPNN model compared favorably with other regression and soft computing models.

3.2 BPNN METHODOLOGY

The ANN structure consists of one or more layers of interconnected neurons or nodes. Each link connecting each neuron has an associated weight. The "learning" paradigm in the commonly used back-propagation (BP) algorithm (Rumelhart et al., 1986), which involves presenting examples of input and output patterns and subsequently adjusting the connecting weights so as to reduce the errors between the actual and the target output values. The iterative modification of the weights is carried out using the gradient descent approach and training is stopped once the errors have been reduced to some acceptable level. The ability of the trained ANN model to generalize the correct input–output response is performed in the testing phase and involves presenting the trained neural network with a separate set of data that have never been used during the training process.

A three-layer, feed-forward neural network topology shown in Fig. 3.2 is adopted in this study. As shown in Fig. 3.2, the BP algorithm involves two phases of data flow. In the first phase, the input data are presented forward from the input to output layer and produces an actual output. In the second phase, the error between the target values and actual values are propagated backwards from the output layer to the previous layers and the connection weights are updated to reduce the errors between the actual output values and the target output values. No effort is made to keep track of the characteristics of the input and output variables. The network is first trained using the training data set. The objective of the network training is to map the inputs to the output by determining the optimal connection weights and biases through the BP procedure. The number of hidden neurons is typically determined through a trial-and-error process; normally the smallest number of neurons that yields satisfactory results (judged by the network performance in terms of the coefficient of determination R^2 of the testing data set) is selected. In the present study, a MATLAB-based BP algorithm BPNN with the Levenberg–Marquardt (LM) algorithm (Demuth and Beale, 2003) was adopted for neural network modeling.

3.3 EVALUATION CRITERIA

The following performance measures were used to assess the performance of the BPNN model: coefficient of determination, R^2; coefficient of correlation, r; root mean squared error, RMSE; mean average error, MAE; relative root mean squared error, RRMSE; and performance index, ρ. Table 3.1 shows the definitions of these performance measures.

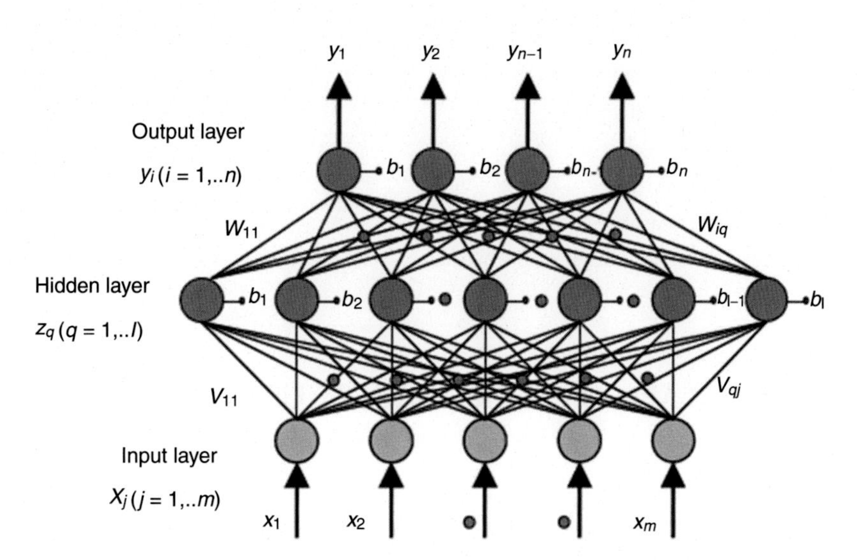

FIGURE 3.2 **BPNN architecture used in this study.** Abbreviation: *BPNN*, back-propagation neural network.

TABLE 3.1 Summary of Performance Measures

Measure	Calculation
Coefficient of determination (R^2)	$R^2 = 1 - \frac{\frac{1}{n}\sum_{i=1}^{n}(Y_i - \bar{Y})^2}{\frac{1}{n}\sum_{i=1}^{n}(y_i - \bar{Y})^2}$
Coefficient of correlation (r)	$r = \frac{\sum_{i=1}^{N}(Y_i - \bar{Y})(y_i - \bar{Y})}{\sqrt{\sum_{i=1}^{N}(Y_i - \bar{Y})^2}\sqrt{\sum_{i=1}^{N}(y_i - \bar{Y})^2}}$
Root mean squared error ($RMSE$)	$RMSE = \sqrt{\frac{1}{N}\sum_{i=1}^{N}(Y_i - y_i)^2}$
Mean average error (MAE)	$MAE = \frac{1}{N}\sum_{i=1}^{N} abs(Y_i - y_i)$
Relative root mean squared error ($RRMSE$)	$RRMSE = \frac{\sqrt{\frac{1}{N}\sum_{i=1}^{N}(Y_i - y_i)^2}}{\frac{1}{N}\sum_{i=1}^{N} y_i} \times 100$
Performance index (ρ)	$\rho = \frac{RRMSE}{1+r}$

$\bar{y}$ is the mean of the target values of y_i; $\bar{Y}$ is the mean of the predicted Y_i; N denotes the number of data points in the used set, training set, testing set or the overall set.

TABLE 3.2 Statistics of Parameters Used for ANN Model Development

Parameters	Parameter description	Min.	Max.	Mean	S.D.
Inputs					
σ'_{mean} (kPa)	Initial effective mean confining pressure	40.0	400.0	103.2	50.8
D_r (%)	Initial relative density after consolidation	−44.5	105.1	51.6	29.8
FC (%)	Percentage of fines content	0	100	18.8	24.2
C_u	Coefficient of uniformity	1.52	28.12	4.14	6.09
D_{50} (mm)	Mean grain size	0.03	0.46	0.21	0.12
Outputs					
Log(W) (J/m^3)	Logarithm of capacity energy	2.48	4.54	3.27	0.42

Min. denotes the minimum value while Max. represents the maximum value; Mean is the average value and S.D. denotes the standard deviation.

3.4 DEVELOPMENT OF CAPACITY ENERGY MODEL USING BPNN

3.4.1 The Database

The database used for BPNN modeling comprises of a total of 405 tests. A summary of the laboratory tests as well as the parameter statistics are listed in Table 3.2. The specific details of the 405 tests, including the test type, values of parameters and the failure mode can be found in Baziar et al. (2011).

Of the 405 data sets, 301 were randomly selected as the training patterns while the remaining 104 were used for testing purposes. As the previously proposed regression and soft computing models did not indicate the specific information of the training and testing patterns, for performance comparison, the criterion of data pattern selection used in this study was based on ensuring that the statistical properties including the mean and standard deviations of the training and testing subsets were similar to each other.

3.4.2 BPNN Model Results

The optimal BPNN model for prediction of capacity energy adopts 11 hidden neurons. The predictions are shown in Fig. 3.3 along with the performance statistics (the coefficient of determination, R^2; the coefficient of correlation, R; the RMSE, the MAE, RRMSE, and performance index, ρ) for the training and testing patterns. It is obvious that the BPNN model has been able to learn the complicated relationship between the capacity energy and the soil initial parameters. The relative errors (defined as the ratio of the difference between the BPNN predicted and the target logarithmic value of capacity energy log(W) divided by the target value, in percentage) for the training and testing patterns are plotted in Fig. 3.4. It is obvious that all of the BPNN estimations of the data patterns fell within ±20% of the target values and most of the predictions were within ±10% of the target values.

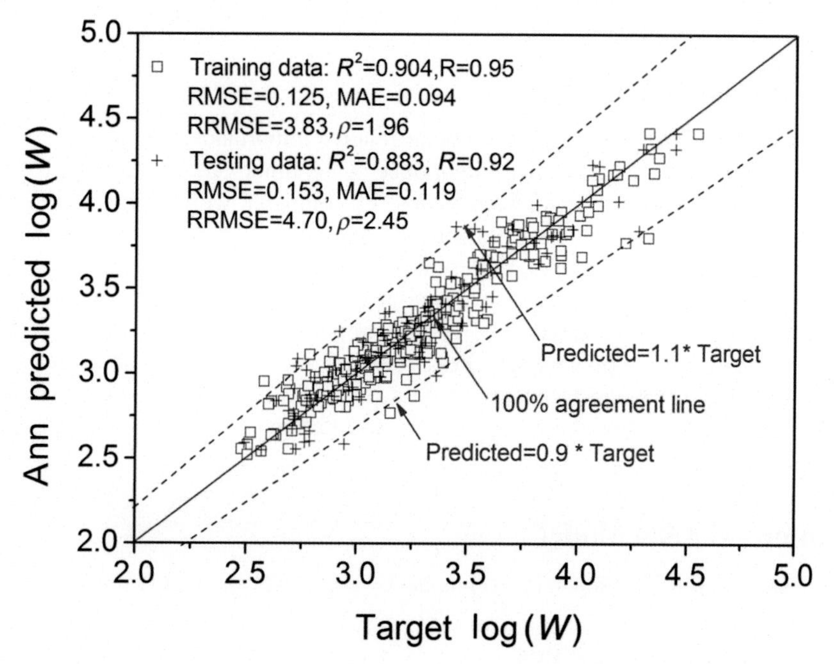

FIGURE 3.3 Comparison between target and BPNN predicted log(*W*). Abbreviation: *BPNN*, back-propagation neural network.

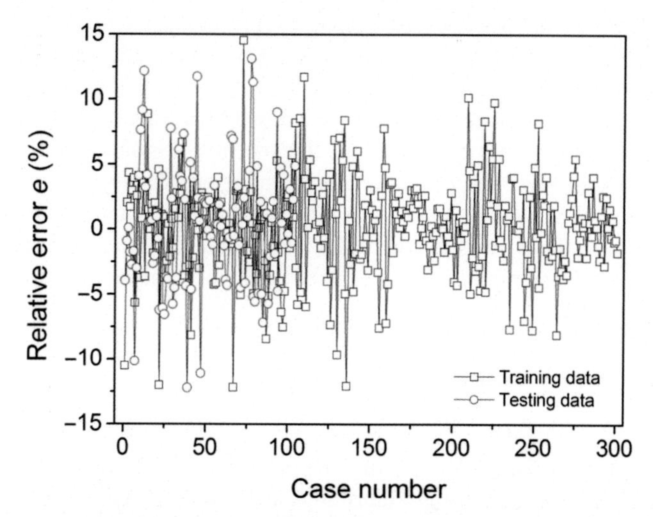

FIGURE 3.4 Variation of the relative errors obtained from the BPNN model. Abbreviation: *BPNN*, back-propagation neural network.

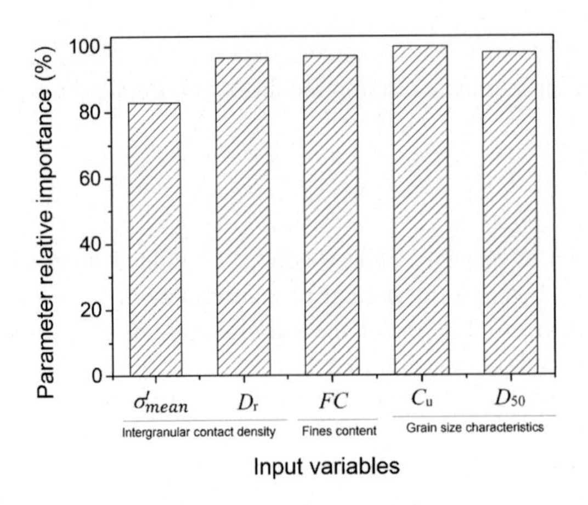

FIGURE 3.5 **Relative importance of the input variables in the BPNN model.** Abbreviation: *BPNN*, back-propagation neural network.

3.4.3 Parameter Relative Importance

The developed ANN model can also provide the parameter relative importance. The procedures for it are provided in Appendix A. Fig. 3.5 gives the plot of the relative importance of the input variables for the BPNN model. The results indicate that the capacity energy Log(W) is more sensitive to C_u compared with the initial mean effective stress σ'_{mean} , which is consistent with the parametric study of Baziar and Jafarian (2007). According to Baziar and Jafarian (2007) and Alavi and Gandomi (2012), as σ'_{mean} and D_r represent the initial density of the soils, they can be categorized into one group referred to as intergranular contact density. Similarly, C_u and D_{50} are grain size distribution parameters and have been grouped as the grain size characteristics or textural properties. Fines Content (FC) is individually considered as a category controlling the potential of pore water pressure buildup. In terms of the three categories, the total relative importance values for intergranular contact density, textural properties, and fines content are 179.28%, 97.06%, and 197.93%, respectively. This confirms the finding by Baziar and Jafarian (2007) that the grain size characteristic is the most influential category.

3.4.4 Easy-to-Interpret ANN Model

The mathematical expression for the developed log(W) model by the BPNN analysis is shown in Appendix B.

3.5 PERFORMANCE COMPARISONS

The overall performance statistics of the models obtained by ANN, LGP, MEP, standard GP, adaptive NF inference system ANFIS, the MARS, and the conventional MLR-based equations are summarized in Table 3.3. It is clearly observed that the BPNN model is able to predict the capacity energy reasonably well compared with other regression and soft computing algorithms. It is not possible to assess which method is more accurate or reasonable since the training and testing patterns used for model development for each method are different. The simple MLR model proposed by Baziar and Jafarian (2007) is not as accurate as other models since it only comprised the linear terms of inputs and thus failed to capture the nonlinear relationships involving a multitude of variables with interaction among each other. The more advanced algorithms such as GP, ANN, AIFIS, LGP, MARS, and MEP are capable of extracting the functional underlying relationships based on the training cases. However, for most of these methods, there is a lack of interpretation as to physical meaning of the generalized weight and bias values. Nevertheless, all these models listed in Table 3.3 give comparably accurate estimations and can be used for cross-validating each other.

3.6 SUMMARY AND CONCLUSION

In the present study, a new estimation model for capacity energy assessment of liquefaction potential of sands and sandy soils is proposed using BPNNs. Comparisons indicate that the proposed BPNN model estimations of the capacity energy are comparable with other models developed using algorithms such as MLR, GP, AIFIS, LGP, MARS, and MEP in ac-

TABLE 3.3 Summary of Performance Measures of the Energy-Based Models for Liquefaction Assessment

Model	No. of data sets	Performances				References
		R^2	R	RMSE	MAE	
MLR	284	0.65	–	0.262	0.213	Baziar and Jafarian (2007)
ANN	284	0.90	–	0.138	0.104	Baziar and Jafarian (2007)
GP	399	0.88	–	0.140	0.109	Baziar et al. (2011)
LGP	301	–	0.87	0.224	0.178	Alavi and Gandomi (2012)
MEP	301	–	0.86	0.233	0.187	Alavi and Gandomi (2012)
GP	301	–	0.81	0.274	0.219	Alavi and Gandomi (2012)
ANFIS	302	0.87	–	0.181	–	Cabalar et al. (2012)
MARS	302	0.88	0.94	0.182	0.155	Zhang et al. (2015)
ANN	405	0.90	0.94	0.132	0.101	This study

"–" indicates that this performance statistics is not provided in the reference.

curacy. The interpretability of the developed BPNN model is discussed and procedures to assess the parameter relative importance are also illustrated.

It should be noted that since the built BPNN model is of a data-driven nature, thus interpolations between the design input variables are more accurate and reliable than extrapolations. Consequently, it is not recommended that the model be applied for values of input parameters beyond the specific ranges in this study.

APPENDIX A PROCEDURES FOR PARTITIONING OF ANN WEIGHTS FOR LOG(*W*) MODEL

This appendix details the procedure for partitioning the connection weights to determine the relative importance of the various inputs using the method by Garson (1991). The method essentially involves partitioning the hidden-output connection weights of each hidden neuron into components associated with each input neuron.

The neural network works with three input neurons, five hidden neurons and one output neuron with the connection weights as shown below:

Table A.1 ANN Weights and Outputs

Hidden neurons	Weights					Output
	Input 1	Input 2	Input 3	Input 4	Input 5	
Hidden 1	−2.4050	−58.1072	1.1758	−4.5727	3.1914	30.7345
Hidden 2	−3.2781	−5.5557	3.5087	−0.2865	−7.4184	7.2321
Hidden 3	−2.2428	−5.0519	−7.8583	3.7089	7.1053	0.4255
Hidden 4	2.0226	5.5300	11.1548	−9.4000	−8.5669	−4.0999
Hidden 5	−14.3763	2.1677	2.1050	−8.5897	−4.4347	1.0071
Hidden 6	−13.2545	−14.4353	−15.4396	−7.3395	−13.8746	−1.1994
Hidden 7	8.0682	1.4211	4.0740	−1.2327	−0.2626	−3.5311
Hidden 8	44.5397	38.6194	4.4275	−12.9313	−48.9027	18.0423
Hidden 9	−0.5771	−0.5889	−3.6987	8.9530	−0.1810	3.8047
Hidden 10	0.4141	2.4533	16.5609	7.5493	−7.5459	7.9102
Hidden 11	−1.6976	0.9095	4.9397	−20.7657	20.1904	−1.9086

The computation process is as follows:

1. For each hidden neuron i, multiply the absolute value of the hidden-output layer connection weight by the absolute value of the hidden-input layer connection weight. Do this for each input variable j. The following products P_{i-j} are obtained:

Table A.2 P_{i-j} results

	P_{i-j}				
Hidden neurons	**Input 1**	**Input 2**	**Input 3**	**Input 4**	**Input 5**
Hidden 1	73.9165	1785.8957	36.1376	140.5396	98.0861
Hidden 2	23.7075	40.1794	25.3753	2.0720	53.6506
Hidden 3	0.9543	2.1496	3.3437	1.5781	3.0233
Hidden 4	8.2925	22.6724	45.7336	38.5391	35.1234
Hidden 5	14.4784	2.1831	2.1199	8.6507	4.4662
Hidden 6	15.8974	17.3137	18.5183	8.8030	16.6412
Hidden 7	28.4896	5.0180	14.3857	4.3528	0.9273
Hidden 8	803.5986	696.7828	79.8823	233.3104	882.3172
Hidden 9	2.1957	2.2406	14.0724	34.0635	0.6887
Hidden 10	3.2756	19.4061	131.0000	59.7165	59.6896
Hidden 11	3.2400	1.7359	9.4279	39.6334	38.5354

2. For each hidden neuron, divide P_{i-j} by the sum for all the input variables to obtain Q_{i-j}. For example for hidden 1,

$$Q_{1-1} = \frac{P_{1-1}}{P_{1-1} + P_{1-2} + P_{1-3} + P_{1-4} + P_{1-5}} = 0.0304$$

3. For each input neuron, sum the product S_j formed from the previous computation of Q_{i-j}. For example,

$$S_1 = Q_{1-1} + Q_{2-1} + Q_{3-1} + Q_{4-1} + Q_{5-1} + Q_{6-1} + Q_{7-1} + Q_{8-1} + Q_{9-1} + Q_{10-1} + Q_{11-1}.$$

Table A.3 Q_{i-j} Results

	Q_{i-j}				
Hidden neurons	**Input 1**	**Input 2**	**Input 3**	**Input 4**	**Input 5**
Hidden 1	0.0346	0.8367	0.0169	0.0658	0.0460
Hidden 2	0.1635	0.2771	0.1750	0.0143	0.3700
Hidden 3	0.0864	0.1945	0.3026	0.1428	0.2736
Hidden 4	0.0552	0.1508	0.3042	0.2563	0.2336
Hidden 5	0.4539	0.0684	0.0665	0.2712	0.1400
Hidden 6	0.2060	0.2243	0.2400	0.1141	0.2156
Hidden 7	0.5358	0.0944	0.2705	0.0819	0.0174
Hidden 8	0.2981	0.2585	0.0296	0.0865	0.3273
Hidden 9	0.0412	0.0421	0.2642	0.6396	0.0129
Hidden 10	0.0120	0.0711	0.4797	0.2187	0.2186
Hidden 11	0.0350	0.0188	0.1018	0.4281	0.4163
Sum	$S_1 = 1.9216$	$S_2 = 2.2366$	$S_3 = 2.2511$	$S_4 = 2.3193$	$S_5 = 2.2714$

4. Divide S_j by max(S_j) for all the input variables. Expressed as a percentage, this gives the relative importance of all output weights attributable to the given input variable. For example, for the input neuron 1, the relative importance (%) is equal to $S_1 \times 100 / \max(S_1, S_2, S_3, S_4, S_5) = 82.85$.

Table A.4 **Relative Importance Results**

Relative importance (%)	Input 1	Input 2	Input 3	Input 4	Input 5
	82.85	96.43	97.06	100.00	97.93

APPENDIX B CALCULATION OF ANN OUTPUT LOG(*W*) MODEL

The transfer functions used for ANN output for Log(*W*) are "logsig' transfer function for hidden layer to output layer and "tansig' transfer function for output layer to target. The calculation process of ANN output for Log(*W*) is elaborated in detail as follows:

From the connection weights for a trained neuron network, it is possible to develop a mathematical equation relating the input parameters and the single output parameter *Y* using

$$Y = f_{sig}\left\{b_0 + \sum_{k=1}^{h}\left[w_k f_{sig}\left(b_{hk} + \sum_{i=1}^{m} w_{ik} X_i\right)\right]\right\} \tag{B.1}$$

in which b_0 is the bias at the output layer, ω_k is the weight connection between neuron k of the hidden layer and the single output neuron, b_{hk} is the bias at neuron k of the hidden layer (k = 1,h), ω_{ik} is the weight connection between input variable i (i = 1, m) and neuron k of the hidden layer, X_i is the input parameter i, and f_{sig} is the sigmoid (logsig &tansig) transfer function.

Using the connection weights of the trained neural network, the following steps can be followed to mathematically express the ANN model:

Step1: Normalize the input values for σ'_{mean}, D_r, FC, C_u, and D_{50} linearly using

$$x_{norm} = 2(x_{actual} - x_{min})/(x_{max} - x_{min}) - 1$$

Let the actual $\sigma'_{mean} = X_{1a}$ and the normalized $\sigma'_{mean} = X_1$

$$X_1 = -1 + 2*(X1_a - 40.0)/(400.0 - 40.0) \tag{B.2}$$

Let the actual $D_r = X_{2a}$ and the normalized $D_r = X_2$

$$X_2 = -1 + 2*(X_{2a} - (-44.5))/(105.1 - (-44.5)) \tag{B.3}$$

Let the actual $FC = X_{3a}$ and the normalized $FC = X_3$

$$X_3 = -1 + 2*(X_{3a} - 0)/(100 - 0) \tag{B.4}$$

Let the actual $C_u = X_{4a}$ and the normalized $C_u = X_4$

$$X_4 = -1 + 2*(X_{4a} - 1.52)/(28.12 - 1.52) \tag{B.5}$$

Let the actual $D_{50} = X_{5a}$ and the normalized $D_{50} = X_5$

$$X_5 = -1 + 2*(X_{5a} - 0.03)/(0.46 - 0.03) \tag{B.6}$$

Step2: Calculate the normalized value (Y_1) using the following expressions:

$$A_1 = 30.7345 - 2.4050\text{logsig}(X_1) - 58.1072\text{logsig}(X_2) + 1.1758\text{logsig}(X_3) - 4.5727\text{logsig}(X_4) + 3.1914\text{logsig}(X_5) \tag{B.7}$$

$$A_2 = 7.2321 - 3.2781\text{logsig}(X_1) - 5.5557\text{logsig}(X_2) + 3.5087\text{logsig}(X_3) - 0.2865\text{logsig}(X_4) - 7.4184\text{logsig}(X_5) \tag{B.8}$$

$$A_3 = 0.4255 - 2.2428\text{logsig}(X_1) - 5.0519\text{logsig}(X_2) - 7.8583\text{logsig}(X_3) + 3.7089\text{logsig}(X_4) + 7.1053\text{logsig}(X_5) \tag{B.9}$$

$$A_4 = -4.0999 + 2.0226\text{logsig}(X_1) + 5.5300\text{logsig}(X_2) + 11.1548\text{logsig}(X_3) - 9.4000\text{logsig}(X_4) - 8.5669\text{logsig}(X_5) \tag{B.10}$$

$$A_5 = 1.0071 - 14.3763\text{logsig}(X_1) + 2.1677\text{logsig}(X_2) + 2.1050\text{logsig}(X_3) - 8.5897\text{logsig}(X_4) - 4.4347\text{logsig}(X_5) \quad (B.11)$$

$$A_6 = -1.1994 - 13.2545\text{logsig}(X_1) - 14.4353\text{logsig}(X_2) - 15.4396\text{logsig}(X_3) - 7.3395\text{logsig}(X_4) - 13.8746\text{logsig}(X_5) \quad (B.12)$$

$$A_7 = -3.5311 + 8.0682\text{logsig}(X_1) + 1.4211\text{logsig}(X_2) + 4.0740\text{logsig}(X_3) - 1.2327\text{logsig}(X_4) - 0.2626\text{logsig}(X_5) \quad (B.13)$$

$$A_8 = 18.0423 + 44.5397\text{logsig}(X_1) + 38.6194\text{logsig}(X_2) + 4.4275\text{logsig}(X_3) - 12.9313\text{logsig}(X_4) - 48.9027\text{logsig}(X_5) \quad (B.14)$$

$$A_9 = 3.8047 - 0.5771\text{logsig}(X_1) - 0.5889\text{logsig}(X_2) - 3.6987\text{logsig}(X_3) + 8.9530\text{logsig}(X_4) - 0.1810\text{logsig}(X_5) \quad (B.15)$$

$$A_{10} = 7.9102 + 0.4141\text{logsig}(X_1) + 2.4533\text{logsig}(X_2) + 16.5609\text{logsig}(X_3) + 7.5493\text{logsig}(X_4) - 7.5459\text{logsig}(X_5) \quad (B.16)$$

$$A_{11} = -1.9086 - 1.6976\text{logsig}(X_1) + 0.9095\text{logsig}(X_2) + 4.9397\text{logsig}(X_3) - 20.7657\text{logsig}(X_4) + 20.1904\text{logsig}(X_5) \quad (B.17)$$

$$B_1 = -0.2432 \times \tanh(A_1) \quad (B.18)$$

$$B_2 = -0.8242 \times \tanh(A_2) \quad (B.19)$$

$$B_3 = -5.3022 \times \tanh(A_3) \quad (B.20)$$

$$B_4 = -5.2137 \times \tanh(A_4) \quad (B.21)$$

$$B_5 = -15.9344 \times \tanh(A_5) \quad (B.22)$$

$$B_6 = -1.8513 \times \tanh(A_6) \quad (B.23)$$

$$B_7 = -19.5763 \times \tanh(A_7) \quad (B.24)$$

$$B_8 = 0.2974 \times \tanh(A_8) \quad (B.25)$$

$$B_9 = -3.3429 \times \tanh(A_9) \quad (B.26)$$

$$B_{10} = 1.0301 \times \tanh(A_{10}) \quad (B.27)$$

$$B_{11} = -2.4767 \times \tanh(A_{11}) \quad (B.28)$$

$$C_1 = 26.768 + B_1 + B_2 + B_3 + B_4 + B_5 + B_6 + B_7 + B_8 + B_9 + B_{10} + B_{11} \quad (B.29)$$

$$Y_1 = C_1 \quad (B.30)$$

Step3: De-normalize the output to obtain Log(W)

$$\text{Log}(W) = 2.48 + (4.54 - 2.48) \times (Y_1 + 1)/2 \quad (B.31)$$

Note: logsig(x) = 1/(1 + exp($-x$)) while tanh(x) = 2/(1+exp($-2x$))–1

ACKNOWLEDGMENTS

The authors would like to express their appreciation to Baziar and Jafarian (2007), Baziar et al. (2011) for compiling the comprehensive database for liquefaction assessment.

REFERENCES

Alavi, A.H., Gandomi, A.H., 2012. Energy-based numerical models for assessment of soil liquefaction. Geoscience Frontiers 3 (4), 541–555.

Baziar, M.H., Jafarian, Y., 2007. Assessment of liquefaction triggering using strain energy concept and ANN model: capacity energy. Soil Dynamics and Earthquake Engineering 27 (12), 1056–1072.

Baziar, M.H., Jafarian, Y., Shahnazari, H., Movahed, V., Tutunchian, M.A., 2011. Prediction of strain energy-based liquefaction resistance of sand-silt mixtures: an evolutionary approach. Computers & Geosciences 37 (11), 1883–1893.

Boulanger, R.W., Idriss, I.M., 2012. Probabilistic standard penetration test-based liquefaction-triggering procedure. Journal of Geotechnical and Geoenvironmental Engineering 138, 1185–1195.

Cabalar, A.F., Cevik, A., Gokceoglu, C., 2012. Some applications of adaptive neuro-fuzzy inference system (ANFIS) in geotechnical engineering. Computers and Geotechnics 40, 14–33.

Chen, Y.R., Hsieh, S.C., Chen, J.W., Shih, C.C., 2005. Energy-based probabilistic evaluation of soil liquefaction. Soil Dynamics and Earthquake Engineering 25 (1), 55–68.

Davis, R.O., Berrill, J.B., 1982. Energy dissipation and seismic liquefaction in sands. Earthquake Engineering & Structural Dynamics 10, 59–68.

Demuth, H., Beale, M., 2003. Neural Network Toolbox for MATLAB-user Guide Version 4.1. The Math Works Inc, Natick, Massachusetts, U.S.A, 153–155.

Dief, H.M., Figueroa, J.L., 2001. Liquefaction assessment by the energy method through centrifuge modeling. In: Zeng, X.W. (Ed.), In: Proceedings of the NSF International Workshop on Earthquake Simulation in Geotechnical Engineering. CWRU, Cleveland, OH, pp. 279–283.

Dobry, R., Ladd, R.S., Yokel, F.Y., Chung, R.M., Powell, D., 1982. Prediction of Pore Water Pressure Build-up and Liquefaction of Sands During Earthquakes by the Cyclic Strain Method. National Bureau of Standards, US Department of Commerce, US Governmental Printing Office, Building Science Series, Washington, DC, 79–82.

Figueroa, J.L., Saada, A.S., Liang, L., Dahisaria, M.N., 1994. Evaluation of soil liquefaction by energy principles. Journal of Geotechnical and Geoenvironmental Engineering 20 (9), 1554–1569.

Garson, G.D., 1991. Interpreting neural-network connection weights. AI Expert 6 (7), 47–51.

Green, R.A., 2001. Energy-based evaluation and remediation of liquefiable soils. PhD dissertation. Virginia Polytechnic Institute and State University, Blacksburg, VA.

Ishihara, K., Yasuda, S., 1975. Sand liquefaction in hollow cylinder torsion under irregular excitation. Soils and Foundations 15 (1), 45–59.

Juang, C.H., Rosowsky, D.V., Tang, W.H., 1999. Reliability-based method for assessing liquefaction potential of soils. Journal of Geotechnical and Geoenvironmental Engineering 125, 684–689.

Juang, C.H., Chen, C.J., Jiang, T., 2001. Probabilistic framework for liquefaction potential by shear wave velocity. Journal of Geotechnical and Geoenvironmental Engineering 127, 670–678.

Juang, C.H., Ching, J., Luo, Z., Ku, C.S., 2012. New models for probability of liquefaction using standard penetration tests based on an updated database of case histories. Engineering Geology 133–134, 85–93.

Liang, L., 1995. Development of an energy method for evaluating the liquefaction potential of a soil deposit. PhD thesis. Department of Civil Engineering, Case Western Reserve University, Cleveland, OH.

Moss, R.E.S., Seed, R.B., Kayen, R.E., Stewart, J.P., Der Kiureghian, A., Cetin, K.O., 2006. CPT-based probabilistic and deterministic assessment of in situ seismic soil liquefaction potential. Journal of Geotechnical and Geoenvironmental Engineering 132, 1032–1051.

Nemat-Nasser, S., Shokooh, A., 1979. A unified approach to densification and liquefaction of cohesionless sand in cyclic shearing. Canadian Geotechnical Journal 16 (4), 659–678.

Ostadan, F., Deng, N., Arango, I., 1996. Energy-Based Method for Liquefaction Potential Evaluation, Phase I. Feasibility Study. U.S. Department of Commerce, Technology Administration, National Institute of Standards and Technology, Building and Fire Research Laboratory, Gaithersburg, Maryland, U.S.A, 43–45.

Rumelhart, D.E., Hinton, G.E., Williams RJ, 1986. Learning internal representation by error propagation. Rumelhart, D.E., Mcclelland, J.L. (Eds.), Parallel Distributed Processing, 1, MIT Press, Cambridge, MA, pp. 318–362.

Seed, H.B., 1980. Closure to soil liquefaction and cyclic mobility evaluation for level ground during earthquakes. Journal of Geotechnical and Geoenvironmental Engineering-ASCE 106 (GT6), 724.

Seed, H.B., Idriss, I.M., 1971. Simplified procedure for evaluating soil liquefaction potential. Soil Mechanics and Foundation Engineering 97 (9), 1249–1273.

Seed, H.B., Idriss, I.M., Makdisi, F., Banerjee, N., 1975. Representation of irregular stress time histories by equivalent uniform stress series in liquefaction analyses. Report No. UCB/EERC-75/29. Earthquake Engineering Research Centre, U.C. Berkeley.

Whitman, R.V., 1971. Resistance of soil to liquefaction and settlement. Soils and Foundations 11 (4), 59–68.

Zhang, W.G., Goh, A.T.C., Zhang, Y.M., Chen, Y.M., Xiao, Y., 2015. Assessment of soil liquefaction based on capacity energy concept and multivariate adaptive regression splines. Engineering Geology 188, 29–37.

Chapter 4

Recent Earthquakes and Volcanic Activities in Kyushu Island, Japan

Hakim Saibi
United Arab Emirates University, Al-Ain, UAE

4.1 INTRODUCTION

In this chapter, we will present two famous earthquakes: the 2005 West off Fukuoka Prefecture is considered as the largest recorded earthquake from interpolate earthquakes at the junction of the SW-Japan Arc and the Ryukyu Arc, and the 2016 Kumamoto earthquake is known for its successive high-intensity earthquakes occurring in less than 1 week. Also we will show the major volcanoes existing in Kyushu Island with their characteristics and volcanic activities.

Japan lies in a zone of extreme crustal instability with the subduction of the Pacific plate and the Philippine Sea Plate. Japan is also a part of the Ring of Fire, where many earthquakes and volcanic eruptions occur. This specific geographic-crustal location of Japan makes it a place of a large number of natural hazards such as strong earthquakes, for example, the 2011 Great East Japan earthquake, M_{JMA} 9 (magnitude determined by Japan Meteorological Agency, JMA) with 40 m high tides of tsunami, big volcanic eruptions, tsunamis, and landslides.

Japan's volcanoes are part of five volcanic arcs. The arcs meet at a triple junction on Honshu Island. Black "teeth" mark the subduction zone with the "teeth" on the overriding plate (Fig. 4.1). The Pacific Plate subducts under the Philippine Plate and forms the Izu-Bonic Arc. The subduction of the Pacific Plate under the Eurasian Plate forms the Northeast Honshu Arc and the Kurile Arc. All these arcs are very active and place of many volcanic activities and earthquakes.

4.2 RECENT EARTHQUAKES IN KYUSHU ISLAND, SOUTHWESTERN JAPAN

4.2.1 The 2005 West Off Fukuoka Prefecture Earthquake

Fukuoka city, capital city of Kyushu Island, southwest of Japan (Fig. 4.2), was struck by a strong earthquake on March 20, 2005 (M_{JMA}7.0). The hypocenters distributed at depths ranging from 2 to 18 km (Fig. 4.3) occurred in the offshore region of the city, which was followed by an M_{JMA}5.8 aftershock on April 20. This strong interplate earthquake caused many casualities and damages to the buildings.

Many seismological and geodetic studies (Shimizu et al., 2006; Uehira et al., 2006) have determined that the earthquake occurred along the extension of the Kego fault, with a length of about 25 km, which runs from the northwest to the southeast under the Sea of Genkai.

The Fukuoka area was investigated by microgravity surveys by the team of Geothermics Laboratory of Kyushu University (Japan) for more than two decades using Scintrex gravimeters in order to understand the fault structures in Fukuoka city. Fig. 4.4 shows the residual Bouguer gravity map of Fukuoka city. The residual gravity is obtained after removing the regional Bouguer gravity component and it helps to study local geological structures such as faults. The residual Bouguer gravity values range between −7.7 and 9.6 mGal, increasing in the northern and western regions of the map area. Two low residual Bouguer gravity anomalies can be seen in the central part of the area extending in the NW–SE direction. This region may reflect a small basin elongated in the NW–SE direction.

Integrating Disaster Science and Management. http://dx.doi.org/10.1016/B978-0-12-812056-9.00004-X

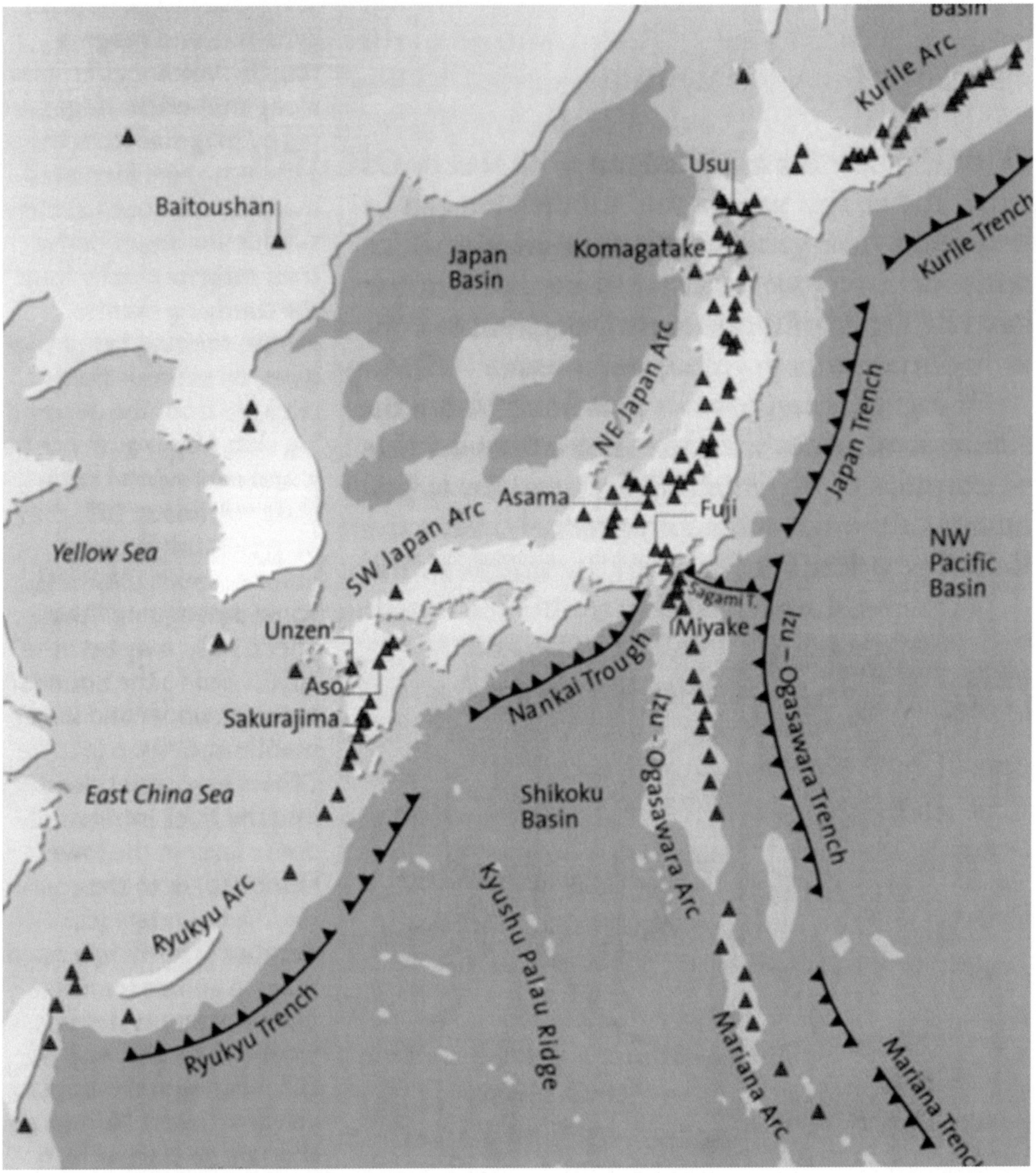

FIGURE 4.1 **Distribution of active volcanoes in and around Japan.**

Saibi et al. (2008) analyzed the measured microgravity data and could detect the Kego fault location using some gravity gradient filtering techniques, such as the horizontal gradient (HG), tilt derivative (TDR), and Euler deconvolution methods. The TDR method has the advantage of responding well to both shallow and deep sources and the map of TDR recognizes the horizontal location and extent of sources. The location of the maximum HG may be used as an indicator of the locations of edges of the source. The Euler solutions give the depths of the sources. Fig. 4.5 illustrates the HG anomaly over fault structure in different modes.

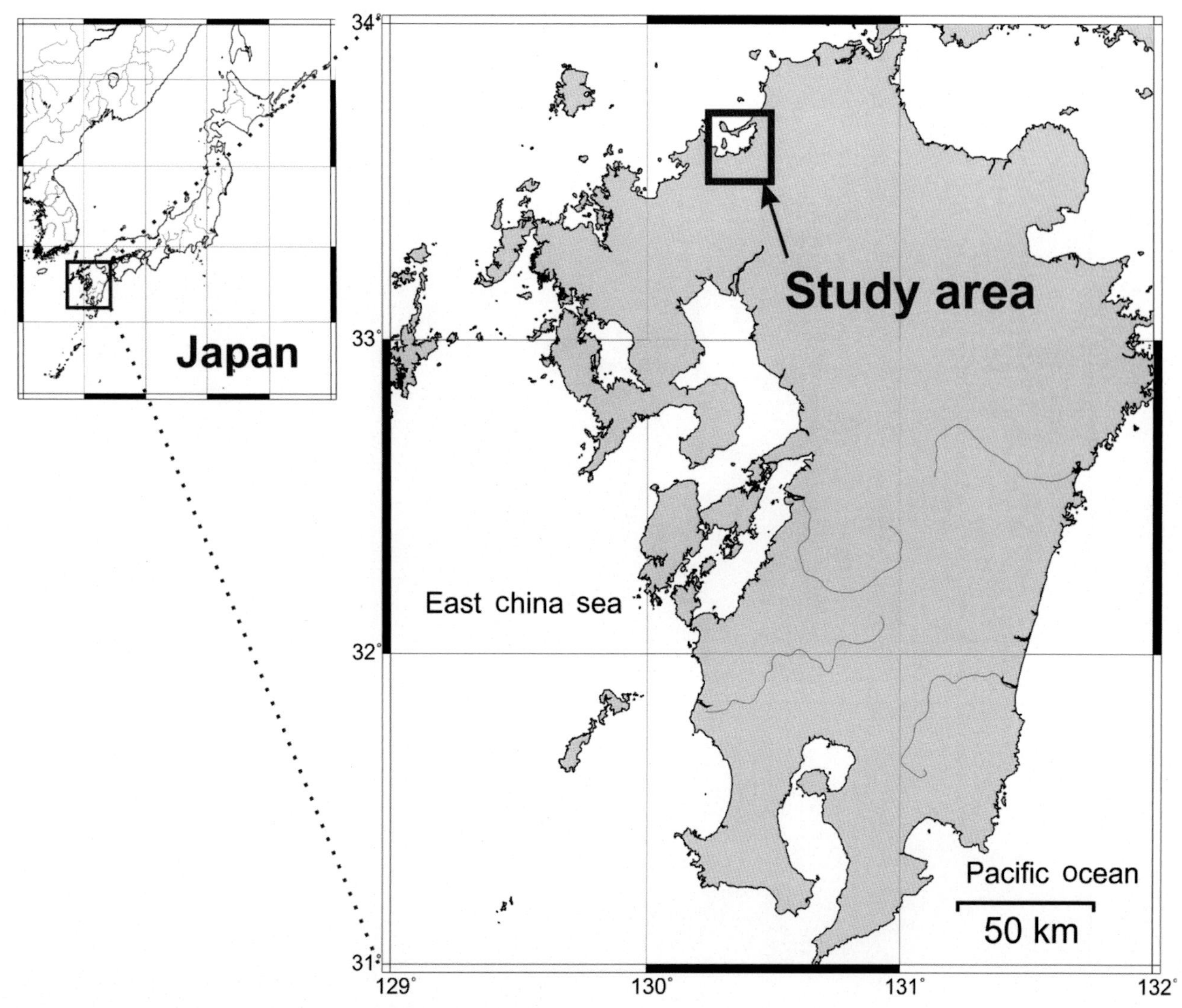

FIGURE 4.2 **Location of the Fukuoka city in Kyushu Island, Japan.**

Fig. 4.6 shows the detected faults from the gravity data. The Fukuoka city is dissected by mainly NW–SE faults as presented by the rose diagram. The possible location of the Kego fault is also presented. Fig. 4.7 shows some photographs from the damages caused by the 2005 West off Fukuoka earthquake and field measurements after the main shock by Kyushu University, Geothermics Lab. Staff members.

4.2.2 The 2016 Kumamoto Earthquake

On April 15–16, 2016, a sequence of strong earthquakes occurred in Kumamoto south of Fukuoka city on the island of Kyushu. A magnitude of $M_{\mathrm{JMA}}7.3$ was recorded on 16th (Japan Standard Time) and a foreshock earthquake of M_{JMA} 6.5 (http://www.jma.go.jp/jma/en/2016_Kumamoto_Earthquake/2016_Kumamoto_Earthquake.html) at a depth of about 10 km. The strong and sudden earthquakes caused more than 60 fatalities and damages to roads and collapse of homes. More than 40,000 people were evacuated to safe places due to the disaster. The earthquakes were located along the Futagawa–Hinagu fault zone (Hashimoto et al., 2017).

The recent development of remote-sensing and radar satellite imaging helped the scientists to understand and study the ground deformation generated by strong earthquake at few mm scale from satellites located at more than 830 km above ground level using InSAR (INterferometric Synthetic Aperture Radar) technology (Figs. 4.8and 4.9).

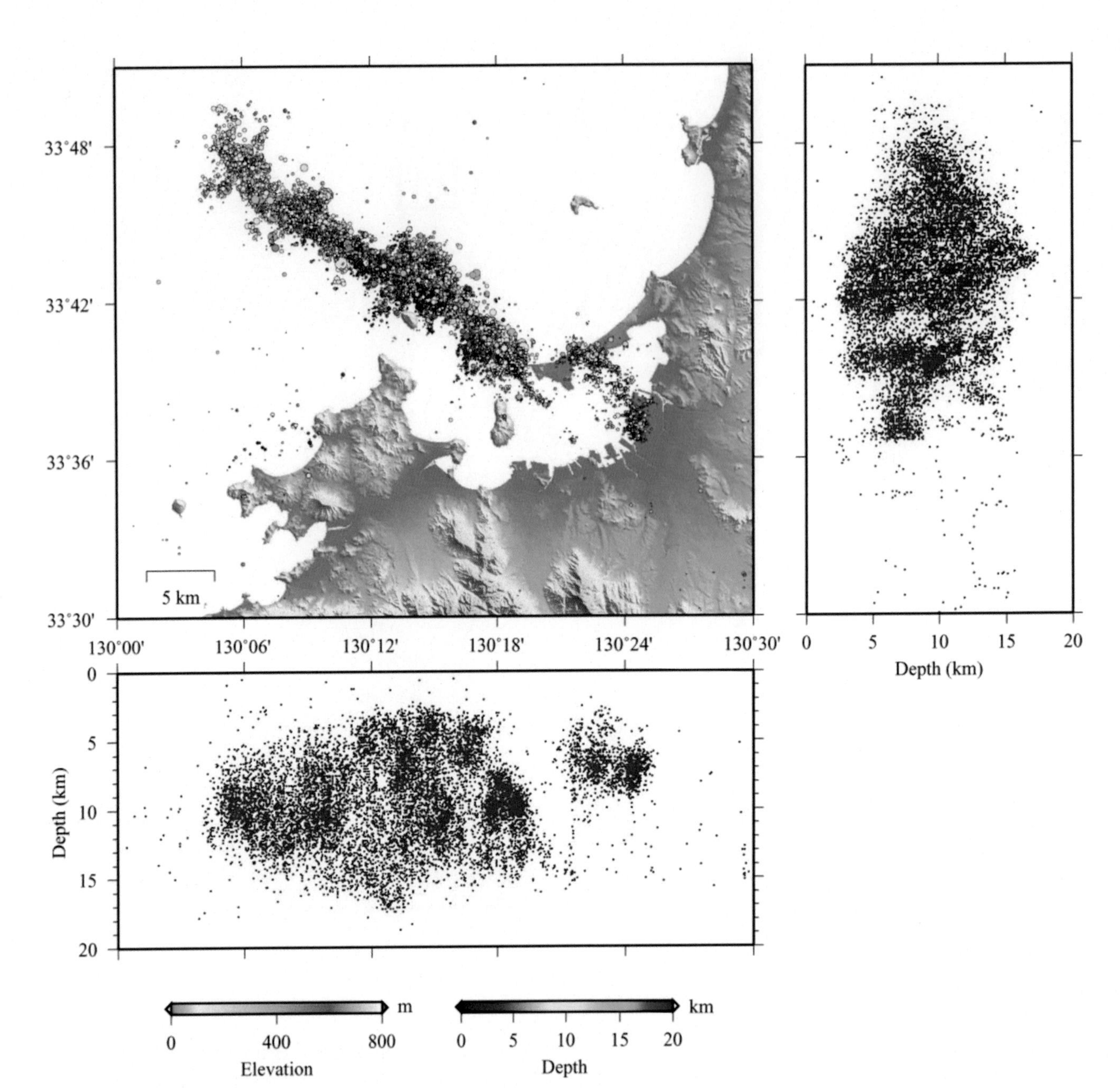

FIGURE 4.3 **Location of the hypocenters and depths after the main shock on March 20, 2005, of the 2005 west off Fukuoka earthquake (Nishijima, 2005).**

Kobayashi (2017) studied the InSAR images of the 2016 Kumamoto earthquake and found a ground displacement of more than 15 cm west of Hinagu fault zone, which played a right lateral fault motion. Fukahata and Hashimoto (2016) calculated the slip of the two main active faults of the 2016 Kumamoto earthquake using a nonlinear inversion of InSAR data and found a slip of 2.4 and 5 m for Hinagu and Futagawa faults, respectively. Fig. 4.10 shows the crustal deformation map after the 2016 Kumamoto earthquake which reached 1 m of deformation after the quake (Geospatial Information Authority of Japan, 2017a).

Fig. 4.11 shows the landslides caused by the earthquake and subsequent rainfall, etc. from the aerial photographs taken by the Geographical Survey Institute after the Kumamoto earthquake. The landslide collapse area is large and roughly 1 ha (football field) or more and small landslides represent an area of roughly 0.1 ha.

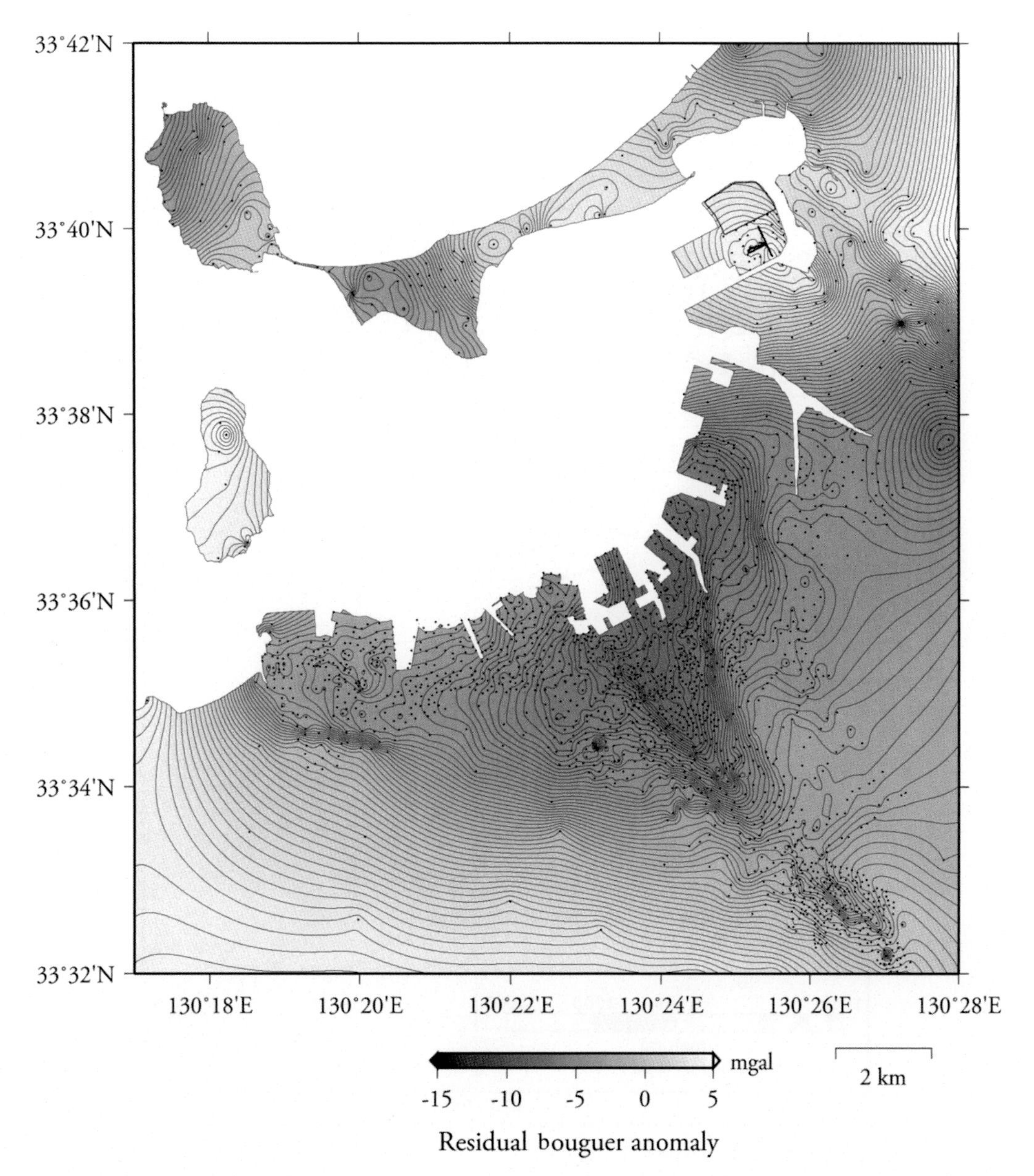

FIGURE 4.4 **Residual Bouguer gravity map of the Fukuoka area. Black points indicate the locations of the gravity stations (Nishijima et al., 2010).**

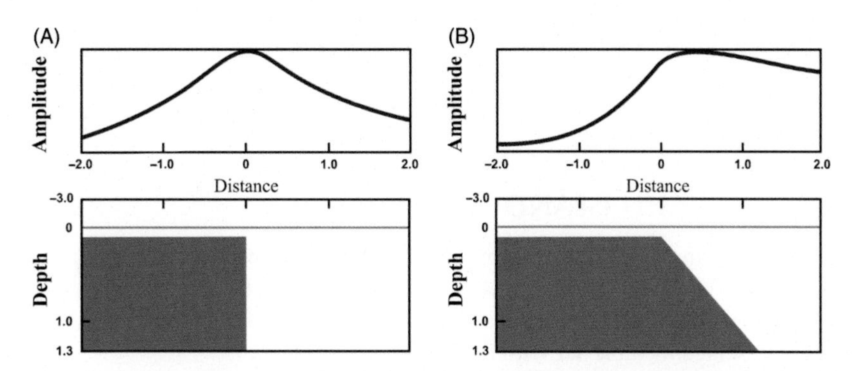

FIGURE 4.5 **Horizontal gradient magnitude over two models: a vertical contact (A) and a dipping contact (B) (Phillips, 2000).**

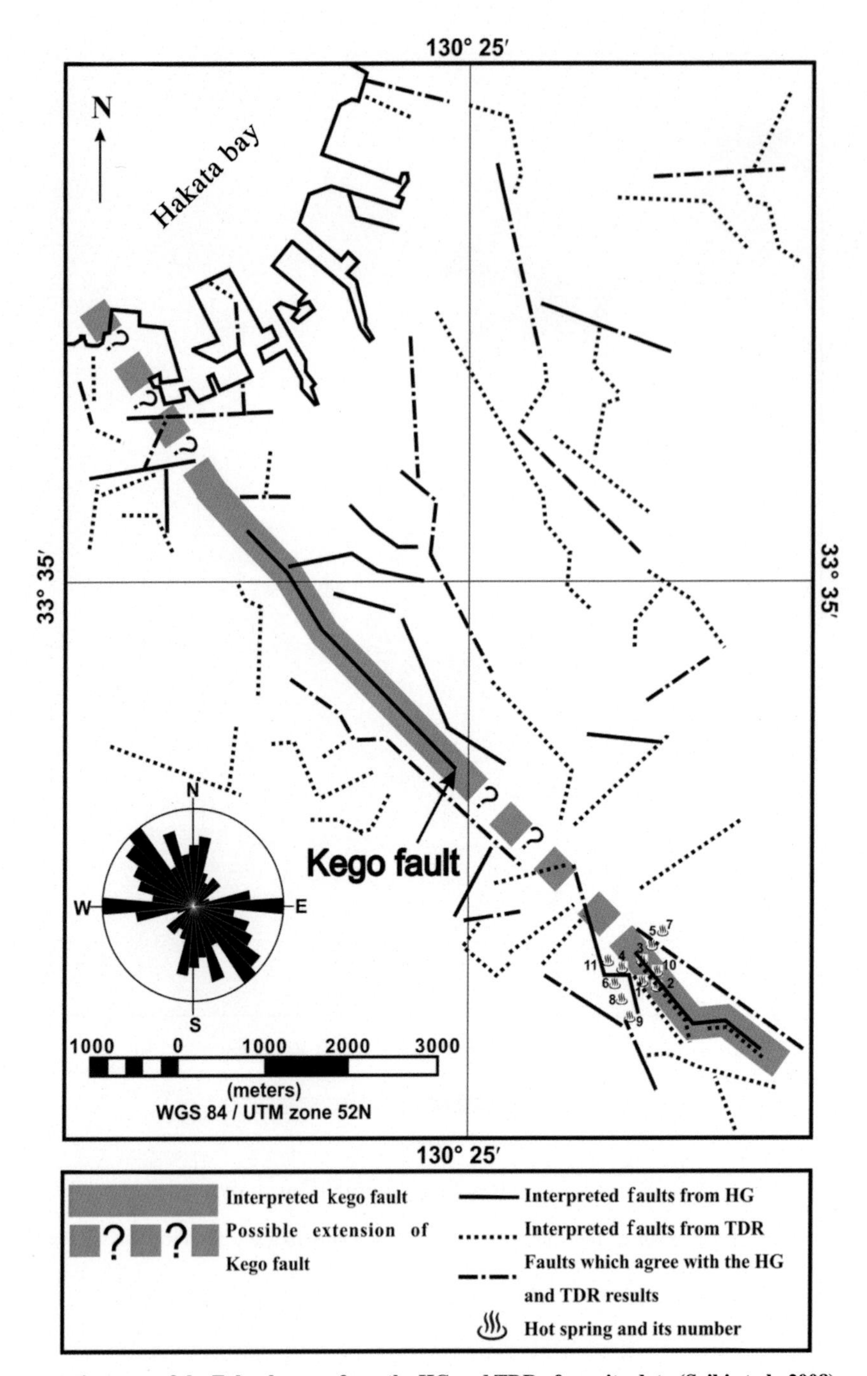

FIGURE 4.6 **Fault interpretation map of the Fukuoka area from the HG and TDR of gravity data (Saibi et al., 2008).**

4.3 VOLCANOES IN KYUSHU ISLAND

Volcanic eruptions pose a severe threat to life and property on a global scale. In order to mitigate risks, it is essential to assess hazards of volcanic activity by evaluating results obtained from detailed monitoring and quantification of active volcanoe dynamics using geophysical, geochemical, and geodetic techniques. Kyushu island is known for its volcanoes. The main volcanoes are represented in Table 4.1 with their characteristics.

FIGURE 4.7 **Photographs of damages caused by the 2005 West off Fukuoka earthquake and geophysical monitoring surveys.**

4.3.1 Unzen Volcano

Unzen volcano is an active volcano located in Nagasaki Prefecture near Shimabara city (Fig. 4.12). Unzen is composed of several overlapping stratovolcanoes. The highest peak of Unzen volcano is named Fugen-dake at 1359 m. In 1991, a large eruption generated a pyroclastic flow which killed 43 people.

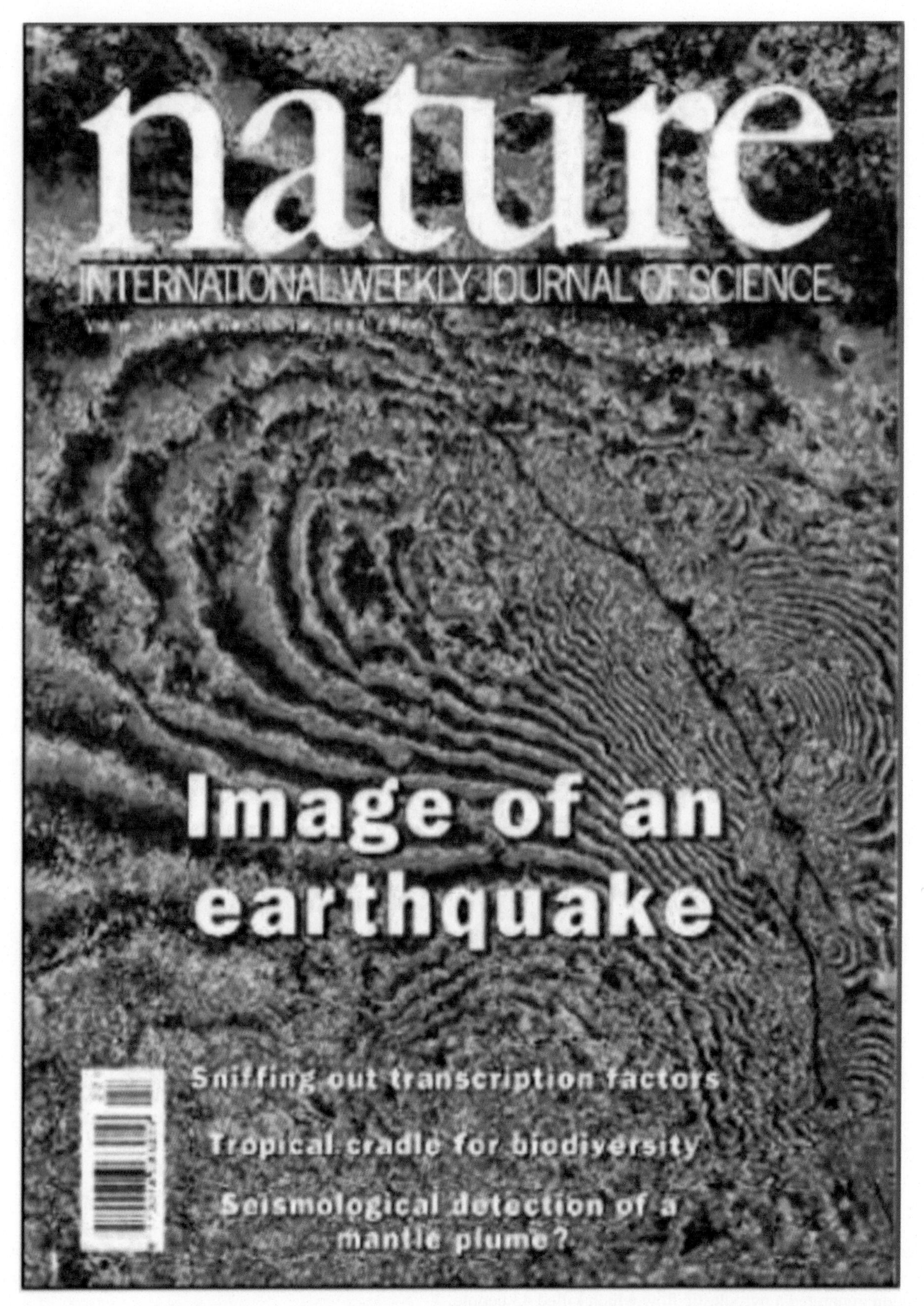

FIGURE 4.8 **Photograph of the front cover of *Nature* showing the new application of InSAR for studying earthquakes way back in 1993.** Since that year, more InSAR examples were generated to resolve the presence of atmospheric artifacts and phase decorrelation problems.

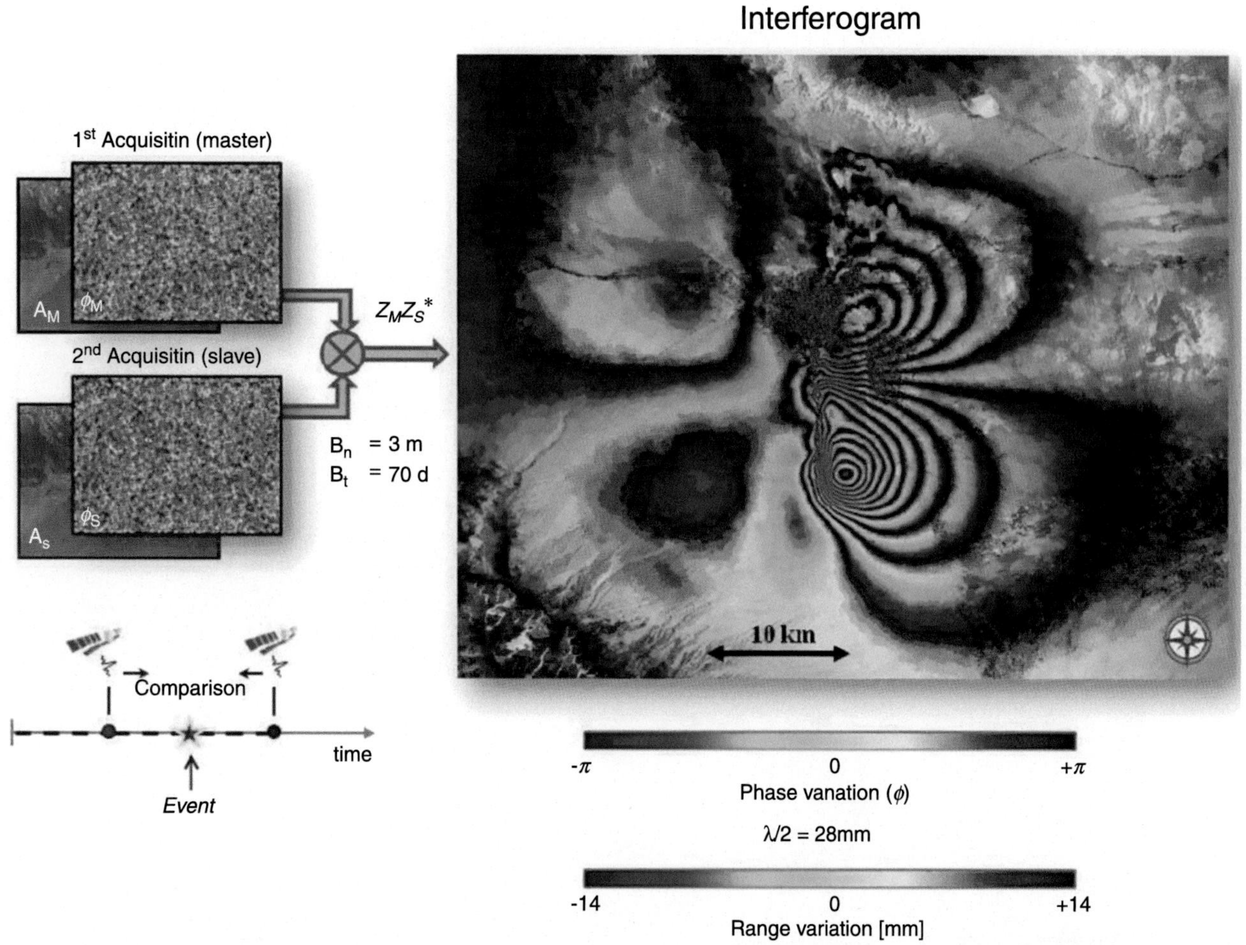

FIGURE 4.9 Example on how we can get an interferogram from radar images of two satellites (master and slave) for studying local deformation generated by earthquakes. *Source*: Ferretti (2015).

The prediction of eruption for volcanoes after long periods of dormancy is difficult because of the shortage of enough recent observation and monitoring. In the last several decades, there have been serious eruptions at some dormant volcanoes such as Unzen volcano. In May 20, 1991, a new lava dome appeared in the crater of the summit region of Unzen volcano (Fig. 4.13).

4.3.2 Kuju Volcano

The earliest magmatic activity of Kuju volcano (Fig. 4.14) was the andesite eruption about 1600 years ago (Kamata and Kobayashi, 1997). After a few hundred years of dormancy, Kuju volcano came to life with an ash eruption on October 11, 1995. Kuju volcano has three types of eruptions: Pyroclastic eruption (interval of several 10,000 years), magmatic eruption (formation of domes, 1000–2000 years), and phreatic eruption (10 years to 100 years).

4.3.3 Monitoring and Studying Activity of Volcanoes

Ground deformation using InSAR or global positioning system techniques is an important geodetic observable that provides information on ground–surface deformation during the time interval spanned by the image acquisitions. The emplacement of magma in the crust is potentially associated with ground deformation, by volume increases related to processes such as crustal anatexis, injection of magma into host rock or into an established chamber, hydrothermal excitation, and magma crystallization and degassing. Time-lapse microgravity surveys provide vital information on mass changes/redistribution and movement of fluids/gases beneath a volcano (Fig. 4.15).

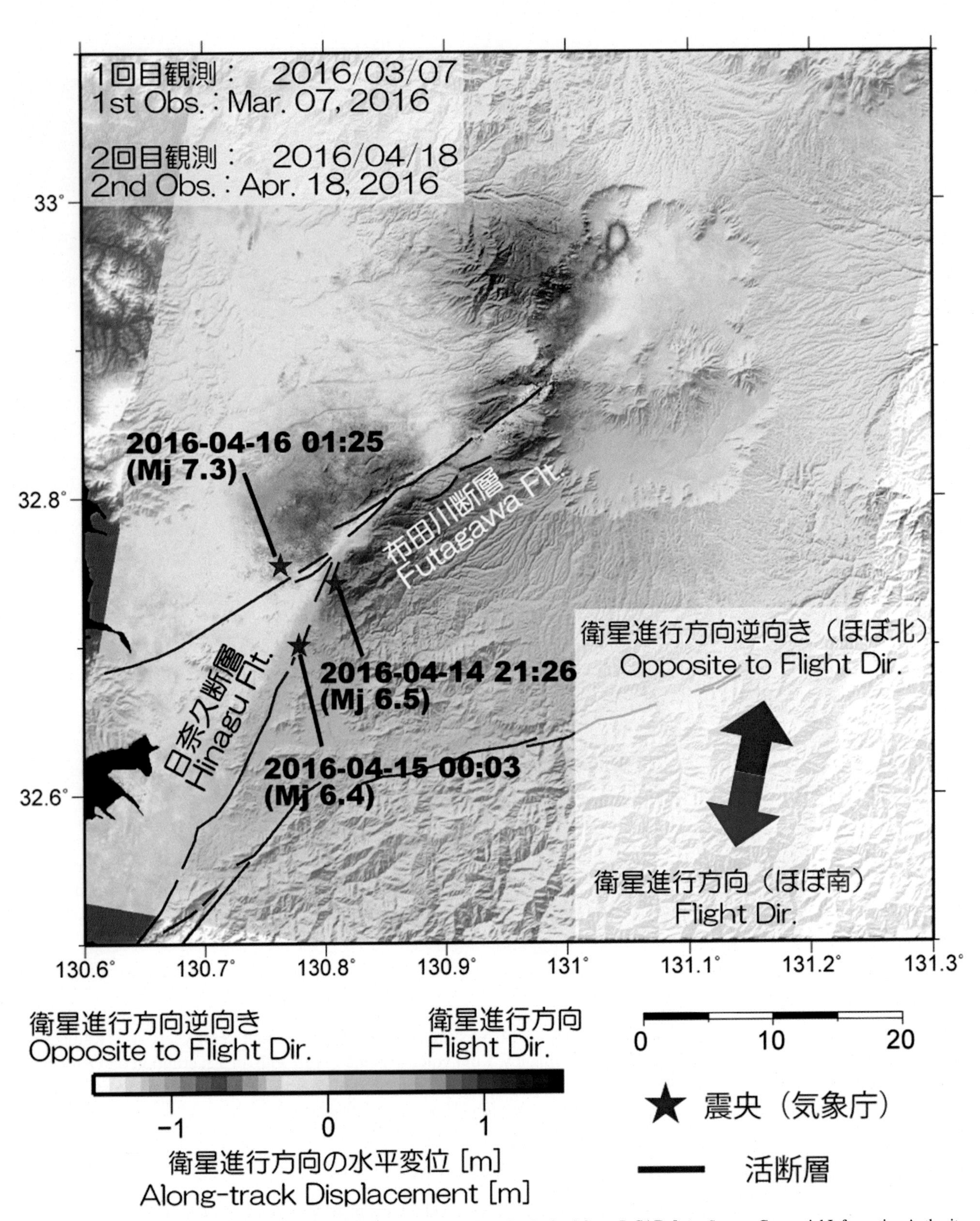

FIGURE 4.10 **Crustal deformation map of the 2016 Kumamoto earthquake derived from InSAR data.** *Source*: Geospatial Information Authority of Japan (2017a) (http://www.gsi.go.jp/cais/topic160428-index-e.html). ALOS-2/PALSAR-2: "Analysis by GSI from ALOS raw data of JAXA".

FIGURE 4.11 Aerial photograph showing the landslides after the 2016 Kumamoto earthquake (photograph taken on April 16). *Source*: Geospatial Information Authority of Japan website (http://www.gsi.go.jp/BOUSAI/H27-kumamoto-earthquake-index.html).

TABLE 4.1 Volcanoes of Kyushu Island, Japan (Volcano Discovery, 2017)

Volcano Name	Type	Location	Activity
Aso	Caldera	Central Kyushu	Active with frequent ash eruptions (0.3, 0.2, 0.15 Ma, 90,000 years ago)
Fukue-jima	Shield volcanoes	Fukue island	Last eruption 3000 years ago
Ibusuku	Calderas	South of Kyushu	Last activity few thousand years ago
Kirishima	Shield	South of Kyushu	Erupted in 2011
Kuju	Stratovolcanoes	Central Kyushu	Erupted in 1995–96
Sakurajima	Stratovolcano	South Kyushu	Younger volcano, constant activity, and one of the most active volcanoes in the world
Sumiyoshi-ike	Maars	South Kyushu	4550 BC, 5050 BC
Tsurumi	Lava domes	Eastern Kyushu	867 AD, 771 AD, 200 BC ± 50 years
Unzen	Stratovolcano	Western Kyushu	Many eruptions from 1990 to 1995

Source: https://www.volcanodiscovery.com.

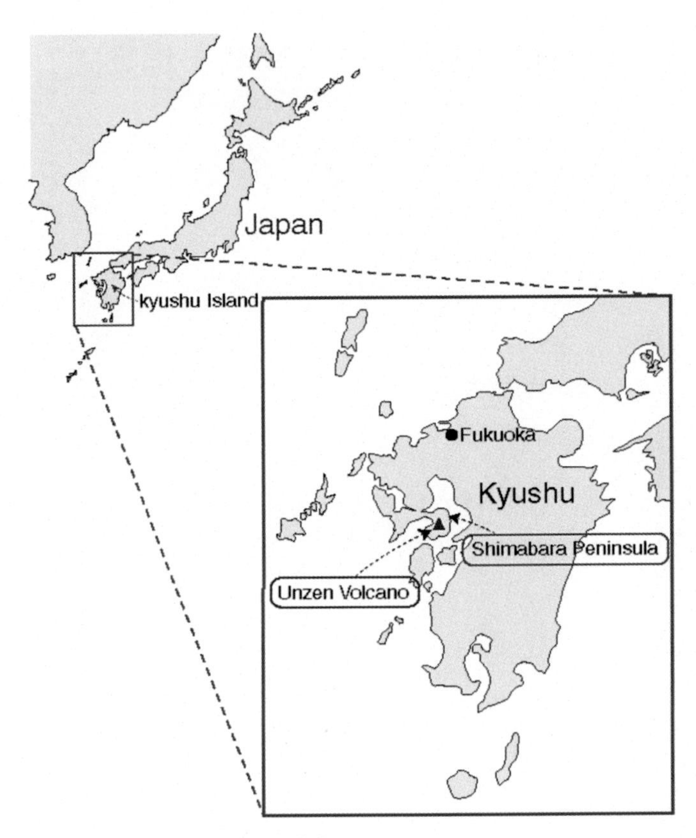

FIGURE 4.12 **Location of Unzen volcano in Kyushu Island, south Japan.**

FIGURE 4.13 **Photograph of Unzen volcano.**

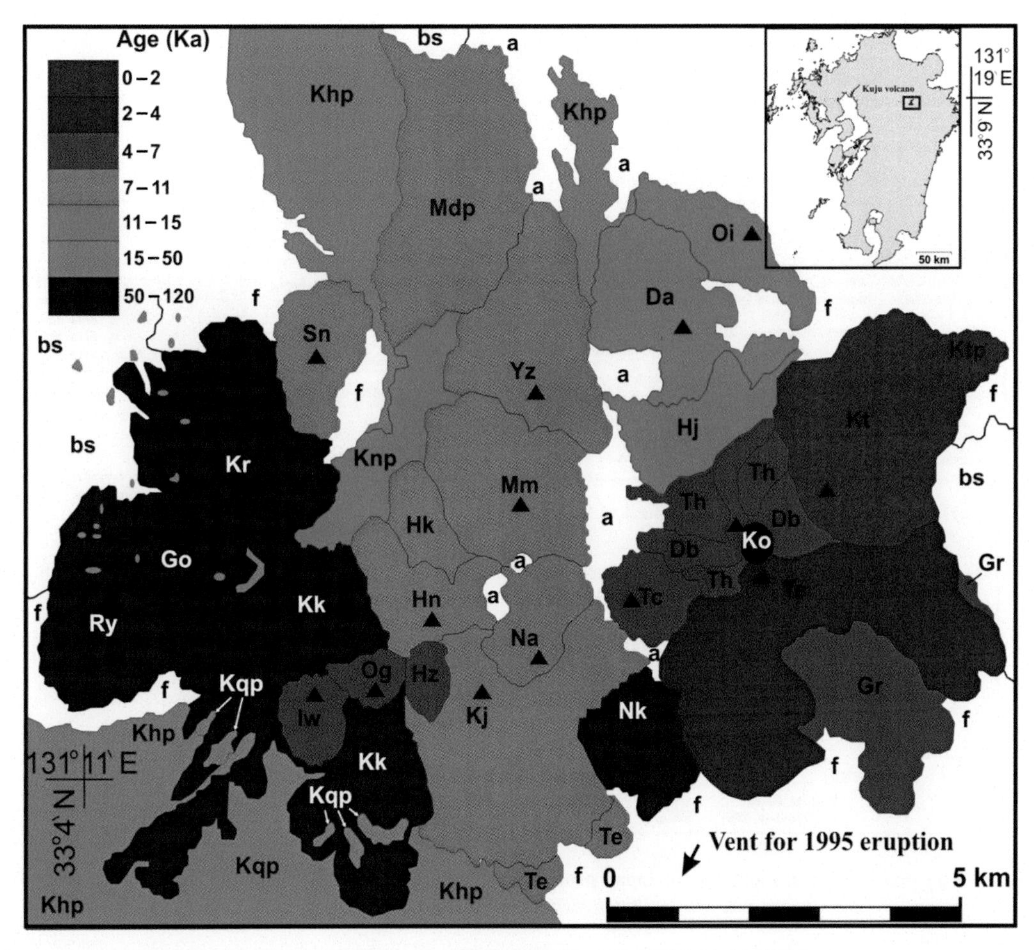

FIGURE 4.14 **Location and geologic map of Kuju volcano, modified from** Kamata and Kobayashi (1997).

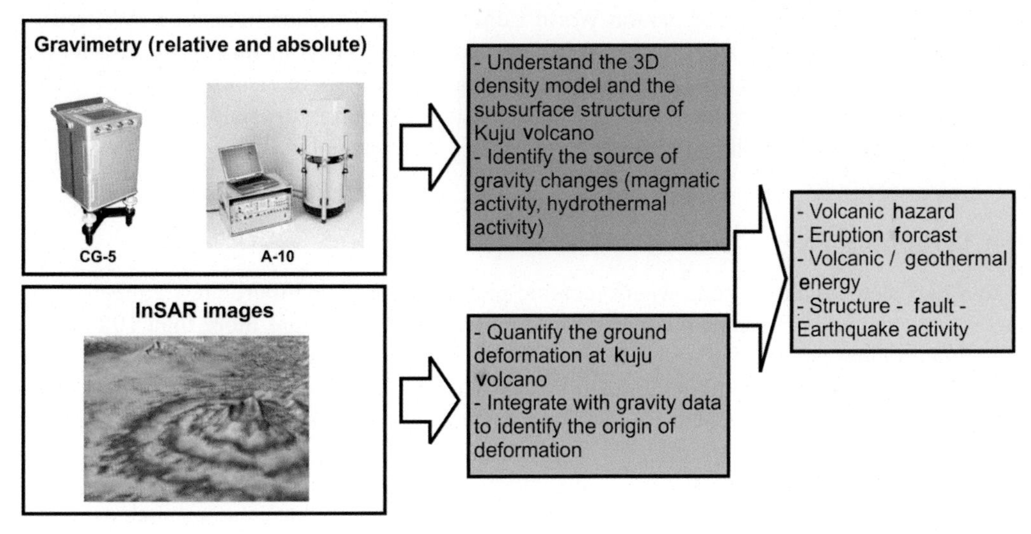

FIGURE 4.15 **Microgravity and InSAR as advanced technological tools in monitoring volcanic activities.**

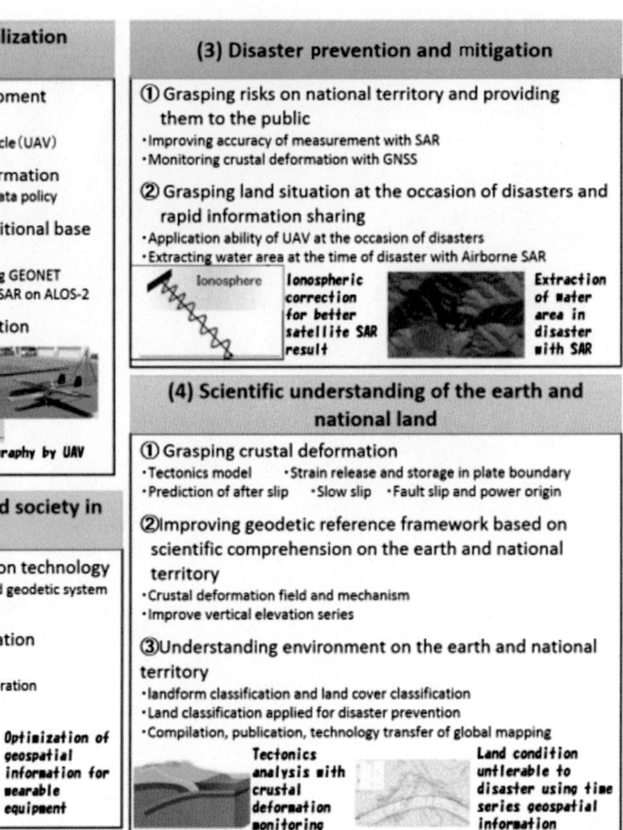

FIGURE 4.16 **Disaster prevention and mitigation in Japan proposed by the GSI of Japan** Geospatial Information Authority of Japan (2017b). *Source*: Geospatial Information Authority of Japan website (http://www.gsi.go.jp/ENGLISH/page_e30091.html).

4.4 DISASTER PREVENTION AND MITIGATION IN JAPAN

The Japanese government (Cabinet Office Japan, 2015) published the 53rd edition of the White Paper on Disaster Management, which highlights the 3rd United Nation World Conference on Disaster Reduction (WCDRR) held in Japan in March 2015. It provides information on the recent disaster statistics, disaster management, disaster policies, and the initiatives implemented by government and local communities on disaster risk reduction actions.

Japanese research institutes and universities are also promoting research in the field of disaster mitigation and prevention. As an example, the GSI developed a strategy in this field as presented in Fig. 4.16.

4.5 CONCLUSIONS

Japan is situated in a plate boundary zone where tectonic plates converge and there are frequent strong earthquakes in coastal areas that can create big tsunamis or offshore earthquakes. There are more than 100 active volcanoes in Japan, and 20% of earthquakes rated magnitude 6 or more are located in Japan. Japan is also susceptible to many other natural disasters such as tsunamis, typhoons, and landslides. The number of victims is much lower compared to other countries due to the culture of risk prevention, structure earthquake-resistant, and high-technology communication systems.

Scientifically, it is still difficult to predict when an earthquake will strike or when a volcano will erupt. However, a continuous geoscientific monitoring of faults and volcanoes using the advanced technology such as remote-sensing (InSAR), time-lapse geophysical monitoring, volcano gas monitoring, and seismicity monitoring will surely help to mitigate further earthquakes and volcanic eruption disaster.

REFERENCES

Cabinet Office Japan, 2015. White paper disaster management in Japan. http://www.bousai.go.jp/kaigirep/hakusho/pdf/WP2015_DM_Full_Version.pdf.

Ferretti, A., 2015. Satellite InSAR data: reservoir monitoring from space. EAGE Lecture. The University of Tokyo, Japan.

Fukahata, Y., Hashimoto, M., 2016. Simultaneous estimation of the dip angles and slip distribution on the faults of the 2016 Kumamoto earthquake through a weak nonlinear inversion of InSAR data. Earth, Planets and Space 68, 204, https://doi.org/10.1186/s40623-016-0580-4.

Geospatial Information Authority of Japan, 2017a. http://www.gsi.go.jp/cais/topic160428-index-e.html (accessed 07.17).

Geospatial Information Authority of Japan, 2017b. http://www.gsi.go.jp/ENGLISH/page_e30091.html (accessed 07.17).

Hashimoto, M., Savage, M., Nishimura, T., Horikawa, H., Tsutsumi, H., 2017. Special issue "2016 Kumamoto earthquake sequence and its impact on earthquake science and hazard assessment". Earth, Planets and Space 69, 98, https://doi.org/10.1186/s40623-017-0682-7.

Kamata, H., Kobayashi, T., 1997. The eruptive rate and history of Kuju volcano in Japan during the past 15,000 years. J. Volcanol. Geotherm. Res. 76 (1–2), 163–171.

Kobayashi, T., 2017. Earthquake rupture properties of the 2016 Kumamoto earthquake foreshocks (M_j 6.5 and M_j 6.4) revealed by conventional and multiple-aperture InSAR. Earth, Planets and Space 69, 7, https://doi.org/10.1186/s40623-016-0594-y.

Nishijima, J., 2005. Personal communication. Kyushu University.

Nishijima, J., Fujimitsu, Y., Fukui, Y., 2010. Densed gravity survey around the Kego fault, Fukuoka city. Chikyu Mon. 32 (4), 251–258.

Phillips, J.D., 2000. Locating magnetic contacts: a comparison of the horizontal gradient, analytic signal, and local wavenumber methods. 70th Annual International Meeting, SEG, Expanded Abstracts, 402–405.

Saibi, H., Nishijima, J., Hirano, T., Fujimitsu, Y., Ehara, S., 2008. Relation between structure and low-temperature geothermal systems in Fukuoka city, southwestern Japan. Earth, Planets and Space 60, 821–826.

Shimizu, H., Takahashi, H., Okada, T., kanazawa, T., Iio, Y., Miyamachi, H., Matsushima, T., Ichiyanagi, M., Uchida, N., Iwasaki, T., Katao, H., Goto, K., Matsumoto, S., Hirata, N., Nakao, S., Uehira, K., Shinohara, M., Yakiwara, H., Kame, N., Urabe, T., Matsuwo, N., Yamada, T., Watanabe, A., Nakahigashi, K., Enescu, B., Uchida, K., Hashimoto, S., Hirano, S., Yagi, T., Kohno, Y., Ueno, T., Saito, M., Hori, M., 2006. Aftershock seismicity and fault structure of the 2005 west off Fukuoka prefecture earthquake (M_{JMA}7.0) derived from urgent joint observations. Earth, Planets and Space 58 (12), 1599–1604.

Uehira, K., Yamada, T., Shinohara, M., Nakahigashi, K., Miyamachi, H., Iio, Y., Okada, T., Takahashi, H., Matsuwo, N., Uchida, K., Kanazawa, T., Shimizu, H., 2006. Precise aftershock distribution of the 2005 west off Fukuoka prefecture earthquake ($M_j = 7.0$) using a dense onshore and offshore seismic network. Earth, Planets and Space 58 (12), 1605–1610.

Volcano Discovery, 2017. https://www.volcanodiscovery.com (accessed 07.17).

Chapter 5

Winter Storms

Cüneyt Çalişkan
Emergency Aid and Disaster Management School of Health, Çanakkale Onsekiz Mart University, Çanakkale, Turkey

5.1 OVERVIEW

"Storm" means a disorder about the weather, and most weather events can result in a storm (Doswell, 2015). Winter storm is a natural meteorological event which causes a change in life by disturbing or suspending the normal flow of daily life (Guha-Sapir et al., 2012; EM-DAT, 2017a). In the past, winter storms used to result in socioeconomic problems such as wars (Pfister et al., 2010) and today, they may cause significant troubles in the performance of public services like education, health, telecommunication, transportation, and entertainment (Wightman et al., 2010). Serious public security threats that might arise due to the long-term suspension of such services can only be overcome by disaster continuum practices (Alexander, 2002).

Many of the weather-sourced hazards come out due to extreme temperatures, rains, or strong winds. When combined with temperatures below the freezing point, strong winds can cause instant heat loss in the human body with wind-chill effect. Long-term exposure to cold weather can cause life-threatening frostbite and hypothermia (Ross, 2002). Behavioral adaptation is the best way to adapt to cold weather. Heat loss can be slowed down or prevented by avoiding contact with cold surfaces or avoiding swallowing cold substances, getting protected from wind and rain, wearing protective clothing, or by moving to a physically hotter area (Wightman et al., 2010). Winter storms are various events significantly damaging agricultural and industrial services, interrupting ecological systems (Patz et al., 2000), influencing the mental health, increasing the amount of substance abuse (Yun et al., 2010), significantly affecting vulnerable groups (Cutter et al., 2003), preventing water, land, and air transportation, causing the closure of schools and public and business areas, interrupting daily life activities of families and isolating them, etc. (Winter).

In this chapter, the effects of winter storms on the society will be revealed out first. Secondly, the disaster continuum will be demonstrated within the framework of damage reduction and recovery stages. Finally, the public health perspective will be reflected for the protection of the public against the effects and results of winter storms and for the improvement of the current conditions. The list is given in Box 5.1 (National Oceanic and Atmospheric Administration, 2017).

5.2 THE SIGNIFICANCE OF WINTER STORMS

Winter storms are rare natural events that are exclusively experienced in America, Asia, and Europe. More than 3500 people have lost their lives in 63 winter storms since 2000 and more than 80 million people have been affected (EM-DAT, 2017b). In the United States, winter storms are among the significant weather events following the flood and violent storms (National Centers for Environmental Information, 2017a). Winter storms make up 11 of the 106 Disaster Statements of the Presidency declared in 2016 by the Federal Emergency Management Agency (FEMA) concerning climatic disasters (Federal Emergency Management Agency). In the logs of the period between 1980 and 2016, the US National Climatic Data Center reported 203 climatic disasters, each of which caused over $1 billion of economic damage. Fourteen winter storms among these have killed more than 1000 people since 1980 (Table 5.1) (National Centers for Environmental Information, 2017a).

A big winter storm hit the United States and lasted from February 14–20, 2015. It caused $3.1 billion damage and killed more than 30 people. The storm and the resultant cold wave affected several places in eastern and northeastern states. Boston city was particularly affected by this storm due to the load stress on the roofs caused by the accumulation of a great amount of snow and also due to the blocking of the transportation corridors. Direct loss of Massachusetts only extended beyond $1 billion. Many motor vehicle collisions (MVCs), hypothermia, and carbon monoxide cases caused many deaths in Tennessee. Flights were canceled, the highway was closed to traffic, schools and workplaces were closed, and electricity was cut off. Forest fires emerged, and houses were evacuated (National Centers for Environmental Information, 2017b).

In October 2013, a strong low-pressure system brought powerful rain and wind to the United Kingdom, France, Belgium, Denmark, and Germany. Winds blowing at 97–113 km/h (60–70 miles/h) speed knocked over many trees and caused chaos

Integrating Disaster Science and Management. http://dx.doi.org/10.1016/B978-0-12-812056-9.00005-1

BOX 5.1 Winter Storms Definitions

Avalanche—A mass of snow, rock, and/or ice falling down a mountain or incline. In practice, it usually refers to the snow avalanche. In the United States, the term snow slide is commonly used to mean a snow avalanche.
Blizzard—A blizzard means that the following conditions are expected to prevail for a period of 3 h or longer: sustained wind or frequent gusts to 35 miles an hour or greater; and considerable falling and/or blowing snow (i.e. reducing visibility frequently to less than 1/4 mile).
Blizzard Warning—Issued for winter storms with sustained or frequent winds of 35 mph or higher with considerable falling and/or blowing snow that frequently reduces visibility to 1/4 of a mile or less. These conditions are expected to prevail for a minimum of 3 h.
Blowing Snow—Blowing snow is wind-driven snow that reduces surface visibility. Blowing snow can be falling snow or snow that has already accumulated but is picked up and blown by strong winds. Blowing snow is usually accompanied by drifting snow.
Blowing Snow Advisory—Issued when wind-driven snow reduces surface visibility, possibly, hampering traveling. Blowing snow may be falling snow or snow that has already accumulated but is picked up and blown by strong winds.
Coastal Flooding—Flooding which occurs when water is driven onto land from an adjacent body of water. This generally occurs when there are significant storms, such as tropical and extratropical cyclones.
Conduction—Flow of heat in response to a temperature gradient within an object or between objects that are in physical contact.
Convection—Generally, transport of heat and moisture by the movement of a fluid.
Cyclone—A large-scale circulation of winds around a central region of low atmospheric pressure, counterclockwise in the Northern Hemisphere, clockwise in the Southern Hemisphere.
Drifting Ice—In hydrological terms, pieces of floating ice moving under the action of wind and/ or currents.
Drifting Snow—Drifting snow is an uneven distribution of snowfall/snow depth caused by strong surface winds. Drifting snow may occur during or after a snowfall. Drifting snow is usually associated with blowing snow.
Drizzle—Precipitation consisting of numerous minute droplets of water less than 0.5 mm (500 μm) in diameter.
Evaporation—The process of a liquid changing into a vapor or gas, usually water in meteorology.
Flood—Any high flow, overflow, or inundation by water which causes or threatens damage.
Freeze—A freeze is when the surface air temperature is expected to be 32 °F or below over a widespread area for a climatologically significant period of time.
Freezing Rain—Rain that falls as a liquid but freezes into glaze upon contact with the ground.
Freezing Rain Advisory—Issued when freezing rain or freezing drizzle is forecast but a significant accumulation is not expected. However, even small amounts of freezing rain or freezing drizzle may cause significant travel problems.
Frostbite—Human tissue damage caused by exposure to intense cold.
Heavy Snow—Snowfall accumulating to 10.2 cm or more in depth in 12 h or less or snowfall accumulating to 15.2 cm or more in depth in 24 h or less.
Heavy Snow Warning—Issued by the National Weather Service when snowfall of 6 inches (15 cm) or more in 12 h or 8 inches (20 cm) or more in 24 h is imminent or occurring. These criteria are specific for the Midwest and may vary regionally.
Hypothermia—A rapid, progressive mental and physical collapse that accompanies the lowering of body temperature.
Ice Fog—A suspension of numerous minute ice crystals in the air or water droplets at temperatures below 0 °C, based at the Earth's surface, which reduces horizontal visibility. Also called ice-crystal fog, frozen fog, frost fog, frost flakes, air hoar, rime fog, pogonip.
Ice Jam—In hydrological terms, a stationary accumulation that restricts or blocks streamflow.
Ice Storm—An ice storm is used to describe occasions when damaging accumulations of ice are expected during freezing rain situations. Significant accumulations of ice pull down trees and utility lines resulting in loss of power and communication. These accumulations of ice make walking and driving extremely dangerous. Significant ice accumulations are usually accumulations of 1/4″ or greater.
Ice Storm Warning—This product is issued by the National Weather Service when freezing rain produces a significant and possibly damaging accumulation of ice. The criteria for this warning varies from state to state, but typically will be issued any time more than 1/4″ of ice is expected to accumulate in an area.
Lake Effect Snow—Snow showers that are created when cold, dry air passes over a large warmer lake, such as one of the Great Lakes, and picks up moisture and heat.
Nor'easter—A strong low-pressure system that affects the Mid Atlantic and New England States. It can form over land or over the coastal waters. These winter weather events are notorious for producing heavy snow, rain, and tremendous waves that crash onto Atlantic beaches, often causing beach erosion and structural damage. Wind gusts associated with these storms can exceed hurricane force in intensity. A nor'easter gets its name from the continuously strong northeasterly winds blowing in from the ocean ahead of the storm and over the coastal areas.
Rain—Precipitation that falls to earth in drops more than 0.5 mm in diameter.
Sleet—Sleet is defined as pellets of ice composed of frozen or mostly frozen raindrops or refrozen partially melted snowflakes. These pellets of ice usually bounce after hitting the ground or other hard surfaces. Heavy sleet is a relatively rare event defined as an accumulation of ice pellets covering the ground to a depth of 12.7 mm or more.

BOX 5.1 Winter Storms Definitions—(*Cont.*)

Sleet Warning—Issued when accumulation of sleet in excess of 12.7 mm is expected; this is a relatively rare scenario. Usually issued as a winter storm warning for heavy sleet.
Snow—Precipitation in the form of ice crystals, mainly of intricately branched, hexagonal form and often agglomerated into snowflakes, formed directly from the freezing (deposition) of the water vapor in the air.
Snow Flurries—Snow flurries are an intermittent light snowfall of short duration (generally light snow showers) with no measurable accumulation (trace category).
Snow Shower—A snow shower is a short duration of moderate snowfall. Some accumulation is possible.
Snow Squall—A snow squall is an intense, but limited duration, period of moderate to heavy snowfall, accompanied by strong, gusty surface winds, and possibly lightning (generally moderate-to-heavy snow showers). Snow accumulation may be significant.
Wind Chill—Reference to the Wind Chill Factor; increased wind speeds accelerate heat loss from exposed skin and the wind chill is a measure of this effect.
Winter Storm Warning—This product is issued by the National Weather Service when a winter storm is producing or is forecast to produce heavy snow or significant ice accumulations. The criteria for this warning can vary from place to place.
Winter Storm Watch—This product is issued by the National Weather Service when there is a potential for heavy snow or significant ice accumulations, usually at least 24–36 h in advance. The criteria for this watch can vary from place to place.
Winter Weather Advisory—This product is issued by the National Weather Service when a low pressure system produces a combination of winter weather (snow, freezing rain, sleet, etc.) that present a hazard, but does not meet warning criteria.

Source: This information is in the public domain. It is retrieved from the National Weather Service Glossary (National Oceanic and Atmospheric Administration, 2017).

TABLE 5.1 Billion-Dollar Events to Affect the United States From 1980 to 2016 (CPI-Adjusted)

Disaster type	Number of events	Percent frequency (%)	CPI-adjusted losses (in billion dollars)	Total losses (%)	Average event cost (in billion dollars)	Deaths
Drought	24	11.8	223.8	19.1	9.3	2993[a]
Flooding	26	12.8	110.7	9.4	4.3	515
Freeze	7	3.4	25.3	2.2	3.6	162
Severe Storm	83	40.9	180.1	15.3	2.2	1546
Tropical Cyclone	35	17.2	560.1	47.7	16.0	3210
Wildfire	14	6.9	33.0	2.8	2.4	184
Winter Storm	14	6.9	41.3	3.5	3.0	1013
All Disasters	203	100.0	1174.3	100.0	5.8	9623

[a]*CPI, Consumer price index.*
Source: This information is in the public domain. It is retrieved from the National Oceanic and Atmospheric Administration (National Centers for Environmental Information, 2017a).

in the transportation line in the South United Kingdom. 625,000 houses in the United Kingdom and around 1.2 million people in Denmark were deprived of electricity. The damage due to the winter storm was approximately $1.35 billion in West and North Europe. In the same year in Asia, around 1000 ships got stuck in ice in the Bohai Sea between North China and North Korea due to extreme cold (LeComte, 2014).

5.3 FORMATION MECHANISM AND PROPERTIES OF WINTER STORMS

Many of the impacts of winter storms are associated with precipitation. This precipitation can occur as rain, snow, sleet, freezing rain, or ice pellets (Eather, 2017; Gibson and Stewart, 2007). When they are accompanied with low temperatures and winds aloft, several hazards might arise (Winter Storm, 2017a). Three common components: humidity, lifting, and cold weather take part in the formation of these winds. First, humidity forms with the vaporizing of water resources, the primary of which are seas and oceans. Humid clouds can cause rain. Secondly, humidity must be lifted up in the air for humid clouds to turn into rain. This can also happen when air climbs up from hills or from a lake. Thirdly, the temperature at ground level and inside cloud can be below the freezing point that might create snow or ice. Thus, humid clouds can precipitate as snow or ice pellets (Clements and Casani, 2016).

Winter storms that happened throughout the history were given names according to the place or time they hit or the theater they damaged. In 2012, the Weather Channel announced the winter storms to be named in a systematic way. The latest classification of this controversial issue (Wikipedia, 2017a) was published for the 2016–2017 period (Winter Storm, 2016). Still, winter storms can be classified according to the type of rain or storm (Eather, 2017). As each climatic event has a unique ecological, organizational, and sociological context, the common properties learned (Donahue and Tuohy, 2006) from a particular event are the following:

1. A winter storm can be a moderate degree snow that lasts for a few days or a blinding snow storm that lasts for a few days. Some winter storms can be so big that they can affect a few states, whereas some others can affect a single community only (Winter Storm, 2017a).
2. Local people of extremely cold regions can be more prepared for the potential risks of winter storms. For example, infrastructural systems they prepare can resist against extremely cold weather conditions for weeks and even for months, whereas the water pipes in milder regions might not have sufficient insulation and thus can blow out in a single night (Clements and Casani, 2016).
3. Though perceptions and definitions might vary from one region to another, vulnerable population is a common factor. Too young or too old people are affected by low temperature as well as hot temperature regardless of local norms or extremities. Regional infrastructures developed on the basis of these temperature norms have great significance (Clements and Casani, 2016). Regarding social perceptions, postevents of the Katrina storm that hit New Orleans in the United States showed that, women and youth were affected by climatic events due to their limited adaptive capacities and sociocultural discrimination (Overton, 2014).
4. Winds aloft accompanied by cold weather conditions pose significant threats. They might accelerate heat loss in the human body, thus can cause the wind-chill effect. In ice storms, winds can cause trees, telecommunication, and power lines to hit each other and fall down, can cause floods in coastal regions, suspension of transportation, deathly MVCs on roads, a complete chaos of the urban life, and slowing of emergency services. When excessive amounts of ice or snow get accumulated, weaker structures might collapse or people might slip, fall down, and injure themselves during the cleaning works of these, or carbon monoxide poisoning might happen during the use of fuel oil-sourced generators or propane heaters as alternative power sources when electricity is cut off (Broder et al., 2005; Clements and Casani, 2016; Piercefield et al., 2011; Wightman et al., 2010). The US National Weather Service define winter storms as "Deceptive Killers" because most cases of death are stated to be related to storm winds (Flinn, 2017). Therefore, it is necessary to define and characterize the vulnerable populations and conditions regarding fatal or nonfatal risk factors that might be related to winter storms in order to better organize the activities for public health and health communication (Iqbal et al., 2012).

Winter storms (Zielinski, 2002) are also important for public and private institutions in order to get prepared for the potential effects of snowfall (Kocin and Uccellini, 2004; Squires et al., 2014) or ice (Spia-index.com, 2017) severity categories, to respond to such effects and to deliver help to the populations under risk. Zielinski (2002) developed a five-level potential effect index for winter storms. The index was prepared based on the following parameters: the intensity of winter storms, atmospheric pressure, time period, speed, and rain. Though the index might vary depending on the regions and the social structures, it might be helpful in the prediction of the effect of the storm.

5.4 MITIGATION

Disaster mitigation measures can be defined as hazard prevention or risk mitigation (Disaster Prevention and Mitigation, 2017). Efficient mitigation measures can prevent a repetition of the damage and break the continuum for disaster damage and reconstruction (FEMA, 2016). Thus, mitigation can be considered to be the keystone for disaster management. Although the remaining three components (preparation, response, and recovery) respond to hazards, mitigation measures try to prevent the risk of threat before the disaster occurs and to eliminate or mitigate its results (Coppola, 2011). For example, FEMA and several programs covering all natural threats are implemented. General objectives of these cover the attempts for mitigation: (1) to reduce the loss of life and property and create safer societies, (2) to promote fast recovery of people after a flood or any other disaster, and (3) to minimize the financial effects of disasters at Federal Treasury, States, Tribes, and Regional government levels (FEMA, 2016). According to FEMA, an efficient mitigation program has three basic components: risk management, mitigation, and insurance (FEMA, 2016). Here, mitigation attempts will be discussed in detail within the framework of winter storms.

5.4.1 Risk Management

Risk management is the first priority in reducing vulnerability against natural threats (Rice and Spence, 2016). In order to reduce or eliminate the risks of winter storms, efficient strategies must be defined (FEMA, 2016). The following general objectives can be used when reviewing the appropriate options for preventing or reducing the effect of winter storm risks (Coppola, 2011).

1. It is not yet possible to *reduce the risk* of the occurrence of winter storms. It is technically impossible to intervene in winter storms like blizzards, ice storms, lake-effect storms, or nor'easters. On the other hand, the damages caused by such can be prevented. Removing unfixed tools and equipment from house gardens, adjusting the roof's capacity to bear snow overloads, and preferring not to drive in bad weather conditions (Datla et al., 2013) can be listed among the examples of such measures.
2. It is aimed to reduce potential damage of winter storms on human beings, structures, or the environment, in other words, to *reduce the results of risks*. Mechanisms that enable the strengthening of structures against potential damage of strong winds, wind shelters for exposed population, and activity and action limiting regulations in highly risky regions can significantly reduce the results of winter storms.
3. The damage risks of winter storms are so high that, though the results can be reduced, they cannot be acceptable. Total *risk avoidance* for winter storms can generally be achieved by evacuating all the people from the area under risk. For example, in the winter storm that hit North Carolina in January 2017, hundreds of people were evacuated (Winter Storm, 2017b). Yet, this is a difficult process and each event can generate its unique results. If avoidance is not an option, public authorities can instruct the society and guide them to perform useful actions (Sellnow and Sellnow, 2010). For example, they can instruct the people to avoid driving during winter storms (Rice and Spence, 2016).
4. Extremely hot weather and strong winds must be present for winter storms to occur. These are unique properties for particular geographical regions and ***risk acceptance*** of each region is different. Sociocultural patterns or cost–benefit analyses, etc. can affect certain risk reduction measures negatively.
5. In winter storms, *risk transfer, sharing, and spreading* are the ultimate damage reduction goal. Here, it is aimed to transfer risks between people within the scope of insurance and with international reassurance in order to minimize material losses of the victims of the disaster.

5.4.2 Mitigation

Mitigation is the elimination or mitigation of long-term risks over current and future building area. Damage reduction measures in winter storms include land usage planning, adoption of sound building practices (FEMA, 2016), improvement of socioeconomic status, early notice capacity building, cultural practices (Greenough et al., 2001), individual preparedness, and other activities intended for a life of property risks through various granting programs (FEMA, 2016). Damage and effect reduction measures for winter storms are classified into two basic categories: structured and unstructured measures (Greenough et al., 2001).

1. Structured mitigation is defined as a risk reduction attempt realized through building or modification of the physical environment with the implementation of engineering solutions. In other words, these are the practices where the human being tries to dominate nature.
2. Unstructured mitigation is defined as a measure that reduces risk by modifying the human behaviors or natural processes without the need for the use of engineering structures. In other words, these are the practices where the human being tries to adapt to nature.

5.4.2.1 Structured Mitigation

Structured mitigation measures include construction, engineering, or other mechanic restoration measures intended for the reduction of the risks or effects of winter storms. Any measure related to winter storms is an effect reduction practice and examples will be given for particular items in each category. General structured mitigation measures are as follows:

1. Techniques for the protection of buildings and infrastrutures (FEMA, 2013):
   a. *Adding building insulation to walls and attics.*
   b. *As buildings are modified, using new technology to create or increase structural stability.*

   c. *Retrofitting public buildings to withstand snow loads and prevent roof collapse.* Flat roofs are particularly vulnerable to the risk of collapse and water damage. On the other hand, metal roof covering systems, steel decks, and plates on the beams have the ability to withstand heavy loads or rains and to save. The overload effect caused by rain or water is due to water "accumulation" in lower areas of the roof. Thus, snow, ice, and rain add severe load to the roofs and cause them to collapse. Roof collapsing can cause serious injuries and loss of lives (Preventing, 2012). For example, 57 people died when the roof of a Bazaar collapsed in 2006 in Moscow due to severe snowing (Moscow, 2016).
2. Techniques for the protection of power lines (FEMA, 2013):
   a. *Establishing standards for all utilities regarding tree pruning around lines.*
   b. *Using designed-failure mode for power line design to allow lines to fall or fail in small sections rather than as a complete system to enable faster restoration.*
   c. *Installing redundancies and loopfeeds.*
   d. *Burying overhead power lines.* This is the burying of overhead electricity or telecommunication lines. This is generally done in for esthetical purposes in cities (Fig. 5.1) (Editorials, 2016). This also prevents power cuts in lines during strong winds, lightning storms, intense snow, or ice storms (Wikipedia, 2017b).
3. Strategies intended for the reduction of effects on the roads (FEMA, 2013):
   a. *Planning for and maintaining adequate road- and debris-clearing capabilities.*
   b. *Using snow fences or "living snow fences" (e.g. rows of trees or other vegetation) to limit blowing and drifting of snow over critical roadway segments.*
   c. *Installing roadway-heating technology to prevent ice/snow buildup.* Snow melting systems (hydronic, heat pipe, and electrical technologies) (Minsk, 1999) facilitate the flow of traffic safety during winter storms (Fig. 5.1). Moreover, they can supersede other snow melting operations (snow-cleaning vehicles and salt, sand, or other chemical deicers) and protect asphalt (Wikipedia, 2017c).

5.4.2.2 Unstructured Mitigation

Unstructured mitigation measures include activities intended for the reduction of the risks or effects of winter storms generally without the need for the use of engineering practices. Any measure related to winter storms is an effect reduction practice, and examples will be given for particular items in each category. General unstructured damage reduction measures are the following:

1. Consciousness-raising activities concerning winter storms (FEMA, 2013):
   a. *Informing the public about severe winter weather impacts.*
   b. *Producing and distributing family and traveler emergency preparedness information about severe winter weather hazards.*
   c. *Including safety strategies for severe weather in driver education classes and materials.*
   d. *Encouraging homeowners to install carbon monoxide monitors and alarms.*

FIGURE 5.1 Under-floor heating in the street. This photograph is in the public domain. Retrieved from Wikipedia at: https://en.wikipedia.org/wiki/Snowmelt_system#/media/File:%22Under-floor_heating%22_in_the_street.jpg.

BOX 5.2 Preparations of Ambulance Staff for Severe Winter Conditions, Çanakkale, Turkey

Çanakkale is located in the Southwest of Marmara region in Turkey and it possesses land both in Asia and Europe continents. It owns the Dardanelles Strait connecting the Mediterranean Sea and the Black Sea. Çanakkale has 12 towns in total. Temperatures go down starting from the coast toward the higher interior parts and there are always strong winds in the Dardanelles Strait.

As the ambulance staff mostly work outside, they are often exposed to cold weather. Their deficiencies regarding their personal protection against the cold might cause a public health problem with regards to access to urgent health response. In this context, measures taken by 167 ambulance staff were evaluated within the scope of personal measures and measures taken for the house and vehicle. Important findings of the research are given below:

1. 57.5% of the participants were following the news about weather conditions in cold weather. According to their statement, 8.4% of the participants had traffic accident in cold weather, 17.4% of them got injured, 9% of them had hypothermia, and 8.4% of them had frostbite.
2. With respect to personal measures, more than half of the participants stated that they wore waterproof boots and coat, gloves, scarf, and beret with ear pads.
3. With respect to measures taken at home in cold weather, more than half of the participants stated that they had an alternative heat source, 3-days food ready-to-eat food reserve and wall thermometer before the winter came; fewer than half of the participants stated that they had first-aid kit, fire extinguisher, operative smoke, and carbon monoxide sensor detectors. Again, fewer than half of the participants stated that they had their doors and windows insulation and chimney checks every year and that, the outer pipes of their houses were insulated against frost.
4. The majority of the participants who had cars stated that they regularly had their cars maintained before the winter came.

The participants who participated in the research were found to have deficiencies with respect to their personal measures and the measures they took for their houses and vehicles. The findings of the study emphasized the need for the training of health personnel about personal measures and measures for their houses and vehicles regardless of their occupations.

Source: Çalışkan et al. (2014).

  e. *Educating citizens that all fuel-burning equipment should be vented to the outside.*
  f. *Informing urgent operation staff of departments such as fire stations, police stations, and emergency services about the impacts of severe winter storms.* Box 5.2 gives information about the winter storm preparations of ambulance personnel of a city. Although this study is about the preparation activities, it informs the ambulance staff about the unstructured risk reduction measures for winter storms.

2. Activities concerning vulnerable populations (FEMA, 2013):
   a. *Identifying specific at-risk populations that may be exceptionally vulnerable in the event of long-term power outages.*
   b. *Organizing outreach to vulnerable populations, including establishing and promoting accessible heating centers in the community.* For instance, Wisconsin State authorities conducted a study about the special groups in the community as part of the general community study for long-term power cuts. The study aimed to ensure life safety of special groups during long-term power cuts. Relevant data was collected within this framework and planning was made for the delivery of appropriate home healthcare, shelter, and transfer services to these special groups when necessary (Long, 2010).

5.4.3 Insurance

Insurance spreads the financial load of the purchasable hazard on the disaster area, thus relieves the victims of disaster from their financial burden. Insurance is a concept that refers to sharing of the financial results of the disaster by the people included in the insurance pool, not only by the victims affected. Thus, financial risk of a hazard is transferred, shared, and spread among the people through insurance. Insurance is generally considered for the risks of flood, earthquake, and fire.

In the United States, winter storms caused $2.1 billion insured loss in 2015 and 15 winter storms caused $11.9 billion insured loss from 1985 up to 2015 (MunichRE-NatCatSERVICE, 2016a). In Europe, winter storms caused $3.1 billion insured loss in 2010 and 10 winter storms caused $36.1 billion insured loss from 1980 up to 2015 (MunichRE-NatCatSERVICE, 2016b). Insurance studied about the Kyrill (2017) winter storm is given in Box 5.3.

5.5 RECOVERY

The impacts of climatic events on health depend on the vulnerabilities of the natural environment and local people and their capacities for recovery (Greenough et al., 2001). On the other hand, even the best damage reduction measures can only

BOX 5.3 Kyrill Winter Storms, European 2007

In 2007, the winter storm named Kyrill hit the UK and many European countries including Germany, France, Netherlands, Denmark, and Austria with a speed of up to 200 km/h. The storm caused suspension of transportation and power cuts, floods, and damages in buildings and 47 people died in this storm (Kyrill, 2017).

In Europe, Kyrill met two of the three criteria for extreme loss which are intensity, size, and location. The only criterion Kyrill did not meet was: intensity (Kyrill, 2017). Kyrill caused $10 billion economic damage. Only $5.8 billion of this amount was insured (Munichre, 2016).

Average demand concerning winter storms in Europe is relatively lower compared to other threats and happens around $1500. A big winter storm will cause great losses. Therefore, a well-organized insurance penetration in Europe can contribute to relatively higher insured losses.

Source: Tatge, 2017 (Kyrill, 2017).

mitigate such impacts. Reconstruction, reorganization, and repair process of the item affected by the disaster and refunctioning of such item maybe even better than its former state is called "recovery". The recovery process can start long before the occurrence of a disaster, within the framework of the emergency management system including predisaster planning, mitigation, and preparation activities.

US Department of Home Security developed eight principles in 2011, which maximize the chance for recovery if applied properly. Key recovery skills serve as a guide for disaster authorities. Key skills are the critical criteria that are required to be applied by the whole society for the achievement of the National Preparedness Objective. These provide a common vocabulary defining the key functions to be developed and implemented in the entire society for national preparedness. Key recovery skills will be analyzed within the framework of winter storms: planning, public information and warning, operational coordination, economic recovery, health and social services, housing, infrastructure systems, and natural and cultural resources (FEMA, 2011).

5.5.1 Planning

Planning is the execution of a systematic process in which the entire society is included for the development of applicable strategic, operational, and/or tactical approaches. A well-organized and comprehensive predisaster and postdisaster local, regional, and national recovery process must be implemented. In a region that is exposed to winter storms, efficient recovery activities are performed through preparation activities. These activities form the ground for the recovery process. Predisaster and postdisaster recovery planning is significant for the communities to gain flexibility and reach recovery objectives (FEMA, 2011). The responsibility for disaster preparation starts with the individual and then, it combines with the wider responsibilities of the society and the local state (FEMA, 2011; Koçak et al., 2015). Individuals from all strata must be included in the community planning process. Tasks of these individuals must be communicated to them through various ways before the disaster comes out. Communication of the tasks is also important for the facilitation of postdisaster recovery. Furthermore, these tasks support the implementation of the priorities and policies and adaptation to decision-making processes before the disaster (FEMA, 2011). During Kentucky Ice storm that occurred in 2009, the water sector was exposed to overloading. Preparation, response, and recovery activities of the water sector are shown in Box 5.4.

5.5.2 Public Information and Warning

Public health measures can be taken by emphasizing the predisaster risk communication and by adapting the interventions to racial, ethnic, and linguistic minorities (Iqbal et al., 2012). In this sense, the main objective of public training is to prevent the results of an event before they occur. US Center for Disease Control and Prevention prepared a guide about winter storms for the public healthcare personnel, shareholders, and the primary healthcare personnel. Here, better understanding and management of winter storms was aimed (Rhodes and Kailar, 2005; National Disaster Education Coalition, 2004). Public messages about any risk or threat must be clear, coherent, and accessible and where necessary, must deliver coordinated, instant, reliable, and applicable information to the entire community in a culturally and linguistically appropriate manner. Public information messages can respond to the expectations of the society. The messages must be short, clear, and accurate and must be formed in different languages to address to the individuals talking different languages in such a society. The place, time, and duration of winter storms can be determined today, thanks to the advanced technical mechanisms used in the weather forecast. In parallel with this, the data collected can be communicated to the public through radio, television (FEMA, 2011), or Internet (Bonnan-White et al., 2014) to warn them so that they get prepared. On the other hand, a

BOX 5.4 Example of Water Sector Impacts and Response to a Winter Storm—Kentucky 2009 Ice Storm

A winter storm hit Kentucky in January 2009 and it caused the longest power cut in its history. Despite the fact that the storm was announced to be coming before it came out, the storm was stronger than expected and there appeared significant impacts on the water sector. Many water services organized by Kentucky Public Service Commission (PSC) were affected by the ice storm and more than 32,000 consumers were unable to access water service at the certain point of the storm. Hickory Water Area in Kentucky-Graves lost its entire water services during the storm. Though there was a water reservoir that could be sufficient for 48 h, consumers could not be delivered water from the reservoir due to power cut and the lack of a spare power generator. Several services were repaired one day after the ice storm with the support of a significant amount of electricity suppliers. PSC gave some recommendations about how to get prepared for the future events in water and wastewater facilities. These recommendations include an information exchange regarding the establishment of call centers for potential service suspensions and organization of a common assistance network.

Source: Kentucky Public Service Commission, 2009 (The Kentucky Public, 2009).

research conducted has revealed out that despite the online information about power cuts, the local people would like to be informed by phone as well (Palm, 2009).

5.5.3 Operational Coordination

Operational coordination means the establishment and continuity of an integrated and coordinated operational structure and process which includes the appropriate participation of all key shareholders and which supports the usage of basic skills (FEMA, 2011). Operational coordination supports the cooperation, information exchange, and efficient use of resources by private and civil society organizations at all levels of the society and the state (Spence et al., 2011). Cooperation must start with the available local resources and must be more comprehensive where necessary. Cooperation between health security or communication network services is an example of such that is intended for the performance of immediate recovery activities to minimize the damage of the people living in a region where the winter storms hit.

5.5.4 Economic Recovery

New job and employment opportunities must be developed in order to recover economic and commercial activities (including food and agriculture) and to contribute to a society with economic welfare. Economic recovery is realized by several complex and interdependent components. The public sector, nonprofit organizations, and the private sector contribute to this process by undertaking the costs after the disaster and by taking active steps for the reconstruction of the local economy. In an economic recovery, the private sector meets the needs for reconstruction and actively participates in recovery planning studies. The individuals, civil society organizations, and governments generally need a temporary economic recovery after an event and there is the tendency to transfer the resources to trivial needs as time goes by (FEMA, 2011).

5.5.5 Health and Social Services

In an event of threat, it is necessary to protect and improve the welfare of the entire society and to strengthen the skills, flexibility, independence, and soundness of health and social services. Therefore, timely restoration of health (hospitals, dialysis centers, etc.) and social (child care, family aids, etc.) services for the recovery of the society. This can be achieved with the cooperation of all partners and shareholders in the affected area (FEMA, 2011).

In addition to having the capacity to directly or indirectly affect the human health, winter storms also cast negative impacts on economic, social, and health services. For example, the ice storm that hit Montreal in 1998 had temporary impacts on critical infrastructure (suspension of transportation, restriction of emergency services, damages to trees, and personal properties) (Henson et al., 2007). Immediate recovery and reconstruction of public services can help the society regain its welfare.

5.5.6 Housing

Housing solutions that efficiently support the needs of the entire society contribute to the sustainability and flexibility of such a society. During winter storms, structures that are not sound enough might get repairable or irreparable damages (FEMA, 2011). Still, people build even more houses in areas that are not resistant against winter storms (Upfront, 2007). Therefore, temporary or permanent housing problems occur after disasters. Both types of housing incur expensive processes. The core ability regarding housing means that public authorities and planners focus on sufficient amount of cost-efficient housings that are globally accessible and they implement sound housing solutions that efficiently support the needs of the entire society and that contributes to sustainability and flexibility. Labor power, material, time, design, supervision, and financial resources are the basic elements underlying such solutions (FEMA, 2011).

5.5.7 Infrastructure Systems

This recovery process aims to make critical infrastructure functions consistent and minimize health and security threats. The objective of the recovery process is to prevent the problems that might occur after the disasters and to respond to the needs of the society. Vulnerable points in infrastructure systems can be detected and strengthened through public–private sector cooperation (FEMA, 2011).

Transportation and power mechanism infrastructures are particularly damaged during winter storms. Heavy snowing, ice accumulation, and strong winds can cause delay or suspension in transportation, can damage trees and power lines, and can cause the collapse of roofs and other structural damages (Heneka and Ruck, 2008). Normally, ice and snow removal equipment and other necessary materials are made available in regions that are exposed to winter storms. During the monitoring of a winter storm, restoration activities such as the cleaning of the transportation ways and making them operational can be delayed for 1–2 days (OCIA, 2014).

5.5.8 Natural and Cultural Resources

This recovery process includes the practices and organizations that cover the appropriate protection, preservation, rehabilitation, and restoration of natural and cultural resources in a manner that is suitable for their historical properties. The primary element of this process is preservation. Natural and cultural resources are unique and mostly vulnerable by nature. Even the best restoration activity may not be the same as the original. Therefore, the entire society must be informed and encouraged in this regard. Participation of the entire society to the preservation of resources and relevant information exchange must be encouraged regarding the identification and usage of such resources and for the empowerment of common objectives (FEMA, 2011). For example, sand (Duxbury, 1999) and pebble (Stive et al., 2002) coasts act as a natural coastal block. They temporarily protect the mainland from the destruction of winter storms by distributing the high energies of big waves to protect the coast landscape, ecosystems, human societies, and the infrastructures on coasts with high energy (Brooks et al., 2017).

5.6 WINTER STORMS AND PUBLIC HEALTH OPERATION: MITIGATION AND RECOVERY

According to Winslow (1920), public health is "the science and art of preventing disease, prolonging life, and promoting health through the organized efforts and informed choices of society, organizations, public and private communities, and individuals." Today, we can define the strategic objectives of public health as 5P (prevention, protection, promotion, and population based on preparedness). Preparedness concept, which has recently been included in the definition, emerged with the emergency cases threatening the public health in the 21st century. These threats have become a significant focus point of public health works at national, regional, and local levels. Public health threats can occur for numerous reasons: bioterrorism, natural infectious diseases, terrorist activities, industrial or transportation accidents, and climatological disasters (ASPE). Weather event generally create threats that harm the public infrastructure and preventive public health services that bring about new and common burdens on the society, harm critical response infrastructure, and lead to compromises in response capacity (Hunter et al., 2016). The impacts of climatological threats on public health can be mitigated with vaccination, infectious disease control measures (Çalışkan and Özcebe, 2013; Noji, 2005;), and environmental health activities (Güler, 2012; Noji, 2005). All activities performed for the prevention of winter storms can be considered within the framework of 5Ps (Noji, 2005). These are given in detail below (ASPE):

1. Prevention (individual- and community-focused) is the stage where the spreading of infectious diseases is prevented.
2. Protection (policies/regulations, enforcement) is the protection of the society against environmental threats and injuries.
3. Promotion (voluntary, education, and advocacy) is the promotion of preventive healthcare including mental health, lifelong healthy habits, and recovery.
4. Population-based (communication, groups) activities are organized public activities intended for the continuity of the health of each individual.
5. Preparedness is the ability to respond to natural and human-sourced disasters.

TABLE 5.2 Winter Storms Disaster Health Framework

	Disaster cycle	Health services	Public health				Essential public health services	Health services								
								Home			Vehicle			Workplace		
											Pre-	During	Post-	Pre-	During	Post-
Postdisaster and disease	Recovery	Tertiary (Rehabilitative)	Prevention (Tersiyer)		Protection		Assessment	Is the reconstruction plan flexible?	Can the plan be applied in a modular manner?	Can the plan be updated against the new threats?						
							Policy development	Does the reconstruction plan respond to the needs of the society?	Is the cooperation of shareholders suitable during the activation of the plan?	Is the plan being adjusted to the laws?						
							Assurance	Is the reconstruction plan disseminated in the society?	Is cooperation with other shareholders is enabled during the implementation of the plan?	Are the outcomes of the plan being monitored and assessed?						
	Response	Sekonder (Therapeutic)	Prevention (Sekonder)				Assessment									
							Policy development									
							Assurance									
Predisaster and disease	Preparedness	Primary (Protector)	Prevention (Primordial/Primery)	Promotion		Preparedness	Assessment									
							Policy development									
							Assurance									
	Mitigation						Assessment	Is there a plan for prevention of the threat or for risk mitigation?	Can the plan be applied in case of threat?	Is the plan updated depending on needs?						
							Policy development	Is there a plan that is acceptable at international level?	Can the plan be highly activated in case of threat?	Is the repeatability of the plan being enabled through updates?						
							Assurance	Is the national plan being activated and disseminated?	Was the plan activated in case of threat?	Are the outcomes of the plan being monitored and assessed?						

As we can classify disasters into two stages as the pre- and postpublic threatening events (Altıntaş, 2015; Eryılmaz, 2007), we can also classify the applications and practices requiring health service intervention (Öztek et al., 2015) as pre- and postdisease stages. Thus, the components in the areas of disaster and health can be adjusted to each other. Pre event health services address to primary and preventive mitigation and preparation stages of the disaster continuum. Secondary health-care services refer to the response stage of the disaster continuum. Third-level rehabilitation services refer to the recovery stage of the disaster continuum. So, as the 5Ps of public health are naturally included in health services, they can be adjusted to the components of the disaster continuum as well. The basic functions of public health, such as the assessment, policy development, and application functions are also included in each stage of health service and disaster continuum (Table 5.2).

Although the first 3Ps of public health (prevention, protection, and promotion) look like they are specific to the health sector, they can be adjusted to the modern disaster continuum, indeed. Early detection of threats is possible through prevention and damages can be minimized. Prevention is divided into two: primordial and primary (AFMC). Primordial prevention can help the performance of population-based activities against winter storms and primary prevention can help the performance of activities that directly aim at the disaster. Although winter storms action planning is a population-based activity, cleaning of the ice and snow that has accumulated on the roof is a measure against the risk of collapse. Promotion deals with the improvement of public health. With protection activities, life losses, injuries, financial, power, spirit and time losses, and loss of the country reputation can be prevented (Maya and Çalışkan, 2016). Population-based activities are the activities performed for the continuity of health and preparedness that cover the requirements for being prepared against emergency cases and disasters.

Basic public services take place at each stage of the disaster management continuum. Assessment stage is the monitoring and inquiring of the events affecting health; policy development is the stage including informing of public, health problems, labor power, assessment of health services, and inquiring of new solutions (Cdc.Gov, 2014). These components can also be adjusted to every health threat. Table 5.2 presents a sample planning developed within the scope of the principles of public health during winter storms within the framework of the recovery and mitigation stages of the disaster continuum. In literature, disaster preparation can be classified into categories as pre-, during, and postdisaster activities (Ready.gov, 2017). Winter storms are also classified into periods as such. Winter storms can have various impacts depending on the human environment and applications and procedure of each region are unique. We can classify these areas as house, car, exterior environment, or workplace. Table 5.2 gives the example addressing to house environment only.

REFERENCES

AFMC, Basic concepts in prevention, surveillance, and health promotion, AFMC Primer on Population Health, Association of Faculties of Medicine of Canada Content, California, USA, http://phprimer.afmc.ca/Part1-TheoryThinkingAboutHealth/Chapter4BasicConceptsInPreventionSurveillanceAndHealthPromotion/Thestagesofprevention

Alexander, D., 2002. Principles of Emergency Planning and Management. Oxford University Press, New York, NY.

Altıntaş, K.H., 2015. Disasters and disaster medicine. In: Güler, Ç., Akın, L. (Eds.), Public Health: Basic Knowledge. third ed. Hacettepe Üniversity Press, Ankara, pp. 1106–1129, (in Turkish).

ASPE. Public health promotion and protection, disease prevention, and emergency preparedness. In: HHS Strategic Plan, Fiscal Years 2007–2012 (Strategic Plan). U.S. Department of Health & Human Services. https://aspe.hhs.gov/report/hhs-strategic-plan-fiscal-years-2007–2012-strategic-plan/chapter-3-strategic-goal-2-public-health-promotion-and-protection-disease-prevention-and-emergency-preparedness.

Bonnan-White, J., Shulman, J., Bielecke, A., 2014. Snow tweets: emergency information dissemination in a US county during 2014 winter storms. PLoS Curr. 6. doi: 10.1371/currents.dis.100a212f4973b612e2c896e4cdc91a36.

Broder, J., Mehrotra, A., Tintinalli, J., 2005. Injuries from the 2002 North Carolina ice storm, and strategies for prevention. Inj. 36 (1), 21–26. doi: 10.1016/j.injury.2004.08.007.

Brooks, S.M., Spencer, T., Christie, E.K., 2017. Storm impacts and shoreline recovery: mechanisms and controls in the southern North Sea. Geomorphol. 283, 48–60. doi: 10.1016/j.geomorph.2017.01.007.

Çalışkan, C., Özcebe, H., 2013. Epidemics of infectious diseases in disasters and control measures of them [Afetlerde enfeksiyon hastali{dotless}klari{dotless} salgi{dotless}nlari{dotless} ve kontrol önlemleri]. TAF Prev. Med. Bull. 12 (5)doi: 10.5455/pmb1-1344684524.

Çalışkan, C., Algan, A., Koçak, H., Biçer, B.K., Şengelen, M., Çakir, B., 2014. Preparations for severe winter conditions by emergency health personnel in Turkey. Disaster Med. Public Health Preparedness 8, 170–173. doi: 10.1017/dmp.2014.28.

Cdc.Gov, 2014. The 10 essential public health services. Centers for disease control and prevention. http://www.cdc.gov/nphpsp/essentialservices.html (accessed 28.02.17).

Clements, B.W., Casani, J.A.P., 2016. Winter storms, second ed. Disasters and Public Health: Planning and Response. Elsevierdoi: 10.1016/B978-0-12-801980-1.00020-9, 471–487.

Coppola, D.P., 2011. Mitigation. Introduction to International Disaster Management. Elsevierdoi: 10.1016/B978-0-12-382174-4.00004-5, 209–250.

Cutter, S.L., Boruff, B.J., Shirley, W.L., 2003. Social vulnerability to environmental hazards. Soc. Sci. Q. 84 (2), 242–261. doi: 10.1111/1540-6237.8402002.

Datla, S., Sahu, P., Roh, H.-J., Sharma, S., 2013. A comprehensive analysis of the association of highway traffic with winter weather conditions. Procedia—Soc. Behav. Sci. 104, 497–506. doi: 10.1016/j.sbspro.2013.11.143.

Disaster Prevention and Mitigation, 2017. Public safety Canada. https://www.publicsafety.gc.ca/cnt/mrgnc-mngmnt/dsstr-prvntn-mtgtn/index-en.aspx (accessed 08.02.17).

Donahue, A.K., Tuohy, R.V., 2006. Lessons we don't learn: a study of the lessons of disasters, why we repeat them, and how we can learn them. Homeland Secur. Affairs 2 (2), 1–28, https://www.hsajorg/articles/167.

Doswell, C.A., 2015. Mesoscale meteorology | severe storms. Encyclopedia of Atmospheric Sciences. Elsevier, Electronic books. https://doi.org/10.1016/B978-0-12-382225-3.00366-2, 361–368.

Duxbury, 1999. Duxbury comprehensive plan. Massachusetts. http://www.town.duxbury.ma.us/public_documents/duxburyMA_Planning/Comprehensive Plan 1999/a991206 Cover Pages Duxbury Final Report.pdf.

Eather, 2017. Know the difference between types of winter precipitation. The weather channel. https://weather.com/safety/winter/news/types-winter-precipitation-20120423#/1 (accessed 04.02.17).

Editorials, 2016. Burying overhead power lines. *The Japan Times*. http://www.japantimes.co.jp/opinion/2016/11/06/editorials/burying-overhead-power-lines/#.WJ8YsFWLTIU (accessed 11.02.17).

EM-DAT, 2017a. General classification. http://www.emdat.be/classification (accessed 31.01.17).

EM-DAT, 2017b. Disaster profiles. http://www.emdat.be/disaster_profiles/index.html (accessed 31.01.17).

Eryılmaz, M., 2007. Introduction to disaster. In: Eryılmaz, M., Dizer, U. (Eds.), Disaster Medicine. second ed. Unsal Press, Ankara, pp. 77–85, (in Turkish).

Federal Emergency Management Agency. Disaster Declarations by Year. https://www.fema.gov/disasters/grid/year.

FEMA, 2011. National disaster recovery framework. http://www.fema.gov/media-library-data/20130726-1820-25045-5325/508_ndrf.pdf.

FEMA, 2013. Mitigation ideas: a resource for reducing risk to natural hazards. Retrieved from https://www.fema.gov/media-library/assets/documents/12318.

FEMA, 2016. FEMA's federal insurance and mitigation administration fact sheet. https://www.fema.gov/media-library/assets/documents/12318.

Flinn, R., 2017. About winter storms. Pennsylvania emergency management agency. http://www.pema.pa.gov/planningandpreparedness/readypa/Pages/Winter-Storms.aspx#.WJslaVWLTIX (accessed 08.02.17).

Gibson, S.R., Stewart, R.E., 2007. Observations of ice pellets during a winter storm. Atmos. Res. 85 (1), 64–76. doi: 10.1016/j.atmosres.2006.11.004.

Greenough, G., McGeehin, M., Bernard, S.M., Trtanj, J., Riad, J., Engelberg, D., May 2001. The potential impacts of climate variability and change on health impacts of extreme weather events in the United States. Environ. Health Perspect. 109, 191. doi: 10.2307/3435009.

Guha-Sapir, D., Vos, F., Below, R., Ponserre, S., 2012. Annual Disaster Statistical Review 2011: The Numbers and Trends. Centre for Research on the Epidemiology of Disasters, Brussels.

Güler, Ç., 2012. Environmental health measures in disasters. In: Güler, Ç. (Ed.), Environmental Health: The Environment and Ecological Connections. Yazıt Press, Ankara, pp. 1287–1310, (in Turkish).

Heneka, P., Ruck, B., 2008. A damage model for the assessment of storm damage to buildings. Eng. Struct. 30 (12), 3603–3609. doi: 10.1016/j.engstruct.2008.06.005.

Henson, W., Stewart, R., Kochtubajda, B., 2007. On the precipitation and related features of the 1998 ice storm in the Montréal area. Atmo. Res. 83 (1), 36–54. doi: 10.1016/j.atmosres.2006.03.006.

Hunter, M.D., Hunter, J.C., Yang, J.E., Crawley, A.W., Aragón, T.J., 2016. Public health system response to extreme weather events. J. Public Health Manage. Pract. 22 (1), E1–E10. doi: 10.1097/PHH.0000000000000204.

Iqbal, S., Clower, J.H., Hernandez, S.A., Damon, S.A., Yip, F.Y., 2012. A review of disaster-related carbon monoxide poisoning: surveillance, epidemiology, and opportunities for prevention. Am. J. Public Health 102 (10), 1957–1963. doi: 10.2105/AJPH.2012.300674.

Koçak, H., Çaliskan, C., Kaya, E., Yavuz, Ö., Altintas, K.H., 2015. Determination of individual preparation behaviors of emergency health services personnel towards disasters. J. Acute Dis. 4 (3), 180–185. doi: 10.1016/j.joad.2015.04.004.

Kocin, P.J., Uccellini, L.W., 2004. Supplement to a snowfall impact scale derived from northeast storm snowfall distributions. Bull. Am. Meteorol. Soc. 85 (2), 194–1194. doi: 10.1175/BAMS-85-2-Kocin.

Kyrill, T.Y., 2017. The winter storm that walloped most of Europe. AIR. http://www.air-worldwide.com/Blog/Kyrill,-the-Winter-Storm-That-Walloped-Most-of-Europe/ (accessed 13.02.17).

LeComte, D., 2014. International weather highlights 2013: super Typhoon Haiyan, super heat in Australia and China, a long winter in Europe. Weatherwise 67 (3), 20–27. doi: 10.1080/00431672.2014.899800.

Long, 2010. Long-term power outage preparedness in Wisconsin. http://emergencymanagement.wi.gov/training/ltpo_docs/LTPO Preparedness in Wisconsin_web version_plain cover.pdf.

Maya, İ., Çalışkan, C., 2016. Evaluating disaster education and training programs at the level of undergraduate degree in the world and Turkey sample. Eur. J. Turk. Stud. 11 (9), 579–604. doi: 10.7827/TurkishStudies.9761.

Minsk, L., 1999. Heated bridge technology. Washington, DC. https://www.fhwa.dot.gov/publications/research/infrastructure/bridge/99158/99158.pdf.

Moscow, 2016. Moscow roof collapse kills 56. http://english.cri.cn/706/2006/02/23/53@54574.htm (accessed 11.02.17).

MunichRE-NatCatSERVICE, 2016a. Loss events in the U.S. 1980–2015: 10 costliest winter storms/damages ordered by overall losses. https://www.munichre.com/en/reinsurance/business/non-life/natcatservice/significant-natural-catastrophes/index.html.

MunichRE-NatCatSERVICE, 2016b. Loss events in Europe 1980–2015: 10 costliest winter storm events ordered by overall losses. https://www.munichre.com/en/reinsurance/business/non-life/natcatservice/significant-natural-catastrophes/index.html.

National Centers for Environmental Information, 2017a. Billion-dollar weather and climate disasters. https://www.ncdc.noaa.gov/billions/summary-stats (accessed 30.01.17).

National Centers for Environmental Information, 2017b. State of the climate: national overview for February 2015. http://www.ncdc.noaa.gov/sotc/national/201502 (accessed 31.01.17).

National Disaster Education Coalition, 2004. Winter Storms. Washington, DC. http://www.disastereducation.org/library/public_2004/Winter_Storms.pdf.

National Oceanic and Atmospheric Administration, 2017. National weather service glossary. http://w1.weather.gov/glossary/index.php?letter=w (accessed 03.02.17).

Noji, E.K., 2005. Public health issues in disasters. Crit. Care Med. 33 (Supplement), S29–S33. doi: 10.1097/01.CCM.0000151064.98207.9C.

OCIA, 2014. Critical infrastructure security and resilience note: winter storms and critical infrastructure. http://www.npstc.org/download.jsp?tableId=37&column=217&id=3277&file=OCIA_Winter_Storms_and_Critical_Infrastructure_141215.pdf.

Overton, L.R.-A., September 2014. From vulnerability to resilience: an exploration of gender performance art and how it has enabled young women's empowerment in post-hurricane, New Orleans. Procedia Econ. Finance 18, 214–221. doi: 10.1016/S2212-5671(14)00933-2.

Öztek, Z., Üner, S., Eren, N., 2015. Health services and health management. In: Güler, Ç., Akın, L. (Eds.), Public Heallth: Basic Knowledge. third ed. Hacettepe University Press, Ankara, pp. 1480–1522, Turkish.

Palm, J., 2009. Emergency management in the Swedish electricity grid from a household perspective. J. Contingencies Crisis Manag. 17 (1), 55–63. doi: 10.1111/j.1468-5973.2009.00557.x.

Patz, J.A., Engelberg, D., Last, J., 2000. The effects of changing weather on public health. Annu. Rev. Public Health 21 (1), 271–307. doi: 10.1146/annurev.publhealth.21.1.271.

Pfister, C., Garnier, E., Alcoforado, M.-J., et al., 2010. The meteorological framework and the cultural memory of three severe winter-storms in early eighteenth-century Europe. Clim. Change 101 (1), 281–310. doi: 10.1007/s10584-009-9784-y.

Piercefield, E., Wendling, T., Archer, P., Mallonee, S., 2011. Winter storm-related injuries in Oklahoma, January 2007. J. Safety Res. 42 (1), 27–32. doi: 10.1016/j.jsr.201011.004.

Preventing, 2012. Preventing roof collapse due to snow. http://www.harleysvillegroup.com/images/ITK/1112/Z-1618.pdf.

Ready.gov, 2017. Prepare for emergencies. https://www.ready.gov/prepare-for-emergencies (accessed 28.02.17).

Rhodes, B., Kailar, R. 2005. The public health information network messaging system. https://www.cdc.gov/phin/tools/phinms/documents/white-paper--on-securing-phinms.pdf.

Rice, R.G., Spence, P.R., 2016. Thor visits Lexington: exploration of the knowledge-sharing gap and risk management learning in social media during multiple winter storms. Comput. Hum. Behav. 65, 612–618. doi: 10.1016/j.chb.2016.05.088.

Ross, S., 2002. In: Witherick, M. (Ed.), Severe Weather. second ed. Nelson Thornes Ltd., Cheltenham, UK.

Sellnow, T., Sellnow, D., 2010. The instructional dynamics of risk and crisis communication: distinguishing instructional messages from dialogue. Rev. Commun. 10 (2), 112–126. doi: 10.1080/15358590903402200.

Spence, P.R., McIntyre, J.J., Lachlan, K.A., Savage, M.E., Seeger, M.W., 2011. Serving the public interest in a crisis: does local radio meet the public interest? J. Contingencies Crisis Manag. 19 (4), 227–232. doi: 10.1111/j.1468-5973.2011.00650.x.

Spia-index.com, 2017. What is the Sperry-Piltz ice accumulation index? http://www.spia-index.com/ (accessed 07.03.17).

Squires, M.F., Lawrimore, J.H., Heim, R.R., Robinson, D.A., Gerbush, M.R., Estilow, T.W., 2014. The regional snowfall index. Bull. Am. Meteorol. Soc. 95 (12), 1835–1848. doi: 10.1175/BAMS-D-13-00101.1.

Stive, M.J., Aarninkhof, S.G., Hamm, L., et al., 2002. Variability of shore and shoreline evolution. Coastal Eng. 47 (2), 211–235. doi: 10.1016/S0378-3839(02)00126-6.

The Kentucky Public, 2009. The Kentucky public service commission report on the September 2008 wind storm and the January 2009 ice storm. Kentucky. https://psc.ky.gov/IkeIce/Report.pdf.

Upfront, 2007. Winter storm damage in US doubles in 50 years. New Sci. 196 (2632), 6. doi: 10.1016/S0262-4079(07)62996-0.

Wightman, M.J., Fenno, A.J., Dice, H.W., 2010. Winter storms. In: Kristi, K.L., Carl, S.H. (Eds.), Disaster Medicine: Comprehensive Principles and Practices. Cambridge University Press, New York, NY, pp. 586–608.

Wikipedia, 2017a. Winter storm naming in the United States. https://en.wikipedia.org/wiki/Winter_storm_naming_in_the_United_States (accessed 04.02.17).

Wikipedia, 2017b. Undergrounding. https://en.wikipedia.org/wiki/Undergrounding (accessed 11.02.17).

Wikipedia, 2017c. Snowmelt system. https://en.wikipedia.org/wiki/Snowmelt_system (accessed 11.02.17).

Winslow, C.-E.A., 1920. The untilled fields of public health. Sci. 51 (1306), 23–33. doi: 10.1126/science.51.1306.23.

Winter Weather Preparedness. Illinois, US. http://ready.illinois.gov.

Winter Storm, 2016. Winter storm names for 2016–2017 revealed. The weather channel. https://weather.com/storms/winter/news/winter-storm-names-2016-2017 (accessed 04.02.17).

Winter Storm, 2017a. The disaster center. http://www.disastercenter.com/guide/winter.html (accessed 04.02.17).

Winter Storm, 2017b. Winter storm in Northern California becomes "Serious situation" as widespread flooding. http://ktla.com/2017/01/09/winter-storm-in-northern-california-becomes-serious-situation-as-widespread-flooding-evacuations-occur/ (accessed 08.02.17).

Yun, K., Lurie, N., Hyde, P.S., 2010. Moving mental health into the disaster-preparedness spotlight. N. Engl. J. Med. 363 (13), 1193–1195. doi: 10.1056/NEJMp1008304.

Zielinski, G.A., 2002. A classification scheme for winter storms in the Eastern and Central United States with an emphasis on Nor'easters. Bull. Am. Meteorol. Soc. 83 (1), 37–51. doi: 10.1175/1520-0477(2002)083<0037:ACSFWS>2.3.CO;2.

Chapter 6

Spatial Association Between Forest Fires Incidence and Socioeconomic Vulnerability in Portugal, at Municipal Level

A.N. Nunes, L. Lourenço
University of Coimbra, Coimbra, Portugal

6.1 INTRODUCTION

Fire plays an important role in certain forest ecosystems, especially not only in the Mediterranean basin, but also in other Mediterranean-type of areas of the world (Ganteaume et al., 2013; Gill, 2006; Keeley and Keeley, 1988). Therefore, fire has been a key tool used by humans for several thousands of years and a vital component in ecosystem dynamics (Nunes et al., 2016; Pausas and Ramon-Vallejo, 1999; San-Miguel-Ayanz et al., 2012). Uncontrolled fires cause, however, large environmental and economic damages, especially in the Mediterranean region (San-Miguel-Ayanz et al., 2012). Nowadays, wildfires rank top of all European forest problems (Barbati et al., 2010), affecting landscape, wildlife, vegetation, soils, water, and air quality, in addition to the human wellbeing (Cerdà and Lasanta, 2005; Certini, 2005; DeBano et al., 1998; Finlay et al., 2012; Malkinson et al., 2011; Miranda et al., 2008; Novara et al., 2013).

Portugal has the highest relative burnt area of all Southern European countries, between 1980 and 2015 (JRC, 2016). Therefore, several studies have been addressed to the drivers behind wildfires in Portuguese territory, linking them mainly with climate/weather conditions (Carvalho et al., 2008; Lourenço and Gonçalves, 1990; Pereira et al., 2005; Trigo et al., 2006;) and changes in the landscape mosaic, as a consequence of agricultural abandonment and a marked increase in land covered by shrubs, grass, and other light vegetation that is very prone to fire (Bajocco and Ricotta, 2008; Carmo et al., 2011; Catry et al., 2009; Moreira et al., 2009; Moreira et al., 2011; Nunes et al., 2005, 2016; Nunes, 2012; Oliveira et al., 2012, 2014). The association between social and economic vulnerability and wildfire incidence, particularly, in terms of burnt area, has received less attention.

Indeed, vulnerability assessment has received, at the international level, growing attention and is considered a crucial component for understanding and reducing the negative impacts of natural and manmade hazards that hamper sustainable development (Birkmann, 2006; Cutter, 2015; UNISDR, 2009, 2015; Tedim, 2012; Kovats et al., 2014). According to the International Strategy for Disaster Risk Reduction (Birkmann, 2006), vulnerability included social, economic, physical, and environmental components and was considered a tool and a precondition for risk assessment. This approach distinguished the potential capacity (available resources) of a community or element to deal with the impacts from the application of this capacity (used resources) by the society. Subsequently, vulnerability was defined as the physical, social, economic, and environmental conditions which increase the susceptibility of a community to the impact of hazards (UNISDR, 2009). According to the International Panel on Climate Change (Kovats et al., 2014), vulnerability reflects the propensity of the exposed elements to be adversely affected, whereas exposure refers specifically to the presence of people, infrastructure, and other assets in hazardous areas.

The concept of place vulnerability, introduced by Cutter (1996), incorporates the concepts of biophysical exposure, social response, and the geographic context in which these issues emerge. Prior and Eriksen (2013) advocated that what

Integrating Disaster Science and Management. http://dx.doi.org/10.1016/B978-0-12-812056-9.00006-3

makes wildfires disastrous is their social impact, and most wildfire studies have ignored this social element, instead being enclosed within ecology or environmental research. As Wigtil et al. (2016) state, wildfire incidence and impacts can vary with social status.

Therefore, some studies have examined the spatial distribution of wildfire potential and its outcomes against selected social and ecological conditions (Chuvieco et al., 2014; Galiana-Martin and Karlsson, 2012; Haas et al., 2013; Wigtil et al., 2016). However, most of these studies examined wildfire impacts on human settlements without considering how heterogeneous social and economic conditions can affect hazard outcomes (Wigtil et al., 2016).

In fact, social vulnerability, in terms of low socioeconomic status of residents, has the effect of exacerbating community risk to wild land fire occurrence and devastation because socially vulnerable populations are generally less able to either mitigate wildland fire risk or recover from such events (Blaikie et al., 1994; Cutter et al., 2000; Evans et al., 2007; Gaither et al., 2011; Lynn and Gerlitz, 2006; Mercer and Prestemon, 2005; Poudyal et al., 2012; Prior and Eriksen, 2013; Wigtil et al., 2016). For instance, Cutter et al. (2000) consider that socially vulnerable groups such as the elderly, lower income, racial minorities, and women are more likely to be exposed to a larger number of hazards and/or be less able to recover from disasters (e.g. wildfire, hurricanes, chemical spills) than wealthier, more able-bodied individuals, and communities. Along similar lines, Mercer and Prestemon (2005) found a positive association between poverty and the area of wildland burned and wildland fire intensity, suggesting that once wildland fires are ignited, poorer communities have fewer resources to extinguish the fire. Morrow (1999) and Lynn and Gerlitz (2006) also argue that poor communities are less able to absorb the effects of natural disasters. Poudyal et al. (2012) defend that forest fires occurrence is a type of disaster, which is a function happening not only because of exposure to biophysical hazards, that is, fire-prone woodlands, but more importantly because of the sensitivity of social groups to hazards.

In this context, understanding social issues that contribute to social vulnerability—social, economic, and political factors—involving wildfire vulnerability and management (Cortner and Field, 2007; McCaffrey et al., 2013) is necessary to address persistent economic, structural, and human life losses from wildfires as well as rising wildfire management costs (Bracmort, 2014; Gall et al., 2011; Hoover and Bracmort, 2015). Poudyal et al. (2012) stress that particular attention, in terms of wildfire mitigation and adaptation, should focus on places where socially marginal populations intersect with higher wildfire risk because these populations have the added vulnerabilities of lower capacity; yet, poor and working-class communities may be less likely than upper-income communities to participate in wildfire protection programs.

Cutter et al. (2000) consider, moreover, that a key component of any vulnerability assessment is the acquisition of systematic baseline data, particularly, at the local level. Pearson et al. (2008) also argue that the degree of vulnerability depends on the temporal and spatial, socioeconomic, and biophysical factors which interact differently according to the context. Thus, local forms of spatial analysis have recently gained in prominence (Mennis, 2006). Geographically weighted regression (GWR) is a local spatial statistical technique used to analyze spatial nonstationarity, defined as when the measurement of relationships among variables differs from location to location (Fotheringham et al., 2002). In fact, GWR improves the predictive performance of classical linear regression by considering regression as a spatial nonstationary process and by capturing the spatial variability of wildfire driving factors but also determining and quantifying their contribution and errors (Rodrigues et al., 2014; Nunes et al., 2016). Thus, several authors have used this regression procedure in wildfire analysis (Chuvieco et al., 2012; Koutsias et al., 2005; Martínez-Fernández and Koutsias, 2011; Martínez-Fernández et al., 2013; Nunes et al., 2016; Oliveira et al., 2014; Rodrigues et al., 2014).

Based on the assumption that the association between wildland fire incidence and socioeconomic vulnerability varied geographically, the main goals of this study were: (i) to analyze the spatial patterns of wildfires on a municipal level; (ii) to identify the most critical social and economic variables associated with spatial incidence of burnt areas, by comparing the performance of classical linear regression and GWR modeling; and (iii) to map spatial variation in the relationships between social and economic vulnerability and wildfire incidence in order to identify spatial clusters.

6.2 MATERIALS AND METHODS

6.2.1 Study Area

Mainland Portugal is composed of 278 municipalities and is geographically located in the extreme west coast of Europe, in the Iberian Peninsula. It covers an area of 89,015 km^2. Major social and economic changes have affected the entire Portuguese territory, mainly in the last five decades. Since 1960, the population declined in 65% of the Portuguese municipalities, with figures varying from 71% to 0.2%, whilst 35% recorded an increase of 0.3%–675%. The northern and central inland municipalities, as well as those in the southern region, were the most affected by depopulation, leading to the abandonment of agricultural lands and a reduction in the size of herds and the amount of forest fuels consumed by grazing and gathering firewood (Moreira et al., 2011; Nunes, 2012; Nunes et al., 2016). Mass emigration to other European countries and to coastal areas within Portugal explains this declining population and the parallel trend of an aging population. Conversely, the population in the coastal municipalities augmented during the period 1960–2011 and these areas nowadays have the highest concentrations, densities, and urban populations, as well as younger residents (Bandeira et al., 2014). The dichotomy inland/coastal is notorious in mainland Portugal.

Depopulation and aging are becoming a dominant component of peripheral rural areas especially in more remote regions, devoted to an increasing isolation and precarious conditions, more vulnerable to poverty and to the risk of social exclusion. Problems in access to services represent another common problem of more peripheral rural areas.

Conversely, innovation and modernization initiatives of the different sectors are mostly concentrated in coastal urban areas, where general population seeks better opportunities, thus producing a high concentration of people around major cities. Two clusters in the coastal region are, however, well defined, corresponding to the metropolitan areas of Lisboa and Porto, in which the population is much more concentered (with the highest densities), younger, and more qualified. The population of these areas also experience the highest per capita income, particularly, in the Lisboa Metropolitan area (Fig. 6.1).

6.2.2 Data Collection

Aiming at examining the association between wildfire incidence and socioeconomic vulnerability in mainland Portugal, firstly, an indicator of wildland fire incidence was identified and after it were acknowledged the variables related with social and economic vulnerability. In this study, municipality was the geographical unit that performed the best according to our assessment and given the specific goals and time period of the study. The unit of municipality was also selected for the analysis because geography approximates community groupings.

In order to compare the incidence of wildfires among municipalities, it was adopted a burnt area index, for the period of 2000–15, once it results from the relationship between the burnt area and the total surface area of the municipality. Burnt area can be understood as a result of effectiveness of fire prevention, detection, firefighting resources and infrastructures, and effectiveness in firefighting efforts (Lex and Goldammer, 2001). The burnt area database was compiled using data from the Instituto da Conservação da Natureza e das Florestas (Portuguese National Institute for Nature Conservation and Forestry) (ICNF), the Portuguese Government Forestry Service.

Spatial autocorrelation in the geographical distribution of burnt area index was evaluated using a global Moran's *I* test. The Moran's *I* test measures the degree to which burnt area pattern is clustered, dispersed, or randomly distributed across the municipalities by computing the deviation from the mean for each tract (Anselin et al., 2006). Moran's *I* coefficient between

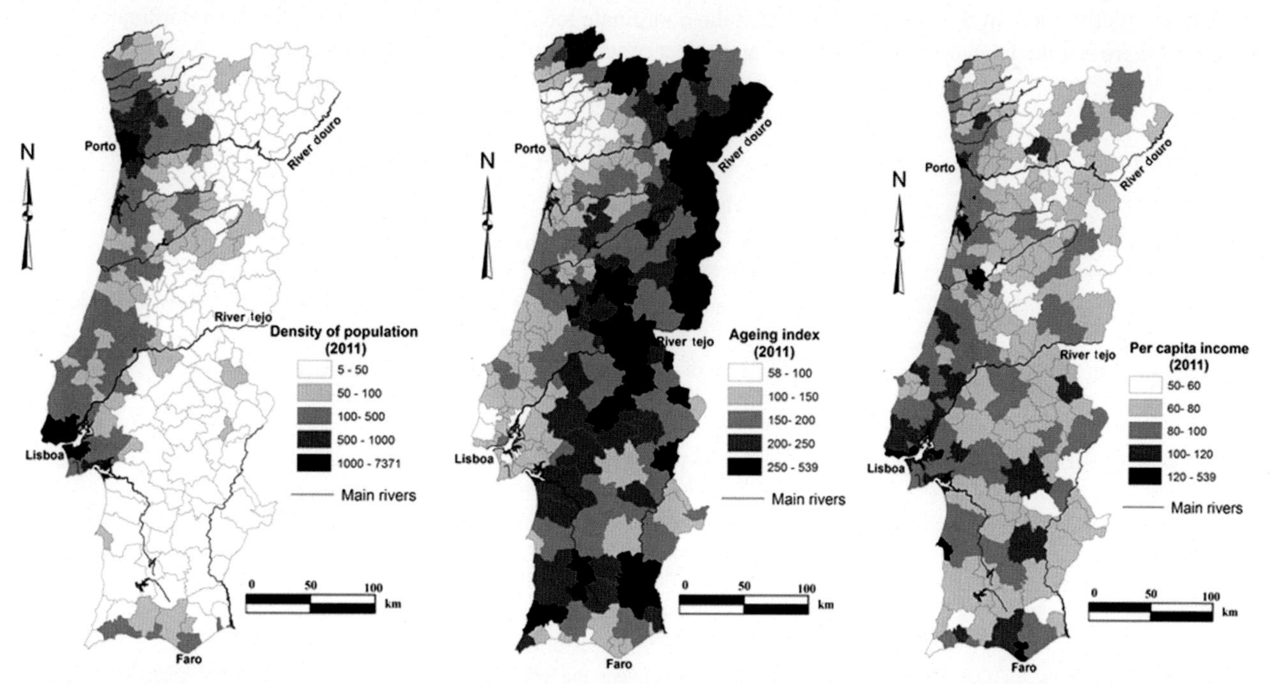

FIGURE 6.1 **Spatial distribution of population density (people per km^2), ageing index (in %), and per capita income (in Euros), at municipal level, in mainland Portugal.**

0 and 1 indicates positive spatial autocorrelation or clustering, negative coefficients between 0 and 1 indicate dispersion or a dissimilar neighboring values, and those near 0 indicate weak or no spatial autocorrelation (Fortin and Dale, 2005).

The selection of the variables to be integrated in the social and economic vulnerability assessment was based on extensive literature review and on the possibility to obtain the required data. Data for socioeconomic vulnerability were obtained from the census, provided by Statistics Portugal. Pearson correlation coefficients were then applied to assess the presence of collinearity among the variables. A correlation coefficient threshold between predictor variables of $|r| > 0.7$ ($P < 0.05$) was classified as a suitable indicator for the point where collinearity begins to severely distort model estimation and subsequent prediction (Dormann et al., 2013). As a result, only the variables identified in Table 6.1 are considered in the statistical analysis at a municipal level.

Since each of the variables was measured on a different scale, it was necessary to standardize each as index value. We used the following formula, with values ranging between 0 and 1:

$$\text{Norm}(\text{var}\,0-1) = \text{var}(\text{value}) - \min(\text{var}) / \max(\text{var}) - \min(\text{var})$$

With:

Norm(var 0–1) being the resulting normalized value in the scale 0–1;
var(value), the original value of the variable obtained for selected unit (municipality);
min(var), the lowest value of that variable recorded at national level;
max(var), the highest value of that variable recorded at national level.

6.2.3 Model

Simple linear regression (SLR) and GWRs were used to examine the most significant association between social and economic vulnerability and the percentage of burnt area in relation to the municipality's area. Model estimation began with the analysis of the linear relationship between wildfire incidence and each variable selected for socioeconomic vulnerability. SLR model assumes that the studied relationship is stationary, that is, the estimated parameters do not vary. To test the hypothesis that the relationship varies spatially, we use GWR. GWR estimates local parameter values, using the below equation (Fotheringham et al., 2002):

$$y_i = \beta_0(u_i, v_i) + \sum_{j=1}^{k} \beta_j(u_i, v_i)x_{ij} + \varepsilon_i$$

where y_i is the value of the outcome variable at the coordinate location i where (u_i, v_i) denotes the coordinates of i, and β_0 and β_j represent the local estimated intercept and effect of variable j for location i, respectively.

TABLE 6.1 Burnt Area Index and List of Variables Used in the Socioeconomic Vulnerability

Variable	Description	Unit
Burnt area index (BAI) 2000–15	Relationship between the annual burnt area and the total surface of the municipality	%
Density of population (DP) 2001–11	The intensity of settlement expressed as the ratio between (total) population and surface (land) area	People per km^2
Aging index (AI) 2001–11	The ratio of the number of elderly persons of an age when they are generally economically inactive (aged 65 and over) to the number of young persons (from 0 to 14)	%
Unemployment rate (UR) 2001–11	Represents the unemployed persons as a percentage of the civilian labor force	%
Per capita income (PCI) 2001–11	Measures the average income earned per person in a given area (city, region, country, etc.) in a specified year. It is calculated by dividing the area's total income by its total population	Euros
Primary sector (PS) 1999–2009	Fraction of the population employed in the agriculture and forestry sector	%
Density of small livestock (DSL) 1999–2009	Relationship between the total number of sheep and goat and the total surface of the municipality	Number of heads per km^2

In fact, global regression methods may be inadequate for large geographical areas such as the Portuguese territory, due to the use of stationary coefficients for the whole study area, perhaps covering local interactions within the factors. Thus, several authors (Koutsias et al., 2005, 2010; Nunes et al., 2016; Oliveira et al., 2014;) observed that the explanatory power of classical linear regression increased considerably after varying relationships were assumed instead of constant ones, using GWR.

There are a number of different goodness-of-fit measures when GWR are applied. R^2 measures the proportion of the variation in the dependent variable which is accounted for by the variation in the model. Values closer to 1 indicate that the model has a better predictive performance. However, its values can be influenced by the number of the variables which are in the model, that is, increasing the number of variables will never decrease the R^2. In this context, the adjusted R^2 is a preferable measure since it contains some adjustment for the number of variables in the model (Fotheringham et al., 2002).

A slightly different measure of goodness-of-fit is provided by the Akaike information criterion (AIC). Unlike the R^2, the AIC is not an absolute measure; it is a relative measure and can be used to compare different models which have the same independent variables. It is a measure of the "relative distance" between the model that has been fitted and the unknown "true" model. Thus, it is simply a way of ranking the models and the model with the lowest corrected AIC (AICc) is classified as the "best" out of all the models specified for the data analyzed (Mazerolle, 2004; Snipes and Taylor, 2014). Comparing the GWR AICc value to the simple regression, AICc value is one way of assessing the benefits of moving from a global to a local regression model (GWR) (Desktop, ArcGIS, 10.2, ESRI).

A key phase in the GWR analysis is the bandwidth definition, which helps us to identify the optimal number of neighbors (bandwidth calibration) since regression output will differ significantly according to this parameter value. Fixed and adaptive kernels are the different methods for bandwidth calibration available in any GWR model. The first, the fixed kernel, imposes an equal distance threshold for each regression point, thus the number of neighbors will probably vary from one regression point to another according to the spatial point pattern. In the second, the adaptive kernel, the number of neighbors to be considered for each regression point is stipulated, thus the adaptive kernel changes the distance threshold to fit the number of data points. In general, fixed kernels should be appropriate in a scenario where the point cloud is regularly distributed over space and the adaptive approach is more suitable for spatially clustered patterns. Finally, the optimum distance bandwidth value or optimum number of neighbors can be determined in two ways: by minimizing the square of the residuals (Cleveland, 1979) or by minimizing the AIC (Hurvich et al., 1998; Kupfer and Farris, 2007).

In this study, an adaptive kernel type and a selected optimal bandwidth by minimizing both the square of the residuals and the AICc were applied. Various regression models were determined using different numbers of neighbors and AICc statistics were compared to assess the goodness-of-fit for each model. On the basis of the AICc scores, the number of neighbors selected was 10.

Given that the errors for a well-fitted model are randomly distributed across a study area, the spatial fit of GWR models were examined using Moran's I analysis of the spatial residuals. The residuals from the GWR model were tested for spatial autocorrelation using the global Moran's I statistic to determine whether the assumptions of regression were being met (Griffith and Layne, 1999).

ArcGIS software (10.2, ESRI) was used to compute the results for the global and municipal regression models and Moran's I statistics. All the maps were also generated using ArcGIS software (ArcGIS, ESRI).

6.3 RESULTS

6.3.1 Burned Area Per Municipality

In Portugal, fires are not distributed uniformly throughout the territory; however, they vary greatly, both in terms of the annual ignition density and percentage of burnt area (Nunes et al., 2016). Although average burnt area per municipality was 1.6%, for the period 2000–15, the spatial variability ranges between a maximum of 5.5% and a minimum of 0.0%, with a standard deviation of 1.5% (Table 6.2).

Spatially, the largest percentages of burnt areas are found in the municipalities of northern and central inland Portugal, covering the vast majority of the highland territories. In these municipalities, the average annual burnt area amounts to more than 3.5% of the total municipality surface (Fig. 6.2). In Southern Portugal, two other municipalities (Monchique and Tavira) can be identified as the most affected by fire.

Conversely, the municipal territories that reveal the lower proportion of burned area are located in the coastal areas and in the Alentejo region. These spatially clustered patterns are confirmed by the results of the global Moran's Index, indicating the presence of a statistically significant positive spatial autocorrelation in mainland municipalities (Moran's $I = 0.307$; Z-score: 9.432; P-value 0.000).

TABLE 6.2 Descriptive Statistics for Burnt Area Index at Municipal Level

n	Maximum (%)	Mean (%)	Median (%)	Minimum (%)	Std. Deviation (%)	Variance
278	5.5	1.6	1.2	0.0	1.5	2.1

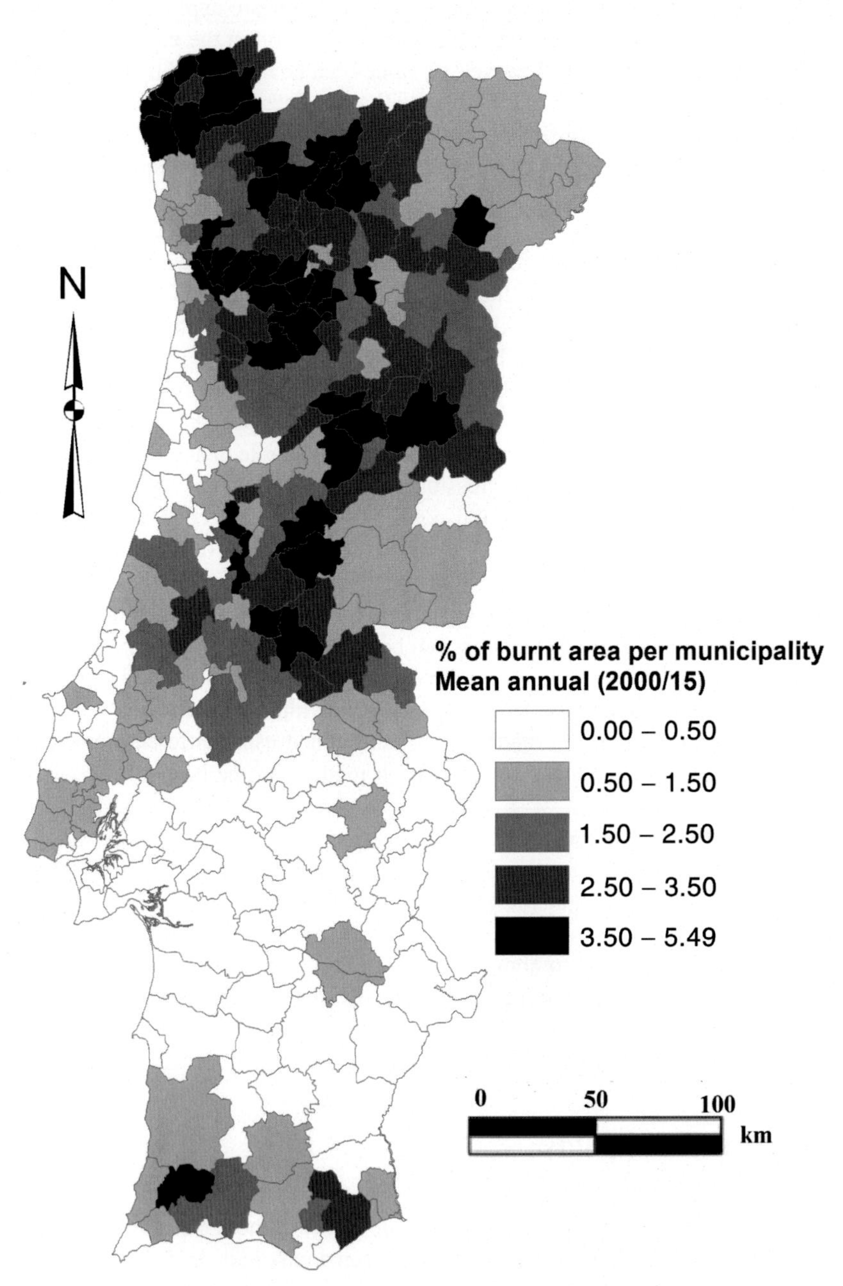

FIGURE 6.2 **Spatial distribution of mean annual burnt area, at municipal level, between 2000 and 2015.**

6.3.2 Association Between Socioeconomic Variables and Burnt Area

In order to find relationships between burnt area index and each variable selected for socioeconomic vulnerability at national level, certain differentiated features can be observed in Table 6.3, as a result of applying SLR and GWRs.

Despite the significant statistical correlations (*P*-value <0.05) obtained by applying the SLR, the *adjusted coefficient of determination* (in %) shows a very *weak* fit in the relationship, since the respective values range from 1% to 8%. The aging index emerges with the best positive correlation (R^2: 0.08), followed by the unemployment rate (R^2: 0.07).

The other variables presented very low associations with the burnt index: negative with the density of population, per capita income, and small livestock density and positive with the primary sector. On the contrary, the results achieved by applying GWR denote a significant improvement in the degree of association between socioeconomic variables and burnt area. The overall R^2 augmented to values ranging from 0.67 to 0.70 showing a high predictive potential for socioeconomic burnt area incidence modeling. Indeed, the GWR results greatly surpass the SLR, the classical regression technique, and enable nonstationary relationships between dependent and predictive variables to be detected. The AICc scores obtained for both methods also suggest that GWR techniques are the most efficient for the data analyzed.

Mapping the local R^2 values to see where GWR predicts well and where it predicts poorly may provide clues about important variables that may be missing from the regression model. In fact, when municipal R^2 values were mapped on the basis of relationships between burnt areas and each of socioeconomic variable, significant differences in the proportion of variance explained by GWR were observed. The analysis of maps shows that the density of small livestock and the unemployment rate include the highest percentage of municipalities in the final class, where the coefficient of variability explained by these variables is over 80% (Fig. 6.3). However, the proportion of municipalities included in the first class, where the coefficient of variability explained is lower than 20%, is very high and ranged from 54% to 63%. These results demonstrate that the spatial incidence of burnt area depends on the combination of multiple variables since the performance of the model is very low, meaning that it is necessary to combine or include different variables for a better explanation of the extent of wildfires.

Analyzing the spatial autocorrelation (Moran's *I*) of the GWR residuals for all the variables, the obtained result demonstrates that the *Z*-scores are negative and the *P*-value is statistically significant (<0.05), thus the spatial distribution of high values and low values in the data set is more spatially dispersed than would be expected if underlying spatial processes were random (Desktop, ArcGIS, 10.2, ESRI).

When considering all the variables together (multivariate models), the multiple linear regression (MLR) selected four variables as statistically significant, increasing the predictive power to an adjusted R^2 of 0.28. Two variables are positively related with burnt area (aging index and unemployment rate), whereas the other two, per capita income and density of small livestock, show a negative *correlation* coefficient. The combination of the same variables in GWR model demonstrates an increase in explained variance, recording an adjusted R^2 of 0.74. At municipal scale, substantial improvements in GWR R^2 can be observed in Fig. 6.4 where the four combined predictive variables explain the large degree of variance in the burnt area observed during the period of 2000–15, since around two-thirds of municipalities were included in the final classes, where the explained variance is over 60% (Table 6.4).

The residuals from the GWR model were tested for spatial autocorrelation using the Moran's *I* statistic, which produced Moran's *I* of -0.09, a *Z*-score of -3.11, and a *P*-value of 0.00, suggesting a tendency toward a *dispersion in* the spatial pattern.

6.4 DISCUSSION AND CONCLUSIONS

The results obtained clearly show a strong spatial association between the incidence of burnt areas and some socioeconomic variables that contribute to *wildfire vulnerability* in mainland Portugal. In general, the results demonstrated that the municipalities with high burnt areas displayed high social and economic vulnerability as a result of the higher aging index and

TABLE 6.3 Coefficients of Correlation Resulting from the Application of SLR and GWR Between % of Burnt Area, Per Municipality, and the Variables Used in the Socioeconomic Vulnerability

Annual Average (n = 278)	Population Density	Aging Index	Unemployment Rate	Per Capita Income	Primary Sector	Small Livestock
SLR adjusted R^2	(−) 0.01***	(+) 0.08***	(+) 0.007***	(−) 0.01**	(+) 0.01**	(−) 0.01***
AICc	64.25	38.93	42.8	61.30	62.15	63.58
GWR adjusted R^2	0.70	0.69	0.69	0.70	0.67	0.70
AICc	−66.85	−34.03	−40.77	−27.1	−25.13	−70.28
Residual squares	2.83	2.83	2.84	2.74	3.16	2.81
Moran's index	−0.10	−0.10	−0.09	−0.08	−0.09	−0.10
Z-score	−2.39	−2.43	−2.15	−1.98	−2.12	−2.43
P-value	0.01	0.01	0.03	0.04	0.03	0.03

Source: (+) positive correlation; (−) negative correlation; Model variable significance ** = 0.05; *** = 0.01.

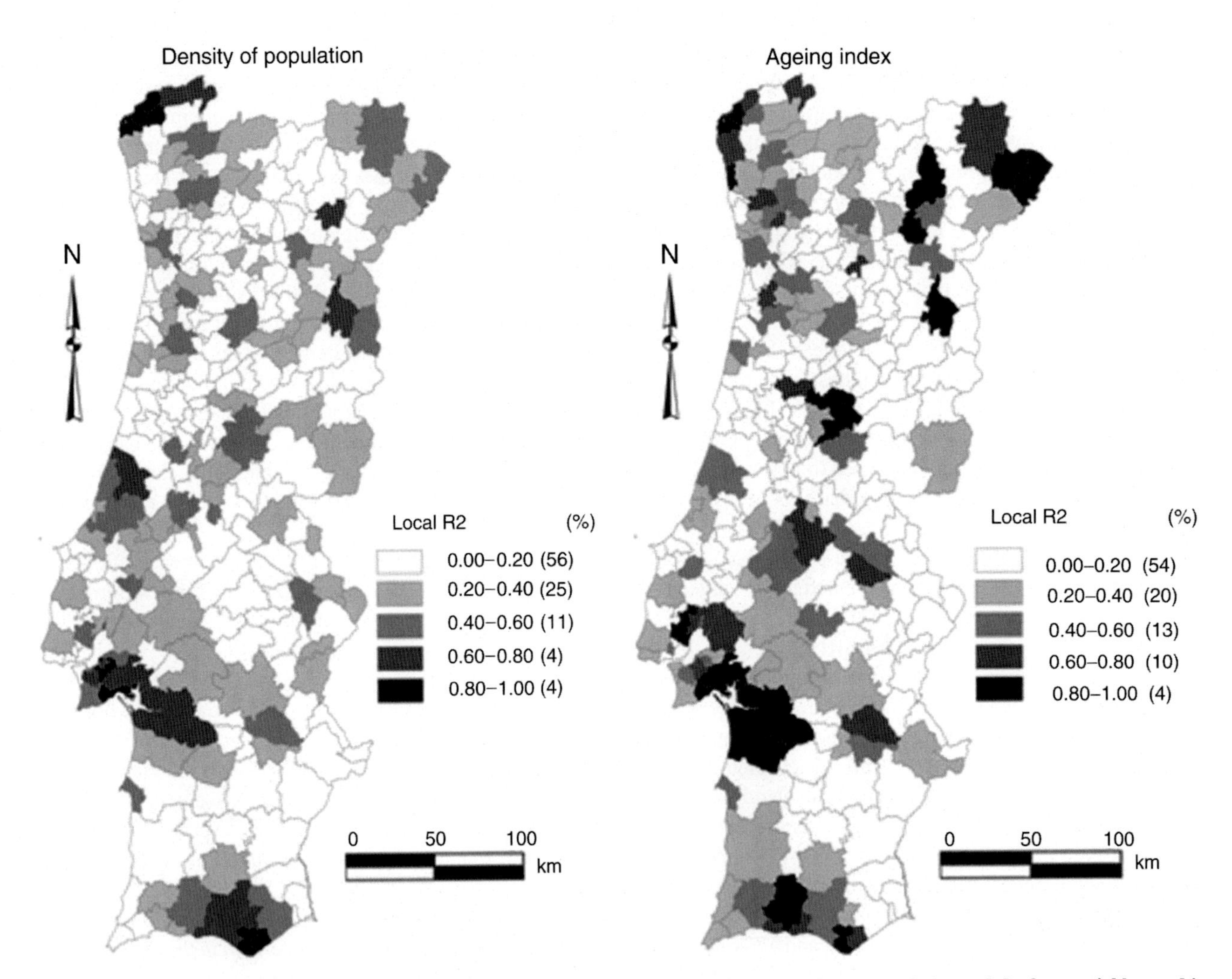

FIGURE 6.3 **Spatial distribution of local R^2 results from the application of GWR between the burnt area index and the four variables used in the socioeconomic vulnerability (the percentage of municipalities is given in parentheses).**

unemployment rates. Conversely, higher-income populations and the prevalence of higher livestock densities, namely sheep and goats, influence negatively on the burnt extension. The combination of these variables can either enhance or detract from a community's ability to mitigate disaster (Cannon, 1994).

Several authors have shown that elderly people become more vulnerable to disasters as they are likely to lack adequate economic resources and physical ability to respond effectively and they are more likely to suffer health consequences, physical harm, and be slower to recover (Birkmann et al., 2013; Bodstein et al., 2014; Cutter et al., 2000). Moreover, in Portugal, the elders are strongly associated with illiteracy levels (Pearson's correlation = 0.77), thus these people are less efficient in obtaining services or information about environmental protection, demonstrate greater exposure to accidents, and are less able to recover from the effects of natural disasters. The characteristics of the population, namely the high proportion of elderly people and few younger people, render these communities more highly sensitive groups to hazards.

A high unemployment rate also seems to lead to the creation of conditions for a greater incidence of burnt areas in certain regions of Portugal. Martínez et al. (2009) and Oliveira et al. (2012) also founded a positive association between unemployment and wildfires' incidence in some European countries and identified the economic difficulties as a possible reason for increasing conflicts that result in more deliberate fires.

Although the fraction of the population employed in the agriculture and forestry sector show a positive association with higher wildfires incidence (Catry et al., 2009; Verdú et al., 2012), at a municipal level, grazing activity seems to have a negative effect in burnt area (Nunes et al., 2016), suggesting that the endurance of traditional Mediterranean silvopasture activities reduces the vulnerability and impact to wildfires and increases the resilience of landscape as to fire.

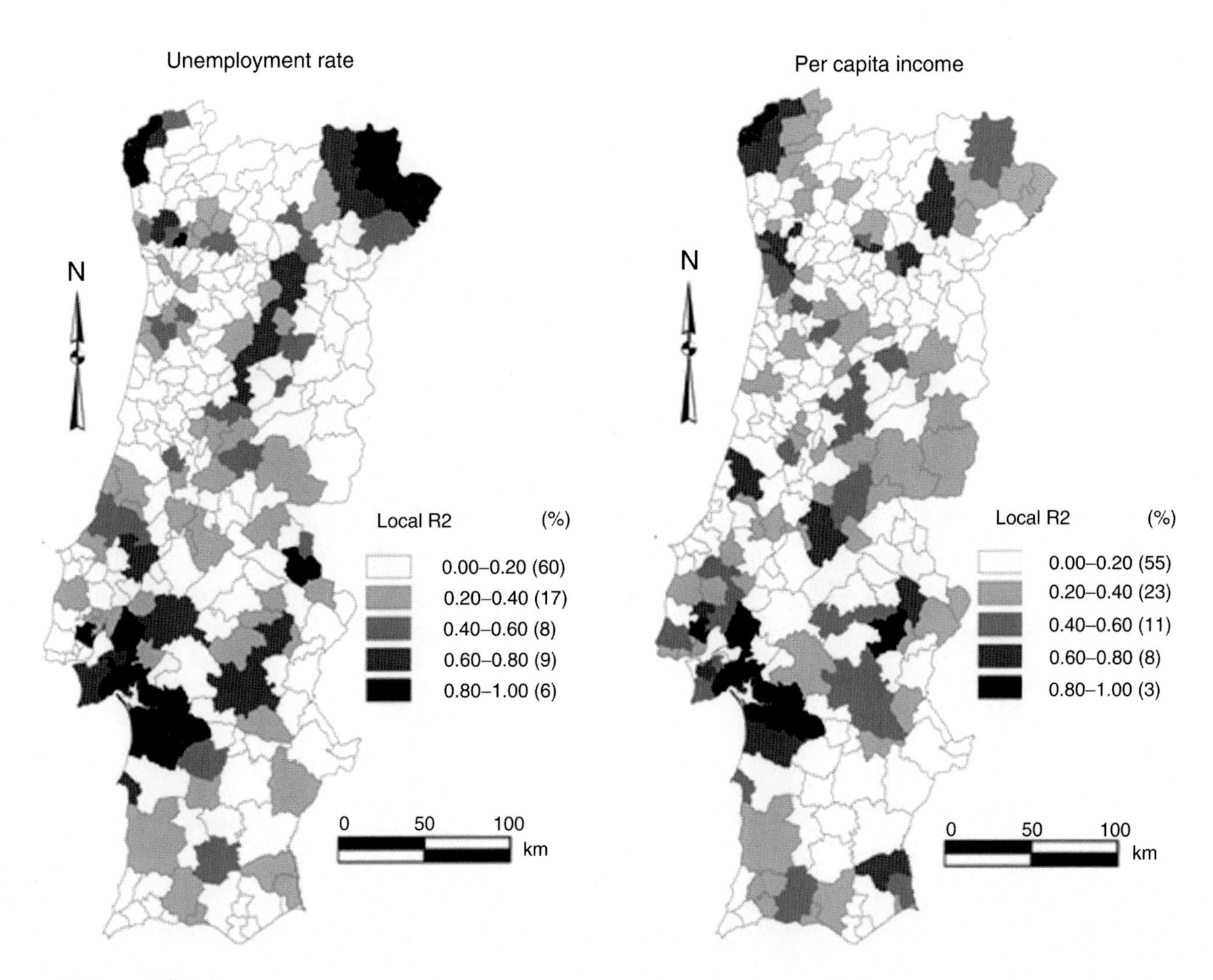

FIGURE 6.3 *(Cont.)*

It is therefore clear that socioeconomic vulnerability influences risk and harms exposure and also affects the capacity to recover from the effects of natural disasters (Blaikie et al., 1994; Bühler et al., 2013; Cutter et al., 2000; Evans et al., 2007; Gaither et al., 2011; Lynn and Gerlitz, 2006; Morrow, 1999). Cutter et al. (2000) strengthen that social conditions including elderly, lower income, higher unemployment rates are more likely to be exposed to a larger number of hazards, can influence an individual's or a community's ability to plan for, cope with, and recover from environmental hazards. Gaither et al. (2011) refer that this perspective places as much emphasis on the social dimensions of disaster, that is, on suspected societal conditions and inequities which may cause some groups to be less prepared for and less able to recover from hazard events.

The overlap between socioeconomic vulnerability in terms of low socioeconomic status of residents and wildfire incidence in Portuguese territory suggests a need to evaluate wildfire management policies with regard to social and economic conditions. However, socially vulnerable communities are generally less engaged in wildfire mitigation programs (Gaither et al., 2011) even when they are exposed to high levels of wildfire risk (Ojerio et al., 2011).

In Portugal, unfortunately, the management policies related to forest fires during the last decades revealed the overemphasized extinction to the detriment of prevention (Mourão and Martinho, 2016) and individual education to the detriment of more comprehensive sustainable development planning. Mateus and Fernandes (2014) estimated that fire presuppression and suppression absorbed 94% of the fire management budget in 2010. Moreover, hazard identification and emergency management applications, as in other parts of the world, mostly consider only aspects of vulnerability concerned with physical exposure. In fact, few studies have incorporated socioeconomic vulnerability measures with wildfire potential to examine communities' adoption of and participation in wildfire mitigation, prevention, and management programs (Gaither et al., 2011; Ojerio et al., 2011; Poudyal et al., 2012). The use of information regarding the economic and social susceptibilities are essential for a holistic vulnerability assessment (Tate, 2011).

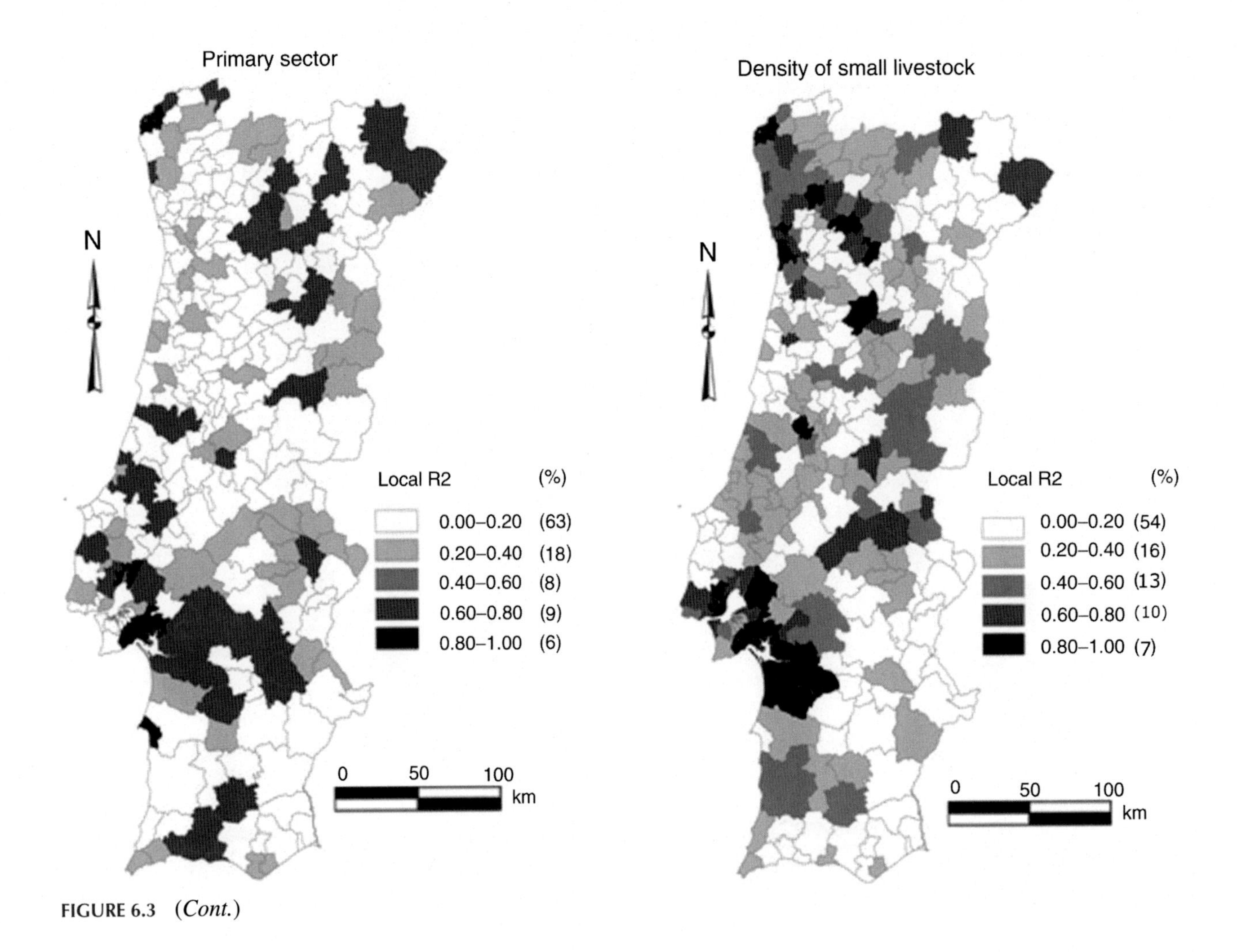

FIGURE 6.3 (*Cont.*)

Thus, wildfire prevention and mitigation effects should give priority to the most vulnerable people and places to safeguard that hazard reduction resources and strategies are rightfully distributed (Collins, 2008; Gaither et al., 2011; Ojerio et al., 2011; Poudyal et al., 2012). Carreiras et al. (2014), when analyzing the policies to deal with wildfire risk in Portugal and Spain, stressed the importance of local characteristics and the stakeholders' involvement in designing effective strategies to reduce fire risk. They argued, for example, that all such solutions have to accommodate changing interests driven, for example, by an aging population.

Various factors contribute to vulnerability and so identifying the many factors that contribute to it in a particular location using one or a few indicators is very difficult. Evaluating the vulnerability of a place also requires a wide-ranging knowledge base, methods, and the perspectives of diverse stakeholders (Adger, 2006; Khan, 2012). A holistic understanding of disaster risk depends on consideration of multiple hazards, social vulnerability, and physical vulnerability (Tate, 2011).

The presented study is exploratory and has several limitations; more studies have to be conducted in order to better capture the complexity of socioeconomic vulnerability. We cannot be sure that we have identified the most relevant variables and processes that determine socioeconomic vulnerability or if all the processes that influence vulnerability can even be measured. The selected models, the SLR versus the GWR, demonstrated that the GWR performance is higher, corroborating the existence of spatial variations in the association between the burnt area and the vulnerability regarding socioeconomic variables. However, some caution should be paid when the size of the kernel bandwidth was selected and to the possible collinearity in the local regression coefficients, which may limit interpretation of their distributional patterns (Wheeler and Tiefelsdorf, 2005).

On the other hand, the development of a strategy based on territorial characteristics and social and economic objectives should improve the participatory process and involve the local communities in land management. In fact, an effective forest

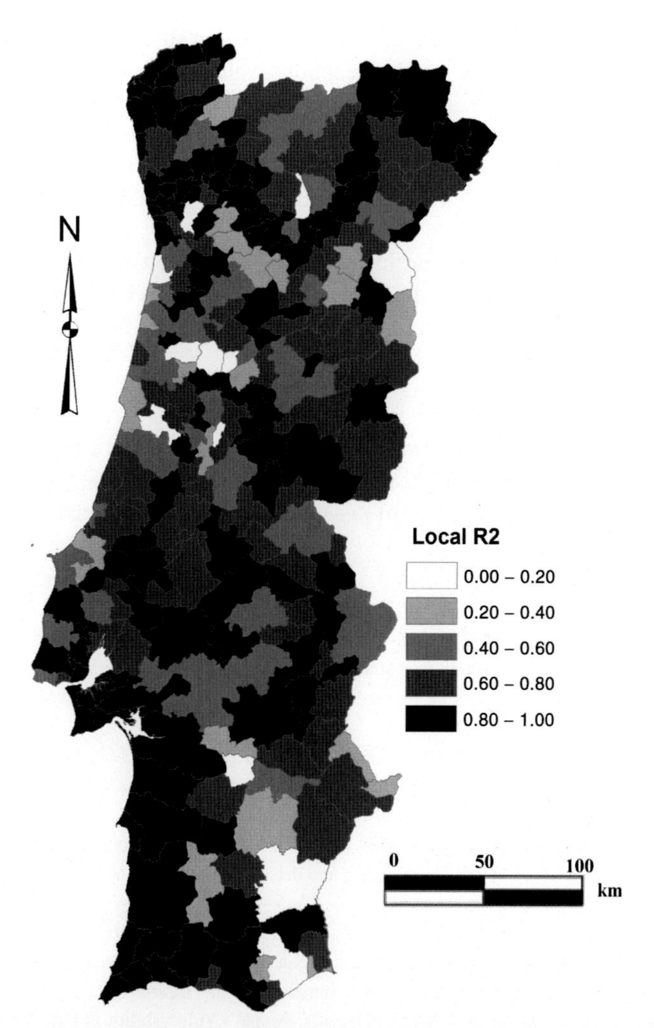

FIGURE 6.4 **Spatial distribution of local R^2 results from the application of GWR between the burnt area index and four socioeconomic variables: aging index, unemployment rate, income per capita, and small livestock density.**

TABLE 6.4 Adjusted R^2 Resulting from the Application of MLR and GWR Between the Burnt Area Index and Four Socioeconomic Variables: Aging Index, Unemployment Rate, Per Capita Income, and Small Livestock Density

SLR Adjusted R^2: 0.28	GWR Adjusted R^2: 0. 74
(+) Aging index ***	Aging index
(+) Unemployment rate ***	Unemployment rate
(−) Per capita income ***	Per capita income
(−) Small livestock density***	Small livestock density
AICc: −37.89	AICc: 422.49
	Residual squares: 0.26
	Moran's *I*: −0.09
	Z-score: −3.11
	P-value: 0.00

Source: (+) positive correlation; (−) negative correlation; Model variable significance *** = 0.01.

resources management and planning for hazard mitigation and adaptation should include an active and collaborative stakeholder participation in order to be consistent with local perceptions, values, needs, and expectations. It is essential, therefore, to establish a framework in order to meet the needs and the interests of local stakeholders and involve them in the community participation and implementation of strategies for an effective forest resources management. In fact, stakeholder participation is a key issue and also a key *challenge* in planning, implementation, and monitoring of *wildfire management,* since primary stakeholders' participation in forest resources management is affected by social and economic factors, such as age, gender, education, or resident status.

Even though this study can be understood as a starting point to promote better knowledge of the complex socioeconomic vulnerability interactions, the obtained results could be useful to support a more adequate fire management for the prevention phase and an important tool to communicate complex issues from science to policymakers or to the general public.

ACKNOWLEDGMENTS

This study was funded by the Centre for Studies in Geography and Spatial Planning (CEGOT), which is co-financed by the European Regional Development Fund (ERDF) through the COMPETE 2020—Operational Programme Competitiveness and Internationalization (POCI) and national funds by FCT under the POCI-01-0145-FEDER-006891 project (FCT Ref: UID/GEO 04084/2013).

REFERENCES

Adger, W.N., 2006. Vulnerability. Global Environ. Change 16, 268–281.

Anselin, L., Syabri, I., Kho, Y., 2006. GeoDa: an introduction to spatial data analysis. Geog. Anal. 38, 5–22.

Bajocco, S., Ricotta, C., 2008. Evidence of selective burning in Sardinia (Italy): which land cover classes do wildfires prefer? Landscape Ecol. 23, 241–248, http://dx.doi.org/10.1007/s10980-007-9176-5.

Bandeira, M.L., Azevedo, A.B., Gomes, C.S., Tomé, L.P., Mendes, M.F., Baptista, M.I., Moreira, M.J.G., 2014. Dinâmicas demográficas e envelhecimento da população portuguesa, 1950–2011: evolução e perspectivas, Fundação Francisco Manuel dos Santos (in Portuguese). https://www.ffms.pt/upload/docs/dinamicas-demograficas-e-envelhecimento-da-populac_efe8FbqdjUGZx3LduUIzgg.pdf.

Barbati, A., Arianoutsou, M., Corona, P., De Las Heras, J., Fernandes, P., Moreira, F., Papageorgiou, K., Vallejo, R., Xanthopoulos, G., 2010. Post-fire forest management in southern Europe: a COST action for gathering and disseminating scientific knowledge. iForest—Biogeosciences and Forestry 3, 5–7. doi: 10.3832/ifor0523-003.

Birkmann, J., 2006. Measuring Vulnerability to Natural Hazards: Towards Disaster Resilient Societies. United Nations Publications, New York, NY.

Birkmann, J., Cardona, O.D., Carreño, M.L., Barbat, A.H., Pelling, M., Schneiderbauer, S., Kienberger, S., Keiler, M., Alexander, D., Zeil, P., Welle, T., 2013. Framing vulnerability, risk and societal responses: the MOVE framework. Nat. Hazards 67, 193–211.

Blaikie, P., Cannon, T., Davies, I., Wisner, B., 1994. At Risk: Natural Hazards, People's Vulnerability, & Disaster. Routledge, London, UK.

Bodstein, A., Lima, V.V., Abreu de, A.M., 2014. The vulnerability of the elderly in disasters: the need for an effective resilience policy. Ambiente & Sociedade 17 (2), 157–174, Available from: http://www.scielo.br/scielo.php?script=sci_arttext&pid=S1414753X2014000200011&lng=en&nrm=iso.

Bracmort, K., 2014. Wildfire protection in the wildland–urban interface. Congressional research service report RS21880. http://fas.org/sgp/crs/misc/RS21880.pdf (accessed 11.03.16).

Bühler, M.D., de Torres Curth, M., Garibaldi, L.A., 2013. Demography and socioeconomic vulnerability influence fire occurrence in Bariloche (Argentina). Landscape and Urban Planning 110, 64–73.

Cannon, T., 1994. Vulnerability analysis and the explanation of "natural" disasters. In: Varley, A. (Ed.), Disasters, Development, and Environment. Wiley, New York, NY, pp. 13–30.

Carmo, M., Moreira, F., Casimiro, P., Vaz, P., 2011. Land use and topography influences on wildfire occurrence in Northern Portugal. Landscape Urban Plan 100, 169–176. doi: 10.1016/j.landurbplan.2010.11.017.

Carreiras, M., Ferreira, A.J.D., Valente, S., Fleskens, L., Gonzales-Pelayo, Ó., Rubio, J.L., Stoof, C.R., Coelho, C.O.A., Ferreira, C.S.S., Ritsema, C.J., 2014. Comparative analysis of policies to deal with wildfire risk. Land Degradation & Development 25, 92–103. doi: 10.1002/ldr.2271.

Carvalho, A., Flannigan, M.D., Logan, K., Miranda, A.I., Borrego, C., 2008. Fire activity in Portugal and its relationship to weather and the Canadian Fire Weather Index System. Int. J. Wildland Fire 17, 328–338.

Catry, F.X., Rego, F.C., Bacão, F.L., Moreira, F., 2009. Modeling and mapping wildfire ignition risk in Portugal. Int. J. Wildland Fire 18 (8), 921–931. doi: 10.1071/WF07123.

Cerdà, A., Lasanta, T., 2005. Long-term erosional responses after fire in the Central Spanish Pyrenees: 1. Water and sediment yield. Catena 60 (1), 59–80. doi: 10.1016/j.catena.2004.09.006.

Certini, G., 2005. Effects of fire on properties of forest soils: a review. Oecologia 143, 1–10. doi: 10.1007/s00442-004-1788-8.

Chuvieco, E., Aguado, I., Jurdao, S., Pettinari, M.L., Yebra, M., Salas, J., Hantson, S., de la Riva, J., Ibarra, P., Rodrigues, M., Echeverría, M., Azqueta, D., Román, M.V., Bastarrika, A., Martínez, S., Recondo, C., Zapico, E., Martínez-Vega, F.J., 2012. Integrating geospatial information into fire risk assessment. Int. J. Wildland Fire 23 (5), 606–619. doi: 10.1071/WF12052.

Chuvieco, E., Martinez, S., Roman, M.V., Hantson, S., Pettinari, M.L., 2014. Integration of ecological and socio-economic factors to assess global vulnerability to wildfire. Global Ecol. Biogeogr. 23 (2), 245–258.

Cleveland, W.S., 1979. Robust locally weighted regression and smoothing scatterplots. J. Am. Stat. Assoc. 74, 829–836.

Collins, T., 2008. What influences hazard mitigation? Household decision making about wildfire risks in Arizona's white mountains. Prof. Geogrepher 60, 508–526.

Cortner, H.J., Field, D.R., 2007. Foreword: synthesis and collaboration. In: Daniel, T.C., Carroll, M.S., Moseley, C., Raish, C. (Eds.), People, Fire, and Forests: A Synthesis of Wildfire Social Science. Oregon State University Press, Corvallis, OR, pp. 7–14.

Cutter, S.L., 1996. Vulnerability to environmental hazards. Prog. Hum. Geogr. 20 (4), 529–539.

Cutter, S.L., 2015. Pool knowledge to stem losses from disasters. Nature 522, 7–9.

Cutter, S.L., Mitchell, J.T., Scott, M.S., 2000. Revealing the vulnerability of people and places: a case study of Georgetown, South Carolina. Annals of the Association of American Geographers 90 (4), 713–737.

DeBano, L.F., Neary, D.G., Folliott, P.F., 1998. Fire's Effects on Ecosystems. Wiley, New York, NY, p. 333.

Dormann, C.F., Elith, J., Bacher, S., Buchmann, C., Carl, G., Carré, G., Marquéz, J.R.G., Gruber, B., Lafourcade, B., Leitão, P.J., Münkemüller, T., McClean, C., Osborne, P.E., Reineking, B., Schröder, B., Skidmore, A.K., Zurell, D., Lautenbach, S., 2013. Collinearity: a review of methods to deal with it and a simulation study evaluating their performance. Ecography 36, 27–46.

Evans, A., DeBonis, M., Krasilovsky, E., Melton, M., 2007. Measuring community capacity for protection from wildfire. Forest Guild Research Paper.

Finlay, S.E., Moffat, A., Gazzard, R., Baker, D., Murray, V., 2012. Health impacts of wildfires. PLoS Currents Disasters 2 (1), Edition 1.

Fortin, M.-J., Dale, M.R.T., 2005. Spatial Analysis: A Guide for Ecologists. Cambridge University Press, Cambridge, UK.

Fotheringham, A.S., Brunsdon, C., Charlton, M.E., 2002. Geographically Weighted Regression: The Analysis of Spatially Varying Relationships. Wiley, Chichester, UK.

Gaither, C.J., Poudyal, N., Goodrick, S., Bowker, J.M., Malone, S., Gan, J., 2011. Wildland fire risk and social vulnerability in the Southeastern United States: an exploratory spatial data analysis approach. For. Policy Econ. 13, 24–36.

Galiana-Martin, L., Karlsson, O., 2012. Development of a methodology for the assessment of vulnerability related to wildland fires using a multi-criteria evaluation. Geog. Res. 50, 304–319.

Gall, M., Borden, K.A., Emrich, C.T., Cutter, S.L., 2011. The unsustainable trend of natural hazard losses in the United States. Sustainability 3 (12), 2157–2181.

Ganteaume, A., Camia, A., Jappiot, M., San-Miguel-Ayanz, J., Long-Fournel, M., Lampin, C., 2013. A review of the main driving factors of forest fire ignition over Europe. Environ. Manage. 51, 651–662.

Gill, N., 2006. What is the problem? Usefulness, the cultural turn, and social research for natural resource management. Aus. Geogr. 37, 5–17.

Griffith, D.A., Layne, L.J., 1999. A Casebook for Spatial Statistical Data Analysis: A Compilation of Analyses of Different Thematic Data Sets. Oxford University Press, New York.

Haas, J.R., Calkin, D.E., Thompson, M.P., 2013. A national approach for integrating wildfire simulation modeling into wildland–urban interface risk assessments within the United States. Landscape Urban Plann. 119, 44–53.

Hoover, K., Bracmort, K., 2015. Wildfire management: federal funding and related statistics. Congressional research service report RS43077. Wildfire potential and social vulnerability. Int. J. Wildland Fire. http://nationalaglawcenter.org/wp-content/uploads/assets/crs/R43077.pdf (accessed 11.12.16).

Hurvich, C.M., Jeffrey, S.S., Chih-Ling, T., 1998. Smoothing parameter selection in nonparametric regression using an improved Akaike information criterion. Journal of the Royal Statistical Society: Series B (Statistical Methodology) 60 (2), 271–293.

JRC, 2016. Forest fires in Europe, Middle East and North Africa 2015. Technical report no. 15, EUR 28148 EN.

Keeley, J.E., Keeley, S.C., 1988. Chaparral. In: Barbour, M.G., Billings, W.D. (Eds.), North American Terrestrial Vegetation. Cambridge University Press, New York, NY, pp. 165–207.

Khan, S., 2012. Vulnerability assessments and their planning implications: a case study of the Hutt Valley, New Zealand. Natural Hazards 64 (2), 1587–1607.

Koutsias, N., Chuvieco, E., Allgöwer, B., Martínez, J., 2005. Modelling wildland fire occurrence in Southern Europe by geographically weighted regression approach. In: De la Riva, J., Chuvieco, E.P. (Eds.), 5th International Workshop on Remote Sensing and GIS Applications to Forest Fire Management: Fire Effects Assessment. Universidad de Zaragoza-EARSeL, Zaragoza, pp. 57–60.

Koutsias, N., Martínez-Fernández, J., Allgöwer, B., 2010. Do factors causing wildfires vary in space? Evidence from geographically weighted regression. GIScience & Remote Sens. 47, 1548–1603.

Kovats, R.S., Valentini, R., Bouwer, L.M., Georgopoulou, E., Jacob, D., Martin, E., Rounsevell, M., Soussana, J.-F., 2014. Europe. In: Barros, V.R., Field, C.B., Dokken, D.J., Mastrandre, M.D. (Eds.), Climate Change 2014 Impacts, Adaptation, and Vulnerability. Part B: Regional Aspects. Contribution of Working Group II to the Fifth Assessment Report of the Intergovernmental Panel on Climate Change. Cambridge University Press, Cambridge, UK and New York, NY, pp. 1267–1326.

Kupfer, J.A., Farris, C.A., 2007. Incorporating spatial non-stationarity of regression coefficients into predictive vegetation models. Landscape Ecol. 22 (6), 837–852, http://link.springer.com/10.1007/s10980-006-9058-2.

Lex, P., Goldammer, J.G., 2001. Fire situation in Germany. Int. For. Fire News 24, 22–30.

Lourenço, L., Gonçalves, A.B., 1990. As situações meteorológicas e a eclosão-propagação dos grandes incêndios florestais registados durante 1989 no Centro de Portugal. II Congresso Florestal Nacional, Porto, Portugal, pp. 755–763.

Lynn, K., Gerlitz, W., 2006. Mapping the relationship between wildfire and poverty. In: Andrews, P.L., Butler, B.W. (comps.), Fuels Management-How to Measure Success: Conference Proceedings. 28–30 March 2006, Portland, OR. Proceedings RMRS-P-41. U.S. Department of Agriculture, Forest Service, Rocky Mountain Research Station, Fort Collins, CO, pp. 401–415.

Malkinson, D., Wittenberg, L., Beeri, O., Barzilai, R., 2011. Effects of repeated fires on the structure, composition, and dynamics of Mediterranean maquis: short- and long-term perspectives. Ecosystems 14, 478–488.

Martínez, J., Vega-Garcia, C., Chuvieco, E., 2009. Human-caused wildfire risk rating for prevention planning in Spain. J. Environ. Econ. Manage. 90, 1241–1252.

Martínez-Fernández, J., Koutsias, N., 2011. Modelling fire occurrence factors in Spain. National trends and local variations. In: San Miguel, J., Camia, A., Gita, I., Oliveira, S. (Eds.), Advances in Remote Sensing and GIS Applications in Forest Fire Management: From Local to Global Assessments. JRC Scientific and Technical Reports. Publications Office of the European Union, Stressa, Italy, pp. 203–208, October 20–21.

Martínez-Fernández, J., Chuvieco, E., Koutsias, N., 2013. Modelling long-term fire occurrence factors in Spain by accounting for local variations with geographically weighted regression. Nat. Hazard. Earth Syst. Sci. 13, 311–327.

Mateus, P., Fernandes, P.M., 2014. Forest fires in Portugal: dynamics, causes and policies. Reboredo, F. (Ed.), Forest Context and Policies in Portugal, Present and Future Challenges. World Forests, 19, Springer International Publishing, Switzerland, pp. 219–236.

Mazerolle, 2004. APPENDIX 1: Making sense out of Akaike's Information Criterion (AIC): its use and interpretation in model selection and inference from ecological data. In: Mazzarolle, M.J. (Ed.), Mouvements et reproduction des amphibiens en tourbières perturbèes. PhD Thesis, Universitè Laval.

McCaffrey, S., Toman, E., Stidham, M., Shindler, B., 2013. Social science research related to wildfire management: an overview of recent findings and future research needs. Int. J. Wildland Fire 22, 15–24.

Mennis, J.L., 2006. Mapping the results of geographically weighted regression. The Cartographic Journal 43 (2), 171–179.

Mercer, E., Prestemon, J.P., 2005. Comparing production function models for wildfire risk analysis in the wildland–urban interface. For. Policy Econ. 7 (5), 782–795.

Miranda, A.L., Monteiro, A., Martins, V., Carvalho, A., Schaap, M., Builtjes, P., Borrego, C., 2008. Forest fires impact on air quality over Portugal. Air Pollution Modeling and Its Application XIX. NATO Science for Peace and Security Series C: Environmental Security. Springer, Dordrecht, 190-198.

Moreira, F., Vaz, P., Catry, F., Silva, J.S., 2009. Regional variations in wildfire preference for land cover types in Portugal: implications for landscape management to minimise fire hazard. Int. J. Wildland Fire 18, 563–574.

Moreira, F., Viedma, O., Arianoutsou, M., Curt, T., Koutsias, N., Rigolot, E., Barbati, A., Corona, P., Vaz, P., Xanthopoulos, G., Mouillot, F., Bilgili, E., 2011. Landscape—wildfire interactions in southern Europe: implications for landscape management. J. Environ. Econ. Manage. 92, 2389–2402.

Morrow, B.H., 1999. Identifying mapping community vulnerability. Disasters 23 (1), 1–18.

Mourão, P.R., Martinho, V.D., 2016. Discussing structural breaks in the Portuguese regulation on forest fires—an economic approach. Land Use Policy 54, 460–478.

Novara, A., Gristina, L., Rühl, J., Pasta, S., D'Angelo, G., La Mantia, T., Pereira, P., 2013. Grassland fire effect on soil organic carbon reservoirs in a semiarid environment. Solid Earth 4 (2), 381.

Nunes, A.N., 2012. Regional variability and driving forces behind forest fires in Portugal an overview of the last three decades (1980–2009). Appl. Geogr. 34, 576–586. doi: 10.1016/j.apgeog 2012.03.002.

Nunes, M.C.S., Vasconcelos, M.J., Pereira, J.M.C., Dasgupta, N., Alldredge, R.J., Rego, F.C., 2005. Land cover type and fire in Portugal: do fires burn land cover selectively? Landscape Ecol. 20, 661–673.

Nunes, A.N., Lourenço, L., Meira, A.C., 2016. Exploring spatial patterns and drivers of forest fires in Portugal (1980–2014). Sci. Total Environ. 573, 1190–1202.

Ojerio, R., Moseley, C., Lynn, K., Bania, N., 2011. Limited involvement of socially vulnerable populations in federal programs to mitigate wildfire risk in Arizona. Nat. Hazard.. (1), 28–36.

Oliveira, S., Oehler, F., San-Miguel-Ayanz, J., Camia, A., Pereira, J.M.C., 2012. Modeling spatial patterns of fire occurrence in Mediterranean Europe using multiple regression and random forest. J. Environ. Econ. Manage. 117–129.

Oliveira, S., Pereira, J.M.C., San-Miguel-Ayanz, J., Lourenço, L., 2014. Exploring the spatial patterns of fire density in Southern Europe using geographically weighted regression. Applied Geography 51, 143–157.

Pausas, J., Ramon-Vallejo, R., 1999. The role of fire in European Mediterranean ecosystems. In: Chuvieco, E. (Ed.), Remote Sensing of Large Wildfires in the European Mediterranean Basin. Springer, Berlin, Germany, pp. 3–16.

Pearson, L., Nelson, R., Crimp, S., Langridge, J., 2008. Climate change vulnerability assessment: review of agricultural productivity. CSIRO Climate Adaptation Flagship Working Paper No. 1.

Pereira, M.G., Trigo, R.M., DaCamara, C.C., Pereira, J.M.C., Leite, S.M., 2005. Synoptic patterns associated with large summer forest fires in Portugal. Agricultural and Forest Meteorology 129, 11–25.

Poudyal, N.C., Johnson-Gaither, C., Goodrick, S., Bowker, J.M., Gan, J., 2012. Locating spatial variation in the association between wildland fire risk and social vulnerability across six southern states. Environ. Manage. 623–635.

Prior, T., Eriksen, C., 2013. Wildfire preparedness community cohesion and social–ecological systems. Global Environmental Change 23 (6), 1575–1586.

Rodrigues, M., de la Riva, J., Fotheringham, S., 2014. Modeling the spatial variation of the explanatory factors of human-caused wildfires in Spain using geographically weighted logistic regression. Applied Geography 48, 52–63.

San-Miguel-Ayanz, J., Schulte, E., Schmuck, G., Camia, A., Strobl, P., Liberta, G., Giovando, C., Boca, R., Sedano, F., Kempeneers, P., McInerney, D., Withmore, C., Oliveira, S.S., Rodrigues, M., Durrant, T., Corti, P., Oehler, F., Vilar, L., Amatulli, G., 2012. Comprehensive monitoring of wildfires in Europe: the European forest fire information system (EFFIS). In: Tiefenbacher, J. (Ed.), Approaches to Managing Disaster—Assessing Hazards Emergencies and Disaster Impacts, Available from: http://ec.europa.eu/environment/forests/pdf/InTech.pdf/ (assessed 21.01.16).

Snipes, S., Taylor, D.C., 2014. Model selection and Akaike Information Criteria: An example from wine ratings and prices. Wine Economics and Policy 3(1), 3–9.

Tate, E., 2011. Indices of social vulnerability to hazards: model uncertainty and sensitivity. PhD dissertation. Department of Geography, University of South Carolina.

Tedim, F., 2012. Enhance wildfire risk management in Portugal: the relevance of vulnerability assessment. In: Charles, C. (Ed.), Wildfire and Community: Facilitating Preparedness and Resilience. Thomas Publisher, Illinois, USA, pp. 66–84.

Trigo, R.M., Pereira, J.M.C., Pereira, M.G., Mota, B., Calado, T.J., Dacamara, C.C., Santo, F.E., 2006. Atmospheric conditions associated with the exceptional fire season of 2003 in Portugal. Int. J. Climatology 26, 1741–1757.

UNISDR, 2009. UNISDR Terminology on Disaster Risk Reduction United Nations International Strategy for Disaster Reduction (UNISDR), Geneva, Switzerland.

UNISDR, 2015. UNISDR Sendai Framework for Disaster Risk Reduction 2015–2030, United Nations Office for Disaster Risk Reduction (UNISDR).

Verdú, F., Salas, J., Vega-García, C., 2012. A multivariate analysis of biophysical factors and forest fires in Spain, 1991–2005. Int. J. Wildland Fire 21 (5), 498–509.

Wheeler, D., Tiefelsdorf, M., 2005. Multicollinearity and correlation among local regression coefficients in geographically weighted regression. J. Geog. Syst. 7 (2), 161–187.

Wigtil, G., Hammer, R.B., Kline, J.D., Mockrin, M.H., Stewart, S.I., Roper, D., Radeloff, V.C., 2016. Places where wildfire potential and social vulnerability coincide in the coterminous United States. Int. J. Wildland Fire. doi: 10.1071/WF15109.

FURTHER READING

Goldammer, J.G., Lex, P., 2001. Fire situation in Germany. FRA global forest fire assessment 1990–2000. Forest Resources Assessment Programme, Working Paper 55. FAO, Rome, pp. 326–335.

Haque, C.E., Etkin, D., 2007. People and community as constituent parts of hazards: the significance of societal dimensions in hazards analysis. Nat. Hazards 41, 271–282.

Rego, F.C., 1992. Land use changes wildfires. In: Teller, A., Mathy, P., Jeffers, J.N.R. (Eds.), Responses of Forest Ecosystems to Environmental Changes. Elsevier Applied Sciences, London, UK, pp. 367–373.

Solangaarachchi, D., Griffin, A.L., Doherty, M.D., 2012. Social vulnerability in the context of bushfire risk at the urban–bush interface in Sydney: a case study of the Blue Mountains and Ku-ring-gai local council areas. Nat. Hazards 64 (2), 1873–1898.

Chapter 7

Landslide Risk Assessment in Parts of the Darjeeling Himalayas, India

Shraban Sarkar
Department of Geography, Cooch Behar Panchanan Barma University, Cooch Behar, West Bengal, India

7.1 INTRODUCTION

Landslides are considered to be one of the most dangerous natural hazards worldwide for the reason that they cause human fatalities and damage to properties. Landslide occurrence is a common phenomenon after intense rainfall in the mountainous areas (Jia et al., 2012). Since landslide is a natural phenomenon and quite uncertain in nature, it is difficult to control its occurrence. However, an advanced prediction of landslide event may reduce the potential loss of life and property. As a consequence, the susceptibility or likelihood analysis for landslides is becoming an important task for earth scientists and spatial modelers.

When the susceptibility map is combined with the temporal and magnitude information of landslide events, this can be converted into a landslide hazard map (Martha et al., 2013). The resultant landslide hazard map can be further incorporated into a socio-economic framework to evaluate the potential social, or economic damages or losses, that is, risk, and the evaluation system is known as risk assessment (Miller, 1988). Varnes and IAEG Commission on Landslides and Other Mass Movements on Slopes (1984) define the landslide risk as, "the expected number of lives lost, persons injured, damage to properties and disruption of economic activity" due to landslides for a given area and reference period. Risk assessment quantifies the vulnerability and cost of elements at risk. Risk assessment is the ultimate goal of landslide study that helps to mitigate the problem (Liu et al., 2012) and proper land use management (Lateltin et al., 2005).

A number of studies have been done in the field of landslide risk assessment and adaptive methods (Cardinali et al., 2002; Dai et al., 2002; Espizua and Bengochea, 2002; Finlay et al., 1999; Michael-Leiba et al., 2000) and it is revealed that landslide risk may be assessed by qualitative, semiquantitative, and quantitative estimation approaches. In the qualitative assessment, the risk may be determined by experts' knowledge (Cardinali et al., 2002; Petley, 1998; cf. Crozier and Glade, 2005). In semiquantitative approach, weights were assigned to slope instability factors (Chowdhury and Flentje, 2003), and finally, the risk was classified with qualitative implications (Anbalagan and Singh, 1996; Hadmoko et al., 2010; Kanungo et al., 2008). Whereas in the quantitative method, the risk was analyzed by hazard, the vulnerability of elements at risk, and consequences, and expressed in numerical forms in many ways (Carrara et al., 1991; Catani et al., 2005; Ghosh et al., 2012; Zêzere et al., 2008).

Risk is the combination of landslide hazard and vulnerability (Cardinali et al., 2002; Martha et al., 2013; van Westen et al., 2006) and is expressed as

$$\textit{Landslide risk} = \textit{Landslide hazard} \times \textit{Vulnerability} \times \textit{Elements at risk}$$

The vulnerability may be defined as the level of potential damage or degree of loss of elements at risk, subjected to a landslide occurrence of a given intensity (Fell, 1994; Wong et al., 1997; cf. Kanungo et al., 2008). The vulnerability was presented as probability, which varies from 0 (i.e. nothing destroyed) to 1 (i.e. all destroyed). However, vulnerability assessment is a complex issue which is rarely considered in an appropriate and thoughtful manner (Crozier and Glade, 2005; van Westen et al., 2006).

Many components constitute elements at risk and can be categorized as people, structure, and infrastructure (dams and power plants, residential buildings, commercial buildings, industries, livestock farms, cemeteries, etc.), public services (schools, hospitals, and others), transportation network (railways, roads), and mining. Among these elements, it is challenging to calculate the risk of human life because it is not a static substance. The evaluation method of expected damages or losses of the above elements exposed at risk varies with the scale of the study (Abella, 2008), and is difficult to assess the vulnerability of all elements at the regional level.

Integrating Disaster Science and Management. http://dx.doi.org/10.1016/B978-0-12-812056-9.00007-5

Many studies have been successfully carried out to evaluate the vulnerability of buildings, population, and roads in India (Anbalagan and Singh, 1996; Ghosh et al., 2012; Jaiswal et al., 2011; Martha et al., 2013). Owing to lack of information, high-resolution stereoscopic data have been used to identify buildings in some cases over Himalayas (Martha and Kumar, 2013; Martha et al., 2013). Das et al. (2011) assessed the vulnerability of vehicles in the Garhwal Himalayas, whereas the probability of damages of roads is illustrated by Kanungo et al. (2008) and Ghosh et al. (2012) in the Darjeeling Himalayas.

The present study area has limited information regarding past landslide events, consequences, details of elements at risk, their condition, and values in price. Lack of such data creates the problem of detailed risk estimation of most of the developing countries. A limited number of studies in landslide hazard and risk have been conducted in such data-scarce region at a medium scale. Under these circumstances to estimate the potential losses of elements at risk, that is, buildings and living population in buildings, and roads, due to future landslide occurrences, the present study has adopted quantitative and semiquantitative estimation approaches.

7.2 STUDY AREA

The present study has focused on Kalimpong town and surrounding hills (26°58′N to 27°11′N and 88°26′E to 88°39′E), which is part of the Darjeeling Himalayas (Fig. 7.1). The areal extent of the study area is about 330 km^2 and is characterized by rugged mountainous terrain with a steep slope. The climate is classified as Cwb (subtropical highland climate) according to Köppen's climate classification with the mean annual rainfall and relative humidity being 2754 mm and 76%, respectively. June to September are the wettest months and receive heavy rain due to the southwest monsoon (Regional Sericulture Research Station, Kalimpong, 2010).

Geologically, the area is covered by Darjeeling Gneiss and Daling Series of metamorphic rocks, which can be further classified into slate, chlorite sericite schist, chlorite quartz schist, golden silvery mica schist, garnetiferous mica schist, and the coarse-grained gneiss (Pawde and Saha, 1982). Tectonically, the area falls under the Himalayan fold-thrust belt, where highly crystalline rocks of Darjeeling Gneiss, thrust over the low-grade metasedimentary rocks of Daling Series along the Main Central Thrust (MCT) (Ghosh et al., 2011; Searle and Szule, 2005). Occurrence of shallow earthquakes around the area is the witness of tectonic activeness of the region (Joshi et al., 2010).

The Kalimpong town hosts around 49,403 populations with compacted settlement and the entire Kalimpong district consists of population density 239 persons/km^2 (Census of India, 2011). The settlements were interconnected by state highway (SH)-12 and minor roads. National highway (NH)-10 is the lifeline of Sikkim that connects with remaining India passing through the region.

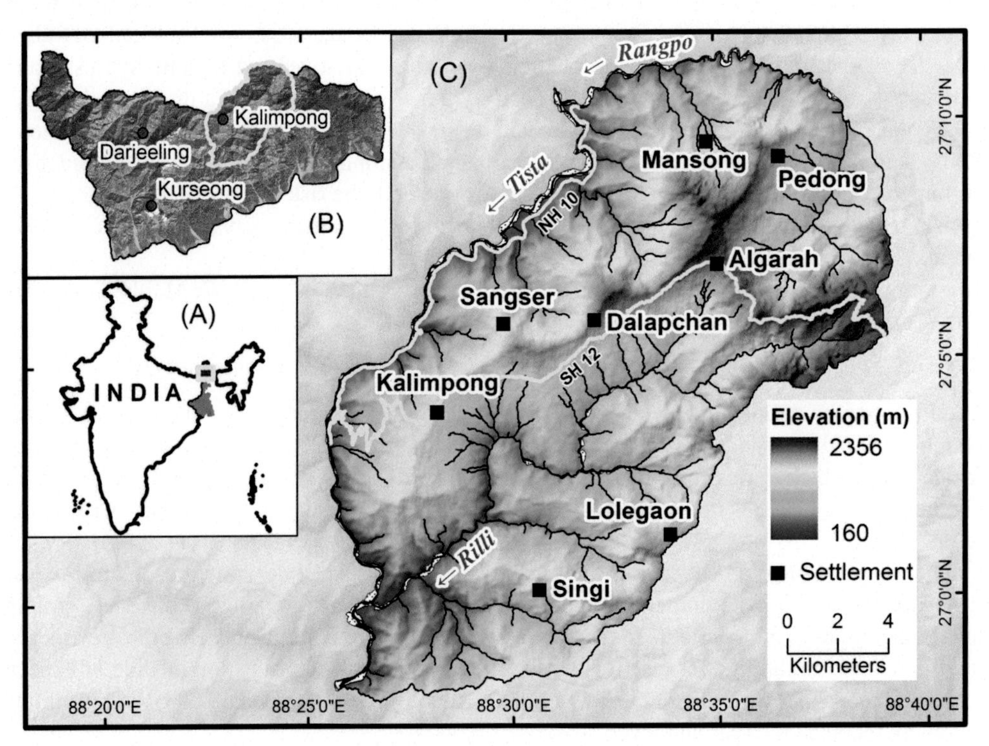

FIGURE 7.1 **Location map.** (A) India, (B) Darjeeling Himalayas, and (C) study area with major settlements and road network.

7.3 MATERIALS AND METHODS

Landslide occurrence is a function of various geo-environmental factors, that is, topography, geomorphology, bedrock geology, and land use (Cruden, 1993; Hansen, 1984; Varnes, 1978; cf. Zhou et al., 2002). Analyzing the relationship of landslide occurrence with these factors helps to understand the nature of failure and to predict vulnerability in advance.

Since landslide risk assessment is the ultimate step of landslide studies, to assess the landslide risk, the present study follows these steps: (a) the landslide susceptibility has been assessed by finding the relationship of existing landslide scars and 16 landslide causing geo-environmental factors, (b) landslide hazard has been evaluated by combining the spatial, temporal, and magnitude probabilities of landslide events, and finally (c) landslide risk to buildings, population, and roads for the 10-year return period has been estimated.

To make the susceptibility map, all the thematic factor maps were derived from the satellite images, Google Earth (® Google Inc.) images, geological map, laboratory analysis, and field survey. Whereas a set of multitemporal remote-sensing images were used to make the landslide inventory map, which was further used to compute the temporal probability of landslide events and its magnitude.

Since the study area has no detailed information regarding building structure at large scale, a high-resolution stereoscopic Cartosat-1 (2007) satellite image along with Resourcesat-1 Mx (2005) and Google Earth (® Google Inc.) images (2009 and 2011) were used to manually extract the settlement units (SUs) and building footprints. Census data of 2001 and 2011 have been used to project the population of each household first and then estimate for each identified building through statistical techniques (Goerlich and Cantarino, 2013; Lung et al., 2013). Finally, counted structure and projected population in each household were validated by limited field verification.

Thick evergreen forest makes it difficult to identify the road network from satellite images only. Therefore, along with satellite images, Google Earth (® Google Inc.) images, GPS tracking, and field verification have been conducted to make final road network map.

Details of data sets and variables extracted from them are given in Table 7.1.

7.3.1 Landslide Susceptibility and Hazard Assessment

To prepare the landslide susceptibility map, all the thematic factor maps and the map of the existing landslide were first converted into a raster grid and then the relationship between the landslide causing factors and existing landslides were calculated by bivariate-based information value method. The degree of susceptibility was represented by classifying the map into three zones, that is, high, moderate, and low. Landslides occurred (2022 scars) by the year 2009 were used to calibrate the prediction, whereas 183 landslide scars which occurred during 2010–2011 were used to validate the prediction.

TABLE 7.1 Details of Data Used in the Study

Data Type	Source	Year	Spatial Resolution	Variables Extracted/Purpose
Declassified satellite image	KH-9	1974, 1975, 1979	20–30 ft	Landslide inventory map
Satellite image	Landsat ETM+	2000	15 m	Landslide inventory map
	Resourcesat-1 LISS IV Mx	2005	5.8 m	Land use and landslide inventory map
	Cartosat-1 (Stereo)	2007	2.5 m	Topographical and hydrological parameters, and landslide inventory map
Google Earth image	Google Inc.	2009, 2011		Landslide inventory map, number of buildings, and road network
Geological map	GSI			Lithology and structure
Census data	Census of India	2001, 2011		Density of population, number of buildings, and administrative map
Field survey		2008–2012		Field verification of land use and landslides

After preparing the landslide susceptibility map (spatial probability), landslide hazard map has been prepared by combining the temporal and magnitude probabilities of landslide events. The temporal probability has been computed using Poisson probability distribution for 37 years (1975–2011) of landslide events. And the landslide area statistics has been put on probability density function, fitted with inverse-gamma distribution, and cumulative distribution function to calculate the magnitude probability of landslides.

7.3.2 Landslide Risk Assessment

By considering the SUs as a basic map unit, risk or probable loss of elements at risk (e.g. building and population) has been estimated by exposure-based approach proposed and used by Lee and Jones (2004) and Ghosh et al. (2012). The adopted calculating method for probable loss at each SU (L_{SU}) can be expressed as

$$L_{SU} = Cell_{Aff_{SU}} \times Cell_{Pr\,ob_R}$$

In the above equation, $Cell_{Aff_{SU}} = P(S) \times Cell_{N_{SU}}$ and $Cell_{Pr\,ob_R} = N_{Elm} / Cell_{N_{SU}}$, $Cell_{Aff_{SU}}$ is the number of affected pixels in a SU, $P(S)$ is the spatial probability [$P(S)$ = Total landslide area in a susceptibility zone/Total area of the susceptibility zone], $Cell_{N_{SU}}$ is the total number of pixels in a SU, $Cell_{Pr\,ob_R}$ is the probability of pixels occupied by an element at risk (e.g. buildings) in a SU, and N_{Elm} is the number of elements at risk (e.g. buildings) within a SU.

Since population is a dynamic substance, it is very difficult to assess the daily movement of population, leaving or worked in buildings over a large area. Therefore, the risk of population has roughly been estimated through damages of buildings (Papathoma-Köhle et al., 2007).

To estimate the probable risk to roads, the present study has implemented the concept of "danger pixel" approach suggested by Kanungo et al. (2008). Danger pixels (pixels occupying the high and moderate susceptible zone) have been identified along the 10-m buffer of roads in both sides. Taking into account the limitation of available information, the present study is guided by indirect cost estimation approach (Zêzere et al., 2007) to evaluate the economic loss to reconstruct the fully or partially damaged roads. Risk to vehicles in the roads and persons inside the vehicles have not been considered in the present research.

7.4 RESULTS AND DISCUSSION

7.4.1 Risk to Buildings and Population

The density of buildings gradually decreases with distance from the center of the town toward downslope. In the present study, a total of 17,343 building footprints within 90 SU (22.14 km^2 in area) was identified (Fig. 7.2A). Out of total buildings, 58% (9984) were in urban area and well constructed by concrete and generally 2–3 tiers. Whereas the remaining buildings were found in rural areas and distributed sparsely. Rural buildings were not well built and were constructed using light materials like bamboo, timber, and bricks with metallic shades. These structures generally consist of a bedroom, a kitchen, and a cattle house. The population estimation shows that a total of 81,942 persons were living in above 90 SU (Fig. 7.2B).

Based on exposure analysis, risk to buildings that are likely to be affected with maximum temporal probability of 10-year return period (T_{10}) is presented in Fig. 7.3. Depending on the variation of $P(S)$, which varies with landslide densities in each susceptibility zone, risk has been simulated for three scenarios, that is, minimum, maximum, and average spatial probability. The calculated values have been categorized into three zones, that is, high, moderate, and low, by natural break classification method to make the comparison between three simulations. Results (Fig. 7.3 and Table 7.2) reveal that within a specified period, the probability of damage or loss of elements at risk depends upon likelihood of landslides. In minimum spatial probability [$P(S)$], almost the entire area comes under low-risk zone. Whereas a steep scarp in upper catchment of Tumthang khola (Mangchu and Icha forest) was mentioned as a moderate-risk zone and 357 buildings containing 767 persons will likely to be affected here in future. The same scarp land was considered high-risk zone in maximum and average $P(S)$. The southern and western parts of the Kalimpong town falls under high-risk zone and moderate-risk zone as per maximum and average $P(S)$, respectively. However, it is clearly noticed that the central part of the Kalimpong town is always in low-risk zone, but the vulnerability of elements at risk is gradually higher toward downslope.

Previous studies on landslide risk assessment in Nilgiri Hills (Jaiswal et al., 2011), Garhwal Himalayas (Anbalagan and Singh, 1996; Das et al., 2011; Martha et al., 2013), and in the Darjeeling Himalayas (Kanungo et al., 2008; Ghosh et al., 2012) agreed that potential loss of elements at risk was directly linked with landslide susceptibility or spatial probability,

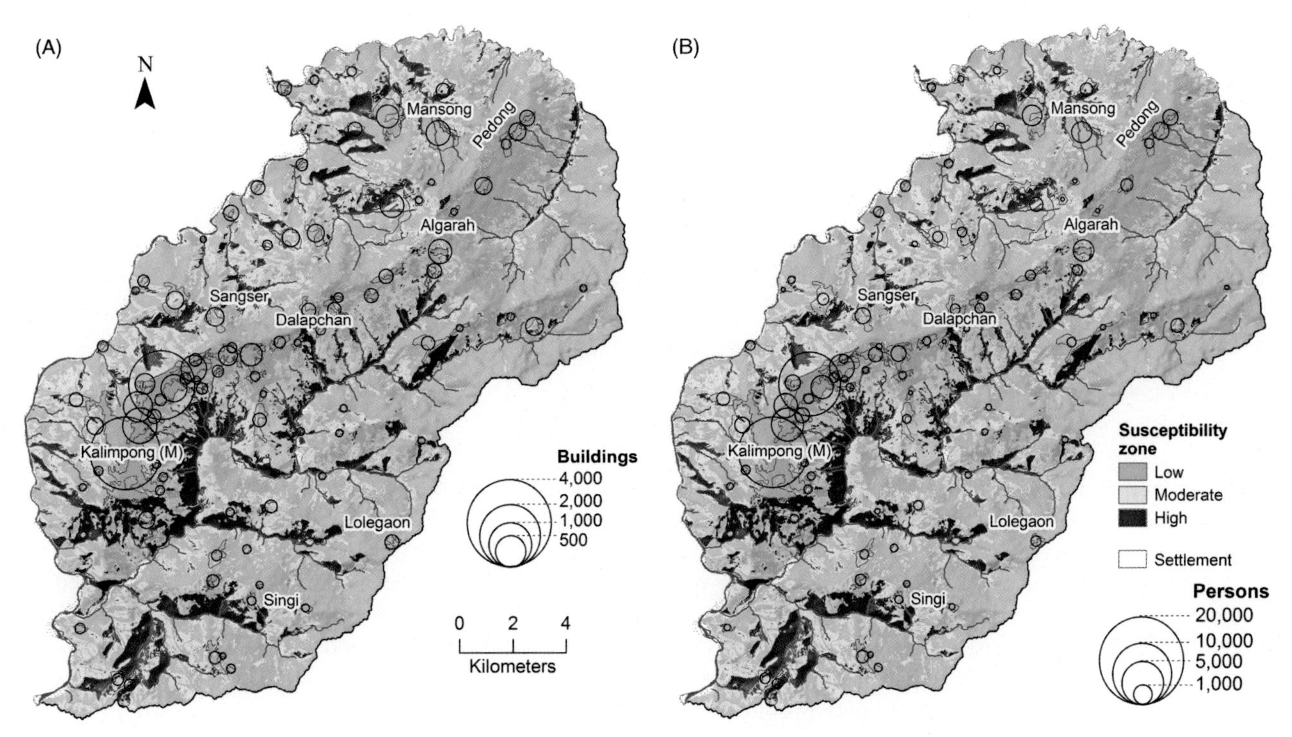

FIGURE 7.2 Elements at risk for (A) buildings and (B) population, overlaying landslide susceptibility maps.

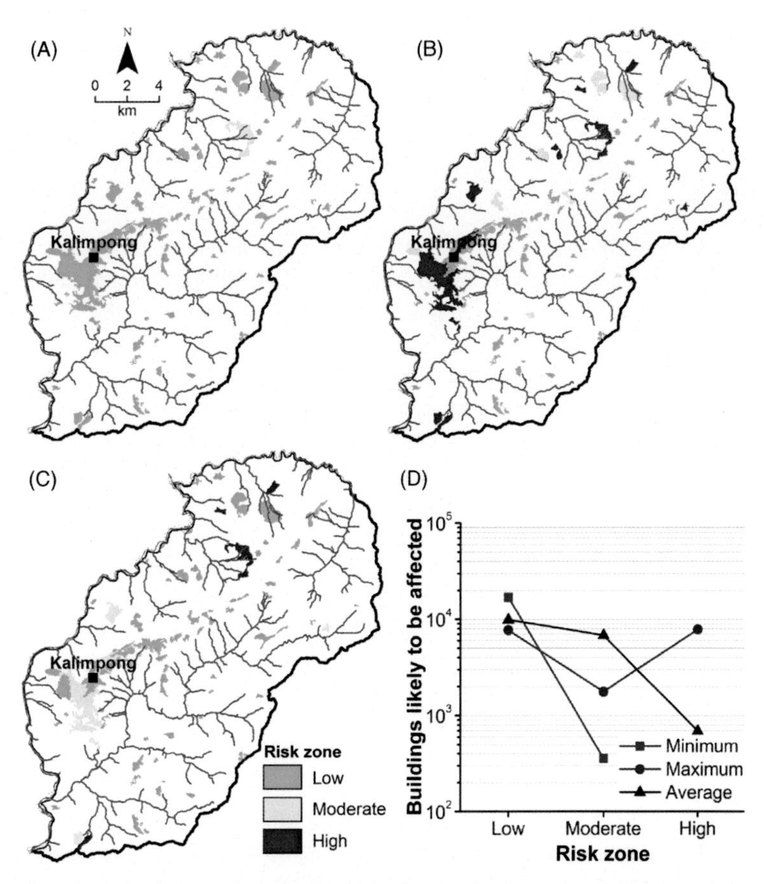

FIGURE 7.3 Risk analysis for buildings at maximum T_{10}: (A) in minimum spatial probability, (B) in maximum spatial probability, (C) in average spatial probability, and (D) the number of buildings likely to be affected in each risk zone.

TABLE 7.2 Areal Coverage of Risk Zones Along with Buildings and Persons Likely to be Affected by Future Landslide Events at Maximum T_{10}

Spatial Probability	Risk Zone	Area (km^2)	Buildings	Persons
Minimum	High	0	0	0
	Moderate	1.25	357	767
Maximum	High	9.38	7858	40,941
	Moderate	4.04	1770	7530
Average	High	2.07	691	2306
	Moderate	6.02	6798	37,266

temporal probability, and intensity of the events. A similar conclusion has been drawn in the present research too, although types of the events (Ghosh et al., 2012) have not been separately incorporated here due to lack of detailed landslide inventory map.

Owing to complexity and difficulty, vulnerability of elements at risk have not been computed here and assumed as 1 (Ghosh et al., 2012). Therefore, it only estimates the number of affected elements at risk rather than the number of destroyed. In many studies, risks are expressed in terms of cost (Catani et al., 2005; Remondo et al., 2005; Sterlacchini et al., 2007), but cost is a dynamic component and it is difficult to estimate the values of buildings at a large area. Therefore, in the present study, the risk to buildings in terms of cost has not been expressed.

Another simplified assumption was adopted here to estimate the persons likely to be affected. The vulnerability of human depends upon house structure, time of landslide occurrence, occupation structure of people, and many others (Papathoma-Köhle et al., 2007). Ghosh et al. (2012) in the Darjeeling Himalayas has estimated that the persons likely to be affected by day and nighttime and observed that persons in the night are more vulnerable. However, in the present study, it is assumed that people will live in their houses during the landslide events. Although landsliding is a natural process, it need not follow all the adaptive assumptions, but such kinds of estimation may be reliable and useful in data-poor large Himalayan terrain.

7.4.2 Risk to Roads

The area has good road network and consists of NH-10, SH-12, metalled, and unmetalled roads. The 22.92-km long NH-10 passing through the area along the banks of river Tista from Gangtok to Siliguri, which have maximum vehicular movement. The second important road is 41.65-km long SH-12, connecting Tista bazar in the West to Lava in the East, passing through the Kalimpong town (Fig. 7.4).

Results estimated for landslide risk for each kind of roads have separately been presented in Fig. 7.5 and Table 7.3. Visual inspection of the risk maps shows that the road segment is higher risk-prone where it passes through gullies at steep side slope of hills and on concave plan curvature. It is found that SH-12 may highly get affected at 17 segments which cover 1.83-km road length and most of the places are in between Kalimpong and Algarah. A spatial concentration has been given to NH-10 (Fig. 7.6), because almost every year during monsoon the road is affected in many places and has disrupted the communication. In Fig. 7.6, probable risk of NH-10 has been linked with landslide hazard, for landslide area, $A_L \geq 1000\ m^2$ at 5-, 10-, 20-, and 40-year return periods. The probability of hazard along the NH may be maximum 39% likelihood of landslides for a 40-year return period, where 28 segments covering 5.44-km road may get affected. The significant part of the road segments likely to be affected is crossing over small streams or steep gullies through small culverts. These channels are ephemeral in nature but during rain, channels receive a tremendous amount of water from its upper catchment. This enormous amount of water flow down over a steep slope along with bringing boulders which washout the road and culverts. Field inspection during rainy season is corroborating in such findings. The identified maximum risk-prone segments also follow up the higher probability of hazards.

In the present study, it is assumed that the average cost (in INR) for reconstruction of the damaged roads per km is, 200 lakhs for NH, 150 lakhs for SH, 70 lakhs for metalled roads, and 40 lakhs for unmetalled roads. Based on the above values and calculated risk, expected loss (in INR) to reconstruct the damaged roads has been estimated in Table 7.3. The estimation reveals that within a year INR 4210 lakhs (65,88,650 USD) may be lost because of damage to roads caused by future landslide events.

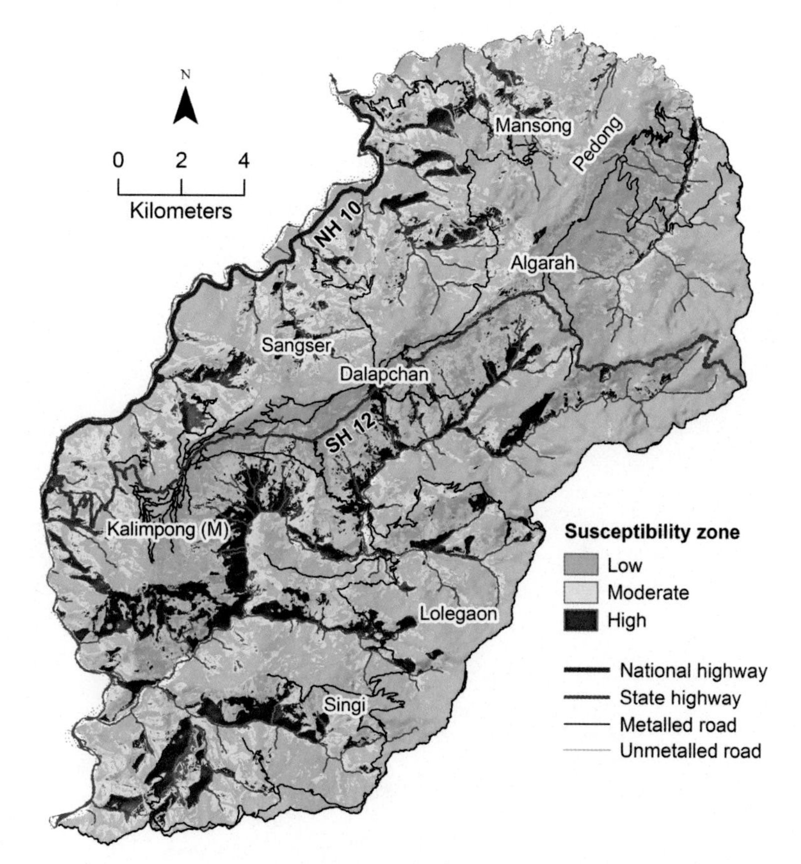

FIGURE 7.4 Elements at risk for roads, overlaying landslide susceptibility map.

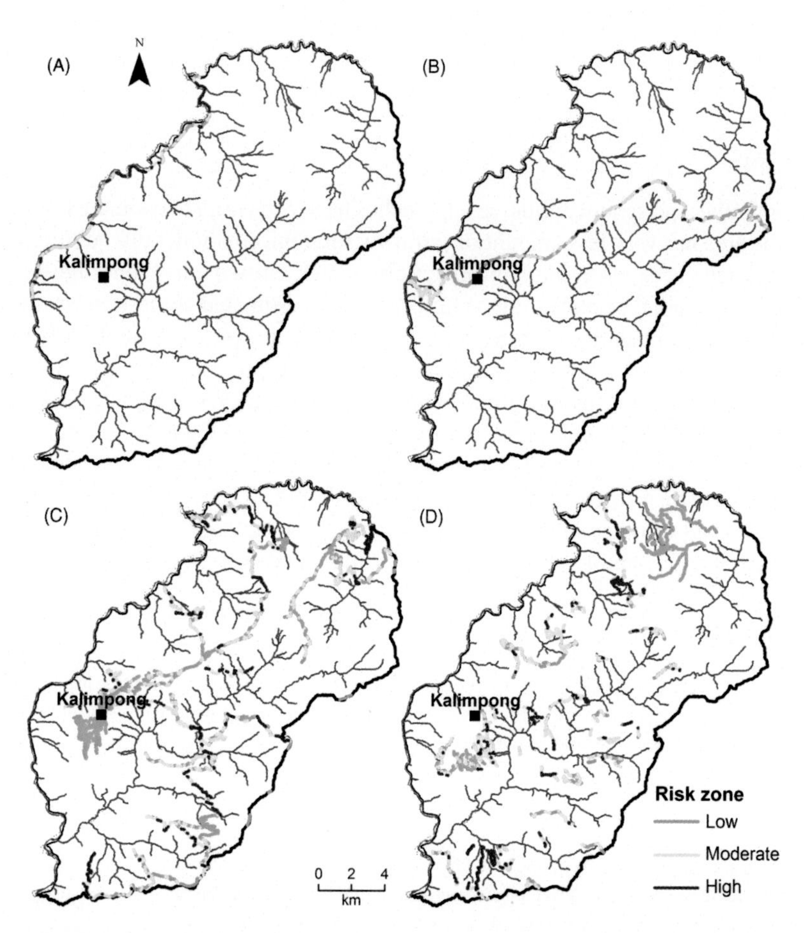

FIGURE 7.5 Risk analysis for roads: (A) national highway, (B) state highway, (C) metalled roads, and (D) unmetalled roads likely to be affected in each risk zone.

TABLE 7.3 Different Types of Roads Likely to be Affected by Landslide and Expected Loss Due to Future Landslide Events

	Road Type and Length (km)			
Risk Zone	NH	SH	Metalled Road	Unmetalled Road
Low	4.57	23.20	113.85	68.63
Moderate	12.91	16.62	82.80	71.54
High	5.44	1.83	27.06	23.82
Total length	22.92	41.65	223.71	163.98
Expected loss (lakh INR)	1088	275	1894	953

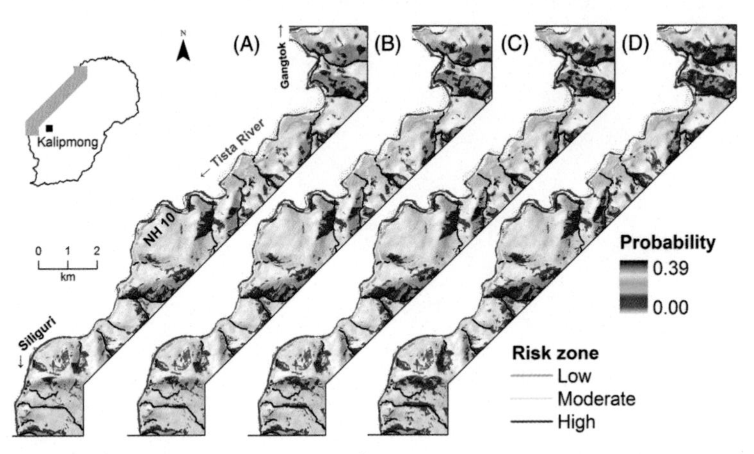

FIGURE 7.6 Landslide hazard and risk for landslide area, $A_L \geq 1000$ m^2, along national highway (NH)-10, (A) 5-year, (B) 10-year, (C) 20-year, and (D) 40-year return periods scenario.

7.5 CONCLUSIONS

In the present study, quantitative and semiquantitative approaches have been used to estimate the potential loss due to future landslide events. Since the area has very less amount of infrastructural information, only buildings, the population residing in the buildings, and roads have been taken into consideration for risk assessment. From high-resolution satellite images, first 90 SUs have been marked from the entire study area. Then risk is estimated at these SUs within a specified period (10 years) with maximum temporal probability at maximum, minimum, and average spatial probability. It is found that the Kalimpong town is in the low-risk zone; however, villages toward downslope may be affected in future.

Semiquantitative danger pixel approach has estimated the potential damage to the road. It is found that 5.44 km NH and 1.83 km SH are passing through the high-risk-prone zone. So, these segments likely to be affected by landslides in future.

Landslide risk map help us to implement landslide management program to avoid any potential loss of life and property. The high-risk-prone zone should be restricted for construction and should be given priority to immediate remedial measures by construction of retaining wall or other site-specific engineering structures. Whereas, in the moderate risk-prone zone, long-term programs can be taken by land-use planning, reforestation, and biotechnical measures.

ACKNOWLEDGMENTS

This article is a part of author's Ph.D. research carried out by the financial support of University Grants Commission, New Delhi, and has been submitted at Banaras Hindu University, Varanasi. The author is grateful to Dr. Archana K. Roy (IIPS, Mumbai) and Prof. K. N. Prudhvi Raju (BHU, Varanasi) for there constructive comments and suggestions.

REFERENCES

Abella, E.A., 2008. Multi-scale landslide risk assessment in Cuba. PhD thesis, ITC Printing Department, Enschede, the Netherlands.

Anbalagan, R., Singh, B., 1996. Landslide hazard and risk assessment mapping of mountainous terrains: a case study from Kumaun Himalaya, India. Eng. Geol. 42, 237–246.

Cardinali, M., Reichenbach, P., Guzzetti, F., Ardizzone, F., Antonini, G., Galli, M., et al., 2002. A geomorphological approach to the estimation of landslide hazards and risks in Umbria, Central Italy. Nat. Hazards Earth Syst. Sci. 2, 57–72.

Carrara, A., Cardinali, M., Detti, R., Guzzetti, F., Pasqui, V., Reichenbach, P., 1991. GIS techniques and statistical models in evaluating landslide hazard. Earth Surf. Processes Landforms 16, 427–445.

Catani, F., Casagli, N., Ermini, L., Righini, G., Menduni, G., 2005. Landslide hazard and risk mapping at catchment scale. Landslides 2, 329–342. doi: 10.1007/s10346-005-0021-0.

Census of India, 2011. Village and Town Wise Primary Census Abstract (PCA), District Census Handbook – Darjiling, Series 20, Part XII-B, 2011. Directorate of Census Operations, West Bengal.

Chowdhury, R., Flentje, P., 2003. Role of slope reliability analysis in landslide risk management. Bull. Eng. Geol. Environ. 62, 41–46.

Crozier, M.J., Glade, T., 2005. Landslide hazard and risk: issues, concepts and approach. In: Glade, T., Anderson, M., Crozier, M.J. (Eds.), Landslide Hazard and Risk. John Wiley & Sons, Chichester, UK, pp. 1–40.

Cruden, D.M., 1993. A simple definition of a landslide. Bull. Assoc. Eng. Geol. Environ. 43, 27–29.

Dai, F.C., Lee, C.F., Ngai, Y.Y., 2002. Landslide risk assessment and management: an overview. Eng. Geol. 64, 65–87.

Das, I., Kumar, G., Stein, A., Bagchi, A., Dadhwal, V.K., 2011. Stochastic landslide vulnerability modeling in space and time in a part of the Northern Himalayas, India. Environ. Monit. Assess. 178, 25–37. doi: 10.1007/s10661-010-1668-0.

Espizua, L.E., Bengochea, J.D., 2002. Landslide hazard and risk zonation mapping in the Río Grande Basin, Central Andes of Mendoza, Argentina. Mt. Res. Dev. 22 (2), 177–185.

Fell, R., 1994. Landslide risk assessment and acceptable risk. Can. Geotech. J. 31, 261–272.

Finlay, P.J., Mostyn, G.R., Fell, R., 1999. Landslide risk assessment: prediction of travel distance. Can. Geotech. J. 36, 556–562.

Ghosh, S., Carranza, E.J., van Westen, C.J., Jetten, V.G., Bhattacharya, D.N., 2011. Selecting and weighting spatial predictors for empirical modeling of landslide susceptibility in the Darjeeling Himalayas (India). Geomorphology 131, 35–56.

Ghosh, S., van Westen, C.J., Carranza, E.J., Jetten, V.G., 2012. Integrating spatial, temporal and magnitude probabilities for medium-scale landslide risk analysis in Darjeeling Himalayas, India. Landslides 9, 371–384.

Goerlich, F.J., Cantarino, I., 2013. A population density grid for Spain. Int. J. Geogr. Inf. Sci., org/10.1080/13658816.2013.799283.

Hadmoko, D.S., Lavigne, F., Sartohadi, J., Hadi, P., Winaryo, 2010. Landslide hazard and risk assessment and their application in risk management and landuse planning in Eastern flank of Menoreh Mountains, Yogyakarta Province, Indonesia. Nat. Hazards 54, 623–642, 10.1007/s11069-009-9490-0.

Hansen, M.J., 1984. Strategies for classification of landslides. In: Brunsden, D., Prior, D.B. (Eds.), Slope Instability. John Wiley & Sons, Chichester, UK, pp. 1–25.

Jaiswal, P., vanWesten, C.J., Jetten, V., 2011. Quantitative estimation of landslide risk from rapid debris slides on natural slopes in the Nilgiri hills, India. Nat. Hazards Earth Syst. Sci. 11, 1723–1743. doi: 10.5194/nhess-11-1723-2011.

Jia, N., Mitani, Y., Xie, M., Djamaluddin, I., 2012. Shallow landslide hazard assessment using a three-dimensional deterministic model in a mountainous area. Comput. Geotech. 45, 1–10. doi: 10.1016/j.compgeo.2012.04.007.

Joshi, K.C., Sengupta, S., Kandpal, G.C., 2010. Macroseismic study of 20th May 2007 Sikkim earthquake—its seismotectonic implications for the region. J. Geol. Soc. India 75, 383–392.

Kanungo, D.P., Arora, M.K., Gupta, R.P., Sarkar, S., 2008. Landslide risk assessment using concepts of danger pixels and fuzzy set theory in Darjeeling Himalayas. Landslides 5, 407–416.

Lateltin, O., Haemmig, C., Raetzo, H., Bonnard, C., 2005. Landslide risk management in Switzerland. Landslides 2, 313–320. doi: 10.1007/s10346-005-0018-8.

Lee, E.M., Jones, D.C., 2004. Landslide Risk Assessment. Tilford, London, UK.

Liu, C.-N., Dong, J.-J., Chen, C.-J., Lee, W.-F., 2012. Typical landslides and related mechanisms in Ali Mountain highway induced by typhoon Morakot: perspectives from engineering geology. Landslides 9, 239–254. doi: 10.1007/s10346-011-0298-0.

Lung, T., Lübker, T., Ngochoch, J.K., Schaab, G., 2013. Human population distribution modellingat regional level using very high resolution satellite imagery. Appl. Geogr. 41, 36–45, org/10.1016/j.apgeog.2013.03.002.

Martha, T.R., Kumar, K.V., 2013. September, 2012 landslide events in Okhimath India—an assessment of landslide consequences using very high resolution satellite data. Landslides 10, 469–479.

Martha, T.R., van Westen, C.J., Kerle, N., Jetten, V., Kumar, K.V., 2013. Landslide hazard and risk assessment using semi-automatically created landslide inventories. Geomorphology 184, 139–150.

Michael-Leiba, M., Baynes, F., Scott, G., 2000. Quantitative landslide risk assessment of Cairns, Australia. In: Bromhead, E.N., Dixon, N., Ibsen, M.-L. (Eds.), Landslides in Research, Theory and Practice, Proceedings of the 8th International Symposium on Landslidespp. 1059–1064.

Miller, S.M., 1988. A temporal model for landslide risk based on historical precipitation. Math.Geol. 20 (5), 529–542.

Papathoma-Köhle, M., Neuhäuser, B., Ratzinger, K., Wenzel, H., Dominey-Howes, D., 2007. Elements at risk as a framework for assessing the vulnerability of communities to landslides. Nat. Hazards Earth Syst. Sci. 7, 765–779.

Pawde, M.B., Saha, S.S., 1982. Geology of the Darjeeling Himalaya. Miscellaneous Publication No. 41, Part II. Geological Survey of India, 50–55.

Petley, D.N., 1998. Geomorphological mapping for hazard assessment in a neotectonic terrain. Geogr. J. 164 (2), 183–201.

Regional Sericulture Research Station, 2010. Weather Data, Kalimpong, West Bengal, India (unpublished).

Remondo, J., Bonachea, J., Cendrero, A., 2005. A statistical approach to landslide risk modelling at basin scale: from landslide susceptibility to quantitative risk assessment. Landslides 2, 321–328. doi: 10.1007/s10346-005-0016-x.

Searle, M.P., Szule, A.G., 2005. Channel flow and ductile extrusion of the high Himalayan slab–the Kanchenjunga–Darjeeling profile, Sikkim Himalaya. J. Asian Earth Sci. 25, 173–185.

Sterlacchini, S., Frigerio, S., Giacomelli, P., Brambilla, M., 2007. Landslide risk analysis: a multi-disciplinary methodological approach. Nat. Hazards Earth Syst. Sci. 7, 657–675.

van Westen, C.J., van Asch, T.W., Soeters, R., 2006. Landslide hazard and risk zonation—why is it still so difficult? Bull. Eng. Geol. Environ. 65, 167–184.

Varnes, D.J., 1978. Slope movements: types and processes. In: Schuster, R.L., Krizek, R.J. (Eds.), Landslide Analysis and Control, National Academy of Sciences, Transportation Research Board Special Report, 176, Washington, DC, 11–33.

Varnes, D.J., 1984. IAEG Commission on Landslides and Other Mass Movements on Slopes. Landslide Hazard Zonation: A Review of Principles and Practice. The UNESCO Press, Paris, pp. 1–63.

Wong, H.N., Ho, K., Chan, Y.C., 1997. Assessment of consequence of landslides. In: Cruden, D., Fell, R. (Eds.), Landslide Risk Assessment. Balkema, Rotterdam, pp. 111–149.

Zêzere, J.L., Oliveira, S.C., Garcia, R.A., Reis, E., 2007. Landslide risk analysis in the area North of Lisbon (Portugal): evaluation of direct and indirect costs resulting from a motorway disruption by slope movements. Landslides 4, 123–136. doi: 10.1007/s10346-006-0070-z.

Zêzere, J.L., Garcia, R.A., Oliveira, S.C., Reis, E., 2008. Probabilistic landslide risk analysis considering direct costs in the area north of Lisbon (Portugal). Geomorphology 94 (3–4), 467–495.

Zhou, C.H., Lee, C.F., Li, J., Xu, Z.W., 2002. On the spatial relationship between landslides and causative factors on Lantau Island, Hong Kong. Geomorphology 43, 197–207.

Chapter 8

Forest and Land Fires in Indonesia: Assessment and Mitigation

Lailan Syaufina
Bogor Agricultural University, Bogor, Indonesia

8.1 FOREST AND LAND FIRES IN INDONESIA

Fire has been used as a tool in traditional farming in Indonesia for centuries as it is the easiest way to clear a land. Fires have occurred in the forests of Kalimantan at least since the 17th century (Bowen et al., 2001). Fire is considered endemic in tropical rain forest as described by historical records and as evidenced by charcoal in soil profiles, but rare with return intervals of hundreds to thousands of years (Cochrane, 2003). The preindependence era was marked by some regulations (ordinance) that control forest fires. The regulations consist of Java and Madura Forest Ordinance 1927 (Chapter 20), *Provinciale Bosverordening Midden Java* (Chapter 14), *Rijkolad-Soerakarta Ongko* 11 (1939). Meanwhile, the postindependence era was marked by forest fires on a large scale (Bowen et al., 2001). Large forest fire was experienced by Indonesia in 1982–83 when about 3.6 million ha tropical rain forest in East Kalimantan burned out. The fire event seemed to shock the world as tropical rain forest is known to be humid—always wet—and evergreen forest can also be ablazed. Since then, a great attention and concern on forest fire including impacts of the fire have been moving forward and so did research on fire impacts and fire assessment as well as fire mitigation. The fire recurred in 1987, 1991, 1994, and 1997–98 since then forest and land fires occur every year until the most recent fire occurrence in the year 2015 as indicated by yearly hotspot distribution. Fig. 8.1 shows that forest and land fires fluctuated annually according to certain patterns as indicated by the hotspot. During the 1997–2015 period, the highest hotspot was found in 1997 and the lowest one was found in 2014. The hotspot is increased periodically, such as in the years 2002, 2004, 2006, 2009, 2012, 2013, 2014, and 2015. The incidence of major fires occurred in conjunction with extreme weather event of *El Nino*, which caused a long drought period in Indonesia.

In the last two decades, forest and land fires in Indonesia reached the peak in 1997–98 which burned about 10 and 11.7 million ha of forest and land (Asian Development Bank, 1999; Tacconi, 2003) and about 75 million people were exposed to haze pollution and fire with a total economic loss of about US $3.5–9.7 billions (Barber and Schweithelm, 2000; Ruitenbeek, 1999; Tacconi, 2003). Almost all ASEAN member states were blanketed by transboundary haze as the impact of forest and land fires in Indonesia that was dominated by fire in peatland.

As an indicator of forest and land fires, the hotspot was also found in almost all provinces of Indonesia, particularly, in Sumatra and Kalimantan (Fig. 8.2). During the period 1997–2013, the highest hotspot was found in Riau province, followed by Central Kalimantan, West Kalimantan, South Sumatra, and East Kalimantan.

One of the most obvious environmental issue in Indonesia in the last few decades is peatland fire, which occur almost annually in Sumatra and Kalimantan particularly. Peatland fire is characterized by fuel types and burning patterns, which comprise surface fire and ground fire. When peatland surface is relatively dry but water table likely to prevent fire spread, surface fire predominates the peatland area. On the other hand, when the water table is low and peat moisture content is less than the critical peat moisture content of about 110% (Frandsen, 1997) to 117.39% (Syaufina et al., 2004a), peat fire is classified as ground fire and may predominate the peatland area. The surface fire is dominated by flaming combustion, while the peat fire (ground fire) is dominated by smoldering process, where fire is kept at a very low burning rate from some decimeters to tens of meters per day (Artsybashev, 1983) or for weeks with a burning rate of less than 1.5 g per square meter per hour or about 0.025 cm soil layer decreasing per hour (Chandler et al., 1983). Peatland fire can be classified into three classes, namely weak, moderate, and high fire severity with burning peat depth <25, 25–50, and >50 cm, respectively (Artsybashev, 1983).

Different from fire in temperate region where natural fire possibly happens, almost 100% of forest and land fires in Indonesia caused by human activities, intentionally and unintentionally. Lightning is an important cause of fire in the

Integrating Disaster Science and Management. http://dx.doi.org/10.1016/B978-0-12-812056-9.00008-7

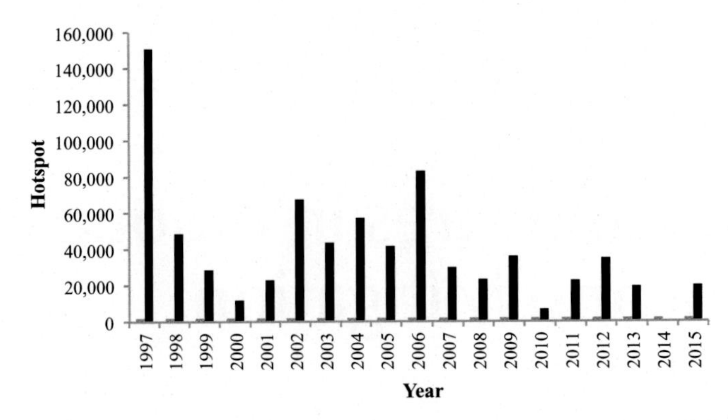

FIGURE 8.1 **Yearly distribution of hotspot as an indicator for forest and land fire in Indonesia in the period of 1997–2015 (data analyzed from Ministry of Environment and Forestry, Indonesia).**

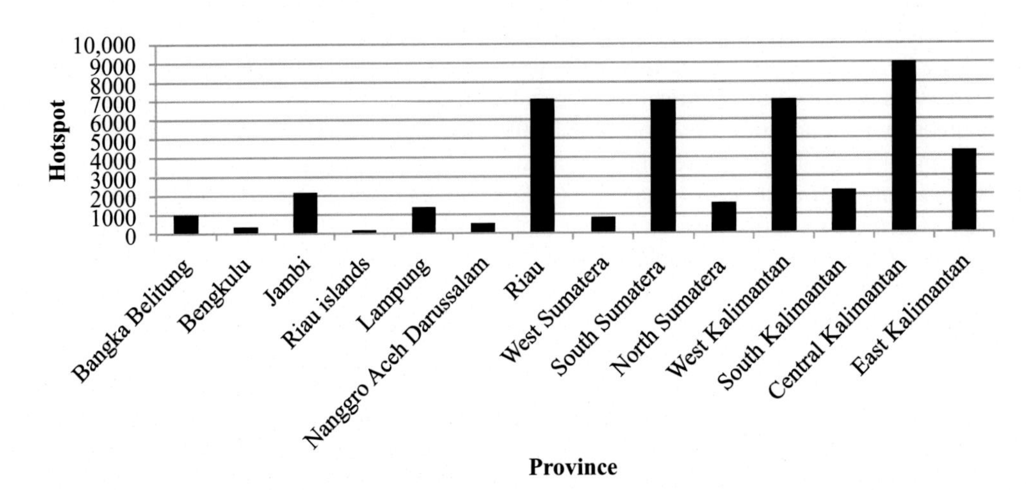

FIGURE 8.2 **Hotspot distribution in various provinces in Indonesia in the period of 1997–2013 (data analyzed from Ministry of Forestry, Indonesia).**

temperate region, which may not continue combustion process as it occurs altogether with rain that suppresses ignition process. Many studies revealed that human activities have been the most important causes of fire in Indonesia, either intentionally or unintentionally (Applegate et al., 2001; Colfer, 2002; Dennis et al., 2005; Goldammer and Seibert, 1990; Suyanto, 2000; Tomich et al., 1998).

The majority of the 1997–98 fires were set by plantation companies and by small-scale farmers to clear land. In some areas with land tenure conflicts and arson attacks. Fires quickly raged out of control due to poor logging practices, which had resulted in increased fuel loads, and drought conditions exacerbated by the 1997 El Niño climatic oscillation (EEP, 1998; Meijaard and Dennis, 1997; Schweithelm and Glover, 1999). It corresponds to the explanation of Secretariat of the Convention on Biological Diversity (2001) that in the last two decades the increased commercial exploitation of forests, apart from degrading forests and making them more fire-prone, has also led to an increased use of fire in clearing tropical forests for agricultural expansion and plantations.

A comprehensive study conducted by Center for International Forestry Research (CIFOR)/ICRAF (Applegate et al., 2001) on forest fire occurrences in 1997–98 at 10 study areas of 6 provinces in Sumatra and Kalimantan, Lampung, Jambi, South Sumatra, Riau, West Kalimantan, and East Kalimantan, indicates that the forest and land fires in Indonesia were caused by human activities. Direct causes of forest and land fires were as follows: (a) fire used in land clearing; (b) fire used as a weapon in land tenure conflict; (c) unintentional fire spread; and (d) fire used for the extraction of natural resources. Meanwhile, indirect causes of fire found in the study include (a) land tenure, (b) land uses allocation, (c) economic incentives/disincentives, (d) forest and land degradation, (e) impacts of population characteristic changes, and (f) lack of institutional capacity. The study results are still relevant with current forest fire and land fire occurrences.

8.2 INDONESIAN PEATLAND AT A GLANCE

Peatland plays a very important role in global environment balance, including climate regulation, water management, biodiversity conservation, forest product, carbon storage and carbon sequestration, and community livelihood. Peatland area in Indonesia ranks first in the world for tropical peatland and ranks fourth after Canada, Russia, and the USA for general peatland area. Tropical peatland covers about 10%–12% of the world's total peatland area, though it stores about 191 Gt C (Page and Rieley, 1998) or about one-third from the total carbon storage in peatland. Assuming that with an average peat depth of 5 m, tropical peatland ecosystem stores 2500 ton C/ha, compared to the average of 1200 ton C/ha in peatland, in general (Diemont et al., 1997). The latest data from the Ministry of Agriculture of Indonesia indicated that peatland area in Indonesia covering 14.9 million ha is distributed in three main islands of Sumatra, Kalimantan, and Papua (BBSDLP, 2011). The peatland has high biodiversity, including tree species of Jelutung (*Dyera custulata*), Ramin (*Gonystylus bancanus*), Meranti (*Shorea* spp.), and Kempas (*Kompassia malaccensis*), and wildlife species of *sinyulong* crocodile (*Tomistoma schlegelii*), Sumatran tiger (*Panthera tigris sumatrae*), honey bear (*Helarctos malayanus*), tapir (*Tapirus indicus*), and arowana fish.

Peat is formed by decomposition and accumulation process of plant materials that grow on the land which is influenced by dry and wet season period. Peat formation occurs in a long period with the formation rate of about 1 mm per year (Charman, 2002), which means 1 m deep peat needs 1000 years to form. Peat in Sumatra and Kalimantan is sometimes more than 10 m deep. Based on the decomposition level, peat is classified as *fibrist* (one-third decomposed), *saprist* (two-thirds decomposed), and *hemist* (between *fibrist* and *saprist*).

Different from temperate peat which originated from homogenous materials such as spaghnum and other shrubs species, tropical peat has a high variation as originated from various vegetation species (Andriesse, 1988). On the other hand, tropical peatland is very sensitive to changes, hence it needs careful and wise management in converting peat swamp forest (Wösten et al., 2008).

As a unique ecosystem, peatland act as a sponge which absorbs excessive water during rainy season and releases water and flows to other places during dry season to prevent drought in the adjacent areas. On the other hand, peatland is a sensitive ecosystem to environmental changes, especially climate change and hydrology. In the last few decades, peatland changed in Indonesia, Malaysia, and Thailand. Indonesia peatland which originally formed as peat swamp forest has been developed and converted into various land uses such as forest plantation, oil palm plantation, agriculture farming, settlement, and other development purposes that cause disturbance of peatland ecosystem function as triggered by irreversible drying characteristic of peat. Hence, fire can easily burn the peatland area, starting on the peat surface.

Changes in peatland cover from swamp forest to agriculture, plantation, or settlement may cause decreasing water level in the peat. As a consequence, the decreasing water level may increase the risk of forest and land fires due to dry peat condition (Susilo et al., 2013) and may decrease peat moisture content as an indicator for fire risk (Syaufina et al., 2004b). Moreover, drainage peatland leads to high fire risk in dry season and may accelerate CO_2 emission. About 30% of CO_2 emissions come from land use, land-use change, and forestry (LULUCF) has contributed to huge peatland degradation (Couwenberg et al., 2010; Hooijer et al., 2006;).

8.3 IMPACTS OF FOREST AND LAND FIRES IN INDONESIA

There are still few in-depth studies on the effect of fires on tropical rain forest biodiversity. Table 8.1 (Syaufina, 2015) indicates that studies on forest fire impacts on biodiversity and human dimension vary in magnitude and types. One can draw on a number of case studies that took place after the fires in 1982–83 and 1997–98 in Indonesia. Intensive researches were conducted in relation to the large forest fire events. The 1982–83 fire episode promotes research on fire impacts on biodiversity, particularly, in East Kalimantan, where the largest fire event occurred in the region. Forest fire event in Kutai National Park, East Kalimantan, resulted in a widespread mortality of reptiles and amphibians (Leighton, 1984; MacKinnon et al., 1996). Fruit-eating birds such as hornbills declined dramatically and only insectivorous birds, such as woodpeckers, were common due to the abundance of wood-eating insects.

The postfire assessment of the 1982–83 in Indonesia fires provided some qualitative assessments on regeneration potential of forests affected by fire (Schindele et al., 1989). The degree of expected recovery of the forest depended on the intensity of burning. For the undisturbed primary forest, it was stated that full recovery of the forest could be expected within a few years (Schindele et al., 1989). However, in the disturbed forests, the prognosis for recovery in the presence of fire was not positive. In lightly disturbed burnt forest, the potential for recovery was good, but not without the help of rehabilitation methods. In moderately disturbed burnt forest, it was unlikely that there would be timber production for at least 70 years and in heavily disturbed forest it would take hundreds of years to return to a typical rainforest ecosystem in the absence of

TABLE 8.1 Forest Fire Impacts on Ecosystem and Human Dimension (Syaufina, 2015)

No.	Research Aspect	Location	Brief Description	Source
A.	**Impacts on Biodiversity**			
1.	Aboveground biomass	East Kalimantan	Loss of aboveground biomass induced by the 1998 fire in felled and un-felled stands was 102 and 147 t ha-1, respectively. Fire disturbance has decreased biomass of forests in East Kalimantan and increased the difficulty of biomass recovery	Toma et al. (1999)
2.	Forest succession	East Kalimantan	The forest with heavy degree of damage undergoes secondary succession process. In light and moderate damage forest, changes in seedling species composition took place	Sumardi (1999)
3.	Forest succession	East Kalimantan	4 and 13 months after fire, vegetation and their types change vigorously. There were six types of recovered vegetation such as Macaranga, woody shrubs, herb, fern, grass, and climber types. Fern, climber, and grass types seemed to become competition process. The macaranga, woody shrubs, and herbaceous types are playing more facilitation	Kobayashi et al. (1999)
4.	Forest succession	East Kalimantan	At 11 years after fire, the total basal area of trees in severely burnt forest was still much smaller than that of in lightly burnt forest area. Lightly burnt forest area was dominated by primary (climax) species, while severely burnt forest area was dominated by secondary (pioneer) species	Oka, Ng. P. (1999)
5.	Succession	West Java	*Leucas lavandulaefolia*, *Paspalum conjugatum* Linn., and *Melastoma malabathricum* Linn. are that in burned area. Those species have reproductive modes by seed or spores, which have fire adaptive traits. Some species have disappeared after fire, such as *Amorphophalus variabilis* Bl, *Lastonia cilora*, *Piper aduncum* Linn., *Demosdium triquetrum* (L.).DC., *Erigeron sumatranensis* Retz., *Lasianthus purpureus* Bl., *Peperomia pellucid*, *Urena lobata* Linn., *Ageratum conyzoides*, and *Ottochloa nodosa*	Rahardjo (2002)
6.	Succession	North Sumatra	Forest fire in pine (*Pinus merkusii* jungh et de Vriese) plantation Aek Nauli, North Sumatra decreased the number of species by 27.5%, consist of 33.3% of tree species (in seedling stage), and 22.7% of shrub species. However, pine seedlings were found in burned area (180 individuals/ha) 4.5 times higher compared with that of in unburned area (40 individuals/ha)	Pangaribuan (2003)
7.	Forest ecosystem	East Kalimantan	Burned forests had reduced canopy and ground cover, lower tree species richness and diversity, and higher canopy tree, seedling, and sapling mortality than unburned forests. Species richness in peat swamp forests at Tanjung Puting was reduced by 59% by the fires and in two lowland dipterocarp forests by 24% (Sungai Wain) and 57% (Kutai). Resurveying of transects at Tanjung Puting 8 months after the first survey showed that burned forests suffer higher tree mortality and further species loss in the months following fires. In addition, species richness was lower in forests that had been logged prior to burning than in forests that had been undisturbed before the fires. The long-term ecological effects of burning on forest diversity, structure, and species composition are discussed, and the conservation implications of the high fire hazard in Indonesia are considered	Yeager et al. (2003)
8.	Forest ecosystem	East Kalimantan	The burnt subplots were dominated by pioneer or secondary tree species, such as *Mallotus* spp., *Macaranga* spp., *Ficus* spp. and *Vernonia arborea*. Local distribution of some indicator species (such as primary tree species: *Pholidocarpus majadum*, *Diospyros* spp., *Eusideroxylon zwageri* and species of Dipterocarpaceae; pioneer or secondary tree species *Vernonia arborea*, *Macaranga* spp., *Mallotus* spp., *Ficus uncinulata*, *Piper aduncum*, *Peronema canescens*) within the plot were figured	Simbolon (2005)
9.	Soil microbes	East Kalimantan	Fire has decreased Actinomycetes, bacteria, and fungi. After fire the species of decomposers (*Trichoderma* sp., *Mucor* sp. and *Penicillium* sp.) disappear	Iskandar and Nurhiftiani (1999)

10.	Soil microbes	Riau	Certain bacteria grow significantly after fire, such as phosphate-solvent bacteria. It is triggered by the increasing phosphate content in peat after fire in Riau, Indonesia	Wiratama (2010)
11.	Soil macrofauna	West Java	Species richness index has decreased from 7.9662 in unburned area to 3.2018 in burned area, in which about 14 insecta ordo were found in unburned area compared to 8 ordo found in burned area	Abidin (2005)
12.	Soil macrofauna	West Java	Fire caused macrofauna families loss including: Thripidae, Tetranichidae, Ellateridae, Staphylinidae, Scarabaidae, Julidae, Polydesmidae, Blattidae, Oedemeridae,Cercopidae, Mantidae, Tenebrionidae, Acrididae, Reduviidae, Scydmaenidae, Oxyopidae, Salticidae, and Tetragnathidae. Moreover, Ordos loss after fire including Blattaria, Homoptera, Mantodea, and Thysanoptera. On the contrary, there were new families found after fire, namely Linyphiidae, Podoridae, Sminthuridae, Zetorchestidae, Ephylomatidae, Rhysotritiidae, Phytoseiidae, Argasidae, Veigaiidae, and Nitidulidae	Buliyansih et al. (2007)
13.	Wildlife	East Kalimantan	As of 1997 before fire, there were 21 individuals of gibbon (*Hylobates muelleri*) from 6 different groups. Habitat degradation due to 1998 fire has caused disappearance of 6 individuals including 2 infants and disruption of three families. Home ranges of surviving groups have extended more than twice as large as they were	Oka, T. (1999)
14.	Wildlife	East Kalimantan	The Wanariset Orangutan Reintrosuction Project reintroduced 82 ex-captive orangutans to free forest life in Sungai Wain Protection Forest from 1992 to 1997. The census of orangutan nests throughout the unburned forest 2.5–4.5 after fires were extinguished indicated a total of number of more than 13–17 orangutans remaining in the forest. It was suggested that reintroduced orangutans left Sungai Wain forest or died under combined effects of drought and fire	Russon and Susilo (1999)
15.	Insects	East Kalimantan	Of the surveyed families, Bostrychiade, which can attack dry wood was the least affected by burning. The next most tolerant family was Cerambycidae, Carabidae, Platypodidae, and Scolytidae were damaged by the fires and had not yet recovered by a half of year after fire. The number of Cerambycidae species that can inhabit the upper parts of trees increased after fire in a high tree stand	Makihara et al. (1999)
B.	**Human dimensions**			
1.	Fire causes	Indonesia	Fire causes in Indonesia are mostly due to human activities related to: shifting cultivation, land clearing for forest plantation, and estate crops, logging	Saharjo (1999)
2.	Fire causes	East Kalimantan	People living around and inside Sungai Wein Protection Forest practice slash and burned to clear the land. They have no alternative to clear the land in a cheap and simple way	Sukmajaya (1999)
3.	Health impacts	Indonesia	Smoke haze come from forest fire/biomass burning that contain various components that disturb human health in the form of gas and particle. Among the gas components are CO, SO_2, NO_2, and aldehydes. Other several compounds such as O_3, CO_2, and hydrocarbons may also provide bad impacts on the lungs. Decreasing air quality to a very dangerous level of health includes respiratory disease and acute respiratory infection	Faisal et al. (2012)
4.	Health impacts	Jambi Province	Forest fire occurrence in Batanghari District, Jambi Province in 2008 showed an increase in air quality parameters such as PM10, SO_2, CO and O_3 although still below of the standard. Data obtained were respiratory disease with a prevalence of 55.9% and pneumonia with a prevalence of 7.35%	Perwitasari and Sukana (2012)
5.	Health impacts	Central Kalimantan	Air quality in Palangka Raya, Central Kalimantan, was rated as 'unhealthy/very unhealthy/dangerous' on 81% of days from September–November 2006 and, in October 2006, 30 of 31 days were 'dangerous' (Board for the Control of Environmental Impacts in Palangka Raya Area, 2006)	Harisson et al. (2009)

fire. In terms of the ecological impact of fires in tropical rain forests, the replacement of vast areas of forest with pyrophytic grasslands is probably the most negative impact (Secretariat of the Convention on Biological Diversity, 2001).

Different from the first large fire in 1982–83, the 1997–98 fire episode has promoted two important aspects of study, namely fire impacts on biodiversity and fire impacts on human dimensions. In Borneo, the orangutan (*Pongo pygmaeus*) suffered a 33% decline in its population because of the 1997–98 forest fires (Schindele et al., 1989). Sumatra reported fire damage in the Bukit Barisan Selatan National Park during the fires of 1997 (Rijksen and Meijaard, 1999). The loss of fruit trees reduced the fruit availability to a large number of omnivorous species, such as primates and squirrels, Sun Bear (*Ursus malayanus*) and civets as well as ungulates such as mouse deer (*Tragulus* sp.) and muntjac (*Muntiacus* sp.). The reduction in densities of ground squirrels and tree shrews suggested that rodent densities, in general, declined which adversely affected the food supply for small carnivores such as the leopard cat (*Prionailurus bengalensis*). The destruction of tree cavities affected birds and mammals such as tarsiers, bats, and lemurs. Finally, the extensive fires destroyed the leaf litter and its associated arthropod community, further reducing food availability for omnivores and carnivores (Kinnaird and O'Brien, 1998).

Similar to fire event in 1982–83, most studies of fire event in 1997–98 were also conducted in East Kalimantan. Forest condition in the region has been in a vulnerable stage to fire as logged over areas were scattered in the whole province, besides degraded peatland dominating the region. Dryland forest fuel and peat fuel were available and became more flammable which were intensified by extreme weather conditions. On the other hand, part of the heart of tropical rain forest Borneo located in East Kalimantan wherein rich biodiversity in terms of flora and fauna was possibly found. Therefore, East Kalimantan has been a focus of study on fire impacts on biodiversity. The studies included: fire impacts on ecosystem, succession results, wildlife, insects, and soil microbes. However, there are several studies conducted in other regions such as Java and Sumatra, which related with fire impacts on forest succession, soil macrofauna, and soil microbes.

A review of fire impacts to the tropical forest ecosystem noted that studies on the topic for South East Asia are still limited (Syaufina and Ainuddin, 2011). Fire effects on tropical forest biodiversity varies from low-to-high magnitude. On the one hand, direct effects of fire to vegetation may kill plants and cause injury and indirect effects of fire to vegetation include open wounds, which attract pest and disease attack. On the other hand, fire alters forest structure and composition. The magnitude of fire effects on tropical forest biodiversity is influenced by several factors, namely fire intensity, fire severity, soil types, postfire precipitation, and burned area (Syaufina and Ainuddin, 2011). 1997–98 fire episode has altered the focus of fire impact studies. There are two main reasons behind the changes, namely (1) the fire occurrence was not limited to forest area anymore but also vast land area outside the forest and (2) the fire occurrence was dominated by smoldering peatland fire, which contributes to the transboundary haze pollution and hence human health and carbon emission/global warming issue. Since then, fire studies have significant change in direction.

Compared to the study of fire impacts on biodiversity, study of fire impacts on human dimensions is lesser in number and magnitude. Bad impacts of transboundary haze pollution produced by peat-dominated fire have promoted researches on fire impacts on human health. Peat mainly burns with a nonflaming processes and is generally a very low-intensity combustion process. It produces high emissions of particulate matter, carbon monoxide (CO), and other compounds of incomplete combustion (Schwela et al., 1999), which makes it particularly detrimental to respiratory health (Goh et al., 1999). Smoke haze that come from forest fire/biomass burning contain various components that disturb human health in the form of gas and particle. Among the gas components are CO, sulfur dioxide (SO_2), nitrogen dioxide (NO_2), and aldehydes. Other several compounds such as ozone (O_3), carbon dioxide (CO_2), and hydrocarbons may also provide bad impacts on the lungs (Faisal et al., 2012).

In contrast to the great impacts, studies on fire impacts on human health in Indonesia are still limited, especially, on the statistics of people affected and air quality. It seems that there are no in-depth studies conducted on these aspects. It may imply less awareness of people on the dangers of the fire impacts on human health. In consequence, human-induced fires are still the environmental problem in Indonesia.

The impacts of peatland fire, in particular, has caused serious impacts on all aspects of ecology, economy, and social. Transboundary haze pollution has been a tremendous phenomenon in the ASEAN region produced by peatland fire in 1997–98 fire episode, which burned an area estimated between 10 and 11.7 million ha (Artsybashev, 1983; Syaufina et al., 2004a), the number of people affected by smoke haze and fire were 75 million, and the total economic loss to the region was as much as US $3.5–9.7 billion (Chandler et al., 1983; Syaufina et al., 2004a). These fires and the resulting haze imposed immense economic, public health, and ecological costs across Southeast Asia. Economic estimates suggest that the cumulative costs were US $9.2 billion (Barber and Schweithelm, 2000). This figure includes firefighting costs, losses of agricultural and plantation crops, short-term health damages, losses in tourism and transportation revenues, and losses in the forestry sector. It does not include long-term health costs, reduction in ecological services, or biodiversity losses.

8.4 FIRE ASSESSMENT

Weather and climate play a very important role in forest fire occurrences. They determine fuel availability and fire season period and severity (Chandler et al., 1983). On the other hand, fire risk is defined as the possibility of a fire starting as determined by the presence and any causative agent activity (FAO, 1986). Fire risk is also defined as the union of two components of fire hazard as the fuel and its susceptibility to burn and fire ignition as the presence of external causes (both anthropogenic and natural) (Chuvieco et al., 2003). Fire assessment needs determination of fire hazard as prediction of fixed or varied environmental factors (fuel, weather, topography) and fire ignition factors. In tropical areas like Indonesia, the most important weather element that influences forest and fire is rainfall as it directly determines fuel moisture content and the fire season (Syaufina et al., 2004b). Forest and land fires in Kalimantan and Riau province occur during dry season when there is no rain for at least 3 days consecutively or when daily precipitation is less than 30 mm. Therefore, rainfall has been used widely in fire assessment (Boer et al., 2010).

Fire assessment model has been developed tremendously since the 1997–98 fire episode. Spatial analyses are used to integrate fire hazard and fire ignition factors. Forest fire forecasts can be determined by using a precipitation accumulation for 2 months prior to fire occurrence and by using Monte Carlo simulation (Prasasti et al., 2014). Efforts to anticipate and address fire risk should be carried out as early as possible, that is, 2 months in advance if the probability of fire risk had exceeded the value of 40%. The study built fire risk prediction model based on CMORPH rainfall data which is useful to overcome limitations on observational data.

Grid-based geographical information system (GIS) with support of multicriteria analysis (MCA) can be used to map fire-prone areas/fire danger in peat swamp forest. Variables tested include land use, road network, slope, aspect, soil, vegetation, and altitude. The use of GIS in fire assessment was used in several areas, such as Central Kalimantan (Samsuri et al., 2012), Rawa Aopa National Park South East Sulawesi (Sugiarto et al., 2013), and West Kalimantan (Kayoman, 2010). In general, data used for fire risk map include topographic map, GCM (temperature, rainfall maps), landsat image, and Bing (land use map, river map, road map, fire area map), hotspot from MODIS image, and socioeconomic data as model development parameters.

Remote sensing as a satellite-based technology plays a very important role in forest and land fires assessment, which include fire detection, fire impact assessment, early warning system, fire management plan, and postfire rehabilitation plan. Furthermore, satellite data have been used to monitor biomass burning at regional and global scale for more than two decades using algorithms that detect the location of active fires at the time of satellite overpass and in the last decade using burned area algorithms that map directly the spatial extent of the area affected by fires such as the MODIS Burned Area Product (MCD45). Forest fire research is one of many appropriate GIS applications. The diversity of factors that affect the beginning and spreading of a forest fire dictates the use of an integrated analysis approach. Considering the intrinsic dynamism of this phenomenon, remote-sensing imagery is also very valuable for these kinds of studies. It provides a quick evaluation of the vegetation status as well as a survey on the effects of fire upon the environment. However, information on the extent and magnitude of the technology application is still limited (Syaufina et al., 2016). Challenges in the research topics for the future include effective Early warning system, fire-prone area mapping, information system–hotspot accuracy, fire impacts on biodiversity, and emission factor combining satellite-based. Ground studies would achieve optimum results and play an important role in minimizing forest and land fire occurrences as well as their impacts in Indonesia.

The government of Indonesia has developed a forest fire warning system to predict fire occurrence for the whole Indonesia region, which is based on weather elements. The system is published daily by Meteorological, Climatological, and Geophysical Agency at http://www.bmkg.go.id/cuaca/kebakaran-hutan.bmkg?w=1&u=1&p= as shown in Fig. 8.3. Any information of forest and land fires in Indonesia can also be accessed through the web site managed by the Ministry of Environment and Forestry at http://sipongi.menlhk.go.id/home/main. The information covers hotspot distribution map, graphs, and other fire control activities.

8.5 FIRE MITIGATION

8.5.1 Forest- and Land Fire-related Policies

The Government of Indonesia realizes that forest and land fires is important environmental issues to be solved. In 2014, the Government merged the Ministry of Forestry and Ministry of Environment into Ministry of Environment and Forestry. With this merger, management of forest and land fires is expected to be more efficient and effective as it is under one main institution compared to previous status when forest and land fires were under separate ministries. In the policy aspect, the Government has released laws and regulation on forest and land fires from central level and regional or local level as well.

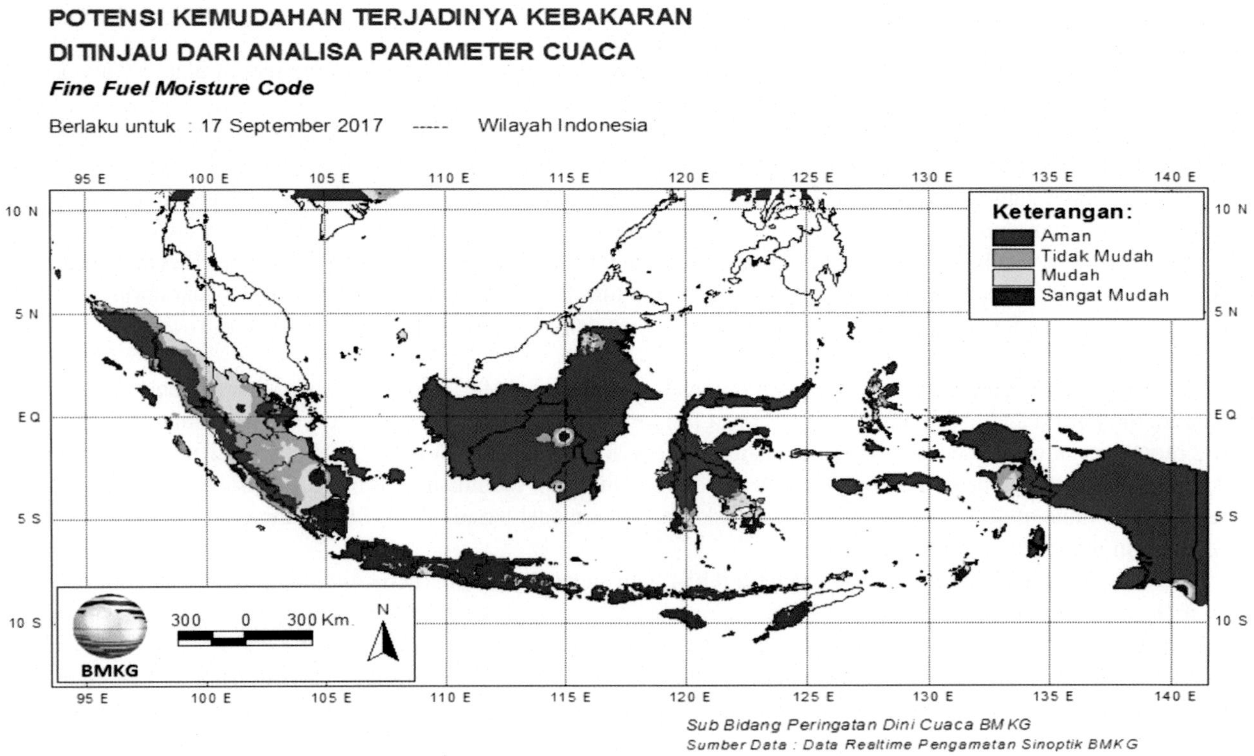

FIGURE 8.3 **Daily fire risk map of Indonesia released by meteorological, climatological, and geophysical agency (BMKG).**

1. Law No. 41 Year 1999 on Forestry
   Chapter 50 verse 3 point (d) of this Law states that everybody is forbidden to burn forest. In the explanation it said that "everybody" means individuals, institutions, or firms/companies. Limited forest burning or controlled burning is only permitted for any special objective or on unavoidable conditions, that is, forest fire control, pest and diseases control, and habitat management for wildlife. The controlled burning needs to obtain permission from the authority.
2. Government Regulation (PP) No. 4 Year 2001 on Environmental Pollution and Degradation Control Related with Forest and Land Fires
   The regulation covers the scope of forest and land fire control and environmental pollution due to forest and/or land fire, including prevention, suppression, rehabilitation, and monitoring. It also covers standard criteria of environmental degradation and environmental pollution at national as well as provincial levels. Chapter 11 of the regulation states that anybody is forbidden to burn forest and land. Any permit holder or any company has an obligation to prevent and suppress forest and land fires and postfire rehabilitation.
   In terms of authority, the Ministry of Forestry needs to coordinate cross/transboundary forest and land fire suppression. At the provincial level, the governor takes up the responsibility in coordinating district/city cross-boundary forest and land fire suppression. The Governor may form and appoint the authority to control forest and land fire occurrence in their region. In a district/city level, the head of district/major takes up the responsibility in controlling forest and land fires in their authority.
3. Government Regulation (PP) No. 45 Year 2004 on Forest Protection
   According to this regulation, forest fire control is all effort conducted to protect forest from fire, which covers prevention, suppression, and postfire activities. It is forbidden to burn the forest as stated in Law No. 41 Year 1999. Limited forest burning or controlled burning only permitted for a special objective or on unavoidable conditions, that is, forest fire control, pest and diseases control, and habitat management for wildlife. The controlled burning needs to obtain permission from the authority (Ministry of Forestry). Responsibility of forest fire control is at different levels based on the location of forest fire occurrences. The Government also established fire brigades from national level, provincial level, district/city level, and village/community level.
4. Law No. 24 Year 2007 on Disaster Management
   According to this Law, natural forest and land fires is included in natural disaster. Meanwhile, man-induced fire is classified as nonnatural disaster. Disaster management is a series of efforts covering determination of disaster risk

development policy, disaster prevention, emergency response, and rehabilitation. The government in national as well as regional levels is responsible for disaster management. This Law also establishes the National Board for Disaster Management as described in Chapter 10 verse (1) which consist of (a) an advisory for disaster management and (b) an organizer for disaster management. At the regional level, disaster management is under Regional Board for Disaster Management.

5. Ministry of Forestry Regulation No. P.12/Menhut/Year 2009 on Forest Fire Control
It contains guidance in forest fire control to be implemented effectively and efficiently. The scope of the regulation includes fire prevention, suppression, and postfire activities. Chapter 5 stated that forest fire prevention in national level covers
    1. Forest fire risk mapping;
    2. Developing forest fire information system;
    3. Partnership with community;
    4. Standardize forest fire control equipment;
    5. Programming forest fire control extension and campaign; and
    6. Arranging training structure of forest fire control

    By merging the two ministries into one, Ministry of Environment and Forestry, all regulation deal with forest fire may also be applicable to land fire. On a daily basis, forest and land fire control in Indonesia is under coordination of Directorate of Forest and Land Fires Control.
6. Law No. 18/2004 on Plantation
This Law was initiated by the agricultural sector, especially addressed to plantation activities. Chapter 25 (1) states that any plantation company has an obligation to maintain sustainability of environmental function. Moreover, Chapter 26 clearly states about prohibition of land clearing using fire for plantation. To strengthen the implementation of the Laws, the Ministry of Agriculture releases regulation of Ministry of Agriculture Regulation No.14/Permentan/PL.110/2/2009 on the Guidelines of Peatland Uses for Oil Palm Cultivation in which zero burning is a compulsory for any land preparation for oil palm plantation establishment.
Zero burning is a land clearing activity without burning for forestry, plantation, agriculture, transmigration, and mining. Vegetation debris from slashing and cutting are not burned but are piled on certain places among planting rows, chopped and spread on the soil surface, or are used for other purposes like furniture, fire wood, charcoal briquette, and compost.
7. Regulation No. 71 Year 2014 on Peatland Ecosystem Protection and Management
The ASEAN Peatland Management Strategy (APMS) was first launched in 2006 to be an important reference for the ASEAN member states in managing their peatlands. The strategy was formulated based on the ASEAN Peatland Management Initiative (APMI), which is based on the concern of transboundary haze pollution problem in the ASEAN member states due to mismanagement of peatland area. Each member state is called for implementing the APMS, in policy aspect as well as technical aspect.
On the other hand, the ministerial meeting of the ASEAN member states on environment in June 2002 agreed to adopt ASEAN Agreement on Transboundary Haze Pollution (AATHP) which needs to be followed by every single country adopting it. Indonesia was the last member state to adopt the agreement in 2014 by Law No. 26 year 2014 on the Adoption of AATHP.
In line with the APMS and AATHP, the Government of Indonesia has released policy on peatland management in Government Regulation No. 71 Year 2014 on Peatland Ecosystem Protection and Management which was revised by the Government improved the Regulation by Government Regulation (PP) No. 57/2016 on Changing of the Government Regulation No. 71/2014 on Protection and Management of Peatland Ecosystem. There are three main important points in the regulation which indicate that the Government of Indonesia has seriously taken peatland into account in sector development and environmental management, namely (a) using the word "peatland ecosystem" as the Government view not only to use peatland as an object but to use peatland as an ecosystem which influences each other in forming balance, stability, and productivity; (b) inclusion term of peatland hydrological unit which means that the peatland ecosystem is located between two rivers, between river and sea, and/or in a swampy area; and (c) peatland ecosystem in Indonesia is classified into two functions: protection function and cultivation function. The implementation of the Laws is supported by the establishment of National Board of Peatland Restoration.

8.5.2 Integrated Forest and Land Fires Prevention Patrols

High degree of commitment of the Government of Indonesia is also indicated by implementation of Integrated Forest and Land Fire Prevention Patrols initiated by Ministry of Environment and Forestry (MoEF) which was launched in 2016 in eight fire-prone provinces in Sumatra and Kalimantan. The patrol team consists of: fire brigade of MoEF, army, police, and villagers. The patrol is conducted daily in targeted villages (about 731 villages within the 8 fire-prone provinces), which prioritize on early detection, data updated, the presence of fire guard on the site, and synergy among institutions and village community. The patrol team provides socialization on fire prevention to community and do early detection of village-based fire risk assessment every day. The patrol's results are reported tierly to village task force, district/province task forces to central government through communication network of smartphone applications of *whatsapp* in order to be monitored and followed up in shorter time period especially when forest and land fires occur. It is expected that the patrol may support fire prevention and early fire suppression in site level effectively in order to minimize forest and land fires in Indonesia.

8.5.3 Good practices in forest and land fires mitigation

Having used fire for centuries, traditional farmers in Indonesia have their own local wisdom in preparing their agricultural land which is environmentally friendly. Studies show that traditional farmers implement controlled burning techniques to prepare the land, which considers weather factor, dryness of fuel, safety, and impacts. Controlled burning practices are still implemented by Dayak community in Kalimantan, Baduy community in West Java, local community near National Parks in Jambi, local community in Sulawesi, local community in Papua, and local community in Nusa Tenggara.

In 2010–14, Indonesia participated in the ASEAN Peatland Forest Project funded by IFAD/GEF together with other ASEAN countries. One of the main objectives of the project is the implementation of sustainable peatland management related to minimizing forest and land fires. The activities which were focused in Riau, West Kalimantan, and Central Kalimantan provinces covered:

(a) Enhancement awareness on sustainable peatland management.
(b) Formulation policies related sustainable peatland management.
(c) Development of partnership program among government, private sectors, community in sustainable peatland management.
(d) Development of demo sites and pilot sites for sustainable peatland management.

Pilot sites in this project including zero burning and controlled burning practices are described as follows:

1. Pilot site of pineapple farming at Riau province
   Pilot site locations: Mumugo village, Rokan Hilir District, Pelintung and Guntung villages, Dumai City, Sepahat, and Tanjung Leban villages, Bengkalis District, 4 ha per each village at community private lands. The pilot sites were prepared by zero burning land clearing using manual slashing and herbicide application. Pineapple seedlings planted in the sites were about 10,000 seedling per ha. The pilot sites were established as incentives for Fire Awareness Community Masyarakat Peduli Api group for their voluntary efforts in supporting fire prevention and suppression in each region. So far, hundreds of MPAs have been formed in fire-prone villages in Indonesia by several institutions, government as well as private sectors and communities. They are trained and involved in fire control activities and other community empowerment.
2. Controlled burning in land preparation for agriculture at West Kalimantan Province
   Pilot site location: Rasau Jaya village, West Kalimantan Province. Pilot sites were established by farmer groups in land preparation using controlled burning. Normally, the farmers implement open burning for land preparation after slashing shrubs and grasses. The farmers changed the habit of open burning by controlled burning where land slashing debris was compiled and burnt in a drum. Ash and charcoal from the burning process were then used as a fertilizer for agricultural crops, such as corn. Therefore, land productivity may increase and environmental impacts can be minimized.
3. Zero burning land preparation for degraded peatland rehabilitation at Central Kalimantan Province
   Pilot site location: Jabiren village, Jabiren Raya subdistrict, Pulang Pisau District, Central Kalimantan. The cleared land was planted with 4000 seedlings of Gaharu (*Aquilaria microcarpa*) and 3000 seedlings of Jelutung (*Dyera polyphylla*) in 6 ha land. The tree stand was mixed with other agricultural crops such as corn and other horticultural crops. The mixed crops planting system is called agroforestry.

REFERENCES

Abidin, Z., 2005. Study on forest fire effects on soil biota using Forest health Monitoring at Hunting Park of Masigit Kareumbi, Sumedang (in Indonesian). Department of Forest Management, faculty of Forestry IPB. In: Syaufina, L., Ainuddin, A.N. (Eds.), 2011 Impacts of fire on SouthEast Asia tropical forests biodiversity: A Review. Asian. J. Plant Sci., 10(4), p. 238244.

Andriesse, J.P., 1988. Nature and management of tropical peat soils. FAO Soils Bulletin 59. Food and Agriculture Organization of the United Nations, Rome, Italy, 165 pp.

Applegate, G., Chokkalingam, U., Suyanto, S., 2001. The Underlying Causes and Impact of Fires in Southeast Asia. Final report. Center for International Forestry Research, Jakarta.

Artsybashev, E.S., 1983. In: Pandit, V., Badaya Trans, K. (Eds.), Forest Fires and Their Control. Oxonian Press, New Delhi, India, p. 160.

Asian Development Bank, 1999. Causes, extent, impact and costs of 1997/1998 fires and drought. Final report, Annex 1 and 2. Planning for fire prevention and drought management project. Asian Development Bank TA 2999-INO. Jakarta, Indonesia.

Barber, C., Schweithelm, J., 2000. Trial by fire: forest fires and forest policy in Indonesia's era of crisis and reform. Report of World Resources Institute, Forest Frontiers Initiative, in collaboration with WWF Indonesia and Telapak Indonesia Foundation.

BBSDLP [Balai Besar Sumberdaya Lahan Pertanian]. 2011. Peta Sebaran Lahan Gambut Indonesia. Kementerian Pertanian RI.

Boer, R., Ardiansyah, M., Prasasti, I., Syaufina, L., Siddiki, R., 2010. Analisis Hubungan antara Jumlah Titik - titik Panas (Hotspot) dengan Luas Kebakaran Hutan dan Curah Hujan. Prosiding Pertemuan Ilmiah Tahunan XVII dan Kongres Mapin V: Teknologi Geospasial untuk Ketahanan Pangan dan Pembangunan Berkelanjutan, IPB International Convention Centre. Bogor.

Bowen, M.R., Bompard, J.M., Anderson, I.P., Guizol, P., Gouyon, A., 2001. Anthropogenic fires in Indonesia: a view from Sumatra. In: Peter, E., Radojevic, M. (Eds.), Forest Fires and Regional Haze in Southeast Asia. Nova Science Publishers, Huntington, NY, pp. 41–66.

Buliyansih, A., Syaufina, L., Haneda, N.F., Agustus 2007. Keanekaragaman arthropoda tanah di hutan pendidikan Gunung Walat. Media Konservasi 12 (2), 57–66.

Chandler, C., Cheney, P., Thomas, P., Trabaud, L., Williams, D., 1983. Fire in Forestry. Wiley, Canada, p. 450.

Charman, D., 2002. Peatlands and Environmental Change. Wiley, UK.

Chuvieco, E., Agaudo, I., Cocero, D., Riano, D., 2003. Design of an empirical index to estimate fuel moisture content from NOAA-AVHRR analysis in forest fire danger studies. Int. J. Remote Sens. 24, 1621–1637. doi: 10.1080/01431160210144660.

Cochrane, M.A., 2003. Fire science for rainforests. Nature 421 (27), 913–919.

Colfer, C.J.P., 2002. Ten propositions to explain Kalimantan's fires—a view from the field. In: Colfer, C.J.P., Resosudarmo, I.A.P. (Eds.), Which Way Forward? Forests, Policy and People in Indonesia. Resources for the Future, An RFF Book, Washington, DC/Bogor, Indonesia, pp. 309–324.

Couwenberg, J., Dommain, R., Joosten, H., 2010. Greenhouse gas fluxes from tropical peatlands in South-east Asia. Global Change Biol. 16 (6), 1715–1732.

Dennis, R.A., Mayer, J., Applegate, G., Chokkalingam, U., Colfer, C.J.P., Kurniawan, I., Lachowski, H., Maus, P., Permana, R.P., Ruchiat, Y., Stolle, F., Suyanto, Tomich, T.P., 2005. Fire, people and pixels: Linking social science and remote sensing to understand underlying causes and impacts of fires in Indonesia. Human Ecology 33, 465–504.

Diemont, W.H., Nabuurs, G.J., Rieley, J.O., Rijksen, H.D., 1997. Climate change and management of tropical peatlands as a carbon reservoir. In: Rieley, J.O., Page, S.E. (Eds.), Biodiversity and Sustainability of Tropical Peatlands. Samara Publishing, Cardigan, UK, pp. 363–368.

EEP, 1998. Forest and Land Fires in Indonesia: Impacts, Factors and Evaluation. State Ministry for Environment, Jakarta, Indonesia.

Faisal, F., Yunus, F., Harahap, F., 2012. Impacts of forest fire on respiratory (in Indonesian). CDK-189 39 (1), 2012.

FAO, 1986. Wild land fire management terminology. Report number 70, FAO Forestry Paper, Roma, M-99. ISBN: 92-5-0024207.

Frandsen, W.H., 1997. Ignition probability of organic soils. Can. J. For. Res. 27, 1471–1477.

Goh, K.T., Schwela, D.H., Goldammer, J.G., Simpson, O., 1999. Health guidelines for vegetation fire events. Background Papers. Published on behalf of UNEP, WHO, and WMO. Institute of Environmental Epidemiology, Ministry of the Environment, Singapore. Namic Printers, Singapore, 498 pp.

Goldammer, J.G., Seibert, B., 1990. The impacts of droughts and forest fires on tropical lowland rain forest of East Kalimantan. Goldammer, J.G. (Ed.), Fire in the tropical biota. Ecosystem Processes and Global Challenges Ecological Studies, 84, Springer-Verlag, Berlin, Germany, pp. 11–31.

Harisson, M.E., Page, S.E., Limin, S., August 2009. The global impacts of Indonesian forest fire. Biologist 56 (3), 156–163.

Hooijer, A., Silvius, M., Wösten, H., Page, S., 2006. PEAT-CO_2, assessment of CO_2 emissions from drained Peatlands in SE Asia. Delft Hydraulics Report Q3943, Delft, The Netherlands, 36 pp.

Iskandar, E., Nurhiftiani, I., 1999. Soil microbial status before and after fire of research plot of Bukit Soeharto Education Forest, East Kalimantan. In: Suhartoyo, H., Toma, T. (Eds.), In the Proceedings 3rd International Symposium on Asian Tropical Forest Management. Pusrehut Special Publication No. 8 (1999) Tropical Forest Research Center, Mulawarman University and Japan International Cooperation Agency. Samarinda, 20–23 September, 1999.

Kayoman, L., 2010. Pemodelan Spasial Resiko Kebakaran Hutan dan Lahan Di Provinsi Kalimantan Barat. Available from: http://mobile.repository.ipb.ac.id/handle/123456789/41137?show=full.

Kinnaird, M.F., O'Brien, T.G., 1998. Ecological effects of wildfire on lowland rainforest in Sumatra. Conserv. Biol. 12, 954–956.

Kobayashi, S., Sutisna, M., Delmy, A., Toma, T., 1999. Initial phase of secondary succession at the burnt logged-over forest in Bukit Soeharto, East Kalimantan, Indonesia – Which vegetation types are facilitation or competition process? In: Suhartoyo, H., Toma, T. (Eds.), In the Proceedings 3rd International Symposium on Asian Tropical Forest Management. Pusrehut Special Publication No. 8 (1999) Tropical Forest Research Center, Mulawarman University and Japan International Cooperation Agency. Samarinda, 20–23 September, 1999.

Leighton, M., 1984. The El Nino Southern Oscillation Event in Southeast Asia: Effects of Drought and Fire inTropical Forest in Eastern Borneo. WWF-US, Washington, DC, 32pp.

MacKinnon, K., Hatta, G., Halim, H., Mangalik, A., 1996. The Ecology of Kalimantan. Periplus Editions, Singapore.

Makihara, H., Kinuura, H., Yahiro, K., Soeyamto, C., 1999. Effects of forest fires on various Coleopterous insects in a tropical rain forest of East Kalimantan. In: Suhartoyo, H., Toma, T. (Eds.), In the Proceedings 3rd International Symposium on Asian Tropical Forest Management. Pusrehut Special Publication No. 8 (1999) Tropical Forest Research Center, Mulawarman University and Japan International Cooperation Agency. Samarinda, 20–23 September, 1999.

Meijaard, E., Dennis, R., 1997. Forest Fires and Indonesia: Bibliography and Background Information. World Wildlife Fund, Amsterdam, The Netherlands.

Oka, Ng.P., 1999. Do burnt tropical rain forests need rehabilitation? In: Suhartoyo, H., Toma, T. (Eds.), In the Proceedings 3rd International Symposium on Asian Tropical Forest Management. Pusrehut Special Publication No. 8 (1999) Tropical Forest Research Center, Mulawarman University and Japan International Cooperation Agency. Samarinda, 20–23 September, 1999.

Oka, T., 1999. Effects of the forest fire 1998 on the family of gibbon – a family in need is a family indeed. In: Suhartoyo, H., Toma, T. (Eds.), In the Proceedings 3rd International Symposium on Asian Tropical Forest Management. Pusrehut Special Publication No. 8 (1999) Tropical Forest Research Center, Mulawarman University and Japan International Cooperation Agency. Samarinda, 20–23 September, 1999.

Page, S.E., Rieley, J.O., 1998. Tropical Peatlands : a review of their natural resources functions with particular reference to Southeast Asia. Int. Peat J. 8, 95–106.

Pangaribuan, H.F.O., 2003. A study on fire ecosystem at Pinus merkusii Jungh. Et de Vriese stand of Aek Nauli, North Sumatera. Department of Forest Management, Faculty of Forestry, IPB. In: Syaufina, L., Ainuddin, A.N. (Eds.), 2011 Impacts of fire on SouthEast Asia tropical forests biodiversity: A Review. Asian. J. Plant. Sci., 10, pp. 238–244, 4.

Prasasti, I., Boer, R., Syaufina, L., June 2014. Application of CMORPH data for forest/land fire risk prediction model in central Kalimantan. Int. J. Remote Sens. Earth. Sci. 11 (1), 41–54.

Perwitasari, D., Sukana, B., Juni 2012. Gambaran kebakaran hutan dengan kejadian penyakit ISPA dan pneumonia di Kabupaten Batanghari Provinsi Jambi tahun 2008. J. Ekol. Kesehatan 11 (2), 148–158.

Rahardjo, S., 2002. Species composition and fire understorey adaptation in burned area under Pinus merkusii Jungh et de Vriese stand (a case study in Gunung Walat Educational Forest, Sukabumi (in Indonesian). Graduate School. IPB. In: Syaufina, L., Ainuddin, A.N. (Eds.), 2011 Impacts of fire on SouthEast Asia tropical forests biodiversity: A Review. Asian. J. Plant. Sci., 10, pp. 238–244, 4.

Rijksen, H.D., Meijaard, E., 1999. Our Vanishing Relative: The Status of Wild Orangutans at the Close of the Twentieth Century. Kluwer Academic Publications, Dordrecht, The Netherlands.

Ruitenbeek, J., 1999. Indonesia. In: David Glover, Timothy Jessup (Eds.), Indonesia's Fires Haze: The Cost of Catastrophe. ISEAS, Singapore.

Russon, A.E., Susilo, A., 1999. The effects of drought and fire on orangutans reintroduced into Sungai Wain Forest, East Kalimantan. In: Suhartoyo, H., Toma, T. (Eds.), In the Proceedings 3rd International Symposium on Asian Tropical Forest Management. Pusrehut Special Publication No. 8 (1999) Tropical Forest Research Center, Mulawarman University and Japan International Cooperation Agency. Samarinda, 20–23 September, 1999.

Saharjo, B.H., 1999. The role of human activities in indonesian forest fire problem. In: Suhartoyo, H., Toma, T. (Eds.), In the Proceedings 3rd International Symposium on Asian Tropical Forest Management. Pusrehut Special Publication No. 8 (1999) Tropical Forest Research Center, Mulawarman University and Japan International Cooperation Agency. Samarinda, 20–23 September, 1999.

Samsuri, S., Jaya, I.N.S., Syaufina, L., 2012. Model Spasial Tingkat Kerawanan Kebakaran Hutan dan Lahan (Studi Kasus Propinsi Kalimantan Tengah); Foresta Indonesian Journal of Forestry Faperta USU - Persatuan Peneliti Kehutanan Sumut; ISSN: 2089–9890; Vol. 1; No. 1; Maret; Hal. 12–18.

Schindele, W., Thoma, W., Panzer, K. 1989. Investigation of the steps needed to rehabilitate the areas of east Kalimantan seriously affected by fire. The Forest Fire 1982/83 in East Kalimantan. Part I: The fire, the effects, the damage and the technical solutions.

Schweithelm, J., Glover, D., 1999. Causes and impacts of the fires. In: Glover, D., Jessup, T. (Eds.), In Indonesia's Forest Fires and Haze: The Cost of Catastrophe. Institute of Southeast Asian Studies, Singapore, pp. 1–13.

Schwela, D.H., Goldammer J.G., Morawska, L.H., Simpson, O., 1999. Health Guidelines for Vegetation Fire Events. Guideline document. Published on behalf of UNEP, WHO, and WMO. Institute of Environmental Epidemiology, Ministry of the Environment, Singapore. Double Six Press, Singapore, 291 pp.

Secretariat of the Convention on Biological Diversity, 2001. Impacts of human-caused fires on biodiversity and ecosystem functioning, and their causes in tropical, temperate and boreal forest biomes. CBD Technical Series no. 5, Montreal, SCBD, 42 p.

Simbolon, H., April 2005. Dinamika hutan Dipterokarp campuran Wanariset Semboja Kalimantan Timur setelah tiga kali kebakaran tahun 1980-2003. Biodiversitas 6 (2), 133–137.

Sugiarto, D.P., Gandasasmita, K., Syaufina, dan L., Juni 2013. Analisis Risiko Kebakaran Hutan dan Lahan di Taman Nasional Rawa Aopa Watumohai dengan Pemanfaatan Pemodelan Spasial. Majalah Ilmiah Globë 15 (1), 68–76.

Sukmajaya, E.W., 1999. The community living around and inside Sungai Wain Protection Forest and their relationship with forest fires. In: Suhartoyo, H., Toma, T. (Eds.), In the Proceedings 3rd International Symposium on Asian Tropical Forest Management. Pusrehut Special Publication No. 8 (1999) Tropical Forest Research Center, Mulawarman University and Japan International Cooperation Agency. Samarinda, 20–23 September, 1999.

Sumardi, 1999. Forest fire, the damage and correlation with regeneration. In: Suhartoyo, H., Toma, T. (Eds.), In the Proceedings 3rd International Symposium on Asian Tropical Forest Management. Pusrehut Special Publication No. 8 (1999) Tropical Forest Research Center, Mulawarman University and Japan International Cooperation Agency. Samarinda, 20–23 September, 1999.

Susilo, G., Yamamotoa, K., Imaia, T., 2013. Modeling groundwater level fluctuation in the tropical peatland areas under the effect of El Nino. Procedia. Environ. Sci. 17, 119–128.

Suyanto, S., 2000. Fire, Deforestation and Land Tenure in the North-Eastern Fringes of Bukit Bukit Barisan Selatan National Park, Lampung. Center for International Forestry Research (CIFOR), Bogor, Indonesia, http://www.cifor.cgiarorg/fire/pdf/pdf71.pdf.

Syaufina, L., 2015. Research status on the relationship between forest fire and biodiversity and human dimension in Indonesia. Paper Presented in the National Seminar on Promoting Research on Forest and Land Fires Mitigation, Adaptation and Impact to Human and Biodiversity. SEAMEO BIOTROP, Bogor, Indonesia, 28 April, 2015.

Syaufina, L., Ainuddin, A.N., 2011. Impacts of fire on south east Asia tropical forests biodiversity: a review. Asian. J. Plant. Sci. 10 (4), 238–244.

Syaufina, L., Saharjo, B.H., Tiryana, T., 2004. The estimation of greenhouse gases emission of peat fire. Working Paper No. 04. Environmental Research Center. Bogor Agricultural University. Bogor.

Syaufina, L., Nuruddin, A.A., Basharuddin, J., See, L.F., Yusof, M.R.M., January-June 2004b. The effects of climatic variations on peat swamp forest conditions and fire behaviour. Trop. For. Manage. J. 10 (2).

Syaufina, L., Sitanggang, I.S., Erman, L.M., 2016. Challenges in Satellite-based Research on Forest and Land Fires in Indonesia: Frequent Item Set Approach. Procedia. Environ. Sci. 33, 324–331.

Tacconi, L., 2003. Kebakaran hutan di Indonesia: Penyebab, Biaya dan Implikasi Kebijakan. CIFOR. pp. Vi + 28.

Toma, T., Matius, P., Sutisna, M., 1999. Fire and human impacts on aboveground biomass of lowland Dipterocarp Forests in East Kalimantan. In: Suhartoyo, H., Toma, T. (Eds.), In the Proceedings 3rd International Symposium on Asian Tropical Forest Management. Pusrehut Special Publication No. 8 (1999) Tropical Forest Research Center, Mulawarman University and Japan International Cooperation Agency. Samarinda, 20–23 September, 1999.

Tomich, T.P., Fagi, A.M., de Foresta, H., Michon, G., Murdiyarso, D., Stolle, F., Van, M., Noordwijk, January–March 1998. Indonesia's fires: smoke as a problem, smoke as a symptom. Agroforestry Today, 4–7.

Wiratama, A., 2010. Exploration of Potential Bacteria as Biofertilizer Used in Burned and Unburned Peat Soil from Riau (in Indonesian). Department of Silviculture, Faculty of Forestry IPB. In: Syaufina, L., Ainuddin, A.N. (Eds.), Impacts of fire on SouthEast Asia tropical forests biodiversity: A Review. Asian. J. Plant. Sci., 10, pp. 238–244, 4.

Wösten, J.H.M., Clymans, E., Page, S.E., Rieley, J.O., Limin, S.H., 2008. Peat water interrelationships in a tropical Peatland ecosystem in Southeast Asia. Catena 73, 212–224.

Yeager, C.P., Marshall, A.J., Stickler, C.M., Colin, A., Chapman, A., 2003. Effects of fires on peat swamp and lowland dipterocarp forests in Kalimantan, Indonesia. Tropical biodiversity 8 (1), 121–138.

Chapter 9

Lessons From Tsunami Recovery Towards Guidelines of Housing Provision in Malaysia

Ruhizal Roosli*,**

*Universiti Sains Malaysia, School of Housing, Building and Planning, Malaysia; **Northumbria University of Newcastle, United Kingdom

9.1 INTRODUCTION

The tsunami in the year 2004 was the worst disaster in Malaysia (Foong et al., 2006). From 1968 to 2004, Malaysia had experienced 39 disasters. About 49% was contributed by natural disasters. After 2004, no major catastrophe occurred than the disaster caused by the acceptance of abundant rainfall. Heavy rain (primary disaster) that caused floods and landslides (secondary disaster) dominated most of those natural disasters (Ismail, 2003). Poorly controlled land use, design of buildings, maintenance of equipment and machinery, and attitudes of personnel in regulatory compliance are inevitably tended to the potential of disaster (secondary disaster) (Roosli and Collins, 2016). Moreover, failures in regulation and compliance were identified as the key vulnerability and disaster causes in Malaysia. Aini et al. (2007) found that organizational error and regulatory failures were the main types of hidden error that contributed significantly to the disasters with 53.6% and 37%, respectively. Unfortunately, local experience in Malaysia suggested by recent academic work proposed that no work has been done to understand community participation in reconstruction housing program after disaster (Aini et al., 2006).

Malaysia still does not have a specific guideline to build housing after a disaster especially in disaster-prone areas (Roosli and O'Brien, 2011). Currently, the only related reference is the Policy and Mechanism on National Disaster and Relief Management to determine the kind of action to be taken to minimize the effects of the disasters. In this policy, there is no specific requirement on permanent postdisaster housing provision. In general, allocation for physical postdisaster recovery is under infrastructure development which is a responsibility of the Public Works Department (NSC, 2011). In general, the legislation on urban planning in Malaysia such as Planning Acts and Building By-Law states that the local authorities have the authority to approve plans for public and private buildings.

At the international level, many guidelines have been prepared that are found suitable for postdisaster housing related to tsunami recovery works. However, many postdisaster programs in developing countries are not successful (UN-HABITAT, 2001). Some of the main basic texts that provide the foundation for the response of the international community and aid organizations in humanitarian emergencies as mentioned by Corsellis and Vitale (2005) are the *UNHCR Handbook for Emergencies* and *Sphere Humanitarian Charter and Minimum Standards in Disaster Response*. Which guidelines can be adapted that best describes the situation in Malaysia? This is an important question to ask, and further discussion needs to be done in order to enhance the current implementation of postdisaster program.

This chapter is a review of recent national disaster mechanism experiences in the housing sector. This chapter is an effort to promoting resilient housing, safety and security, and secure tenure in a prone area. Key lessons will emerge from the review process and analysis. These inputs will then have influence to the content that will be developed and presented as guidelines. These key lessons are perhaps the best practical (operational and technical) guidelines compared to other international cases to be adapted to the national situations. An overall objective is to support humanitarian responses to disaster and conflicts for resilience house construction to tsunami-prone area.

9.2 THE URGENCY OF POSTDISASTER HOUSING PROVISION AFTER THE 2004 TSUNAMI IN MALAYSIA

After the tsunami, many houses were damaged and destroyed. Families already stricken by poverty were unable to reconstruct their houses and leave it to the authorities to look after the matter. The authorities also have a number of obstacles in providing this housing (Foong et al., 2006). The reasons include a lack of relevant national and state policies and action

Integrating Disaster Science and Management. http://dx.doi.org/10.1016/B978-0-12-812056-9.00009-9

plans; existence of regulations on urban planning and environment which have not been adjusted to manage flood; slow response to flood disasters due to lack of capacity and resources; and lack of public awareness on flood variability and flood-induced hazard mitigation (due to tsunami) (Soti and Herard, 2012).

Even if the country has abundance of rules and regulations in providing housing, it is only applicable to the formal type of development or housing (Kennedy et al., 2008). Many numbers of housing in the affected area were identified as informal housing especially in villages that were constructed free from obligation to these rules and regulations. As a result, it is not surprising when many houses hit by the floods due to Tsunami were washed away. In addition, some of the projects initiated and monitored by responsible agencies also fail to protect disaster victims' rights.

Postdisaster housing is put up after disasters by governmental and/or private institutions. There are two types of housing built after a disaster: one is principally a shelter put up for immediate relief purposes and the other one is more permanent housing with long-term settlement purposes (Jha, 2010). Permanent housing, which is the last stage of housing recovery, basically aims to be a final solution after disasters to provide housing individually which would fulfil the needs of the inhabitants in relatively much longer period of time. They aim not only to serve as housing units or only basic protection, but also to satisfy all necessary requirements regularly. Victims who previously lived in squatter homes would be provided a new settlement and permanent houses from the government. This is an effort not only to satisfy the victims, but also to speed up the process of housing provision by the government.

For Malaysia, the existence of guidelines for the construction of houses in tsunami-prone areas is necessary to at least mitigate the effects of destruction caused by this type of disaster and making preparations to face small-scale flooding or otherwise. Universiti Sains Malaysia has developed the Management Model Disaster Risk for Sustainable Development (DRM-SD) through the Centre for Global Sustainability Studies (CGSSUSM) as a result from previous disaster occasions. The model (Fig. 9.1) contains the cycle that include prevention (prevention), readiness (preparedness), action (response), and recovery (recovery). These four concepts are bound by Governance (governance) which focuses on events before, during, and after a disaster. This model connects all DRM-cycle components and explains the importance of a progressive risk reduction as the best option for disaster management in local governance and country towards sustainability.

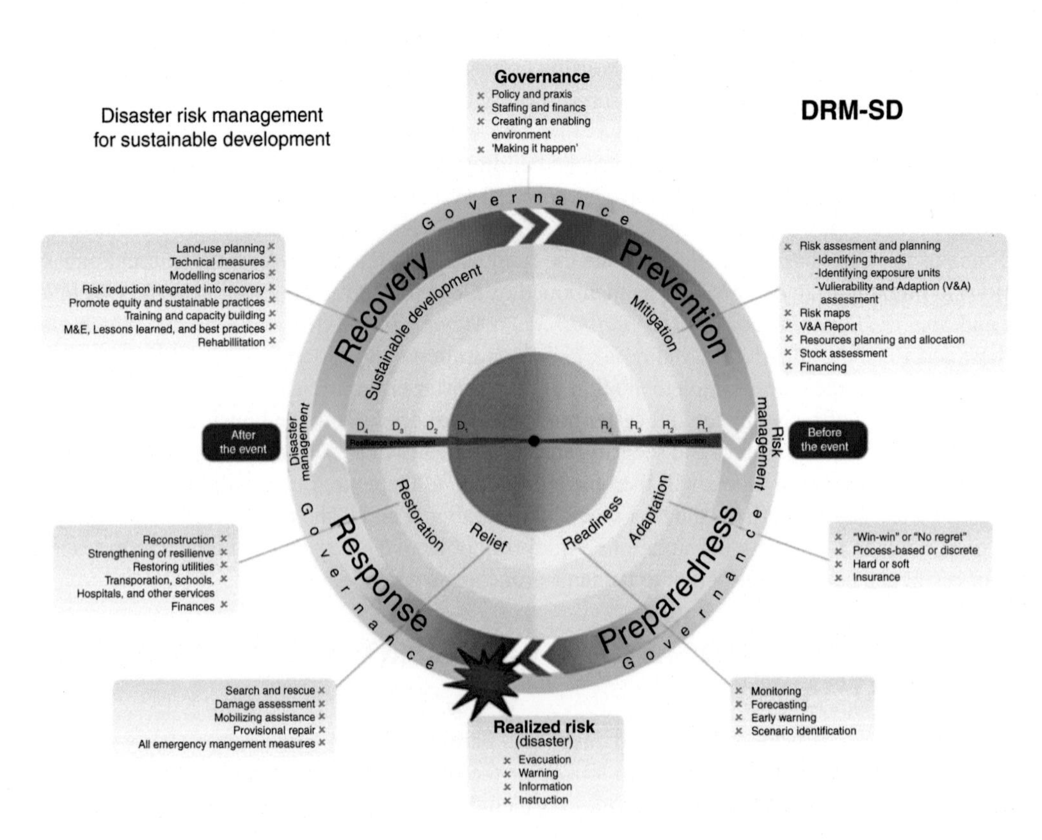

FIGURE 9.1 **Disaster risk management model for sustainable development, DRM-SD (CGSS, 2013).**

9.3 ISSUES WITH THE BUILT ENVIRONMENT AND SUSTAINABLE RECONSTRUCTION PROCESS

The built environment refers, in general terms, to human settlements, buildings, and infrastructure (transport, energy, water, waste, and related services) (RICS, 2009). The sector includes the commercial property and construction industries and the built environment and related professions.

The term "built environment professional" includes those refer to as "practitioners", primarily concerned with providing technical support services—consultation and briefing, design, planning, project management and implementation, technical investigations including monitoring and evaluation studies (Lloyd-Jones, 2009). They may be employed directly by a client or indirectly through a contractor.

Built environment professionals may also be concerned with designing and implementing policy, standards, and regulation of the built environment—factors that are critical in reducing the risks from hazards—or are exclusively or partly involved with training, professional education, and research. The professions include land surveyors, planners, administrators, and land tenure specialists who are concerned with sectors such as housing and land issues that are particularly highlighted in any postdisaster situation (Lloyd-Jones, 2009).

Sustainable reconstruction offers the chance to improve the quality of buildings, the environment, and living conditions in disaster-affected regions. However, conventional reconstruction efforts often failed because of a one-sided approach, for example, one that focuses only on technical or construction aspects (Barakat, 2003). There were cases where houses were constructed but without the necessary infrastructure, water supply, and sanitation, because of one-dimensional attitudes and, among other challenges, institutional constraints, bureaucracies, etc. Often, conventional reconstruction neglects important social and livelihoods issues which result in a poorer economic situation for beneficiaries with interrupted social relations.

It is important to integrate the principles of sustainability strategically from the earliest stages of reconstruction in order to avoid major failures during reconstruction (Schneider, 2012). The key principles are:

1. Learn from experiences, which dealt with effective and efficient reconstruction, and from traditional building technologies which survived disasters;
2. Establish and maintain a well-functioning project-management process;
3. Ensure local participation in decision-making processes;
4. Anchor the project in the local context;
5. Coordinate with other donors to identify potential synergies;
6. Determine communication and knowledge-sharing strategy;
7. Develop a risk strategy;
8. Conduct regular monitoring and evaluation (M&E);
9. Choose the lifespan of houses to be built;
10. Provide adequate temporary shelters;
11. Consider reusing and recycling temporary housing components for permanent houses to be built in the future;
12. Consider the overall development concerns and priorities of your organization;
13. Follow principles of bio-climatic and adaptable design.

9.3.1 Postdisaster Housing

Emergency housing or temporary accommodation refers to disaster-affected families' provisional place to stay at the earlier stage of disaster until they get permanent housing (Corsellis and Vitale, 2005). It is actually the period of physical, social, and emotional recovery in the rehabilitation phase. Unfortunately, rehabilitation is often overlooked by related agencies to the rights of disaster victims. Corsellis and Vitale (2005) suggested that there are two simple steps of approach (other than complicated) in providing emergency housing that are the approach to satisfy the individual family and then the community as a whole. These approaches acquire ambitious plans of community reconstruction, disaster victims' participation, guided self-help construction, and holistic measures of development (Lizarralde, 2002).

Postdisaster reconstruction theories emphasize the response according to phases concerning social aspects of the housing reconstruction process. They note that housing is an industrialized product that is provided to the affected community. However, most cases of postdisaster reconstruction revealed unachievable project missions and dissatisfaction results within the last three decades because disaster community exclude the fact that housing the victims after disaster is also the work of providing houses to the community as a whole (Lizarralde, 2002). The rights to housing, however, are preserved in the international legislations to the rights of adequate housing.

"Shelter after Disaster: Guidelines for Assistance" is the first guide on shelter and housing inspired from diverse 1970s disasters' experience (UN/OCHA, 2010). Unfortunately, very limited supplementary work has been done in order to enhance the use and content (UN/OCHA, 2010). The "transitional settlement: displaced populations" draft was then produced as a result of the revised version between 2002 and 2004 in order to inspire new ideas in risk reduction response (Corsellis and Vitale, 2005). According to this guide, the term "emergency management" typically means a major focus on the preparedness and response phases of disasters.

The use of "disaster plans" frequently refers to the full range of activities from mitigation through recovery. However, the summary guidelines that include entries on the shelter sector and that are published by most operational organizations (international specification) do not have a significant body of literature to refer to and found to have repetition of a few key guidelines (Crawford, 2002). Consequently, the operations guidelines was introduced in order to avoid confusion in emergency housing over the meanings of commonly used terms such as "emergency shelter", "temporary shelter", "temporary housing", "permanent shelter", "dwellings", "housing", "building", "recovery", and "reconstruction" as you can see in Fig. 9.2.

These terms are commonly used from transitional settlement phase (tent, temporary shelter) to permanent housing reconstruction phase (from the period of disaster impact to project accomplishment) in order to find an easier situation to describe, support, and integrate its contribution for a wider response.

According to El-Masari (1997), design of a house should be based on a comprehensive analysis of the physical conditions of the building in relation to probable disaster(s). Shape, height, building materials, construction techniques, and space arrangements of the building should be improved and modified by applying appropriate strengthening measures. All of these could be undertaken by the help of regularization and upgrading process.

9.4 CONFLICT IN THE DISASTER RECOVERY PROCESS OF HOUSING RECONSTRUCTION

Housing reconstruction after disaster is a crucial issue because of physical, social, psychological, and environmental effects (Davis, 1981). An improved strategy is crucial to accelerate the reconstruction process for upgrading condition of human settlements. Community participation is a key term to understanding the community's social needs (Turner, 1977). Community participation is seen by some as a way for stakeholders to influence development by contributing to project design, influencing public choices, and holding public institutions accountable for the goods and services they provide. Some view

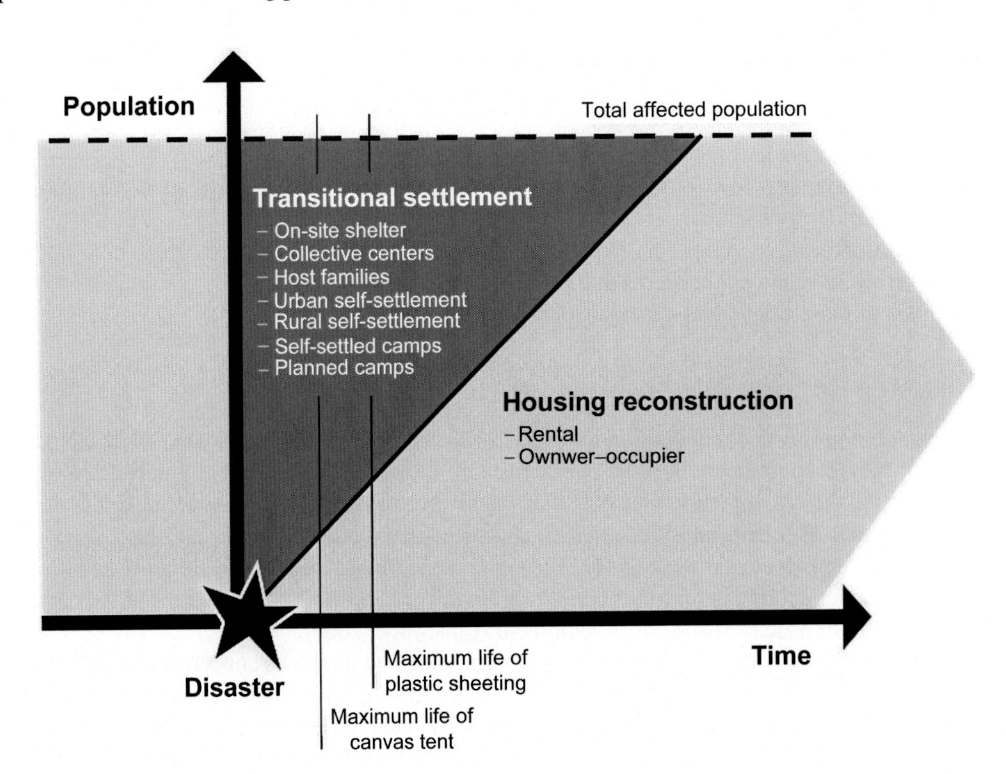

FIGURE 9.2 **Postconflict and in complex emergencies from emergency shelter to transitional settlement to permanent housing (Corsellis and Vitale, 2005).**

participation as the direct engagement of affected populations in the project cycle—assessment, design, implementation, monitoring, and evaluation—in a variety of forms. Still others consider participation as an operating philosophy that puts affected populations at the heart of humanitarian and development activities as social actors with insights, competencies, energy, and ideas of their own (Turner, 1972).

Reconstruction begins at the community level. A good reconstruction strategy engages communities and helps people work together to rebuild their housing, their lives, and their livelihoods. A very strong commitment and leadership from the top are needed to implement a bottom-up approach, because pressure is strong in an emergency to provide rapid, top-town, autocratic solutions. Engagement of the community may bring out different preferences and expectations, so agencies involved in reconstruction must be open to altering their preconceived vision of the reconstruction process. Numerous methods exist for community participation, but they need to be adapted to the context, and nearly all require facilitation and other forms of support.

Presently, shelter needs in Malaysia after disasters cause different habitation modes, which can be summarized in these alternatives:

1. Moving to public shelters;
2. Moving to shelters of friends and relatives;
3. Moving to a second undamaged house or rent a house;
4. Organization of camps or tent shelters beside the damaged buildings (Price et al., 2000).

Relocation of affected population to new and safe sites can be an effective tool to reduce the probable secondary disaster and to create a resilient community. As argued by Arslan and Unlu (2007), relocation always has effects on the behavior of disaster victims by means of changing their daily activities and livelihood. They prefer not to be relocated and stay close to their home and if possible try to rebuild their shelter (Foong et al., 2006). Capability of the people is also being questioned in building their own house according to situational needs without reference to building codes. Housing should be built according to rules and regulations, controlled by authority through inspection, incentive, and punitive measures (Burgess, 1978).

With regard to issues concerning who should play a major part in reconstruction process, a majority of the affected population and regulators usually seek minimal government involvement in order to avoid steep taxes and more importantly avoid bureaucracy that would complicate the situation (e.g. legal actions and mass media attentions) (Turner, 1977). The only way to comply with the specific requirement in providing temporary accommodation is by establishing collaboration between emergency reliefs and mechanism in rehabilitation and to implement planned programs (Comerio et al., 1996). This productive solution and atmosphere enable stakeholders to adopt appropriate recovery strategy and meets the need of the population. Thus, scholars suggested that the approach in housing resettlement should be based on the reason why houses are provided (Burgess, 1978; Turner, 1977).

However, there is a boundary between involvements of victims in postrecovery work (Arslan and Unlu, 2007). Commonly, most of the victims were in a traumatic case even if they had no injuries or damaged buildings (Mulwanda, 1992). The state of this current situation limits active participation from victims due to low capability and readiness physically and mentally. They must know their own capability and readiness before they can get involved in the decision-making process because mitigation effort demands consideration to legal framework (Mulwanda, 1992). Disaster victims must take rational action that facilitates government to establish self-reliant communities so as to reduce their vulnerability to natural disasters.

Secondary disasters (collateral disaster) often cause far more damage and problems than a primary disaster (a primary disaster such as an earthquake that causes or brings in its wake one or more disaster such as a fire or tsunami) (Comerio et al., 1996). Often the pressures of time, limited contingency budget, and people's needs counteract the demands of quality and suitability (Morago, 2005). Undesirable housing conditions partly or entirely due to noncompliance with building codes and other related legislation are considered as potential risk of moderate or serious harm to the health and safety of the occupiers (or others) of the residential premises such as:

1. Physical injury (burns, wound, and serious bodily injury);
2. Psychological distress (sadness, frustration, anxiety, and a number of other negative mood states);
3. Emotional and psychological trauma;
4. Psychosocial effects as a result of secondary disaster (Levine et al., 2007).

The ability of the built environment to withstand the impacts of natural forces plays a direct role in determining the casualties and economic costs of disasters. Disaster-resistant construction of buildings and infrastructure is an essential component of local resiliency (Cernea and McDowell, 2000). Any postdisaster recovery framework or policy in providing housing must come with specific rules and regulation such as Housing Act, Building Codes, and Law of Property Act. Such rules and regulations will give authority or local authority powers to intervene where housing conditions are unacceptable to reduce the risk of human casualties (Imrie, 2007). Engineering codes, standards, and practices have been acknowledged and practiced to prevent natural hazards. However, investigations after disasters have revealed shortcomings in construction techniques, code enforcement, and the behavior of structures under stress (Mileti, 1999).

On the other hand, it is also important to examine information related to people who are going to live in the buildings because lessons can be learned on the involvement of local residents (Spence, 2009). Local residents have knowledge of the place they live in. It would be crucial to establish timely contact with local residents and use the information to guide the rescue effort and fulfill what local residents expect to receive. Without understanding local needs, it would cause not only confusions, but also waste of resources (Xiulan, 2008). The affected population must be strongly represented on the body that directs recovery. They should be incorporated, coordinated, and extended as part of the recovery planning process. Participatory planning must involve people from the start of the planning process. It is simply not enough to ask them for their opinion of a plan that has already been drawn rather listen to communities concerned as input to recovery plans especially related to housing resettlement or reconstruction.

At this juncture, the question is whether the development of the affected community is free from potential risk with only authority or local authority playing their part by complying with rules and regulations in providing housing. Arslan and Unlu (2007) argued that assessment of needs should be based on the expressed priorities of affected communities because recovery is not just in physical appearance in order to restore normalcy. Therefore, the stakeholders should invest and give serious consideration in justice and human rights as part of an attempt to restore peace and promote democracy and understanding in postdisaster recovery period.

9.4.1 Dilemma in Assistance to Postdisaster Housing

Many of the difficulties encountered by agencies' shelter and housing work are typical issues that these agencies experience when working in the shelter and housing sector. Shelter and housing work is the "least successful form of aid when compared to other humanitarian intervention sectors" (ALNAP, 2003). While some of the problems noted by ALNAP are generic to all humanitarian projects, others are more specific to the shelter and housing sector (Table 9.1).

Housing provision is always a complex relief effort, leading to rehabilitation divide and confusion as to objectives and responsibilities (ALNAP, 2003). Other than authorities, a few donors and nongovernmental organizations (NGOs) are willing to fund housing work outside of an emergency context and a postdisaster emergency presents one of the few opportunities available for upgrading the quality of vulnerable housing. Government is responsible to ensure safe housing for their citizens. However, action in the emergency of recovery period becomes unorganized by all parties/agencies especially at disaster-affected area. In many cases, the role of government in rebuilding houses becomes muddled.

Scholars suggested that the approach in housing resettlement should be based on the reason why houses are provided (Burgess, 1978; Turner, 1977). The understanding concerning housing as a product and the process of achieving those products is vital mainly in terms of temporary shelter before reconstruction of permanent housing (Davis, 1987). There are always arguments between the term "housing" compared to "houses". In English, the word "housing" can be used as a noun or as a verb. When used as a noun, housing describes a commodity or product. "The verb 'to house' describes the process or activity of housing" (Turner, 1977, p. 151). John F.C. Turner has been the most influential writer about housing in the developing world in the postwar period. The "empowerment" (participation) of local people

TABLE 9.1 Difficulties in Postdisaster Reconstruction

Common Agency Difficulties in Shelter and Housing Projects	Common Agency Difficulties in All Humanitarian Assistance Projects
Inadequate understanding of land title and tenure issues	Inadequate understanding of issues related to social processes and their relationship to humanitarian interventions
Pressure to allocate funding resources in a highly visible manner to meet political demands of donors and host governments	Low levels of preparedness by agencies
The inconsistency of housing: where some members of affected populations receive a substantially large improvement to their assets, while others in the same communities may not	Inadequate selection and training of staff
	Delays in the implementing a response
	Slow disbursement of project funds to field operations

Source: ALNAP (2003).

is significant and should be sufficient in the process of redevelopment (Turner, 1977). Housing should be treated as a process rather than simply a product. "There is a difference between a concern with houses and a concern with housing, whereas houses are cultural and social objects and consumables, housing is a cultural process and a social activity" (Turner, 1977, p. 58).

However, agencies suggested that states utilize local solutions and procurement instead of prefabricated or imported shelter with the use of international guidelines as reference (Crawford, 2002). The biggest problem is the bureaucracy's ability at national level to convince officials of the importance of international standards in the international context (ProVention Consortium, 2004). Officials usually want something more definitive from government in terms of clarification (dissemination of information) to the importance of international specification (standards) that is accepted and endorsed (based on previous empirical experiences) on a world-wide basis (Davis, 1987). Users of standards also want something that can make outcome distinctions based on a unanimous understanding (official announcement or endorsement) over the use of international specifications in a national context because they will give some assurance that their efforts, resources, and trusts will provide positive results.

However, each disaster situation is unique and requires distinctive (different) proceedings. The dilemma in disaster response is not only limited to producing a well-planned settlement based on the present legal infrastructure, but also a solution for people to "bounce back" in the direction of normal life or even better (Corsellis and Vitale, 2005). Thus, Johnson (2002) introduced two specific considerations in order to provide a solution for temporary accommodation:

1. The potential of the particular community's human and financial resources;
2. The possibility of project durability (hazard resistant and constructive livelihood).

Efforts in relief assistance should focus on the approach to utilize optimum resources and sustain project durability. Generally, national relief programs (disaster plan) come with a complete legal infrastructure inspired by national and international experiences (Corsellis and Vitale, 2005). From this notion, the only way to comply with the specific requirement in providing temporary accommodation is by establishing collaboration between emergency reliefs and mechanism in rehabilitation and to implement planned programs. This productive solution and atmosphere enables stakeholders to adopt appropriate recovery strategy and meets the need of the population. Thus, scholars suggested that the approach in housing resettlement should be based on the reason why houses are provided (Burgess, 1978; Turner, 1977).

9.4.2 Relief Coordination

Coordination always involves an organized direction of activities before, when disaster occurs, and after the disaster. However, weak coordination is considered as a barrier in emergency management and was found as a major issue in disaster response (Johnston, 2004). Coordination requires efficiency in disaster planning, effectiveness in relief works, and knowledge exchange among relief workers and victims. The present standard of coordination is considered insufficient due to inadequate disaster resources (e.g. managing the flow of goods, information, and finances) that remain as a key principle in disaster operation (ProVention Consortium, 2004).

Simultaneously, scholars argued that the degree of relief coordination by many international governmental and nongovernmental organizations is not sufficient to effectively meet the needs of disaster victims. Affected countries as well as the governmental agencies and nongovernmental organizations still lack a strong relationship and networking especially with other international communities (Quarantelli, 1984). Without national and international coordination, delays in disaster response may occur due to repetitions of effort and limited exchange of information and technology.

Normally, the contacts among relief workers are often difficult because information is generally lacking in the scene of disasters (Quarantelli, 1984). The loss of power and phones lines and destruction of transportation networks limit communication between humanitarian organizations and victims and among relief workers themselves. Thus, there is always a dilemma in local, national, and international governmental and nongovernmental agencies at the disaster scene concerning abundance of relief efforts by aid agencies (Johnston, 2004). It is often unknown that resources are available and even the involvement and contribution of suppliers is unpredictable. This creates many redundancies and duplicated efforts and materials. This scenario is chaotic and creates difficulties in order to make any changes and improvement in relief works (Johnston, 2004). This confusion happens because of a failure to communicate and understand their own strengths, resources, roles, and responsibilities in disaster works.

Meanwhile, coordination in relief works is also ineffective due to different objectives and values from NGOs in providing relief. As argued by Minear and Weiss (1995), international organizations have to consider their own programs in disaster relief. They pose themselves as international relief bodies but always give priority in pursuing their own objectives, for example, promoting their own brand and commercial agendas. As a result, the humanitarian works turn to self-interest in the name of relief works. The real relief workers from local and national organizations unanimously disagree with the nature of this adopted emergency approach (Minear and Weiss, 1995).

The relief organizations also struggle when it comes to budget allocations in running any humanitarian works (Minear and Weiss, 1995). Without financial support, any humanitarian work meets with a dead end. Funds in these cases depend upon charities and donations and organization reserves. In some cases, organizations seem even keener in collecting funds and donations rather than encouraging humanitarian works. They put their name within the international system rather than the international relief network. Consequently, disaster victims are the real victims here not only from natural or technological behavior, but also from human behavior. Thus, scholars suggested that there is a need for more "collaborative problem solving" in the response of institutional disaster risk reduction (Mileti, 1999). This collaboration (national and international) should concentrate on consultation and understands roles and responsibilities between agencies involved (Minear and Weiss, 1995).

This collaboration will also encourage exchange of technologies (e.g. rescue equipment, early warning systems, and building constructions and materials) in order to reduce immediate and future impacts by reducing the vulnerability of the people and societies involved. Unfortunately, lack of opportunities exists among relief workers, especially government officers, to utilize knowledge in technology transfer due to distractions from other components of a postdisaster reconstruction project (Barakat, 2003). Mileti (1999) noted that organizations that work in disaster relief and postconflict reconstruction focused on nonhousing sectors as the first step to livelihood such as governance issues, donations, budget, disease control, violence and armed conflict, and social capital because housing is the last resort after all considerations are resolved.

On the other hand, as argued by Christoplos (2006), shelter/housing is the primary protection for disaster victims from hazards in the scene of disaster. Unfortunately, officials underestimate specific challenges at the disaster scene because in any postdisaster planning and recovery, reconstruction of infrastructure and housing is the main indicator to measure the outcome of risk reduction (Christoplos, 2006). Compliance with regulations (e.g. building codes and planning acts) and technology requirements is the key to higher hazards resistance. For example, an effective design of earthquake-resistant buildings, especially to the enactment of effective seismic-design regulations, would have safer protection even if still remained seismically vulnerable in earthquake regions (Mileti, 1999).

9.4.3 Postdisaster Housing Financing Models

Postdisaster housing financing models are as (1) outright gift, (2) partial support, and (3) loans (Barakat, 2003). There are advantages and disadvantages of these three models which are presented in Table 9.2. These three models of postdisaster-housing as follows:

1. *The contractor model:* Sometimes housing reconstruction programs are contracted to professional construction companies. Large-scale contracted constructions may have disadvantages. For instance, specific housing needs of individual communities may not be met, and diversity within the community may not be taken into consideration. On the other hand, large numbers of houses with standard specifications can be constructed relatively quickly using staff with technical expertise, employing specialist skills. This model is also appropriate when target groups lack the skills and resources to undertake the construction work themselves.
2. *The self-build model:* This model, which is also called self-help or owner-driven, enables the communities to undertake construction works of their houses themselves. The model is possible when labor is available, housing design is relatively simple, communities have a tradition of self-build, and there is strict time limit. Postdisaster reconstruction work can be set up on a family self-help basis or as a joint community reconstruction program. In some instances, food for work is also included as a part of the program.
3. *Cooperative reconstruction:* This model, which is an alternative to self-build model, focuses on mobilizing a community to undertake reconstruction program together. It requires a high level of community involvement and cooperation. In this model, materials are provided for the whole community as a whole, rather than for individual families. Agencies can control the process and make sure that community members are benefiting from the program equally.

The study conducted by Barakat (2003) revealed that there were three different ways of locating permanent housing in the area, which are:

1. Constructing houses on original lots of the previous ones,
2. Constructing a new settlement close to the old one, and
3. Constructing new settlement far from the old one.

TABLE 9.2 Advantages and Disadvantages of Various Housing Reconstruction Finance Options

Finance Option	Description	Advantages	Disadvantages
Outright gift	Beneficiaries are given houses on the basis of meeting certain conditions of entitlement. The recipient has no obligation to repay the cost of the house	Removes the need to set up a system to recuperate costs. Allow recipients to use their assets to meet other needs	Encourages dependency and undermines local coping mechanisms. Bypasses and thus weakens local institutions. Is often imposed solution. The assisting agency cannot recuperate money for new projects. Number of houses provided is limited
Partial contribution through self-help	Beneficiaries may receive building material and/or technical advice, and/or a partial grant. They build their own house, usually on a communal basis or by contracting local builders	Removes the need to set up a system to recuperate costs. Allows recipients to use their assets to meet other needs. Increases involvement and participation by the recipients	As with the outright gift, this option can undermine both local capacity to cope and local institutions. Materials provided may not meet the requirements or aspirations of the recipients. Time spent on building may conflict with other priorities of the recipients, such as income generation, which may be a vital element in family recovery
Loans	There are many variations of loan programs. The most common for reconstruction is the long-term loan. Some loans may be without interest, while others apply normal interest rates	People without resources are able to rebuild their homes and repay the loan over time. Recipients have freedom to build a house according to their own choice. Encourages independence and sustainability	May encourage renters to become owners. Credit systems may not exist and so may need to be set up. Loans may be a significant financial burden for recipients, especially if they have no previous experience of credit systems. Loan systems are costly to administer. Many financial institutions favor only the most credit-worthy people and they demand the creditor's house as a guarantee

Source: Barakat (2003).

The failure of so many of these projects means that huge amounts of resources—both financial and human power—are being misallocated. Key factors that are affecting project success in the area are:

1. *Responding to the local needs:* The overriding principle of any project must be that it deals with the needs of the people who have been affected by the disaster. Often programs are implemented without first consulting with local populations about what they feel their needs are. Without a clear understanding of the needs of the affected population, the relevance of the program will be limited.
2. *Understanding the situation dynamics:* This is especially important in postconflict circumstances, but relates to all postdisaster situations. The situation in which a project is going to be implemented is not a static thing, it is always changing. There will be many factors affecting the project—both directly and indirectly—its design, implementation, and outcomes. Mapping out all of these specifics—what the issues are, who the main actors are, what power dynamics exist between them, and so forth—is a crucial exercise when planning any postdisaster project.
3. *Misallocation of resources:* Resources in postdisaster situations are precious, and it is therefore serious when funds and manpower are diverted into projects that do not succeed. Many postdisaster projects become extremely expensive, much more so than is necessary. The result is a lack of resources for other important postdisaster relief and rehabilitation projects.
4. *Short-termism:* Another factor affecting the success of projects is negligence in placing the specific project in the broader context of postdisaster rehabilitation. Often, in postdisaster situations, the focus of the project can be very

specific, for example, the provision of shelter or the rebuilding of roads. However, these specific issues are connected in the broader environment of the postdisaster situation and the long-term development of the area. To put it another way, a project cannot exist in a vacuum. It will affect, and be affected by, the broader context of postdisaster rehabilitation. The role of local authorities in the long-term development and rehabilitation of their communities after disaster is central. The pressure on them to enact short-term solutions to deal with the effects of the disaster need to be tempered with an understanding of the longer-term issues and strategies.

5. *Dependency versus capacity:* This point leads to the fifth factor affecting project success in postdisaster situations. The issue of building capacity rather than dependency is particularly acute in postdisaster cases. Many projects bring important resources (funding or expertise) necessary for the emergency relief phase after the disaster; however, once the emergency need has been met, the funding or expertise is gone. This perpetuates a dependence of the affected communities on aid and relief projects, rather than fostering the development of local capacity so that the affected community may become self-sufficient. Working with local partners to improve capacity not only benefits the affected communities, but also lessens the burden on aid organizations working in the region.
6. *Accountability:* The issue of accountability is closely linked with the first factor affecting project success—addressing the needs of the local communities—however, it does bear separate examination. When projects are not connected to the area in which they operate, that is, they are not staffed by local people, they do not use local resources, etc.—they also do not have to be accountable to the local populations regarding their impact. This distance helps to perpetuate the cycle of failed projects, as the lessons from each project are not passed on to the next. The view of the affected groups as passive recipients of aid also impacts the level of accountability, as they are not seen as partners to whom the project must answer with regard to its success or failure. Accountability to the local population engenders a feeling of ownership of the project, which is a key factor for project sustainability.
7. *Quality assessment:* This final factor is most directly connected to the solutions examined in this resource guide. Without built-in assessment mechanisms, these projects cannot learn from their past mistakes and determine where they have gone wrong. In addition, this disconnects the web of postdisaster projects generally—not just from the local groups—but from other projects, as they cannot learn from other project mistakes if the lessons are not being examined and recorded.

9.5 REVIEWS OF INTERNATIONAL GUIDELINES IN SHELTER/HOUSING SECTOR

Postdisaster housing reconstruction is a process that is the interaction of complex social, technological, and economic factors and actions. The process of postdisaster housing reconstruction is comprised of four different periods: The predisaster, immediate relief, rehabilitation, and reconstruction periods. The actions and measures defined in the process also fall into four categories: policy-making, organization, implementation, and evaluation and follow-up (IFRCRCS, 2012; UNDRO, 1982). Actions related to policy-making and various actions about organization are realized in the predisaster period, and the remaining actions are realized in the postdisaster phases. The analysis would be a first step for realizing a more precise organization plan which omits the frequent mistakes for the implementations in Malaysia.

Some of the main reference was used in providing housing after a disaster during the reconstruction process are:

1. Shelter a2.117 mmfter Disaster, Davis, Ian (1978) Oxford Polytechnic Press, Oxford.
2. Guidelines for Postdisaster Housing by Oxfam
3. The Sphere Handbook: Humanitarian Charter and Minimum Standards in Disaster Response, Sphere Project Office, Geneva
4. Safer Homes, Stronger Communities: A Handbook for Reconstructing After Natural Disasters, World Bank
5. Sustainable Reconstruction in Disaster-Affected Countries: Practical Guidelines, United Nations Environment Programme and Skat—Swiss Resource Centre and Consultancies for Development
6. Handbook on Housing and Community Reconstruction, The World Bank, Washington, DC
7. After the Tsunami—Sustainable Reconstruction Guidelines for South-East Asia, Swiss Resource Centre and Consultancies for Development—in partnership with Sustainable Building and Climate Initiative (SBCI) of UNEP—United Nations Environment Program
8. Reduce Tsunami Risk: Strategies for Urban Planning and Guidelines for Construction Design, ADPC (Asian Disaster Preparedness Center), Bangkok
9. Guidelines for Housing Development in Coastal Sri Lanka, Ministry of Housing and Construction, Colombo
10. Guidelines for Reducing Flood Losses, United Nation
11. A People's Guide to Building Damages and Disaster Safe Construction, UNNATI, Ahmedabad

12. Shelter After Disaster, Oxford Polytechnic Press, Oxford
13. Guidelines for Reconstruction of Building in Cyclone Affected Areas in Orissa, HUDCO (Housing and Urban Development Corporation Ltd), New Delhi
14. Housing Reconstruction After Conflict and Disaster, Humanitarian Practice Network, London
15. A Practical Guide to the Construction of Low Cost Typhoon-resistant Housing, BSHF, Coalville
16. Technical Guidelines for Resilience House Construction to Climate Change (2015), UN-Habitat Worldwide
17. Handbook on Design and Construction of Housing for Flood-prone Rural Areas of Bangladesh, ADPC (Asian Disaster Preparedness Center), Bangkok

9.5.1 General Guiding Principles

The content of Housing and Community Reconstruction Handbook Guiding Principles (Jha, 2010) are:

1. Do not just reconstruct houses, reactivate communities;
2. Put owners in charge of reconstruction and address needs of tenants and squatters;
3. Provide an effective organizational structure;
4. Use reconstruction to rethink the future and conserve the past;
5. Collaborate with communities rather than just inviting their participation;
6. Promote civil society engagement consistent with reconstruction policy;
7. Use assessment and monitoring to improve reconstruction outcomes;
8. Use reconstruction to mobilize disaster risk management policy reform;
9. Manage financial resources and stabilize family finances;
10. Avoid relocation or mitigate all its impacts;
11. Avoid sacrificing hard-won policies to facilitate reconstruction;
12. Establish environmental sustainability as a reconstruction objective.

Some of the very major references in the Postdisaster Housing Provision were adapted from *General Guiding Principles for Housing and Community Reconstruction Handbook* (Jha, 2010). The *Sphere Handbook* outlined key dilemmas and challenges (summarized in Table 9.3A) faced by humanitarian agencies concerning issues emerged in providing accommodation in disaster scene. This handbook is the most considerable because the operational organization (the "Sphere Project") keeps updating from time to time (Crawford, 2002). "InterAction" from the United States of America memberships and the Steering Committee for Humanitarian Response (e.g. Care International, Caritas Internationalis, the International Committee of the Red Cross, the International Federation of Red Cross, and Red Crescent Societies) constituted a project called Sphere Project in 1997 for setting the standards in this guideline that contribute to an operational framework for accountability in disaster assistance efforts.

> *Sphere is based on two core beliefs: first, that all possible steps should be taken to alleviate human suffering arising out of calamity and conflict, and second, that those affected by disaster have a right to life with dignity and therefore a right to assistance. Sphere is three things: a handbook, a broad process of collaboration and an expression of commitment to quality and accountability.*
>
> (Sphere Project, 2004, p. 5)

This "Humanitarian Charter and Minimum Standards" is based on the principles and provisions of international humanitarian laws, international human rights laws, refugee laws, and the Code of Conduct for the International Red Cross and Red Crescent Movement and Nongovernmental Organizations (NGOs) in Disaster Relief. This handbook identified minimum standards to be attained in disaster assistance, in each of five key sectors (water supply and sanitation, nutrition, food aid, shelter, and health services). Table 9.3A, Table 9.3B and Table 9.3C describes comparison of the guideline reviews.

These guidelines describe the core principles that govern humanitarian action and restore the right of populations affected by disaster and live with dignity. Other guideline contents are considered as being repetitive of a limited number of basic principles, not making coherent sense and consistent progress over time, being difficult to source and generally outdated (Crawford, 2002).

The key early text published on emergency shelter, Ian Davis' *Shelter After Disaster*, appeared in 1978. Later, Oxfam, UNDRO, UNHCR, UNICEF, and other major agencies began to focus on shelter especially the transition from immediate/temporary/transitional to permanent housing. By 2000, the Sphere Project had brought together many of the field guidelines, research, and the operational experience of practitioners in an attempt to set minimum standards for the aid community and disaster victims.

TABLE 9.3A Guideline Reviews of SPHERE

Publisher	SPHERE		
Title	British Red Cross Annual Review 2000 plus 2000 Trustees' Report and Accounts		
Date	2000		
Description	This guide is set out as a series of sections each specifying minimum standards and offering key indicators and guidance notes. It refers too many of the key texts discussed in this report and works through detailed steps to carry out, for example, a needs assessment. Effective disaster response is set out as a process and much emphasis is put on not operating in isolation from other over-arching sectors like logistics		
Common standards	1: Participation (e.g. local capacity, communication, and transparency and sustainability) 2: Initial assessment (e.g. checklists, assessment team, and sources of information) 3: Response (e.g. meeting the minimum standards, capacity, and expertise) 4: Targeting (e.g. targeting mechanisms and monitoring errors of exclusion and inclusion) 5: Monitoring (e.g. use of monitoring information and information sharing)		
Shelter and settlement standard	1: Strategic planning (e.g. return to home; land and building ownership; and natural hazards) 2: Physical planning (e.g. access to shelter locations; and access and emergency escape) 3: Covered living space (e.g. safety and privacy; duration, and climate context) 4: Design (e.g. materials and construction; and ventilation and participatory design) 5: Construction (e.g. sourcing of shelter materials; and labor and construction standards) 6: Environmental impact (e.g. sustainability and the management of environmental resources)		
Key advise	1: Standard, indicators, and guidance notes 2: 3.5–4.5 m^2 per person in a shelter 3: Kit list of utensils, tools, and soap 4: 45 m^2 per person in a camp 5: Minimum space between dwellings and site gradient 6: 1 km^2 of forest could serve up to 500 people 7: Annual fuel wood consumption 600–900 kg per person		
Policy	Need	Implementation	Response
	Three scenarios: people stay at home; people are displaced and stay in host communities; and people are displaced and stay in clusters	N/A	N/A
Process	Standard	Indicator	Guidance
Priorities	Analysis (needs assessment), monitoring, evaluation, and consultation	Site selection, planning, security, and environmental concerns	Housing, living quarters, clothing standard, household items, livelihood support, and environmental concerns
Provision	N/A	Human resources, competent staff, and local capacity	N/A

Source: Crawford (2002).

From all the tables, there is a need to have a consolidated institutional assistance in housing provision, especially to assist less-developed area with their own capacities if possible to reduce hazards and vulnerability as well as strategies designed to protect the environment and to improve economic growth, levels of education, and living conditions of the entire population. Once the relevant authorities are unable and/or unwilling to fulfill their responsibilities, they are obliged to allow humanitarian organizations to provide humanitarian assistance and protection as stated in the Humanitarian Charter and Minimum Standards (Sphere Project, 2004).

Most scholars conclude that development is the solution in solving disaster problems and that the level of satisfaction in development and level of satisfaction to meet certain needs of disaster victims are related. The United Nations also looks at any development as an important objective in relief assistance. However, the reduction of disasters in any development comes along with both benefits and flaws. The benefit of any legitimate development might be seen as guided by international legislations. Due to the provision of development and reconstruction, compliance with the law is the only way to deliver better enforcement techniques. Unfortunately, uncertainty in an international legal context and lawmaking is also contributed to a dilemma in relief assistance especially to the framework convention on victims' rights to adequate housing.

TABLE 9.3B Guideline Reviews of OXFAM

Publisher	**OXFAM**		
Title	The Oxfam Handbook of Development and Relief, Vol II		
Date	1995		
Description	Although Oxfam have led the way in developing innovative shelter strategy over the last two decades, very little space is dedicated to this subject in their 1995 handbook It is worth noting, however, that other Oxfam pamphlets cover shelter kits in more detail elsewhere, for example Water, Sanitation and Shelter Packs, Oxfam Public Health Team, Oxfam 1996 This handbook again precedes much of the important recent work on shelter		
Common standards	N/A		
Shelter and settlement standard	N/A		
Key advise	• Site selection criteria • Provide public health specialists • Purchase locally or regionally • Choose low-cost measures • Roofing treated as crucial		
Policy	Need	Implementation	Response
	Assessing need for Settlements Housing Temporary shelter Food storage Housing reconstruction	Determining response Public infrastructure Do not undermine local capacity Future mitigation (sustainable, traditional techniques)	Via local NGOs Local/regional purchase Low-cost measures Safety Local investment
Process	Baseline data, impact of events, present situation, probable evolution, conclusions for action	N/A	Core dwellings initially phased implementation
Priorities	N/A	Site selection	N/A
Provision	N/A	Public Health Adviser Existing materials Makeshift shelters Roofing	Technical expertise Tents from stockpiles (advantages and disadvantages of each type) Training (building practice)

Source: Crawford (2002).

9.6 CHALLENGES OF POLICY IMPLEMENTATION

Decisions as regard to any policies is known as the "implementation" process (Pressman and Wildavsky, 1973). Pressman and Wildavsky (1973) introduced two perspectives as the sensible approach for an ideal administration and policy implementation. First, the manager and other superiors apply the "top-down perspective" that encourage them to implement policies rather than providing written plans. Simultaneously, the wave of enforcement should move backwards that involves the lowest level of organization known as "the bottom-up perspective" (Yates, 1977). This "street-level bureaucrats" approach requires more understanding to the real situation and which policy needs to be put in place. As suggested by Johnston (2004), the main aspect to measure the level of success in policy implementation must consider local control rather than hierarchy. Therefore, a number of policy involvements in political systems are apparently challenging. The multiplicities of rules that govern a regulatory framework consist of civil and criminal laws, regulatory statutes, and codes of conduct that administer the practicing bodies (Johnston, 2004). Foremost, regulators focus on the outcomes as a result of the accomplishment of regulatory aims. This is the type of governance concerning the culture of compliance with prescriptive rules. However, regulators must be encouraged to allow some flexibility in order to achieve greater outcomes.

Flexibility in regulatory practice creates a growing uncertainty regarding the state of designing or understanding policy implementation due to conflict in dealing with crime accusation, misconduct, and dishonesty (Johnston, 2004). The conflict is the subject of political issues and not an activity caused by technical measures. The issues are always in relation to responsibility and the risk to misconceptions of "political contests" (preference adaptation). Even the political pressure

TABLE 9.3C Guideline Reviews of UNHCR

Publisher	**UNHCR**		
Title	The UNHCR Handbook for Emergencies		
Date	1999		
Description	A standard text which sets out the policy and objectives of the UNHCR. It gives comprehensive advice but no prescribed technical solutions. The handbook annexes give more details on the materials and equipment available It refers the reader to specialist UN departments for expert advice		
Common standards	N/A		
Shelter and settlement standard	N/A		
Key advise	30–45 m^2 per person in a camp Maximum camp population 20,000 (90 ha) Standards for services and infrastructure Module descriptions 3.5 m^2 per person in tropical climate shelter 4.5–5.5 m^2 per person in cold climates or urban settings Handbook also gives detailed annexes on supplies and transport		
Policy	Need	Implementation	Response
	Long-term planning, bottom-up Three scenarios: dispersed settlement, mass shelter in public buildings, camps	Avoid high population density and large emergency settlements	Technical support Refugee participation Shelter
Process	Site assessment Needs assessment	Selection Planning, Shelter Prioritize, obtain maps and information, identify flaws, make estimates, assess different layouts	N/A
Priorities	Criteria for site selection: water supply, size, land use, land rights, security, protection, topography, drainage, soil conditions, accessibility, climate, health, vegetation	Contingency planning	Sanitation, water supply, roads, fire prevention, administrative, and communal services Shelter standards
Provision	Decentralized, community-based approach, bottom-up services, infrastructure, modular planning, environmental considerations, gender considerations	Information for planning Expertise and personnel	Different shelter types, plastic sheeting, tents, prefabricated shelters, cold climate shelters Reception and transit camps Public buildings and communal facilities

Source: Crawford (2002).

is a nature of any industry (Yates, 1977); it still depends on how well the organization encounters the issue of regulatory techniques in this political challenge.

Johnston (2004) found out that there are three options in order to come across the failures in administration:

1. Failure in implementation due to disruption during the process of implementation. At this stage, the failure occurs either intentionally or accidentally depending on the official's determination to comply with circulated instruction or law.
2. Failure at design stage that involves program formulation. Some programs are simply "crippled at birth" because the policy itself was not competent.
3. Stage of failure to achieve organizational target in implementing policies because in some cases organizational and regulatory aims are different. For example, different institutions carry different targets in humanitarian works (e.g. political interest, gender discrimination, and profit base) and deflect the main target in certain development projects.

Efforts in relief are more difficult compared to other normal planning (e.g. higher education, labors) because implementing disaster planning and recovery must consider immediate action, interaction, and coordination rather than planning a perfect design.

9.7 GOVERNMENT POLICY ON POSTDISASTER HOUSING PROVISION IN MALAYSIA

Moin (2007) noted that any physical development must comply with the MNSC Directive 20 alongside other national legal frameworks in development process such as:

1. National Land Code 1965 (KTN), Housing Development Act (Control & Licensing) 1966 [Act 118], Local Government Act 1976 [Act 171], Urban and Regional Planning Act 1976 [Act 172], and Road, Drainage & Building Act 1974 [Act 133].
2. National Housing Policy, 2011, in order to provide with adequate, accessible, high quality and affordable housing in order to increase higher standard of living and sustainable.
3. Other state land development requirements: Land Conservation Act; Environmental Quality Act 1974; Local Government Act 1976; Road, Drainage and Building Act; Occupational Safety and Health Act; Uniform Building By-Laws; Town and Country Planning Act 1976 (Act 172); Infectious Disease Act; Road Transportation Act; Fire and Safety Act, and other related acts.

Executive order in the MNSC Directive 20 by the Prime Minister is the standard operational procedure to comply with for all departments involved in disaster management. Even if the complete version of the MNSC Directive 20 is restricted, the contents circulated are clear to all departments in the Mechanism of Disaster Management in Malaysia. The MNSC Directive 20 specifies in writing what should be done when disaster strikes (Fig. 9.3), when to use certain clauses of it, and where responsibility lies. This directive includes objectives, scope of areas, stages of the process, responsibility, and review of implication at the end to make sure that the procedure continues to be useful, relevant, and up to date (Moin, 2007). The Malaysia National Security Council (MNSC) Directive 20 clearly stated guidelines on the management of disasters including the responsibilities and functions of various agencies within the scope of national and international legislations.

The reconstruction process, which has the following organizational set up (Diagram 9.1) giving an overview of the partners in it, depends on whether the project will be located in a new area or not and whether the beneficiaries want to implement official design or custom-made designs. It was found that the process was initiated and controlled by (1) the Ministry of Prime Minister Department (*Unit Pemulihan Pasca Banjir* and National Security Council as moderator) and (2) Public Works Department at the federal level (and working hand in hand with the state committee (1—state secretary and state director of economic planning unit; 2—land office representative, district officers, and related officers). Later, private firms, builders, other designers, and the beneficiaries were all participated in the project. At the state level, the private firms appointed, who designed houses, communicated only with the state committee (Diagram 9.2). The various partners in the house building process each have a specific role to play from federal to district level (Diagram 9.3).

9.7.1 Posttsunami Housing Reconstruction in Malaysia

Taking the opportunity to redevelop the affected area after the tsunami disaster, the Kedah State Government has drafted a proposal for a New Town Development Plan to relocate the scattered villages along the shore. A minimum of six villages along the coastal area were relocated to the new residential zones (Taman Permatang Katong). Meanwhile, in Penang, a minimum of four villages along the coastal area were relocated to the new residential zones (Rumah Pangsa Masjid Terapung).

As a result of redevelopment process and local considerations, two types of permanent housing for the tsunami victims adapted the concept of the "Rumah Mesra Rakyat" Housing Scheme, a low-cost flat (Rumah Pangsa Masjid Terapung as you can see in Fig. 9.4) and a low-cost bungalow (detached house) with an approximate total floor area of 100 m^2 (Taman Permatang Katong as you can see in Fig. 9.5). Each of these houses consisted of a living/dining, a kitchen, 3 bedrooms, and 2 bathrooms. Prior to the tsunami disaster, this special housing scheme was initially executed to help low-income families living in the suburban areas in Malaysia to secure decent housing with subsidies under the national budget. The construction of the RMR (People Friendly Home) Housing for the tsunami victims is fully financed by the Tsunami Fund and is solely developed by the SPNB (National Housing Corporation). Affected families who wanted access to these RMR houses had to make application to the respective state governments for both the land and the house. A loan repayment scheme with subsidies from the Tsunami Fund and the National Budget will be offered.

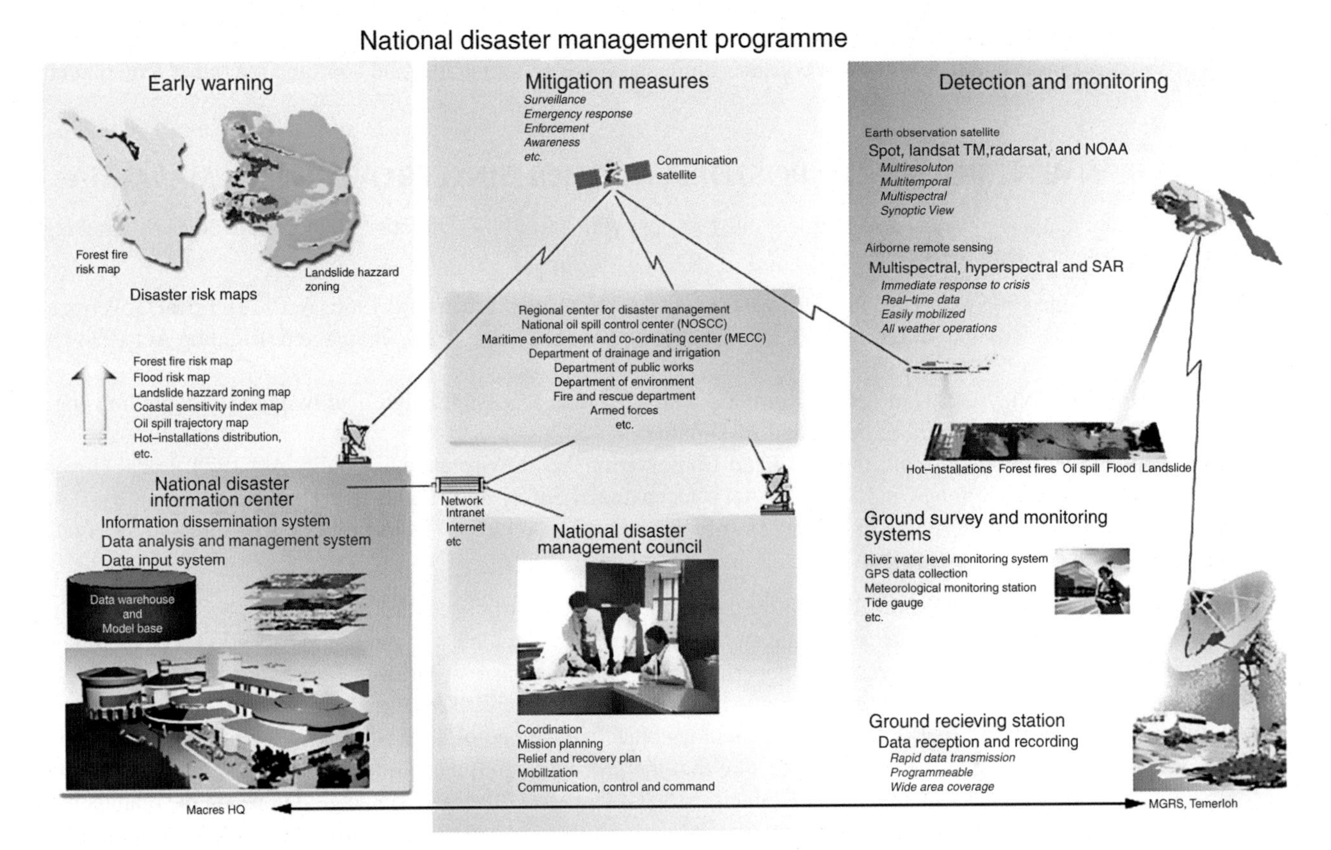

FIGURE 9.3 **Disaster management in Malaysia (Moin, 2007).**

9.8 THE NEED FOR NEW NATIONAL HOUSING DISASTER STRATEGIES

The international disaster community has long time ago agreed to implement the approach of "building back safer", instead of "building back better" that is emphasized even more strongly the notion of disaster preparedness for safe living in buildings which are resistant to future hazards (Kennedy et al., 2008). These reconstruction processes (Fig. 9.6) are an overtime processes and offer the opportunity of achieving sustainable development in the disaster management process (UN/OCHA, 2010). Increasing sustainability is achieved through improved predisaster conditions and reduced risk and vulnerability.

Disaster is a serious disruption of the functioning of society, causing widespread human, material, or environmental losses which exceed the ability of the affected people to cope using its own resources. Disasters are often classified according to their cause whether natural or man-made (DHA/IDNDR, 1992). A disaster is any occurrence that causes damage, ecological disruption, loss of human life, or deterioration of health and health services on a scale sufficient to warrant an extraordinary response from outside the affected community or area (World Health Organization, 1995).

In Malaysia, according to the MNSC Directive 20, for the purpose of this directive, disaster is defined as an incident which occur in a sudden manner and complex in its nature and that causes losses of lives, damages to property or natural environment, and bring a deep effect to local activities. Such an incident needs a management that involves extensive, resources, equipment, skills, and manpower from many agencies with an effective coordination, which is possibly demanding a complex action and would take a long time (NSC, 1997).

According to other international guidelines, authorities should mobilize the affected communities in the target locations to form housing reconstruction committees in order to conduct hazard mapping, identify safe relocation sites, and select families to receive housing assistance (Crawford, 2002). Later, the beneficiaries contracted their own construction company or workers through public bidding. These construction companies or workers have been identified and listed by the authorities. They received comprehensive training and details on construction of the houses designed previously suitable for disaster prone area.

One of the most crucial decisions to be made in postdisaster housing is whether to rebuild damaged houses in their existing locations or resettle disaster-affected families to new sites (Bastable, 2003). Key factors involved in the decision to resettle families in any reconstruction process are:

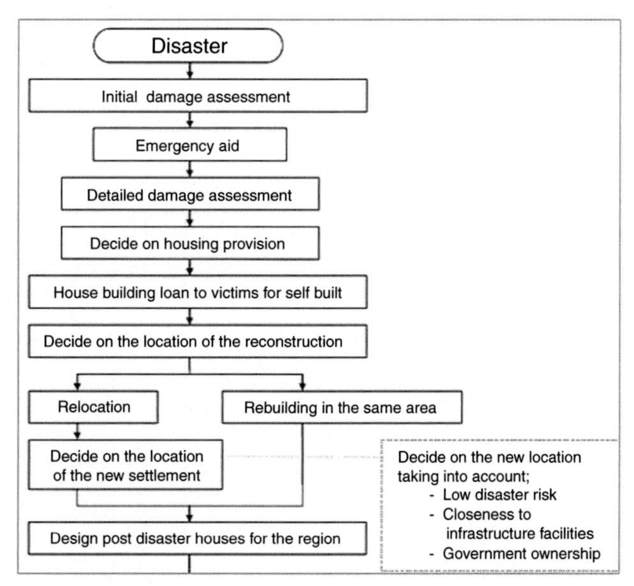

DIAGRAM 9.1 **Flow chart of the reconstruction project.**

1. concerns over safety and vulnerability of disaster-affected communities to future hazards;
2. efficient use of scant municipal resources in delivering land, building materials, utility connections, and other services to multiple affected families;
3. a high-profile way for local authorities to demonstrate their response to the disaster event, and
4. willingness of donors to finance new settlements in support of addressing local poverty and vulnerability.

The resources available to local authorities in disaster-prone areas is limited at the best of times and even more so in postdisaster circumstances. The outcome of this writing might give insights into designing and planning the national policy and disaster management framework by restructuring and reorganizing the present National Disaster Management Mechanism in terms of enhancing the coordination of responsibility between and within government bodies in the National Disaster Management Mechanism particularly as an instruction on the provision of housing (Fig. 9.7).

These guidelines provide guidance regarding the key aspects of sustainable reconstruction. Integrating sustainable building principles during reconstruction is key to the principle of "building back safer", that is, not only to improve resilience to natural hazards in the future, but also to ensure that the opportunity is secured to shift toward buildings and structures that are as energy-efficient, low greenhouse gas emitting and climate-mitigating as possible. Those reconstructing processes include choosing and obtaining building materials and technologies, achieving cost effectiveness and affordability, gaining access to information, using environmentally sound and energy-efficient building practices, and winning institutional and community acceptance.

A guideline is intended to be as comprehensive as possible, as it cannot be considered complete and does not represent a scientific study of sustainable reconstruction practices. Nor, on the other hand, does it provide readymade solutions for construction projects, each of which differs according to locations, budgets, and other conditions.

Although the focus of this guidelines is on housing, the reader will likely find some of the information applicable to other types of buildings (health facilities, schools, public buildings, etc.). With a more clear planning and well-planned housing delivery process, sustainable reconstruction is designed to provide an integrated framework for action by all parties involved.

9.9 LESSONS LEARNED

It is important to understand the interrelated nature of postdisaster issues that need to be adhered. Action on any front alone is not likely to work. Learning from the previous cases in Malaysia in relation to the international environment, the component of the following issues as suggested in the National Disaster Housing Strategy (FEMA, 2009) were identified as aspects of sustainable reconstruction in Malaysia as:

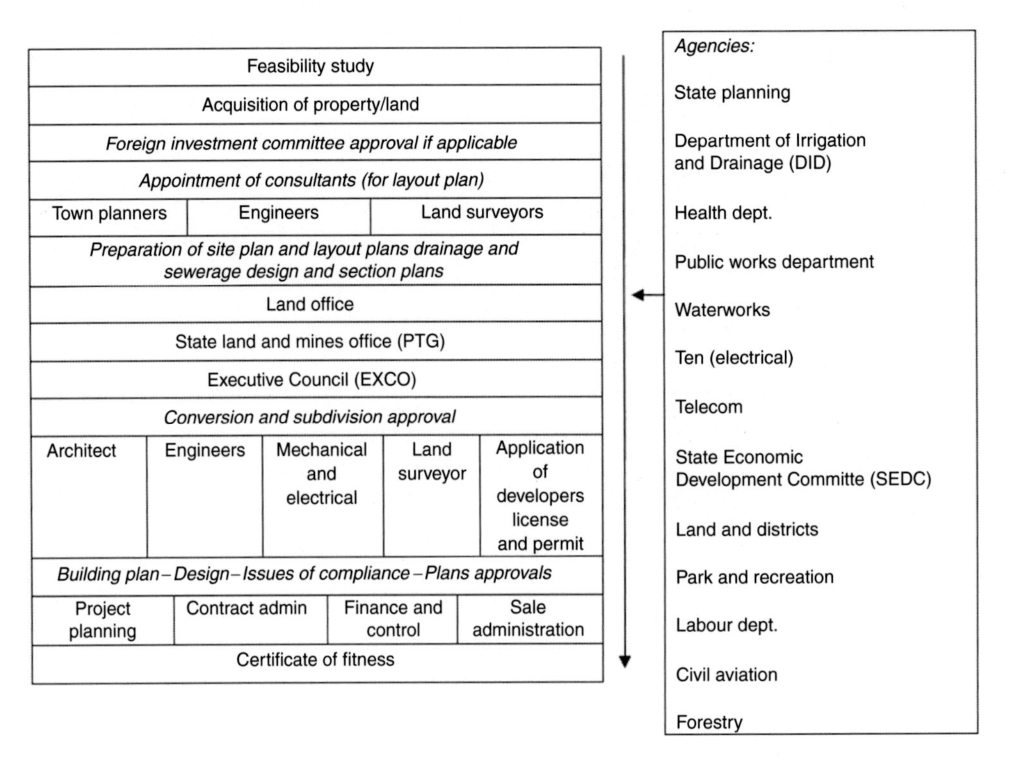

DIAGRAM 9.2 **Land development process in Malaysia (Tan, 2001).**

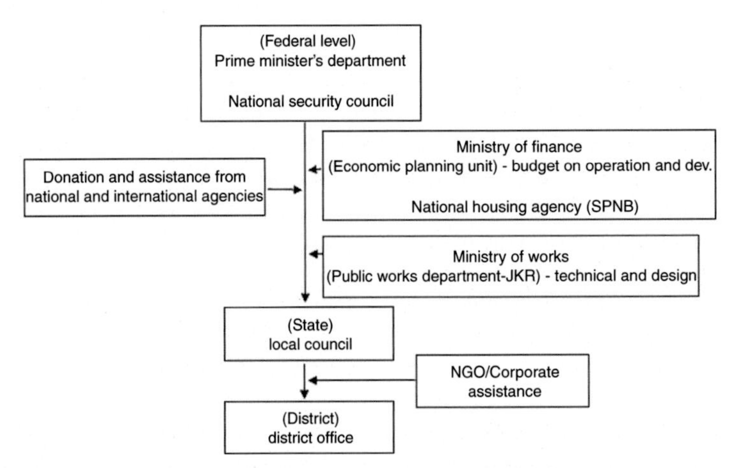

DIAGRAM 9.3 **The Tsunami Postdisaster housing provision in Penang and Kedah, Malaysia.**

1. Institutional:

   Difficulties in effective project management and monitoring were the most common problem areas where reconstruction housing work "gets stuck". It was also found that postdisaster housing work by the authorities is frequently criticized for not sticking within project time and funding constraints, and requiring high levels of management support to keep the project on track toward realizing objectives (input has reached the target group and the client and beneficiaries are satisfied with the outputs).

   Inadequate interaction among the organizations involved in the reconstruction project and the beneficiaries lead to a high level of delays. Postdisaster reconstruction works are initiated, controlled, and undertaken by three-tier government authorities. Officials of these authorities claim that there is no previous learning example/comprehensive research for the reconstruction projects; consequently, rehabilitation works are done without understanding user needs, the geography of the area, and readiness of land allocation. Sometimes, private firms are involved in a part of the project, for example, in the reconstruction project in some location, but detailed investigations are not undertaken by these firms either.

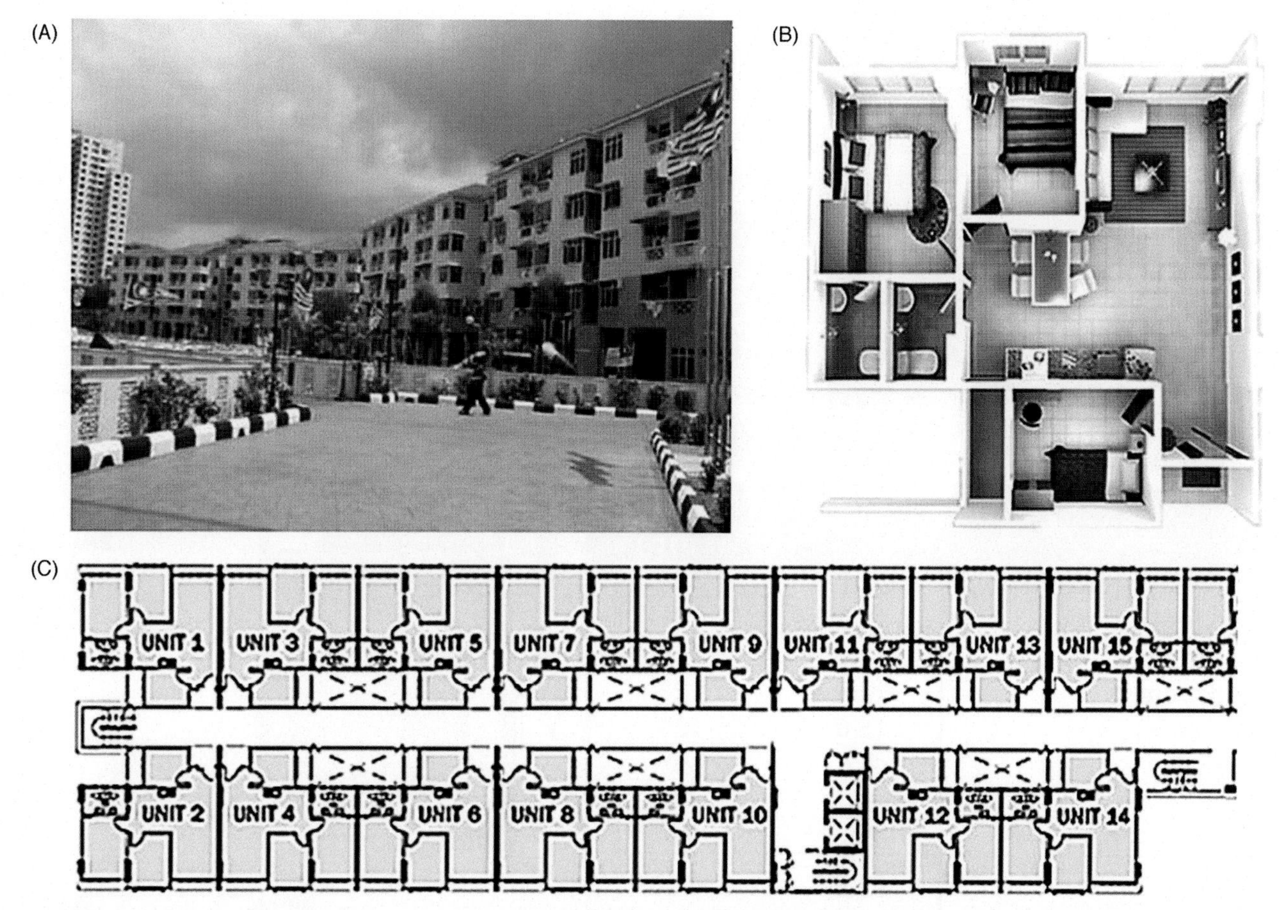

FIGURE 9.4 **New housing scheme for tsunami victims at Rumah Pangsa Masjid Terapung, Jalan Tanjung Bungah (Roosli and O'Brien, 2011).**

2. Technical:

   Technical issues are generally the most easily resolved of all "problem areas" in postdisaster housing and most typically involve local solutions. Other than offering a brief overview of structural and material considerations for disaster-resistant housing, guidelines available do focus on engineering aspects of housing construction. It provides housing designs or models that can be transplanted from JKR approved design to disaster context suitable for flood situation. Local designs and technical expertise has sought out the design issue when considering a shelter or housing intervention even though more studies must be done on postoccupancy impact on the community.

   Other than the design of a house, site selected should (1) be suitable for building housing; (2) not bear high disaster risk; (3) not be exposed to strong wind; (4) be easy to reach; (5) be big enough for facilities, such as mosque and school, required in a village; (6) be close to existing infrastructure facilities in order to make the construction cheaper and easier; and (7) be suitable for the animals owned by the villagers. In addition, the plot selected should be big enough to facilitate the necessary buildings for the house such as store, working area house, etc.

   There is a tendency after a disaster to focus on the tangible costs and to try to employ concrete solutions. Counting the houses destroyed and planning emergency shelter for that number of families may seem like a logical and measurable first step to take when facing a postdisaster situation. However, there are many more intangible issues closely linked to the provision of shelter that must be addressed as published in operational guidelines at international and national levels.

   Most published guidelines by operational organizations do not have a significant body of literature to refer to. The existing international and national shelter/housing policy is, with notable exceptions:

   a. the repetition of a few key guidelines;
   b. inconsistent in which parts of the sector it includes and which it does not, reflecting a lack of clarity in the engagement of organizations with the sector;

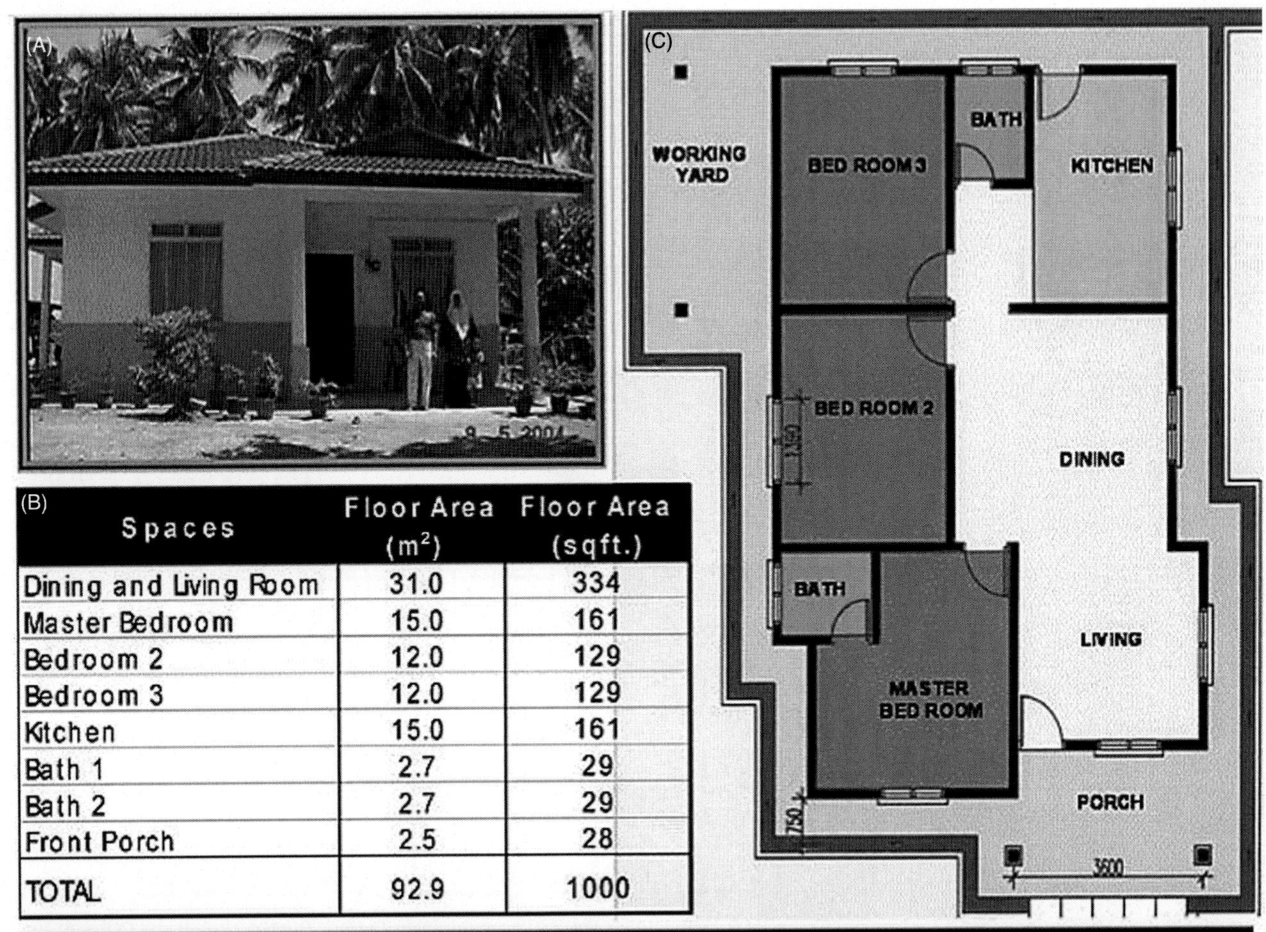

(B)

Spaces	Floor Area (m^2)	Floor Area (sqft.)
Dining and Living Room	31.0	334
Master Bedroom	15.0	161
Bedroom 2	12.0	129
Bedroom 3	12.0	129
Kitchen	15.0	161
Bath 1	2.7	29
Bath 2	2.7	29
Front Porch	2.5	28
TOTAL	92.9	1000

Housing Type (Built up area ~1000sqft.)	Housing Price (RM)	Subsidy by Govt (RM)	Repayment (RM)	Monthly Installment (RM)	Repayment period
On stilts	50,000	16,666	33,334	100	28 years
On ground	40,000	13,333	26,667	100	22 years

FIGURE 9.5 (A) Front view, (B) floor area of 700 sq ft, and (C) layout plan of the block at housing scheme for tsunami victims at Taman Permatang Katong in Kota Kuala Muda (Roosli and O'Brien, 2011).

c. inconsistent in its structure, reflecting the lack of a holistic understanding of the sector and of operational decision-making paths.

In Malaysia, although guidelines and advice are already available, housing sector does not have the same institutional support as other key parts of the emergency aid community such as emergency shelter, food, and cloth. However, attempts to improve the provision of housing in practice have been accompanied by an increasingly sophisticated technique of building construction such as Industrialized Building System and standard designs. This progress has driven investment and research in this building technology area.

3. Environmental and socioeconomic:

Housing is much a process as it is a result of the process. Therefore, one of the most crucial decisions to be made in postdisaster housing is whether to rebuild damaged houses in their existing locations or resettle disaster-affected families to new sites. Understanding of the surrounding areas of the affected area is crucial before any action can be taken. Efforts on evaluation works has to be done thoroughly within the availability of resources. One of the keys to success in housing interventions is ensuring that reconstruction supports community self-reliance in rebuilding after disasters. This involves "building up from the vernacular" by engaging as much as possible with the local ways and means in which houses are typically built. The current guidelines emphasize the importance of understanding and applying solutions based on the local housing context.

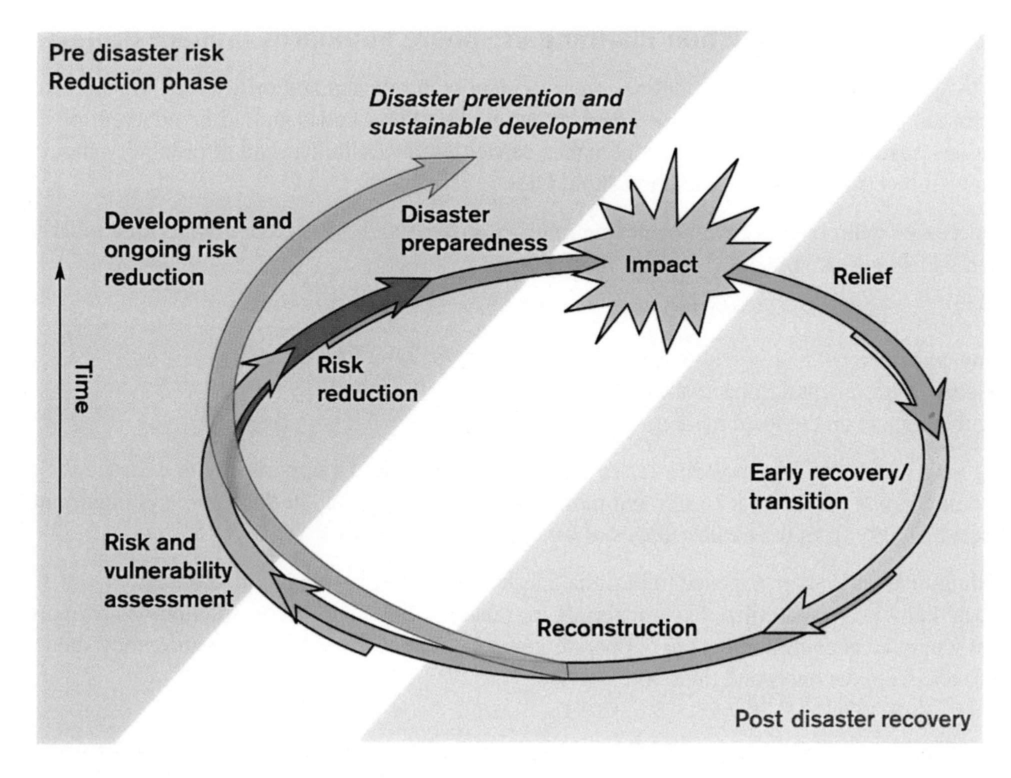

FIGURE 9.6 **Reconstruction phase (Lloyd-Jones, 2009).**

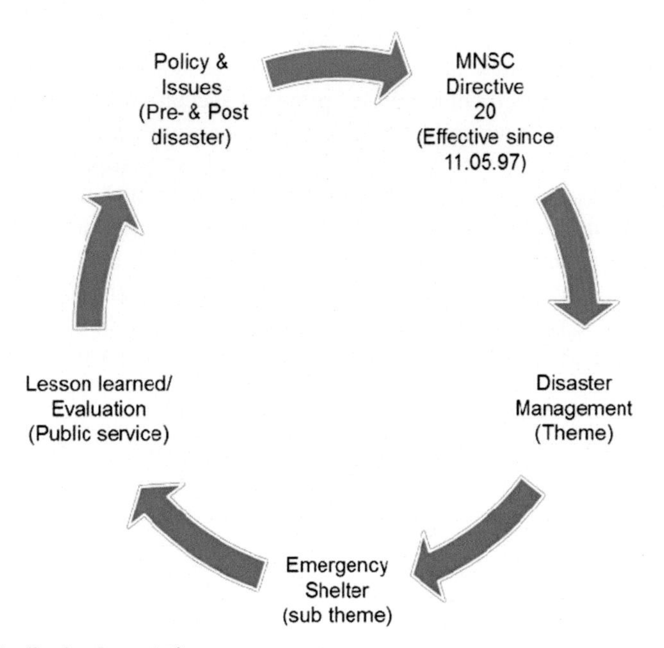

FIGURE 9.7 **Learning process of policy implementation.**

Addressing economic inequalities without widening access to political participation, or conversely, or attempting to "educate" people to change their views of identity with changing the underlying inequalities among groups. Good guideline builds on the first, as it is clear that all of these issues affect, and are affected by, one another. Therefore, a project cannot be designed in a vacuum. Rather, it must be built with an understanding of the entire situation and the different factors involved. The evaluation techniques employed should reflect this and assess both the broader relevance of a project as well as its specific impact.

9.9.1 Sustainable Reconstruction Planning in Sphere Standards (Sphere Project, 2011)

It is important to agree to standards and guidelines on settlement with relevant authorities/line ministries to ensure that key safety and performance requirements are met. Local or national building codes should be adhered to. The local culture, climatic conditions, resources, building and maintenance capacities, accessibility, and affordability should also be kept in mind. Sustainable projects upon their completion should not:

1. put an unnecessary drain on the government line ministries, local authorities, community that inherits the completed project from the perspective of:
   a. staffing levels
   b. maintenance
   c. operating costs
2. have a negative environmental impact, and
3. have a negative impact on the local market.

Developing new settlements in a postdisaster context provides us with an opportunity to enhance the quality of buildings, the environment, and to build back safer and more resilient communities. In this vein, it is important to integrate the principles of sustainability from the earliest stages of settlement in order to:

4. Avoid building in areas that are exposed to hazards, inefficient, and/or not maintainable.
5. Build on local knowledge and utilize local materials for rehabilitation and (re)construction where appropriate.
6. In the event where local communities are to operate/maintain the associated infrastructure, they should be involved in the project cycle from the onset and their voices heard.

9.9.2 Learning from Tohoku Japan Reconstruction Program Allocation (System of Special Zone for Reconstruction)

To accelerate reconstruction and stimulate investment in the affected regions, the Japan authority has established a system of Special Zones for Reconstruction, which offer deregulation and simplified statutory procedures, a variety of tax breaks and financial incentives, and new mechanisms to facilitate land-use restructuring. The system, taking into account requests from the disaster-afflicted communities, provides special arrangements for deregulation and reduced procedures. Assistance in terms of tax, fiscal, and financial arrangements will also be considered. They aim at promoting such measures as land-use restructuring through a unified contact point for multiple authorization processes and seeking a swifter completion of such processes. Furthermore, a legal framework will be introduced with the intention of prompting the introduction of necessary special measures and assistance in which consultation between the national government and disaster-afflicted local governments can take place, reflecting the progress made in developing plans for reconstruction at the regional level (Reconstruction Headquarters, 2011). In total, 52 plans for Special Zones for Reconstruction have been approved, which include those intended for the promotion of town-building through land-use restructuring, renewable energy initiatives to support regional development, and the development of a medical/commercial/industrial cluster.

Designing guidelines requires a holistic consideration as suggested by Doberstein and Stager (2013). Before beginning reconstruction, an attempt should be made to identify all conditions that led to reconstruction of settlement residents' vulnerability to disaster:

1. Conduct a postdisaster vulnerability assessment.
   Before beginning reconstruction, an attempt should be made to identify all conditions that led to reconstruction of settlement residents' vulnerability to disaster.
2. Aim to "build back better".
   With an understanding of the vulnerabilities of settlement residents, reconstruction should seek to reduce vulnerabilities wherever possible, thus creating a more resilient community and one that is aware of the hazards residents face in their daily life.
3. Clearly communicate rezoning decisions.
   Decisions on whether or not to allow reconstruction in hazardous locations should be made as quickly as possible, based on an assessment of ongoing hazard risks. All decisions must be communicated clearly, quickly, and in the correct sequence to surviving residents.

4. Practice community-based disaster management (CBDM) and involve the community in resettlement decisions.
   Settlement residents affected by disaster should help to assess community vulnerabilities and to design strategies and responses that decrease them. If the resettlement of settlement residents is unavoidable (owing to high potential for similar disasters in the future, for instance), every effort should be made to involve them in the selection of a safer relocation site and in housing design, as well as to remain as close as possible to the original community. Long-distance relocation of survivors should be avoided wherever possible, unless this is their collective wish.
5. Provide housing and community design assistance.
   Individual houses and communities should be designed as part of the reconstruction process to withstand better the natural hazards confronting the region. If surviving residents and indeed communities are expected to rebuild their own homes and communities from compensation, design assistance should be offered by government agencies or NGOs to establish capacity among affected communities to rebuild in a more resilient fashion. Homes should be built to the highest standard possible, and existing building codes respected.
6. Enhance tenure security.
   Postdisaster reconstruction should aim to provide secure housing and land title to settlement survivors, especially if reconstruction involves resettlement in new housing sites. If secure tenure did not exist prior to the disaster, further investigation into why these residents were drawn to the area, where they came from, and other factors should be conducted in order to supply adequate formal housing.
7. Ensure effective use of relief and reconstruction aid.
   International aid organizations (if necessary) typically offer significant funding for reconstruction and disaster risk reduction projects, but careful evaluation of the effectiveness of these projects is required to ensure that reconstruction does not re-create or even increase risks. An integrated approach, involving capacity-building projects, community education, early-warning systems, and housing reinforcement may prove to be much more successful in reducing vulnerability than protective structures alone. Both structural and nonstructural postdisaster actions should be taken in conjunction and should not be considered mutually exclusive response options. Aid donors should make countries likely to be affected by natural hazards aware of the disaster response and the risk reduction funding that is available and at the same time promote the message that postdisaster reconstruction should have "vulnerability reduction" as a core guiding principle.

9.9.3 Provisional Guidelines of Postdisaster Permanent Housing

Achievement of immediate objectives in planning for postdisaster program are: has the input reached the target group, and are the client and beneficiaries satisfied with the outputs. However, the success of any new settlement project/program is based on a well-functioning management process. The guidelines are organized according to the typical main steps of a reconstruction project as adapted from Sphere standards (Sphere Project, 2011) and the outcome of this writing. They focus on permanent housing. The main steps are outlined in Table 9.4.

The provisional guideline shows that land, employment, infrastructure, and access to the means of reconstruction as their key priorities. Thus, the needs identified by the affected groups point to more long-term strategies of rebuilding and of the creation of capacity for their involvement in the postdisaster rehabilitation. The involvement of local organizations and government, and the empowering of the affected communities, is much more effective than the transplanting of outside organizations to deal with the problems. This is also closely connected to the concern over the creation of dependency on the donors in the affected society. The building of capacity, through the involvement of the affected groups in their long-term rehabilitation, must be a guiding principle in any postdisaster reconstruction program. Past experience indicates several needs.

First, an institution must be established to collect, organize, and analyze the necessary sector by sector information, from profiling existing baseline to superimposing damage and losses with a unified and comparable approach. The team should be multidisciplinary and inter- institutional, with clearly designated focal points to compile and present the data in a comparable manner, so that they can be summarized and factored into a macroeconomic scenario exercise. Each focal point should have common terms of reference. The global analysts (e.g. macroeconomists, environmental economists, gender experts) will proceed to use the emerging data of damage and losses to: contrast the disaster scenarios to the nondisaster trend; make environmentally related damage and losses visible; and differentiate men and women's impact and roles in the postdisaster process.

Second, a deadline must be established to submit the final report deadlines for the submission of sector data (quantification in a standardized format with agreed common criteria) and accompanying descriptive text must be set depending on

TABLE 9.4 Provisional Guideline of Postdisaster Permanent Housing in Malaysia

N	Steps	Action
Preparation		
1.	Assessment	Select and carry out the necessary assessments: Environmental impact assessment (if necessary) Community assessment Damage assessment Mapping and geospatial information of Kelantan Any other
2.	Set-up	Decide on the objectives Identify the resources (funds, assets, human resources) Establish project structure (office, staff, etc.) Define decision-making, communication, and monitoring procedures
3.	Community	Define criteria for selection Identify and select a community
4.	Partners	Define the other partners Identify roles and responsibilities Make clear contracts
5.	Approach	Select appropriate approaches: Donor- or owner-driven; mixed approach Reconstruction at the same site or relocation
6.	Project definition	Develop a project document, based on environmental, technical, economic, social, and institutional conditions Develop management tools, including an action plan, time schedule, budget, risk strategy, and monitoring plan Clarify what rights the beneficiaries/users will have (to sell, rent) Integrate action plan and budget for training/instruction concerning future maintenance Undertake necessary tenders
7.	Site selection	Assess risks from natural hazards Carry out an environmental impact assessment Check land property ownership and the right to build Ensure location meets the users' needs Identify surrounding settlements/activities Assess stability of soil Assess access to water, sanitation, energy, transport
Planning phase		
8.	Disaster preparedness	Select appropriate measures for disaster preparedness: Vulnerability analysis PASSA—Participatory Approach to Safe Shelter Awareness Community-based risk assessment Contingency plans

9.	Site plan	Check existing master plans Indicate houses, access roads, infrastructure and services, green, recreational, commercial, and religious areas Indicate buffer zones to water catchment, agricultural land, sensitive natural areas Identify required disaster-preparedness measures Identify if any existing structures can be reused/integrated with the new buildings Ensure compliance with zoning and other regulations Explore expectations of future users Maintain social networks
10.	Building design	Select a house shape that suits the climate and culture and that is flood resistant Choose building designs and materials that are energy-efficient, environmentally appropriate, low-cost, and practical
		Select building components (supporting frame, foundation, floors, walls, roof) and technologies according to climate and ensure their resistance to natural hazards Make sure that materials used are environment-friendly, nontoxic, derived from sustainable sources, of good quality and socially accepted Consider reuse or recycling of building material and temporary shelters Design kitchens and stoves to ensure cultural acceptance, hygiene, smoke-less cooking and safety
11.	Infrastructure	Select an appropriate water supply and sanitation system Integrate a sustainable solid-waste-management system Select a sustainable power system that, to the extent that is possible, uses renewable energy sources Opt for access roads with adequate surface and space for extension—establish telecommunication connections
Implementation		
12.	Project management	Establish a team for the implementation Prepare a time line Make use of local sustainable construction technologies Produce the technical documentation Prepare a bill of quantities and detailed budget Undertake tenders Maintain safe, healthy, and socially just working conditions
13.	Quality control	Perform quality control of materials and works Use regular monitoring for control of use of materials, environmental impact, and workplace safety
14.	Environment-friendly site management	Ensure that construction waste is disposed of properly Store fuel and chemicals in contained areas to avoid leakage Minimize transport as far as possible
15.	Material banks	Establish a material bank to facilitate the provision of the needed materials, if appropriate Provide training for masons, laborers, engineers, as needed
16.	Controlled demolition	Consult with the local authority about demolition issues Establish a step-by-step activities plan Sign agreements with the relevant partners
17.	Reuse of debris	Review the local possibilities for reuse of debris Identify the kinds of debris materials which are appropriate for reuse
18.	Maintenance	Design the house for easy and self-evident care and maintenance Ensure all materials can be worked/repaired locally Fully test any and all systems (water, toilets, energy, waste disposal, cleaning, etc.) Provide a checklist of regular actions needed (cleaning of storm-water drains, vegetation control, pest control) Provide training/instructions for cleaning, small repairs, etc. to users and house owners

Source: Adapted from Crawford (2002), Barakat (2003), Bastable (2003), Corsellis et al. (2005), Schneider (2012), and Roosli et al. (2016).

the final deadline. The description will include not only narratives of the events impact on the sector, but also the criteria and assumptions made to establish damage and loss figures.

Third, a deadline must be set for completion of other relevant institutions, and this deadline must be discussed and made compatible with a strategic reconstruction proposal. Caveats as to the accuracy of available data, methodological considerations, and assumptions made must be specifically addressed. The timing for the assessment should be such that, without losing its timeliness, it does not interfere with the ongoing emergency, particularly the search and rescue, although thinking of the future in terms of the needs for the reconstruction process is an immediate task to be pursued since some actions are required to be undertaken promptly, especially those related to providing housing solutions, health and education services, and recovery strategy. Additionally, discussion of the future serves as a therapeutic measure to overcome trauma. The main concern is that it does not interfere with immediate life-saving activities and emergency relief operations.

Each sector team should consult and exchange information with each other to avoid duplication, share data of common interest or of interest in more than one sector, and identify information gaps or lack of information. The sector specialist will not only gather information on baselines and the disasters impact on them (i.e. damage and losses), but on reconstruction needs in the form of sectoral strategic responses. These can be used as inputs to develop an overall reconstruction strategy and possibly project proposals.

The strategic proposal will include a framework for action, based on preexisting policies or development strategies, focusing on adaptation of the latter to the needs for the reconstruction, prioritize and sequence the process, define resource gaps to be filled from government, private and external sources, and profile execution processes in which affected populations and other stakeholders can play key roles in reconstruction.

The function of this provisional guidelines is to provide decision-makers and stakeholders with a quantitative basis to request recovery funding assistance and to design a reconstruction strategy. The quantification, given its sector-by-sector nature, allows for concrete, specific proposals for action in sector or geographic terms. It is a tool for determining priorities (importance vs. urgency) and sequencing (timeline for reconstruction process), that is, to restore livelihood through income and employment, while physical reconstruction of housing, production, and infrastructure proceed.

9.10 CONCLUSION

Postdisaster housing reconstruction is definitely a process. This process is affected by legal, bureaucratic, and social factors as well as by economic and technical factors. Consequently, postdisaster dwelling is the product of this process of relations and it cannot be evaluated independently from this process. In order to comprehend the achievements or failures in a postdisaster housing reconstruction program, the actions in the predisaster, immediate relief, and rehabilitation periods should be appraised as well as the postdisaster dwelling itself. Postdisaster housing involves strategic and tactical decision-making, resembling procurement that is organizing programs of work, allocating resources, initiating and carrying out projects, and sharing responsibilities between the survivors and the experts. Experience shows that predisaster planning is usually inadequate and needs to be updated after the disaster in the light of actual vulnerabilities. No conventional procurement process is possible because there is no clear contracting client, the survivors have few resources and probably no "voice" in decision-making, and resources have to be shared among several options. The case of Malaysia is rather unique because the three-tier administration (level of administration) from federal, state to district requires higher level of commitment due to different political understanding. The tendency in Penang and Kedah was for government, donors, and the media to focus on the number of houses constructed as a measure of achievement. However, the delay in housing provision shows ineffectiveness with a lot of room for improvement is needed. This writing proposes that other organizations such as NGOs, universities, uniform bodies, and/or private firms can be involved in the earlier stage of reconstruction projects as the organizers and/or they can participate in the operations. For long term and permanent housing provision is more appropriate to be handled by governing authorities. In terms of implementations, it was observed that the main problem was the lack of satisfactory actions and policy framework in the predisaster phase. Therefore, although the actions in the postdisaster phases seem to be more satisfactory; the implementations following the flood can hardly be called a success.

REFERENCES

Aini, M.S., Laily, P., Sharifah, A.H., Zuroni, J., Norhasmah, S., 2006. Sustainability knowledge, attitude and practices of Malaysians. WIT Trans. Ecol. Environ. 93, 743–752.

Aini, M.S., Fakhru'l-Razi, A., Daud, M., Adam, M., Kadir, R.A., 2007. Malaysian socio-technical disaster model and operational guide. Available from: www.researchsea.com/html/article.php/aid/2032/cid/6/research/malaysian (accessed 02.03.17).

ALNAP, 2003. Humanitarian Action: Improving Monitoring to Enhance Accountability and Learning. ODI, London.

Arslan, H., Unlu, A., 2007. The evolution of community participation in housing reconstruction projects after Duzce earthquake, Istanbul Technical University. Available from: http://www.grif.umontreal.ca/pages/ARSLAN_%20Hakan.pdf (accessed 01.03.17).

Barakat, S., 2003. Housing reconstruction after conflict and disaster. Humanitarian Practice Network Paper. Publish-on-Demand Ltd., London No. 43.

Bastable, A., 2003. Guidelines for Post-disaster Housing. Oxfam, Oxford.

Burgess, R., 1978. Petty commodity housing or dweller control? A critique of John Turner's views on housing policy. World Dev. 6, 1105–1133.

Cernea, M.M., McDowell, C., 2000. Risks and Reconstruction: Experiences of Resettlers and Refugees. The International Bank for Reconstruction and Development, Washington, DC.

CGSS, 2013. Disaster Risk Management for Sustainable Development (DRM-SD): An Integrated Approach. Centre for Global Sustainability Studies, USM, Penang.

Christoplos, I., 2006. Tsunami recovery impact assessment and monitoring system (TRIAMS). Working Paper. Risk Reduction Indicators, ProVention Consortium.

Comerio, M.C., Landis, J.D., Firpo, C.J., Monzon, J.P., 1996. Residential earthquake recovery: improving California's post-disaster rebuilding policies and programmes. Calif. Policy Semin. 8 (7), 1–11.

Corsellis, T., Vitale, A., 2005. Transitional settlement displaced populations. Available from: www.shelterproject.org (accessed 01.12.16).

Crawford, K., 2002. Existing Guidelines Supporting the Shelter Sector. University of Cambridge, Cambridge.

Davis, I., 1981. Disasters and the Small Dwelling. Pergamon Press, Oxford.

Davis, I., 1987. Developments in the provision of culturally sensitive housing within Seismic Areas (1981–1986). In: Proceedings of Middle East and Mediterranean Regional Conference on Earthen and Low-Strength Masonry Buildings in Seismic Areas. Middle East Technical University, Ankara, pp. 107–115.

DHA/IDNDR, 1992. Internationally agreed glossary of basic terms related to Disaster Management. , United Nations, Geneva.

Doberstein, B., Stager, H., 2013. Towards guidelines for post-disaster vulnerability reduction in informal settlements. Disasters 37 (1), 28–47.

El-Masari, S., 1997. Learning from the people: a fieldwork approach in war-damaged villages in Lebanon. In: Awoyona, A. (Ed.), Reconstruction After Disaster: Issues and Practices. Ashgate, Aldershot, pp. 57–72.

FEMA, 2009. National disaster housing strategy, department of homeland security. Available from: https://www.fema.gov/pdf/emergency/disasterhousing/NDHS-core.pdf (accessed 12.12.16).

Foong, S.L., Shiozaki, Y., Horita, Y., 2006. Evaluation of the reconstruction plans for Tsunami disaster victims in Malaysia. J. Asian Archit. Build. Eng. 5 (2), 293–300.

IFRCRCS, 2012. Post-Disaster Settlement Planning Guidelines. International Federation of Red Cross and Red Crescent Societies, Geneva.

Imrie, B., 2007. The interrelationships between building regulations and Architects' practices. Environ. Plann. B Plann. Des. 34 (5), 925–943.

Ismail, A., 2003. Built environmental and tropical storm in Malaysia. The International Conference on Disaster Management: Disaster Risks and Vulnerabilities Reduction. School of Management, Universiti Utara Malaysia, July 4, 2003.

Jha, A.K., 2010. Safer Homes, Stronger Communities: Handbook for Reconstruction after Natural Disasters. The World Bank, Washington, DC. Available from: www.housingreconstruction.org/housing/home (accessed 12.03.17).

Johnson, C., 2002. What's the big deal about temporary housing? Planning considerations for temporary accommodation after disaster: example of the 1999 Turkish earthquakes. The TIEMS 2002 International Disaster Management Conference. University of Waterloo, May 14–17, 2002.

Johnston, R., 2004. Regulation: Enforcement and Compliance. Australian Institute of Criminology, Canberra.

Kennedy, J., Ashmore, J., Babister, E., Kelman, A., 2008. The meaning of build back better: evidence from post-Tsunami Aceh and Sri Lanka. J. Conting. Crisis Manag. 16 (1), 24–36.

Levine, J.N., Esnard, A.M., Sapat, A., 2007. Population displacement and housing dilemmas due to catastrophic disasters. J. Plann. Lit. 22 (1), 3–15.

Lizarralde, G., 2002. Multiplicity of choice and users' participation in post-disaster reconstruction: The case of the 1999 Colombian earthquake. Conference Proceedings Tiems. Facing the Realities of the New Millennium. Tiems, Waterloo.

Lloyd-Jones, T., 2009. The Built Environment Professions in Disaster Risk Reduction and Response—A Guide for Humanitarian Agencies. Max Lock Centre, University of Westminster, London, UK.

Mileti, D.S., 1999. Disasters by Design: A Reassessment of Natural Hazards in the United States. Joseph Henry Press, Washington, DC.

Minear, L., Weiss, T., 1995. Mercy Under Fire: War and the Global Humanitarian Community. Westview Press, Boulder, CO.

Moin, C., 2007. Disaster mitigation support and management in Malaysia, Prime Minister Department Malaysia, natural disasters: protecting vulnerable communities. IDNDR Proceedings of the Conference. London, October 13–15, 1993.

Morago, L.N., 2005. Forced displacement and human rights: the international legal framework applicable to refugees and internally displaced persons. Available from: http://reliefweb.int/report/world/forced-displacement-and-human-rights (accessed 15.12.16).

Mulwanda, M.P., 1992. Active participants or passive observers? Urban Stud. 29 (1), 89–97.

NSC, 1997. Principle and management mechanism of national disaster relief, Directive No. 20. National Security Council Malaysia, Putrajaya.

NSC, 2011. Standard Operational Procedure—Flood. National Security Council Malaysia, Putrajaya.

Pressman, J.L., Wildavsky, A., 1973. Implementation. University of California Press, Ltd., Los Angeles, CA.

Price, R., Bibee, A., Gonenc, R., Jacobs, S., Konvits, J., 2000. Turkey post-earthquake report. Report. OECD Secretariat, Paris.

ProVention Consortium, 2004. International cooperation on humanitarian assistance in the field of natural disasters, from relief to development. Report of the Secretary-General, Fifty-Ninth Session, Agenda Item 39 (a), A/59/374, Geneva.

Quarantelli, E.L., 1984. Organisational Behaviour in Disasters and Implications for Disaster Planning. National Emergency Training Center, Federal Emergency Management Agency (FEMA), Emmitsburg, MD.

Reconstruction Headquarters, 2011. Reconstruction Headquarters in Response to the Great East Japan Earthquake. Principal Agency of the Government of Japan.
RICS, 2009. The Built Environment Professions in Disaster Risk Reduction and Response A Guide for Humanitarian Agencies. University of Westminster, MLC Press, Westminster.
Roosli, R., Collins, A.E., 2016. Key lessons and guidelines for post-disaster permanent housing provision in Kelantan, Malaysia. Proc. Eng. 145, 1209–1217.
Roosli, R., O'Brien, G., 2011. Social learning in managing disasters in Malaysia. Int. J. Disaster Prev. Manag. 20 (4), 386–397.
Schneider, C., 2012. Sustainable reconstruction in disaster-affected countries practical guidelines. United Nations Environment Programme and Skat—Swiss Resource Centre and Consultancies for Development. Available from: http://staging.unep.org/sbci/pdfs/PracticalGuidelines (accessed 23.12.16).
Soti, B., Herard, D., 2012. Sheltering Literature Review and Annotated Bibliography. Florida International University. Available from: http://digitalcommons.fiu.edu/drr_shelter/ (accessed 21.01.17).
Spence, R., 2009. Estimating shaking-induced casualties and building damage for global earthquake events. Final technical report. Cambridge Architectural Research Ltd., Cambridge.
Sphere Project, 2004. Humanitarian Charter and Minimum Standards in Disaster Response, second ed. The Sphere Project, Geneva.
Sphere Project, 2011. Humanitarian charter and minimum standards in humanitarian response. Geneva. Available from: www.sphereproject.org/content/view/720/200/lang,english (accessed 11.12.17).
Tan, A.L., Project Management in Malaysia—A Comprehensive Approach for Successful Management of Property Development Projects from Inception Until Completion, Kuala Lumpur.
Turner, J.F.C., 1972. Housing as a verb. In: Turner, J., Fichter, R. (Eds.), Freedom to Build: Dweller Control of the Housing Process. Macmillan Company, New York, pp. 148–175.
Turner, J.F.C., 1977. Housing by People: Towards Autonomy in Building Environments. Pantheon Books, New York.
UNDRO, 1982. Shelter after disaster: guidelines for assistance. Office of the United Nations Disaster Relief Coordinator, United Nations, New York.
UN-HABITAT, 2001. Guidelines for the Evaluation of Post Disaster Programmes. UN-HABITAT, Nairobi, Kenya.
UN/OCHA, 2010. Literature Review for Shelter After Disaster. UN/OCHA, Geneva.
World Health Organization, 1995. The world health report 1995: bridging the gaps. World Health Forum 16 (4), 377–385.
Xiulan, Z., 2008. Lessons from Wenchuan Mega earthquake: a social development perspective. In: Proceedings of the International Day for Disaster Reduction. New York, October 13, 2009.
Yates, D., 1977. The Ungovernable City: The Politics of Urban Problems and Policy Making. MIT Press, Cambridge, MA.

FURTHER READING

CGSS, 2015. Draft of Resolution of Kelantan Flood Disaster Management Conference 2015. Solution Framework for Sustainable Development. Universiti Sains Malaysia Main Hall Kelantan, February 14–16, 2015.
Edmond, H.F., Leavy, M., 2012. National Disaster Housing Strategy and Implementation Plan. Nova Science Publisher, Hauppauge, NY.
Ishak, R., Azizi, A., Mohamed, N., 2004. Special report: disaster planning and management. Environmental health unit occupational and environmental health section disease control division. Minister of Health Malaysia, Putrajaya.
Pilon, P.J., 1999. Guidelines for Reducing Flood Losses. United Nation International Strategy for Disaster Reduction, Geneva.
Sauders, G., 2004. Dilemmas and challenges for the shelter sector: lessons learned from the sphere revision process. Overseas Development Institute (ODI) 28 (2), 160–175.

Chapter 10

Drought Prediction With Standardized Precipitation and Evapotranspiration Index and Support Vector Regression Models

Ravinesh C. Deo*, Sancho Salcedo-Sanz, Leopoldo Carro-Calvo**, Beatriz Saavedra-Moreno*,****

**University of Southern Queensland, Springfield, QLD, Australia; **University of Alcalá, Alcala de Henares, Spain*

10.1 INTRODUCTION

Drought is an insidious hazard characterized by below-average precipitation for a short, intense period or for a prolonged period that is sufficient to trigger significant environmental, societal, and economic damage. Developing the capacity for detection, monitoring, and forecasting of drought in near-real-time from weeks, months, to seasons can help mitigate its detrimental impacts. In order to predict future drought-related climatic parameters, two types of models are considered: physically based models, global circulation model (GCM) and the machine learning (ML) or data-driven (computational intelligence) models. GCMs explain the dynamics of physical components of the climate system and global temporal and spatial patterns. However, the underlying equations are complex and must be discretized in space and time. This leads to overparameterization of the variables, equifinality, and model uncertainties (Beven, 2006; Uhlenbrook et al., 1999). While GCMs provide reliable predictions fo r ancillary variables like temperature, less reliable information is obtained for variables that are crucial determinants of drought (e.g. rainfall) (Hudson et al., 2011; Kuligowski and Barros, 1998). By contrast, data-driven models use historical trends in climatic parameters or fluctuations in large-scale climate indices, sea surface temperatures, and other hydrometeorological variables to predict future drought. Such models employ ML algorithms to extract information from climatic parameters that model the linear and nonlinear relationship between predictors (inputs) and predictand (outputs) (Acharya et al., 2013; Deo and Sahin, 2014; Şahin, 2012; Şahin et al., 2014). However, there is emerging evidence that data-driven models can outperform the GCMs in certain spatial regions or hydroclimatic parameters of interest (Abbot and Marohasy, 2012, 2014).

The two most popular data-driven (or ML) models are the artificial neural network (ANN) and support vector regression (SVR) algorithms. Both relate to statistical learning theory which encompasses parameterized functions that are directly derived from the predictor data set. The model learns about relations between various inputs and the projected output, which is trained to generalize the output as an unseen variable related to the input data (Salcedo-Sanz et al., 2014, 2015). The first of these models, ANN is a computational paradigm that mimics the biological structure of the brain (McCulloch and Pitts, 1943). However, the ANN model suffers from difficulty in training the input data, nonoptimal solutions or the inability to produce a global or unique solution with various runs of the model due to differences in initial weights or optimization used (Coulibaly and Evora, 2007; Khan and Coulibaly, 2006). By contrast, the SVR model developed in the early 1990s (Vapnik and Vapnik, 1998) is an advanced ML technique based on the principles of structural risk minimization. This model minimizes the expected error that reduces problems associated with overfitting or parameterizations (Yu et al., 2006). The advantages of SVR models over the ANN are: greater ability to generalize the training data set, fewer parameters needed to design the model, lesser possibility of overfitting, and better performance (Hearst et al., 1998). Consequently, many studies have shown improved performance in problems related to classification (Mukherjee et al., 1997), regression (Dibike et al., 2001; Smola and Schölkopf, 2004), and time-series prediction (Müller et al., 1997, 1999; Tay and Cao, 2001; Thissen et al., 2003). A plethora of literature has applied SVR models in hydrologic studies such as runoff prediction (Asefa et al., 2006; Dibike et al., 2001; Yu and Liong, 2007), flood and streamflow forecasting (Li et al., 2010; Liong and Sivapragasam, 2002; Yu et al., 2006), groundwater monitoring or network designs (Asefa et al., 2004), lake water-level prediction (Asefa et al., 2005; Khan and Coulibaly, 2006), and downscaling of atmospheric parameters for prediction of streamflow (Perera et al., 2011).

Integrating Disaster Science and Management. http://dx.doi.org/10.1016/B978-0-12-812056-9.00010-5

Specifically in the area of climate sciences, the SVR algorithm has been applied in prediction of air temperatures, statistical downscaling of GCM (Radhika and Shashi, 2009; Tripathi et al., 2006), detection of climate change signals, prediction of short-term wind speed, and very short-term air temperatures and rainfall occurrences and amounts (Ortiz-García et al., 2009, 2012, 2014; Salcedo-Sanz et al., 2014, 2015; Sánchez-Monedero et al., 2014). However, for studies focusing on drought, the rainfall time-series has to be converted into a drought index (DI) that is employed as a standardized metric for detecting the onset, termination, severity, duration, or recurrence of drought events. From a practical viewpoint, the prediction of DI is appealing for operational use and decision-making in relation to drought-risk. However, the SVR algorithm has not been applied in prediction of DIs per se, or analysis of future drought in Australia.

The research presented in this work is motivated by a desire to explore the potential of SVR algorithm in predicting the standardized precipitation and evapotranspiration index (SPEI) for nine locations in Australia. While a recent study has applied ANN for predicting SPEI (Deo and Şahin, 2015), this research is the first to apply SVR, which is an improved ML tool based on its better universal approximation theorem. Our decision to develop models for SPEI prediction is based on the premise that drought is a multidimensional phenomenon with interactions between rainfall and other meteorological variables such as evapotranspiration or air temperatures affect the severity of dry conditions (Qian et al., 2011; Song et al., 2014) combined rainfall and evapotranspiration into a composite index (CI) to provide a better assessment of drought compared to the indices solely based on rainfall. In our study, we use the SPEI, which can measure drought on multiple timescales (e.g. 1, 3, and 12 months) and is able to be applied in climatically diverse regions (Vicente-Serrano et al., 2010a, b, 2012a). Unlike the standardized precipitation index (SPI) and the rainfall-decile drought index which are solely based on rainfall (Deo et al., 2009; McKee et al., 1993; Mpelasoka et al., 2008), the SPEI considers combined impacts of rainfall, temperature, and evapotranspiration on the evolution of drought (Beguería et al., 2013; Vicente-Serrano et al., 2012a). Consequently, the SPEI has been applied for drought studies in Europe, China, and the Czech Republic (Beguería et al., 2013; Li et al., 2012; Potop et al., 2012; Vicente-Serrano et al., 2010a,b, 2012a; Yu et al., 2014), including recent works in Australia using ANN models (Dayal et al., 2016a; Deo and Şahin, 2015; Deo et al., 2017), the application of SVR for SPEI-based drought modeling in Australia is a new research step.

The work described here uses the SVR algorithm (Chang and Lin, 2011) for prediction of the monthly SPEI using 12 regressor variables (rainfall, mean, minimum, and maximum air temperatures, and evapotranspiration rates deduced using the Thornthwaite method), the synoptic-scale climate indices related to rainfall variability in the Pacific and Indian Ocean regions (southern oscillation index (SOI), Pacific decadal oscillation (PDO), Indian Ocean dipole (IOD), southern annular mode (SAM)), and sea surface temperatures (Niño 3, 3.4, and 4.0 SSTs), which have previously been used for rainfall predictions (Abbot and Marohasy, 2012, 2014; Deo and Sahin, 2014; Mekanik et al., 2013; Saji et al., 2005). For validating of the model predictions, we compare the mean absolute error (MAE), root-mean square error (RMSE), coefficients of determination including the Nash–Sutcliffe coefficient (Nash and Sutcliffe, 1970), and the Wilmot's index of agreement (Willmott, 1982) calculated for the simulated and observed SPEI together with drought properties in the test period.

The purpose of this investigation is threefold: The first objective consists of developing and applying the SVR model to predict SPEI for the period 1915–2012. A second objective is to conduct a deeper statistical analysis of SVR model performance using key statistical measures. A third objective is to examine the prediction error yield of drought properties in the test period, which also corresponded to the Millennium drought.

The rest of the work is structured as follows: Section 10.2 describes the theoretical basis of the SVR algorithm and the DI (SPEI). In Section 10.3, we detail the hydrometeorological and climate data sets used including the methodology for statistical analysis of model performance assessment. In Section 10.4, we present and discuss the results of our work and in Section 10.5, we arrive at the concluding statements for closing the work.

10.2 THEORETICAL FRAMEWORK

10.2.1 Standardized Precipitation and Evapotranspiration Index

To develop SVR models for drought prediction, we first calculate the SPEI using the freely available R program (Beguería and Vicente-Serrano, 2013) which can be downloaded from http://cran.r-project.org/web/packages/SPEI/index.html. Here, we present a brief description of the computational approach of the DI but for a complete theory and its comparison with other indices, the readers should refer to works of Vicente-Serrano et al. (2010a,b, 2011a,b, 2012a,b). For all stations under consideration, the potential evapotranspiration (PET), which estimates the amount of evaporation and transpiration that is expected to occur if sufficient water was available, was determined using rainfall and temperature data.

In the literature, there are three common methods (Thornthwaite, Hargreaves, and the Penman–Monteith) for calculating PET (Allen et al., 1994; Hargreaves, 1994; Thornthwaite, 1948). The Hargreaves method requires mean solar radiation but if such data are not available, they can be estimated from the station latitude and month of the year. On the other hand, the Penman–Monteith method is the most data-expensive technique as it requires incoming solar radiation or alternatively the percent cloud cover, saturation water pressures, dew point temperature or humidity, and atmospheric surface pressure in order to determine a psychometric constant for the PET (Vicente-Serrano et al., 2010a, b, 2012a). In this study, we use the Thornthwaite method which is the simplest of all and the least data-intensive approach. When only the monthly data is available, as it is in our study, the Thornthwaite (1948) method is used:

$$PET = 16K\left(\frac{10T}{I}\right)^{m} \tag{10.1}$$

where T is the monthly mean temperature (°C), I is the heat index calculated as the sum of 12 monthly index values i, the latter being derived from mean monthly temperature:

$$i = 16K\left(\frac{T}{5}\right)^{1.514} \tag{10.2}$$

and m is deduced empirically ($m = 6.75 \times 10^{-5}I^3 + 7.75 \times 10^{-7}I^2 + 1.79 \times 10^{-2}I + 0.492$), K is a correction coefficient computed as a function of the latitude and month:

$$K = \left(\frac{N}{12}\right)\left(\frac{NDM}{30}\right) \tag{10.3}$$

NDM is the sum of days of the month and N is the maximum number of sun hours calculated using

$$\mathrm{N} = \left(\frac{24}{\pi}\right)\mathrm{ws} \tag{10.4}$$

and ω_S = hourly angle of sun rising ($\omega_S = \arccos(-\tan\varphi \tan\delta)$), φ = latitude in radians, $\delta = 0.4093sen\left(\frac{2\pi J}{365} - 1.405\right)$ is the solar declination in radians and J is the average Julian day of the month.

The next step is to determine the surplus/deficit of water ($D_i = PCN_i - PET_i$) as the difference between precipitation (PCN) and PET. D_i values are aggregated at different timescales, which is similar to the procedure of the SPI. The difference $D_{k\,i,j}$ in a given month j and year i depends on the chosen timescale k. For example, the accumulated difference for 1 month in a particular year i with a 12-month timescale is

$$X_{i,j}^{k} = \sum_{l=13-k+j}^{12} D_{i-1,l} + \sum_{l=1}^{j} D_{i,l} \quad \text{if } j < k \tag{10.5}$$

$$X_{i,j}^{k} = \sum_{i=j-k+1}^{j} D_{i,l} \quad \text{if } \; j \geq k \tag{10.6}$$

where $D_{i,l}$ is the PCN–PET difference in the first month of year i (in mm).

Unlike the SPI, where probability distribution of gamma family (two-parameter gamma or three-parameter Pearson III) is used, SPEI utilizes a three-parameter distribution. Out of common distributions (Pearson III, log-normal, and general extreme value), the log–logistic distribution, $f(x)$, is the most appropriate for standardizing the D series (Vicente-Serrano et al., 2010a):

$$f(x) = \frac{\beta}{\alpha}\left(\frac{x-\gamma}{\alpha}\right)^{\beta-1}\left[1+\left(\frac{x-\gamma}{\alpha}\right)^{\beta}\right]^{-2} \tag{10.7}$$

where α, β, and $\gamma$are the scale, shape, and origin parameters, respectively, for D values in the range ($\gamma > D < \infty$), calculated using the L-moment approach (Ahmad et al., 1988; Singh et al., 1993), where

$$\beta = \frac{2w_1 - w_0}{6w_1 - w_0 - 6w_2}, \quad \alpha = \frac{(w_0 - 2w_1)\beta}{\Gamma(1+1/\beta)\Gamma(1-1/\beta)}, \quad \gamma = w_0 - \alpha\Gamma\left(\frac{1+1}{\beta}\right)\Gamma\left(\frac{1-1}{\beta}\right) \tag{10.8}$$

and $T(\beta)$ is the gamma function of β and probability-weighted moments are given by

$$w_s = \frac{1}{N}\sum_{i=1}^{N}(1-Fi)^s D_i \tag{10.9}$$

where F_i is the frequency estimator calculated following the approach of Hosking (1990):

$$F_i = \frac{i-0.35}{N} \tag{10.10}$$

where i is the range of observations arranged in increasing order and N is the number of data points.

After the parameters of log–logistic distribution are determined, the probability distribution function of D series is calculated as

$$F(x) = \left[1+\left(\frac{\alpha}{x-\gamma}\right)^{\beta}\right]^{-1} \tag{10.11}$$

Finally, the distribution $F(x)$ is used to calculate the SPEI following the classical approximation of Abramowitz and Stegun (1972):

$$SPEI = W - \frac{C_0 + C_1 W + C_2 W^2}{1 + d_1 W + d_2 W^2 + d_3 W^3} \tag{10.12}$$

where $W = \sqrt{-2\ln(P)}$ for $P \leq 0.5$ and P is the probability of exceeding a determined D value, $P = 1 - F(x)$ and $C_0 = 2.515517$, $C_1 = 0.802853$, $C_2 = 0.010328$, $d_1 = 1.432788$, $d_2 = 0.189269$, and $d_3 = 0.001308$. Note that if $P > 0.5$, then it is replaced by $1 - P$ and the sign of the SPEI is reversed. For drought analysis, the grading is similar to the SPI with drought categories as: extreme (SPEI ≤ -2.0), $-2.0 \leq$ SPEI < -1.5 (severe), and $-1.5 \leq$ SPEI < -1.0 (moderate).

SPEI is a time-varying signal where significantly low (high) rainfall will result in negative (positive) values of the DI. In Fig. 10.1, we demonstrate the practicality of the SPEI for detecting the drought onset and termination dates and quantification of its properties. Here, the SPEI and the corresponding monthly precipitation signal for Bathurst (located in the New South Wales) for the severe drought event of 1982–83. Following the run-sum approach of Yevjevich (1967), the onset month of drought (i.e. t_{onset}) was identified for a monthly SPEI was negative (i.e. rainfall conditions were lower rainfall than the normal) and termination month (t_{end}) for the monthly SPEI was in positive again (i.e. rainfall conditions were higher rainfall than the normal).

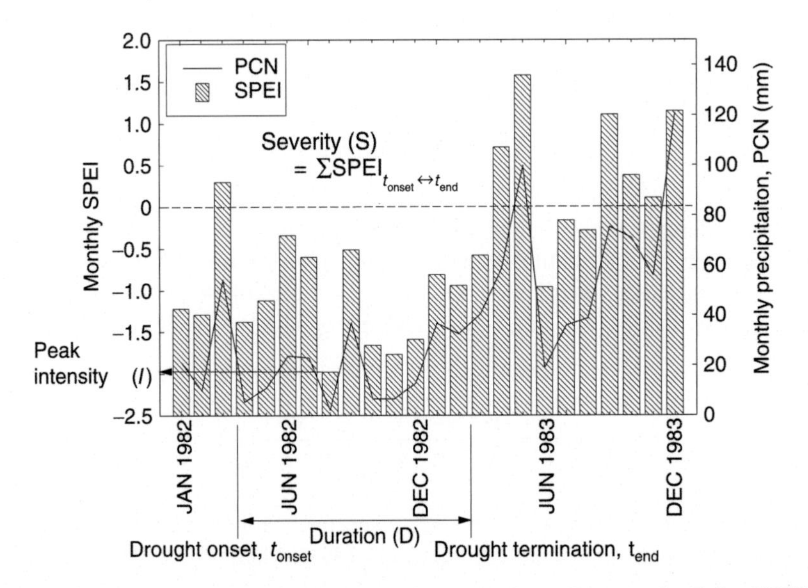

FIGURE 10.1 **Definition of drought properties by monthly Standardized Precipitation and Evaporation Index (SPEI) compared with the corresponding precipitation data (PCN) for Bathurst, New South Wales (station ID 63005).** Note that this data is for the severe drought period of 1982/83, the onset month is identified when the value of the is less than 0, drought severity is the sum of consecutive months with negative SPEI and the duration of drought is the time span over the range *t*onset and *t*end.

Accordingly, the severity (S) was the accumulated negative SPEI, peak drought intensity (I) as the minimum of the SPEI between t_{onset} and t_{ends}, and the drought duration was the sum of all consecutive months between t_{onset} and t_{end}. For Bathurst, the first segment of drought commenced in April 1982 and was terminated in February 1983. The monthly precipitation fluctuated between 6.2 and 99.8 mm, which was reflected in the corresponding SPEI which acquired values between −1.98 and 1.58, respectively. Accordingly, the peak intensity of drought recorded between t_{onset} and t_{end} attained a minimum SPEI of −1.98 (in August 1983) with very low precipitation (6.2 mm). Accordingly, the drought duration was 11 months and severity (accumulated SPEI) was −13.28.

10.2.2 Support Vector Regression

For predicting SPEI, we used C implementation of the SVR model, as described in Chang and Lin (2011) (http://www.csie.ntu.edu.tw/cjlin/libsvm/). SVR is an appealing algorithm for a large variety of regression-based problems since it does not only take into account the error approximation to the input data, but also the generalization of the prediction model. That is, its capability to improve the prediction of the model when a new data set is evaluated by it. SVR firmly grounded in the framework of statistical learning (or VC) theory, which has been developed over the last three decades by earlier works of Vapnik (2000) and Vapnik and Kotz (2006). The VC theory characterizes properties of learning machines which enable them to generalize well to the unseen data. Although there are several versions of SVR model, the classical model, ε-SVR, described in detail in Smola and Schölkopf (2004) used in a number of applications in Science and Engineering (Salcedo-Sanz et al., 2014, 2015) has been considered in this research work (Fig. 10.2).

Given a set of training vectors based on predictor data set denoted by $\Gamma = \{(x_i, y_i), i = 1, \ldots, l\}$, the ε-SVR algorithm for regression problem aims to train a prediction model of the form $y(x) = f(x) + b = w^T\phi(x) + b$, whereby the goal is to minimize a general risk function:

$$R[f] = \frac{1}{2}\|w\|^2 + C\sum_{i=1}^{l} L(y_i, f(x)) \tag{10.13}$$

where w controls the smoothness of the model, $\phi(x)$ is a function of projection of the input space to the feature space, b is a parameter of bias, x_i is a feature vector of the input space with dimension N, y_i is the output value to be estimated, and $L(y_i, f(x))$ is the loss function selected. In this work, we utilize the L1-SVRr (L1 SVR model), which is characterized by an ε-insensitive loss function (Smola and Schölkopf, 2004):

$$L(y_i, f(x)) = |yi - f(x_i)|_\varepsilon \tag{10.14}$$

In order to train this SVR model, it is necessary to solve the following optimization problem (Smola and Schölkopf, 2004):

$$\min\left(\frac{1}{2}\|w\|^2 + C\sum_{i=1}^{l}(\zeta_i + \zeta_i^*)\right) \tag{10.15}$$

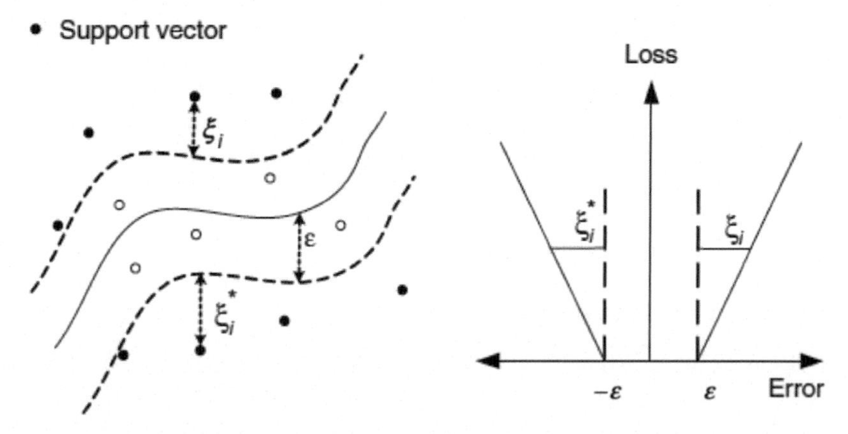

FIGURE 10.2 The concept of support vector regression (SVR) and the ε-insensitive loss function.

subject to

$$y_i - w^T\phi(x_i) - b \leq \varepsilon + \zeta_i,\ i = 1,\ldots,l \quad (10.16)$$

$$-y_i + w^T\phi(x_i) + b \leq \varepsilon + \zeta_i^*,\ i = 1,\ldots,l \quad (10.17)$$

$$\zeta_i, \zeta_i^*,\ i = 1,\ldots,l \quad (10.18)$$

The objective function given in Eqs. (10.15)–(10.18) minimizes the complexity (i.e. the magnitude of w) of the monthly SPEI estimator (i.e. the estimator will tend to be flat if no other considerations are imposed), leading to regularization of the solution, and penalizes errors in estimation that lie outside an ε-tube (goodness of fit) (Li et al., 2010). In other words, for any (absolute) error smaller than $\varepsilon, \xi i = \xi_i^* = 0$. The constant $C > 0$ trades off the importance between the complexity of f and the amount to which deviations larger than ε are tolerated. The dual form of this optimization problem is usually obtained through the minimization of the Lagrange function, constructed from the objective function and the problem constraints. In this case, the dual form of the optimization problem is the following:

$$\max\left(-\frac{1}{2}\sum_{i,j=1}^{l}(\alpha_i + \alpha_i^*)(\alpha_j + \alpha_j^*)K(x_i, x_j) - \varepsilon\sum_{i=1}^{l}(\alpha_i + \alpha_i^*) + \sum_{i=1}^{l} y_i(\alpha_i - \alpha_i^*)\right) \quad (10.19)$$

$$\sum_{i=1}^{l}(\alpha_i - \alpha_i^*) = 0 \quad (10.20)$$

$$\alpha_i, \alpha_i^* \in [0, C] \quad (10.21)$$

In addition to these constraints, the Karush–Kuhn–Tucker conditions must be fulfilled, and also the bias variable, b, must be obtained. The interested reader can consult the extensive work of Smola and Schölkopf (2004) for reference. In the dual formulation of the problem, the function $K(x_i, x_j)$ is the kernel matrix, which is formed by the evaluation of a kernel function, equivalent to the scalar product $(\phi(x_i), \phi(x_j))$. A usual election for this kernel function is a Gaussian of the form

$$K[x_i, x_j] = \exp\left(-\gamma.\|x_i - x_j\|^2\right) \quad (10.22)$$

The final form of the function $f(x)$ depends on the Lagrange multipliers α_i, α_j as follows:

$$f(x) = \sum_{i=1}^{l}(\alpha_i - \alpha_j^*)K(x_i, x) \quad (10.23)$$

In this way, it is possible to obtain an SVR model by means of the training of a quadratic problem for a given hyperparameters C, ϵ and γ. However, obtaining these parameters is not a simple procedure, the implementation of search algorithms is being necessary to obtain the optimal ones or the estimation of them (Ortiz-García et al., 2009). In this case, a grid search has been implemented in order to obtain the optimal parameters of the SVR. Moreover, according to Thissen et al. (2003), the SVM algorithm is characterized by: (1) a global optimal solution which will be found, (2) the result is a general solution avoiding overtraining, (3) the solution is sparse and only a limited set of training points contribute to this solution, and (4) nonlinear solutions can be calculated efficiently because of the usage of inner products.

10.3 MATERIALS AND METHODOLOGY

10.3.1 Study Area and SVR Input Data

For prediction of the SPEI using the ε-SVR algorithm, we utilize 12 regressors that described the geographic and climatic attributes of the study region. Tables 10.1 and 10.2 show the geographic and climatic properties and Fig. 10.3 plots a spatial map. Based on the climatological statistics averaged over 1915–2012, the rainfall, temperature, and evapotranspiration are found to be discernible for each station. The driest and the wettest stations have a mean rainfall of 164.7 mm year^{-1} (Maree)

TABLE 10.1 The Geographical Details and Climatic Characteristics of Data Stations in This Study

				Annual climatological statistics (Rainfall, PCN mm year^{-1}; Temperature, T °C; Evapotranspiration, ET mm)						
Station name	**ID**	**Latitude–longitude**	**Eleva-tion (m)**	**Mean PCN**	**Min PCN**	**Max PCN**	**Mean Temp**	**Min Temp**	**Max Temp**	**Mean ET**
Bathurst (NSW)	63005	149.56°E -33.43°S	713.0	646.5	214.2	1028.5	13.2	6.2	20.1	52.4
Moruya (NSW)	69018	150.15°E -35.91°S	150.2	975.1	472.0	1822.2	15.7	11.3	20.1	58.7
Denil-iquin (NSW)	74128	144.95°E -35.56°S	94.0	402.60	167.7	803.9	15.8	8.8	22.7	58.3
Yamba (NSW)	58012	153.36°E -29.43°S	153.4	1483.2	679.0	2716.8	19.61	15.44	23.77	72.7
Gabo island (VIC)	84016	149.91°E -37.57°S	149.9	923.5	562.9	1537.0	15.0	12.0	18.0	57.3
Wilsons (VIC)	85096	146.42°E -39.13°S	146.4	1061.5	659.1	1490.4	13.9	11.7	16.1	54.8
Palm-erville (QLD)	28004	144.08°E –16.00°S	144.1	697.6	360.6	1028.3	16.9	18.5	20.4	120.1
Merre-din (WA)	10092	118.28°E -31.48°S	315.0	332.8	177.1	565.0	22.8	10.9	24.8	66.3
Marree (SA)	17031	29.65°S –138.06°E	50.0	164.70	39.3	408.7	13.5	11.8	14.7	408.7

TABLE 10.2 The Regressors (Input Parameters) Used in the SVR Models

Meteorological parameters	
Monthly mean precipitation (mm)	PCN
Monthly mean air temperature (°C)	T_{mean}
Monthly maximum air temperature (°C)	T_{max}
Monthly minimum air temperature (°C)	T_{min}
Evapotranspiration (computed by the Thornthwaite method) (mm)	PET
Climate mode indices	
Southern oscillation index	SOI
Pacific decadal oscillation	PDO
Southern annular mode	SAM
Indian Ocean Dipole	IOD
Sea surface temperatures (SST)	
Niño 3.4 SST (5°N–5°S,170°W–120°W)	N3.4
Niño 4.0 SST (5°N–5°S, 160°E–150°W)	N4.0
Niño 3.0 SST (5°N–5°S,150°W–90°W)	N3.0

and 1483.2 mm year^{-1} (Yamba), respectively. The averaged minimum PCN for Maree is approximately 39.3 mm year^{-1} and that for Yamba is 679.0 mm year^{-1}, whereas the averaged maximum PCN is 408.7 mm year^{-1} and 2716.8 mm year^{-1}, respectively. In terms of the mean temperatures, the generally warmer stations are Merredin (max temp. 24.8 °C), Yamba (23.8 °C), Deniliquin (22.7 °C) with Bathurst, Moruya, and Palmerville exhibiting average temperature of 20 °C. It is

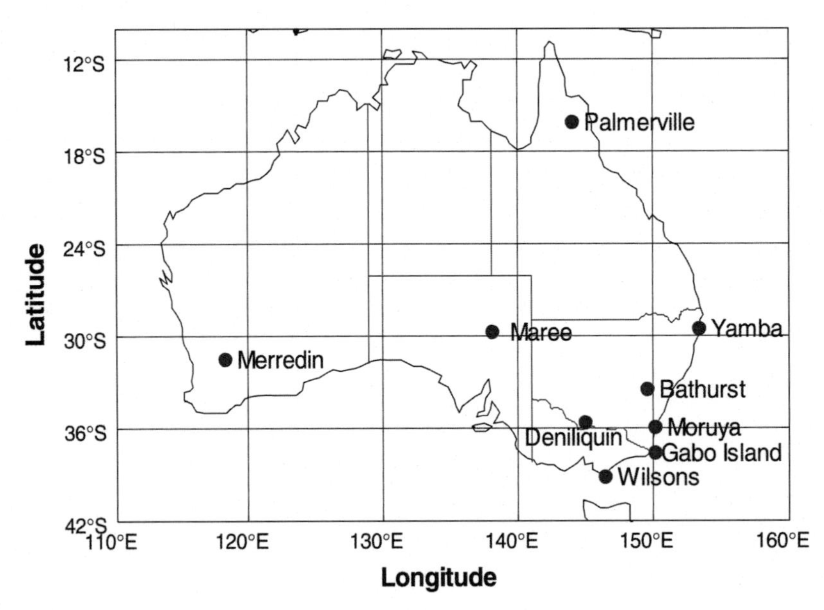

FIGURE 10.3 **A spatial map of the nine locations considered in this study.**

noteworthy that Maree is the coolest of all stations, having a mean air temperature of approximately 14.7 °C. However, the annual averaged evapotranspiration rates for Maree is the highest (ET $\approx$ 408.7 mm year^{-1}) compared to all other stations.

For the development of SVR models, the predictive variables were the hydrometeorological variables (rainfall, maximum temperature, minimum temperature, mean temperature, and evapotranspiration estimated by Thornthwaite method). Additionally, the synoptic-scale climate drivers (SOI, PDO, IOD, and SAM) and sea surface temperatures (Niño 3.0, 3.4, and 4.0) were also used. Except for the index of SAM for which available data from 1948 was used, all others had time-series data from 1915 to 2012. Rainfall data were obtained from the archives of the Australian Bureau of Meteorology (ftp://ftp.bom.gov.au/anon/home/ncc/www/change/HQdailyR) and the temperature data were obtained from the Australian Climate Observations Reference Network–Surface Air Temperature (ACORN-SAT) (http://www.bom.gov.au/climate/change/acorn-sat/). Both were collated from hourly in situ observations at daily time step (Della-Marta et al., 2004; Jones et al., 2009; Lavery et al., 1997; Menne and Williams, 2009). The data quality has been checked using standard normal homogeneity tests, where raw values were adjusted for inhomogeneities caused by factors such as station relocations, instrumental errors, and adverse exposures to the measurement site (Alexandersson, 1986; Torok and Nicholls, 1996). The process also detected and removed gross single-day errors in data records. Rather than making inhomogeneity adjustments in the mean values, daily records were adjusted for discontinuities at the 5, 10, …, 90, 95 percentiles. Missing data were deduced by generating artificial rainfall based on cumulative rainfall distributions (Haylock and Nicholls, 2000). Consequently, both data sets have since been used extensively for climate change studies in Australia (Alexander et al., 2006; Suppiah and Hennessy, 1998).

The sources of supplementary training data used as regression covariate are: the SOI and the IOD index acquired from the Australian Bureau of Meteorology (Trenberth, 1984), the PDO index acquired from the Joint Institute of the Study of the Atmosphere and Ocean (JISAO) (Mantua et al., 1997; Zhang et al., 1997), and the index of SAM acquired from the British Antarctic Survey database (Marshall, 2003). SOI is typically calculated using Troup's method where pressure differences between Tahiti and Darwin are used. PDO is determined using the UKMO Historical SST data set for 1900–81 and the Reynolds Optimally Interpolated SST (V1) (Morid et al., 2007) for January 1982–December 2001 and OI.v2 SST fields from January 2002 onwards. The sea surface temperatures comprise the Niño 3.0 (5°N–5°S, 150°W–90°W), Niño 3.4 (5°N–5°S, 170°W–120°W), and Niño 4.0 (5°N–5°S, 160°W–150°W) indices acquired from the National Centre for Atmospheric Research (NCAR) CGDs Climate Analysis Section (Hurrell and Trenberth, 1999; Trenberth, 1997).

We incorporated these regression covariates for prediction of the DI following the general consensus that synoptic-scale climate drivers are known to moderate drought events in Australia. For example, the positive and the negative phases of the SOI are known to impact rainfall variability in the eastern region. IOD, which represents the coupled ocean and atmosphere phenomenon in the equatorial Indian Ocean, affects drought in the southern half of the continent (Ashok et al.,

2003; Saji et al., 1999, 2005). Likewise, the index of SAM, which represents strength and position of cold front and mid-latitude storm systems, is a significant driver of rainfall variability in the southeast region (Hendon et al., 2007). The effect of SSTs on the evolution of drought in Australia is well known. Considering the potential impact of synoptic-scale climate drivers and SSTs on evolution of the DI, we used these data sets in training of our SVR models. Our approach was consistent with application of these data sets in various ML approaches used for rainfall forecasting (Abbot and Marohasy, 2012, 2014; Deo and Sahin, 2014; Mekanik et al., 2013; Morid et al., 2007) and for prediction of the effective DI in Iran and South Africa (Morid et al., 2007; Masinde, 2013).

10.3.2 Statistical Evaluation of SVR Model Performance

In evaluation of the model performance, we employ five prediction score metrics calculated from the test data set: (1) RMSE, (2) MAE, (3) the coefficient of determination, R^2 (Paulescu et al., 2011; Ulgen and Hepbasli, 2002), (4) the Wilmort's index of agreement, d (Acharya et al., 2013; Willmott, 1981, 1982), and (5) the Nash–Sutcliffe coefficient of efficiency (E) (Krause et al., 2005; Nash and Sutcliffe, 1970). The mathematical formulas are as follows:

$$RMSE = \sqrt{\frac{1}{N}\sum_{i=1}^{n}\left(SPEI_{p_i} - SPEI_{o_i}\right)_t^2} \tag{10.24}$$

$$MAE = \frac{1}{N}\sum_{i=1}^{n}\left|\left(SPEI_{p_i} - SPEI_{o_i}\right)_t\right| \tag{10.25}$$

$$R^2 = \left(\frac{\sum_{i=1}^{n}\left(SPEI_{o,i} - \overline{SPEI}_{o,i}\right)\left(SPEI_{p,i} - \overline{SPEI}_{p,i}\right)}{\sqrt{\sum_{i=1}^{n}\left(SPEI_{o,i} - \overline{SPEI}_{o,i}\right)^2}\sqrt{\sum_{i=1}^{n}\left(SPEI_{p,i} - \overline{SPEI}_{p,i}\right)^2}}\right)^2 \tag{10.26}$$

$$d = 1 - \left[\frac{\sum_{i=1}^{N}(\mathrm{SPEI}_{p,i} - \mathrm{SPEI}_{o,i})^2}{\sum_{i=1}^{N}(|\mathrm{SPEI}_{p,i} - \overline{\mathrm{SPEI}_o}| + |\mathrm{SPEI}_{o,i} - \overline{\mathrm{SPEI}_o}|)^2}\right], \quad 0 \le d \le 1 \tag{10.27}$$

$$E = 1 - \left[\frac{\sum_{i=1}^{N}(\mathrm{SPEI}_{o,i} - \mathrm{SPEI}_{p,i})^2}{\sum_{i=1}^{N}(\mathrm{SPEI}_{o,i} - \overline{\mathrm{SPEI}_o})^2}\right], \quad 0 \le E \le 1 \tag{10.28}$$

where the $SPEI_{p_i}$ and the $SPEI_{o_i}$ are the predicted and the observed values of the monthly SPEI in test period t (test slice), respectively, i is the month of the test data set, and N (=144) is the length (number of samples in the test set) in period t (2001–2012).

When comparing the model performance by RMSE and MAE, the values must be as small as possible to reflect smallest deviations of the predictions from observations. However, the MAE is less sensitive to extreme values than the RMSE (Fox, 1981). For the best model, the R^2 which is determined by the scatter plot of observed and predicted SPEI is expected to be close to unity. Likewise, the d and E should be unity for the perfect fit (Krause et al., 2005). A disadvantage of the coefficient of determination and the Nash–Sutcliffe efficiency is that the differences between the observed and predicted SPEI values are calculated as squared values. Consequently, the larger values in time-series are overestimated, whereas the smaller values are neglected (Legates and McCabe, 1999). This insensitivity is overcome using the Wilmort's Index of Agreement (Willmott, 1981) so that the ratio of the mean square error and potential error yield is considered for the SVR model performance (Willmott, 1984).

10.4 RESULTS AND DISCUSSION

In order to compare the simulated value of SPEI by SVR algorithm with observed SPEI, the scatter plots of the variable have been prepared (Fig. 10.3). According to the degree of scatter between the simulated and observed value of SPEI, the prediction skill of the SVR model for all stations considered are found to be highly discernible. A linear fit model of the $SPEI_p$ versus $SPEI_o$ shows the best predictions for the three stations (Bathurst, Gabo Island, and Wilsons) and the worst predictions for Maree and Merredin. In order to provide better statistical comparison of the goodness of fit, a simple linear regression model of the form

$$SPEI_P = mSPEI_o + C \tag{10.29}$$

has been applied. Here, the performance of the SVR model performance are assessed by the gradient (m), regression coefficient (R), maximum deviation of the predicted SPEI from the linear trend line (maxDev), and the y-intercept. According to this first-order performance measures, if m and R are close to unity and C is zero, then the predicted and the observed SPEI are in parity with each other (i.e. a perfect fit of simulated and observed data is demonstrated). As listed in Table 10.1, the lowest three gradients are 0.674 (Maree), 0.757 (Merredin), and 0.896 (Deniliquin) and the highest three gradients are 0.985 (Wilsons), 0.973 (Gabo Island), 0.944 (Yamba), and 0.935 (Bathurst). This shows that the performance of SVR algorithm is the worst for Maree, which is in agreement with the smallest value of the linear regression coefficient (0.723) and the correspondingly large deviation of predicted SPEI from the linear trend line (maxDev $\approx$ 1.05). Also, a relatively large maxDev (1.030) is obtained for Palmerville; however, this is likely due to outliers in the simulated data as regression coefficient and gradients for the same station are relatively large. Overall, it is construed that good prediction skill of the SVR model has been demonstrated for majority of the stations used in this investigation.

In order to validate statistically the model performance in the test period, we utilize prediction score metrics defined by Eqs. (10.21)–(10.28) to determine the skill of SVR model in simulating the SPEI. Tables 10.3 and 10.4 display the time-averaged MAE, RMSE, and coefficient of determination including Willmott's index of agreement and Nash–Sutcliffe coefficient of efficiency. In agreement with previous results, the worst performance is obtained for Maree with RMSE (0.711), MAE (0.522), and R^2 (0.523), and the slightly better performance for Merredin with MAE (0.359), RMSE (2.516), and R^2 (0.830). The low performance of SVR model for these stations can also be confirmed by the Willmott's index (0.876 for Maree and 0.938 for Merredin) and the Nash–Sutcliffe coefficient (0.643 for Maree and 0.805 for Merredin). Out of all nine stations considered, the best performance is obtained for Gabo Island, with the smallest MAE and RMSE (0.093 and 0.134) and the largest value of R^2 (0.984). For this case, the Willmott's index and the Nash–Sutcliffe coefficient are very close to unity (0.996 and 0.982, respectively). The model predictions for Bathurst are slightly worse than Gabo Island with an MAE value larger by about 22%, RMSE by 14% larger, and the R^2, d, and E lower by only 1% each than Gabo Island. The performance for Wilsons, Yamba, and Moruya is also relatively good, which is evidenced by the prediction metrics of 0.130–0.135 (MAE), 0.170–0.196 (RMSE), and 0.965–0.973 (R^2), Willmott's index (0.990–0.993), and the Nash–Sutcliffe coefficient (0.964–0.973) (Fig. 10.4).

TABLE 10.3 The SVR Model Performances Based on Linear Regression Equation ($SPEI_p = mSPEI_o + C$) of Observed Value of the Standardized Precipitation and Evapotranspiration Index ($SPEI_o$) With the Predicted Values ($SPEI_p$) From 2001 to 2012 for Each Site

	Station ID	Bathurst 63005	Gabo 84016	Moruya 69018	Palmerville 28004	Wilson 85096	Merredin 10092	Deniliquin 74128	Maree 17031	Yamba 58012
Linear regression parameters	M	0.935	0.973	0.950	0.904	0.985	0.757	0.896	0.674	0.944
	r	0.989	0.992	0.982	0.951	0.986	0.911	0.967	0.723	0.981
	maxDev	0.468	0.555	0.403	1.030	0.577	0.873	0.673	1.050	0.557
	C	−0.0135	0.0348	−0.0101	0.00797	−0.0415	0.0357	−0.0118	0.0646	0.0392

The regression coefficient (r) and maximum deviation (maxDev) of predicted index from the modeled index is also shown.

TABLE 10.4 Quantitative Measure of the SVR Model Performance Based on the Prediction Score Metrics Over the Tested Data Set (2001–2012)

Performance Measures					
Locations	**MAE**	**RMSE**	R^2	*d*	*E*
Bathurst	0.114	0.154	0.978	0.994	0.977
Gabo Island	*0.093*	*0.134*	*0.984*	*0.996*	*0.982*
Moruya	0.135	0.175	0.965	0.991	0.964
Palmerville	0.233	0.297	0.905	0.977	0.967
Wilson	0.130	0.170	0.973	0.993	0.971
Merredin	0.359	2.516	0.830	0.938	0.805
Deniliquin	0.195	0.276	0.935	0.979	0.925
Maree	*0.522*	*0.711*	*0.523*	*0.876*	*0.643*
Yamba	0.135	0.196	0.962	0.990	0.963

Key measures of model performance are: root mean square error (RMSE), mean absolute error (MAE), coefficient of determination (R^2), the Wilmort's index of agreement (*d*), and the Nash-Sutcliffe coefficient of efficiency (*E*).
The italicized metrics show the most accurate performance.

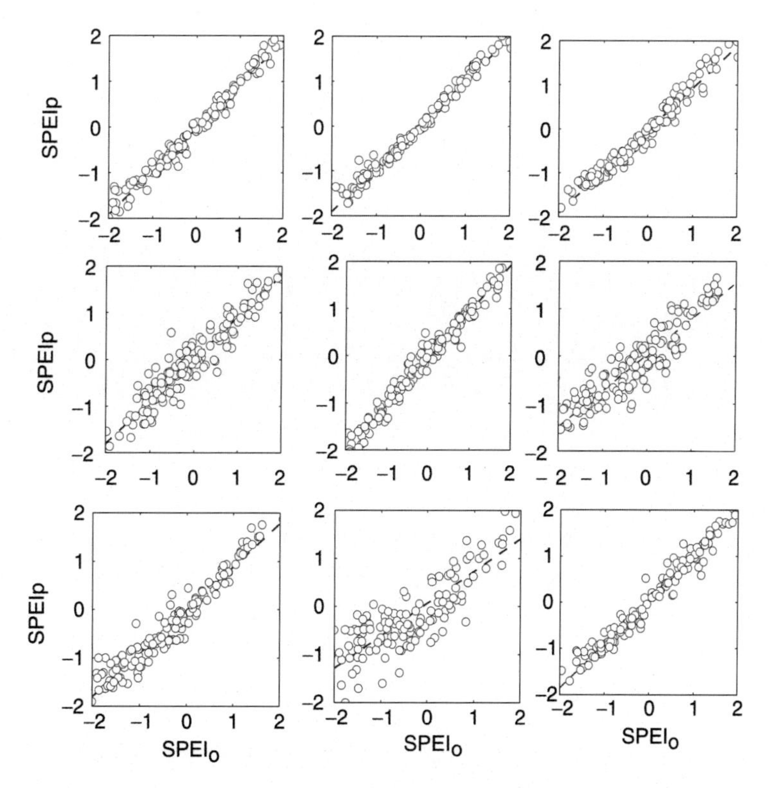

FIGURE 10. 4 Scatter plot of the predicted Standardized Precipitation and Evaporation Index (SPEIp) and the observed Standardized Precipitation and Evapotranspiration Index (SPEIo) based on the SVR algorithm in the test period (2001–12). A linear regression fit of the form SPEIp = mSPEIo + C is displayed. (A) Bathurst, (B) Gabo Island, (C) Moruya, (D) Palmerville, (E) Wilsons, (F) Merredin, (G) Deniliquin, (H) Maree, (I) Yamba.

During this assessment of the SVR model, we are also interested in checking in greater detail the model prediction error encountered per month for each station considered. Fig. 10.5 plots the prediction error (PE) for each station over the tested period (2001–2012). The time series of the magnitude of PE is used to assess visually how close the observed SPEI values are compared to the simulated SPEI. Overall, there appears to be very good agreement between the observed and the predicted DI, with majority of the simulated SPEI within an error bound of ±0.25 (as depicted by the blue boundary lines in

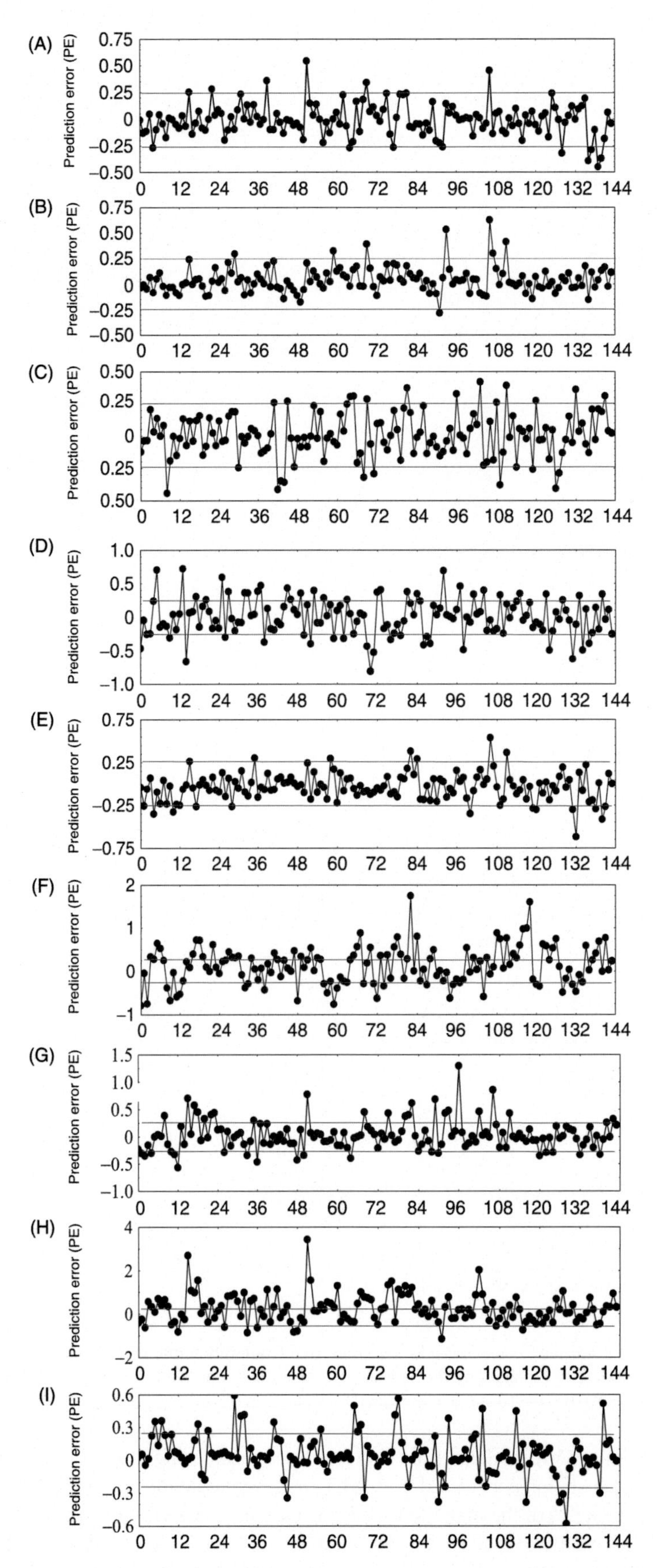

FIGURE 10.5 The SVR model prediction error (PE) yield deduced as the difference between the observed and the simulated SPEI in test period (2000–12). (A) Bathurst, (B) Gabo Island, (C) Moruya, (D) Palmerville, (E) Wilsons, (F) Merredin, (G) Deniliquin, (H) Maree, (I) Yamba. The blue line shows PE bound of ± 0.25.

each panel). In order to analyze the frequency of absolute values of model error for each station in the test period, we show histograms for each station and the all-station totals (Fig. 10.6). Here, the numbers on each bar (of 0.1 bin width) show the monthly count of simulated SPEI over the 144 months in the test data set. The highest frequency of error in the smallest bin (i.e. PE = ±0.1) is evident for Gabo Island (91) followed by Bathurst (80), Yamba (79), Wilsons (76), and Moruya (68). By contrast, the frequency for Maree and Merredin with PE = ±0.1 is very low (29 and 26 out of the 144 months tested, respectively). When considered for all stations (Fig. 10.5), a total of 538 errors are within ±0.1, 290 within ±0.1 and ±0.2, and 162 within ±0.2 and ±0.3.

It is also interesting to compare the relative percentage of model simulations in various error brackets. Thus, we divide simulated errors into the range, $0 \leq |PE| < 0.25$, $0.25 \leq |PE| < 0.50$, $0.50 \leq |PE| < 0.75$, $0.75 \leq |PE| < 1.0$, and $|PE| > 1$,

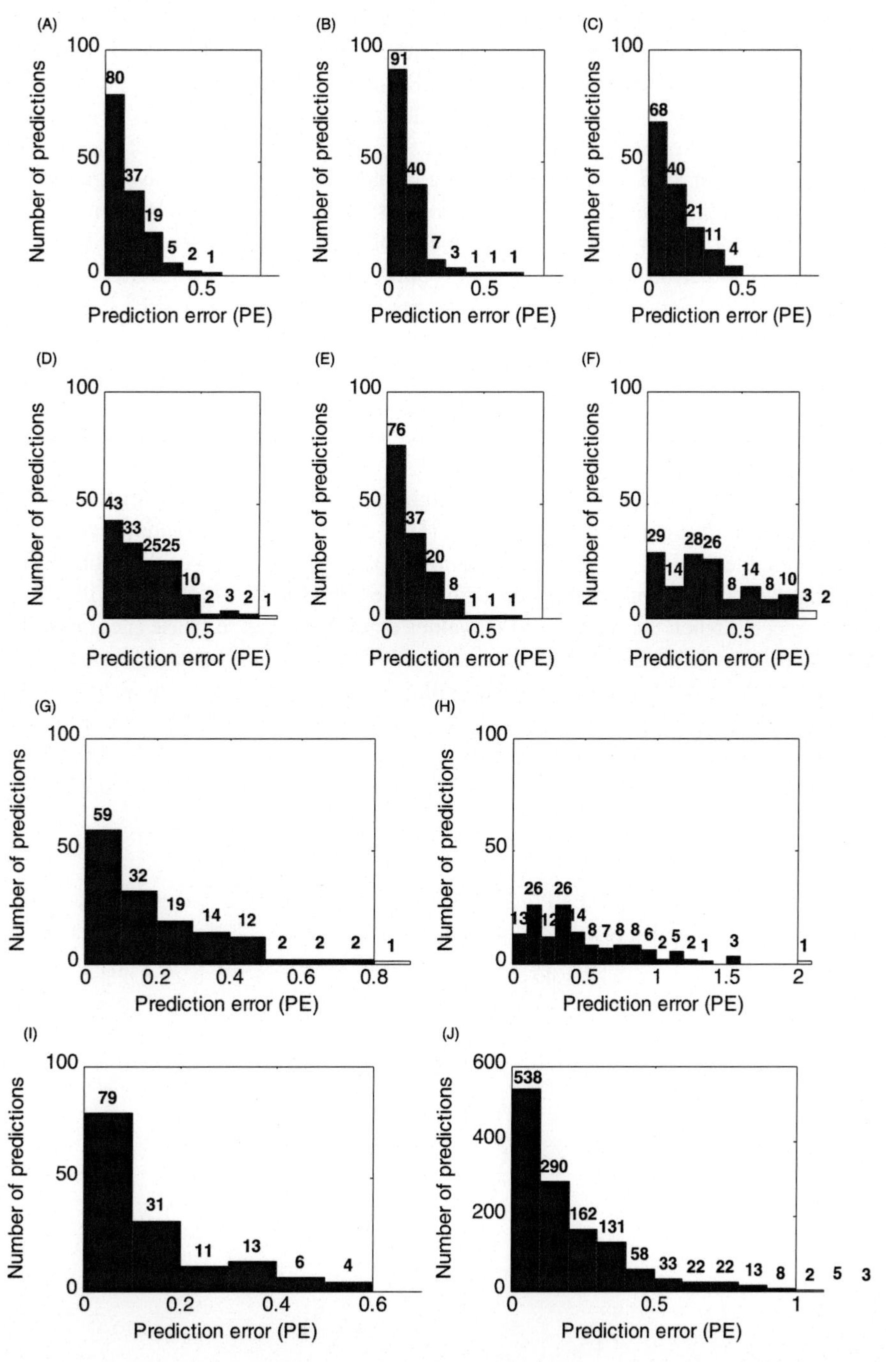

FIGURE 10.6 **A histogram of the frequency distribution versus the magnitude of the prediction errors (PE) in the tested period (2000–12).** (A) Bathurst, (B) Gabo Island, (C) Moruya, (D) Palmerville, (E) Wilsons, (F) Merredin, (G) Deniliquin, (H) Maree and (I) Yamba. In panel (J) the PE values for all stations have been pooled together and each subplot shows the number of predictions in the respective bin (of width = 0.1).

which is illustrated in Table 10.5. Note that the first error bracket contains all of the PE values within ±0.25, the second error brackets within ±0.50, the third error bracket within ±0.75, and so on. It is worth mentioning that the frequency of errors in each error bracket is highly discernible and reflects widely disparate predictions for each case. That is, all the simulations for Moruya, about 94.4% of simulations for Gabo Island, and 89.6% for Bathurst have error magnitude less than 0.25. Similarly, about 84.7% of errors for Wilsons and 81.9% of errors for Yamba have errors less than 0.25. Clearly, the significantly large frequency of model error in smallest error bound certifies excellent performance of the SVR model for the five stations.

When the PE is analyzed for Maree and Merredin, the magnitudes are significantly worse: with approximately 33.3%–36.1% in the lowest bracket, 29.9%–36.8% in the range $0.25 \leq |PE| < 0.50$, 13.9%–18.1% in the range $0.50 \leq |PE| < 0.75$, and approximately 11.1%–1.4% in the largest bracket with $|PE| > 1$. By contrast, none of the predicted errors for the five stations (Bathurst, Moruya, Yamba, Gabo Island, and Palmerville) exceed a magnitude of 1. The same deduction also holds true for the error bracket $0.75 \leq |PE| < 1$, although for the case of Palmerville a very small frequency (≈0.7%) of errors are recorded in this category. Indeed, this reaffirms that the SVR model performs relatively poorly for Maree and Merredin compared to the other stations, in agreement with the results depicted in Fig. 10.3 and Table 10.1. Overall, by all-station averaged values, the SVR model performs very well with over 91% of all errors in the ±0.5 range (Table 10.5).

In Fig. 10.7, a deeper analysis of SVR model performance is undertaken using time-averaged values of mean, standard deviation, minimum, and maximum values of the PE. Here, the first panel is used to show the mean PE whose sign reflects instances of underpredictions (for positive) or overpredictions (for negative) for a given station; second panel shows the standard deviation of the predicted error, while the third and fourth panels show the minimum and maximum values of the PEs. In accordance with the notably large scatter between the predicted and observed SPEI (Fig. 10.4), the predictions for Maree and Merredin exhibit the largest error (0.125–0.234), while those for Gabo Island, Yamba, and Deniliquin are 17%–20% lower. In the case of underpredictions, the SVR model for Wilson yields a value of −0.039. Importantly, the smallest PE in the tested period is evident for Bathurst (−0.011) and Moruya and Palmerville (0.002). Consistent with larger mean errors, the standard deviations for Maree and Merredin is dramatically larger than all other stations considered and so are the minimum and maximum PEs.

The performance of SVR model in prediction of the drought properties is evaluated for the test data set (2001–2012), which also coincided with mega-drought of the Millennium (Dijk et al., 2013; Ummenhofer et al., 2009). In this context, we are interested in deducing the drought severity (*S*), duration (*D*), and intensity (*I*) of drought based on the observed and simulated SPEI following the run-sum approach proposed by Yevjevich (1967), which is also illustrated in Fig. 10.3. Note that a drought month is detected when the monthly SPEI values enter the negative phase (i.e. rainfall conditions are lower than normal), the drought severity is the accumulated negative SPEI, duration is the total number of months with negative SPEI, and drought intensity is the minimum of the SPEI within the drought period. The properties of observed and predicted drought events are exemplified as boxplots (Figs. 10.8–10.10). The ends of the boxplot lie between the lower quartile P25 (25th percentile) and upper quartile P75 (75th percentile), with the second quartile P50 (50th percentile) as the median. Two horizontal lines (known as whisker) extend from the top and bottom of the box. The bottom whisker extends from

TABLE 10.5 Cumulative Frequency (in %) of the Prediction Error (PE) in the Specified ± Error Bound Determined From the Simulated and Observed SPEI in the Tested Period (2001–2012)

Error bound	$\|PE\| > 1$	$0.75 \leq \|PE\| < 1$	$0.50 \leq \|PE\| < 0.75$	$0.25 \leq \|PE\| < 0.50$	$0 \leq \|PE\| < 0.25$
Bathurst	0	0	0.7	10.4	89.6
Gabo	0	0	1.4	4.2	94.4
Moruya	0	0	0.0	0.0	100.0
Palmerville	0	0.7	5.6	32.6	61.8
Wilsons	0	0.0	1.4	14.6	84.7
Merredin	1.4	7.6	18.1	36.8	36.1
Deniliquin	0.7	1.4	3.5	24.3	70.1
Maree	11.1	11.8	13.9	29.9	33.3
Yamba	0	0	2.1	15.3	81.9
All-station	1.5	2.4	5.2	18.7	72.5

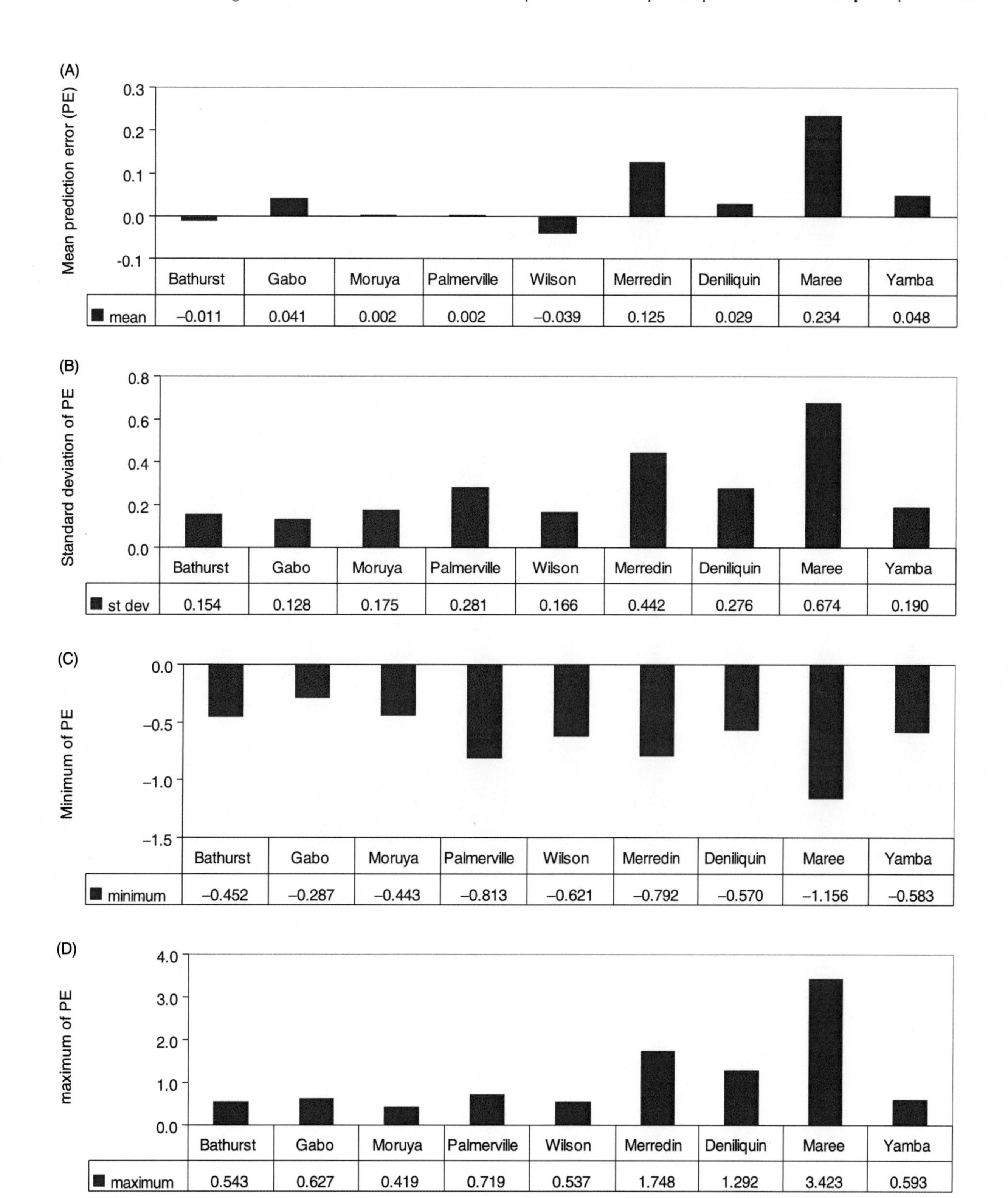

FIGURE 10.7 The time-averaged values of: (A) mean, (B) standard deviation, (C) minimum, (D) maximum values of the model prediction error (PE) in the tested period (2001–12). The table includes the actual values obtained for each station.

P25 to the smallest nonoutlier in the data set, whereas the other one goes from P75 to the largest nonoutlier. It is noticeable that the medians of predicted and observed drought parameters are very close for all stations, except Maree and Merredin (Figs. 10.8–10.10F and H). However, the upper and lower quartile values as well as the maximum and minimum predicted and observed properties of drought are discernible for all stations considered.

In order to analyze closely the spread of the simulated and observed *S*, *I*, and *D*, Tables 10.6 and 10.7 show the percentage differences in P25, P50, and P75. For the case of *S*, the five stations (Bathurst, Gabo Island, Moruya, Palmerville, and Palmerville) exhibit less than 5% difference in the lower quartiles. However, for the median of drought severity, the differences are slightly higher (−6% for Moruya and 12% for Moruya) although for the upper quartile the difference is elevated by about 42% (Palmerville), 27% (Gabo Island), 12% (Bathurst), and 6% for Wilsons. Interestingly, the upper quartile seems to be overestimated by the SVR algorithm for all these cases. For the case of Yamba and Deniliquin, the median of *S* is significantly lower for predicted value (−15% and −17%, respectively), although the predicted lower quartile is somewhat 8%–9% smaller in magnitude than the observed value. It is thus ascertained that the simulated spread of the drought severity parameter exhibits the greatest deviation from the observed value for the case of Merredin and Maree. Specifically, for the case of Merredin, the simulated lower quartile of *S* is approximately 39% smaller, the median is 36% smaller, and the upper quartile is 13% smaller than the observed value. This result also holds true for Maree, where the

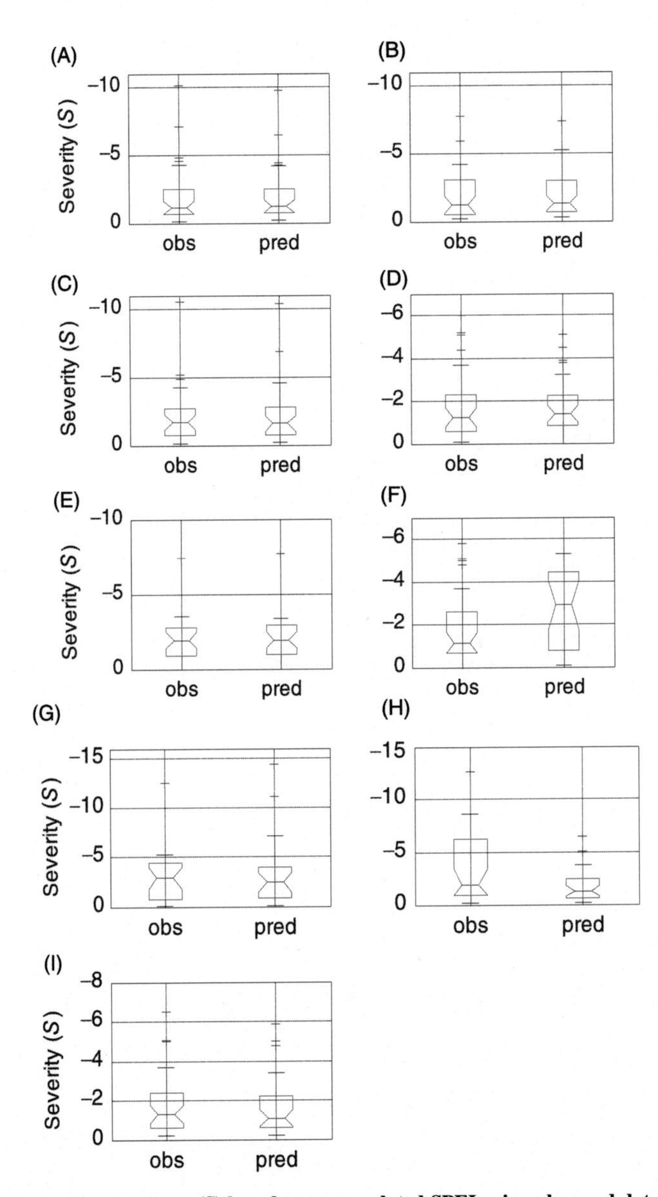

FIGURE 10.8 Boxplot of drought severity parameter (S) based on accumulated SPEI using observed data (obs) and predictions (pred) using the SVR models for the tested period following the approach of Yevjevich et al. (1967). (A) Bathurst, (B) Gabo Island, (C) Moruya, (D) Palmerville, (E) Wilsons, (F) Merredin, (G) Deniliquin, (H) Maree and (I) Yamba.

simulated lower quartile is approximately 62% smaller, the median is 32% smaller, and the upper quartile is 39% smaller than the observed value.

In terms of the spread of the simulated intensity of drought, the largest discrepancy is found for Maree and Merredin, which is internally consistent with relatively poor performance of the SVR model for these cases. Specifically, the SVR model underpredicts the lower quartile, median, and upper quartile value of the intensity of drought event for Merredin by 29%–33%, whereas for Marree the underpredictions are between 23% and 37%. Although the difference in lower quartile and median of S for Bathurst, Gabo Island, Moruya, and Wilsons is less than 10%, the same parameters of I are widely disparate, and range between 1% and 64%. When the statistics of spread is considered for the duration of drought, there appears to be an insignificant difference in the lower quartile and the medians for all stations. By contrast, the upper quartile has been overpredicted by 33% (Bathurst), 13% (Deniliquin), 8% (Moruya), and 7% (Merredin), and underpredicted by approximately 25% for Marree.

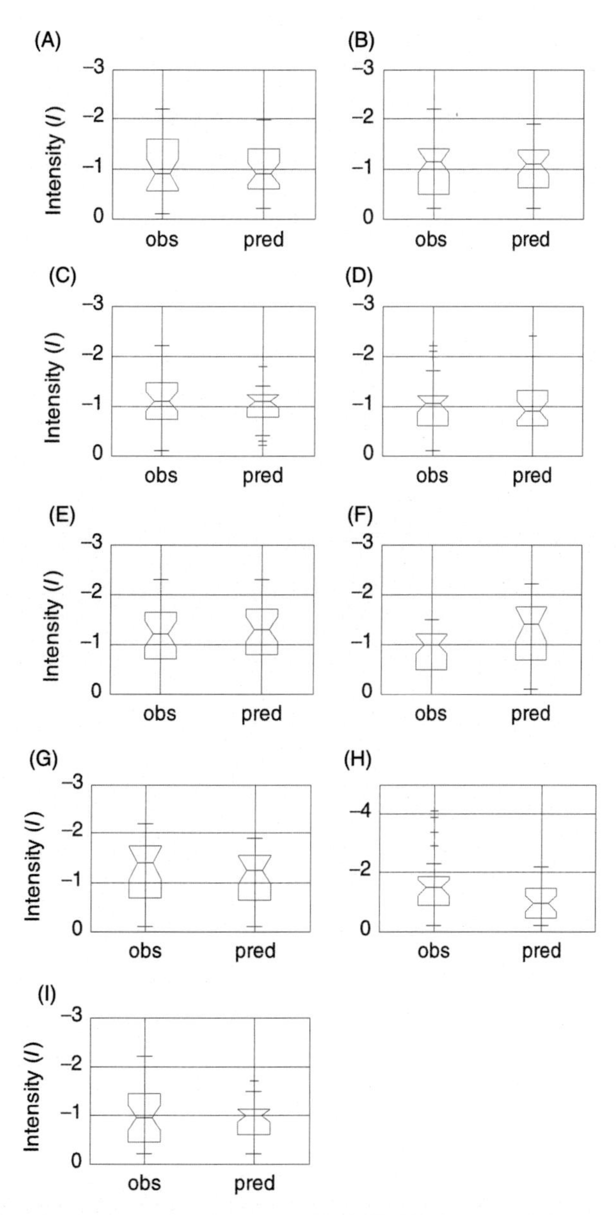

FIGURE 10.9 Boxplot of drought intensity parameter (I) based on minimum SPEI using observed data (obs) and predictions (pred) using SVR models for testing period 2000–12 following the approach of Kim et al. (2011). (A) Bathurst, (B) Gabo Island, (C) Moruya, (D) Palmerville, (E) Wilsons, (F) Merredin, (G) Deniliquin, (H) Maree and (I) Yamba.

A conclusive argument is made by checking the differences in *S*, *I*, and *D* averaged over the test period. Fig. 10.11 displays the column graph of drought parameters calculated from observed and simulated SPEI. On average, the parameter *S* for all but Wilsons and the parameter *I* for all but Wilsons and Palmerville have been underpredicted by the model. Among all of the underpredicted values, the discrepancy in *S* for Maree far exceeds all other stations ($\approx -38.3\%$), followed by Merredin ($\approx -27.5\%$), and Deniliquin, Yamba, and Palmerville ($\approx -8\%$). Although the time-averaged difference in simulated intensity for Maree and Merredin has been similarly large (-30.5% to -47.4%), the same parameter for Deniliquin also differs by -38.3% followed by Yamba (-19.2%), Moruya (-16.2%), and Gabo Island (-13.2%). Finally, it is evidenced that there is less than 5% difference in the prediction of drought duration for all stations except Maree for which the simulated duration is approximately 16% lower in magnitude. This contrasts the severity and intensity where modest differences in the two properties are evident.

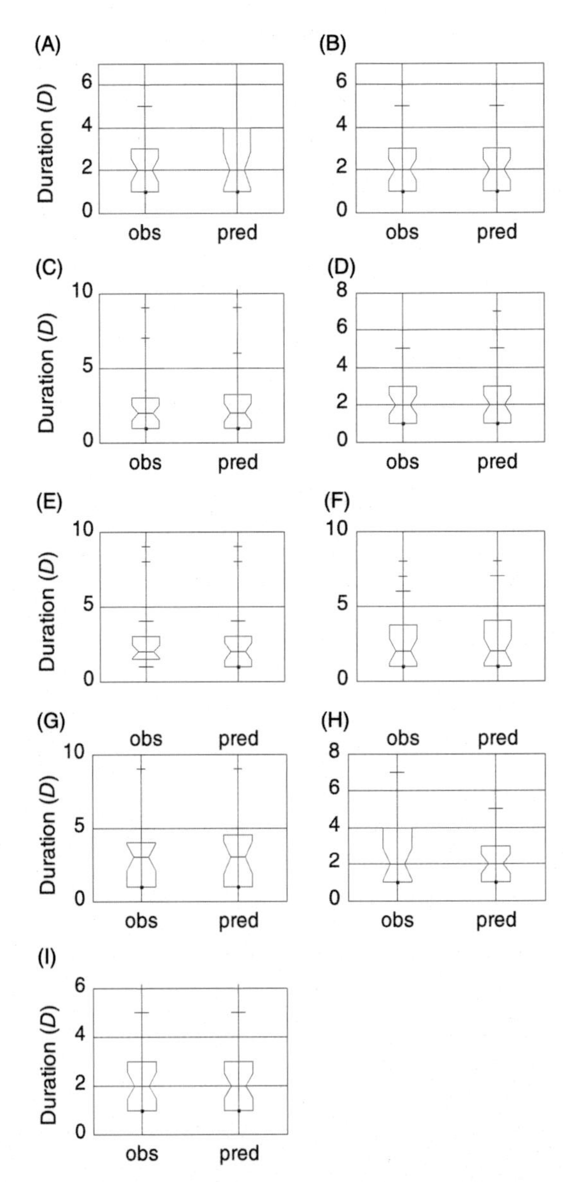

FIGURE 10.10 **Boxplot of drought duration parameter (D) (months) based on sum of consecutive months with negative SPEI using observed data (obs) and predictions (pred) using SVR models for tested period following the approach of Kim et al. (2011).** (A) Bathurst, (B) Gabo Island, (C) Moruya, (D) Palmerville, (E) Wilsons, (F) Merredin, (G) Deniliquin, (H) Maree and (I) Yamba.

TABLE 10.6 Differences in the Statistics of the Spread of Drought Properties (*S*, *I*, and *D*) Represented as Lower Quartiles (P25), Median (P50) and Upper Quartiles (P75) Based on the Predicted and Observed SPEI in the Tested Data Set

Station	Percentage difference in *S*		
	P25	P50	P75
Bathurst	0	4	12
Gabo Island	−2	4	27
Moruya	3	−6	4
Palmerville	−1	12	42
Wilsons	4	0	6
Merredin	−39	−36	−13
Deniliquin	−9	−17	9
Maree	−62	−32	−39
Yamba	−8	−15	0
Station	Percentage difference in *I*		
	P25	P50	P75
Bathurst	−13	64	−33
Gabo Island	−1	24	−4
Moruya	−18	10	7
Palmerville	11	−14	0
Wilsons	3	8	14
Merredin	−33	−29	−29
Deniliquin	−11	−14	−4
Maree	−23	−37	−50
Yamba	−22	5	33
Station	Percentage difference in *D*		
	P25	P50	P75
Bathurst	0	0	33
Gabo Island	0	0	0
Moruya	0	0	8
Palmerville	0	0	0
Wilsons	0	0	0
Merredin	0	0	7
Deniliquin	0	0	13
Maree	0	0	−25
Yamba	0	0	0

10.5 CONCLUSION

In this work, we apply an ML algorithm known as SVR in a problem of predicting the SPEI, using hydrometeorological predictors in the period 1915–2012, for candidate stations in Australia. SVR model has been trained using rainfall, mean, maximum, and minimum temperatures, supplemented by the synoptic-scale climate mode indices (SOI, PDO, IOD, SAM), and sea surface temperatures (Niño 3.0, 3.4, and 4.0) as regressors (inputs) to generate the SPEI as the regressand. In developing SVR models for predicting the SPEI, approximately 88% of the input data (1915–2000) has been used for training the algorithm and 12% of the input data (2001–2012) has been used for testing the model output.

TABLE 10.7 The Prediction Performance of the SVR Model for Quantifying Properties of Drought (*S*, *I*, and *D*) Over the Tested Data Set Based on the Accumulated Negative SPEI Following the Run-Sum Approach of Yevjevich (1967)

Differences in Predicted and Observed Property of Drought (%)			
Location	**Severity (*S*)**	**Peak intensity (*I*)**	**Duration (*D*)**
Bathurst	−2.5	−8.6	1.3
Gabo Island	−5.3	−13.2	2.3
Moruya	−1.1	−16.7	2.2
Palmerville	−7.8	6.7	2.4
Wilson	3.7	2.0	−3.6
Merredin	−27.5	−30.5	2.3
Deniliquin	−7.0	−38.3	2.2
Maree	−38.3	−47.4	−16.1
Yamba	−7.4	−19.2	−2.6

In order to validate how well the SVR model encapsulated the nonlinear relationships between predictors and the objective variable, the performance was assessed using prediction metrics such as the spread, MAE, RMSE, and the coefficient of determination including the Wilmort's index of agreement and the Nash–Sutcliffe coefficient of efficiency. The primary findings of this study can be enumerated as follows:

1. Based on the scatter plot analysis of the test data, the gradient representing the 1:1 ratio of the agreement between predicted and observed SPEI was between 0.674 and 0.973 and the linear regression coefficient was between 0.723 and 0.992, and the maximum deviation was between 0.403 and 1.050. Thus, the worst prediction was for Maree and Merredin, and the best for Gabo Island.
2. In terms of quantitative statistical measures, the model performed well for the tested data set in predicting the monthly SPEI for all stations considered except for Maree and Merredin. The time-averaged MAE encountered ranged from 0.093 to 0.522, RMSE of 0.134 to 0.711, and coefficient of determination of 0.523 to 0.985. Moreover, the Wilmort's index of agreement and the Nash–Sutcliffe coefficient of efficiency were close to unity for all stations except Maree which had a magnitude of 0.876 and 0.643 and Merredin with a magnitude of 0.938 and 0.805, respectively.
3. Based on the SPEI in the tested data set, the frequency of PEs greater than ±0.75 was very small for majority of the stations considered. However, as expected, for the case of Maree, the SVR model performed relatively poorly, as evidenced by approximately 11.8% of all simulations with PE in the range $0.75 \leq |PE| < 1$ and approximately 11.1% in the range $|PE| > 1$. Similarly, for the case of Merredin, the frequency of errors in the larger brackets ($0.75 \leq |PE| < 1$ and $|PE| > 1$) was approximately 7.6% and 1.4%, respectively. Notwithstanding this, more than 91% of all predicted errors had magnitudes less than ±0.25 when considered by the all-station average for the region.
4. To examine the performance of the SVR model in predicting drought events over the tested period, the parameters of drought severity, duration, and peak intensity were assessed. Apart from the case of Maree and Merredin, the error encountered in quantifying drought severity was less than 10%. However, for simulation of drought intensity, a relatively large error was registered for all stations except Wilsons, Palmerville, and Bathurst. Again, the largest difference in the simulated and observed severity and intensity of drought was obtained for Merredin (−27.5%, −30.5%) and Maree (−38.3%, −37.4%), respectively. While the severity and intensity were underpredicted by the SVR model for 7 out of 9 stations considered, the duration was overpredicted for all stations except Wilsons, Maree, and Yamba.

In synopsis, we ratiocinate that the SVR model is highly efficient in the prediction of the DI for majority of the stations considered in the study region. However, its performance in geographically diverse regions appeared to be different, which perhaps reflects the different role of regressors used in training of the SVR model. Finally, the SVR algorithm is certified to be an appealing ML tool for prediction of drought parameters and may yield information that is beneficial for areas related to water resource management, water use and planning, sustainable agriculture, hydrology, and hydrologic engineering.

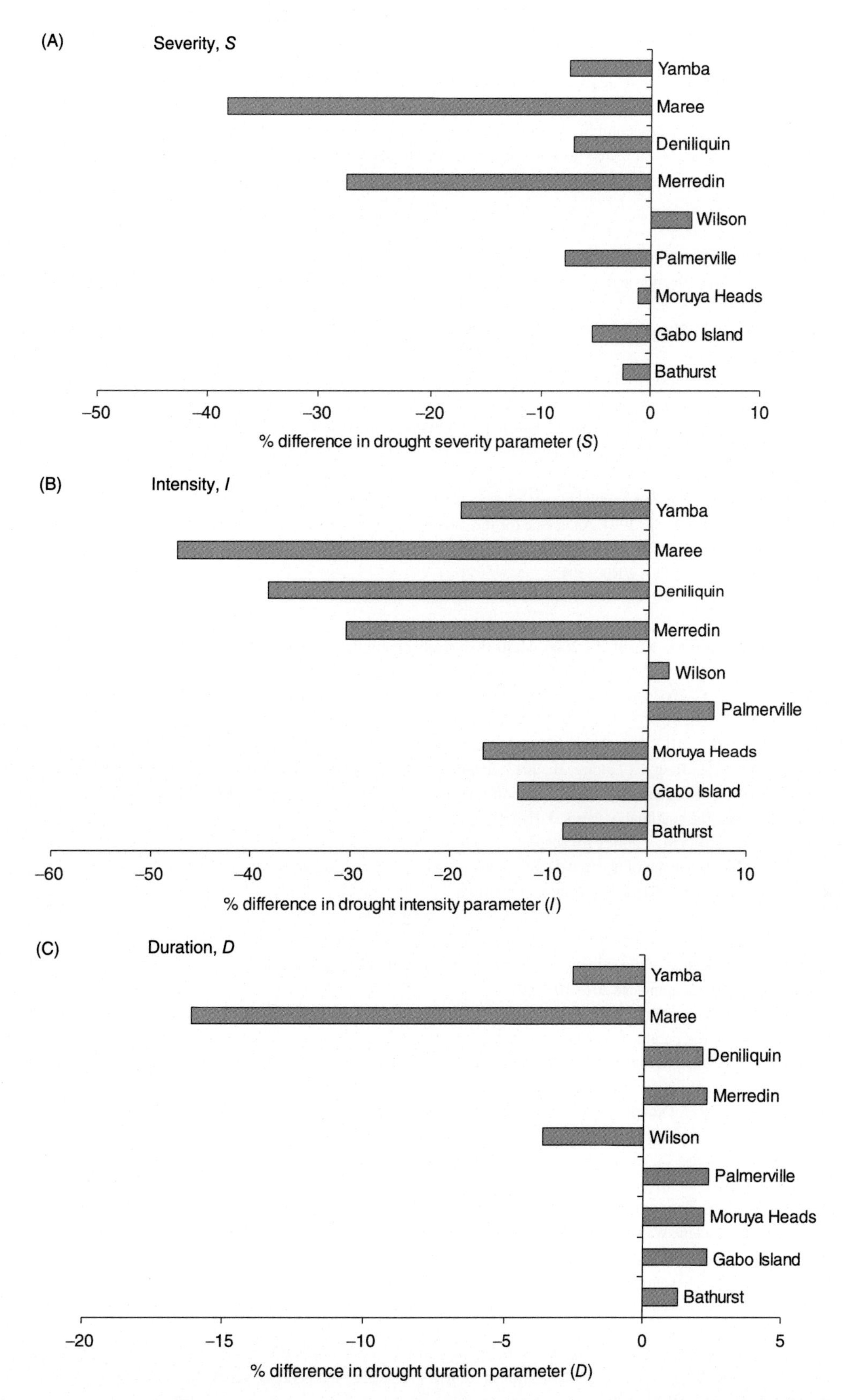

FIGURE 10.11 **Column graphs of differences in parameters representing the drought severity (S), intensity (I) and duration (D) deduced from the observed and the simulated SPEI by SVR models in testing period 2000–12.**

ACKNOWLEDGMENTS

The rainfall, temperature, and SOI data were acquired from Australian Bureau of Meteorology, the PDO data by the Joint Institute of the Study of the Atmosphere and Ocean (JISAO), and SAM from the British Antarctic Survey Database, all of which are greatly acknowledged. School of Agricultural, Computational and Environmental Sciences (University of Southern Queensland) supported Dr R.C. Deo for research time allocation to collaborate with Prof. S Salcedo-Sanz and his research team (Universidad de Alcala, Spain).

REFERENCES

Abbot, J., Marohasy, J., 2012. Application of artificial neural networks to rainfall forecasting in Queensland, Australia. Adv. Atmos. Sci. 29 (4), 717–730.

Abbot, J., Marohasy, J., 2014. Input selection and optimisation for monthly rainfall forecasting in Queensland, Australia, using artificial neural networks. Atmos. Res. 138, 166–178.

Abramowitz, M., Stegun, I.A., 1972. Handbook of Mathematical Functions: With Formulas, Graphs, and Mathematical Tables. Courier Dover Publications, New York, 1046 pp.

Acharya, N., Shrivastava, N., Panigrahi, B., Mohanty, U., 2013. Development of an artificial neural network based multi-model ensemble to estimate the northeast monsoon rainfall over south peninsular India: an application of extreme learning machine. Clim. Dyn. 43 (5), 1303–1310.

Ahmad, M., Sinclair, C., Werritty, A., 1988. Log-logistic flood frequency analysis. J. Hydrol. 98 (3), 205–224.

Alexander, L., et al., 2006. Global observed changes in daily climate extremes of temperature and precipitation. J. Geophys. Res. Atmos. 111 (D05109). doi: 10.1029/2005JD006290.

Alexandersson, H., 1986. A homogeneity test applied to precipitation data. Int. J. Climatol. 6 (6), 661–675.

Allen, R., Smith, M., Pereira, L., Perrier, A., 1994. An update for the calculation of reference evapotranspiration. ICID Bull. 43 (2), 35–92.

Asefa, T., Kemblowski, M.W., Urroz, G., McKee, M., Khalil, A., 2004. Support vectors-based groundwater head observation networks design. Water Resour. Res. 40 (11). doi: 10.1029/2004WR003304.

Asefa, T., Kemblowski, M., Lall, U., Urroz, G., 2005. Support vector machines for nonlinear state space reconstruction: application to the Great Salt Lake time series. Water Resour. Res. 41 (12).

Asefa, T., Kemblowski, M., McKee, M., Khalil, A., 2006. Multi-time scale stream flow predictions: the support vector machines approach. J. Hydrol. 318 (1), 7–16.

Ashok, K., Guan, Z., Yamagata, T., 2003. Influence of the Indian Ocean Dipole on the Australian winter rainfall. Geophys. Res. Lett. 30 (15). doi: 10.1029/2003GL017926.

Beguería, S., Vicente-Serrano, S.M., 2013. Calculation of the standardised precipitation-evapotranspiration index. CRAN (http://sac.csic.es/spei). 1–16.

Beguería, S., Vicente-Serrano, S.M., Reig, F., Latorre, B., 2013. Standardized precipitation evapotranspiration index (SPEI) revisited: parameter fitting, evapotranspiration models, tools, datasets and drought monitoring. Int. J. Climatol. 34 (10), 3001–3023.

Beven, K., 2006. A manifesto for the equifinality thesis. J. Hydrol. 320 (1), 18–36.

Chang, C.-C., Lin, C.-J., 2011. LIBSVM: a library for support vector machines. ACM Trans. Intell. Syst. Technol. (TIST) 2 (3), 27.

Coulibaly, P., Evora, N., 2007. Comparison of neural network methods for infilling missing daily weather records. J. Hydrol. 341 (1), 27–41.

Dayal, K., Deo Ravinesh, C., Apan, A., 2016a. Application of hybrid artificial neural network algorithms for the prediction of standardized precipitation index. IEEE TENCON 2016—Technologies for Smart Nation, IEEE. Singapore.

Della-Marta, P., Collins, D., Braganza, K., 2004. Updating Australia's high-quality annual temperature dataset. Aust. Meteorol. Mag. 53 (2),75–93.

Deo, R.C., Sahin, M., 2014. Application of the extreme learning machine algorithm for the prediction of monthly effective drought index in eastern Australia. Atmos. Res. doi: 10.1016/j.atmosres.2014.10.016.

Deo, R.C., ahin, M., 2015. Application of the artificial neural network model for prediction of monthly standardized precipitation and evapotranspiration index using hydrometeorological parameters and climate indices in eastern Australia. Atmos. Res. 161–162, 65 1.

Deo, R.C., et al., 2009. Impact of historical land cover change on daily indices of climate extremes including droughts in eastern Australia. Geophys. Res. Lett. 36 (8),1–5.

Deo, R.C., Kisi, O., Singh, V.P., 2017. Drought forecasting in Eastern Australia using multivariate adaptive regression spline, least square support vector machine and M5Tree model. Atmos. Res. 184, 149–175.

Dibike, Y.B., Velickov, S., Solomatine, D., Abbott, M.B., 2001. Model induction with support vector machines: introduction and applications. J. Comput. Civil Eng. 15 (3), 208–216.

Dijk, A.I., et al., 2013. The millennium drought in Southeast Australia (2001–2009): natural and human causes and implications for water resources, ecosystems, economy, and society. Water Resour. Res. 49 (2), 1040–1057.

Fox, D.G., 1981. Judging air quality model performance. Bull. Am. Meteorolog. Soc. 62 (5), 599–609.

Hargreaves, G.H., 1994. Defining and using reference evapotranspiration. J. Irrig. Drain. Eng. 120 (6), 1132–1139.

Haylock, M., Nicholls, N., 2000. Trends in extreme rainfall indices for an updated high quality data set for Australia 1910–1998. Int. J. Climatol. 20 (13), 1533–1541.

Hearst, M.A., Dumais, S., Osman, E., Platt, J., Scholkopf, B., 1998. Support vector machines. IEEE Intell. Syst. Appl. 13 (4), 18–28.

Hendon, H.H., Thompson, D.W., Wheeler, M.C., 2007. Australian rainfall and surface temperature variations associated with the Southern Hemisphere annular mode. J. Clim. 20 (11), 2452–2467.

Hosking, J.R., 1990. L-moments: analysis and estimation of distributions using linear combinations of order statistics. J. R. Stat. Soc. Ser. B (Methodol.), 105–124.

Hudson, D., Alves, O., Hendon, H.H., Marshall, A.G., 2011. Bridging the gap between weather and seasonal forecasting: intraseasonal forecasting for Australia. Q. J. R. Meteorol. Soc. 137 (656), 673–689.

Hurrell, J.W., Trenberth, K.E., 1999. Global sea surface temperature analyses: multiple problems and their implications for climate analysis, modeling, and reanalysis. Bulletin of the American Meteorological Society. 80 (12), 2661–2678.

Jones, D.A., Wang, W., Fawcett, R., 2009. High-quality spatial climate data-sets for Australia. Aust. Meteorol. Oceanogr. J. 58 (4), 233.

Khan, M.S., Coulibaly, P., 2006. Application of support vector machine in lake water level prediction. J. Hydrol. Eng. 11 (3), 199–205.

Krause, P., Boyle, D., Bäse, F., 2005. Comparison of different efficiency criteria for hydrological model assessment. Adv. Geosci. 5 (5), 89–97.

Kuligowski, R.J., Barros, A.P., 1998. Experiments in short-term precipitation forecasting using artificial neural networks. Mon. Weather Rev. 126 (2), 470–482.

Lavery, B., Joung, G., Nicholls, N., 1997. An extended high-quality historical rainfall dataset for Australia. Aust. Meteorol. Mag. 46 (1), 27–38.

Legates, D.R., McCabe, G.J., 1999. Evaluating the use of "goodness-of-fit" measures in hydrologic and hydroclimatic model validation. Water Resour. Res. 35 (1), 233–241.

Li, P.H., Kwon, H.H., Sun, L., Lall, U., Kao, J.J., 2010. A modified support vector machine based prediction model on streamflow at the Shihmen Reservoir, Taiwan. Int. J. Climatol. 30 (8), 1256–1268.

Li, W.-G., Yi, X., Hou, M.-T., Chen, H.-L., Chen, Z.-L., 2012. Standardized precipitation evapotranspiration index shows drought trends in China. Chin. J. Eco-Agric. 5, 21.

Liong, S.Y., Sivapragasam, C., 2002. Flood Stage Forecasting with Support Vector Machines. Wiley Online Library.

Mantua, N.J., Hare, S.R., Zhang, Y., Wallace, J.M., Francis, R.C., 1997. A Pacific interdecadal climate oscillation with impacts on salmon production. Bull. Am. Meteorol. Soc. 78 (6), 1069–1079.

Marshall, G.J., 2003. Trends in the Southern Annular Mode from observations and reanalyses. J. Clim. 16 (24), 4134–4143.

Masinde, M., 2013. Artificial neural networks models for predicting effective drought index: factoring effects of rainfall variability. Mitig. Adaptation Strateg. Global Change 19 (8), 1139–1162.

McCulloch, W.S., Pitts, W., 1943. A logical calculus of the ideas immanent in nervous activity. Bull. Math. Biophys. 5 (4), 115–133.

McKee, T.B., Doesken, N.J., Kleist, J., 1993. The relationship of drought frequency and duration to time scales. In: Proceedings of the 8th Conference on Applied Climatology. American Meteorological Society, Boston, MA. pp. 179–183.

Mekanik, F., Imteaz, M., Gato-Trinidad, S., Elmahdi, A., 2013. Multiple regression and Artificial Neural Network for long-term rainfall forecasting using large scale climate modes. J. Hydrol. 503, 11–21.

Menne, M.J., Williams, Jr., C.N., 2009. Homogenization of temperature series via pairwise comparisons. J. Clim. 22 (7), 1700–1717.

Morid, S., Smakhtin, V., Bagherzadeh, K., 2007. Drought forecasting using artificial neural networks and time series of drought indices. Int. J. Climatol. 27 (15), 2103–2111.

Mpelasoka, F., Hennessy, K., Jones, R., Bates, B., 2008. Comparison of suitable drought indices for climate change impacts assessment over Australia towards resource management. Int. J. Climatol. 28 (10), 1283–1292.

Mukherjee, S., Osuna, E., Girosi, F., 1997. Nonlinear prediction of chaotic time series using support vector machines. Neural Networks for Signal Processing [1997] VII. Proceedings of the 1997 IEEE Workshop, IEEE, pp. 511–520.

Müller, K.R., et al., 1997. Predicting time series with support vector machines. Artificial Neural Networks—ICANN'97. Springer, Berlin, Heidelberg, Switzerland, pp. 999–1004.

Müller, K.R., et al., 1999. Using support vector machines for time series prediction. Advances in Kernel Methods—Support Vector Learning. MIT Press, Cambridge, MA, 243–254.

Nash, J., Sutcliffe, J., 1970. River flow forecasting through conceptual models. Part I—A discussion of principles. J. Hydrol. 10 (3), 282–290.

Ortiz-García, E.G., Salcedo-Sanz, S., Pérez-Bellido, Á.M., Portilla-Figueras, J.A., 2009. Improving the training time of support vector regression algorithms through novel hyper-parameters search space reductions. Neurocomputing 72 (16), 3683–3691.

Ortiz-García, E., Salcedo-Sanz, S., Casanova-Mateo, C., Paniagua-Tineo, A., Portilla-Figueras, J., 2012. Accurate local very short-term temperature prediction based on synoptic situation Support Vector Regression banks. Atmos. Res. 107, 1–8.

Ortiz-García, E., Salcedo-Sanz, S., Casanova-Mateo, C., 2014. Accurate precipitation prediction with support vector classifiers: a study including novel predictive variables and observational data. Atmos. Res. 139, 128–136.

Paulescu, M., Tulcan-Paulescu, E., Stefu, N., 2011. A temperature-based model for global solar irradiance and its application to estimate daily irradiation values. Int. J. Energy Res. 35 (6), 520–529.

Perera, B., Sachindra, D., Godoy, W., Barton, A., Huang, F., 2011. Multi-objective planning and operation of water supply systems subject to climate change. World Acad. Sci. Eng. Technol. 60, 571–580.

Potop, V., Boroneanţ, C., Možný, M., Štěpánek, P., Skalák, P., 2012. Observed evolution of drought episodes assessed with the standardized precipitation evapotranspiration index (SPEI) over the Czech Republic. EGU General Assembly.

Qian, W., Shan, X., Zhu, Y., 2011. Ranking regional drought events in China for 1960–2009. Adv. Atmos. Sci. 28, 310–321.

Radhika, Y., Shashi, M., 2009. Atmospheric temperature prediction using support vector machines. Int. J. Comput. Theory Eng. 1 (1), 1793–8201.

Şahin, M., 2012. Modelling of air temperature using remote sensing and artificial neural network in Turkey. Adv. Space Res. 50 (7), 973–985.

Şahin, M., Kaya, Y., Uyar, M., Yıldırım, S., 2014. Application of extreme learning machine for estimating solar radiation from satellite data. Int. J. Energy Res. 38 (2), 205–212.

Saji, N., Goswami, B.N., Vinayachandran, P., Yamagata, T., 1999. A dipole mode in the tropical Indian Ocean. Nature 401 (6751), 360–363.

Saji, N., Ambrizzi, T., Ferraz, S., 2005. Indian Ocean dipole mode events and austral surface air temperature anomalies. Dyn. Atmos. Oceans 39 (1), 87–101.

Salcedo-Sanz, S., Deo, R.C., Carro-Calvo, L., Saavedra-Moreno, B., 2015. http://eprints.usq.edu.au/27227/ Monthly prediction of air temperature in Australia and New Zealand with machine learning algorithms. Theoretical and Applied Climatology. 1-13. doi: 10.1007/s00704-015-1480-4.
Salcedo-Sanz, S., Rojo-Álvarez, J., Martínez-Ramón, M., Camps-Valls, G., 2014. Support vector machines in engineering: an overview. Wiley Interdiscip. Rev. Data Min. Knowl. Discov. 4 (3), 234–267.
Sánchez-Monedero, J., Salcedo-Sanz, S., Gutiérrez, P., Casanova-Mateo, C., Hervás-Martínez, C., 2014. Simultaneous modelling of rainfall occurrence and amount using a hierarchical nominal–ordinal support vector classifier. Engi. Appl. Artif. Intell. 34, 199–207.
Singh, V., Guo, H., Yu, F., 1993. Parameter estimation for 3-parameter log–logistic distribution (LLD3) by Pome. Stochastic Hydrol. Hydraul. 7 (3), 163–177.
Smola, A.J., Schölkopf, B., 2004. A tutorial on support vector regression. Stat. Comput. 14 (3), 199–222.
Song, X., et al., 2014. Spatial–temporal variations of spring drought based on spring-composite index values for the Songnen Plain, Northeast China. Theor. Appl. Climatol. 116 (3–4), 371–384.
Suppiah, R., Hennessy, K.J., 1998. Trends in total rainfall, heavy rain events and number of dry days in Australia 1910–1990. Int. J. Climatol. 18 (10), 1141–1164.
Tay, F.E., Cao, L., 2001. Application of support vector machines in financial time series forecasting. Omega 29 (4), 309–317.
Thissen, U., Van Brakel, R., De Weijer, A., Melssen, W., Buydens, L., 2003. Using support vector machines for time series prediction. Chemom. Intell. Lab. Syst. 69 (1), 35–49.
Thornthwaite, C.W., 1948. An approach toward a rational classification of climate. Geogr. Rev., 55–94.
Torok, S., Nicholls, N., 1996. A historical annual temperature dataset. Aust. Meteorol. Mag. 45 (4).
Trenberth, K.E., 1984. Signal versus noise in the Southern Oscillation. Mon. Weather Rev. 112 (2), 326–332.
Trenberth, K.E., 1997. The definition of El Niño. Bull. Am. Meteorol. Soc. 78 (12), 2771–2777.
Tripathi, S., Srinivas, V., Nanjundiah, R.S., 2006. Downscaling of precipitation for climate change scenarios: a support vector machine approach. J. Hydrol. 330 (3), 621–640.
Uhlenbrook, S., Seibert, J., Leibundgut, C., Rodhe, A., 1999. Prediction uncertainty of conceptual rainfall-runoff models caused by problems in identifying model parameters and structure. Hydrol. Sci. J. 44 (5), 779–797.
Ulgen, K., Hepbasli, A., 2002. Comparison of solar radiation correlations for Izmir, Turkey. Int. J. Energy Res. 26 (5), 413–430.
Ummenhofer, C.C., et al., 2009. What causes Southeast Australia's worst droughts? Geophys. Res. Lett. 36 (4).
Vapnik, V., 2000. The Nature of Statistical Learning Theory. Springer.
Vapnik, V., Kotz, S., 2006. Estimation of Dependences Based on Empirical Data. Springer.
Vapnik, V.N., Vapnik, V., 1998. Statistical Learning Theory. Wiley, New York.
Vicente-Serrano, S.M., Beguería, S., López-Moreno, J.I., 2010a. A multiscalar drought index sensitive to global warming: the standardized precipitation evapotranspiration index. J. Clim. 23 (7), 1696–1718.
Vicente-Serrano, S.M., Beguería, S., López-Moreno, J.I., Angulo, M., El Kenawy, A., 2010b. A new global 0.5 gridded dataset (1901–2006) of a multiscalar drought index: comparison with current drought index datasets based on the Palmer drought severity index. J. Hydrometeorol. 11 (4), 1033–1043.
Vicente-Serrano, S.M., Beguería, S., López-Moreno, J.I., 2011a. Comment on "Characteristics and trends in various forms of the Palmer drought severity index (PDSI) during 1900–2008" by Aiguo Dai. J. Geophys. Res. Atmos. 116 (D19). doi: 10.1029/2011JD016410.
Vicente-Serrano, S.M., et al., 2011b. A multiscalar global evaluation of the impact of ENSO on droughts. J. Geophys. Res. Atmos. 116 (D20). doi: 10.1029/2011JD016039.
Vicente-Serrano, S.M., et al., 2012a. Performance of drought indices for ecological, agricultural, and hydrological applications. Earth Inter. 16 (10), 1–27.
Vicente-Serrano, S.M., Zouber, A., Lasanta, T., Pueyo, Y., 2012b. Dryness is accelerating degradation of vulnerable shrublands in semiarid Mediterranean environments. Ecol. Monogr. 82 (4), 407–428.
Willmott, C.J., 1981. On the validation of models. Phys. Geogr. 2 (2), 184–194.
Willmott, C.J., 1982. Some comments on the evaluation of model performance. Bull. Am. Meteorol. Soc. 63. (11), 1309–1313.
Willmott, C.J., 1984. On the evaluation of model performance in physical geography. Spatial Statistics and Models. Springer, 443–460.
Yevjevich, V., 1967. An objective approach to definitions and investigations of continental hydrologic droughts. Colorado State University, Fort Collins, Colorado.
Yu, X., Liong, S.Y., 2007. Forecasting of hydrologic time series with ridge regression in feature space. J. Hydrol. 332 (3), 290–302.
Yu, P.-S., Chen, S.-T., Chang, I.-F., 2006. Support vector regression for real-time flood stage forecasting. J. Hydrol. 328 (3), 704–716.
Yu, M., Li, Q., Hayes, M.J., Svoboda, M.D., Heim, R.R., 2014. Are droughts becoming more frequent or severe in China based on the standardized precipitation evapotranspiration index: 1951–2010? Int. J. Climatol. 34 (3), 545–558.
Zhang, Y., Wallace, J.M., Battisti, D.S., 1997. ENSO-like interdecadal variability: 1900–93. J. Clim. 10 (5), 1004–1020.

FURTHER READING

Dayal, K., Deo Ravinesh, C., Apan, A., 2016b. Drought modelling based on artificial intelligence and neural network algorithms: a case study in Queensland, Australia. In: Leal Filho, W. (Ed.), Climate Change Adaptation in Pacific Countries: Fostering Resilience and Improving the Quality of Life. Springer, Berlin.
Reynolds, R.W., Smith, T.M., 1994. Improved global sea surface temperature analyses using optimum interpolation. J. Clim. 7 (6), 929–948.

Part II

Recovery and Management

Chapter 11

Earthquake Risk Reduction Efforts in Nepal

Jitendra Bothara*, Jason Ingham, Dmytro Dizhur****
**Miyamoto International NZ Ltd, Christchurch, New Zealand*
***University of Auckland, Auckland, New Zealand*

11.1 INTRODUCTION

Nepal is a mountainous, landlocked country sandwiched between China to the north and India to the south, west, and east and has a population of approximately 29 million (WPR, 2017). The country is situated in the central part of the Himalayas, which extend 2400 km from east to west. Nepal's approximately trapezoidal footprint is 800-km long and 150–250-km wide, for a total area of 147,181 km^2 (NTB, 2017). The country is located between latitudes 26°22′ and 30°27′N, and longitudes 80°04′ and 88°12′E (CBS, 2015). The northern part of the country is a typically rugged terrain comprising high mountain ranges and hills, and the southern part is a continuation of the Gangetic plains (Terai or lowlands). The extreme gradient variation of the terrain along the north–south transect is indicated by an altitude variation from 67 m above sea level in the south to 8848 m at the peak of Mt. Everest. Along the north–south transect, Nepal can be broadly divided into five physiographic regions: (1) High Himalayas, (2) High Mountains, (3) Middle Mountains, (4) Siwalik (High Hills), and (5) Terai (lowland) (see Fig. 11.1).

Approximately 83% of Nepal is comprised of high mountains and hills to the north, with the remaining 17% being Terai. Due to extreme altitude variation and numerous hills, mountains, and deep river valleys, many parts of the country are inaccessible by vehicular transport. This inaccessibility has resulted in the isolation of some communities from the rest of the country (Bothara and Sharpe, 2003).

As a consequence of Nepal's geographical location, the country is situated in a high earthquake hazard region. The major source of earthquakes in Nepal and the overall Himalayan region is subduction of the Indo-Australian Plate beneath the Eurasian plate, with Nepal being situated on the two plate boundaries. The country's high seismicity is related to the movement of tectonic plates across the Himalayas, which has caused the development of several active faults. Seismic hazard mapping and risk assessment undertaken as part of the Nepal Building Code Development Project (MHPP, 1994a) mapped 92 active faults traversing Nepal.

The majority of the country's building stock consists of unreinforced masonry (URM) buildings typically constructed using locally sourced stone, brick, and mud. In recent years, reinforced concrete (RC) frame buildings with masonry infill have become prominent. Irrespective of the construction materials or structural system, the majority of buildings lack basic earthquake-resilient design and construction and are highly vulnerable to earthquake-induced shaking. Similar to the country's building stock, the design and construction of much of Nepal's infrastructure and lifelines do not account for seismic effects. This mix of a highly vulnerable building stock and infrastructure in an environment with high seismic hazard and a community with low awareness of earthquake safety measures has resulted in significantly high seismic risk to the population of Nepal. This risk has been substantiated by various studies (Chaulagain et al., 2015; JICA and MOHA, 2002) and was further highlighted by the 2015 Nepal earthquake sequence. Consequently, Nepal was ranked as the 11th most earthquake-vulnerable country in the world (UNDP, 2009). In terms of seismic vulnerability ranking, Kathmandu Valley was ranked first among 20 cities assessed globally (GESI, 2001).

Nepal is currently undergoing constitutional changes and has adopted a federal system and new constitution that came into effect in September 2015, replacing the Interim Constitution of 2007. In accordance with this new constitution, Nepal is divided into seven provinces and 744 village committees (gaunpalika), metropolitan areas, submetropolitan areas, and municipalities (Republica, 2015; Wikipedia, 2017). The study reported here is based on the earlier administrative system in place at the time of the 2015 Nepal earthquake sequence, where the country was divided into 75 districts, with district-level administration of disaster risk management (DRM) activities and the Ministry of Home Affairs (MOHA) as the nodal point. The new constitutional system will require changes in the relevant acts, regulations, and guidelines for DRR and disaster response.

Rapid population growth and unsustainable development practices in Nepal continuously increase the exposure and vulnerabilities of local communities to earthquake disasters (UNDP, 2015). In this context, Nepal has adopted many preemptive policies and actions to reduce disaster risk. It is generally accepted that seismic risk in Nepal cannot be brought to

Integrating Disaster Science and Management. http://dx.doi.org/10.1016/B978-0-12-812056-9.00011-7

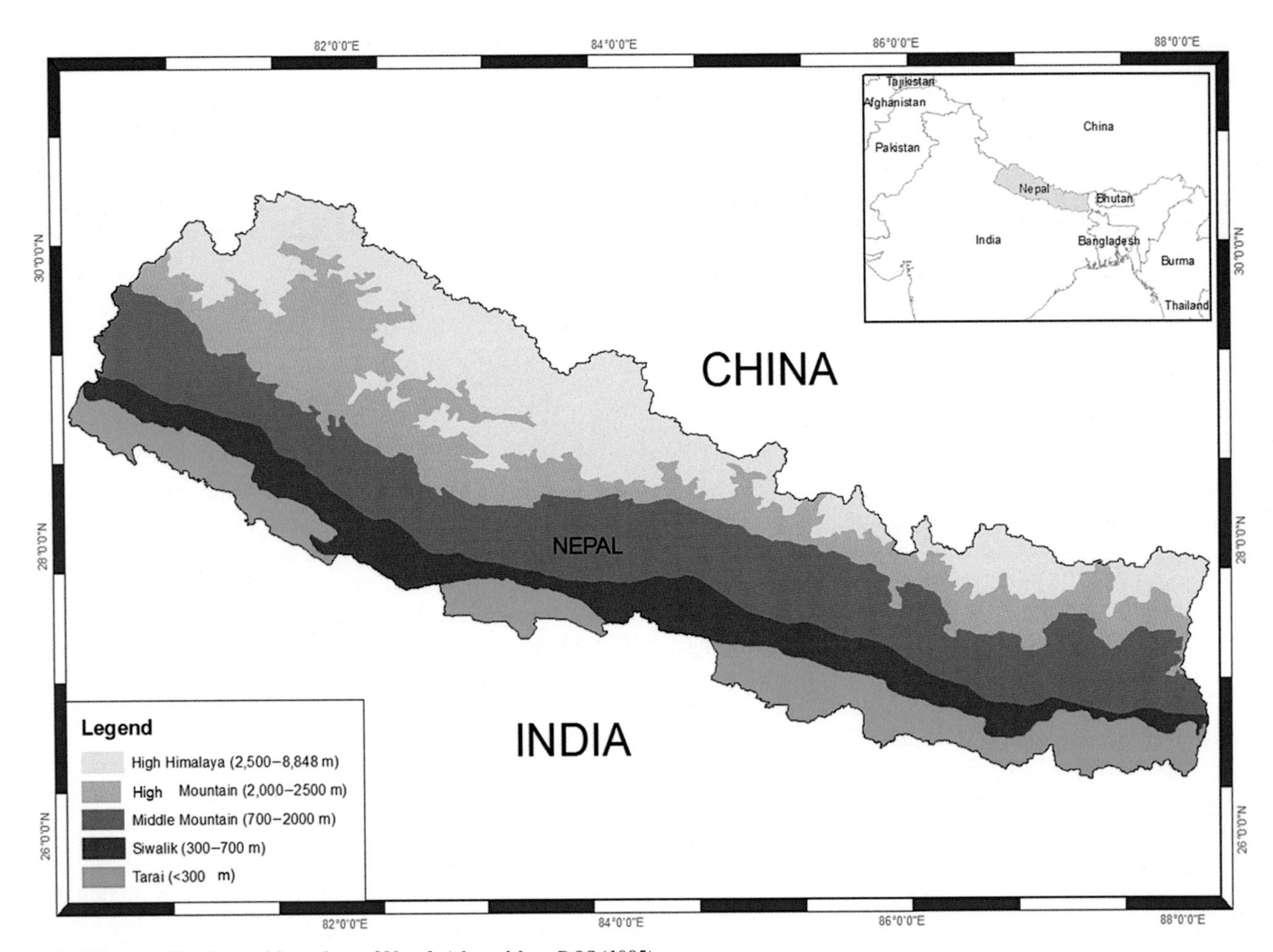

FIGURE 11.1 Physiographic regions of Nepal. *Adapted from DOS (1985).*

a level equivalent to that of developed nations due to the current state of the economy, various sociocultural and technological factors, and inaccessibility to adequate construction materials in remote areas. However, by adopting a balanced and pragmatic approach to EDRR, seismic risk can certainly be reduced.

Earthquakes are naturally occurring phenomena, with the overall imposed risk dependent on the circumstances of individuals, households, and societies, including belief systems, culture, religion, and other local conditions. Many circumstances arise from diverse micro- to macrolevel political, social, economic, and environmental processes (IRDR and ICSU, 2017), which may be more pronounced due to the microcultural environment and isolation of communities due to inaccessibility. Within this context, the importance of positively altering belief systems and cultural values to bring constructive transformation and to reduce the vulnerability of communities has been recognized (IFRC, 2014; Mercer et al., 2014). The encouragement of community-based EDRR via policy intervention is essential to achieve this transformation. Concurrently, efforts are required to more efficiently use resources and to increase earthquake risk reduction awareness and enforcement through public policy. A top-down approach driven by policy can provide a legal basis and generate demand for seismic safety, whereas a bottom-up approach via activities such as raising awareness and undertaking training at the community level enables sustainability of the EDRR process due to an increased understanding of risk. To successfully undertake EDRR in Nepal, it is recognized that top-down and bottom-up processes must be coordinated on all levels (MHPP, 1994b).

Until the late 1980s, Nepal was largely isolated from the rest of the world, with limited access to information and internationally recognized knowledge. Changes in Nepal's political system during the early 1990s, initiatives undertaken by the United Nations (UN) and other international organizations, and the information revolution provided Nepal with exposure to the rest of the world and greater access to vital information. Internationally trained Nepalis are increasingly returning to Nepal to engage with the private and other nongovernmental sector. This influx of knowledge has resulted in a significant transformation in the country's social structure. Multiple EDRR programs were instituted following the 1988 East Nepal earthquake. These programs have helped to bring an enhanced understanding of risk and positive change regarding earthquake risk reduction and have provided risk mitigation options. Rather than fatalistically ignoring earthquake disasters or considering them "God's will", as was common previously, earthquakes are increasingly considered to be disasters that can be prepared for and mitigated. The approaches taken by various institutes to raise earthquake awareness, investments made by the government and international agencies, the changed policy environment, and the government's prioritization of EDRR since the early 1990s appear to have significantly contributed to general risk awareness and efforts at risk reduction. As suggested by Burton et al. (1993), the impacts from natural hazards such as earthquakes can be reduced if people make preimpact hazard adjustments. Overall, it is evident that Nepal is heading toward engagement in earthquake risk reduction although not to the fullest capacity.

The study reported here was focused on efforts made in terms of preparedness, improvement of building stock, and a positive change in the general Nepali psyche. Initiatives adopted by various institutions in Nepal are reviewed, and field observations are reported. The 2015 Nepal earthquake sequence provided an opportunity to evaluate the effect of these earthquake risk reduction initiatives.

11.2 NEPAL'S SEISMICITY AND PREVIOUS EARTHQUAKES

11.2.1 Location and Seismic Risk

The high seismicity of Nepal is related to its location in the vicinity of the boundary between the Indo-Australian Plate and the Tibetan Plateau on the Eurasian Plate (see Fig. 11.2). Most of the country is located on the Indo-Australian Plate, with recorded plate movement of 35–38 mm/year in the northeast (Jouanne et al., 2004). The Himalayas constitutes the northern boundary of the Indo-Australian Plate, with the Chaman Fault serving as the boundary to the west and the Sagaing Fault being the boundary to the east (Molnar and Dayem, 2010). The Himalayas were formed by continuous subduction followed by collision of the Indian and the Eurasian plates for 50 million years (DeCelles et al., 2014; Patriat and Achache, 1984; Rowley, 1996). As a result

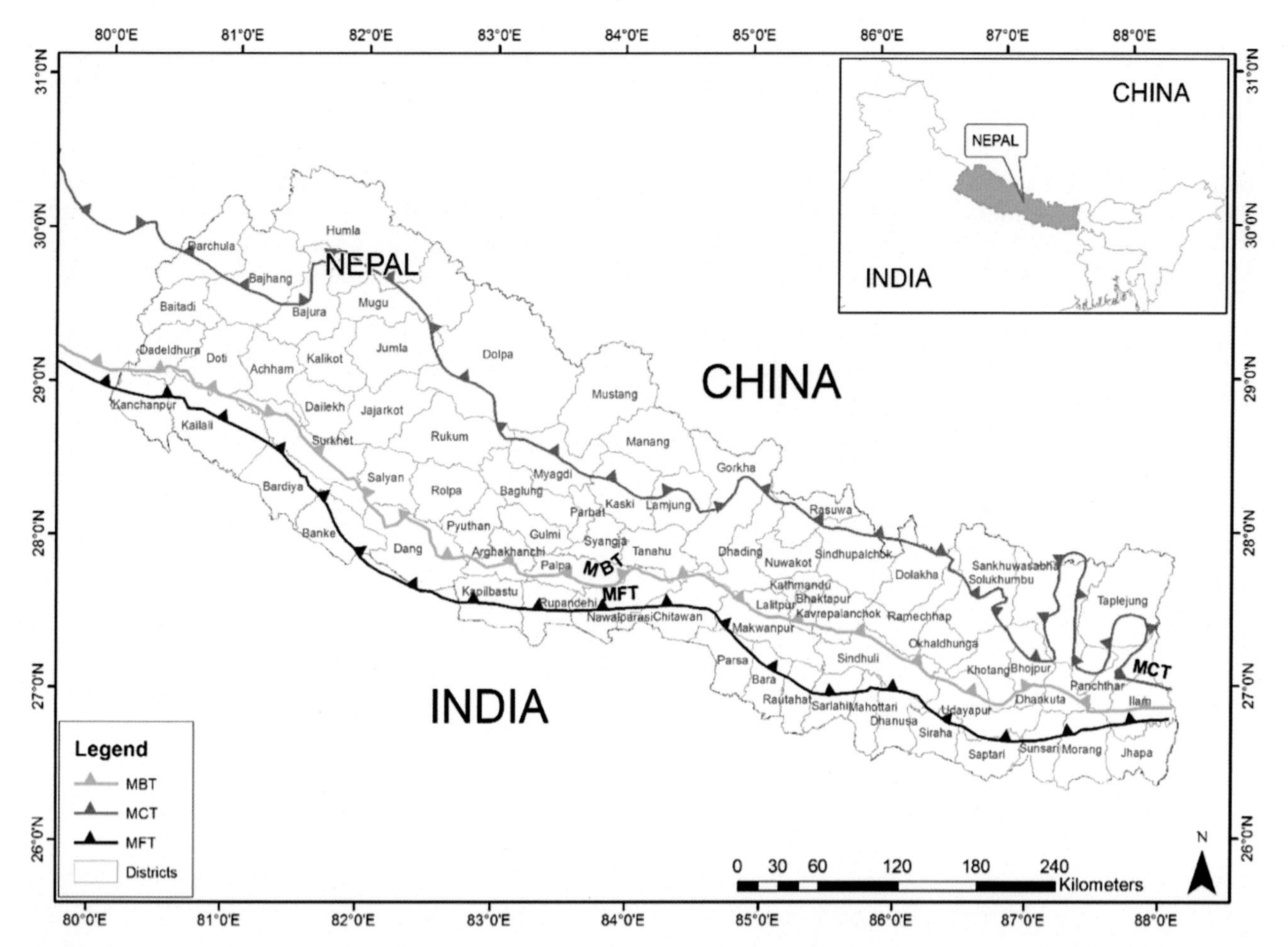

FIGURE 11.2 **Location of mega faults in the context of Nepal—main frontal thrust (MFT), main boundary thrust (MBT), and main central thrust (MCT).** *Adapted from Prajapati et al. (2017).*

of this plate movement, the Himalayan arc experienced a large number of destructive earthquakes in the past. A seismic hazard study conducted by the Global Seismic Hazard Assessment Program (GSHAP) concluded that a peak ground acceleration (PGA) in excess of 0.5 g is expected in Nepal (GSHAP, 1999) for a 475-year return period earthquake.

11.2.2 Major Earthquakes Pre-2015

Nepal has experienced a large number of strong historic earthquakes. The brief summary of documented earthquakes provided herein is split into events that occurred prior to the catastrophic event in 1934 and subsequent events up to 2015. Researchers have reported that large earthquakes with a moment magnitude (M_w) of 7.5 or more occurred in Nepal in 1100, 1505, 1555, 1724, 1803, 1833, and 1897 (Bilham, 2004; Bilham, 2009; Rana and Lall, 2013; Sapkota et al., 2013; Srivastava et al., 2013). Table 11.1 provides a summary of all major earthquake events.

As shown in Fig. 11.3, several events of M_w 6.0 or greater occurred in Nepal over the last century. The great Nepal–Bihar earthquake of January 15, 1934, caused extensive loss of life and property in Nepal and India, with a modified Mercalli intensity (MMI) up to X (10) recorded over an area of approximately 4500 km^2. This earthquake severely damaged buildings in the Kathmandu Valley and other parts of Nepal and resulted in approximately 15,700 deaths, including 8519 fatalities in Nepal (Rana and Lall, 2013). The August 21, 1988, East Nepal earthquake of magnitude M_w 6.8 (Bilham et al., 2001) was centered in Udaypur in eastern Nepal. This earthquake caused 721 deaths in Nepal, destroyed an estimated 60,000 houses, and caused widespread damage to infrastructure (Thapa, 1989).

11.2.3 2015 Earthquake

The M_w 7.8 earthquake of April 25, 2015, and subsequent aftershocks is the most destructive earthquake series in Nepal's recorded history. The 2015 Nepal earthquake sequence affected an area of approximately 22,000 km^2 across 14 districts, resulting in approximately 9000 deaths and 23,000 injuries, with more than 880,000 houses affected (GON, 2015). It is fortunate that the April 25, 2015, earthquake occurred on a Saturday afternoon during school holidays and that the M_w 7.3 aftershock on May 12, 2015, struck in the afternoon during school holidays. The casualty numbers could have been much worse had the earthquake struck at night or during a weekday when children were in classrooms and the general population was at home or work.

Total direct financial losses to Nepal as a result of the 2015 Nepal earthquake sequence are estimated to be NR 700 billion (≈US $7 billion). The private housing sector was the most severely affected sector and accounted for 50% of total direct financial losses, followed by the tourism industry and the Department of Environment and Forestry (NPC, 2015). In Fig. 11.4, the earthquake-affected areas are classified by the damage level for the purpose of prioritizing rescue and relief operations (NPC, 2015). Classification of the areas is based on the effects of the earthquake experienced in individual dis-

TABLE 11.1 Paleoseismic Events and Historic Earthquakes in and Around NEPAL (MMI Greater Than VII) (Bilham, 2004; Bilham, 2009; Bilham et al., 2001; Rana and Lall, 2013; Srivastava et al., 2013)

Date	Information source	Intensity (MMI)	Magnitude (M_w)
11000 BC	Seismite	≥XI	
1100 AD	Trench	≥X	~8.5
1255 AD	Historic and Trench	X	
1408 AD	Historic and Trench	X	>8.5
1505 AD	Historic	≥VII	8.2–8.8
1681 AD	Historic	IX	
1767 AD	Historic	≥VII	
1810 AD	Historic	IX	
1833 AD	Historic	IX–X	7.61
1833 AD	Historic	IX	~7
1833 AD	Historic	VIII	~7
1866 AD	Historic	VIII	7.6
1934 AD	Instrumental	X	>8.0

Abbreviations: *MMI*, modified Mercalli intensity; M_w, earthquake moment magnitude.

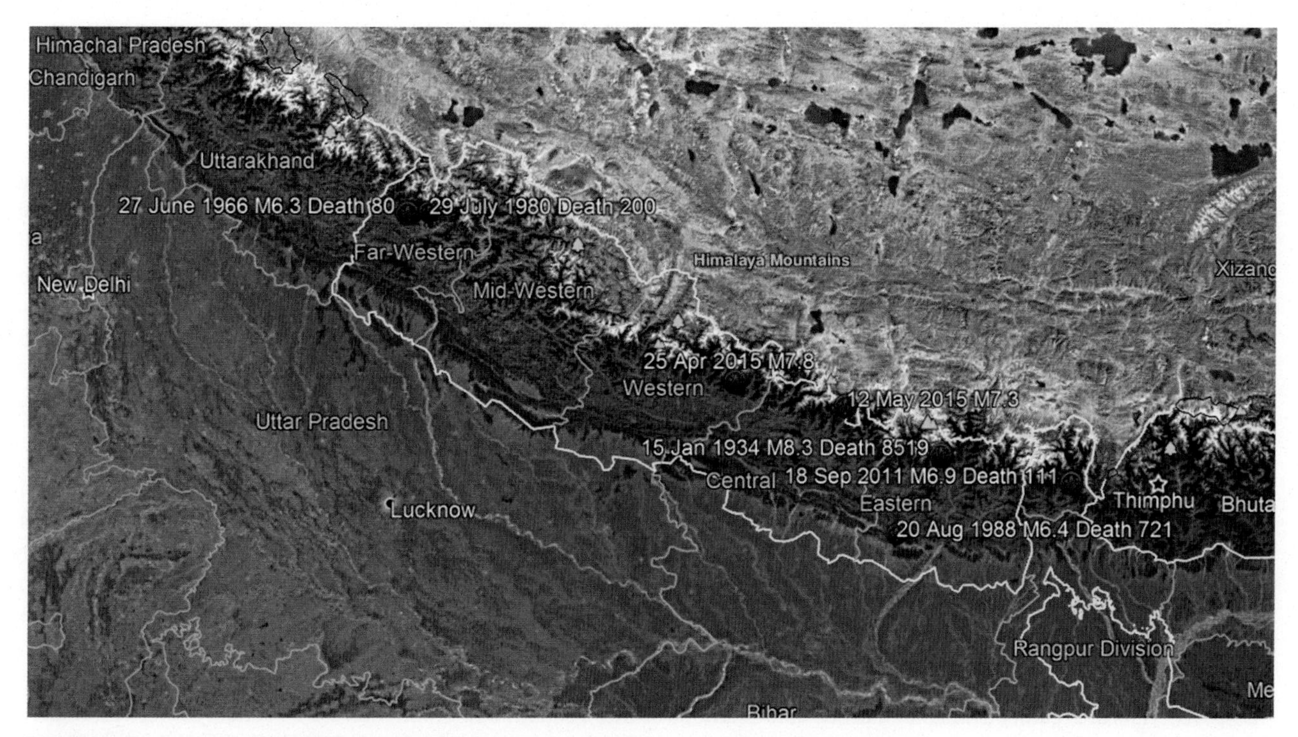

FIGURE 11.3 **Location and magnitude of post-1934 Nepal earthquakes.** *Adapted from Google Earth.*

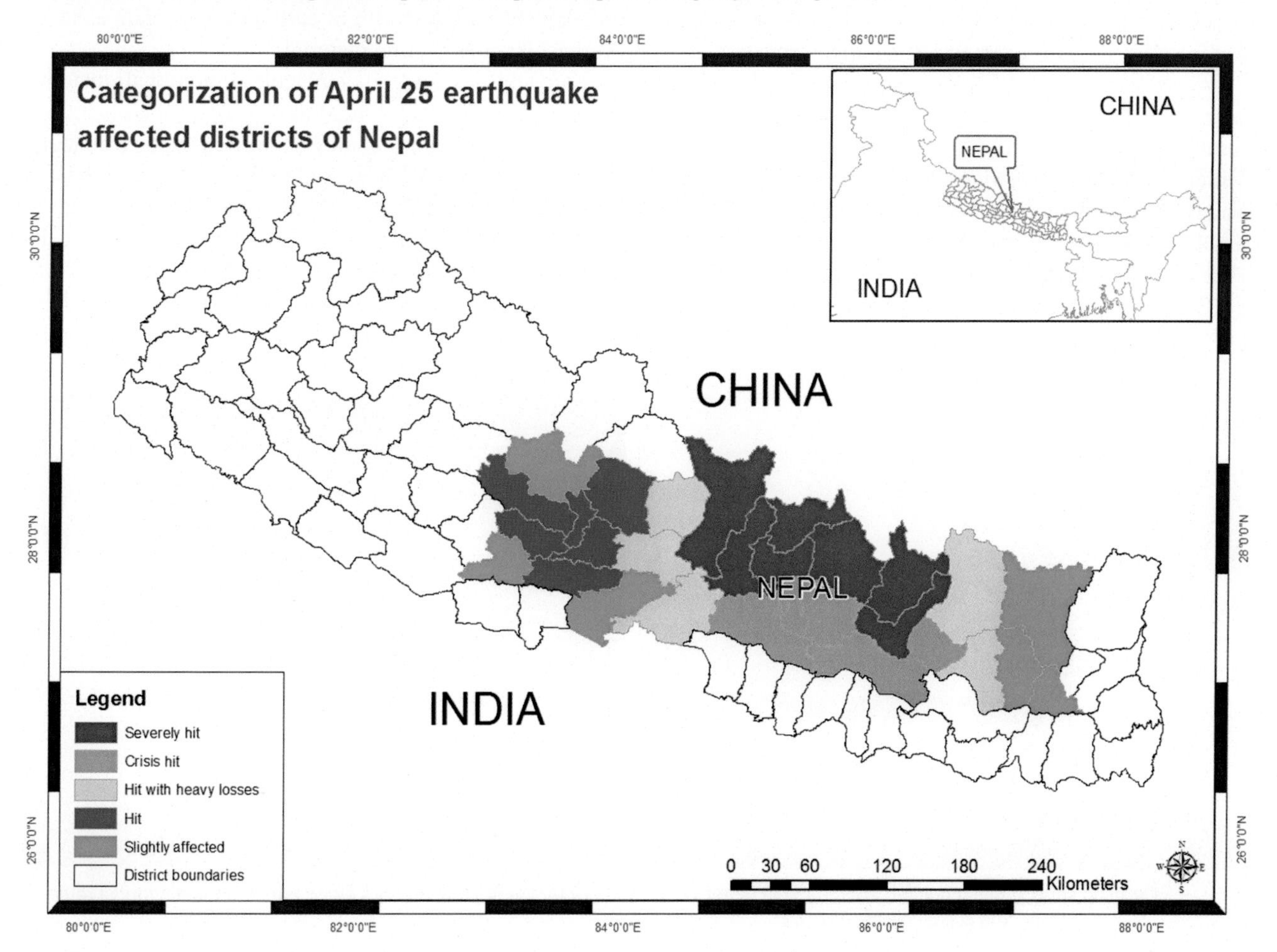

FIGURE 11.4 **Areas in Nepal affected by the 2015 Nepal earthquake sequence.** *Adapted from NPC (2015).*

tricts. Later for prioritizing reconstruction, 14 "severely hit" and "crisis hit" districts were reclassified as "most affected", and another 17 "hit with severe losses", "hit", and "slightly affected" were reclassified as "affected" districts.

11.3 FACTORS CONTRIBUTING TO NEPAL'S EARTHQUAKE VULNERABILITY

As evidenced by the 2015 and earlier earthquakes, Nepal's buildings are highly vulnerable to earthquake-induced shaking and have been a major source of casualties and economic losses. A combination of poor building materials, deficient construction practices, and poor compliance with existing standards and regulations have led to a highly vulnerable building stock in Nepal. The situation is further exacerbated by a limited awareness and understanding of building materials and skills, seismic resilient construction training for craftsperson and professionals, and effective control and monitoring systems, as well as the relative isolation of many parts of the country. Significant focus must be placed on improving the seismic resilience of buildings to address this vulnerability and hence the subsequent sections are directed at issues related to the country's building stock.

11.3.1 Inadequate Building Materials

Building construction in Nepal is mainly governed by local availability of and access to construction materials and technology. Most of the country is inaccessible by vehicular land transport because of its rugged topography and the remoteness of its mountainous regions, posing logistical challenges for transporting building materials (Fig. 11.5). Common building materials in Nepal are stone, fired or unfired clay brick, mud, steel, concrete, bamboo, and timber. The use of these construction materials is mainly based on their local availability and the availability of funds. In hilly and mountainous regions, stone is the most common building material because of its local availability and low cost compared to clay bricks. Similarly, most houses in the Terai and valleys are constructed of clay brick with mud mortar, adobe, timber, and bamboo, as these materials are readily available. Various combinations of these construction materials are commonly found throughout other areas of Nepal. These vernacular buildings are highly vulnerable to earthquake shaking because of their weak construction materials, a lack of integrity between various structural elements, and a dearth of earthquake-resistant elements. Refer to Fig. 11.6 for examples of buildings constructed using vernacular construction materials.

As shown in Fig. 11.7, more than 70% of residential buildings are constructed from vernacular (traditional) materials, with more than 42% constructed using loadbearing mud walls or masonry units laid in mud mortar. In most cash-strapped rural areas, these are generally the only materials available for building construction. A study conducted by the Kathmandu Valley Earthquake Risk Management Project (KVERMP) in 2000 revealed that 63% of school buildings in the Kathmandu Valley were constructed from stone or clay brick with mud mortar or adobe (NSET, 2000). Such mud-based buildings are highly vulnerable to earthquake shaking as observed following the 2015 Nepal earthquake sequence, in which more than 70% of the buildings that suffered damage or destruction were vernacular buildings (Bothara et al., 2016).

Cement and steel construction materials were first introduced to Nepal in the 1950s. The uptake of these materials is growing rapidly for construction of loadbearing masonry and RC frame buildings (see Fig. 11.8 for building typology

(A) Carrying sand in mid-mountain area that has better access and is financially better off

(B) Dragging reinforcing bars along a mid-mountain tourist route

FIGURE 11.5 **Transportation of construction materials in remote areas.**

(A)

A stone residential building with mud mortar masonry on the ground floor and dry stacked stone masonry on the upper storey, Middle Mountains

(B)

A stone school building with mud mortar masonry, Middle Mountains

(C)

A hotel building constructed using mixed materials and structural systems (RC frame and stone on the bottom storey and RC frame and timber on the upper storey), High Mountains

(D)

A mixed masonry residential building (adobe and fired clay brick with mud mortar), Middle Mountains

(E)

Typical clay brick masonry residential building with mud mortar and timber floor and roof structure, Kathmandu Valley

(F)

A mixed-use masonry building (stone in mud on the bottom storeys and clay brick in cement mortar on the top storey with timber floors and RC slab roof), Middle Mountains

(G)

A masonry residential building with adobe on the bottom storey, clay brick with cement mortar on the top two storeys, and an RC slab roof, Kathmandu Valley

(H)

A timber house in the Terai

(I)

A residential building constructed from timber, wattle, and daub, Terai

FIGURE 11.6 **Typical examples of vernacular building types in Nepal.**

examples). However, their use remains limited to areas accessible by vehicular land transport and locations with a strong economy (Bothara and Sharpe, 2003; Bothara et al., 2000). As illustrated during the 2015 Nepal earthquake sequence, building vulnerability has not been reduced significantly despite the growing use of cement and steel in the construction industry, mainly due to inappropriate use of construction materials and noncompliance with earthquake-resistant construction techniques (Dizhur et al., 2016; Gautam and Chaulagain, 2016).

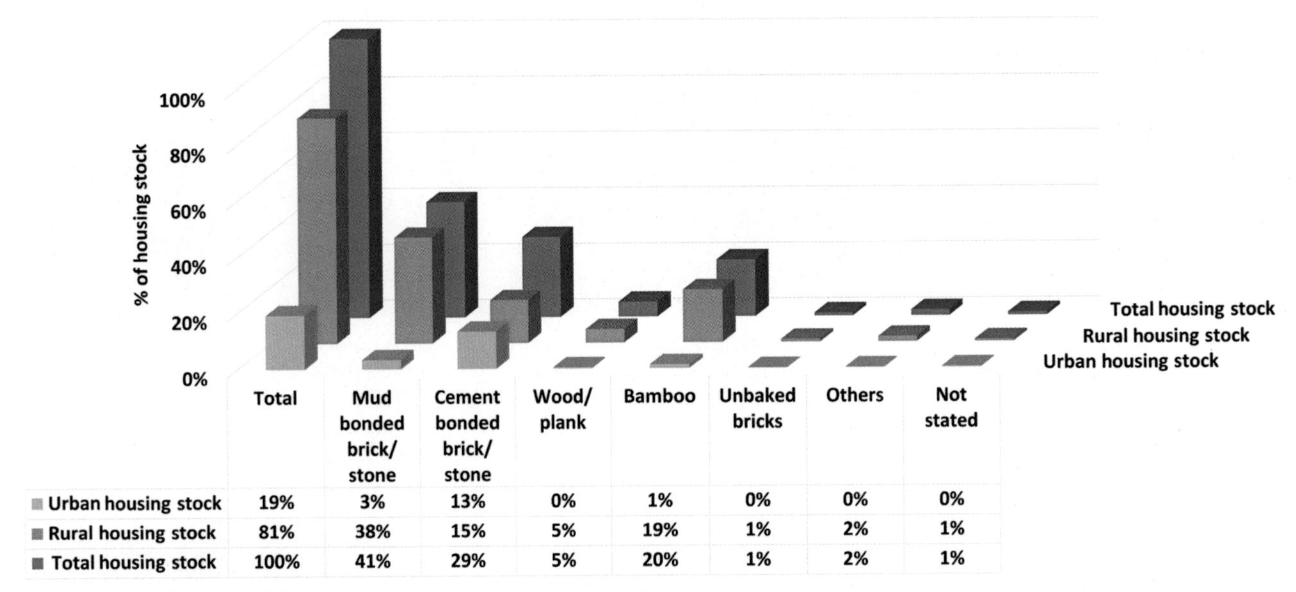

	Total	Mud bonded brick/ stone	Cement bonded brick/ stone	Wood/ plank	Bamboo	Unbaked bricks	Others	Not stated
Urban housing stock	19%	3%	13%	0%	1%	0%	0%	0%
Rural housing stock	81%	38%	15%	5%	19%	1%	2%	1%
Total housing stock	100%	41%	29%	5%	20%	1%	2%	1%

FIGURE 11.7 **Breakdown of residential building typology by walling materials.** *Adapted from CBS (2012); numbers have been rounded off.*

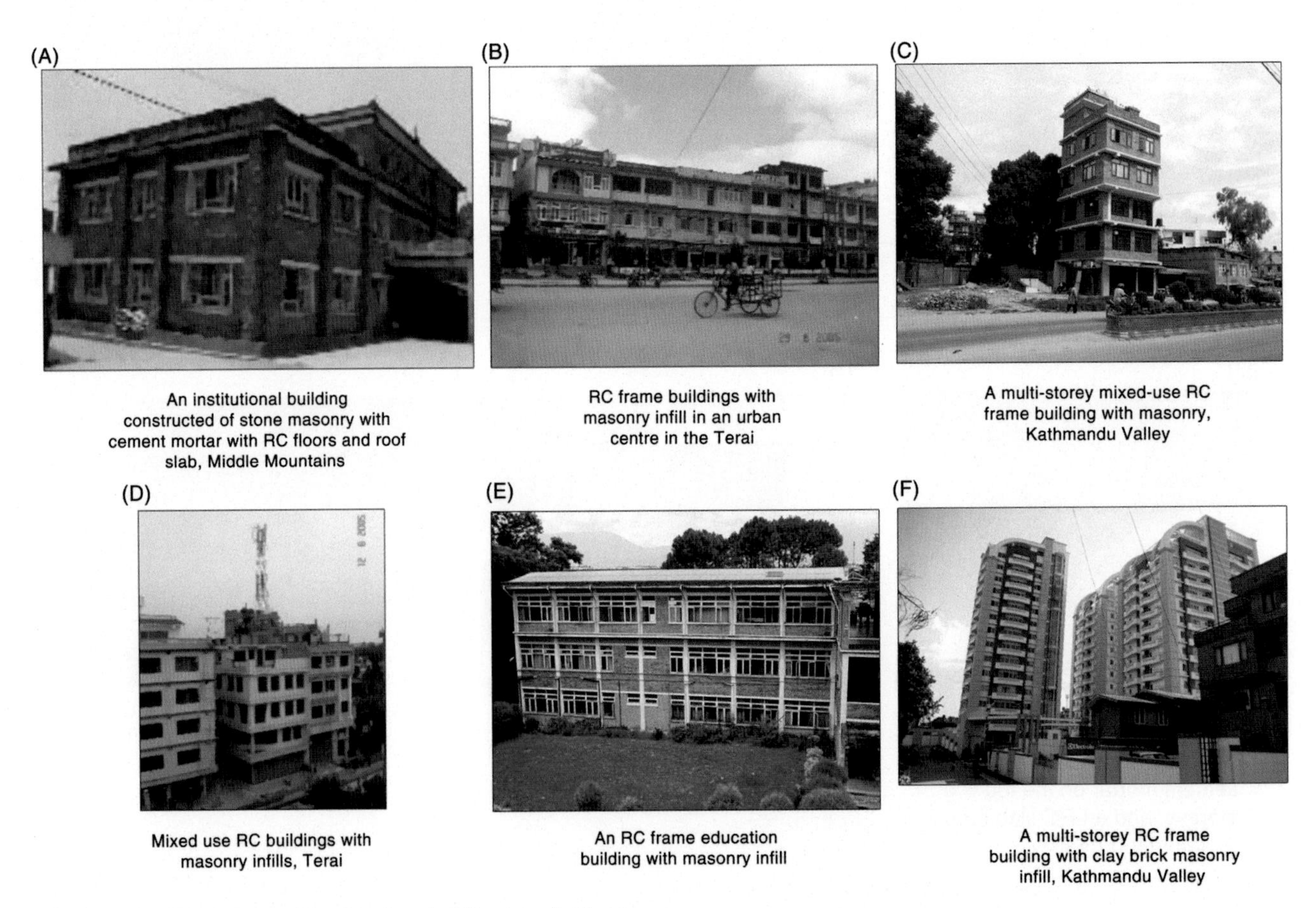

FIGURE 11.8 **Typical examples of modern building types in Nepal.**

11.3.2 Poor Construction and Compliance

A combination of poor building materials, deficient construction practices (Bothara et al., 2000), and a lack of compliance with existing standards and regulations for earthquake-resistant construction (a few common examples are shown in Fig. 11.9) have led to a building stock that is seismically highly vulnerable to earthquakes. Poor construction techniques and lack of compliance affect building quality constructed using vernacular as well as modern materials. In many cases,

(A)

Typical stone masonry wall without bond stones (photograph taken 2017)

(B)

A brick URM building with cement mortar and keys for return walls (photograph taken 1998)

(C)

Insufficient splice length to transfer forces from the upper storey (once constructed) to the lower storey (photograph taken 2006)

(D)

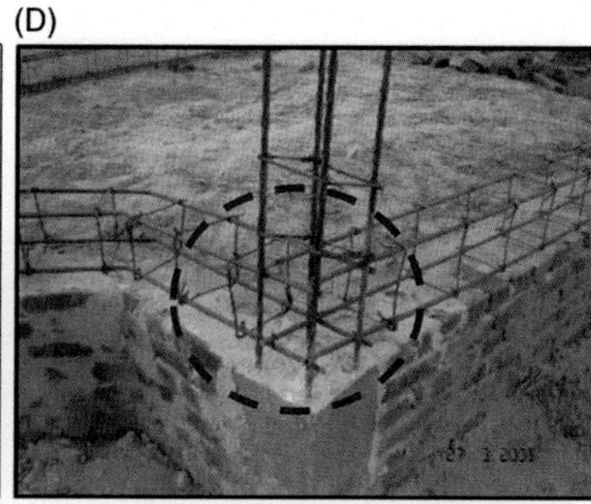

Beam-column joint; note the lack of stirrups in the joint and insufficient anchorage of beams bars in the column (photograph taken 2006)

FIGURE 11.9 **Common deficiencies of building construction practices.**

the use of modern construction materials has increased rather than decreased vulnerability. Anhorn et al. (2015) argued "a profound loss in seismic culture", as manifested through a decline in local building practices, due to a preference to new modern materials in Nepal. For example, it is common to observe stone masonry walls that are constructed without bond stones (see Fig. 11.9A). A lack of bond stones between wythes could lead to delamination of walls during earthquake shaking, leading to a building's collapse. It is common to observe brick keys along a vertical plane at wall corners, which minimizes the integrity between face-loaded and return walls (see Fig. 11.9B). Similarly, deficient splice length and deficient anchorage of beam rebars in columns are few examples (Fig. 11.9C and D) of defective detailing in RC buildings.

In response to the death and destruction caused by the 1988 East Nepal earthquake, the Nepal Building Code Development Project (NBCDP) was established. This project was responsible for preparing various building codes, standards, and guidelines to improve the seismic safety of buildings in Nepal and to develop the legislative framework for their implementation. The outputs from the NBCDP were only promulgated in 2004. A study conducted in 2014 by Manandhar (2015) revealed that over 50% of households surveyed in two districts of Nepal were unaware of the existence of these building codes. The study also showed that within the Kathmandu Valley, more than 35% of new houseowners did not consult engineers before constructing their building.

A study conducted in 2000 by Dixit (2008) revealed that the majority of buildings in both urban and rural areas of Nepal were constructed by their owners, who mostly employed an artisan to direct operations (NSET, 2000). The resulting buildings are characterized by a high degree of informality and highly varied decisions regarding materials and design. Traditional artisans or craftspeople without formal training often play a pivotal role in overall construction activity, and owners rely on them heavily for all types of building-related advice. This scenario has led to the construction of a large number of earthquake-prone buildings in the country.

Building consent is mandatory (in principle) in urban areas of Nepal, but there was no requirement for the submission of structural designs and drawings for all types of buildings. The territorial authorities generally suffer from poor institutional and technical capacity to implement building strength-related provisions (Parajuli et al., 2000; WEF, 2015), although a few authorities have begun to require the consent of structural design (Gautam and Chaulagain, 2016). In addition, there is no current legal framework or compliance system in Nepal for improving the seismic performance of the existing building stock.

11.3.3 Increased Use of Marginal Land

Nepal is experiencing tremendous demographic growth, from 8.5 million to 29 million people over the last 65 years (Countrymeters, 2017). Approximately 17% of the population lives in urban areas of Nepal, and since the 1970s, the urban population has grown at a rate of about 6% (Muzzini and Aparicio, 2013). The increasing population and rapid urban growth combined with unplanned settlements have led to increased human population in areas traditionally barred from settlement, such as river basins, steep slopes, and areas with soft soil (Fig. 11.10A–C), all of which are locations prone to liquefaction, land and rock slides, and shaking amplification (Bothara and Karmacharaya, 2015). Findings from various studies (Anhorn et al., 2015; JICA and MOHA, 2002; Joshi et al., 2013) and observations made in different parts of the country show similar scenarios in rural areas, small and large towns, and cities in all physiographic regions.

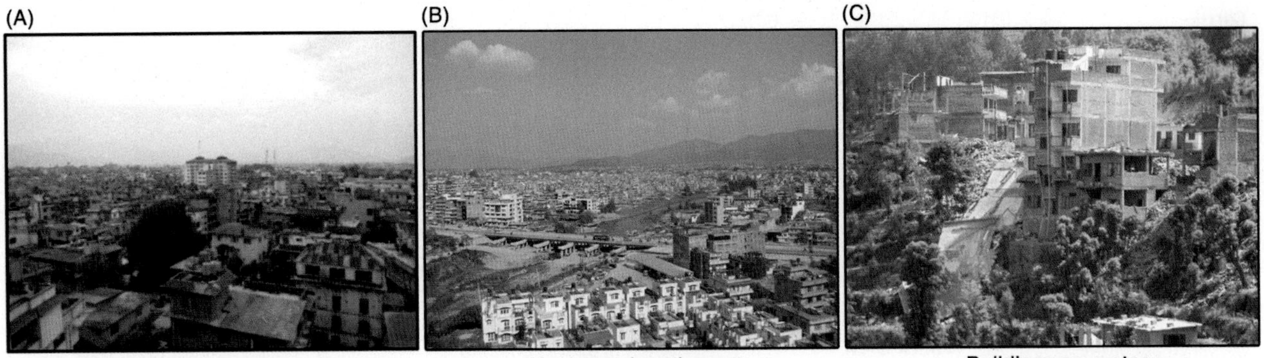

(A) Haphazard city settlement in Kathmandu (2016) (B) Haphazard settlement along a river in Kathmandu Valley (2016) (C) Buildings on a steep slope, Irkhu, Sindhupalchowk, (2015)

FIGURE 11.10 **Marginalization of land and settlements.**

(A) Subdivision of house; note one structure below second floor and multiple structures above the second floor, Kathmandu Valley (B) Small land package with a multi-storey building, Kathmandu Valley (C) Mixed building typologies in a single block without a seismic gap, Kathmandu Valley

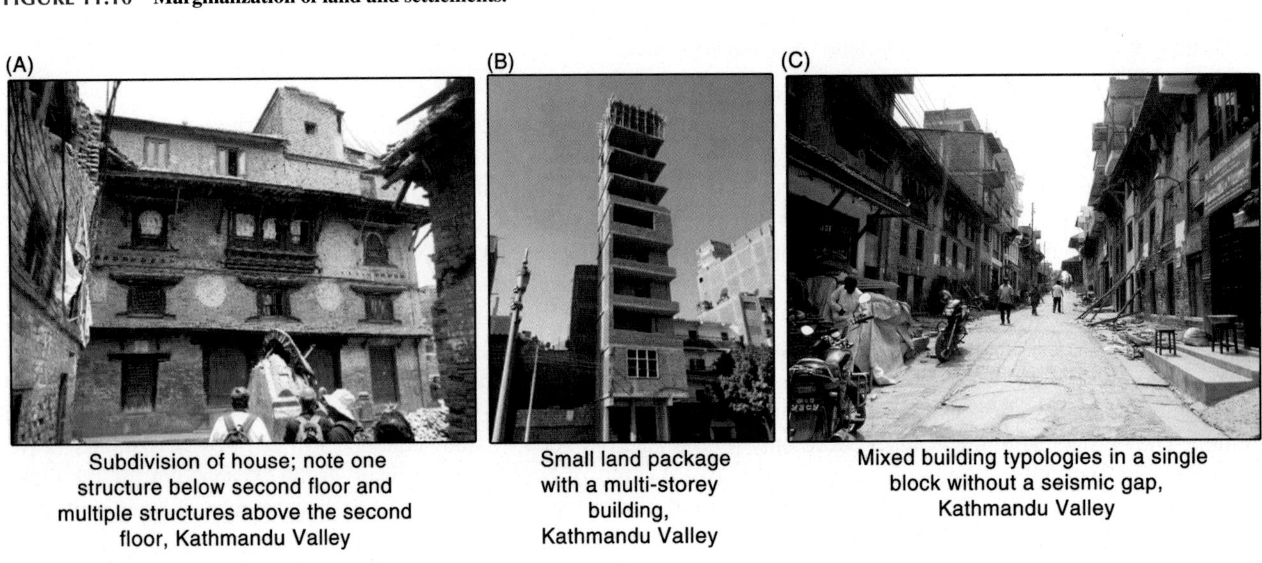

FIGURE 11.11 **Land fragmentation issues.**

Population growth, increasing land prices, and inheritance practices have led to the fragmentation of land into small packages. Based on Nepal's cultural and legal practices, all male children have an equal right to inheritance from their parents, meaning that land is equally subdivided among male children. This practice often results in the subdivision of houses into parts (Fig. 11.11A). Over time, one sibling may decide to demolish his inheritance and construct a new house, potentially leading to high seismic vulnerability of the entire house. Increasing land prices have further exacerbated the situation, forcing the construction of tall, narrow buildings (Fig. 11.11B). With increasing urbanization, population density is also rapidly increasing. In a few areas in the Kathmandu Valley, density up to 1000 people per hectare has been reported (EMI, 2010). Increased population density has further increased the earthquake vulnerability of the population.

The country is also experiencing a rapid change in building typologies. It is a common practice to replace demolished older buildings in a block with a modern building typology, without a seismic separation gap (Fig. 11.11C) between the new and old buildings. The interaction between various building typologies during seismic shaking creates a highly complex situation and potentially increases the vulnerability of the buildings.

11.3.4 Limited Appreciation of Seismic Risk

Past studies (Tuladhar et al., 2014, 2015; Wesson, 1998) and observations, including anecdotal evidence and first-hand experience by the authors, indicate a limited perception of earthquake risk among the public as well as among policy- and decision-makers. Although Nepal suffered a major earthquake in 1934 that resulted in approximately 9000 deaths, there are limited indications of significant enhancement of earthquake safety since then. Nepal is currently marred by many deficient policies, which has impeded progress to improve seismic safety within the country. Although most of Nepal's fatalities

TABLE 11.2 Household Income in Nepal (Average of Per Capita in Real 1995/1996 Nepali Rupees)

Area	Real Mean per capita expenditure (NRs/year)		Total change
	1995/1996	2003/2004	
Nepal	7,191	10,129	41%
Urban	13,115	19,601	49%
Rural	6,753	8,484	26%

US $1 ≈ Nepal Rupees 100 in 2017.
Source: Dillion et al. (2011).

and financial losses during earthquakes have been the result of buildings with insufficient earthquake resilience, relevant authorities have taken limited initiatives to improve this deficiency. This is an even more significant issue in rural areas because 83% of Nepal's population is rural (CBS, 2015).

11.3.5 Poor Social Indicators

The low-income level of many Nepalis has been a major contributor to the country's vulnerability to earthquake disasters. Studies have shown that economically disadvantaged individuals are more vulnerable to natural disasters than their economically advantaged counterparts (UNISDR, 2009). Poverty in Nepal is deeply entrenched and complex in nature, with 31% of Nepalese living below the poverty line (Panthhe and McCutcheon, 2015), most of whom reside in rural areas (ADB, 2013; Rigg et al., 2016). Table 11.2, which presents the per capita income of Nepal, shows that income in rural areas is significantly lower than in urban areas.

Studies (UNDP, 2014) have shown that understanding of vulnerabilities increases with increasing human development index. In 2011, the literacy rate among Nepal's rural population was 62%, compared to 82% in the urban population (CBS, 2015). A study published by Panthhe and McCutcheon (2015) illustrated that education level in the urban areas is significantly higher than that in rural areas. The urban population is more exposed to information on seismic safety than the rural population. The rural population is far less economically privileged than their counterparts in the urban areas. These social indicators suggest that there is less access to, understanding of, and ability to utilize resources and information in rural areas. Despite this dismal scenario, there are limited policy directives and EDRR programs in rural areas to protect the rural population from potential earthquake disasters. This deficiency contributed to the extensive damage and destruction caused by the 2015 Nepal earthquake sequence in rural areas.

11.3.6 Limited Training and Knowledge Transfer

In Nepal, engineers and technicians are typically required to complete tertiary education in order to work in the construction industry, whereas craftspeople acquire their knowledge by the passing down of skills through generations. The engineering education system in Nepal is mainly focused on building construction using modern materials, and seismic design is not part of the regular engineering curriculum (although some education on this subject has been included recently). As a result, most graduate civil engineers are unable to provide meaningful input on the seismic resilience of buildings constructed with even modern materials. In contrast, Fig. 11.7 indicates that more than 70% of the current Nepalese building stock is constructed using vernacular materials. This scenario is expected to continue in the future due to socioeconomic reasons, resulting in the construction of additional earthquake vulnerable buildings (UNESCO, 2010).

More than 82% and 92% of buildings, respectively, in urban and rural areas of Nepal are produced by craftsmen who have no training on earthquake-resistant construction and no access to engineering resources. Although this scenario has changed slightly in recent times, any investment made for the formal training of craftspeople/masons is very limited. Fig. 11.12 shows that despite craftsmen being responsible for constructing the large majority of the building stock in Nepal without any engineering resources, investment into their training is insignificant. The limited training that is available is often short and overly structured, which is not useful given the craftsmen's generally low education levels and lack of exposure to formal education (Bothara, 2003). This scenario significantly impedes improvement of the seismic resilience of the building stock in Nepal.

Nepal is currently experiencing a major socioeconomic change, which has impacted the labor market. This change is due to large-scale foreign migration and resulting remittances (Manandhar, 2015), increasing migration toward cities, an improved economy within Nepal, and better education. As a result, there has been less adoption of traditional construction

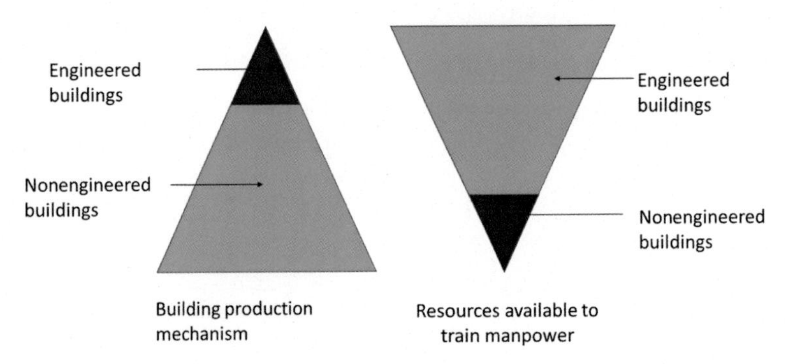

FIGURE 11.12 **Investment versus building procurement mechanism (indicative).**

professions and a lack of transfer of related knowledge to the next generation. While there is no existing research on the impact of outward migration from rural areas on the rural construction industry, it has been observed that this type of outward migration has significantly reduced the productive-age population (age group between 20 and 40 years old) in the rural areas. This migration has significantly hindered reconstruction efforts in these areas after the 2015 Nepal earthquake sequence.

11.4 GLOBAL INITIATIVES FOR DRM

Since the late 1980s, the UN has led a concerted global initiative for DRR. This initiative has ranged from policy creation to implementation, following the UN's realization of the increasing impacts of disasters on human life and the intrinsic relationship between sustainable development and disasters. These initiatives, as discussed in the following sections, have promoted a culture of disaster prevention, which has encouraged a shift from post- to predisaster initiatives. They have also resulted in the development of a proactive policy framework and the initiation of DRM activities in Nepal. Fig. 11.13 presents a brief summary of the relationships between international DRR initiatives and the Nepali DRR policy framework formulation, some of which are discussed below.

11.4.1 International Strategies and the Nepali Response

In 1989, the UN General Assembly (in resolution A/RES/44/236) launched a far-reaching global undertaking to reduce human loss and the impacts of disasters (UN, 1989) by declaring the International Decade of Natural Disaster Reduction (IDNDR) program for 1990–1999. The IDNDR was the first concerted international effort aimed at disaster management

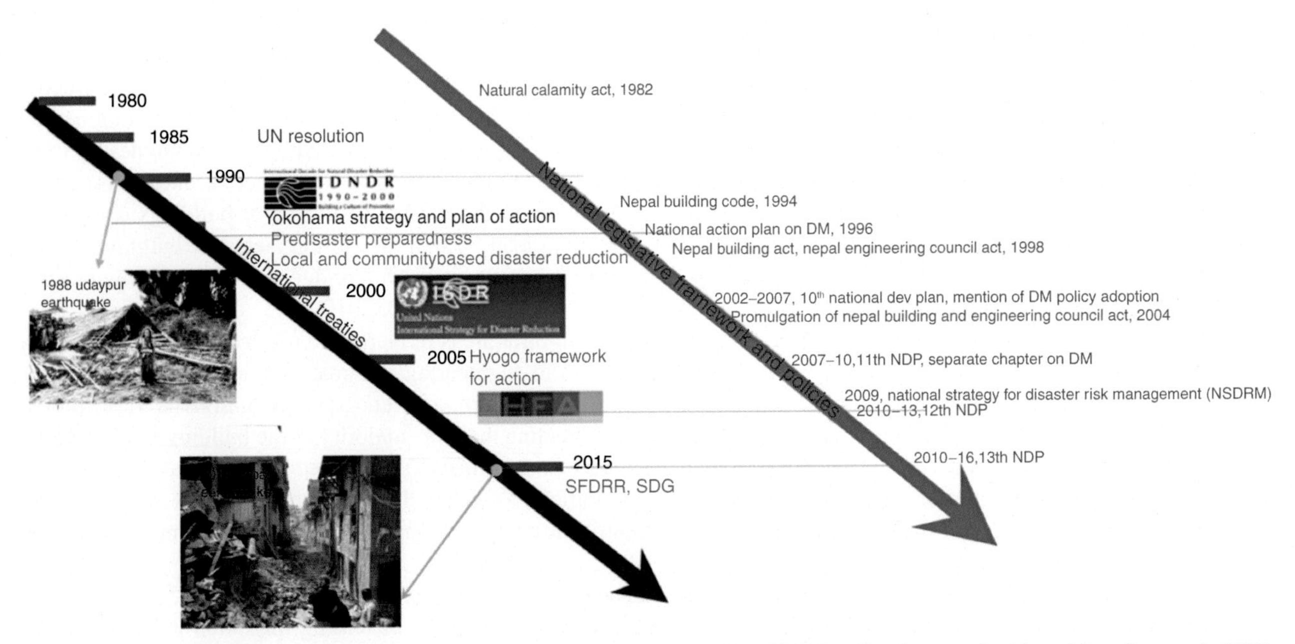

FIGURE 11.13 **Relationships between major international DRR initiatives and Nepali DRR policy framework.** *Adapted from Shaw et al. (2016).*

on the global level. In response to the UN call, Nepal constituted a National IDNDR committee, which developed an action plan with a focus on predisaster management. This action plan identified priority item groups, activities, and executing agencies and established a time line for these actions (MOHA, 1996).

Following on from the UN resolution, as part of the World Conference on Natural Disaster Reduction in 1994, an action plan was developed with an extensive agenda that provided a new direction for disaster management. The action plan set guidelines for action on prevention, preparedness, and mitigation of disaster risk, with a focus on the least developed countries, including landlocked and island nations. The strategy focused on strengthening national capacities and legislative frameworks through a participatory approach and recognition of local knowledge and traditional methods and encouraged a shift from postdisaster response to disaster prevention and management (IDNDR, 1994).

The next milestone in global initiatives for disaster management was the International Strategy for Disaster Reduction (ISDR), which was a UN-sponsored global drive aimed at building disaster-resilient communities and was based on experiences with the IDNDR program. The ISDR embodied principles articulated in a number of major documents adopted during the 1990s, in particular, the "Yokohama Strategy for a Safer World: Guidelines for Natural Disaster Prevention, Preparedness and Mitigation" (UNISDR, 1994). The goal of the ISDR was to promote disaster reduction as an integral component of sustainable development. This strategy had an effect in Nepal, where a network of UN agencies, intergovernmental groups, and nongovernmental or civil society organizations teamed up for a coordinated effort for DRR following the ISDR principles.

In the early 2000s, the Hyogo Framework for Action (HFA) (ISDR, 2005), which was a 10-year action plan, was endorsed by the UN General Assembly in resolution A/RES/60/195 following the 2005 World Disaster Reduction Conference. This action plan recognized an intrinsic relationship between disaster reduction, sustainable development, and poverty eradication. The HFA declaration also placed equal emphasis on formulating appropriate national policies and building capacity at the community level. In response to the HFA, the Government of Nepal adopted the National Strategy for Disaster Risk Management (NSDRM) in 2009, which outlined a vision of developing Nepal as a disaster-resilient community.

The most recent plan is the Sendai Framework for Disaster Risk Reduction (SFDRR) 2015–2030, which was adopted in Sendai, Japan, on March 18, 2015 (UN, 2015) and builds on disaster management experience gained in the previous decades. The plan prioritizes the understanding of disaster risk, strengthening governance to manage disaster risk, investing in DRR for resilience, enhancing disaster preparedness for effective response, and "Build Back Better" in recovery, rehabilitation, and reconstruction. The effects of SFDRR on DRR in Nepal are yet to be observed.

11.4.2 National Legislative Framework and Policies

11.4.2.1 National Calamity (Relief) Act of 1982

Nepal has a relatively short history of managing predisaster efforts in a coherent and systematic way. The country's legal framework for disaster management was put in place in 1982 with the promulgation of the Natural Calamity (Relief) Act. This act allocated the responsibility of preparing for and responding to disasters to the government and provided an administrative structure for disaster management in the country (Pokharel, 2015). Other legislative and policy-level initiatives followed, some of which are discussed below.

To address the shift from post- to predisaster risk reduction efforts, the Government of Nepal has replaced the National Calamity (Relief) Act of 1982 by a comprehensive Disaster Risk Reduction and Management Act, 2017.

11.4.2.2 Relevant Acts and Codes

Until 1994, Nepal had no formal regulations or documents setting out requirements or best practices for achieving the satisfactory seismic performance of buildings. Prior to this point, it was common to use Indian standards for the design of buildings and infrastructure.

As previously mentioned, following the 1988 East Nepal earthquake, the NBCDP was established with funding from the UN Centre for Human Settlements Program (UNCHS(UN-Habitat)) (HMG, 1994) to prepare a building code for the country. Planning for a national building code was guided by the need to tailor it to the local economic climate, lack of trained manpower, low seismic risk awareness levels, and limited accessibility to technology and quality construction materials. The NBCDP project also recognized that enforcement of stringent rules or any standards requiring full engineering input or even the requirement for professional advice for the design of small buildings would be inappropriate for the time and context (Bothara et al., 2000).

The NBCDP proposed a strategy for progressive safety improvements, which involved gradual progression from nonengineered to preengineered to engineered building design and construction. The strategy also allowed the design and construction of buildings following the latest research and international standards. Accordingly, the National Building Code included several construction standards and guidelines on preengineered or nonengineered construction, including mud-based buildings (MHPP, 1994b). From a compliance perspective, the standards were divided into four groups: (1) international state of the art; (2) professionally engineered structures; (3) rules of thumb; and (4) guidelines for rural construction.

The NBCDP also developed a management plan to implement the building code and drafted the Engineering Council Act and Building Act for its effective enforcement. However, due to various reasons, these acts and the building code were not promulgated until 2004.

The Building Act empowered the Department of Urban Development and Building Construction (DUDBC) to provide directives, set building compliance standards, and develop and promote safer building construction practices. The act required the formation of a Building Council and empowered DUDBC to review the design of buildings exceeding a certain size and occupancy and grant approval accordingly (HMG, 1998a). However, the act's focus was on urban areas and much was left to the discretion of territorial authorities, resulting in a lack of effective enforcement and compliance.

The Engineering Council Act (HMG, 1998b) and Council Regulation (HMG, 2000) were also enacted in 2004. This act enabled the formation of a professional engineering body, named the Nepal Engineering Council. However, the act does not encompass a competence-based registration system or professional liability, which has limited the meaningful development of engineering as a competent profession in Nepal. After the 2015 Nepal earthquake sequence, the Nepal Engineering Council has initiated a process to implement a competence-based registration system and develop related guidance.

Another relevant piece of legislation is the Local Self-Governance Act (HMG, 1999). Although this act is not directly concerned with disaster-related issues, it does provide territorial authorities with the ability to formulate and implement plans and programs in their districts (e.g. the incorporation of earthquake risk reduction activities). The Local Self-Governance Act also empowers municipalities to monitor and control building construction in their territories, including consent for building activities. It should be noted that this empowerment is only applicable in municipal (urban) areas, and the act is unclear regarding what needs to be submitted for the design documents required for building consent.

Although the building consent process remains a revenue-generating tool for territorial authorities, a few municipalities have recently begun considering structural engineering design when granting building consent (Gautam and Chaulagain, 2016). However, most municipalities lack the capability to incorporate structural engineering design and consequently face logistical challenges.

11.4.2.3 1996 National Action Plan on Disaster Management

In the 1990s, in response to the UN General Assembly's decision A/RES/44/236, Nepal formulated a national disaster mitigation program and developed a draft National Action Plan on Disaster Management. This action plan was updated following the Yokohama conference to incorporate the Yokohama strategy and make it more pragmatic (MOHA, 1996). The plan demonstrates a paradigm shift from rescue and relief operations to disaster management, which was a shift that has had a crucial impact on macrolevel development planning.

Nepal's 1996 action plan identified four key areas: preparedness, response, reconstruction and rehabilitation, and mitigation. The plan outlined priority items, activities, and timelines for their completion and indicated the agencies responsible for executing these actions. In all the identified areas, the plan acknowledged actions for long-term EDRR (MOHA, 1996).

11.4.2.4 2000s: Relevant National Development Plans

Over time, the policy environment in Nepal for DRR has become more comprehensive and conducive. Nepal's National Development Plans (NDPs) provide policy directives for general development activities. The 10th NDP (2002–2007) (NPC, 2002) was the first plan to acknowledge DRR as a priority, which was further reinforced in subsequent plans.

The 10th NDP emphasized policy formulation for DRR and strengthened institutional mechanisms, risk assessment, and information collection and dissemination. Although little was achieved during the period the plan addressed (Sainju, 2015), the plan did lay a foundation for future programs. Building on increasing awareness and understanding of disasters, the 11th NDP (NPC, 2007) emphasized a shift of focus from postdisaster response to preparedness, prevention, and mitigation as well as the need for change in prevailing policies related to disaster mitigation. The 12th NDP (2010–2013) (NPC, 2010) further expanded provisions on DRM and acknowledged the linkage between sustainable development and DRR. The 12th NDP also adopted a specific policy on DRR and outlined the roles of international nongovernmental organizations (INGOs), nongovernmental organizations (NGOs), community-based organizations (CBOs), and the private sector in DRR. More importantly, the 12th NDP emphasized the inclusion of stakeholders in developing DRR strategies. The 13th NDP (2013–2015) (NPC, 2013) acknowledged DRR as one of the components of sustainable development and focused on bringing disaster risk into the mainstream development process.

11.4.2.5 2009: National Strategy for Disaster Risk Management

The NSDRM program was developed after the Government of Nepal recognized the need for a meaningful and integrated document based on the HFA referred to earlier. The program strategy was to guide and ensure effective disaster management (MOHA, 2009) and was framed around the five priority areas of the HFA. The strategy identified 29 activities for action and proposed a new institutional arrangement for DRM, which entailed the formation of a National Disaster Management Council to be chaired by the Prime Minister (this has not yet been developed).

11.4.2.6 2011: Nepal Risk Reduction Consortium

The Nepal Risk Reduction Consortium (NRRC) is probably the most integrated project to support seismic resilience in the country. This consortium was formed in 2009 by the Government of Nepal and a group of international organizations working to promote the UN's ISDR. The NRRC was launched in 2011 to develop a long-term Disaster Risk Reduction Action Plan based on the NSDRM. It is a unique arrangement that unites aid and development agencies (e.g. the International Federation of Red Cross and Red Crescent Societies, United Nations Development Program (UNDP), United States Agency for International Development (USAID), United Kingdom Department for International Development (DFID) and various NGOs) and financial institutions (e.g. World Bank and Asian Development Bank) in partnership with the Government of Nepal to reduce Nepal's vulnerability to natural disasters. Based on the HFA and NSDRM, the NRRC identifies five flagship priorities for sustainable DRM and immediate action: (1) school and hospital safety; (2) emergency preparedness and response; (3) flood risk management; (4) integrated community-based DRM; and (5) policy/institutional support for DRM (GON, 2013).

11.5 EFFORTS TO REDUCE EARTHQUAKE DISASTER RISK

11.5.1 Kathmandu Valley Earthquake Risk Management Project

More concerted efforts for earthquake risk reduction in Nepal began with the Kathmandu Valley Earthquake Risk Management Project (KVERMP), which received core funding from the Office of Foreign Disaster Assistance of USAID (Dixit et al., 2000). This project brought together stakeholders such as national government agencies, municipal governments, professional societies, academic institutions, schools, and international agencies present in Kathmandu Valley. As part of the project, efforts were initiated on earthquake preparedness and the evaluation of Kathmandu Valley's earthquake risk, and an action plan was developed to mitigate the risk. Risk was presented in laypeople's language (Fig. 11.14A), the School Earthquake Safety Program (SESP) was initiated (described later in this section), mason training programs were developed, and a community-based approach for EDRR was adopted. After the conclusion of the KVERMP, the National Society for Earthquake Technology-Nepal (NSET) extended its programs to other parts of Nepal (NSET, 2017). The success of KVERMP can be attributed to its down-to-earth approach, understanding of local issues, community-based approach, and the presentation of scientific knowledge in simple terms.

11.5.2 Awareness Raising Activities

Significant efforts have been made in Nepal (mostly in urban areas) in recent years to raise public awareness of seismic risk and how its effects can be mitigated. These efforts consider safety as a sociocultural issue, which requires a bottom-up approach to achieve effective DRR. To remind people of impending seismic risk, the Government of Nepal has declared January 15 (the day of the 1934 earthquake) as Earthquake Safety Day. The government, various institutions, and territorial authorities have engaged in innovative efforts for public awareness such as earthquake safety exhibitions; activities to commemorate Earthquake Safety Day; shaking table demonstrations; an earthquake parade; publication and distribution of brochures, pictorial books, illustrated story books, and pamphlets; radio and television programs; street dramas; and dissemination of news and bulletin articles (Fig. 11.14). These efforts provide an example of public–private partnership supported by international agencies.

11.5.3 School Earthquake Safety Program

In response to the high risk faced by schools in the Kathmandu Valley, in 1998, NSET initiated the SESP under the KVERMP. The basic assumption was that schools serve as a community's social and intellectual center to propagate new ideas. The program not only allowed for the strengthening of school buildings and their nonstructural components, but also provided an opportunity to interact and work together with the government, territorial authorities, and communities. In

(A)

Earthquake scenario of Kathmandu Valley (Source: NSET)

(B)

Preparedness for earthquake safety (Source: NSET)

(C)

Street drama for raising awareness of earthquake safety (Source: NSET)

(D)

Brochure on improved construction practices

(E)

'Shock Table' demonstration to convince people of the effectiveness of earthquake-resistant construction (Source: NSET)

(F)

A panoramic view of Earthquake Safety Day (Source: NSET)

FIGURE 11.14 **Public awareness activities and publications.**

addition, the program provided an opportunity to train local craftspeople, raise awareness, develop and disseminate technology, implement the building code, and, most importantly, to influence the next generation (Bothara et al., 2002; Bothara et al., 2004).

In the late 1990s, the SESP culminated in strengthening many school buildings (Fig. 11.15) using cost-effective measures (Bothara et al., 2004). The program was later extended to include earthquake preparedness training for students and teachers (Fig. 11.15C), preparation of earthquake emergency response plans, first aid training, demonstrations of light search and rescue operations, and inclusion of DRM in the school curricula (Dixit et al., 2013; MOHA, 2016). As part of the NRRC's flagship program, these activities have now been extended to other parts of the country. The Government of Nepal has included the retrofit of school buildings in its annual plan, and this work has been completed in many areas of the country (MOHA, 2015).

11.5.4 Hospital Safety Improvements

Studies prior to the 2015 Nepal earthquake sequence (Dixit et al., 2014; MOH, 2003; WHO, 2002) showed that the structural and nonstructural components and contents of most hospital buildings in Nepal faced high seismic risk and it was

(A)

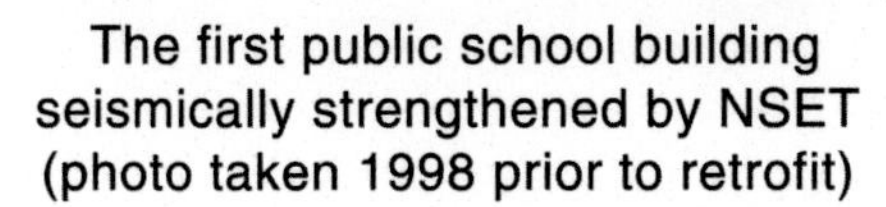

The first public school building seismically strengthened by NSET (photo taken 1998 prior to retrofit)

(B)

The same school building as in (A)

(C)

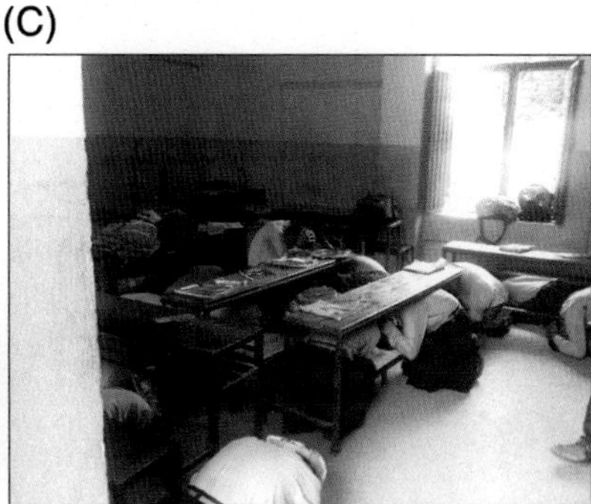

School earthquake drill

FIGURE 11.15 **Seismic retrofitting of school building.** *Source: NSET.*

concluded that these hospitals were expected to be inoperable during even medium earthquake shaking due to nonstructural and content damage. Accordingly, preparedness programs were implemented by the Government of Nepal in public and private hospitals (MOHA, 2016). As a result of related interventions, hospitals performed better during the 2015 Nepal earthquake sequence than would otherwise have been the case (Maharjan et al., 2015).

11.5.5 Human Resource Development

11.5.5.1 Professional Training

In recent decades, Nepal has made significant progress in training in the areas of engineering and DRM. In the mid-1980s, the country's universities offered courses only for a Bachelor of Engineering, whereas many universities now offer post-graduate courses in structural, earthquake, and geotechnical engineering as well as DRM. Courses on DRM are also now available at the bachelor's level. Various NGOs and INGOs also frequently conduct DRM training, including aspects of earthquake engineering for people of different levels such as disaster managers, engineers, and bureaucrats (Fig. 11.16). These changes in professional training have resulted in more professionals in the country and increased public accessibility of their services. The increase in academic activities and number of professionals has also enhanced the exchange of ideas, thoughts, and research findings through conferences and seminars, which in turn has helped EDRR (Chamlagain, 2009).

11.5.5.2 Craftsperson Training

Although craftsperson training in Nepal began after the 1988 East Nepal earthquake to boost reconstruction efforts, in reality the managed and sustained training efforts did not get underway until 1998 when the hands-on training programs

(A)

Formal classroom training

(B)

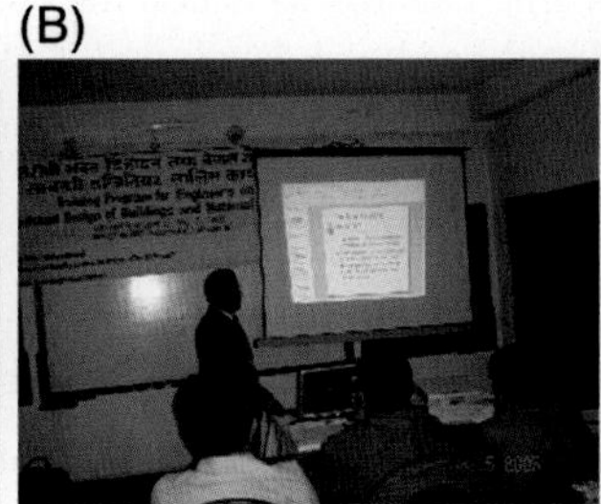

Engineer training

(C)

Mason training on the effect of pre-soaking bricks on mortar and masonry strength (1998) (Bothara, J. K., 2003)

(D)

On-site training: learning ductile detailing skills

FIGURE 11.16 **Training on earthquake-resistant design and construction.**

(A) (B)

Search and rescue training First responder's training

FIGURE 11.17 **Light search and rescue training and first responder training.** *Source: NSET*

were started as part of SESP. Recognizing the socioeconomic climate and craftspeople's potential influence in the community for implementing and sustaining earthquake-resistant construction, NSET took a pragmatic approach to training. This aspect of SESP was a community-based effort in which local craftspeople were encouraged to work toward the goal of training, which included developing a mechanism to access potential houseowners. The training covered both classroom and hands-on sessions (Fig. 11.16). Informal, flexible course materials were designed to address local issues and ensure that the training was interactive (Bothara, 2003). In the last two decades, many institutes, NGOs, and CBOs have started to deliver similar training activities. These organizations have assessed rural areas with increased activity following the 2015 earthquake sequence and have offered training for trainer courses to civil engineers and junior engineers to develop a trickle-down effect. However, there is currently no accreditation or quality control mechanism in place for such trainings.

11.5.5.3 Postearthquake Response Training

NSET conducts Medical First Responder and Collapsed Structure Search and Rescue (CSSR) training in collaboration with the MOHA, Government of Nepal, to implement the Program for Enhancement for Emergency Response (PEER). The program has trained responders from the Nepal Army, Nepal Police, and Nepal Red Cross Society, who are mandated with first responder tasks. Several other institutions also provide end-user training on basic emergency response, community search and rescue (Fig. 11.17), first aid, and damage assessment (Dixit et al., 2015; Jimee et al., 2012; KU, 2010).

11.5.6 Upgrading and Strengthening the Seismological Network

The Department of Mines and Geology (DMG) is an arm of the Government of Nepal that began microseismic monitoring in 1978 with the installation of a short-period vertical seismic station. The number of stations was gradually increased to create the National Seismological Network, which comprised 21 short-period seismic stations in 1998, with an additional seven accelerometer stations added in 2012. These stations provide uniform coverage of the Lesser Himalaya and Sub-Himalayan terrain. As part of the National Seismological Network, 29 GPS stations have also installed to monitor crustal shortening due to the continuous movement of the Indian Plate toward the north (NSC, 2012). The data acquired through this network has helped increase the understanding of the seismic hazards to which the country is exposed. As part of the National Seismological Network a seismic hazard map of Nepal has been developed, historical earthquakes and seismic gaps have been researched, and an attenuation model has been developed.

11.6 OBSERVED CHANGES

11.6.1 Greater Appreciation of Seismic Risk

Until the late 20th century, the basic understanding of disaster risk in Nepal was very limited. Efforts to raise earthquake awareness included providing access to knowledge on disaster mitigation in nonscientific language as well as fighting the prevailing culture of fatalism and ignorance. However, this situation has significantly improved in the last two decades.

Upreti et al. (2012) compared observations from similar research conducted in 1998 and 2009 regarding earthquake awareness. The findings reveal a shift in positive attitude toward earthquake safety (Fig. 11.18) and a marked increase in the percentage of people who consider earthquakes as an issue deserving their attention (from 54% in 1998 to 68% in 2009). However, as part of a more recent study conducted by Manandhar (2015) in two districts of Nepal, it was concluded that messages about the existence of building codes have not reached to at least 50% of the surveyed population.

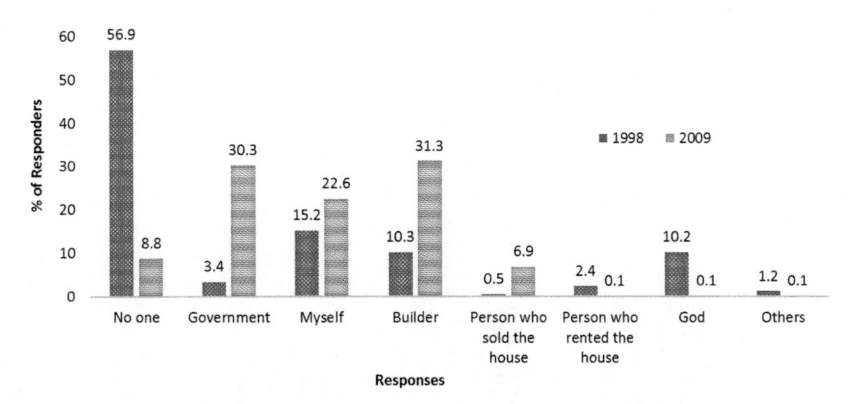

FIGURE 11.18 **Comparative analysis of responses to the query "whom would you blame if an earthquake collapses your building?".** *Adapted from Upreti et al. (2012)*

11.6.2 Guidelines for Building Assessment and Retrofit

Guidelines for building assessment and retrofitting (DUDBC, 2011) were developed by DUDBC in response to the 1996 National Action Plan. Engineers in Nepal have been provided some training in these guidelines and are aware of the processes and products required for building assessment. Retrofitting has also become a household term in Nepal, thanks to the SESP and other follow-up programs that have helped raise awareness in local communities.

The guidelines were developed following international documents, particularly those formulated by the United States of America's Federal Emergency Management Agency (FEMA) and Applied Technology Council (ATC). While the guidelines lack a consistent approach or a focus on local building types at this stage, they are an important first step and, with increasing research activities, could be updated in the future.

11.6.3 Improved Construction Practices

There have been noticeable advances in engineering and building practices over the last two decades. These advances have occurred mainly in urban areas and, to a lesser degree, in remote rural areas. Advances include improved construction quality, the incorporation of earthquake-resistant features such as bands in URM buildings, and seismic detailing of RC building elements (Fig. 11.19). It can be argued that greater accessibility to trained manpower and information, better construction materials, and the improved economic conditions have been crucial to these advances. The training of masons has proved valuable, even in areas where engineers are unavailable, as masons have been able to implement some earthquake-resilient features in buildings (Fig. 11.19C).

11.6.4 Observations Following the 2015 Nepal Earthquake Sequence

The 2015 Nepal earthquake sequence was a litmus test for the effectiveness of work undertaken over the prior two decades. Sensitization programs, awareness-raising activities, and training played a major role in improved rescue and relief opera-

(A)

A ductile RC frame building under construction

(B)

A confined masonry building under construction; note sill band

(C)

Timber bands in a residential house with a light gable

FIGURE 11.19 **Improved construction practices.**

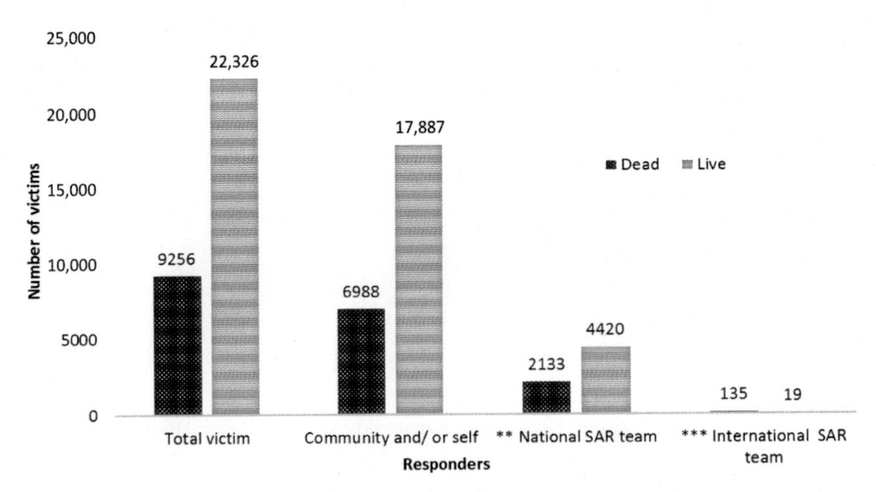

FIGURE 11.20 Victims removed and dead recovered by various search and rescue teams. *Adapted from Dixit et al. (2015). Note: "Nepal Police, July 2, 2015". **Reports by Nepalese Army, Nepal Police and Armed Police Force, June 2015. ***Report by Nepalese Army, June 16, 2015.*

tions. Implementation of the building code and construction guidelines, as well as training of masons and professionals, have also played a significant role in postearthquake recovery and reconstruction.

11.6.4.1 Search and Rescue Contribution

As observed in past disasters, the local community is the first responder. Fig. 11.20 shows the effectiveness of the local community as first responders for search and rescue after the 2015 Nepal earthquake sequence and is a testament to the value of previous training initiatives. Following the earthquake, the local community rescued at least 80% of all those who were rescued. PEER training helped local and national responders to work effectively, either independently or side by side with international Urban Search and Rescue teams (Dixit et al., 2015; U.S. Embassy in Nepal, 2015). Although it would be an exaggeration to say that the effective national contribution to search and rescue was due to PEER training only, the performance of national responders was superb despite logistical challenges.

11.6.4.2 Postearthquake Rapid Visual Assessments

Immediately following the 2015 Nepal earthquake sequence, the Nepal Engineers Association (NEA), DUDBC, NSET, and consulting engineering companies mobilized volunteers or their employees for postearthquake rapid visual assessments (RVAs) of buildings. This mobilization was possible because engineers had either received earthquake preparedness training beforehand or were at least familiar with the process and outcomes (Fig. 11.21A–D). When thousands of engineers arrived to volunteer, NEA mobilized them to undertake RVAs of residential buildings, initially in the Kathmandu Valley and later in other earthquake-affected areas. Similarly, DUDBC and NSET, who are experts in the RVA process, conducted assessments of large buildings, including hospitals, office buildings, and schools, which helped the immediate postearthquake operation of these facilities. The Department of Education also conducted RVAs of school buildings in the earthquake-affected areas and placarded these buildings when necessary (Fig. 11.21D) (Bothara et al., 2017).

11.6.4.3 Resilience of Health Facilities in the Kathmandu Valley

All major tertiary health facilities in the Kathmandu Valley were functional after the 2015 Nepal earthquake sequence (Sharma, 2015), thanks to previous work to improve their seismic resilience. Their continuity of function can also be partly attributed to the low level of earthquake shaking in this area. In a news article after the earthquakes, Kunwar (2015) reported good seismic performance of retrofitted hospital building in the Kathmandu valley. However, similar facilities outside the Kathmandu Valley, particularly in rural areas, became mostly inoperable after the earthquake due to intense shaking, weak buildings, and a lack of preparedness.

11.6.4.4 Overall Greater Seismic Resilience

Improved construction practices and the incorporation of seismic resilience elements led to improved behavior in the 2015 Nepal earthquake sequence (Figs. 11.15B and 11.22) than may have otherwise been expected, with a number of buildings suffering only minor reparable structural damage. However, damage to nonstructural components such as brick partition

(A) Annex III: Rapid Evaluation Form

Rapid Evaluation Safety Assessment Form

Inspection
Inspector ID: ______ Inspection date and time: ______ ☐ AM ☐ PM
Organization: ______ Areas inspected: ☐ Exterior only ☐ Exterior and interior

Building Description
Building Name: ______ Address:
District: ______
Building contact/phone: ______ Municipality/VDC: ______
Approx. "Footprint area" (sq. ft): ______ Ward No: ______ Tole: ______

Type of Construction
☐ Adobe ☐ Stone in mud ☐ Stone in cement ☐ Brick in cement ☐ Wood frame
☐ Bamboo ☐ Brick in mud ☐ Brick in cement ☐ R.C frame ☐ Others: ______

Type of Floor
☐ Flexible ☐ Rigid

Type of Roof
☐ Flexible ☐ Rigid

Primary Occupancy:
☐ Residential ☐ Hospital ☐ Government office ☐ Police station
☐ Educational ☐ Industry ☐ Office Institute ☐ Mix
☐ Commercial ☐ Club ☐ Hotel/Restaurant ☐ Others: ______

Evaluation
Observed Conditions:

	Minor/None	Moderate	Severe
➤Collapsed, partially collapsed, or moved off its foundation	☐	☐	☐
➤Building or any story is out of plumb	☐	☐	☐
➤Damage to primary structural members, cracking of walls, or other signs of distress present	☐	☐	☐
➤Parapet, chimney, or other falling hazard	☐	☐	☐
➤Large fissures in ground, massive ground movement, or slope displacement present	☐	☐	☐
➤Other hazard (Specify) e.g tree, power line etc:	☐	☐	☐

Estimated Building Damage (excluding contents)
☐ None
☐ 0-1%
☐ 1-10%
☐ 10-30%
☐ 30-60%
☐ 60-100%
☐ 100%

Comments: ______

Posting Choose a posting based on the evaluation and team judgment. Severe conditions endangering the overall building are grounds for an Unsafe posting. Localized Severe and overall Moderate conditions may allow a Restricted Use posting. Post INSPECTED placed at main entrance. Post RESTRICTED USE and UNSAFE placards at all entrances.
☐ **INSPECTED** (Green placard) ☐ **RESTRICTED USE** (Yellow placard) ☐ **UNSAFE** (Red placard)
Record any use and entry restrictions exactly as written on placard: ______

Further Actions Check the boxes below only if further actions are needed.
☐ Barricades needed in the following areas:
☐ Detailed evaluation recommended: ☐ Structural ☐ Geotechnical ☐ Other ______
Comments: ______

(B)

Team composed of Nepalese and international engineers with a hospital management team in Kathmandu

(C)

Placard from DUDBC

(D)

School building placarded as "Unsafe"

FIGURE 11.21 Postearthquake RVA and placarding.

walls in RC frame buildings was severe. The 2015 Nepal earthquake sequence also provided a compelling case for investing in retrofitting (WEF, 2015). For example, none of the approximately 300 retrofitted school buildings suffered serious damage or collapsed, even in epicentral areas, despite being constructed of low-strength materials such as stone or brick masonry with mud mortar. Even those school buildings that were located near the epicentral region were in a condition of immediate usability. Almost all of the retrofitted school buildings were used during the earthquake response as emergency shelters, warehouses, health posts, or as safe offices (Dixit et al., 2015).

(A)

Building with earthquake-resilient features in kathmandu valley that survived without damage

(B)

RC frame building with damage limited to infill walls

FIGURE 11.22 **Observed improved performance of building structures.**

11.6.4.5 Postearthquake Reconstruction

Increased awareness of seismic resilience techniques helped promote earthquake-resilient repair and reconstruction after the 2015 Nepal earthquake sequence. Furthermore, because engineers had received previous training or exposed to repair and strengthening techniques, they were able to act immediately (Fig. 11.23).

The post-earthquake reconstruction strategic plan clearly emphasized the "Build Back Better" approach. Accordingly, the Government of Nepal offered financial assistance for reconstruction to individual homeowners with the precondition that their house must be made earthquake-resistant (NPC, 2015). This approach represents a significant improvement in the government's response compared to the 1988 earthquake.

Post-earthquake, a positive change regarding improved construction practices was apparent in earthquake-affected areas. People were aware of earthquake-resilient construction and were making significant efforts to make houses more earthquake-resilient (Fig. 11.24), although they faced many challenges regarding reconstruction (Bothara et al., 2016). Local NGOs, INGOs, CBOs, national and international organizations, and the Government of Nepal have all played key roles in this improved scenario.

11.7 ONGOING ISSUES AND CHALLENGES

The earthquake risk in Nepal is rapidly increasing due to the combined effects of seismic hazard, earthquake vulnerability of the building stock and infrastructure, limited investment, limited political will to implement building-related legislation and codes, fatalism, and a weak economy. Furthermore, there is no legal framework to improve the seismic performance of

(A)

Repair and strengthening of a RC column

(B)

Repair and strengthening of infill wall of a RC frame building

(C)

Stitching of a brick masonry wall

FIGURE 11.23 **Repair and strengthening of buildings damaged by the 2015 earthquake.**

(A)

RC frame building with larger columns and close stirrups in columns (2016)

(B)

Timber bands in a model residential house (2016)

(C)

RC bands and vertical bars in a residential building (2016)

FIGURE 11.24 **Improved construction practices (postearthquake reconstruction).**

the existing building stock despite the vast majority of buildings remaining highly vulnerable to earthquakes. The scaling-up of successful pilot projects and the institutionalization and internalization of achievements is a major challenge for further enhancement and sustainability of EDRR endeavors made thus far.

Many DRR programs in Nepal are driven by international donors. Therefore, the sustainability of EDRR efforts remains doubtful once international support recedes. In addition, these programs generally operate with a limited time frame. For these reasons, the long-term sustainability of EDRR efforts requires continuous investment and a reasonable time frame to cope with changing scenarios and dynamic planning. Although community-based DRM (CBDRM) programs are centered on communities, the decision-making process is controlled by international donors and national experts. As experts and community members perceive risk differently (Laursen, 2015), it is difficult to internalize acquired knowledge until the community actively makes decisions or is involved in the decision-making process. Hence, the challenge is how to involve the community.

The current framework for DRM is focused on urban areas, where a reasonable level of awareness of seismic safety has been initiated. However, there is a need for further improvements. Rural areas, where most of the population lives, have been left far behind and need significant additional input regarding CBDRM (Tuladhar et al., 2014, 2015). The Government of Nepal has proposed initiation of a massive CBDRM program (MOHA, 2016), but CBDRM requires continual interaction, dialogue, egalitarianism, openness, and a mechanism to cope with changing social dynamics. In the current bureaucratic structure in Nepal, the question remains of how to accommodate such an initiative. Furthermore, there remains limited research on issues such as engineering, socio-economic and cultural conditions, earthquake resiliency, and building safety in Nepal in the context of seismic protection.

The effects of earthquake disasters go beyond borders. Hence, regional cooperation, research, and resource sharing are essential for long-term and sustainable EDRR. Education on EDRR is limited and there is no training during engineering education on vernacular building types, which form most of the building stock in Nepal.

Insurance coverage in Nepal is among the lowest in the world (AON, 2015). In the given economy and recognizing the nonengineered vernacular nature of the building stock, procuring of insurance is a challenge. A continual process for dynamic risk assessment is required to reveal the drivers of earthquake risk and the effectiveness of policies focused on reducing risk. The policy framework needs to be responsive to changing dynamics. Safety is a cultural issue and until it becomes a part of daily life, its sustainability will remain questionable. In addition, there is currently no control over the development of haphazard settlements in Nepal.

11.8 RECOMMENDATIONS

The following are recommendations for how Nepal might further improve its earthquake resilience through earthquake risk reduction efforts:

- EDRR should be integrated into development planning, budgeting, implementation, monitoring, and evaluation.
- An alternative funding framework for EDRR is essential so that efforts already made are not discontinued if international funding cannot be sustained.
- Communities should be included in the decision-making process when CBDRM is involved.
- Both top-down and bottom-up processes must work simultaneously. Similarly, the creation of demand for earthquake safety through policy framework, awareness, and meeting these demands through training, market development, and funding must be coordinated for smooth implementation of EDRR.

- The government and other agencies should intensively promote public awareness and earthquake preparedness programs throughout the country.
- Formation of an institution dedicated to EDRR to guide, monitor, and research seismic safety is essential.
- An institutional framework should be developed for action-oriented research on local issues such as seismological and geotechnical research, socioeconomic and cultural issues, and building resiliency and intervention options. The community, NGOs, CBOs, and the private sector should be included in EDRR.
- Existing legislation related to EDRR should be further developed and the building code should be implemented.
- Awareness has been raised to an extent, but the knowledge gained needs to be internalized. The 2015 Nepal earthquake sequence provided an opportunity to disseminate and internalize knowledge due to increased awareness on impending earthquake risk in Nepal.
- Disaster-related insurance policies should be developed.
- A competence-based registration system for engineers and licensing for craftspeople should be implemented.
- Investment should be made in the implementation of EDRR in rural areas.
- The activities of the Nepal DRM Flagship Program should be scaled up, particularly those related to school and hospital safety, emergency preparedness and response capacity, integrated community-based DRR management, and policy and institutional support.
- A legal framework should be developed to improve the seismic performance of the existing building stock.
- Risk-sensitive land-use planning should be developed and implemented.

11.9 CONCLUDING REMARKS

Nepal has made significant improvements in its preparedness for major earthquakes over the last two-and-a-half decades. Global motivation for DRR, investment, and knowledge sharing by international agencies, as well as improved understanding of earthquake disasters and mitigation strategies have played a major role in the country's EDRR efforts. Accordingly, many initiatives have been introduced at both the policy and grassroots levels. Further sustained efforts and investment are needed to capitalize on these initiatives and offset Nepal's growing seismic risk.

ACKNOWLEDGMENTS

The authors acknowledge the contribution of Yogeshwar Parajuli, Santosh Shrestha, Shyam Jnavaly, Bhubaneswori Parajuli, Ramesh Guragain, Hima Shrestha, and others who have directly or indirectly contributed to the development of information provided herein. The editing support provided by Ann Cunningham and Hannah Collins are thankfully acknowledged.

REFERENCES

ADB, 2013. Country poverty analysis (detailed). Nepal Asian Development Bank, Retrieved January 10, 2017. Available from: https://www.adb.org/sites/default/files/linked-documents/cps-nep-2013-2017-pa-detailed.pdf 2013.

Anhorn, J., Lennartz, T., Nüsser, M., 2015. Rapid urban growth and earthquake risk in Musikot, Mid-Western Hills Nepal. Erdkunde 69 (4), 307–325, 0.3112/erdkunde.2015.04.02.

AON, 2015. Nepal earthquake event recap report. Aon Benfield, 2015.

Bilham, R., 2004. Earthquakes in India and the Himalaya. Annals of Geophysics: Tectonics Geodesy and History 47 (2), 839–858.

Bilham, R., 2009. The seismic future of cities. Bulletin of Earthquake Engineering 7 (4), 839–887. doi: 10.1007/s10518-009-9147-0.

Bilham, R., Gaur, V., Molnar, P., 2001. Himalayan seismic hazard. Science 293, 1442–1444.

Bothara, J.K., 2003. Craftsmen: a key for introducing earthquake resistant construction. In: Charlson, A. (Ed.) Earthquake Hazard Centre Newsletter 7 (2), 3–4.

Bothara, J.K., Sharpe, R.D., 2003. Seismic protection in developing countries: where are the gaps in our approach? In: Proceedings of Pacific Conference on Earthquake Engineering. Christchurch, New Zealand. p. 9.

Bothara, J.K., Parajuli, Y.K., Arya, A.S., Sharpe, R.D., 2000. Seismic safety in owner-built buildings. In: Proceedings of 12th World Conference on Earthquake Engineering. Auckland. p. 8.

Bothara, J.K., Dixit, A.M., Nakarmi, M., Pradhanang, S.B., Thapa, R., 2002. Seismic safety of schools in Kathmandu Valley, Nepal: problems and opportunities. In: Proceedings of NZSEE 2002 Conference. New Zealand Society for Eartquake Engineering, Napier, New Zealand.

Bothara, J.K., Pandey, B., Guragain, R., 2004. Seismic retrofitting of low strength masonry non-engineered school buildings. The Bulletin of New Zealand Society for Earthquake Engineering 37 (1), 13–22.

Bothara, J.K., Karmacharya, S., 2015. Creating Safer Cities: Lessons from Christchurch for Kathmandu Valley. In: Commitment - Kathmandu Valley Development: Reflection and Revelations. Kathmandu. Kathmandu Valley Development Authority, Government of Nepal, Nepal, pp. 171–181.

Bothara, J.K., Dhakal, R.P., Dizhur, D., Ingham, J.M., 2016. The challenges of housing reconstruction after the April 2015 Gorkha, Nepal earthquake. Technical Journal of Nepal Engineers' Association, Special Issue on Gorkha Earthquake 2015 XLIII-EC30 (1), 121–134.

Bothara, J., Dizhur, D., Dhakal, R., Ingham, J., 2017. Context and issues during post-earthquake rapid evaluation of buildings after the 2015 Nepal (Gorkha) earthquake. In: Proceedings of Structural Engineering Society New Zealand (Inc.). Structural Engineering Society New Zealand, Wellington, p. 11.

Burton, I., Kates, R., White, G.F., 1993. The Environment as Hazard, 2nd ed. Guilford Press, New York.

CBS, 2012. National population and housing census 2011 (National report), vol. 01. Central Bureau of Statistics, National Planning Commission, Government of Nepal, Kathmandu.

CBS, 2015. Nepal in Figures. Nepal Bureau of Statistics, National Planning Commission Government of Nepal,, Kathmandu, Nepal, Retrieved November 3, 2017. Available from: http://cbs.gov.np/image/data/Publication/Nepal%20in%20Figures%20English%202014.pdf.

Chamlagain, D., 2009. Earthquake scenario and recent efforts towards earthquake risk reduction efforts in Nepal. Journal of South Asia Disaster Studies 2 (1), 57–80, Retrieved from: https://www.researchgate.net/publication/280534382_Earthquake_Scenario_and_Recent_Efforts_Toward_Earthquake_Risk_Reduction_in_Nepal.

Chaulagain, H., Rodrigues, H., Silva, V., Spacone, E., Varum, H., August 2015. Seismic risk assessment and hazard mapping in Nepal. Natural Hazards 78 (1), 583–602. doi: 10.1007/s11069-015-1734-6.

Countrymeters, 2017. Nepal Population. Retrieved July 30, 2017, from Countrymeters: http://countrymeters.info/en/Nepal.

DeCelles, P.G., Kapp, P., Gehrels, G.E., Ding, L., 2014. Paleocene–Eocene foreland basin evolution in the Himalaya of southern Tibet and Nepal: implications for the age of initial India–Asia collision. Tectonics 33, 824–849. doi: 10.1002/2014TC003522.

Dillion, A., Sharma, M., Zhang, X., 2011. Estimating the Impact of Access to Infrastructure and Extension Services Rural Nepal. International Food Policy Research Institute, Wahington, DC, http://dx.doi.org/10.2499/9780896291881.

Dixit, A.M., 2008. Challenges of building code implementation in Nepal. International Symposium 2008 on Earthquake Safe Housing, 28–29 November 2008. United Nations Centre for Regional Development Disaster Management Planning Hyogo Office, Tokyo, Japan, pp. 61–66, Retrieved May 1, 2017, from http://www.preventionweb.net/files/10591_HESITokyoPapers.pdf.

Dixit, A.M., Dwelley-Samant, L.R., Nakarmi, M., Pradhanang, S.B., Tucker, B., 2000. The Kathmandu valley earthquake risk management project: an evaluation. 12th World Conference on Earthquake Engineering. Auckland. .

Dixit, A.M., Yatabe, R., Dahal, R.K., Bhandari, N.P., 2013. Public school earthquake safety program in Nepal. Geomatics Natural Hazards and Risk 5 (4), 293–319. doi: 10.1080/19475705.2013.806363.

Dixit, A.M., Yatabe, R., Guragain, R., Bhandary, N.P., 2014. Non-structural earthquake vulnerability assessment of major hospital buildings in Nepal. Georisk: Assessment and Management of Risk for Engineered Systems and Geohazards 8 (1), 1–13, Retrieved from: http://dx.doi.org/10.1080/17499518. 2013.805629.

Dixit, A.M., Guragain, R., Shrestha, S.N., 2015. Two decades of earthquake risk management actions judged against Gorkha earthquake of Nepal, April 2015. In: Proceedings of New Technologies for Urban Safety of Mega Cities in Asia (USMCA 2015). Kathmandu, Nepal. , Retrieved January 12, 2017, from https://www.researchgate.net/publication/306092563_TWO_DECADES_OF_EARTHQUAKE_RISK_MANAGEMENT_ACTIONS_JUDGED_AGAINST_GORKHA_EARTHQUAKE_OF_NEPAL_APRIL_2015.

Dizhur, D., Dhakal, R.P., Bothara, J.K., Ingham, J.M., 2016. Building typologies and failure modes observed in the 2015 Gorkha (Nepal) earthquake. Bulletin of the New Zealand Society for Earthquake Engineering 49 (2), 211–232.

DOS, 1985. Physiographic regions of Nepal. Department of Survey Government of Nepal, Retrieved July 13, 2017, from http://3.bp.blogspot.com/-UNLug4XqpJ0/ThrnD51cbGI/AAAAAAAAAAk/XTL_X-V85yg/s1600/fig42.jpg.

DUDBC, 2011. Seismic vulnerability evaluation guideline for private and public buildings. Department of Urban Development and Building Construction, Ministry of Urban Development, Government of Nepal, Kathmandu, Nepal.

EMI, 2010. Risk-sensitive land use plan—Kathmandu metropolitan city, Nepal. Quezon City 1101, Earthquakes & Megacities Initiatives (EMI), Philippines. Retrieved from: http://www.emi-megacities.org.

Gautam, D., Chaulagain, H., 2016. Structural performance and associated lessons to be learned from world earthquakes in Nepal after 25 April 2015 (M_w 7.8) Gorkha earthquake. Engineering Failure Analysis 68, 222–243, Retrieved May 16, 2017, from http://dx.doi.org/10.1016/j.engfailanal.2016.06.002.

GESI, 2001. The final report on global earthquake safety initiative (GESI) pilot project. Geohazards International and United Nations Centre for Regional Development, Retrieved January 16, 2017, from http://www.preventionweb.net/files/5573_gesireport.pdf.

GON, 2013. Nepal risk reduction consortium: flagship programme. Government of Nepal, Kathmandu, Retrieved April 1, 2017, from: http://flagship4.nrrc.org.np/sites/default/files/documents/NRRC_Flagship%20Programmes%20%28For%20Web%29_19%20Mar%202013-1.pdf.

GON, 2015. Nepal disaster risk reduction portal. (Nepal disaster risk reduction portal producer) Retrieved January 1, 2017, from: https://web.archive.org/web/20150629024928/http://drrportal.gov.np/incidentreport.

GSHAP, 1999. Global seismic hazard assessment programme. (Global seismic hazard assessment program editor). Retrieved December 10, 2016, from: http://static.seismo.ethzch/GSHAP/eastasia/.

HMG, 1994. A management plan for the introduction of a National Building Code. His Majesty's Government of Nepal, United Nations Development Programme, Kathmandu, Nepal.

HMG, 1998a. The Building Act 2055. His Majesty's Government of Nepal, Kathmandu, Nepal.

HMG, 1998b. Nepal Engineering Council Act 2055. His Majesty's Government of Nepal, Kathmandu, Nepal.

HMG, 1999. Local Self-Governance Act. Local Self-Governance Act. His Majesty's Government of Nepal, Retrieved February 10, 2017, from http://www.undp.org/content/dam/nepal/docs/reports/governance/UNDP_NP_Local%20Self-Governance%20Act%201999,%20MoLJ,HMG.pdf.

HMG, 2000. Nepal Engineering Council Regulation 2057. His Majesty's Government of Nepal, Kathmandu, Nepal.

IDNDR, 1994. Yokohama strategy and plan of action for safer world. In: Proceedings of World Conference on Natural Disaster Reduction. Yokohama Japan. pp. 1–18, Retrieved February 1, 2017, from http://www.unisdr.org/files/8241_doc6841contenido1.pdf.

IFRC, 2014. World disasters report 2014 focus on culture and risk. International Federation of Red Cross and Red Crescent Societies, Geneva, Switzerland, Retrieved May 9, 2017, from http://www.ifrc.org/Global/Documents/Secretariat/201410/WDR%202014.pdf.

IRDR and ICSU, 2017. Issue brief: disaster risk reduction and sustainable development. (Integrated research on disaster risk International Council for Science). Retrieved February 5, 2017, from: http://www.preventionweb.net/: http://www.preventionweb.net/files/35831_35831irdricsubriefdrrsd5b15d1.pdf.

ISDR, 2005. Hyogo framework for action 2005–2015: building the resilience of nations and communities to disasters. International Strategy for Disaster Reduction, Kobe, Japan.

JICA and MOHA, 2002. The study on earthquake disaster mitigation in the Kathmandu Valley, Kingdom of Nepal. Japan International Cooperation Agency (JICA), Ministry of Home Affairs, His Majesty's government of Nepal, Kathmandu. Retrieved from: http://flagship2.nrrc.org.np/sites/default/files/knowledge/JICA%20-%20The%20study%20on%20Earthquake%20Disaster%20Mitigation_Vol1.pdf.

Jimee, G.K., Upadhyay, B., Shrestha, S.N., 2012. Earthquake awareness programs as a key for earthquake preparedness and risk reduction: lessons from Nepal. In: Proceedings of the 15 World Conference on Earthquake Engineering (10). Lisbon. .

Joshi, A., Basnet, S., Dawadi, G.S., Duwal, S., Pandey, K.R., Irwin, D., 2013. Urban Growth Pattern in Kathmandu Valley. Genesis Consultancy (P) Ltd. & Welink Consultants (P) Ltd., Kathmandu, Nepal, Unpublished.

Jouanne, F., Mugnier, J.L., Gamond, J.F., Avouac, J.P., Le Fort, P., Pandey, M.R., Avouac, J.P., April 2004. Current shortening across the Himalayas of Nepal. Geophysical Journal International 157 (1), 1–14. doi: 10.1111/j.1365-246X.02180.x.

KU, 2010. Kathmandu University goes community level training at Sindhupalanchowk and Dolakha District on disaster management (Kathmandu University). Retrieved 11 May, 2017, from Kathmandu University News: http://www.ku.edu.np/news/indexphp?op=ViewArticle&articleId=186&blogId=1.

Kunwar, S.S., 2015. Retrofitted buildings provide a safe space during the earthquake. (08). Retrieved from May 2, 2017, https://www.spotlightnepal.com/2015/05/22/retrofitted-buildings-provide-a-safe-space-during-the-earthquake/.

Laursen, M.R., 2015. Community-based earthquake preparedness in Nepal: a matter of risk perceptions UNISDR. Retrieved May 1, 2017, from https://www.unisdr.org/campaign/resilientcities/assets/documents/privatepages/Community-Based%20Earthquake%20Preparedness%20in%20Nepal%20-%20A%20Matter%20of%20Risk%20Perceptions.pdf.

Maharjan, D.K., Shrestha, H., Guragain, R., 2015. Non-structural vulnerability mitigation in hospital and experience in recent medium intensity earthquake. In: Proceedings of the New Technologies for Urban Safety of Mega Cities in Asia. Kathmandu. p. 10, Retrieved March 15, 2017, from https://www.sheltercluster.org/sites/default/files/docs/d.maharjan_h.shrestha_r_guragain-_non-structural_vulnerability_mitigation_in_hospital_and_experience_in_recent_medium_intensity_earthquake.pdf.

Manandhar, B., 2015. Remittance and earthquake preparedness. Int. J. Disaster Risk Reduct. 15, 52–60, Retrieved from: http://dx.doi.org/10.1016/j.ijdrr.2015.12.003.

Mercer, J., Gaillard, J.C., Crowley, K., Shannon, R., 2014. Culture and disaster risk reduction: lessons and opportunities. In: Fearnley, C., Wilkinson, E., Tillyard, C.J., Edwards, S.J. (Eds.), Natural Hazards and Disaster Risk Reduction: Putting Research Into Practice. Routledge, New York.

MHPP, 1994a. Seismic hazard mapping and risk assessment for Nepal. Nepal Building Code Development Project Nepal, Ministry of Housing and Physical Planning, His Majesty's Government of Nepal, UNCHS, Kathmandu, Nepal. UNDP Project # Nep.88.054-21.03.

MHPP, 1994b. A management plan for the introduction of a National Building Code. Nepal Building Code Development Project Nepal, Ministry of Housing and Physical Planning, His Majesty's Government of Nepal, UNCHS, Kathmandu, Nepal. UNDP Project # Nep.88.054-21.03.

MOH, 2003. Non-structural assessment of hospitals in Nepal. Ministry of health and population (MoH), national society for earthquake technology—Nepal (NSET), World Health Organization (WHO), Kathmandu, Nepal., Retrieved January 14, 2017, from http://www.preventionweb.net/publications/view/1964.

MOHA, 1996. National action plan on disaster management in Nepal. 12. Ministry of home affairs, his Majesty's Government of Nepal, Singh Durbar, Kathmandu, Nepal, Nepal, Retrieved December 22, 2016, from http://www.preventionweb.net/files/30532_nepalnationalactionplandisastermana.pdf.

MOHA, 2009. National strategy for disaster risk management, 2009. Ministry of Home Affairs, Government of Nepal, Kathmandu.

MOHA, 2015. Nepal disaster report 2015. Ministry of Home Affairs, Government of Nepal and DPNet-Nepal, Kathmandu, Nepal, ISBN: 978-9937-0-0324-7.

MOHA, 2016. Disaster risk reduction in Nepal: achievements, challenges and ways forward. National position paper for the Asian Ministrial Conference DRR 2016, p. 12. Ministry of Home Affairs, Government of Nepal, New Delhi, India. Retrieved March 15, 2017, from http://drrportal.gov.np/uploads/document/627.pdf.

Molnar, P., Dayem, K.E., August 2010. Major intracontinental strike-slip faults and contrasts in lithospheric strength. Geosphere 6 (4), 444–467.

Muzzini, E., Aparicio, G., 2013. Urban Growth and Spatial Transition in Nepal: An Initial Assessment. The World Bank, Washington, DC. doi: 10.1596/978-0-8213-9659-9.

NPC, 2002. 10th National Development Plan (2002–2007). National Planning Commission, Government of Nepal, Kathmandu, Nepal.

NPC, 2007. 11th National Development Plan (2007–2010). National Planning Commission, Government of Nepal, Kathmandu, Nepal.

NPC, 2010. 11th National Development Plan (2010–2013). National Planning Commission, Government of Nepal, Kathmandu, Nepal.

NPC, 2013. 13th National Development Plan (2013–2015). National Planning Commission, Government of Nepal, Kathmandu, Nepal.

NPC, 2015. Nepal earthquake 2015: post-disaster needs assessment, Volume A: key findings. National Planning Commission, Government of Nepal, Kathmandu, Nepal.

NSC, 2012. D. o. National Seismological Centre Producer. Retrieved January 14, 2017, from http://www.seismonepal.gov.np/index php?linkId=128.

NSET, 2000. Seismic vulnerability of the school buildings of Kathmandu valley and methods for reducing it. National Society for Earthquake Technology-Nepal, Kathmandu, Nepal.

NSET, 2017. NSET-Home page (National Society for Earthquake Technology-Nepal) Retrieved February 4, 2012, from http://www.nset.org.np: http://www.nset.org.np.

NTB, 2017. Discover Nepal. (Discover Nepal, Producer & Nepal Tourism Board). Retrieved May 6, 2017, from http://www.welcomenepal.com/plan-your-trip/geography.html.

Panthhe, K.P., McCutcheon, A.L., 2015. Rural urban education in Nepal. Int. J. Econ. Res. 6 (1), 30–44, ISSN: 2229-6158.

Parajuli, Y.K., Bothara, J.K., Dixit, A.M., Pradhan, J.P., Sharpe, R., 2000. Nepal building code—need, development philosophy and means of implementation. In: Proceedings of the 12th World Conference on Earthquake Conference. Auckland, New Zealand. p. 7.

Patriat, P., Achache, J., 1984. India–Eurasia collision chronology has implications for crustal shortening and driving mechanism of plates. Nature 311, 615–621, (ISSN 0028-0836).

Pokharel, L.N., 2015. Natural Disaster Relief Act 1982: to manage disaster. Retrieved February 6, 2017, from http://www.nepalromania.com/?p=5892.

Prajapati, S.K., Dadhich, H.K., Chopra, S., 2017. Isoseismal map of the 2015 Nepal earthquake and its relationships with ground-motion parameters, distance and magnitude. J. Asian Earth Sci. 133, 24–37. doi: 10.1016/j.jseaes.2016.07.013.

Rana, B.J., Lall, K., 2013. The Great Earthquake in Nepal (1934 AD). Ratna Pustak Bhandar, Kathmandu, Nepal, ISBN 9789937330152.

Republica, 2015. PM formally announces 744 local units operational. Republica, Kathmandu, Nepal, Retrieved May 7, 2017, from http://www.myrepublica.com/news/16442/.

Rigg, J., Oven, K.J., Basyal, G.K., Lamichhane, R., 2016. Between a rock and a hard place: vulnerability and precarity in rural Nepal. Geoforum 76, 63–74, Retrieved from: http://dx.doi.org/10.1016/j.geoforum.2016.08.014.

Rowley, D.B., 1996. Age of initiation of collision between India and Asia: a review of strategic data. Earth and Planetary Science Letters 145, 1–13.

Sainju, R.S., 2015. Learning workshop on disaster risk management in Nepal (Powerpoint presentation). Retrieved February 7, 2017, from http://dms.nasc.org.np/sites/default/files/documents/RabisSainju.pdf: http://dms.nasc.org.np/.

Sapkota, S.N., Bollinger, L., Klinger, Y., Tapponier, P., Gaudemer, Y., Tiwari, D., 2013. Primary surface ruptures of the great Himalayan earthquakes in 1934 and 1255. Nature Geoscience 6, 71–76.

Sharma, D.C., 2015. Nepal earthquake exposes gaps in disaster preparedness. The Lancet 385 (9980), 1819–1820. doi: 10.1016/S0140-6736(15)60913-8.

Shaw, R., Scheyvens, H., prabhakar, S., Endo, I., 2016. Disaster Risk Reduction for Sustainable Development. Institute of Environmental Strategies (IGES), Kanagawa, Japan.

Srivastava, H.N., Bansal, B., Verma, M., 2013. Largest earthquake in Himalaya: an appraisal. J. Geol. Soc. India 82, 15–22.

Thapa, N., 1989. Bhadau Panch ko Bhookampa 2045 (The Earthquake of 5th Bhadau, 2045 BS, in Nepali). Niranjan Thapa, Kathmandu.

Tuladhar, G., Yatabe, R., Dahal, R.K., Bhandari, N.P., 2014. Knowledge of disaster risk reduction among school students in nepal. Geomatics Natural Hazards and Risk 5 (3), 190–207.

Tuladhar, G., Yatabe, R., Dahal, R.K., Bhandary, N.P., 2015. Disaster risk reduction knowledge of local people in Nepal. Geoenvironmental Disasters 2 (5), 12. doi: 10.1186/s40677-014-0011-4.

UN, 1989. International decade for natural disaster reduction, resolution A/RES/44/236. International decade for natural disaster reduction. United Nations. Retrieved January 22, 2017, from http://www.un.org/documents/ga/res/44/a44r236.htm.

UN, 2015. Sendai framework for disaster risk reduction 2015–2030. United Nations. Retrieved January 31, 2017, from http://www.preventionweb.net/files/43291_sendaiframeworkfordrren.pdf.

UNDP, 2009. Nepal country report—global assessment. ISDR Global Assessment Report on Poverty and Disaster Risk 2009. United Nations Development Programme, Kathmandu, Nepal, Retrieved January 1, 2017, from www.undp.org.np/uploads/publication/201010290938349.

UNDP, 2014. Sustaining human progress: reducing vulnerabilities and building resilience. United Nations Development Programme, Washington, DC, ISBN: 978-92-1-126368-8.

UNDP, 2015. An overview of ten year's progress of implementation of Hyogo Framework for Action (HFA): 2005–2015 in Nepal. Ministry of Home Affairs, Government of Nepal; United Nations Development Programme, Kathmandu.

UNESCO, 2010. Engineering: issues, challenges and opportunities for development. United Nations Educational, Scientific and Cultural Organization, Retrieved from: https://books.google.co.nz/books/about/Engineering.html?id=09i67GgGPCYC&redir_esc=y.

UNISDR, 1994. United Nations International strategy for disaster reduction. Retrieved August 8, 2017, from https://www.unisdr.org/who-we-are/international-strategy-for-disaster-reduction.

UNISDR, 2009. Nepal country report—global assessment of risk. United Nations International strategy for disaster reduction. Retrieved March 2, 2017, from http://www.undp.org/content/dam/nepal/docs/reports/UNDP_NP_Nepal Country Report.pdf.

Upreti, N., Dixit, A.M., Shrestha, S.N., 2012. Raising earthquake awareness in Kathmandu valley: a comparative analysis of achievements during 1999–2009. In: Proceedings of the 15th World Conference on Earthquake Engineering. Lisbon. p. 9.

U.S. Embassy in Nepal, 2015. United States builds capacity of search and rescue and medical first responders. U.S. Embassy in Nepal. Retrieved April 30, 2017, from https://np.usembassy.gov/united-states-builds-capacity-of-search-and-rescue-and-medical-first-responders/.

WEF, 2015. Building resilience in Nepal through public–private partnerships. World Economic Forum, Global Agenda Council on Risk & Resilience. World Economic Forum. Retrieved April 15, 2017, from http://www3.weforum.org/docs/GAC15_Building_Resilience_in_Nepal_report_1510.pdf.

Wesson, E., 1998. Kathmandu Valley Earthquake Risk Perception Study. National Society for Earthquake Technology-Nepal, Kathmandu.

WHO, 2002. A structural vulnerability assessment of hospitals in Kathmandu valley. World Health Organisation, Ministry of Health, Kathmandu, Retrieved March 14, 2017, from http://www.nset.org.np/nset2012/images/publicationfile/20130101161352.pdf.

Wikipedia, 2017. Administrative divisions of Nepal (Wikipedia). Retrieved May 7, 2017, from https://en.wikipedia.org/wiki/Administrative_divisions_of_Nepal.

WPR, 2017. Nepal population 2017. W. P. Review, Producer & World Population Review. Retrieved May 6, 2017, from http://worldpopulationreview.com/countries/nepal-population/.

Chapter 12

Urban Flood Management in Coastal Regions Using Numerical Simulation and Geographic Information System

T.I. Eldho*, P.E. Zope*, A.T. Kulkarni**

*Indian Institute of Technology, Mumbai, India

**Risk Management Solution India Pvt. Ltd., India

12.1 INTRODUCTION

Flooding in urban areas due to heavy rainfall coupled with high tides is a major concern affecting development of coastal cities all over the globe. In the last few decades, in India and other countries, extraordinary rainfall of high intensities resulted in incidents of urban flooding with both increased frequency and magnitude. In cities that affected with flood, this could lead to immense loss of life, property, and livelihoods of its inhabitants. Coastal urban cities are vulnerable to flooding under combined influence of heavy rainfall and high tides.

For the effective coastal urban flood management, we need to understand the flooding phenomena in a particular region due to various influencing factors such as topographic conditions, climatic factors, and tidal influence, through simulation and predictions (Kulkarni et al., 2014b, c). Hence, there is a need of hydraulic flood model capable of incorporating the various components such as overland flow, channel flow, tidal condition, and holding pond effects in the considered area. Due to the complexity of the problem, an integrated approach of numerical modeling coupled with geospatial techniques of remote sensing (RS) and geographic information system (GIS) is required for appropriate flood management (Shahapure et al., 2010).

Increase in population due to migration of people from rural to urban area changes the land-use pattern of the catchment area (Kulkarni et al., 2014a). Land-use land cover (LULC) changes impact the runoff process and increase in flood peaks at downstream of the rivers (Amini et al., 2011). Most of the coastal urban cities like Mumbai, Navi Mumbai, Chennai, etc. in India have limitations on horizontal growth and thus have limitations on infrastructural facilities such as widening of the existing drainage systems (DeFries and Eshleman, 2004; Kulkarni et al., 2014a; Zope et al., 2015, 2016a, b). Coastal urban cities have also limitations on discharge capabilities at main outlet point due to the tidal conditions (Kulkarni et al., 2013; Shahapure et al., 2010; Zope et al., 2012). As such, even a small intensity of rainfall with high tide leads to flooding in low-lying areas (Zope, 2016, Zope et al., 2015). In recent years, due to rapid changes in rainfall pattern and extreme rainfall intensities, major flooding events are taking place all over the world (Zope et al., 2016a). Flooding, being the natural disaster, cannot be avoided; however, the losses occurring due to flooding can be prevented by proper flood mitigation planning. As such, it is necessary to have a proper estimation of flood extent and flood hazard for the different flow conditions so that proper flood evacuation and disaster management plan can be prepared in advance (Zope, 2016; Zope et al., 2015).

In this chapter, various aspects of coastal urban flooding, theoretical developments, and modeling are discussed. The available standard models for coastal flood simulation are briefly discussed. Furthermore, an integrated approach of flood assessment in coastal urban areas using numerical models and geospatial techniques is illustrated. The detailed description and applications of two simulation models, viz. Integrated Flood Assessment Modeling (IFAM) tool and free software packages from Hydrologic Engineering Center (HEC), viz. HEC-HMS and HEC-RAS, are demonstrated. Furthermore, the flood assessments of coastal urban regions are demonstrated using two case studies with integrated approach of numerical simulation and GIS.

Integrating Disaster Science and Management. http://dx.doi.org/10.1016/B978-0-12-812056-9.00012-9

12.2 COASTAL URBAN FLOOD PROBLEMS

Increase in urbanization due to change in LULC as well as increase in extreme rainfall intensity due to climate change and other issues have increased the flooding problems in many of the coastal urban cities. The main effect of urbanization in cities is the drastic change in LULC to increase impervious surface area in terms of built up areas such as building, roads, and other structures (Parkinson and Mark, 2005). Hence, for given rainfall conditions, there is decrease in time of concentration, increase in peak discharge, loss of existing drainage capacity, and flood frequency (Vieux, 2001). The higher runoff rates increase the severity and frequency of flooding at the downstream side of the channel or river. To achieve better planning and designing the storm water drainage system, for a given rainfall conditions, we need to simulate the flow conditions. The urban managers and planners should have proper knowledge of available hydrological and hydraulic models.

In urban hydrology, the major factors that influence rainfall-runoff process include rainfall intensity, evaporation/evapotranspiration, infiltration, overland flow, detention or retention storage, channel flow, and tidal variations (Chow et al., 1988). Generally, the hydrograph generated from the urban catchment is a function of land use, rainfall intensity, duration, shape, and size of the area considered (Singh, 1996). For urban flood management, simulation models are now recognized as main tools that can give important information about flood hazards and risks. Simulation models will provide an insight into the hydrologic processes within the urban catchments, functioning of urban drainage systems, a framework for data analysis, and way of evaluating various developmental plans (American Society of Civil Engineers (ASCE), 1992). Hydrological models play an important role in flood management studies. The urban storm water models should be capable to model the flood flows and predict the flooding areas and its consequences. The urban flood model results summarize the behavior of the urban catchment response as a function of time and space, giving flood plain and hazard maps.

As most of the coastal cities are being surrounded by sea or its creek, the storm water drainage system and discharge at the outlet point are influenced by the tidal variations (Kulkarni et al., 2014a). For most of the cities, while designing the storm water drainage system, there are limitations of outlet levels and depths of the channel due to tidal conditions. In most parts of the world, coastal cities are formed by the reclamation which changes the hydrological characteristics of the natural watershed and soil get saturated due to the groundwater problem. If proper reclamation is not carried out with appropriate drainage system, severe flooding problems may arise (Zope et al., 2015). In coastal urban cities, flooding mostly takes place when high intense rainfall coincides with high tide. Surges and cyclones can also cause severe flooding problems. For example, major coastal cities in India have experienced severe flooding problems and witnessed loss of life and property as well as disruption to traffic and power; among these being Mumbai in 2005, Surat in 2006, Kolkata in 2007 (Guhathakurta et al., 2011), and recently Srinagar and Chennai.

12.3 INTEGRATED URBAN FLOOD SIMULATION

Typically, the urban flood simulation includes number of components which represent various processes such as overland flow, channel flow, tidal flows, effects of storage structure like holding ponds, flood inundation, etc. An integrated flood model combines all these components such that the flood variation can be simulated with respect to space and time (Kulkarni et al., 2014a).

The simulation procedure and important components are briefly described below.

- *Overland flow component:* Here, the runoff is estimated for the given rainfall condition after considering various losses such as infiltration, evaporation, or interception. The overland flow component can be estimated by conceptual/lumped models such as soil conservation curve number (SCS-CN) method, mass balance approach, or by solving the governing equations such as Saint Venant equations (Singh, 1996).
- *Channel flow component:* The overland flow joins the channel and routed through the channel to get the discharge at the given location. The routing can be done by using mass balance approach or by solving governing Saint Venant equations or its simplification (Chow et al., 1988).
- *Tidal effects and flood storage:* In coastal urban areas, the tidal influence is considered through tidal models. The effects of flood storage structures such as holding/detention ponds can be simulated by mass balance-based approach (Shahapure et al., 2010).
- *Flood inundation component:* This component gives the flood impacts on depth-wise and areal-wise, giving the flood hazards. The flood inundation can be directly obtained in 2D/3D models or can be indirectly estimated by mass balance-based approach from 1D models.

In the integrated flood simulation approach, all these components are combined to have a simulation model, which can be used to predict the flood variation for a given urban area (Kulkarni et al., 2013).

To develop an efficient flood simulation model for a given urban area, we need a large number of data sets such as geomorphological features, digital elevation model (DEM), LULC, climate parameters (such as rainfall and temperature), tidal data, etc. To collect and deal with such huge data sets, we need latest data collection and database management tools such as RS and GIS. In the integrated flood simulation, the data set such as DEM, LULC, and land features are obtained from RS, and GIS is used for data manipulation, preparation of various maps, preprocessing of data, and postprocessing of the results to get flood inundation and hazard maps (Vieux, 2001). The numerical/conceptual/lumped model is the simulator of hydrologic/hydraulic process. Thus, computer models with GIS and RS have become integral part of the coastal urban flood simulation.

12.4 URBAN FLOOD SIMULATION MODELS

Generally, the urban simulation models include hydrologic and hydraulic models. The hydrologic models simulate the rainfall to runoff process, considering various losses and provide the quantity of flow with respect to space and time. The input to the hydraulic model is the known flow amount (typically the output of a hydrologic model) and output results are generally the flow height, location, velocity, direction, and pressure at a particular location or time (Chow et al., 1988; Singh, 1996). Number of flood simulation models which can be used to simulate the hydrologic processes in urban catchments have been developed in the last few decades. Some of the commonly used hydrologic and hydraulic models for urban flood simulation are briefly described here.

HEC-HMS Model: Hydrologic Engineering Center's Hydrologic Modeling System (HEC-HMS), developed by the U.S. Army Corps of Engineers, can be used to simulate rainfall–runoff process for dendrite watershed system and can be used as distributed, semidistributed or lumped model, event-based or continuous simulation model (www.hec.usace.army.mil/software/hec-hms/). The inputs to the model include basin parameters and climate variables, whereas the outputs include hydrographs, and peak discharge is being used as input to hydraulic models such as HEC-RAS model.

HEC-RAS Model: This model is used to perform steady and unsteady river flow analysis (www.**hec**.usace.army.mil/.../**hec-ras**/) by solving Saint Venant equations or its simplified forms using finite difference method for unidirectional flow conditions and by solving energy equations (USACE, 2010, 2011). The water surface profiles and water surface extents generated in HEC-RAS are used as inputs for generation of flood plain mapping in ARC-GIS using postprocessing with HEC-GeoRAS.

SWMM Model: Storm Water Management Model (SWMM) developed by the U.S. Environmental Protection Agency (USEPA) has capabilities of simulating single as well as continuous event of urban runoff quantity as well as quality in storm water as well as sewered and combined drainage system. The model can be used for drainage design, flood control strategies, flood plain mapping, and for nonpoint source loading purpose (http://www.epa.gov/ednnrmrl/models/swmm).

STORM: Storage, Treatment, Overflow, Runoff model (STORM) developed by the U.S. Corps of Engineers can be used to simulate runoff and pollutant load (Zoppou, 2001).

PRMS: Precipitation-Runoff modeling system (PRMS) developed by the U.S. Geological survey (USGS) is used to simulate runoff for both daily as well as very small time interval (Wurbs, 1995).

MIKE 11: It is developed by DHI (Danish Hydraulic Institute) and can be used for simulation of river and channels flows, water quality, and sediment transport system of all types of water bodies. It has capabilities of flood plain, flood encroachment, and flood control design strategies and can be used for flood forecasting and resource operation, operation of irrigation and drainage system, tidal, and storm surge studies in rivers and estuaries (http://www.mikebydhi.com/).

IFAM Model: Integrated Flood Assessment Model is a web GIS-based integrated flood model with capabilities of 1D–1D and 1D–2D flood plain model (Kulkarni et al., 2014b, c). Both the web GIS server and the associated hydrological model have been indigenously built at IIT Bombay. The data input to the model is from the client-side through a web browser. The model is capable of simulating 1D overland flow using mass balance approach, 1D diffusion wave-based channel flow model, and quasi-2D raster-based floodplain model. The outputs from the IFAM include: (a) discharge and stage hydrographs; (b) water level profile plot, and (c) flood map animation in the case of flooding in channel.

In this paper, two case studies of coastal urban flood simulation are presented. The first one using the IFAM model and second one using HEC-HMS and HEC-RAS. The details of these models are briefly discussed below.

12.5 FLOOD SIMULATION USING IFAM

The hydrologic processes considered in the IFAM are (Kulkarni et al., 2014a): 1D mass balance-based overland flow; 1D diffusion wave-based channel flow solved using finite element method, and quasi-2D raster-based floodplain model solved using continuity equation. In the mass balance-based overland flow model, it is assumed that gravitational force is the only driving force for overland flow and is modeled using mass balance approach (Shahapure et al., 2010). The overland flow for any of the subareas can be evaluated by the continuity equation as follows:

Inflow – outflow = Increment in storage per unit time.

$$I - qL = \frac{\Delta V}{\Delta t} \tag{12.1a}$$

$$r_e A_c - qL = \frac{\Delta V}{\Delta t} \tag{12.1b}$$

where I is the inflow into the subarea, r_e is the excess rainfall rate, A_e is the subcatchment area considered, q is the overland flow from subarea of the catchment per unit length of the channel in the form of crossflow into the channel, ΔV is the increment in detention storage, Δt is the time step, and L is the length of the channel segment intercepting the overland flow.

Eq. (12.1b) can be written in the following form:

$$K_1 h_{t+\Delta t}^{5/3} + 100 h_{t+\Delta t} = K_2 \tag{12.2}$$

where

$$K_1 = \left[\frac{L S_o^{1/2} \Delta t}{2 n_o A_c}\right], \quad K_2 = 100 h_t + \Delta t \left[\frac{(r_e)_t + (r_e)_{t+\Delta t}}{72}\right] - K_1 h_t^{5/3} \tag{12.3}$$

where S_o is the slope of the overland flow plane which is assumed to be equal to the friction slope and n_o is the Manning's roughness coefficient applicable to the overland flow (s m$^{-1/3}$).

The detailed formulations are available in Shahapure et al. (2010) and Kulkarni et al. (2013, 2014a).

Channel flow is the main form of surface water flow, and all other surface flow processes contribute to it. The continuity and momentum equations considering diffusion wave are

$$\frac{\partial Q}{\partial x} + \frac{\partial A}{\partial t} - q = 0 \tag{12.4}$$

$$\frac{\partial Q}{\partial t} + \frac{\partial (Q^2 / A)}{\partial x} + gA\left(\frac{\partial H}{\partial x} + S_{fc}\right) = 0 \tag{12.5}$$

where Q is the discharge in the channel (m^3/s), A is the area of flow in the channel (m^2), S_{fc} is the friction slope of the channel, g is the acceleration due to gravity, h_c is the depth of flow in the channel, and x, t are the spatial and temporal coordinates. ($\partial H/\partial x$) is the slope of water surface elevation. The diffusion wave formulation for this has been first derived by Hromadka and Yen (1986), which has been solved using Galerkin's weighted residual approach using linear line element with two nodes (Desai et al., 2011). The formulation details are available in Shahapure et al. (2010) and Kulkarni et al. (2013, 2014a). The initial boundary condition for the problem consists of water levels obtained from the steady-state analysis, whereas downstream boundary conditions are in the form of specified tidal stage specified by local authorities.

The moment when channel reaches its bankful stage, water ceases to be contained in the channel and spills over into the floodplains. This has been achieved using a raster-based floodplain model (RBFP). This is nothing but an integrated 1D channel flow model (described earlier) and a 2D floodplain routing model. The water level in each cell (pixel) of the DEM is evaluated based on the adjacent four neighbors. The simplest way to achieve distributed routing of water over floodplain is to treat each cell as storage volume and solve the continuity equation. The change in the cell volume over time is equal to the fluxes into and out of it during the time step. The water level in the central cell is determined based on the water levels in the adjacent cells and their corresponding discharges. The continuity equation can be expressed as follows (Bates and De Roo, 2000; Cunge et al., 1980):

$$\frac{V_{i,j}^{t+\Delta t} - V_{i,j}^{t}}{\Delta t} = Q_{up} + Q_{down} + Q_{left} + Q_{right} \tag{12.6}$$

where $V_{i,j}^{t}$ is the volume of water in the cell of ith row and jth column at time t, Q_{up}, Q_{down}, Q_{left}, and Q_{right} are the flow rates (here flux entering into the cell is considered as positive) from the up, down, left, and right adjacent cells, respectively.

Based on the above equations, an IFAM has been developed (Kulkarni et al., 2013, 2014a). At the beginning, the IFAM model reads channel details, rainfall data, tidal conditions, and DEM raster data. Initially, the model performs steady-state analysis in the channel using initial tidal conditions. At this stage, the number of floodplain cells adjacent to each channel cell in all the four directions is also computed. In case there is any overtopping of the channel, the IFAM will perform flood routing on all such floodplain cells. At this stage, time step increment wise, simulation begins with overland flow which is being computed from effective rainfall falling over the catchment grid. The control then switches to channel model where any lateral discharge occurring from the overland flow grids and/or due to interaction between the channel and raster flood model is incorporated. The node-wise discharge and water levels are computed for each time step. Once the channel water levels are computed, channel overtopping condition (i.e. channel water level > channel bank level) is checked for each time step until the condition is true.

Once the raster flood model is activated, the floodplain routing is carried out in all four directions for each cell using Eq. (12.6) and the water level is updated for each cell. The model then proceeds like-wise till the specified simulation time. The model code has been written in MATLAB and has followed modular approach with each submodel (i.e. function/subroutine) for independent processes like steady-state analysis, DEM preprocessing, overland flow model, dynamic wave channel flow, etc. and called for within the main model in the time loop. The space-wise water level information for all the cells is stored in the form of the matrix of the size of floodplain. For every specified time interval, this matrix is saved as raster (GIS file format) which can then be viewed in ArcMap to view the areal flood extent. Some of the limitations of the IFAM model include: 1D nature of the overland flow simulation, difficulty to handle supercritical flows, requirements of huge data sets such as DEM, soil map, LULC variation, etc. for proper simulation and higher computational time requirements.

12.6 FLOOD SIMULATION USING HEC-HMS AND HEC-RAS

The integrated flood modeling approach using HEC-HMS and HEC-RAS include flood simulation and hazard mapping, for different storm flow conditions and rainfall events (Zope, 2016; Zope et al., 2015, 2016a, b). Flood hydrographs for different flow conditions and land-use conditions can be generated using SCS-CN as a loss method, SCS unit hydrograph method for transformation, and kinematic wave method for flood routing in the HEC-HMS model with integration of HEC-GeoHMS (www.hec.usace.army.mil/software/hec-hms/). Hydraulic modeling can be done by using the HEC-RAS model with integration of HEC-GeoRAS (www.**hec**.usace.army.mil/.../**hec-ras**/). Flood plain maps and flood hazard maps for different storm flow conditions can be developed for various land-use conditions.

For hydrologic modeling, inputs required are proper DEM, LULC map, soil map, initial abstraction map, and curve number maps. For generation of these maps, GIS and RS techniques are used. The main basic important input on which overall results depends is the selection of corrected DEM. The DEM can be obtained from various sources and be corrected using actual survey. Arc GIS can be used to delineate the catchment boundary. To obtain accurate delineation and stream network generation, actual surveyed data along the river alignment can be used. LULC maps can be generated by using Landsat/TM, Landsat/ETM, and IRS P6/L-d satellite images. For meteorological input, intensity duration frequency curves (Chow et al., 1988) for the considered city can be developed for different storm return periods. Basin file for overland flow estimation can be generated using the HEC-GeoHMS model and used as an input in the HEC-HMS model for generation of flood hydrographs and peak discharges for different storm flow conditions. Geometry file can be generated in the HEC-GeoRAS model, an extension of Arc GIS, which has been used as an input in the HEC-RAS model for hydraulic modeling. Water surface extents and profiles generated by simulation of the HEC-RAS model for different flow conditions can be used as an input for developing the flood plain maps for different storm flow conditions as postprocessing process in the HEC-GeoRAS model. Flood depths and flood extent generated in flood plain modeling is used as the main input for flood hazard modeling and flood risk assessment (Zope, 2016; Zope et al., 2015, 2016a, b). Some of the limitations of the HEC-HMS and HEC-RAS models include: 1D nature of the models, requirements of river cross-sectional data, requirements of huge data sets such as DEM, soil map, LULC variation, etc. for proper simulation.

12.7 CASE STUDIES

Here, two case studies of coastal urban catchments are considered for the flood plain simulation. First case study is the catchment of Vashi, Navi Mumbai, India, and the second one is the catchment of Dahisar River, Mumbai, India. In the first case study, the flood plain simulation is carried out by applying the IFAM, developed at IIT Bombay (Kulkarni et al., 2014a, b). In the second case study, flood plain simulation is carried out using the integrated flood modeling approach of HEC-HMS with HEC-GeoHMS and HEC-RAS with HEC-GeoRAS and GIS and RS (Zope et al., 2015, 2016a, b).

12.7.1 Case Study 1: Vashi Coastal Urban Catchment

The IFAM model has been applied to Vashi watershed of Navi Mumbai, India, with an area of 3.05 km^2 (Fig. 12.1) for urban flood assessment. The watershed has been delineated using DEM, existing storm water network plan, and with field data. The dominant flow direction for Vashi is from north to south into the creek. The DEM has been extracted from a pair of stereo images of the Cartosat-1 satellite with a spatial resolution of 2.5 m. For the purpose of floodplain simulation, the DEM has been resampled to 10 m spatial resolution. The land-use map has been extracted from the supervised classification of the satellite imagery of April 13, 2010 obtained from the IRS-LISS IV sensor with a spatial resolution of 5.8 m. The land use for the area includes built-up (78%), water body (5%), vegetation (15%), and marshy land (2%). The overland flow Manning's roughness values considered are: built up (0.015), water body (0.03), vegetation (0.15), and marshy land (0.06) (Vieux, 2001).

The surface runoff is modeled using one-dimensional 90 overland flow subgrids as indicated in Fig. 12.1. The grids are digitized in Arc GIS 10.0 based on the flow direction and major and minor storm water drain network. The database for

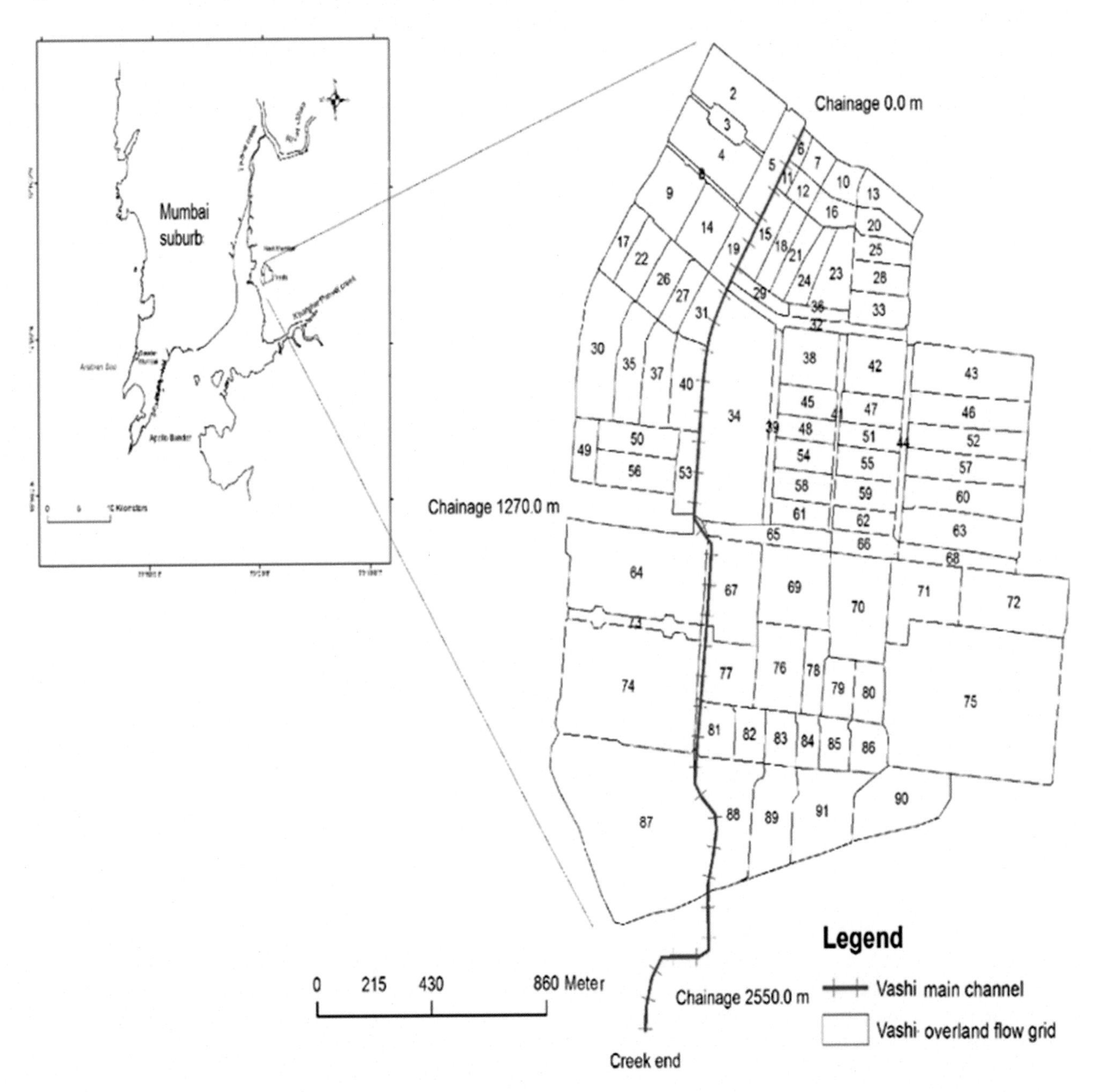

FIGURE 12.1 **Location map of Vashi Watershed with model grid.**

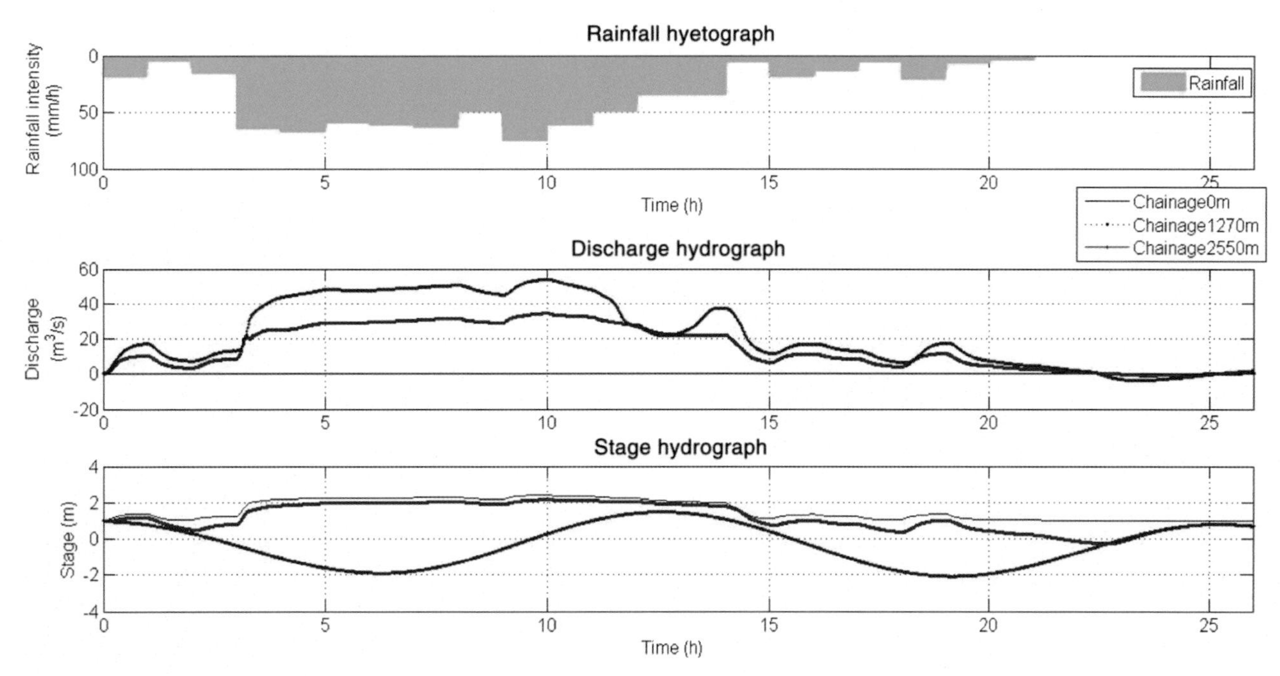

FIGURE 12.2 **Simulated discharge and stage hydrograph of Vashi watershed (July 26, 2005).**

the overland flow elements includes overland element id, area of each element, averaged slope, averaged roughness values, average width of the flow element, and channel node where the corresponding overland flow element drains the water. The Manning's roughness value of 0.03 has been considered for channel.

12.7.1.1 Results and Discussion

The model has been applied to Vashi catchment for an extreme rainfall event of July 26, 2005 with a uniform time step of 2 s. For this event, the rainfall started at 03:00 h of the day during which the tidal stage was in high water condition at 1.72 m (above msl). The event has the peak rainfall intensity of 76 mm/h occurring for over 1-h duration. The simulated discharge and stage hydrographs (at certain chainages) along with the tide creek boundary is shown in Fig. 12.2. The channel overtopping occurred at 3.2 h from the start of simulation. The peak discharge of 54.21 m^3/s occurs at 10 h at the channel end. The simulated flood extent at certain time intervals of the simulation is shown in Fig. 12.3. At the 10th hour of simulation, the flood extent has spread to east at the middle reach of the channel. Subsequently, at the 16th hour, the flood extent starts receding due to decrease in rainfall.

The cumulative rainfall depth for the event is 745 mm, the total surface runoff at the end of the channel is 695.25 mm, and the net water depth on the floodplain surface at the end of the simulation is 42.95 mm. Thus, the total outflow from the channel is 738.20 mm (695.25 + 42.95) which reasonably matches with the cumulative rainfall depth for the event. Infiltration has been neglected as the area is densely urbanized and flow contribution into the channel during high-tide condition has not been explicitly computed for water balance. The average of the mass balance error computed over the floodplain at each time step is 0.037%. Incidentally, the flooded area of Agricultural Produce Market Committee (APMC) market of Vashi, shown in Fig. 12.3, matches with the inundated area simulated by the model. However, these photographs qualitatively validate the simulation results, in the absence of spatial and temporal flood records. Simulation results indicate that wetting and drying has been simulated. Furthermore, the flooding at the southern end is due to tidal variation during the simulation where the channel drains into the creek.

This case study demonstrated the applications of IFAM for flood simulation in coastal areas. The model outputs of flood hydrographs and flood plain maps can be used for flood hazard analysis, flood vulnerability studies, and risk assessments. Furthermore, the model results can be used for flood mitigation and planning of coastal cities.

12.7.2 Case Study 2: Dahisar River Urban Catchment

Mumbai, the capital of Maharashtra and financial hub of India, consists of four major rivers, viz. Mithi River, Oshiwara River, Poisar River, and Dahisar River (Zope, 2016). All these rivers discharge their flow to the Arabian Sea through

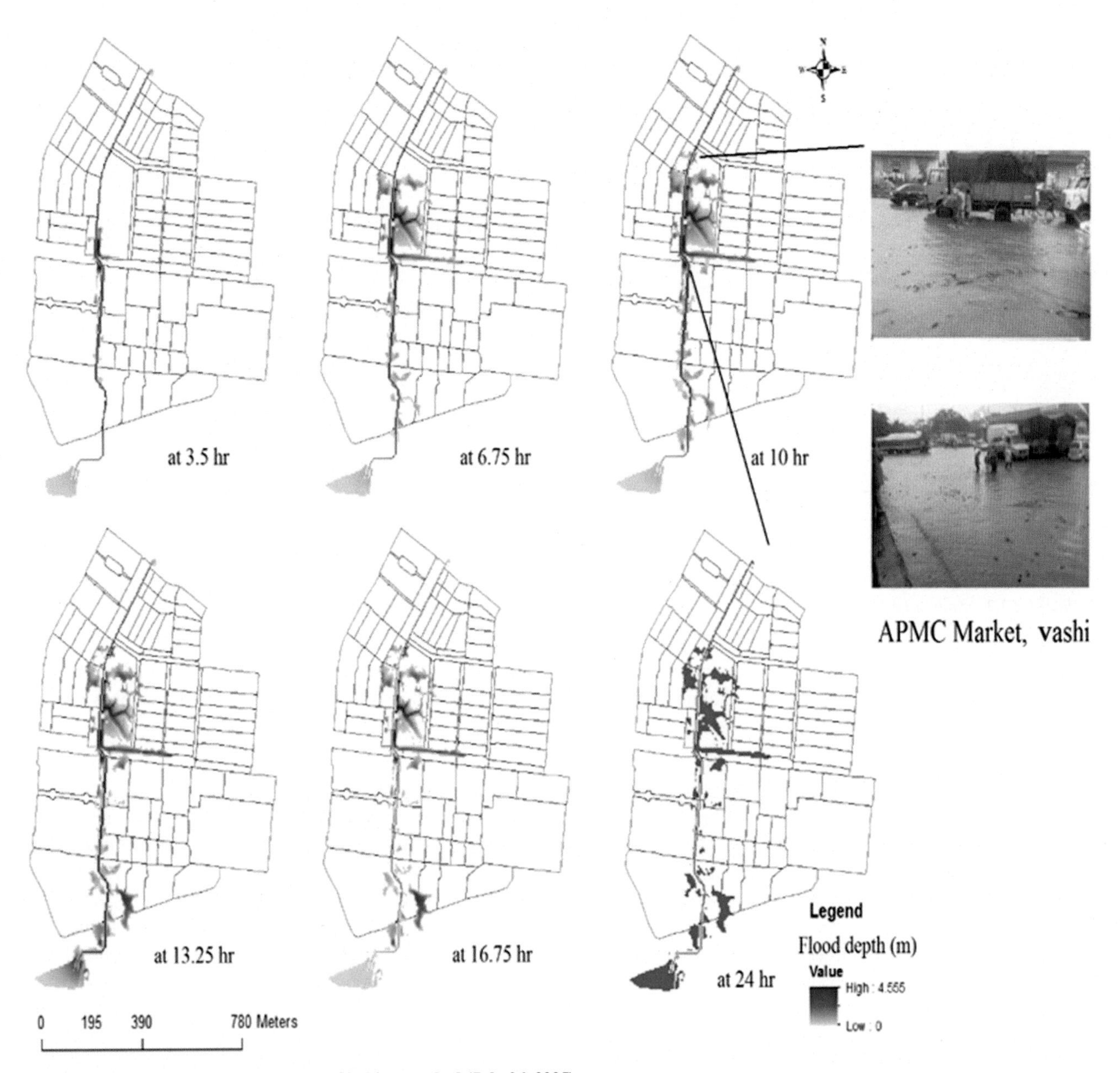

FIGURE 12.3 Simulated flood extent for Vashi watershed (July 26, 2005).

creeks. Dahisar River catchment extends between latitudes 19°8′0″N and 19°18′0″N as well as longitudes 72°50′0″E and 72°57′0″E. Location map of Dahisar River is shown in Fig. 12.4. Dahisar River originates from the spillway of Tulsi Lake (Fig. 12.4). Upstream part of the river falls in national park and downstream part is fully developed and urbanized. The ground elevation varies from 0.35 to 490.35 m with respect to the mean sea level (msl). Upstream part of the river possesses high ground elevation, whereas the downstream part of the river is with less gradient. River discharges its flow to Arabian sea through Manori Creek.

The total length of the river is 18.14 km out of which 9.19 km upstream length falls in the national park area. The total catchment area of the river is 34.29 km^2. Delineation of the watershed has been carried out using the HEC-GeoHMS software, extension of the AEC GIS 9.3.1 (USACE, 2010). To obtain fine sectional data along the flow path of the river, actual surveyed data along the river alignment was collected from the Municipal Corporation and these data points were used to replace the data points from the Shuttle Radar Topographic Mission (SRTM) (htpp://srtm.csi.cgiar.org) DEM. As such, the new corrected SRTM DEM was used for delineation of the catchment boundary and generation of subwatersheds. The corrected DEM is shown in Fig. 12.5.

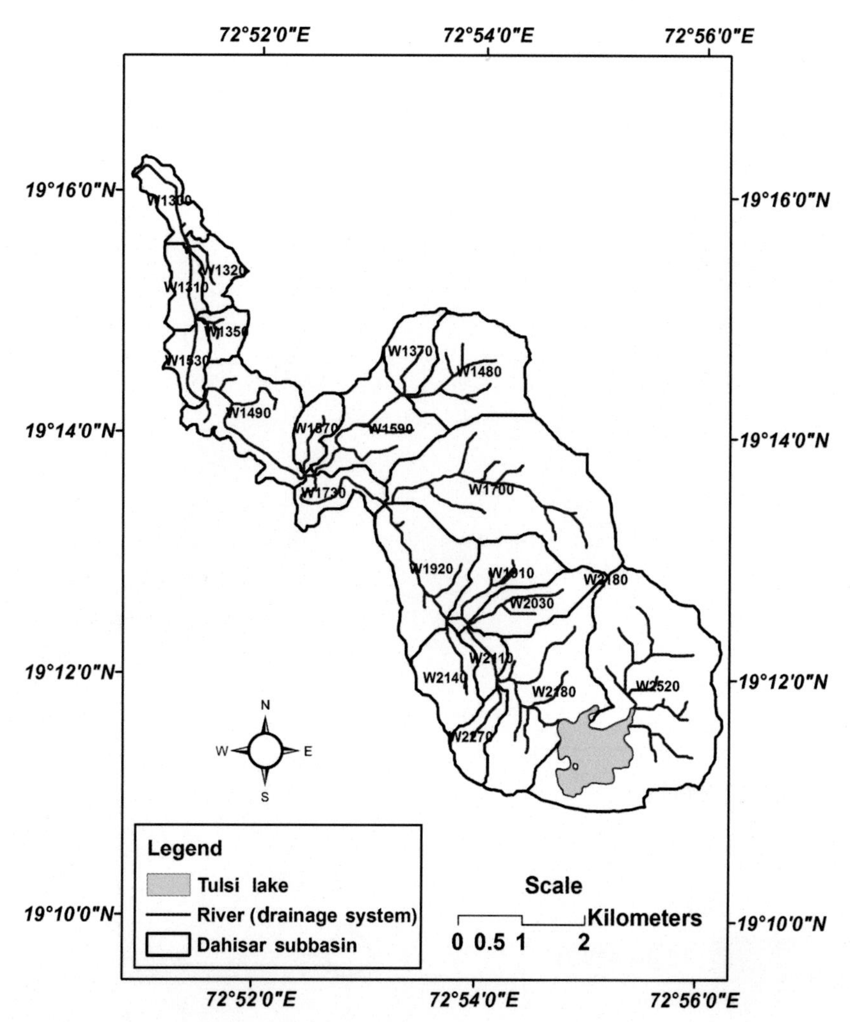

FIGURE 12.4 **Location of Dahisar River catchment with sub catchment.**

The total catchment area of the Dahisar River is 34.29 km^2 and it consists of 20 subcatchments as shown in Fig. 12.4. The satellite image IRS P6/ L-4 of the year 2009 has been used for the extraction of LULC. Supervised classification based on the Gaussian Maximum likelihood has been carried out using the ERDAS Imagine software. LULC classes observed for the analysis are: open land, built-up land, water body, dense vegetation, and less dense vegetation. Manning's n map was prepared by assigning the n values for each land-use class as 0.015, 0.03, 0.1, 0.035, and 0.15 to built-up land, water body, dense vegetation, less dense vegetation, and open land, respectively (Vieux, 2001). Composite curve numbers for each sub-basin has been extracted from the generated curve number map.

Hydrological modeling has been carried out using the HEC-HMS model, and flood hydrographs at each junction as well as outlet were generated (Gul et al., 2010; USACE, 2010). In this study, the SCS-CN method is used as a loss method, the SCS-Unit hydrograph method as a transform method, and the kinematic wave method as a routing method in the HEC-HMS modeling (Zope, 2016; Zope et al., 2015, 2016a, b). Flood hydrographs generated at each junction will be the input to HEC-RAS for hydraulic modeling. Simulation was carried out for 24 h rainfall depth with 10- and 100-year return periods. Flood water depth and water extent profiles have been extracted by using the HEC-RAS model with integration of HEC-GeoRAS for pre- and postprocessing. The geometry prepared with the details such as stream centerline, left and right banks, flow paths, cutlines, and junction points superimposed on triangulated irregular network TIN) is shown in Fig. 12.6. Flood hydrograph at each junction derived from the HEC-HMS model are the input as flow data in HEC-RAS. TIN has been used to extract the attributes of the spatial geometric data. In this study, the upstream boundary condition is normal depth and downstream boundary condition was tidal depth, as the river discharge in Manori creek, Arabian Sea (Zope, 2016; Zope et al., 2015). Flood depth and flood extent profiles are the inputs for generation of the flood inundation map.

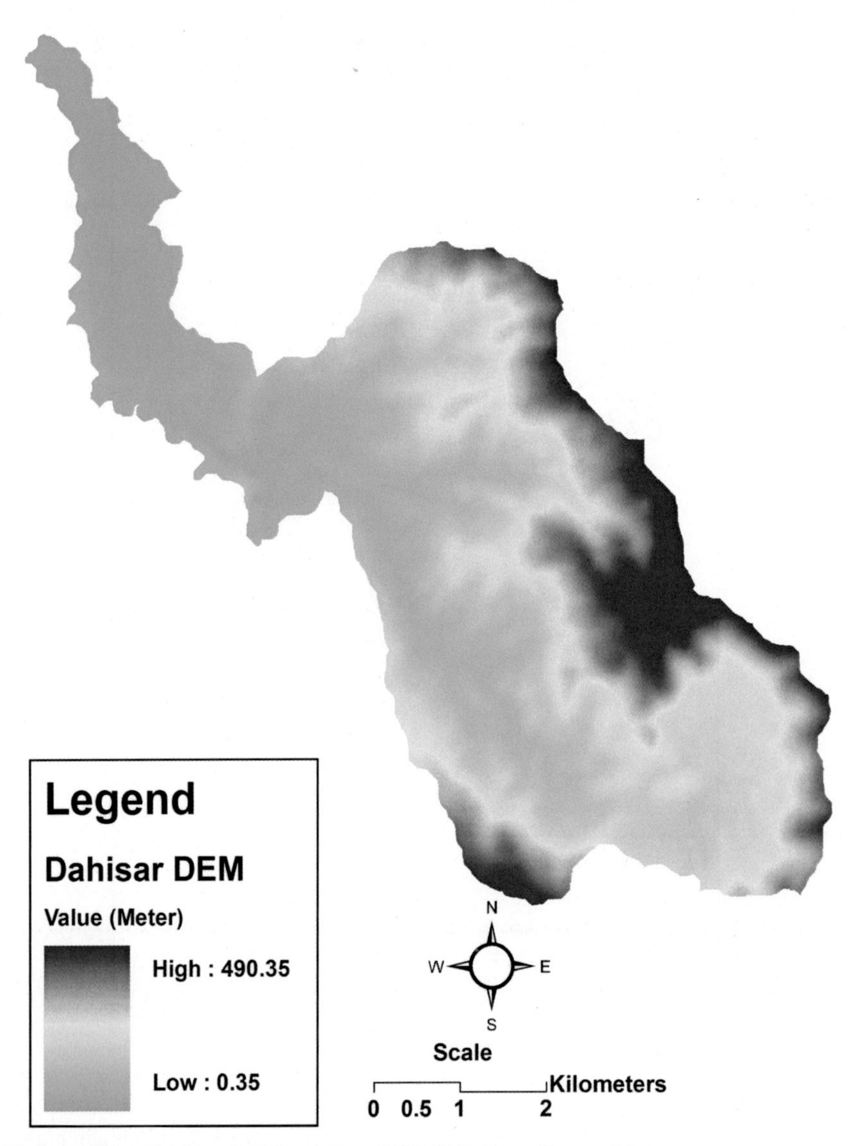

FIGURE 12.5 **DEM of Dahisar River Catchment. Abbreviation: *DEM*, digital elevation model.**

12.7.2.1 Results and Discussion

For generation of the flood inundation and flood hazard maps, the important inputs required are the flood peak discharge and hydrographs at each junction as well as at the outlet of the river. Flood hydrographs generated for 10- and 100-year return periods are shown in Figs. 12.7 and 12.8, respectively. Flood peak discharge at the outlet of the river was observed as 1292.40 and 2778.5 m^3/s for 10- and 100-year return periods, respectively.

The main input essential for the preparation of the flood hazard and risk assessment as well as flood management planning is the generation of flood plain maps for different flow conditions. Flood polygons are generated with the help of TIN, and water surface profiles extracted from HEC-RAS. For generation of flood inundation map, HEC-GeoRAS software, an extension of Arc GIS, has been used (Zope, 2016; Zope et al., 2015). Flood plain maps generated for the 10- and 100-year return period flow conditions are presented in Figs. 12.9 and 12.10, respectively. An aerial extent of the flood inundation area for 10-year return period is 1.84 km^2 and for 100-year return period storm flow condition is 2.55 km^2. Flood inundation maps generated in this study can be used for flood mitigation and disaster management planning for the local municipal corporation.

In this study, the flood hazard map along the river alignment are prepared by using the methodology described by Zope et al. (2016a, b). Raster grid maps of maximum flood extent, slope distance, distance from river, and distance from coast bank are prepared. Standardization of all the maps was done by using Slice tool in Arc GIS by dividing the total area into 100 zones. The flood hazard map was generated by performing weighted overlay analysis (Zope et al., 2016a, b).

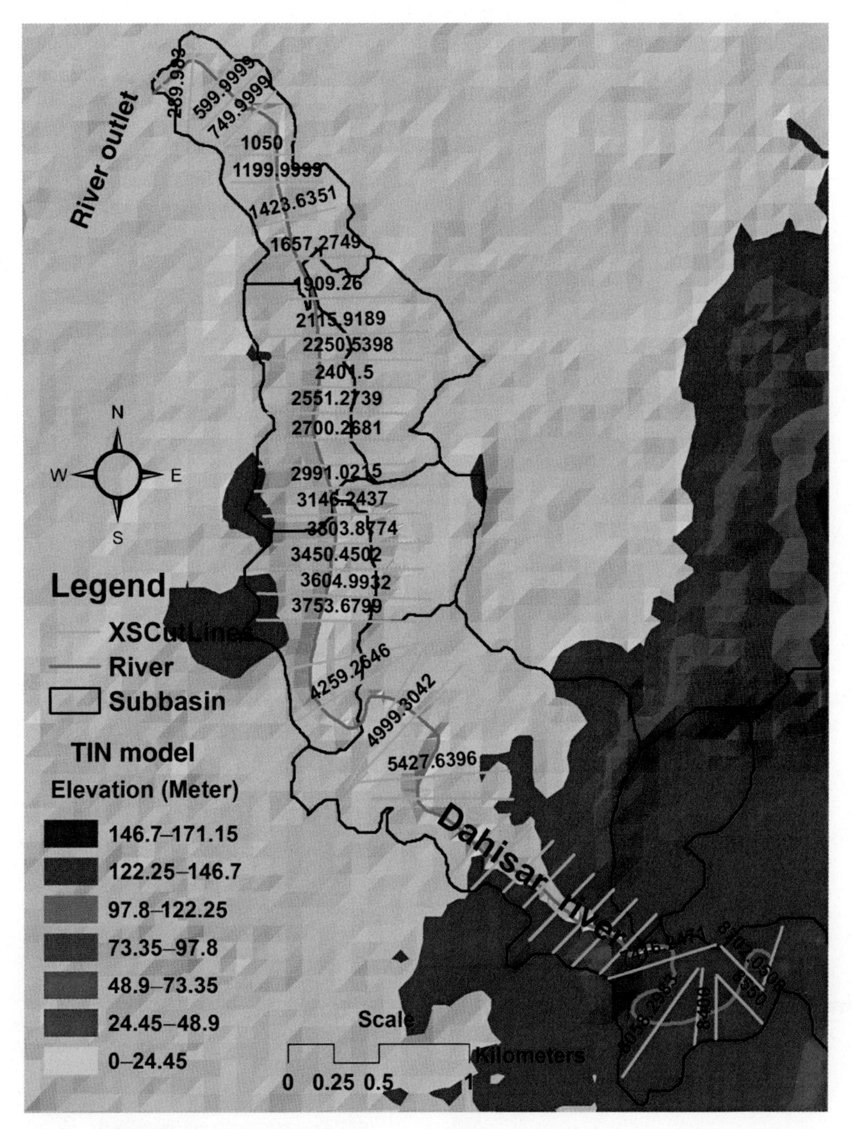

FIGURE 12.6 **HEC-RAS geometry for Dahisar River Catchment.**

The flood hazard has been classified into five categories, viz. very high, high, medium, low, and very low hazards. The flood hazard map generated for 100-year return period flood in this study is presented in Fig. 12.11. The flood hazard map generated in this study can be used as an effective input for flood evacuation planning, flood vulnerability, and risk assessment planning.

12.8 CONCLUDING REMARKS

Urban flooding is a major problem in many parts of the world and is one of the most natural disastrous event which takes place every year, especially in the coastal cities. Urban flood, being a natural disaster, cannot be avoided; however, the losses occurring due to flooding can be prevented by proper flood mitigation planning. As such, it is necessary to have a proper estimation of flood extent and flood hazard for the different flow conditions so that proper flood evacuation and disaster management plan can be prepared in advance. The coastal urban flooding is a complex phenomenon which may occur in various forms such as: urban flooding due to high intensity rainfall; due to inadequate drainage and flooding caused by overtopping in the channels or rivers; flooding due to high tides, etc. In coastal urban cities like Mumbai, mostly severe flood scenarios take place due to combination of surface flooding, channel overtopping, and tidal flooding. For effective coastal urban flood management and mitigation plans, the possible flooding scenario is to be simulated for extreme rainfall events, or various return periods of rainfall and other design scenarios. This chapter discussed the integrated approach of

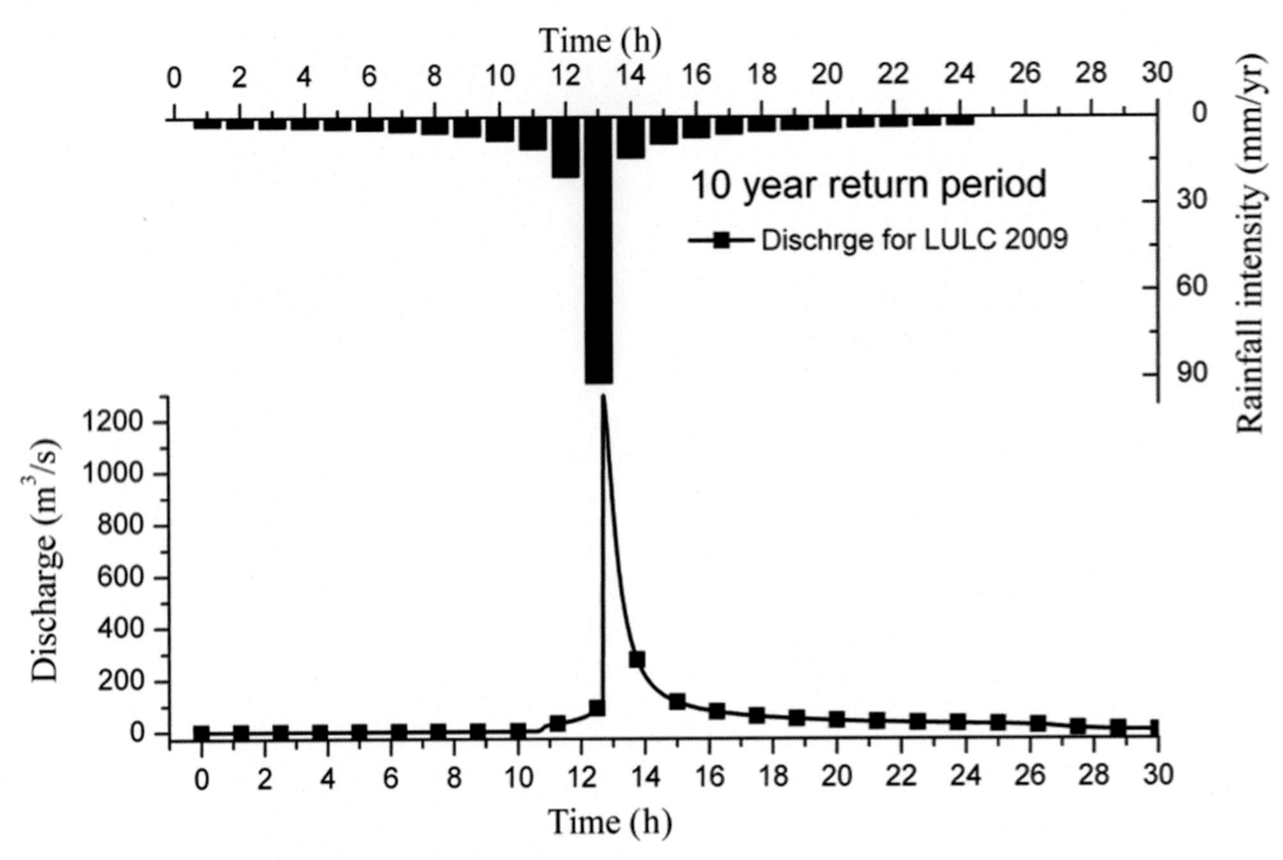

FIGURE 12.7 **Flood hydrograph at outlet for 10-year return period for Dahisar River catchment.**

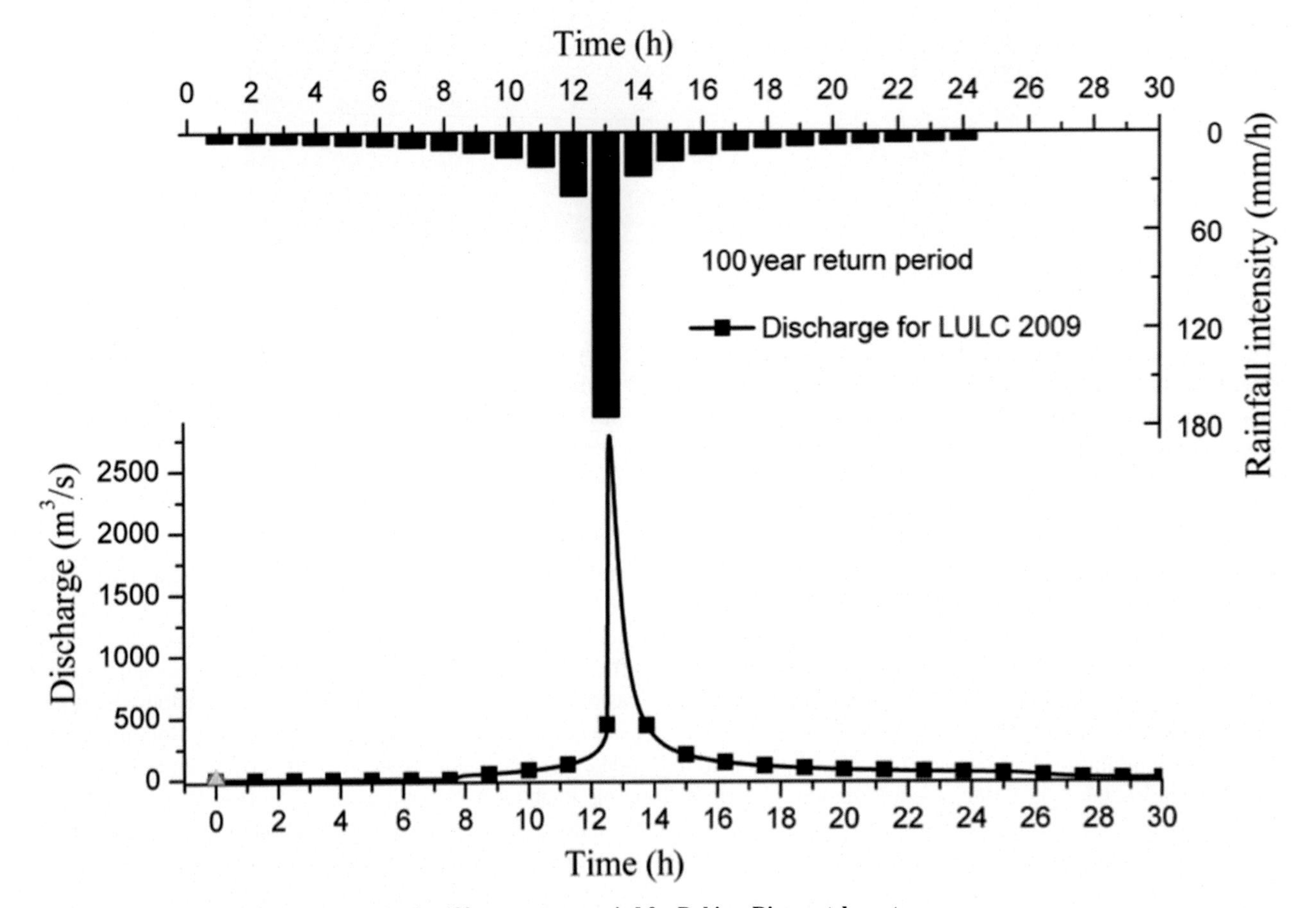

FIGURE 12.8 **Flood hydrograph at outlet for 100-year return period for Dahisar River catchment.**

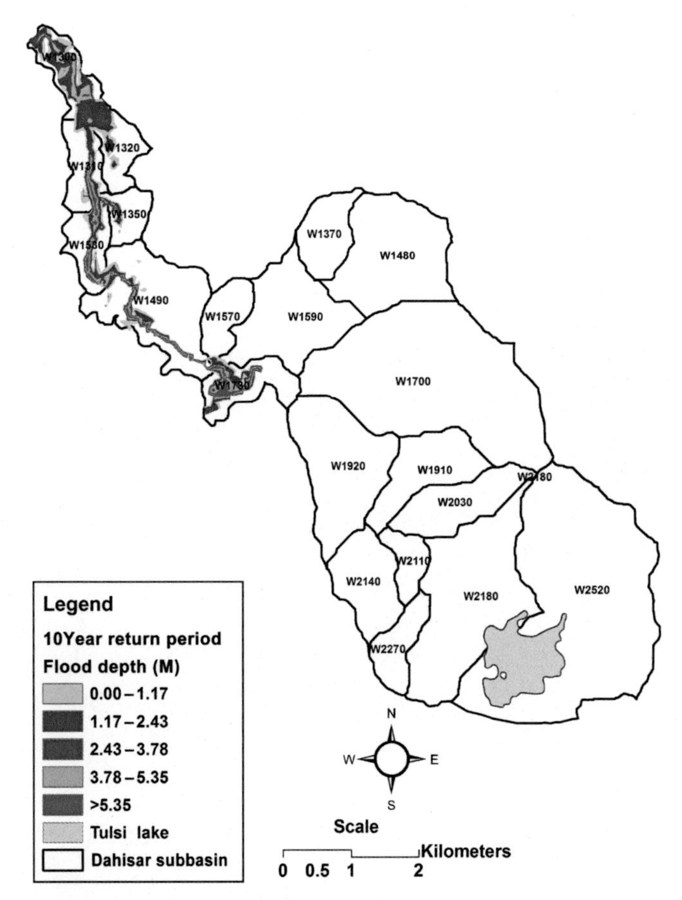

FIGURE 12.9 Flood plain map for 10-year return period for Dahisar River catchment.

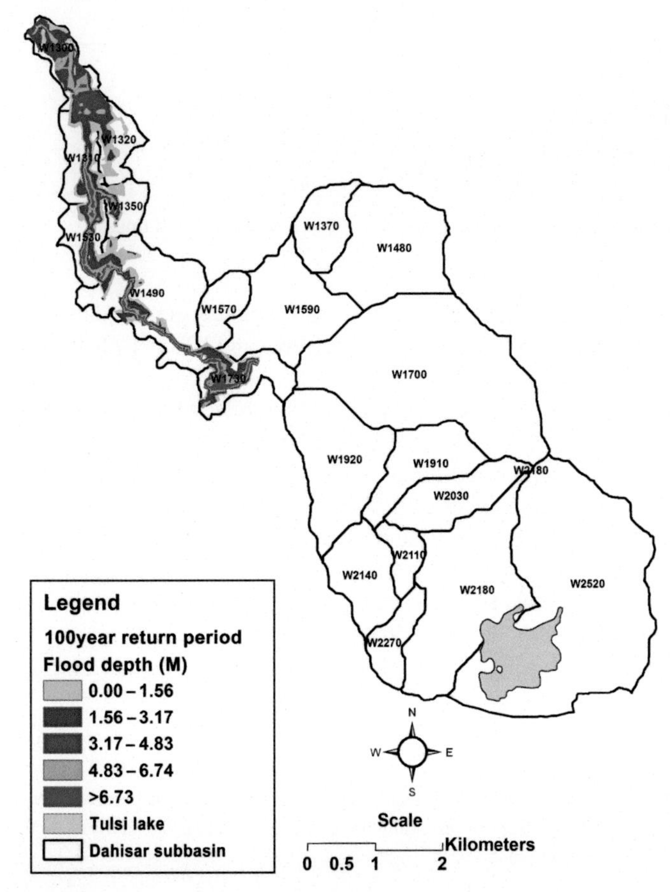

FIGURE 12.10 Flood plain map for 100-year return period for Dahisar River catchment.

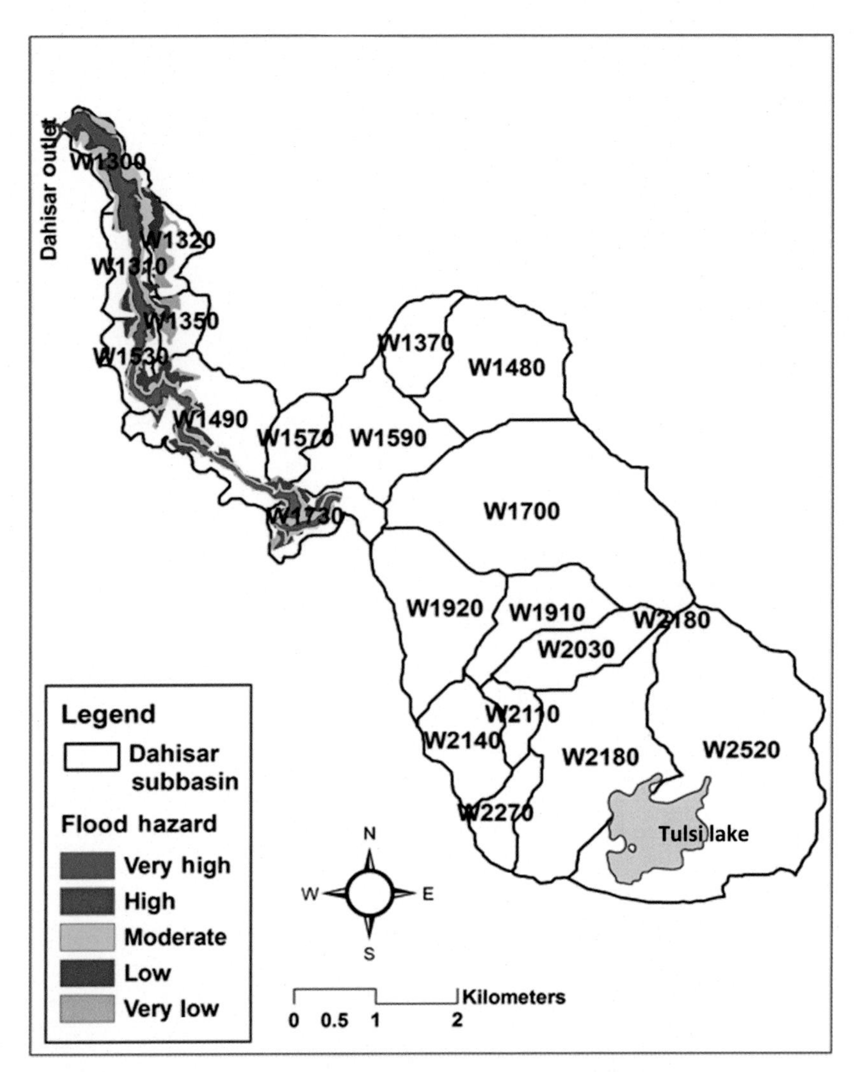

FIGURE 12.11 **Flood hazard map for Dahisar River catchment for 100-year return period flood.**

coastal urban flood simulation using computer models, GIS, and remotely sensed data. Various urban flood models available are discussed briefly. Two of the urban flood models, viz. IFAM and HEC-HMS-RAS, are discussed in detail and their applications are illustrated through two case studies.

The IFAM model is an integrated raster-based flood inundation model where overland flow has been modeled using the 1D mass balance approach, 1D channel flow has been modeled using diffusion wave approximation, and floodplain flow has been modeled using the quasi-2D storage cell-based model. The model has been applied for a coastal urban watershed of Navi Mumbai for an extreme rainfall event of July 26, 2005. The case study demonstrated the application of IFAM model for urban flood simulation. The HEC-HMS-RAS is an open source model which can be effectively used in flood simulation, as demonstrated in the second case study. As demonstrated, both models have good potential to be used in urban flood simulation studies. Flood plain and flood hazard maps generated by these models can be used as flood evacuation planning and flood disaster management planning by the municipal authorities.

ACKNOWLEDGMENTS

The authors acknowledge their sincere gratitude to the Department of Science and Technology (DST), Govt. of India, New Delhi, for sponsoring the present study through the 09DST033 project. The authors thank the engineers of City and Industrial Development Corporation of Maharashtra Limited and Navi Mumbai Municipal Corporation for providing data of the study area. The authors are thankful to the Municipal Corporation of Greater Mumbai for providing actual surveyed data.

REFERENCES

American Society of Civil Engineers (ASCE), 1992. Design and Construction of Urban Storm Water Management Systems. ASCE, New York.

Amini, A., Ali, T., Ghazali, A., Aziz, A., Akib, S., 2011. Impacts of land-use change on stream flows in the Damansara Watershed, Malaysia. Arab. J. Sci. Eng. 36 (5), 713–720.

Bates, P., De Roo, A.P., 2000. A simple raster-based model for flood inundation simulation. J. Hydrol. 236 (1–2), 54–77.

Chow, V., Maidment, D.R., Mays, L.W., 1988. Applied Hydrology. McGraw-Hill, New York.

Cunge, J.A., Holly, F.M., Verwey, A., 1980. Practical Aspects of Computational River Hydraulics. Pitman Advanced Publishing Program, London.

DeFries, R., Eshleman, K.N., 2004. Land use change and hydrologic processes: a major focus for the future. Hydrol. Process 18, 2183–2186.

Desai, Y.M., Eldho, T.I., Shah, A.H., 2011. Finite Element Method with Applications in Engineering. Pearson Education, New Delhi.

Guhathakurta, P., Sreejith, O.P., Menon, P.A., 2011. Impact of climate change on extreme rainfall event and flood risk in India. J. Earth Syst. Sci. 120, 359–373.

Gul, G.O., Harmancroglu, N., Gul, A., 2010. A combined hydrologic and hydraulic modeling approach for testing efficiency of structural flood control measures. Nat. Hazards 54, 245–260. doi: 10.1007/s11069-009-9464-2.

Hromadka, T.V., Yen, C.C., 1986. A diffusion hydrodynamic model (DHM). Adv. Water Resour. 9 (3), 118–170.

Kulkarni, A.T., Eldho, T.I., Rao, E.P., Mohan, B.K., 2013. An Integrated flood simulation model for a coastal urban catchment of Navi Mumbai: a case study. In: Proc. Hydro 2013 International, December 4–6, 2013. IIT Madras, India.

Kulkarni, A.T., Bodke, S.S., Rao, E.P., Eldho, T.I., 2014a. Hydrological impact on change in land use/land cover in an urbanizing catchment of Mumbai: a case study. Indian Society of Hydraulics (ISH) J. Hydraul. Eng. 20 (3), 314–323.

Kulkarni, A.T., Eldho, T.I., Rao, E.P., Mohan, B.K., 2014b. An integrated flood inundation model for simulating heavy rainfall event of coastal urban watershed in Navi Mumbai, India. Nat. Hazards 73, 403–425.

Kulkarni, A.T., Eldho, T.I., Rao, E.P., Mohan, B.K., 2014c. A web GIS based integrated flood assessment modeling tool for coastal urban watersheds. Comput. Geosci. 64, 7–14.

Parkinson, J., Mark, O., 2005. Urban Stormwater Management in Developing Countries. IWA Publishing, London.

Shahapure, S.S., Eldho, T.I., Rao, E.P., 2010. Coastal urban flood simulation using FEM, GIS and remote sensing. Water Resour. Manage. 24 (13), 3615–3640.

Singh, V.P., 1996. Kinematic Wave Modelling in Water Resources. Wiley-Interscience, New York.

USACE, 2010. HEC-RAS River analysis system. Hydraulic Reference Manual Version 4. 1. USACE, Davis, CA, USA.

USACE, 2011. GIS tools for support of HEC-RAS USING ArcGIS. HEC-GeoRAS: User's Manual Version 4.3.93. USACE, Davis, CA, USA, 2011.

Vieux, B.E., 2001. Distributed Hydrologic Modeling Using GIS. Kluwer Academic Publishers, Dordrecht, The Netherlands.

Wurbs, R.A., 1995. Water Management Models—A Guide to Software. Prentice Hall, New York, 29.

Zope, P.E., 2016. Integrated urban flood management with flood models, hazard, vulnerability and risk assessment. Ph.D. Thesis, submitted to Department of Civil Engineering. I.I.T. Bombay, Mumbai, India.

Zope, P.E., Eldho, T.I., Jothiprakash, V., 2012. Spatio-temporal analysis of rainfall for Mumbai city. Int. J. Environ. Res. Dev. 6 (3), 545–553.

Zope, P.E., Eldho, T.I., Jothiprakash, V., 2015. Impacts of urbanization on flooding of coastal urban catchment: a case study of Mumbai city, India. Nat. Hazards 75 (1), 887–908.

Zope, P.E., Eldho, T.I., Jothiprakash, V., 2016a. Impacts of Land use – Land Cover change and Urbanization on Flooding: A Case Study of Oshiwara River Basin in Mumbai, India. *Catena* 145, 142–154.

Zope, P.E., Eldho, T.I., Jothiprakash, V., 2016b. Derivation of rainfall intensity duration frequency curves for Mumbai city, India. J. Water Resour. Prot. 8, 756–765.

Zoppou, C., 2001. Review of urban storm water models. Environ. Model. Software 16, 195–231.

Chapter 13

Probabilistic Analysis Applied to Rock Slope Stability: A Case Study From Northeast Turkey

Nurcihan Ceryan*, Ayhan Kesimal, Sener Ceryan***
**Balikesir University, Balikesir, Turkey*
***Karadeniz Technical University, Trabzon, Turkey*

13.1 INTRODUCTION

Rock slope instability is a major hazard for human activities and often causes economic losses, property damage (maintenance costs), as well as injuries or fatalities. Assessment of the stability of natural or engineering rock slopes has vital prospects to manage the risk of the failure that can develop on these slopes.

The parameters of concern in a typical slope stability assessment are categorized as the ones that describe the physicomechanical characteristics of the rock material, discontinuities, and rock mass (such as cohesion and friction angle), whereas the geometric parameters are those that describe the orientation, location, or the size of the discontinuities. It is known that geological structures and rock weathering have prominent roles in rock slope instability and the presence of tectonic structures might lead to the generation of fractures and finally failure in the rock mass (Donati and Turrini, 2002; Regmi et al., 2014; Vatanpour et al., 2014). In addition, the geometric relationships between the rock slope and discontinuities is the other geometric parameter relating to rock slope stability. Understanding the failure mechanism (mainly governed by the geological structures and the geometry of the slope) is one of the most important concerns in every rock slope stability assessment. The general modes of failure can be categorized as planar, wedge, toppling, and circular. However, in reality, the failure mechanisms are much more complicated. To realistically model such complex behaviors, the slope geometry and the properties of the geological structures need to be determined (Shamekhi, 2014).

In general, the primary objectives of the rock slope stability analyses are (Eberhart, 2003):

- to determine the rock slope stability conditions,
- to investigate potential failure mechanisms,
- to determine the sensitivity/susceptibility of slopes to different triggering mechanisms,
- to test and compare different support and stabilization options, and
- to design optimal excavated slopes in terms of safety, reliability, and economics.

Commonly, the stability of a slope can be expressed in one or more of the following terms (Wyllie and Mah, 2004):

- *Factor of safety, Fs:* Stability is quantified by the limit equilibrium of the slope: Fs > 1 means stable slope.
- *Strain:* Failure is defined by the onset of strains great enough to prevent safe operation of the slope or that the rate of movement exceeds the rate of mining in an open pit.
- *Probability of failure:* Stability is quantified by the probability distribution of difference between resisting and displacing forces (safety margin), which are each expressed as probability distributions.
- *Load and resistance factor design:* Stability is defined by the factored resistance being greater than or equal to the sum of the factored loads.
- *Reliability index:* Reliability is the probability that a slope will be safe, for example, it will survive and not fail under given conditions. It is, therefore, the probability of success Ps which is the probability that the Fs is greater than 1 or that the safety margin (SM) will be greater than zero, for example.

Integrating Disaster Science and Management. http://dx.doi.org/10.1016/B978-0-12-812056-9.00013-0

The rock slope analysis technique is chosen depending on both the site conditions and the potential mode of failure, with careful consideration being given to the varying strengths, weaknesses, and limitations inherent in each methodology (Eberhart, 2003).

Slope stability analysis can be classified into deterministic analysis or probabilistic analysis depending on how uncertainty is incorporated and evaluated. The input for a deterministic analysis is a set of parameters of fixed values (usually at the mean values of the data obtained from site investigations). These types of analyses are based on the calculation of a safety factor that is defined as the ratio of the forces resisting the slide of a rock block over the forces causing the slide. The process of a deterministic analysis is one-off and is implemented by either limit equilibrium analysis or the numerical method. Owing to different sources of uncertainty entailed in such stability problems, the deterministic approach cannot, albeit simple and straightforward, reflect quantified uncertainties explicitly and sufficiently. In recent years, probabilistic approach with the calculation of probability of failure instead of a Fs against failure has become more common (Nilsen, 2000). Probabilistic method takes into consideration the inherent variability and uncertainties in the analysis parameters. In probabilistic analysis, usual parameters related to geometry and unit weights are known as constant parameters. The parameters water pressure, the active frictional angle, and seismic acceleration may, however, vary within wide limits. They do not have a single fixed value. There is no way to predict exactly what the value of one of these parameters will be at any location. Hence, these parameters are described as random variables. When we consider the uncertainties contained in the rock mass and in the slope stability analysis method to be applied, the following important questions in slope reliability analysis must be answered:

- How uncertainties in slope stability analysis can be characterized?
- How the failure probability of a slope can be calculated efficiently both for an individual slip surface and for the slope as a system?
- How the performance information can be used to update the reliability of a slope?
- How to implement reliability-based design (RBD) with incomplete probabilistic information?

Uncertainty can be dealt with in different ways. Parametric and scenario analysis—the assessment of possible ranges of behaviors through variation of input properties and consideration of different conditions—can be applied (from Hammah and Yacoub, 2009). However, statistical simulation is even more powerful. In addition to identify possible outcomes, it can quantify uncertainty and estimate the likelihoods of outcomes occurring. As a result, probabilistic simulation helps engineers to develop more robust and economic designs and solutions. For slope stability, common statistical simulations used in probabilistic calculations are: Monte Carlo Simulation (MCS), First Order Reliability Method (FORM), First Order Second Moment Method (FOSM), Point Estimate Method (PEM), Latin Hypercube Method (LHM), Response Surface Method (RSM), and robustness methods. Although the probabilistic methods take into consideration the inherent variability and uncertainties in the analysis parameters, they include the significant drawbacks and limitations as following (Phoon, 2008).

Probabilistic methods have the ability to include the inherent variation exhibited by almost all parameters that influence slope stability. Also, a probabilistic approach emphasizes the fact that a slope collapse cannot be completely neglected for any choice of slope angle, although the likelihood for failure can be very small.

A probabilistic approach to design is more easily integrated with mine design than a deterministic approach, as it can serve as input to risk analysis for an open pit. Cost–benefit analyses of failures are very attractive but suffer from difficulty associated with estimating all costs involved.

These methods can be applied to almost any type of problems, but a major drawback is the assumption of failure model, commonly based on existing nature. It is widely accepted that great variability and data deficiency exist in natural phenomena.

In the back-analysis of slope failure, the general shape of the sliding surface and volume of the failed mass are known; however, the shear-strength parameters of the sliding surface, the groundwater condition, and the effect of dynamic force are unknown. Based on known information, back-analysis is often carried out to improve an understanding of the unknown parameters at the moment of slope failure. There can be a quantifiable degree of error in the measuring procedures or when an initial estimate of the descriptive statistics of the governing parameters can be made than a probabilistic type of back-analysis is more appropriate. However, posterior mean and standard deviation values of the input parameters, cohesion, and friction angle in the probabilistic back-analysis depend on prior information. In addition, prior information about these parameters such as shear strength value is obtained from the in situ measurements or laboratory tests. There are, however, some limitations and uncertainty in these tests and in situ measurements. Various back-analysis methods have been developed to overcome these limitations and the uncertainty arising from the geological material.

In this study, rock slope stability analysis, especially probability analysis, will be reviewed. In addition, probabilistic slope stability analysis and probabilistic back-analysis were applied to landslides in Araklı-Tasonu quarry NE Turkey. This limestone quarry provides approximately 80% of the raw material needed for the cement plant of Trabzon Cement Factory, which is the region's most important economic enterprise. The material in the quarry was obtained using an uncontrolled blasting technique for approximately 10 years. Then, three landslides occurred on October 3, 2005 (Landslide 1), March 20, 2006 (Landslide 2), and October 19, 2006 (Landslide 3) at the quarry following heavy rainfall. Kesimal et al. (2008) stated that the first plane failure had been strongly influenced by the acceleration of uncontrolled blasting operations, in addition to the heavy rainfall. After landslides, annual production has decreased significantly, and the region's economic output has declined. In addition, new tension cracks on the slopes occurred and the rock masses between the new rock slopes formed by the three landslides (main scarps of the landslide) and the tensile cracks and limited to fractures on the both sides. Considering this condition, it was thought that the new failure can be developed in these slopes. Therefore, to assess the failure mechanism, determine the range of the shear strength mobilized in the failure plane, estimate the groundwater condition at the time of failure, and evaluate the influence of blasting; back-analysis using probabilistic limit equilibrium technique was carried out. And then probabilistic stability analysis of these new slopes named Arc1, Arc2, and Arc3 were performed. In these analysis, Monte Carlo method is used to obtain the probability density function (PDF) of safety factors. For slope stability analysis performed in this study, RocPlane (Version .3.0, Rocscience Inc.) program, which is rented with the grant of the Scientific Research Project of Balikesir University Geological Engineering Department (Project No. 2012/106, Project Manager: Dr. Nurcihan Ceryan), was used.

13.2 BACKGROUND

13.2.1 The Category of the Rock Slope Stability Analysis

The methods used to evaluate rock slope stability are grouped differently by different researchers. According to Eberhart (2003), these methods can be grouped under two headings; first one is conventional methods including kinematic analysis, limit equilibrium analysis, and rockfall simulators and the other one is numerical methods including continuum approach, discontinuum approach (distinct-element method, discontinuous deformation analysis, and particle flow codes), and hybrid approach. The author did not take into account the use of rock mass classification in assessing the rock slope stability.

According to Gao (2015), the rock slope stability can be divided into two groups. The first group is quantitative analysis method including a theory method such as the limit equilibrium method (LEM), the limit state method, and a numerical simulation method. The computation of the LEM based on the force and momentum equilibrium (or the presence of the failure in rock mass such as sliding and toppling) is simple and results in determining its safety factor. However, some realistic assumptions are required, and the reliability of the results is significantly dependent on the accuracy of the assumptions. Because the stability is primarily evaluated by experience, it is highly subjective (Gao, 2015). In the numerical simulation method, the safety factor is generally calculated using the shear strength reduction technique or numerical models that relate existing strength to limit equilibrium strength. Although the computation is intuitive and simple, some computing parameters used in the numerical analysis should be provided beforehand and that the determination of these parameters is very difficult. Because the evaluation of the stability of the rock slope is dependent on some experience, some subjective factors exist (Gao, 2015). The second group can be referred to as the qualitative analysis method including rock engineering system (RES), the rock mass classification, and the engineering analogy methods, and soft computing techniques such as the fuzzy evaluation method, the multicriteria decision-making approach, the support vector machine method, the genetic algorithm method, the artificial neural network method, and colony clustering algorithm. In the qualitative analysis method, the stability of the rock slope is described by the status of the stability instead of the safety factor (from Gao, 2015). However, this method considers the complicated geological factors of the stability of the rock slope. Some uncertain factors of the stability of the rock slope can be reflected in different aspects, and the application is very convenient. Because the main influence factor of the stability of the rock slope is the geological environment, the qualitative analysis method is a practical and extensively applied method (from Gao, 2015).

13.2.2 Deterministic Analysis

Both the conventional and numerical methods are efficient techniques when dealing with deterministic slope stability analysis. In the deterministic approach, the stability of the slope is also described by a single value for the stability factor. Thus, the uncertainty is not explicitly considered. According to its definition, a Fs >1 means that stabilizing forces are greater than driving forces and hence a stable slope. An alternative method in the rock slope stability analysis is partial factor method. According to this method, partial factors for action and materials are applied instead of an overall safety factor (Nilsen, 2000). The partial factor method normally tends to give more conservative design than the approaches which uses deterministic Fs (Nilsen, 2000). Deterministic analysis enjoys a long history of development, and acceptable levels of Fs for various conditions are well established. It has been taken as a routine step for slope stability analysis. However, deterministic analysis uses fixed input parameters and can only cope with the risk of uncertainty by requiring a large Fs value (Qian, 2012; Read and Stacey, 2009). In addition, the same Fs for different failure modes may have different probability of failures (Einstein and Baecher, 1982; Low, 2007). According to Tabba (1984), for Fs between 1.2 and 1.8 (a stable slope), the probability of failure is very sensitive to the degree of uncertainty and the inherent variability in the input parameters. Neglecting many sources of uncertainty and resulting in less reliable output are the main disadvantages of deterministic analysis. On the other hand, it is possible to use deterministic approach by means of a sensitivity analysis. A sensitivity analysis can produce a good qualitative understanding of the factors that are most important for a specific rock slope, but cannot quantify the actual chance of failure.

13.2.3 Uncertainty in the Estimating Rock Slope Stability

An important factor, often neglected in a conventional rock slope stability assessment, is to identify the sources of uncertainty that affect the reliability of the output results (Duncan, 2000). In the estimation of rock slope stability, there are many kinds of uncertainties involved, like the input parameters uncertainty, the calculations uncertainty, and the procedure uncertainty in which the input parameters are transferred into the numerical model. Generally, they can be divided into the following three aspects (Shen, 2012):

The first one is physic uncertainty. This type of uncertainty is an innate property of nature. In slope stability, it is attributed to the natural variability or randomness of some property, such as site topography, site stratigraphy and variability, groundwater level, in situ soil and/or rock characteristics, engineering properties of rock mass, and rock behavior. The natural variability, however, is one of the most important sources of uncertainty, especially when dealing with rock mass and discontinuity network (Baecher and Christian, 2003). The natural or inherent variability is either related to different characteristics of the geological structures at different locations (spatial) or different characteristics of them at a single location, but at different times (temporal) (Shamekhi, 2014). The spatial variability of a common source of uncertainty that is commonly observed in the geometric parameters that describe discontinuities such as orientation and size should be considered in the numerical simulations (Shamekhi, 2014). This kind of uncertainty can be quantified by measurements and statistical estimations, but it is unpredictable and therefore irreducible via collection of more experimental data or using more refined models (Shen, 2012).

The second one is knowledge (or statistical) uncertainty. Parameters such as the friction angle of discontinuity, the uniaxial compressive strength and elasticity modulus of rock materials, the inclination and orientation of discontinuities in a rock mass, and the measured in situ stresses in the rock surrounding an opening do not have a single fixed value but may assume any number of values. There is no way of predicting exactly what the value of one of these parameters will be at any given location. Hence, these parameters are described as random variables. In slope probabilistic analysis, the establishment of the probability distribution of every random variable is a fitting process based on the limited data from measurements or tests (Shen, 2012). Therefore, there are three major subcategories introduced: site characterization uncertainty, model uncertainty, and parameter uncertainty. (1) The lack of knowledge of the governing geometric parameters, (2) the simplifying assumptions in the modeling methods (model uncertainty), and (3) the low level of accuracy in the acquired data (Baecher and Christian, 2003). Human errors in data collection such as field measurements, or lack of data due to the inaccessibility of some of the geological structures can be categorized as knowledge uncertainty. It should be noted that regardless of the low level of knowledge uncertainty in the input parameters, the output results might still suffer from this disadvantage. This is mainly due to the simplifying assumptions that exist in the numerical simulations such as those involved in LEMs (from Shamekhi, 2014).

The last type of uncertainty is simulation uncertainty. Simulation uncertainty is attributed to lack of understanding of physical laws that limits our ability to model the real world (Shen, 2012). An important factor, often neglected in a conventional rock slope stability assessment, is to identify the sources of uncertainty that affect the reliability of the output results (Duncan, 2000). Slope stability analysis and design is a process which simulates the relationship of a set of input random and the output data (e.g. safety factor, reliability index, failure probability, etc.) through some mathematical models (e.g. equations, functions, algorithms, calculation simulation programs, etc.), and these models are constructed based on mathematical mechanical abstract about the real process (Shen, 2012). The uncertainty can exist in the input parameters, the calculations, and the procedure in which the input parameters are transferred into the numerical model. Both the natural variability and knowledge uncertainty can be reduced significantly by incorporating field investigation techniques that are capable of collecting sufficient and representative field data with an acceptable accuracy level. However, the model uncertainty may not be resolved thoroughly only by relying on complicated numerical simulations, which in many cases are also computationally expensive (from Shamekhi, 2014).

13.2.4 Probabilistic Methods

13.2.4.1 Basic Terms

Probabilistic methods have long been used in engineering disciplines with significant success. In the probabilistic approach, in general, each combination of the strength parameters, which may be selected randomly or nonrandomly, creates a different realization of the slope, while the geometry remains the same. By analyzing each realization, a deterministic value for the Fs is obtained. Finally, a probabilistic distribution can be fitted to the values of factor of safety and accordingly, the probability of failure can be estimated (Chiwaye and Stacey, 2010; Miller et al., 2004; Shamekhi, 2014).

To illustrate the methodology, assume that the load and the strength of a structure element, for example, a slope, can be described by two PDFs, respectively, as shown in Fig. 13.1 (Calderon, 2000; Shen, 2012). The strength, or resistance, of the element is termed *R* and the load is denoted *S*. The respective mean and standard deviations of each distribution are denoted m_R and m_S. The mean value of Fs is defined as the distance of the means of resistance force and load. The possibility (or probability) of the slope failure Pf is indicated by the overlap of the PDFs of resistance force and load (Fig. 13.1A). In a purely deterministic approach when using only the mean strength and load, the resulting Fs would be significantly larger

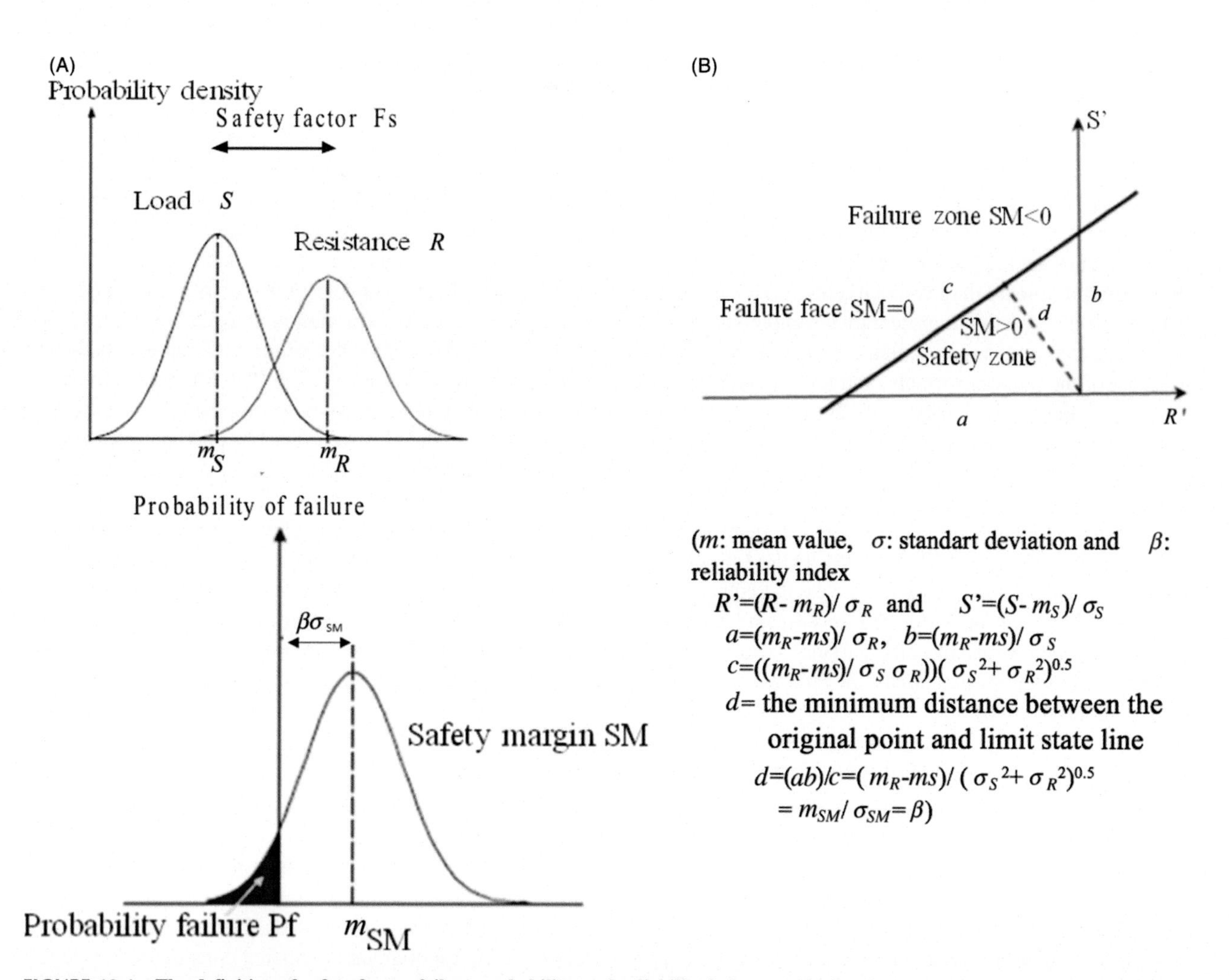

FIGURE 13.1 **The definition of safety factor, failure probability, and reliability index considering the load and the strength of a structure element described by two probability density (after Calderon, 2000; Shen, 2012).**

than unity, implying stable conditions. In estimating the probability that the load exceeds the strength of the construction element, it is common to define an SM as (Fig. 13.1B and C) (Calderon, 2000; Shen, 2012):

$$SM = R - S \quad (13.1)$$

The SM is one type of performance function which is used to determine the probability of failure. The performance function is often denoted $G(x)$ (Harr, 1987):

$$G(x) = R(x) - S(x) \quad (13.2)$$

where x is the collection of random input parameters, which make up the resistance and the load distribution.

The PDF for the SM is illustrated in Fig. 13.1C. In this case, failure occurs when the SM is less than zero. The probability of failure, Pf, is the area under the density function curve for values less than zero, as shown in Fig. 13.1C. Reliability is the probability that a slope will be safe, for example, it will survive and not fail under given conditions. The reliability index β_s can be defined with the help of mean and standard deviations of the performance function $G(X)$, where X represents the probabilistic distribution of the input parameters (Fig. 13.1B). In the above discussion, the state function is assumed as a linear function about two independent normal distributed variables. If the random variables are not normal distributed or independent, they should be transformed into independent normal distributed variables through some means (Ang and Tang, 1984).

For a rock slope, the performance function denoted $G(x)$ can be defined as $G(x) = \text{Fs} - 1$. In this case, the probability of failure, Pf, is the area under the density function curve of Fs or values less than 1. And, the probability of success Ps which is the probability that the Fs greater than 1. In other words, the reliability of a slope is the computed probability that a slope will not fail and is 1 minus the probability of failure:

$$Ps = 1 - Pf \quad (13.3)$$

If the PDF of safety factor is normally distributed, the corresponding reliability index β is defined as (Fell et al., 1988; Tabba, 1984):

$$\beta = (\mu_{Fs} - 1) / \sigma_{Fs} \quad (13.4)$$

where μ_{Fs} is the mean of safety factor and σ_{Fs} is the standard deviation of safety factor.

In general, probabilistic analysis is performed by the following two steps (Tabba, 1984). In the first step, the obtained geotechnical is analyzed to determine the basic statistical parameters (i.e. mean and variance) and the PDF in order to represent and predict the random property of the geological and geotechnical parameters (Calderon, 2000; Tabba, 1984). The PDF is used to model the relative likelihood of a random variable. In cases where it is believed that a given set of measured data represents a set of representative sample values of the variable, and no other information is available, a probability density distribution is representative of the random variable (Tabba, 1984). The mean value of the PDF represents the best estimate of the random variable, and the standard deviation or the coefficient of variance of the PDF represents an assessment of the uncertainty (Park et al., 2001). The second step, risk analysis of slope stability is accomplished using the basic statistical parameters and the probability density distribution developed from the first step. That is, once the probabilistic properties of the input parameters are assumed, the probability of failure can be evaluated by many different risk analysis procedures.

Various methods have been adopted in determining the probability of failure in geotechnical engineering. Some of the available methods are MCS, LHM, FORM, FOSM, PEM, RSM, etc. These methods can be considered generally well known (Christian, 2004). In some cases, the equation for the Fs or for the margin of safety can be solved analytically to give the variance of the factor of safety (from Calderon, 2000). In particular, if the margin or Fs can be expressed as the sum of the random variables, the variance is simply the sum of the variances of each contributing variable. Unfortunately, when the performance function of the Fs contains several variables, this method can be named as direct calculation methods (from Calderon, 2000).

13.2.4.2 Monte Carlo Analysis and Latin Hypercube Sampling

Monte Carlo analysis is a computer-based method of analysis developed in the 1940s that uses statistical sampling techniques to obtain a probabilistic approximation to the solution of a mathematical equation or model (Christian, 2004). Monte Carlo method is a method in which the analyst creates a large number of sets of randomly generated values for the uncertain parameters and computes the performance function for each set. The method calculates the probability failure of slope based on the assumption of the PDF of input random variables (Ang and Tang, 1984; Ross, 1995). The Monto Carlo simulation follows a four-step process. In the first step, for each component random variable being considered, it selects a random value that conforms to the assigned distribution. In the second step, the value of the Fs using the adopted performance function and the output values obtained from step 1 are calculated. Next step includes repeating steps 1 and 2 many times, storing the Fs result from each calculation. During each pass, a random value from the distribution function for each parameter is selected and entered into the calculation. Numerous solutions are obtained by making multiple passes through the program to obtain a solution for each pass. The appropriate number of passes for an analysis is a function of the number of input parameters, the complexity of the modeled situation, and the desired precision of the output. The final result of an MCS is obtaining a probability distribution of the safety factor, probability of failure, and reliability index. The main advantage of the Monte Carlo method lies in the fact that the results from any MCS can be treated using classical statistical methods; thus, the results can be presented in the form of histograms, and methods of statistical estimation and inference are applicable (Subramanyan et al., 2011). The other advantages of this method are: (1) the convergence velocity of simulation is unconcerned with the dimension of random variables; (2) the complexity of the performance function is not related to the simulation procedure; (3) the error of the results is very easy to be determined (Hammah and Yacoub, 2009; Morgan and Henrion, 1990; Shen, 2012). The disadvantage of MCS is that the number of simulations increases substantially with the reduction of failure probability, and the calculation time will be prolonged immensely (Baker and Cornell, 2003; Shen, 2012). In addition, in most applications, the actual relationship between successive points in a sample has no physical significance; hence, the randomness/independence for approximating a uniform distribution is not critical. Moreover, the error of approximating a distribution by a finite sample depends on the equal distribution properties of the sample used for $U(0,1)$ rather than its randomness. Once it is apparent that the uniformity properties are central to the design of sampling techniques, constrained or stratified sampling techniques become appealing (Morgan and Henrion, 1990).

Latin Hypercube Sampling is the most common variance reduction scheme for MCS (Iman et al., 1980; Startzman and Watterbarger, 1985). The sampling method is a technique for improving the efficiency of MCS (Olsson, 2002). The Monte Carlo method randomly selects samples from the valid domain of a variable, which results in an ensemble of numbers without guarantee, whereas the LHM adopts a more systematic sampling approach. It first divides the domain of an input variable into a number of equal-sized bins (Hammah and Yaccoub 2009). It then obtains a random sample from each of those bins. This ensures an ensemble of random numbers that more accurately conforms to the input probability distribution over the domain Latin hypercube sampling is generally recommended over simple random sampling when the model is complex or when time and resource constraints are an issue (Hammah and Yaccoub, 2009; Subramanyan et al., 2011). Morgan and Henrion (1990) stated, "Latin Hypercube sampling seems to be particularly helpful. Although it can introduce slight bias in the estimate of moments, in practice this seems negligible."

13.2.4.3 The Point Estimate Method

The PEM was originally developed by Rosenblueth (1975, 1981). With this method, random variable distributions are represented by two point estimates at ±1 standard deviation from the mean. The analysis is computed for all possible combinations of point estimates, from which statistical results are computed (e.g. mean and standard deviation of output variables). The method applies appropriate weights to each of the point estimates of the response variable to compute moments (Valley and Duff, 2011; Fig. 13.2). The weights can differ for different points. For the slopes, it involves computing the Fs using values of each variable that are one standard deviation above or below the expected value and combining the results to estimate the expected value and standard deviation of the Fs. Hoek (2006) noted that while the PEM technique does not provide a full distribution of the output variable, as do the MCS and LHM, it is very simple to use for problems with relatively few random variables and is useful when general trends are being investigated. When the probability distribution function for the output variable is known, for example, from previous Monte Carlo analyses, the mean and standard deviation values can be used to calculate the complete output distribution. Despite its simplicity and rigorous mathematical basis (Christian and Baecher, 1999, 2002), there are several other disadvantages of the PEM. The main disadvantage of the PEM is that it suffers from the "curse of dimensionality", as the number of random variables increases, the number of point evaluations increases exponentially, significantly increasing computational effort. Harr (1989) and Hong (1998) proposed variations to the PEM which entail fewer runs by locating the evaluation points farther away from the mean, but with the potential disadvantage of evaluating outside the domain of definition for bounded variables. Valley et al. (2010) pointed out that when the output distribution docs not resemble a normal frequency curve, the PEM methodology will adequately estimate mode but fail to account for the full variability, probably due to the close location of the estimation points not allowing for the inclusion of failure modes which manifest further from the mean. In the study performed by Wang and Huang (2012), a Microsoft Excel-based program, RosenPoint, was developed for the Rosenblueth approach.

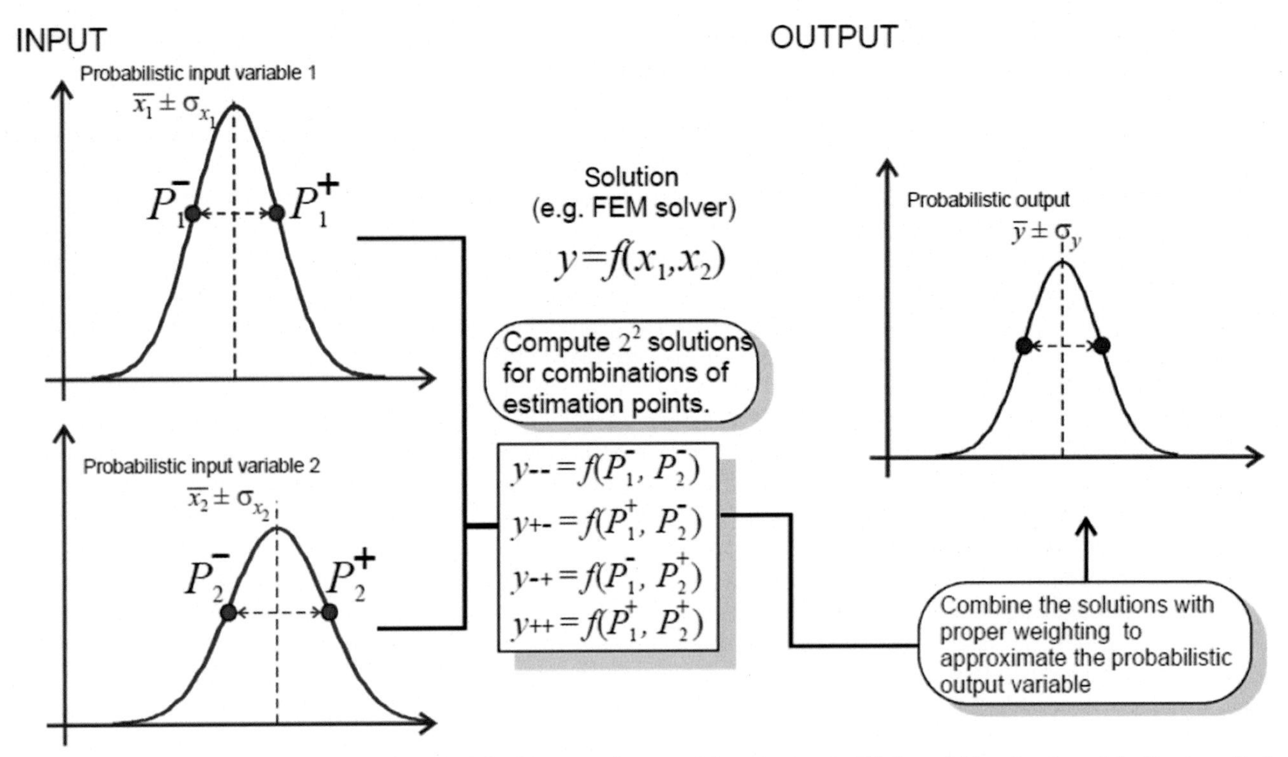

FIGURE 13.2 **Illustration of the computation principle of an approximation of the output probabilistic variable using the point estimate method (in this example, the case with only two probabilistic input variables is assumed (from Valley and Duff, 2011)).**

13.2.4.4 FOSM and SOSM Methods

The primitive probabilistic analysis method, "Second Moment Pattern", produces a linearization of performance function about the mean value of the input random variables (Cornell, 1971). These concerns the first or second order of the performance function based on the first and second moments of random variables, so is called the First Order Second Moment (FOSM) method or the Second Order Second Moment (SOSM) method (Duncan, 2000). The drawback of the FOSM and SOSM methods is that the results depend on the mean value of the input variables at which the partial derivatives of the SM are evaluated (invariance problem). Therefore, these methods are accurate only for linear functions, and the error will be quite large when the degree of nonlinearity of performance function is higher (Baecher and Christian, 2003; Shen, 2012). One of the great advantages of the FOSM method is that it reveals the relative contribution of each variable in a clear and easily tabulated manner (Baecher and Christian, 2003; Morgan and Henrion, 1990). This is very useful in deciding what factors need more investigation and even in revealing some factors whose contribution cannot be reduced by any realistic procedure. Many of the other reliability analysis methods do not provide this information (Baecher and Christian, 2003; Russelli, 2008; Shen, 2012).

All terms in the Taylor series that are of a higher order than the ones are assumed negligibly small and then discarded (Baecher and Christian, 2003). From the remaining first-order terms, the probability that the performance function is less than any given value can be calculated. Considering a performance function SM of n random variables X_i, its Taylor's series expansion about the mean value of the random variables $\mu_1,\ldots,\mu_n$, truncated after first-order terms, gives (from Russelli, 2008):

$$SM\left(X_1,\ldots\ldots X_n\right) = SM\left(\mu_1,\ldots\mu_n\right) + \sum_{i=1}^{n}\left(X_i - \mu_{Xi}\right)\frac{\partial SM}{\partial X_i} \tag{13.5}$$

(Eqs. (13.6) and (13.7), from Russelli, 2008). The derivatives evaluated at $\mu_1,\ldots,\mu_n$ are considered as linearization points. The mean value and the variance of the performance function are given proximately by the following equations:

$$\mu_{SM}\left(x_1,\ldots x_n\right) = SM\left(\mu_{Z1,\ldots Z2}\right) \tag{13.6}$$

$$Var\left[SM\left(X_1,\ldots,X_2\right)\right]=\sum_{i=1}^{n}\left(\frac{\partial SM}{\partial Xi}\right)^2 Var(Xt)+2\sum_{i=1}^{n}\sum_{j=1}^{n}\left(\frac{\partial SM}{Xi}\frac{\partial SM}{Xj}Cov\left(X_iX_j\right)\right) \tag{13.7}$$

If the random variables are uncorrelated, then the term with the covariance drops out. In general, for n random variables, $2n + 1$ calculations are involved. In practice, it is sometimes complicated to evaluate derivatives of nonlinear functions (Russelli, 2008). The usual output of the FOSM method is the reliability index given by Eq. (13.4) for normally distributed variables. For nonnormal PDFs, the reliability index evaluated by the FOSM method is only an approximation. Very refined methods have been developed to convert general PDFs into standard normal distributions (Ang and Tang, 1984; Russelli, 2008).

To avoid the misusc of the FOSM method in probabilistic analyses, one should be referred to its limitations, which are listed below (from Russelli, 2008):

- The accuracy of the FOSM method diminishes as the nonlinearity of a function increases.
- The skewness of the output PDF is not provided.
- The shape of the PDF of the input variables is not taken into account, and the random variables are described using only their mean and standard deviation. In this way, no information about the shape of the PDF of the output is provided, but it has to be assumed. This assumption introduces a source of inaccuracy.
- The FOSM method is applied primarily to problems without spatial correlation among the input variables. With extra calculation effort, the method can be applied for two correlated random variables, but this can be very cumbersome.

The SOSM method represents a slight extension of the FOSM method. Actually, with the SOSM method, it is possible to include the second-order terms of the Taylor's series expansion in the evaluation of the mean value of a performance function (from Russelli, 2008).

13.2.4.5 First-Order Reliability Method

While the FOSM method and its extension by the Rosenblueth PEM are powerful tools that usually give excellent results, they do involve approximations that may not be acceptable (Chiwaye, 2010). One approximation is that the moments of the failure criterion can be estimated accurately enough by starting with the mean values of the variables and extrapolating linearly (Baecher and Christian, 2003). A second is that the form of the distribution of the performance function is known and can be used to compute the probability of failure. The Hasofer–Lind method (Hasofer and Lind, 1974), also well known as FORM, can overcome this difficulty by calculating the derivatives of the SM at a critical point on the failure surface, also called design point. An iterative solution is usually required to find this point, but the process tends to converge very rapidly (Russelli, 2008). FORM was initially derived for use with noncorrelated variables. However, Low and Tang (1997) proposed a method of using the FORM method with correlated variables. The FORM method can also be adapted to be used on variables that are not normally distributed. In Fig. 13.3A, the joint PDF of the random variables R and S is linearized in the design point P. When the variables are nonnormally distributed, then they can be transformed into standard normal variables with zero mean value and standard deviation equal to unity, as described by Ang and Tang (1984) and plotted in the standard normalized space as in Fig. 13.3B (from Russelli, 2008). The shortest distance between the failure point P and the origin of the normalized space is the reliability index β (from Russelli, 2008):

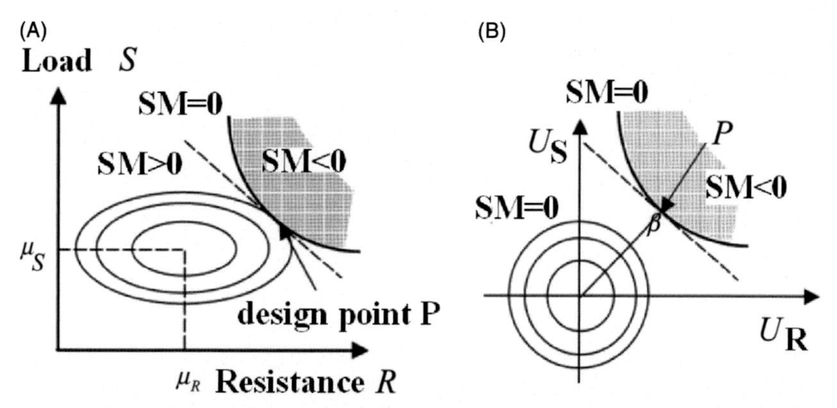

FIGURE 13.3 Linearization of the SM in the design point in the RL-plane (A) and definition of the reliability index for normalized random variables (B) (from Russelli, 2008).

$$\beta = \min_{(SM=0)}\left(\sqrt{U_R^2 + U_L^2}\right) \tag{13.8}$$

In second-order reliability methods, the failure function is not linear as it is in FORM. In SORM, the second-order approximation of the function is established. So, if the limit state function is not linear, it will improve the result by including the second derivate of the failure function when the design value is determined. This is only if the limit state function is smooth; if it, on the other hand, is rough, then the result might also be worse (Schweckendiek, 2006).

The most important advantages of the FORM method are the following (Russelli, 2008):

- The reliability index and the failure probability are independent of the safety format used and they can be evaluated also for nonlinear functions.
- It is a more efficient method for estimating low probability of failure when compared with other approaches, as will be shown in Chapter 6.
- The sensitivity factors give additional information on the influence of the input random variables on the performance function.

The important references for FORM theory and its application and analytical and numerical optimization outlines are Christian et al. (1992, 1994), Vrijling (1997), and Nadım (2006).

13.2.4.6 Response Surface Method

The RSM method originated by Bosx and Wilson (1951) is a collection of statistical and mathematical techniques helpful for developing, improving, and optimizing processes through empirical model building (Kola, 2014). The RSM, an alternative to the MCS method, replaces the numerical model with an approximated less-expensive surrogate model, which can be used to estimate the system response and analyze uncertainty propagation (Li et al., 2011). The basic idea of the RSM is to approximate the implicit limit state function using an equivalent explicit mathematical function of the random variables involved in the limit state function. Because the approximated function is explicit, the FORM can be applied to estimate the probability of failure (Tan et al., 2013). The method generally engages a combination of both computation and visualization. It is based on a group of carefully designed mathematical and statistical experiments which is used to develop an adequate functional relationship between a response of interest (output variable) influenced by several independent variables (input variables). An experiment is a series of tests, called runs, in which changes are made in the input variables in order to identify the reasons for changes in the output response. In general, the structure of the relationship between the inputs and output (response) is unknown but can be truly approximated by the RSM in which the convergence to the real relation improves by a number of smooth functions (from Dadashzadeh, 2015). The use of quadratic response surface models makes the method simpler than standard nonlinear techniques for determining optimal designs (Kola, 2014).

Li et al. (2016) reviewed previous studies on developments and applications of RSMs in different slope reliability problems. According to Wong (1985), the RSM becomes less accurate when the region of interest is expanded, unless more point estimates are employed. Li et al. (2011) proposed a stochastic response surface (SRSM) method for reliability analysis of rock slopes and showed that the accuracy of the proposed method is higher than that for the FORM and is much more efficient than MCS. The basic idea of SRSM is to approximate model inputs and outputs in terms of random variables such as the standard normal variables by a polynomial chaos expansion. The unknown coefficients in the polynomial chaos expansion are determined using a probabilistic collocation method (from Li et al., 2011). According to Li et al. (2011), the major advantage of SRSM is that it allows existing deterministic numerical codes, such as a finite-element analysis code, to be used as a "black box" within the method. SRSM and the accuracy of the proposed SRSM is higher than that of the FORM. Furthermore, it is more efficient than the direct MCS. Stankovic´ et al. (2013) used FORM enforced with RSM to study the stability of an open pit coal mine in Monte Negro to obtain probability of failure with sufficient accuracy, but with various simplifications. Zhang et al. (2013) studied the system reliability of soil slopes with RSM. Tan et al. (2013) studied the application of RSM in slope stability analysis. It is to be noted that in spite of the fact that available discrete element codes like 3DEC requires high run time per model, no study has been performed so far on the application of the RSM using probabilistic analyses in discontinuum medium (from Dadashzadeh, 2015).

13.2.4.7 Robustness Methods

The robustness of a system or component is the degree to which its properties (performances) are not affected by the uncertainties of input variables or uncertainties of environmental conditions. It measures the insensitivity of the system or component properties to parameter variation and uncertainties in environment (Huyse, 2001). Robustness is usually measured with the variance or standard deviation of the performance function $Y = g(X)$. For two designs with the same mean value as shown in Fig. 13.2, Design 1 is more robust than Design 2 since the former has a narrower distribution (lower variance). Both reliability and robustness are critical constituents of high quality (from Huyse, 2001; Zang et al., 2002). However, reliability and robustness are conceptually different. The differences are (Huyse, 2001):

1. Reliability is concerned with the performance distribution at the tails of the PDF, whereas robustness is concerned with the performance distribution around the mean of the performance function.
2. Reliability is more related to safety for the avoidance of extreme catastrophic events, whereas robustness deals with the everyday fluctuations and is more related to the avoidance of quality loss.

Because of the difference between reliability and robustness, there are two different design methodologies dealing with reliability and robustness, namely RBD and robust design. As suggested by the name, RBD makes a design reliable or ensures the probability of failure less than the required level. Robust design makes a design not sensitive to uncertainty or reduces the variations of design performance. At the analysis level, we will first discuss how to assess reliability and robustness for a given design and then at the design level (from Huyse, 2001; Zang et al., 2002).

According to Wang et al. (2015), in a geotechnical RBD, the statistics of the noise factors, including the coefficients of variation (COVs) of the noise factors (mainly referring to uncertain rock properties here in) and the coefficients of correlation between noise factors, are often difficult to ascertain. When these statistics are overestimated or underestimated, the design obtained from the traditional RBD can be cost-inefficient or unsafe. The robust design concept is introduced to the RBD to minimize the effect of the uncertainty associated with the estimated statistics of noise factors (from Wang et al., 2015).

In the proposed Robust Geotechnical Design framework by Wang et al. (2013), the computed failure probability of rock slope is modeled as the response of the system. The variation in the computed failure probability caused by the uncertainty in the estimated variation of rock properties is evaluated using statistical methods. Wang et al. (2013) stated that the robustness for RBD is achieved if the variation of the failure probability (i.e. the system response) can be minimized by manipulating design parameters of the rock slope. However, higher design robustness is often achieved at a higher cost. Thus, a multiobjective optimization considering cost and robustness is needed to select the optimal designs among those in the acceptable design space. Multiobjective optimization does not usually produce a single best design with respect to all design objectives. Rather, the result is often expressed in a "Pareto Front", which is a set of optimal designs that are "non-dominated" by any other designs in all aspects (from Wang et al., 2013). In the said study, the design robustness is achieved by adjusting design parameters (i.e. slope angle and height, and protection measures) that can be controlled by the engineer. The authors indicated that without considering design robustness, traditional RBD methods may produce a least-cost design that was initially shown as adequate by meeting the failure probability requirement but later found inadequate because of an underestimation of the variation of noise factors.

Xu et al. (2014) indicated that the variability of shear characteristics of rock discontinuities is often difficult to ascertain. Thus, even with the RBD approach, which allows for consideration of the uncertainty of input parameters, the design of a rock slope system may be either cost-inefficient (overdesign) or unsafe (underdesign), depending on whether the variation of input parameters is overestimated or underestimated. The authors presented feasible approach to addressing this problem using robust design concept considering the uncertainty about the variation of input parameters being a critical issue in an RBD.

13.2.5 Rock Mass Classification Systems Applied to Slope Stability

One of the common tools used in the evaluation of the stability and design of rock slope is rock mass classification systems. Due to the long-term geological processes, the geometrical and mechanical properties of a rock mass are extremely complex. Moreover, its environment is also complex (e.g. both the groundwater conditions and in situ stress have great variations) (Zheng et al., 2016). To quantify the complex properties of a rock mass based on the past experience, various taxonomies, usually called rock mass classification systems, have been developed (Zheng et al., 2016). Rock mass classification has been applied successfully in tunneling and underground mining. They are also used to estimate the strength and deformation properties of rock masses.

According to Li (2016), rock mass classification systems have increasingly attracted the attention of scholars worldwide because these systems are able to obtain a quantitative experience equation, as well as consider many factors affecting the stability of rock mass. At present, classification systems have become an important tool for quick assessment the stability of rock slope (Li et al., 2016). A number of empirical rock mass rating methods have been proposed for the assessment of slope stability. These include Rock Mass Strength (RMR; Selby, 1980), Slope Mass Rating (SMR; Romana, 1985; Romana et al., 2003), Slope Rock Mass Rating (SRMR; Robertson, 1988), Slope Stability Index (Singh et al., 1986), Chinese Slope Mass Rating (CSMR; Chen, 1995), Rock Slope Deterioration Assessment (Nicholson and Hencher, 1997;

Nicholson, 2000, 2002, 2003, 2004), Slope Stability Probability (SSPC; Hack, 1998; Hack et al., 2003), Modified Slope Stability Probability Classification (Lindsay et al., 2001)Volcanic Rock Face Safety Rating (Singh and Connolly, 2003), Falling Rock Hazard Index (Singh, 2004), Modification of Slope Mass Rating (Tomas, 2008; Tomas et al., 2007, 2012), and an alternative rock mass classification proposed by Pantelidis (2010). Among these classification systems, Romana's classification systems can be considered as a universally used system. SMR is a well-established technique that is appropriate for slope risk assessment and provides quantitative adjustment factors (Salmi and Hosseinzadeh, 2015).

At first, classification systems made for underground excavations were applied to slope stability problems (Hack, 2002). Generally, these systems inherited the main features of the underlying classification system for underground excavations. This caused that some systems have strange components or parameters that are not applicable in slope stability or are missing parameters that are important in slope stability problems (Hack, 2002). For this, some classification systems have been modified for slopes (e.g. the RMS, SMR, SRMR, and CSMR systems comprise modifications of the RMR system). Later on, classification systems were created directly for rock slopes (e.g. SSPC).

The parameter used these classifications can be grouped under three headings. First group includes the parameters in the fundamental Rock Mass Rating System such as the intact rock strength, the Rock Quality Designation index, the condition of discontinuities, the spacing of discontinuities, and the groundwater outflows (Pantelidis, 2009). As known, RMR system (Bieniawski, 1976) was initially developed for underground structures. The second type of the factors relevant to the method of excavation, such as the slope height and dip, the grade of weathering of the rock mass and the dip, and orientation of discontinuities, are not as common as the five previously mentioned; however, they are used in about half of the existing classification systems (Pantelidis, 2009). Other factors appear to be of less importance and refer to the stabilization and protective measures, failure history, stresses that act on the slope, direct disturbance (e.g. human activities), and the condition of slope (overhangs at the slope face, face irregularity, vegetation cover; Pantelidis, 2009). According to Pantelidis (2009), there are uncertainties in the correspondence between the slope stability factor and the rating value.

SSPC system developed by Hack (1996) is based on a three-step approach and on the probabilistic assessment of independently different failure mechanisms in a slope. In this system, the stability is determined in two analyses. The first analysis is the determination of the stability of the slope related to the discontinuities in the rock mass. This analysis is related to the orientation of the discontinuities and the slope. The second analysis determines the stability of the slope in relation to the strength of the rock mass in which the slope is made. This second analysis is independent of the orientation of the discontinuities and the slope.

The other qualitative analysis method is the systems approach, that is, the RES. In the RES approach, the interaction matrix is both the basic analytical tool and also the main presentation technique for characterizing the most important parameters, and their interaction mechanisms, in a rock engineering project (Hudson, 1992). It has also been widely used in other rock mechanics applications such as the general problem of stability of slopes (Budetta et al., 2008; Castaldini et al., 1998; Ceryan and Ceryan, 2008; KhaloKakaie and Zare Naghadehi, 2012a, b; Mazzoccola and Hudson, 1996; Rozos et al., 2008; Shang et al., 2005; Zare Naghadehi et al., 2011, 2013; Zhang et al., 2004). According to the RES methodology, the main objective involves defining the principal causative and triggering factors that are responsible for the manifestation of slope instability phenomena, quantifying their interactions, obtaining their weighted coefficients, and calculating the slope instability index, which refers to the inherent potential instability of each slope (from Gao, 2015).

13.2.6 Conventional Methods of the Rock Slope Stability Analysis

Rock slope analysis with conventional methods is mostly deterministic in nature that provides only single value of Fs. Stead et al. (2006) indicated that conventional rock slope analyses in current practice invariably begin with engineering geological investigations of the discontinuities, leading to kinematic and limit equilibrium stability assessments. In their study, they give a tablet to provide a summary of conventional methods, together with their advantages and limitations. Table 13.1 was modified after Coggan et al. (1998) by the said authors.

To evaluate the stability of a rock slope, the structural condition of the unfavorably oriented discontinuities in the rock slope needs to be evaluated as to whether or not the structural condition results in an instability (Park et al., 2016). This procedure is known as kinematic analysis. Kinematic analyses of discontinuity-controlled rock slope instabilities take into account the comparison of the orientation of discontinuity planes or their intersections with friction angles, slope geometry, and slope orientation. Angular relationships between discontinuities and slope surfaces are applied to determine the potential and modes of failures (Kliche, 1999). Numerous studies have been performed to determine failure modes utilizing stereographic projection technique (Admassu and Shakoor, 2013; Aksoy and Ercanoğlu, 2007; Böhme et al., 2013; Cruden, 1978; Gischig et al., 2011; Gokceoglu et al., 2000; Goodman, 1976; Hocking, 1976; Hoek and Bray, 1981; Lucas, 1980; Markland, 1972; Matherson, 1988; Park and West, 2001; Park et al., 2005; Smith, 2015; Vatanpour et al., 2014;

TABLE 13.1 Conventional Methods of Analysis (After Stead et al., 2006)

Analysis method	Critical input parameters	Advantages	Limitations
Stenographic and kinematic	Critical slope and discontinuity geometry; representative shear strength characteristics	Simple to use and show failure potential. Some methods allow analysis of critical key blocks. Can be used with statistical techniques to indicate probability of failure and associated volumes	Suitable for preliminary design or for noncritical slopes, using mainly joint orientations. Identification of critical joints requires engineering judgments. Must be used with representative discontinuity strength data
Limit equilibrium	Representative geometry, material/joint shear strength, material unit weights, groundwater, and external loading/support conditions	Much software available for different failure modes (planar, circular, wedge, toppling, etc.). Mostly deterministic but some probabilistic analyses in 2D and 3D with multiple materials, reinforcement, and groundwater profiles. Suitable for sensitivity analysis of Fs to most inputs	Fs calculations must assume instability mechanisms and associated determinacy requirements. In situ stress, strains, and intact material failure not considered. Simple probabilistic analyses may not allow for sample/data covariance
Rockfall simulation	Representative slope geometry and surface condition. Rock block sizes, shapes, unit weights, and coefficients of restitution	Practical tool for siting structures and catch fences. Can utilize probabilistic analysis. 2D and 3D codes available	Limited experience in use relative to empirical design charts

Yoon et al., 2002). This procedure should be carried out for the probabilistic analysis as well as the deterministic analysis. Deterministic (conventional) kinematical approach based on the use of most observable discontinuity orientations is the commonly employed method for the preliminary assessments of stability of slopes excavated in jointed rock masses. But, other possible discontinuity orientations distributed around the central clusters may exist and they may also contribute to failures. In addition, deterministic conventional kinematical approach, the stereonet-based kinematic analysis, assumes that a uniform slope orientation and a tightly clustered orientation of the discontinuities exist. However, the slope orientation, represented by the slope aspect and slope angle, cannot be uniform if the slope faces are rugged, as in many natural rock slopes, or if the slope was created by deficient blasting (Park et al., 2016). The effects of such discontinuities and variability of slope orientation on instabilities are assessed by probabilistic kinematical approaches (Tuncay and Ulusay, 2000). The most detailed studies on the probability-based kinematic analysis approach were performed by McMahon (1971), Zanbak (1977), CANMET (1981), and Gokceoglu et al. (2000). McMahon (1971) stated that using the statistical distribution models, the discontinuity contour diagrams in the stereographic projection network can be rearranged. The author has also investigated the possibility of planar slip for different slope angles based on the probable distributions of discontinuities on the stereonet. Gokceoglu et al. (2000) indicated that the orientation values can be normalized, assuming that the orientation values are normally distributed and using the theoretical normal distribution curve to be obtained from the mean and standard deviation of these orientation data.

The stability of each kinematically possible failure mechanism can be determined by limit equilibrium analysis. The deterministic method of limit equilibrium analysis for rock slope stability was introduced by Jaeger (1971) and Kutter (1974), and recent techniques, which are commonly used, were established by Hoek and Bray (1981) and Goodman (1976). Many of the limit equilibrium analyses have been available for more than 40 years and can be considered reliable slope design tools (Hoek et al., 2000). According to Hoek (2009), a well-designed limit equilibrium program is probably the best tool for "what if" type analyses at a conceptual slope design stage or to investigate failures and possible remedial actions. Representative geometry and material characteristics, rock mass shear strength parameters (cohesion and friction), discontinuity shear strength characteristics, reinforcement characteristics, and external support data are the critical input parameters for limit equilibrium analysis. In the said analysis, each parameter is represented by a single mean value. The LEM, the kinetic analysis, is purely a force-based calculation, an effective tool for analyzing forces acting on the slope, predicting the moment of failure when the driving forces exceed the resisting forces. Wyllie and Mah (2004) indicated that the rock can be assumed to be a Mohr–Coulomb material in which the shear strength is expressed in terms of the cohesion and friction angle for all shear type failures. The strength of a rock mass is usually represented by its cohesion and friction angle, with slope failures evaluated using the linear Mohr–Coulomb yield criterion (Li et al., 2012). However, the researchers showed that the yield criteria for rock masses to be nonlinear (Dong-Ping et al., 2016; Douglas, 2002; Hoek and Brown, 1980; Sheorey, 1997; Yudhbir et al., 1983). For this, a linear failure envelope may not be suitable for estimating the rock slope stability (Li et al., 2012). The output of such analysis was a single value for the Fs that was used as the parameter of choice for stability assessment. However, in this approach, the Fs is highly sensitive to the level of uncertainty in the input values and is acceptable when low-scattered input data with a low level of uncertainty is available (Shamekhi, 2014). Due to simplification, and constraints related to the mathematical formulation, the displacements within critical points of the slope cannot be calculated effectively using LEM (Suikkanen, 2014). According to Nilsen (2016), due to the uncertain and variable character of input parameters, the limit equilibrium approach has evident shortcomings for stability analysis of rock slopes, particularly when the partial factor principle is applied. In addition, it is well known that the solution obtained from the LEM is not rigorous, as neither static nor kinematic admissibility conditions are satisfied (Li et al., 2012). Moreover, in order to find a solution, arbitrary assumptions must be made regarding inter-slice forces for a two-dimensional (2D) and inter-column forces for a three-dimensional (3D) case (Li et al., 2012).

An alternative to the FOS approach to rock slope stability is the probabilistic method. This method is based on the calculation of the probability of failure (Pf) of the slope. In general, since the geometry is simple, the failure mechanism is known in advance, and discretization of the geometry is not required, the input parameters, either geometric or physical, can be easily defined stochastically (Ahmadabadi and Poise, 2016; Li et al., 2011; Suikkanen, 2014). In this approach, the input parameters are described as probability distributions rather than point estimates of the values. For this, the inherent variability of each input parameter is characterized by a probabilistic distribution function and the first and second statistical moments (mean and standard deviation). By combining these distributions within the deterministic model used to calculate the FOS, the probability of failure of the slope can be estimated (Chiwaye, 2010). In the limit equilibrium techniques, transferring the variability of the geometric/strength parameters is not considered as a huge challenge. For this, Monte Carlo sampling is commonly used to generate different realizations of the slope based on the inherent variability of the input parameters. However, due to a high level of simplifications in such analysis, the modeling uncertainty is of a concern. In addition, in these techniques, there are the limitations such as assuming the behavior of the materials as linear and not considering the dynamic forces, loading sequences and some parameters in the boundary conditions. These limitations increase the level of modeling uncertainty and restrict the application of conventional methods in more complex mechanisms that are the dominant situations in reality (Stead et al., 2006). To avoid such limitations, numerical methods for the stability analysis of the rock slopes are considered as choices that are more reliable. However, these methods suffer from some other disadvantages, especially when incorporated in a probabilistic analysis (Shamekhi, 2014).

13.2.7 Numerical Methods

Stead et al. (2006) presented how rock slope analyses may be undertaken using three levels of sophistication. According to these authors, Level I analyses include the conventional application of kinematic and limit equilibrium techniques with modifications to include probabilistic techniques, coupling of groundwater simulations, and simplistic treatment of intact fracture and plastic yield. Level II analyses involve the use of continuum and discontinuum numerical methods. In addition to simple translation (Stead et al., 2006). Level II techniques can be applied to complex translational rock slope deformations where step-path failure necessitates degradation and failure of intact rock bridges along basal, rear, and lateral release surfaces (from Stead et al., 2006).

The results of many research showed that limitations equilibrium method has following limitation and disadvantages:

- When movement is detected in a slope, LEMs are not suitable for evaluating the impact of such movements on the overall stability (Corkum and Martin, 2004).
- Moreover, they are limited to simplistic problems in their scope of application, encompassing simple slope geometries and basic loading conditions, and as such, provide little insights into slope failure mechanisms (Desai and Christian, 1977; Eberhart, 2003).
- Many rock slope stability problems involve complexities relating to geometry, material anisotropy, nonlinear behavior, in situ stresses, and the presence of several coupled processes (pore pressures, seismic loading, and etcetera) (Chiwaye, 2010).
- The limit equilibrium solution only identifies the onset of failure, whereas the numerical solution includes the effect of stress redistribution and progressive failure after failure has been initiated. The resulting Fs allows for this weakening effect

- The numerical models can also be used to determine the Fs of a slope in which a number of failure mechanisms can exist simultaneously or where the mechanism of failure may change as progressive failure occurs (Hoek et al., 2000).

Cundall (2002) compared the characteristics of numerical solutions and LEMs in solving the Fs of slopes and concluded that continuum mechanics-based numerical methods have the following advantages: (a) no predefined slip surface is needed; (b) the slip surface can be of any shape; (c) multiple failure surfaces are possible; (d) no statistical assumptions are needed; (e) structures (such as footings, tunnels, etc.) and/or structural elements (such as beams, cables, etc.) and interfaces can be included without concern about compatibility; and (f) kinematics is satisfied.

To deal with these complexities and the other disadvantage of LEMs, numerical methods are used for slope stability analyses. Advances in computing power and the availability of relatively inexpensive commercial numerical modeling codes means that the simulation of potential rock slope failure mechanisms could, and in many cases should, form a standard component of a rock slope investigation (Stead et al., 2006). However, numerical models take much longer times to compute compared to limit equilibrium models (Chiwaye, 2010). Limit equilibrium methods can make thousands of safety factor calculations almost instantaneously but numerical methods require longer times to make just one safety factor calculation (Valdivia and Lorig, 2000). Another drawback with numerical methods is that they are generally not easy to use (Chiwaye, 2010). In the numerical method, the entire slope is divided into elements. Elements are modeled with stress–strain relationships and deformation properties that define how the material behaves. After the stress states and boundary conditions are specified, the numerical method is able to compute the deformation and displacement of a rock mass. It can also compute the FOS of a rock mass by applying the shear strength reduction technique (Chiwaye, 2010; Read and Stacey, 2009). In the method, there is no necessity for predefined failure surface and mode or statistical assumptions. Moreover, multiple failure surfaces can be taken into account (Jing, 2003). The shear strength reduction technique usually will determine a safety factor equal to or slightly less than LEMs (Chiwaye, 2010). Reviews of the numerical method for rock mass were given by Jing and Hudson (2002) and Jing (2003). The numerical method has two major advantages. Firstly, it is capable of computing the deformation and displacement of a rock mass. Secondly, its process of analysis is more rigorous than that of LEM (e.g. the failure surface is sought out during the analysis instead of being preassumed and FS is calculated by the shear strength reduction technique). On the other hand, the numerical method is slow compared with LEM, making it

TABLE 13.2 Numerical Methods of Analysis (After Stead et al., 2006)

Analysis method	Critical input parameters	Advantages	Limitations
Continuum modeling (e.g. finite clement, finite difference)	Representative slope geometry; constitutive criteria (e.g. elastic, elasto-plastic, creep, etc.); groundwater characteristics; shear strength of surfaces; in situ stress state	Allows for material deformation and failure, including complex behavior and mechanisms, in 2D and 3D with coupled modeling of groundwater. Can assess effects of critical parameter variations on instability mechanisms. Can incorporate creep deformation and dynamic analysis. Some programs use imbedded language (e.g. FISH) to allow user to define own functions and subroutines	Users should be well trained, experienced, observe good modeling practice, and be aware of model/software limitations. Input data generally limited and some required inputs are not routinely measured. Sensitivity analyses limited due to run time constraints, but this is rapidly improving
Discontinuum modeling (e.g. distinct element, DDA)	Slope and discontinuity geometry; intact constitutive criteria (elastic, elasto-plastic, etc.); discontinuity stiffness and shear strength; groundwater and in situ stress conditions	Allows for block deformation and movement of blocks relative to each other. Can model complex behavior and mechanisms (combined material and discontinuity behavior, coupled with hydromechanical and dynamic analysis). Able to assess effects of parameter variations on instability. Some programs use imbedded language (e.g. FISH) to allow users to define own functions and subroutines	As above, experienced users needed. General limitations similar to those listed above. Need to simulate representative discontinuity geometry (spacing, persistence, etc.). Limited data on joint properties available (e.g. joint stiffness)

unsuitable for certain types of analysis (such as sensitivity analysis or probabilistic analysis) where stability analysis needs to be repeated many times (Chiwaye, 2010).

According to Eberhart (2003), the most common numerical methods of analysis applied for rock slope stability can be divided into three approaches: continuum, discontinuum, and hybrid modeling. Many numerical methods have been developed and used in the geotechnical engineering. Stead et al. (2006) discussed the advantages and disadvantages of some of the methods (Table 13.2).

These numerical methods may be divided into two approaches: continuum and discontinuum modeling (Eberhart, 2003). Continuum codes assume the material is continuous throughout the body. Continuum modeling is best suited for the analysis of slopes that are comprised of massive, intact rock, weak rocks, and soil-like, or heavily fractured rock masses. The actual rupture surface does not form in the continuum modeling; so, the afterfailure analysis is not possible. Also, the discontinuities inside a rock mass cannot be modeled explicitly except for few major ones (Alzo'ubi, 2016). To overcome some of the shortcomings of the continuum modeling, new approaches have been developed such as introducing new constitutive models and simulating localization of shear bands in the intact material (Alzo'ubi, 2016). Finite-element method, finite-difference method, and finite-boundary method are all continuum methods (Jing, 2003).

If the stability of the rock slope is controlled by movement of joint-bounded blocks and/or intact rock deformation, then the use of discontinuum modeling should be considered (Stead et al., 2006). Discrete Element Method (DEM) and Discrete Fracture Network method are mainly for discontinuum modeling. The modeling methods treat the rock slope as a discontinuous rock mass by considering it as an assemblage of rigid or deformable blocks. It allows the deformation and movement of blocks relative to each other, so it can model complex behavior and mechanisms. The analysis includes sliding and opening of rock discontinuities controlled by the normal and shear stiffness of joints. It requires representative slope and discontinuity geometry, intact constitutive criteria, discontinuity stiffness and shear strength, groundwater characteristics, and the in situ stress state (Chiwaye, 2010; Eberhart, 2003; Stead et al., 2006). Discontinuum modeling constitutes the most commonly applied numerical technique to rock slope stability. Several variations of discrete-elements methodology exit: distinct-element methods, discontinuous deformation analysis, and particle flow codes. The major limitation of discontinuum modeling is that it requires representative discontinuity geometry (spacing, persistence, etc.) along with joint data and properties of each block (Chiwaye, 2010; Eberhart, 2003).

Stead et al. (2006) stated that continuum and discontinuum codes as described above often fail to realistically simulate the progressive failure of rock slopes, particularly the dynamics of kinematic release accompanying complex internal distortion, dilation, and fracture. The importance of developing kinematic release through fracturing in selected mechanisms is a key issue in rock slope analysis that is not addressed by conventional numerical models. Stead et al. (2004) emphasize the need to consider rock slope failures using the principles of fracture mechanics with appropriate consideration of damage, energy, fatigue, and time dependency. These approaches have been developed to address this situation, and hybrid approaches are increasingly being adopted in rock slope analysis. The models formed by hybrid approaches may include combined analyses using limit equilibrium stability analysis and finite-element groundwater flow and stress analysis. These models have been used for a considerable time in underground rock engineering including coupled boundary finite-element and coupled boundary-distinct element solutions. Recent advances include coupled particle flow and finite-difference analyses (Table 13.3). In the study performed by Stead et al. (2006), Level III analyses involve the use of hybrid continuum–discontinuum codes with fracture simulation capabilities (Table 13.3). These codes are applicable to a wide spectrum of rock slope failure modes, but are particularly well suited to complex translation/rotational instabilities where failure requires internal yielding, brittle fracturing, and shearing, in addition to strength degradation along release surfaces (Stead et al., 2006).

There are two other methods which do not follow this classification: Meshless Methods (MM) and Artificial Neural Networks (Bobet, 2010). Jing and Hudson (2002), Jing (2003), and Bobet (2010) have discussed the different numerical methods applied in rock mechanics.

A deterministic approach when used with the numerical methods can only be trusted for a general evaluation of the slope stability. However, due to high level of uncertainty, such results are not reliable to be used in practical designs or decision-making (Shamekhi, 2014). For this reason, the probability approach has also found applications in numerical analysis. Since the geometry and the discretization of a model are independent of its physical parameters, the inherent variability in the strength parameters can be transferred into a numerical model with an affordable computational cost (from Shamekhi, 2014). Many authors have efficiently implemented the probabilistic strength parameters into different numerical models including the FOSM, PEM, the Hasofer–Lind approach (FORM), MCS, and the RSM (Dadashzadeh, 2015; Duncan, 2000; Duzgun and Bhasin, 2009; Griffiths and Fenton, 2004; Griffiths et al., 2009; Wolff, 1996).

The statistical analysis of slope stability, however, is not limited to the inherent variability in the strength parameters. The variability in geometric parameters can have a noticeable impact on the probability of failure. More importantly,

TABLE 13.3 Advantages and Limitations of Advanced/Hybrid Numerical Methods of Analysis (Stead et al., 2006)

Analysis method	Critical input parameters	Advantages	Limitations
Particle flow code (e.g. PFC, ELFEN)	Problem geometry, particle shape, size and size distribution; particle density, bond stiffness, and strength (normal and shear); bonding type and tightness of packing configuration	Ideal for simulating particle flow, but can also simulate behavior of solid material (e.g. intact or jointed rock) through bonded assemblage of particles, most notably the fracturing and disintegration of the bonded assemblage. Dynamic analysis possible, as well as 2D and 3D simulations. Some programs use imbedded language (e.g. FISH) to allow users to define own functions and subroutines	Input parameters are based on micromechanical properties, requiring calibration through simulation of laboratory testing configurations (i.e. to correlate particle bonding properties to Young's modulus, compressive strength, etc.). Particles are rigid and often cylindrical (2D) or spherical (3D). Simulation of brittle fracture not based on physical laws/principles of fracture mechanics
Hybrid finite/discrete element codes (e.g. ELFEN, DDA)	Combination of input parameters listed in Table 13.2 for both continuum and discontinuum standalone models (e.g. elastic, elasto-plastic, etc., for continuum; stiffness, shear strength, etc., for discontinuities); damping factors; tensile strength and fracture energy release rate for fracture simulation	Combines advantages of both continuum and discontinuum methods. Coupled finite-/discrete element models able to simulate intact fracture propagation and fragmentation of jointed and bedded media. Incorporates efficient automatic adaptive re-meshing routines. Dynamic, 2D and 3D analyses possible using wide variety of constitutive models (plastic, viscoplastic, etc.)	Complex problems require high memory capacity. Comparatively little practical experience in use. Requires ongoing calibration and constraints. Yet to be coupled with groundwater

geometric parameters define failure mechanisms. If these parameters are defined deterministically, the uncertainty in the predicted failure mechanism will be very high (from Shamekhi, 2014). Recently, a few studies have attempted to capture the inherent variability of some of the geometric parameters in analyses with the finite-element or DEM (Brideau et al., 2012; Hammah and Yaccoub, 2009; Shamekhi and Tannant, 2015). In these approaches, similar to what have been done for strength parameters, random samples are selected from the probabilistic distribution of the geometric parameters (Shamekhi, 2014). Shamekhi and Tannant (2015) suggested a new methodology for performing rock slope stability analyses that incorporate probabilistic variability in geometric parameters such as joint orientation and trace length in finite-element models of a slope. The said methodology minimizes the number of geometric realizations that need to be considered and constructed in finite-element software.

13.2.8 Soft Computing Methods Applied in Slope Stability

Although the probabilistic approach has been proposed as an objective tool for representing uncertainty in failure model and material characteristics, these methods have many drawbacks and deficiencies (Ceryan, 2016). The probabilistic characteristics of a random variable would be described completely if the form of the distribution function and the associate parameters are specified. In addition, the form of the distribution function may not be known in some practices (Park et al., 2012). The other main drawbacks of the probabilistic methods are that almost all of the present probabilistic methods assume independency between the different rock slope parameters. Park et al. (2012) indicated that, frequently, only the maximum and minimum values for an uncertain parameter can be obtained precisely; therefore, an uncertain parameter can be expressed only with an interval between the minimum and the maximum.

It is known that engineering behavior of a rock mass is controlled by many factors, related to its nature and the environmental conditions. The rock mass is largely discontinuous, anisotropic, inhomogeneous, and nonelastic, and a rock mass is also a fractured porous medium, under complex in situ conditions of stresses, temperature, and fluid pressures, and is under stress and continuously loaded by dynamic movements (Jing, 2003). Most of these factors take effect simultaneously and have complicated interactions with each other in practical engineering (Yang and Zhang, 1997) and the geological conditions for the rock mass may be largely uncertain. So, behavior of a rock mass can be complex and uncertain. Due to these situations, numerical analyses are exact modeling of in situ rock masses, which is still not possible. In addition, they are expensive and time-consuming tasks and always implemented under certain restricted conditions. Considering the deficiencies of probabilistic approach and numerical analyses to slope stability, soft computing methods have been applied to evaluate rock slope stability by many researchers (Ceryan, 2016).

The study performed by Li and Mei (2004) is based on results of the statistical analysis of a large amount of measured data in rock slope engineering, and the fundamental fuzzy model of displacements and deformations of rock slope is established by using the theory of fuzzy probability measures. Aksoy and Ercanog˘lu (2007) suggested a different approach to kinematic analyses of discontinuity-controlled rock slope instabilities. Ferentinou and Sakellariou (2007) applied supervised ANNs using back-propagation learning algorithm for the prediction of slope performance under static and seismic loading. In addition, they applied unsupervised ANNs using the efficient visualization techniques offered by self-organizing maps in lithological classification of unsaturated soils and in classification of dry and wet slopes according status of stability and failure mechanism. Park et al. (2008) indicated the previous studies combined the fuzzy set theory with the approximate method such as PEM or FOSM. Goshtasbi et al. (2008) used a genetic algorithm in a heavily jointed rock mass in order to investigate the critical circular slip surface and modification of slope surface. Park et al. (2012) proposed a useful approach for properly addressing the fuzzy uncertainties caused by incomplete information and for using fuzzy set theory to evaluate the reliability of a slope. In this study, uncertain parameters are taken into account as fuzzy numbers, and uncertainties in the input variables are handled in fuzzy-based MCS and fuzzy-based reliability approach in order to obtain the probability of slope failure. Zare Naghadehi et al. (2013) proposed a new Mine Slope Instability Index (MSII) to assess the stability conditions of slopes in open-pit mining. The approach employs the RES approach to account, in an objective and systematic way, for the complex interactions that exist between parameters in real project. Basarir and Saiang (2013) investigated the applicability of fuzzy systems to RMR basic and adjustment parameters included in the SMR system. In the study, two different rock masses were considered, and SMR values were calculated by considering RMR basic and the adjustment factors and developed fuzzy systems. Hosseini and Gholinejad (2014) investigated the slope stability condition by using fuzzy estimation method based on fuzzy possibility theory. Liu et al. (2014) proposed a new analysis strategy using the concept of "cloud" proposed on the basis of probability theory and fuzzy mathematics to account for the fuzziness and randomness simultaneously. They first utilized the cloud models to generate cloud memberships which demonstrate the degree of each evaluated factor belonging to the five ranked grades. Also, the said author considered the varying contributions of different factors by introducing the weight matrix which is obtained by expert opinions. They then applied the AHP, like in fuzzy methods, to obtain a comprehensive evaluation of slope stability. Kumar et al. (2017) proposed models based on the Minimax Probability Machine (MPM) for the prediction of the stability of epimetamorphic rock slope. To realize the aim, they developed two models: Linear Minimax Probability Machine and Kernelized Minimax Probability Machine. The experimental results given in their study demonstrated that MPM-based approaches are promising tools for the prediction of the stability status of epimetamorphic rock slope.

Because the influence factors of the rock slope stability are numerous and their relations are complex, the clustering idea is extensively used in engineering analogies (Gao, 2015). According to Gao (2015), because the environmental influence factors of the rock slope stability are complex, the clustering problem is a complicated fuzzy random optimization problem, which cannot be adequately solved by traditional methods. The author proposed a new clustering optimization method, namely the abstraction ant colony clustering algorithm to solve this.

13.3 BACK-ANALYSIS

The term back-analysis involves a procedure where different parameters and hypotheses of a trial problem, which can be expressed numerically, are varied in order for the results of the analysis to match a predicted performance as much as possible (Vardakos et al., 2007). Back-analysis problems may be solved in two different ways, defined as inverse and direct approaches (Cividini et al., 1981). Cividini et al. (1981, 1983) give an insightful review of back-analysis principles and aspects, including examples of both direct and inversion methods. In the inverse approach, the mathematical formulation is just the reverse of ordinary stress analysis (Jeon and Yang, 2004). It numerically solves some of the material parameters or loading conditions based on measured displacements. Rapid numerical solution is one of the advantages of the inverse method. However, the number of the measured values should be greater than the number of unknown parameters, so that optimization techniques can be used to determine the unknowns (Jeon and Yang, 2004). The direct approach is based on an iterative procedure correcting the trial values of unknown parameters by minimizing error functions; hence, no formulation of the inverse problem is required (Jeon and Yang, 2004). This method has the advantage that it can be applied to nonlinear problems without having to rely on a complex mathematical background (Gioda and Maier, 1980). Gioda (1985) presented an example of the back-analysis of a geotechnical embankment problem where both the inverse and the direct approach were used. Besides the above-mentioned methods, soft computing methods such as fuzzy, the neural network, or genetic algorithm are also used (Feng et al., 2000; Deng and Lee, 2001; Jeon and Yang, 2004; Lee et al., 2006).

Back-analysis approach has been applied in engineering problems, especially in geotechnical and mining engineering. This approach has been widely applied to identify in situ stress field rock mass deformation modulus and strength parameters, rock mass hydraulic properties, rock mass zoning, boundary conditions, loads acting on the tunnel linings, etc., through direct application of closed-form solutions or numerical methods (Cai and Chen, 1987; Cai et al., 2007; Deng and Nguyen Minh, 2003; Hisatake and Hieda, 2008; Jeon and Yang, 2004; Kaiser et al., 1990; Mello Franco et al., 2002; Oggeri and Oreste, 2012; Okui et al., 1997; Sakurai and Takeuchi, 1983; Sakurai et al., 2003; Tonon et al., 2001; Vardakos et al., 2007; Zhang et al., 2006). In the literature, there are also interesting applications of back-analysis. For example, the study by performed Kamp et al. (2010) examined how intense the earthquake-triggered landsliding was compared to preseismic landsliding and how well the (afterward) generated landslide susceptibility map for 2001 predicted potential risk zones within the region.

In the back-analysis of slope failure, general shape of sliding surface and volume of failed mass are known, while the shear strength parameters of sliding surface are unknown. Based on this information, the knowledge on these parameters are updated which are unknown at the moment of slope failure (Loupasakis and Konstantopoulou, 2007; Zhang et al., 2010a, b). Limit equilibrium techniques are commonly adopted methods due to their simplicity for structurally controlled slopes (Sharifzadeh et al., 2010).

Gioda (1985) points out that a distinction of back-analysis methods can also be made considering deterministic methods and probabilistic approach. The deterministic back-analysis methods for the slope failure under influence of an earthquake or a blasting method can be classified under two headings, pseudo-static (Aydan and Ulusay, 2002; Hack et al., 2007; Mavrouli et al., 2009; Wyllie and Mah, 2001) and dynamic methods based on Newmark's displacement-type analysis (Aydan and Ulusay, 2002; Dong et al., 2009; Whu and Tsai, 2011). In a deterministic method, the slope stability model is usually believed or assumed accurate, and the purpose of back-analysis is to find a set of parameters that would result in the slope failure (Akgun and Kockar, 2004; Aydan and Ulusay, 2002; Harris et al., 2011; Hatzor and Levin, 1997; Sancio, 1981; Sharifzadeh et al., 2010; Sonmez et al., 1998; Styles et al., 2011; Tiwari et al., 2005; Tokashiki and Aydan, 2011; Tuncay and Ulusay, 2001; Tutluoglu et al., 2011; Wesley and Leelaratman, 2001; Whu and Tsai, 2011). But, there are uncertainties in a deterministic back-analysis method (Leroueil and Tavenas, 1981; Duncan and Stark, 1992; Gilbert et al., 1998; Stark and Eid, 1998; Tang et al., 1999; Deschamps and Yankey, 2005). When high precision measurements are available or when the back-analysis model is not highly sensitive to measurement errors, then a deterministic approach can be followed. At a minimum, the back-calculated strength from the methods can be used to verify the strength values from the laboratory measured (Frayssines and Hantz, 2009; Hatzor and Levin, 1997; Sancio, 1981; Tiwari et al., 2005). For example, in the study by performed Hatzor and Levin (1997), the shear failure of large rock slope along a clay-filled bedding in limestone was back-analysis. The shear strength of both the clay infilling and clay-limestone contact was determined using direct shear test, as well as triaxial test. The authors pointed out that shear strength of the clay infilling may be greater than the clay-limestone contact, but the shear strength of the contact is mainly governed by the effective internal friction angle, the cohesion is negligible.

In recent years, the soft computing methods and probabilistic approach are often used in the studies of the landslide susceptibility (Althuwaynee et al., 2012; Cervi et al., 2010; Oh and Pradhan, 2011; Pradhan, 2010a,b,c; Pradhan and Youssef, 2010) and back-analyses for the failure in natural and engineering slope (Zhang et al., 2010a, b). Their probabilistic back-analysis shares the concept presented by Eykhoff (1974) in parameter identification. The importance of the error involved in the measurements is taken into account in their analyses. Gioda and Sakurai (1987) present a survey of back-analysis methods using deterministic and probabilistic approaches and principles with reference to tunneling problems. Ledesma et al. (1996) and Gens et al. (1996) described a minimization procedure along with reliability estimates of the final calculated parameters, coupled with the finite-element method which is similar to that of Eykhoff (1974) and Cividini et al. (1981) that make use of a priori information from prior geological investigations. Luckman et al. (1987) performed Bayesian updating for estimating pore pressures using the first-order reliability formulations. Honjo et al. (1994) employed an extended Bayesian method to back-analyze an embankment on soft clay, which is also based on the assumption that the number of observed data is larger than the number of parameters to be updated. A maximum likelihood approach with extension to Kalman filtering principles was presented and used by Hoshiya and Yoshida (1996). Gilbert et al. (1998) and Chowdhury et al. (2004) used discrete probability distributions in Bayesian updating of uncertain parameters. In their studies, a continuous random variable should be approximated as a discrete variable with sufficient small intervals if accurate reliability evaluation is required. Feng and An (2004) suggested the integration of an evolutionary neural network and finite-element analysis using a genetic algorithm for the problem of a soft rock replacement scheme for a large cavern excavated in alternating hard and soft rock strata. The method of neural networks in back-analysis has also been used by Chua and Goh (2005). In their work, a method termed as Bayesian back-propagation (EBBP) neural network was used via a combination of a genetic algorithm and a gradient descent method to determine the optimal parameters. Finno and Calvello (2005) performed back-analysis of braced excavations using a maximum likelihood type of objective function and local search optimization. Zhang et al. (2010a) proposed two efficient methods based on a system identification approach derived from Bayesian theory for probabilistic back-analysis of slope failures. The probabilistic back-analysis methods are based on the system. The back-analysis method proposed by Zhang et al. (2010b) is formulated based on Bayes' theorem and solved using the Markov chain MCS method with a Metropolis–Hasting algorithm. According to the authors, the method is very flexible as any type of prior distribution can be used. In the above-mentioned study, it is also found that the correlation of cohesion and friction angle of soil does not affect the posterior statistics and the remediation design of the slope significantly, while the type of the prior distribution seems to have much influence on the remediation design. Oggeri and Oreste (2012) investigated the static behavior of a rock mass in a tunnel by a probabilistic approach and the back-analysis. The study clearly shows that both the preliminary estimation of the rock characteristics and the measurement during excavation of tunnel present a level uncertainty that can be described with a probability distribution, and the back-analysis can be developed to the probabilistic type approach. In a probabilistic back-analysis, it is recognized that the slope stability model may not be perfectly accurate, and numerous combinations of slope stability parameters may result in slope failure (Zhang et al., 2010b). There can be a quantifiable degree of error in the measuring procedures or when an initial estimate of the descriptive statistics of the governing parameters can be made, then a probabilistic type of back-analysis is more appropriate (Vardakos, 2007).

13.4 A CASE STUDY: PROBABILISTIC STABILITY ANALYSIS OF ARAKLI-TASONU LANDSLIDES, NE TURKEY

13.4.1 Description of the Study Area

The Tasonu limestone quarry, where the landslides occurred, is one of the biggest limestone quarries in the eastern Black Sea region (northeast Turkey; Fig. 13.4). This limestone quarry provides approximately 80% of the raw material needed for the cement plant in Trabzon (The Trabzon cement factory). The quarry has a pit area measuring ~700 m × 850 m. The material in the quarry has been obtained from uncontrolled blasting technique for a period of roughly 10 years. Three landslides have taken place on October 3, 2005 (Landslide 1), March 20, 2006 (Landslide 2), and October 19, 2006 (Landslide 3) in the quarry after heavy rainfall.

Meteorological records for the last 30 years reveal that the area has received a mean monthly rainfall of 72 mm. The minimum value was 35 mm, measured in July, and the maximum value was 120 mm, recorded in October. The heaviest rainfall occurs between October and January, with a monthly average of 94 mm (Ceryan, 2009).

13.4.2 Geological Setting

The oldest unit in the study area is Campanian and Maastrichtian basaltic and andesitic pyroclastics (Ceryan, 2009). The Kirechane formation of Campanian age unconformable rest on the Caglayan Formation (Fig. 13.5).

A number of dip and dip direction measurements were taken from bedding and joint planes of the Campanian aged Tasonu Formation. These measurements revealed that bedding planes in the rock mass dip toward SE and SW at different locations with inclinations of 9–20°. Its thickness varies between 5 and 180 cm. In the landslides, dip direction and dip of

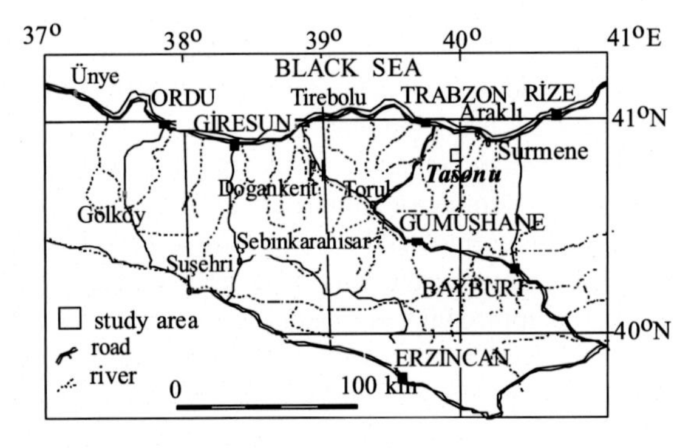

FIGURE 13.4 **Location of the study area.**

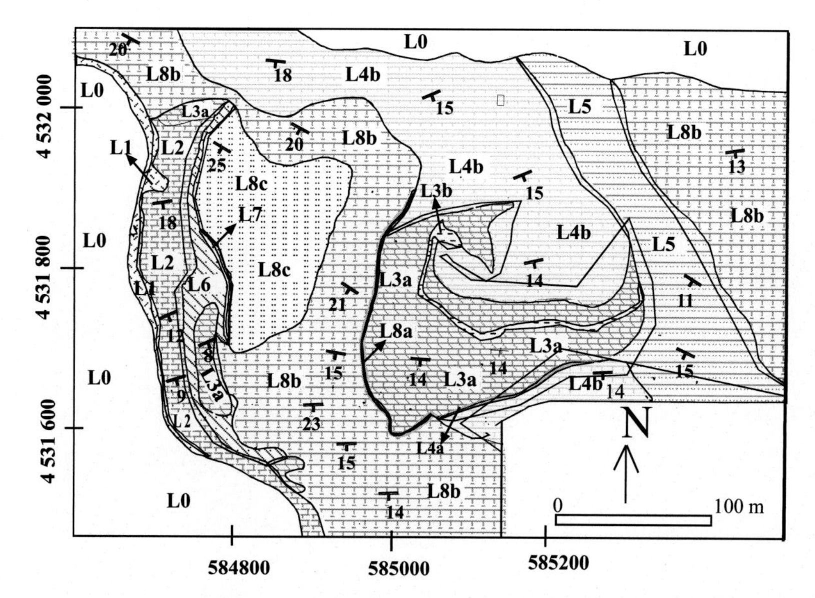

FIGURE 13.5 **The geological map of the Tasonu quarry.** L0: basalt, andesit, and their pyroclastics; L1: volcanic pebbly red tuff; L2: red tuff alternating with white limestone; L3A: limestone with macrofossils and karstic voids; L3B: red tuff; L4: marl red sandy clayey limestone; L5: alternate with sandy limestone clayey limestone and marl; L6: volcanic tuff intercalate with clayey limestone and mar; L7: sandy pebbly limestone; L8B: carbonate cemented sandstone intercalated with clayey limestone and marl; L8A: Lower part of the sandstone contains silicified level; L8C: Interbedded common macrofossilliferous with biotite tuffacous carbonate cemented sandstone and sandy limestone (Ceryan, 2009).

bedding planes are 178–184 and 13–15°, respectively. Excavated slope and strata dip directions in this area are the same. In the Tasonu Formation, two discontinuities set perpendicular to the plane of the bedding were observed. One of them is approximately parallel to the direction of the layer, whereas the other approximately perpendicular to the direction of the layer. These discontinuity surfaces are planar with slightly rough and mostly karsts. Approximately 43% of them are unfilled, 42% of them are clay-filled, and 5% of them are filled with calcite.

13.4.3 Description and Mechanism of the Arakli-Tasonu Landslides

The clayey layers with the same dip direction as the quarry slopes are observed on the bench faces (Fig. 13.6A). In April 2005, these layers of weakness appeared on the slope excavation, and tensile cracks 38–91 m from the excavation faces developed on the northern side of the quarry (Fig. 13.6B). After this date, following the results of Kesimal et al. (2008), production at the quarry continued using blasting. However, excavation using blasting had the least impact in terms of slope stability. After the cracks were noticed, they were marked on the topographic map (Fig. 13.7). Groundwater levels in the cracks and drillings were measured after heavy rains (Kesimal et al., 2005; Ceryan, 2009). The depth of groundwater level measured in the tension cracks approximately parallel to the bedding ranged from 28.8 to 31.3 m, 13,2 to 15.7 m, and 15.3 to 17.8 m, respectively, before Landslides 1, 2, and 3, as shown in Fig. 13.8A–C (Kesimal et al., 2005; Ceryan, 2009).

On October 3, 2005, a large rock mass, about 52 m high, 57 m wide, and with a length along the strike of 50 m, slipped a distance of approximately 25 m along the red-white-colored clay layer dipping 14° into the quarry (Landslide 1; Fig. 13.8A). Then, on March 20, 2006, the large rock mass, about 46 m high, 90 m wide, and with a length along the strike of 158.2 m, slipped a distance of approximately 25 m along the same clay layer. Finally, on October 19, 2006, the other large rock mass, about 48 m high, 77 m wide, and with a length along the strike of 138.3 m, slipped a distance of approximately 25 m along the same clay layer (Landslide 3; Fig. 13.8B). The failure plane was the clay layer in all three failures (Figs. 13.6 and 13.8). This clay layer is inclined to the outside of the slope with 14° and it has appeared to the surface by the excavation made on the slope toe (Fig. 13.6A). The direction of this layer and the direction of the excavation slopes are parallel to each other. The said landslides developed as a planar failure. The landslides resulted in the destruction of two houses, a mosque, and a school.

Shortly after Landslide 3, some tension cracks started to appear at the back of the slope almost parallel to the slope face, and these were followed by additional cracks. The first crack measured about 0.5–1.0 m wide, starting at a distance of 25 m from the newly developed free surface and had a channel approximately 46 m in depth. It quickly filled with debris

FIGURE 13.6 The excavation slopes where landslides occurred (A) and previously developed tension cracks (B).

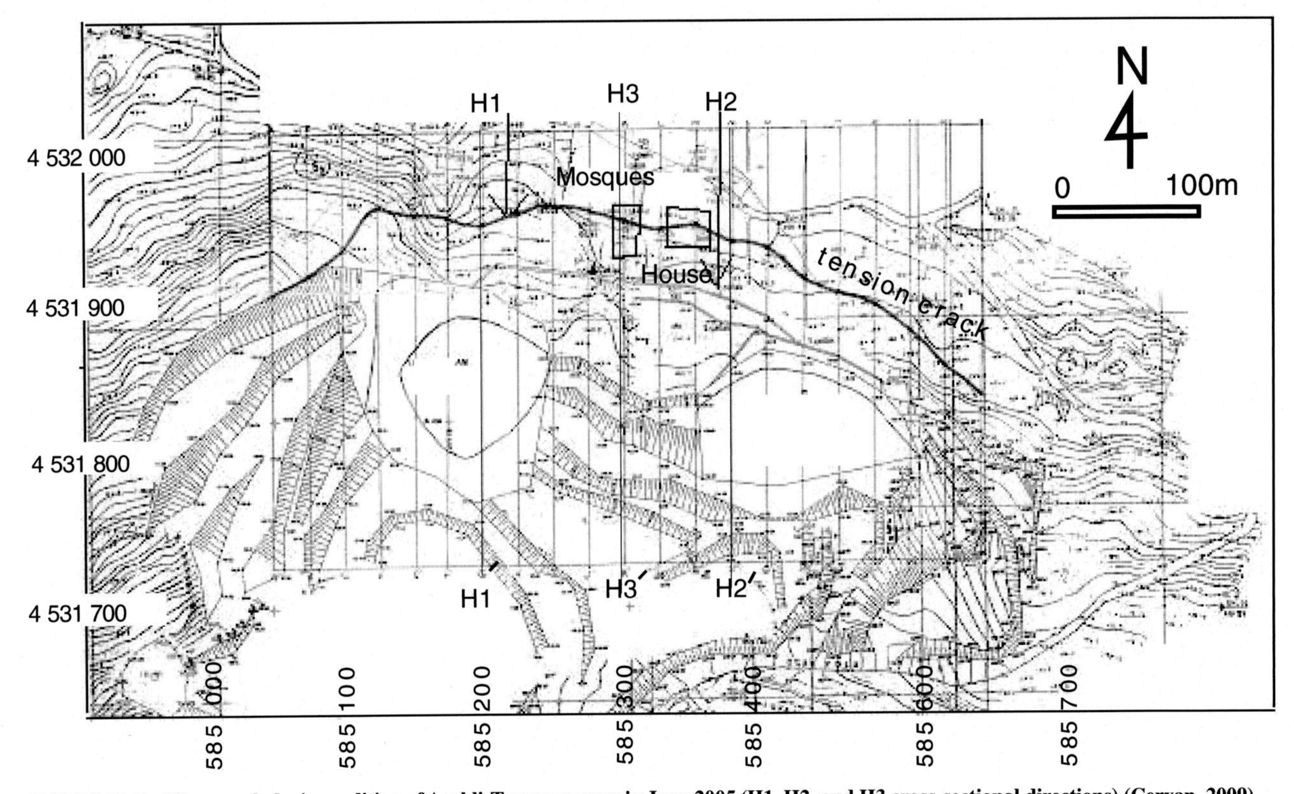

FIGURE 13.7 The morphologic condition of Arakli-Tasonu quarry in June 2005 (H1, H2, and H3 cross-sectional directions) (Ceryan, 2009).

FIGURE 13.8 The landslides occurred on October 3, 2005 (A), March 20, 2006 (C), and October 19, 2006 (B) in Arakli-Tasonu limestone quarry.

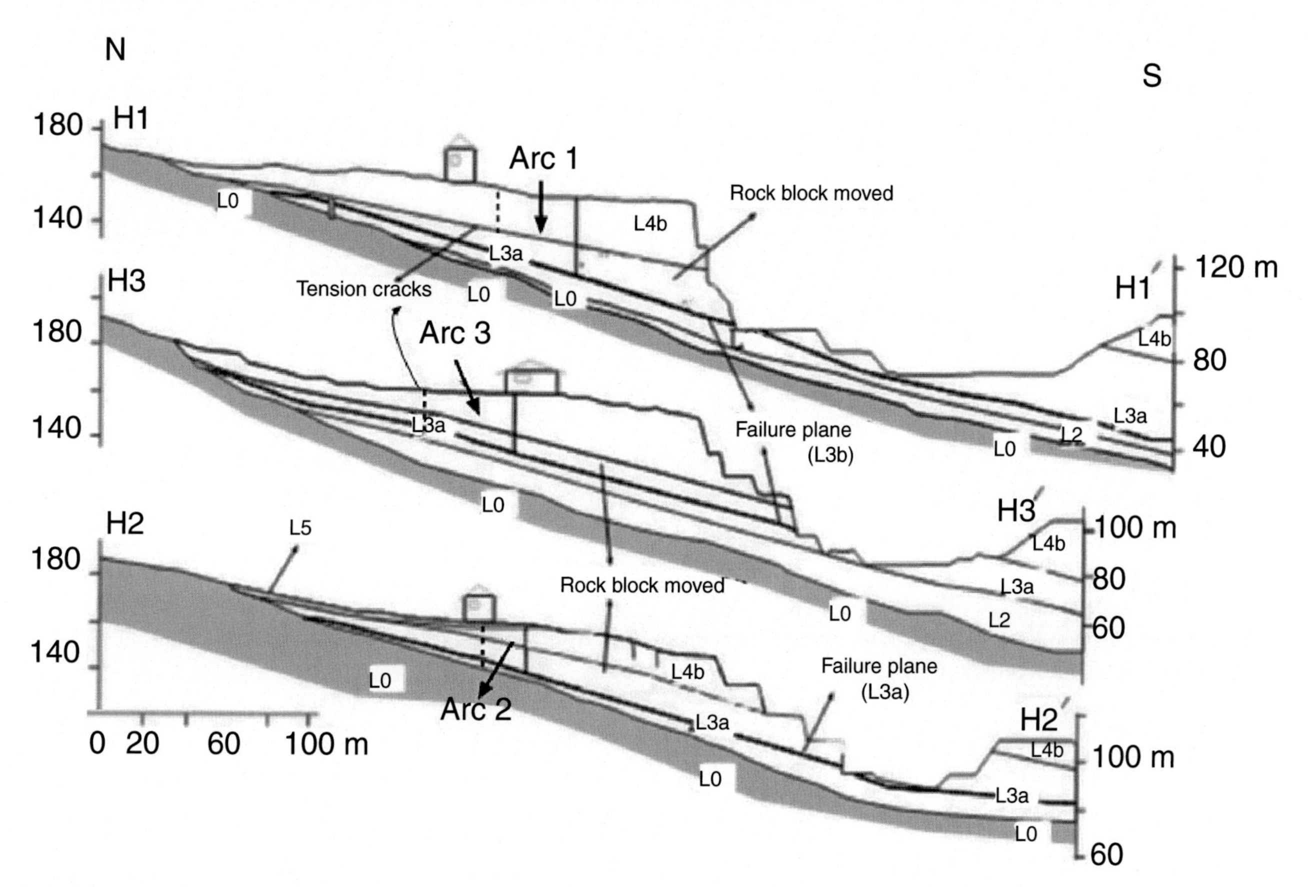

FIGURE 13.9 **The cross-section through the failed slopes.** L0: basalt, andesit, and their piroklastic; L2: red tuff alternate with white limestone; L3a: common macroshelly karstic voided limestone; L4b: red sandy clayey limestone (Ceryan, 2009).

and water. After these slips, a step and a high slope at the back became exposed because the displaced mass at the front and back had become unstable.

The slope had an overall angle of 80° measured from toe to crest, and the first sliding mass consisted of three main units: red sandy, clayey limestone (lithological unit L4b in Fig. 13.5), with a thickness of about 21–31 m (upper unit); macroshelly karstic voided limestone, with a thickness of about 19–36 m (lithological unit L3a in Fig. 13.5); and an overlying weaker zone of clay, which was the failure plane (Figs. 13.5 and 13.9). Units in the other landslides were the same though with different thicknesses. At the foot of the sliding mass, ground heave was observed coupled with axial splitting owing to high compressive stresses at the toe.

Site investigations and eyewitness accounts revealed that the failures had developed in the thick clay layer (Figs. 13.6A and 13.10). This red-white-colored layer was formed by weathering of tuffite. The thickness of the clayey levels that formed the landslide slip plane range from 30 to 115 cm (Figs. 13.10 and 13.11). The surfaces of the limestone layers above and below the sliding surface were planar with a slight roughness. Since the thickness of the clay layer was greater than the amplitude of roughness of the bedding plane, the sliding plane passed through the layer of clay (Figs. 13.10 and 13.11).

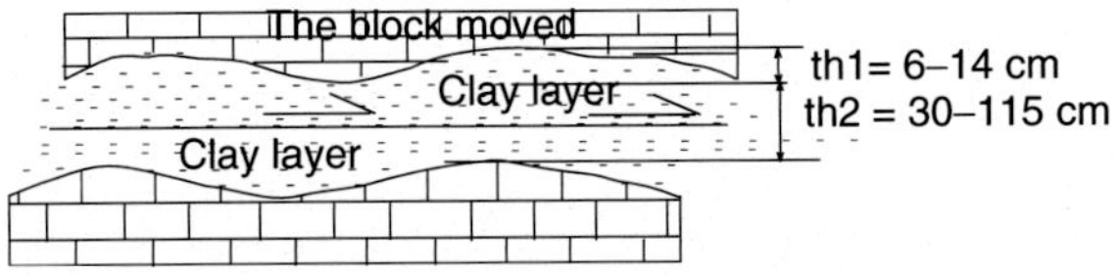

The block moved
Clay layer
Clay layer
th1= 6–14 cm
th2 = 30–115 cm

FIGURE 13.10 The failure plane passing through the clay layer (th1: amplitude of roughness of bedding plane, th2: thickness of clay layer).

FIGURE 13.11 The plane of failures and clayey level forming this plane.

TABLE 13.4 The Result of Sieve Analyses and Physical Properties, Atterberg Limits, and Shear Strength Parameters of the Clay Samples Taken From the Clay Layer

	Max	Min	Mean	Standard deviation
Sand (%)	43	12	31	11.7
Silt (%)	10	3	5	1.9
Clay (%)	82	50	63	12.8
Natural water content (%)	25	9	15	5.4
Liquid limit (%)	94	52	66	12.3
Plastic limit (%)	30	18	24	3.5
Plasticity index (%)	70	24	42	16
Shrinkage limit (RL)	14	10	12	1.3
Specific gravity	2.611	2.534	2.592	
Natural unit weight (kN/m^3)	17.9	15.34	16.51	0.87
Dry unit weight (kN/m^3)	15.9	12.5	14.4	1.25
Saturated unit weight (kN/m^3)	19.7	11.6	18.2	2.28
Effective cohesion (kPa)	24.1	9.2	14.6	5.6
Effective angle of friction (°)	20	9.0	15.3	4.1

13.4.4 Laboratory Testing

Index properties, mineralogical characteristics, and shear-strength properties of the clay layer as the failure plane of the landslides were determined by the laboratory studies. All the samples were labeled and transported to the laboratory daily. Sieve analyses, physical properties, Atterberg limits, and direct shear-strength tests were carried out on total 12 undisturbed clay samples according to ASTM (1994) standards (Table 13.4; Fig. 13.12). In addition, X-ray diffraction analysis was performed on the clay samples to determine the mineralogical composition. X-ray diffractograms of the samples were performed at Black Sea Technical University using a Philips PW-1140 diffractometer.

Grain-size analyses on the clay samples from the clay layer revealed that the clay fraction dominates the material. The size fractions of the samples were: 51%–83% clay, 33%–44% sand, and 3%–10% silt. The values of the liquid limit of the samples varied between 52% and 94%, whereas the values of the samples' plasticity index varied between 24% and 70%. The samples from the clay layer were determined to be the CH group, high-plasticity sandy clay–high-plasticity clay. The results of mineralogical analysis of the clay samples indicated that clay minerals were dominant (77% smectite, 9% illite). The other minerals were calcite (7%) and feldspar plus quartz (5%). The mineralogical composition and field observations confirmed that the clay layer was a weathering product of tuffite.

Considering the failure mechanism of the said landslides, consolidated-drained direct shear-strength tests were carried out on 11 undisturbed specimens taken from the clay layer. Consolidated-drained direct shear tests were carried out by shearing three samples and applying three different normal stresses (Fig. 13.12). The Mohr–Coulomb failure criterion was considered to represent the shear strength of the materials. The linear relations with very high correlation coefficients were obtained as failure envelopes of the clay layer.

13.4.5 Deterministic Back-Analysis

The pseudo-static simple standard calculation of slope stability under influence of an earthquake or a blasting is given Fig. 13.13. Vertical acceleration is often neglected and, generally, only horizontal peak acceleration is considered as illustrated in Fig. 13.10. Back-analyses of a failed slope with the pseudo-static method to find the shear strength parameters of the failure plane were performed by using Eq. (13.9). In deterministic back-analysis, the safety factor (Fs) is accepted as equal to 1 (Eq. 13.9) , and then Eq. (13.10) is derived to obtain mobilized cohesion values during failure (from Hack et al., 2007; Mavrouli et al., 2009; Wyllie and Mah, 2001).

$$Fs = \frac{cL + [W(\cos\alpha - k\sin\alpha) - U - V\sin\alpha]\tan\phi}{[W(\sin\alpha + k\cos\alpha) + V\cos\alpha]} \tag{13.9}$$

$$c = \frac{[W(\sin\alpha + k\cos\alpha) + V\cos\alpha] - [W(\cos\alpha - k\sin\alpha) - U - V\sin\alpha]\tan\phi}{L} \tag{13.10}$$

where W, U, and V are calculated using the following equations:

$$W = A\gamma \tag{13.11}$$

$$\mathrm{V} = 0.5\gamma_w Z_w^2 \tag{13.12}$$

$$\mathrm{U} = 0.5\mathrm{L}\gamma_w Z_w \tag{13.13}$$

where α is the dip angle of the failure plane (degree), c is the cohesion (kPa), ϕ is the friction angle (degree), γ is the unit weight of the rock mass (kN/m^3), γ_w is the unit weight of water (kN/m^3), L is the length of the failure plane (m), A is the basal area of the block (m^2), W is the weight of the sliding block (kN/m), k is the horizontal component of seismic coef-

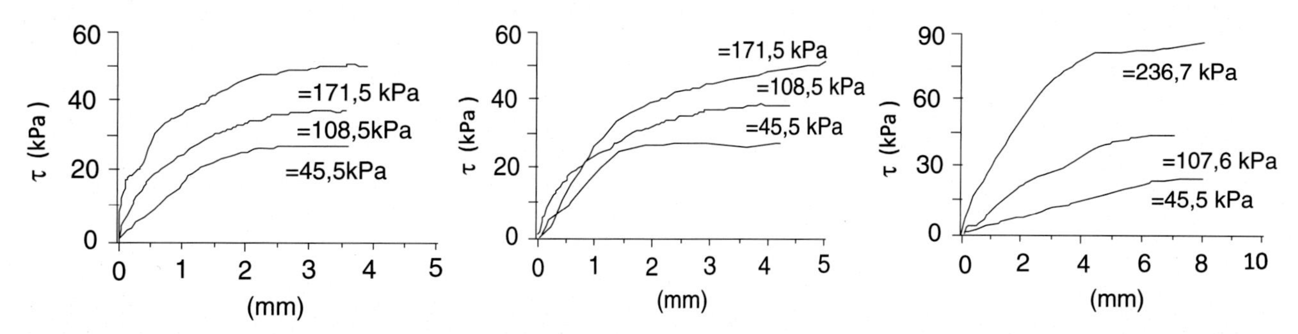

FIGURE 13.12 An example for the stress–strain curves for three direct shear tests of the clay samples from the failure plane of the landslides.

ficient caused by blasting, z_w is the depth of the water in the tension crack (m), U is the water forces acting on the failure plane (kN/m), and V is the water forces acting in the tension crack (kN/m).

Prefailure geometry was estimated from the photographs and the topographical map produced in 2004 before the landslides were occurred (Figs. 13.6 and 13.7). The geometrical characteristics were obtained from the cross-section through the failed slopes. The physical properties and weights of the rock mass blocks used in the analysis were given in Table 13.5.

The inclination of all failure planes was taken as 14° measured during the field studies. As the landslide occurred after a period of heavy rain falls, the variation of water pressure is most likely to be one of the main causes of the landslides. The average value of the depth of the water in the tension crack in the tensile cracks for Landslides 1, 2, and 3 were found to be 6, 8.5, and 8 m, respectively (Ceryan, 2009). During the rainy season, the groundwater level was observed to change up to 2.5 m (Ceryan, 2009; Kesimal et al., 2005) (Table 13.5). Immediately after the development of tension cracks behind the slopes, the excavations have been continued by blasting with minimal seismic effect. According to blast-induced horizontal acceleration values obtained from 73 shots, the average value was considered as 0.013 g (m/s^2) on slopes (Kesimal et al., 2005). Maximum value was 0.035, whereas the minimum value was zero.

Cohesion and internal friction angles were determined by comparison of deterministic back-analyses of the three landslides and the laboratory tests. Then the relation between c and ϕ for each landslide profile was drawn for the comparison with the range of c and ϕ pairs obtained from the direct shear tests (Fig. 13.14).

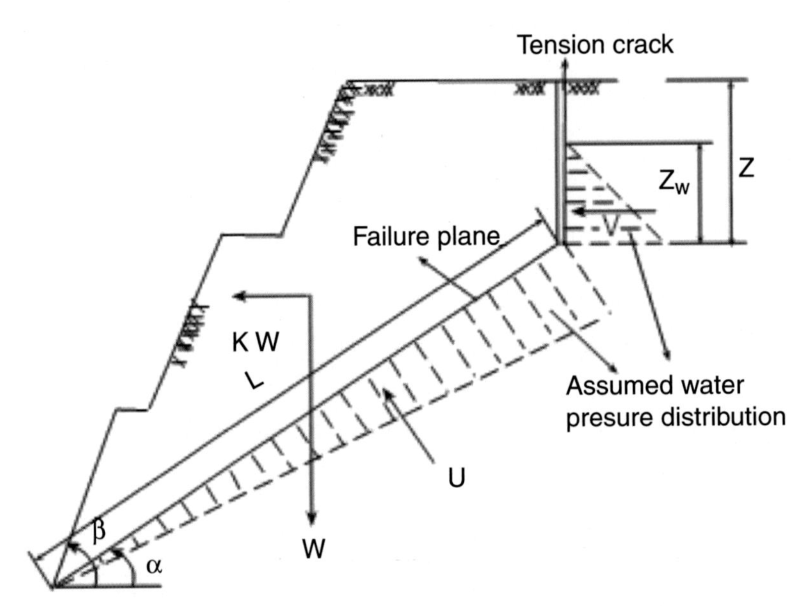

FIGURE 13.13 Limit equilibrium model for pseudo-static stability analysis (W is the weights of sliding block, k is the horizontal component of seismic coefficient caused by blasting, z is the depth of the tension crack, z_w is the depth of the water in the tension crack, L is the length of the failure plane, U is the water forces acting on the failure plane, V is the water forces acting in the tension crack, α is the dip angle of the failure plane) (Hack et al., 2007).

TABLE 13.5 The Geometrical Characteristics, Physical Properties, and Weights of Rock Mass Block Moved and Water Condition in Tension Cracks

	A	γ	α	k	L	z_w	W	$U{*}L$	V
Landslide 1	2808	22.1	14	0.013	80.3	6	62056.8	2409	180
Landslide 2	3088	22.1	14	0.013	158.2	8.5	68244.8	6723.5	361.25
Landslide 3	3551	22.1	14	0.013	138.3	8	78477.1	5532	320

W is the weights of sliding block, k is the horizontal component of seismic coefficient caused by blasting, z is the depth of the tension crack, z_w is the depth of the water in the tension crack, L is the length of the failure plane, U is the water forces acting on the failure plane, V is the water forces acting in the tension crack, and α is the dip angle of the failure plane.

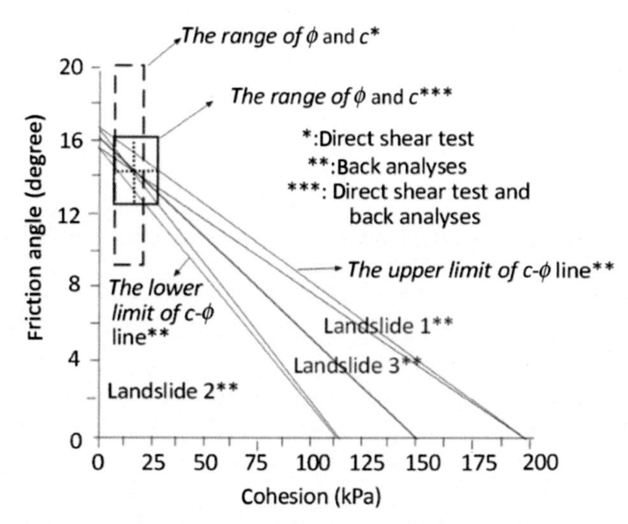

FIGURE 13.14 **The result of the deterministic back-analyses for Arakli-Tasonu landslides.**

13.4.6 Probabilistic Back-Analysis

The probabilistic back-analysis of failed slope profiles in Arakli-Tasonu quarry was performed by using the method proposed by Zhang et al. (2010a). In this method, let $g(\theta,r)$ denote a slope stability model (such as a model based on an LEM), where θ is a vector denoting uncertain input parameters and r is a vector denoting the input parameters without uncertainty. In other words, r is dropped from the slope stability model for simplicity (Zhang et al., 2010a). The uncertain input parameters θ may include both soil strength parameters and pore-water pressure parameters. For simplicity, assume that the prior knowledge on θ can be described by a multivariate normal distribution with a mean of μ_θ and a covariance matrix of C_θ. The objective of probabilistic back-analysis is then to improve the probability distribution of θ based on the observed slope failure information. To quantify the effect of model imperfection, the model uncertainty can be modeled as a random variable, which is defined as (Zhang et al., 2010a):

$$\varepsilon = y - g(\theta) \tag{13.14}$$

where y is the actual Fs and ε is a random variable characterizing the modeling uncertainty. For simplicity, assume ε follows the normal distribution with a mean of μ_ε and a standard deviation of σ_ε.

Let $\mu_{\theta/d}$ and $C_{\theta/d}$ denote the improved mean and covariance matrix of θ, respectively. As a multivariate normal distribution can be fully determined by its mean and covariance matrix, the task in the probabilistic back-analysis is then reduced to determining $\mu_{\theta/d}$ and $C_{\theta/d}$. For a general slope stability model $g(\theta)$ is a point that maximizes the chance to observe the slope failure event and it denotes the most probable combination of parameters that had led to the slope failure event. μ_θ/d can be obtained by minimizing the following misfit function $2S(\theta)$ (Zhang et al., 2010a):

$$2S(\theta) = \frac{\left[g(\theta) + \mu_\varepsilon - 1\right]^T \left[g(\theta) + \mu_\varepsilon - 1\right]}{\sigma_\varepsilon^2} + (\theta - \mu_\theta)^T C_\theta^{-1} (\theta - \mu_\theta) \tag{13.15}$$

The improved covariance matrix of θ, $C_{\theta/d}$, which describes the magnitude of uncertainty in each component of θ as well as the dependence relationships among various components of θ, can be determined as follows (Zhang et al., 2010a):

$$C_{\theta/d} = \left(\frac{G^T G}{\sigma_\varepsilon^2} + C_\theta^{-1} \right)^{-1} \tag{13.16}$$

$$G = \left. \frac{\partial g(\theta)}{\partial \theta} \right|_{\theta = \mu\theta/d} \tag{13.17}$$

where **G** is a row vector representing the sensitivity of $g(\theta)$ with respect to θ at $\mu_{\theta/d}$.

The back-analysis of a failed slope profile based on the above equations is called "probabilistic back-analysis method". Eqs. (13.15)–(13.17) are general and are applicable whether $g(\theta)$ is linear or not. However, when $g(\theta)$ is approximately linear, $\mu_{\theta/d}$ and $C_{\theta/d}$ can be determined analytically with the following equations (13.18)–(13.20):

$$\mu_{\theta/d} = \mu_\theta + C_\theta H^T \left(H C_\theta H^T + \sigma^2 \right)^{-1} \left[1 - g(\mu_\theta) - \mu_\theta \right] \tag{13.18}$$

$$C_{\theta/d} = \left(\frac{H^T H}{\sigma_\varepsilon^2} + C_\theta^{-1} \right)^{-1} \tag{13.19}$$

$$H = \left. \frac{\partial g(\theta)}{\partial \theta} \right|_{\theta=\mu\theta} \tag{13.20}$$

where $g(\mu_\theta)$ = predicted Fs calculated at point μ_θ and $\mathbf{H}$ is the row vector representing the sensitivity of $g(\theta)$ with respect to θ point μ_θ.

The back-analysis based on Eqs. (13.18)–(13.20) is called "simplified probabilistic back-analysis". Compared with the method by optimization, the simplified method is easier to apply, as it does not involve any minimization procedure. The limitation of the simplified method is that it is applicable only when $g(\theta)$ is largely linear. In the two suggested methods by Zhang et al. (2010a), the improved joint probability distribution of the uncertain parameters is approximated by the multivariate normal distribution, and the task of back-analysis is then reduced to the determination of the mean $\mu_{\theta/d}$ and covariance matrix ($C_{\theta/d}$) of the multivariate normal distribution. The formulas to calculate $\mu_{\theta/d}$ and $C_{\theta/d}$ have been summarized in Eqs. (13.15)–(13.17) and Eqs. (13.18)–(13.20), respectively.

The back-analysis of the Arakli-Tasonu landslides by which probabilistic back-analysis method by optimization described above was implemented in three steps as given by (Zhang et al., 2010a):

TABLE 13.6 Probabilistic Back-Analysis of Landslide 2 (Input Parameters are k, z_w, c, and ϕ)

	A	B	C	D	E	F	G	H	I	J	K	L	M
1	c	ø	A	γ	α	k	L	zw	W	U*L	V	F1	F2
2	14,4368737	14,641	3088	22,1	14	0,013	158,2	8,5017	68245	6724,8	361,394	17747,2	17721,4
3					Fs								
4					1,00145								
5	**Prior information**												
6		μθ	σθ		Cθ								
7	c	14,6	5,6		31,36	0	0	0					
8	ϕ	15,25	4,1		0	16,81	0	0					
9	k	0,013	0,0005		0	0	2,5E-07	0					
10	zw	8,5	0,42		0	0	0	0,1764					
11													
12	**Misfit function**												
13		με	σε		Observed Fs			**θ–μ**		error1	error2	2S(θ)	
14		0	0,05		1			-0,1631		0,0008	0,02292731	0,0237734	
15								-0,609					
16								6E-07					
17								0,0017					
18													
19	**Posterior mean & covariance**												
20			**μθ|d**		G					G^T			
21		c	14,43687		0,0089	0,0623	-3,97057	-0,0166		0,0089			
22		ϕ	14,641							0,06228			
23		k	0,013							-3,97057			
24		zw	8,50171							-0,01663			
25			σθ|d		Cθ|d					ρ$_{c\phi}$			
26		c	5,49948		30,244	-4,172	3,96E-06	0,0117		1	-0,68996957	0,0014387	0,00506
27		ϕ	1,0995		-4,172	1,2089	1,48E-05	0,0437			1	0,0269093	0,09468
28		k	0,0005		4E-06	1E-05	2,5E-07	-4E-08				1	-0,000198
29		zw	0,41985		0,0117	0,0437	-4,1E-08	0,1763					1
30													

c is the cohesion (kPa), ϕ is the friction angle (°), γ is the unit weight of the rock mass (kN/m^3), A is the basal area of the block (m^2), α is the dip angle of the failure plane, k is the horizontal component of the seismic coefficient caused by blasting in the length of the failure plane (m), W is the weights of sliding block (kN/m), z is the depth of the tension crack, z_w is the depth of the water in the tension crack, U is the water forces acting on the failure plane (kN/m^2), and V is the water forces acting in the tension crack (kN/m).

TABLE 13.7 Results of the First Probabilistic Back-Analysis (θ_1: Prior Information 1)

The clay layer				The failure plane of						The clay layer		
				Landslide 1		Landslide 2		Landslide 3				
θ_1	μ_θ	σ_θ	v_θ	$\mu_{\theta d}$	$\sigma_{\theta d}$	$\mu_{\theta d}$	$\sigma_{\theta d}$	$\mu_{\theta d}$	$\sigma_{\theta d}$	$\mu_{\theta d}$	$\sigma_{\theta d}$	$v_{\theta d}$
c	14.6	5.6	38.4	14.488	5.572	14.437	5.499	14.473	5.545	14.49	5.608	39
ϕ	15.25	4.1	26.9	14.481	0.840	14.641	1.099	14.604	0.951	14.72	1.042	7.1
k	0.013	0.0005	3.9	0.013	0.0005	0.013	0.0005	0.013	0.0005			
z_w	6.0	0.42	7.0	6.013	0.4199	8.501	0.4199	8.04	0.4199			
	8.5	0.42	4.9									
	8.0	0.42	5.3									

TABLE 13.8 Results of the Second Probabilistic Back-Analysis (θ_2: Prior Information 2)

The clay layer				The failure plane of						The clay layer		
				Landslide 1		Landslide 2		Landslide 3				
θ_2	μ_θ	σ_θ	v_θ	$\mu_{\theta d}$	$\sigma_{\theta d}$	$\mu_{\theta d}$	$\sigma_{\theta d}$	$\mu_{\theta d}$	$\sigma_{\theta d}$	$\mu_{\theta d}$	$\sigma_{\theta d}$	$v_{\theta d}$
c	17.1	6.19	36.2	17.28	5.859	17.37	5.22	17.45	5.59	17.45	5.60	32
ϕ	14.1	1.14	8.1	14.18	0.694	14.16	0.83	14.21	0.76	14.20	0.77	5.4
k	0.013	0.0005	3.9	0.013	0.0005	0.013	0.0005	0.013	0.0005			
z_w	6.0	0.42	7.0	6.013	0.4199	8.501	0.4199		0.4199			
	8.5	0.42	4.9					8.04				
	8.0	0.42	5.3									

1. Select a stability model $g(\theta)$ and identify the uncertain parameters θ;
2. Quantify the knowledge on θ prior to the back-analysis and on the model uncertainty of $g(\theta)$. This step is in fact a process of determining μ_θ, C_θ, μ_ε, and σ_ε; and
3. Improve the probability distribution of θ considering the slope failure event. In this step, $\mu_{\theta/d}$ and $C_{\theta/d}$, which contain the improved knowledge on θ, are calculated using the formulas presented in the previous section.

The back-analysis for Landslide 2 was given as an example for the back-analysis carried out in the study (Table 13.6). In the analysis, the Excel worksheet given in Zhang et al. (2010a) was used. In Table 13.6, shear strength force F_1 and driving force F_2 are the resisting force given as

$$F1 = [W(\cos\alpha - k\sin\alpha) - U - V\sin\alpha]\tan\phi \tag{13.21}$$

$$F2 = [W(\sin\alpha + k\cos\alpha) + V\cos\alpha] \tag{13.22}$$

According to the results of the probabilistic back-analysis with different prior information performed in this study, it can be said that posterior values of the mean and standard deviation of the input parameters in the method given by Zhang et al. (2010a) depend on prior information of these parameters in the probabilistic back-analysis technique. On the other hand, it is known that variation in cutting parameters obtained from the laboratory tests can result from the heterogeneity of the sample, distribution state of the sample, and type of the sampling. Therefore, it can be inferred that usage of only laboratory tests is not suitable as prior information in the probabilistic back-analysis. Considering this condition, two probabilistic back-analyses for all the landslides were carried out in the present study. In the first analysis, the values of c and ϕ taken as input parameters are the values obtained from direct shear test (Table 13.7). These values were named as prior information 1. In the second probabilistic back-analysis, the prior information values of c and ϕ were obtained by comparing the results of direct shear tests and the deterministic back-analysis (Table 13.8). These values were named as prior information 2.

According to the result of consolidated-drained direct shear tests, the values of the internal friction angle (ϕ) of the samples vary between 9° and 20°, while the values of cohesion (c) of the samples vary between 9 and 24 kPa. The mean and standard deviation of ϕ of the clay samples were 15.3° and 4.1°, respectively, whereas the values of c were 14.6 and 5.6 kPa, respectively (prior information 1). The coefficient of variation, ϕ is 26.9, and the value of c is 38.4 (Table 13.4).

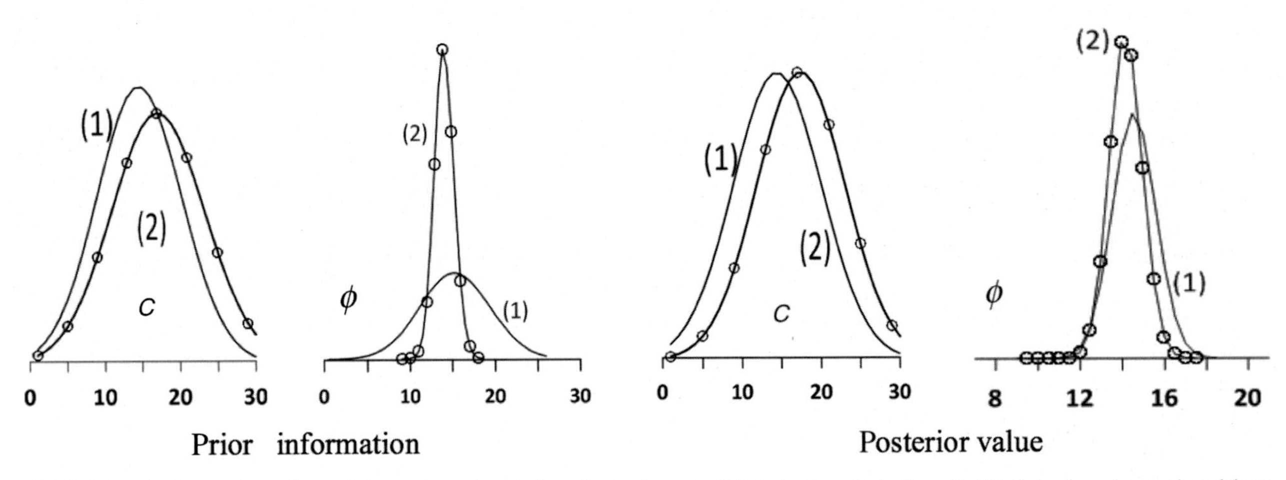

FIGURE 13.15 Comparison of prior information and posterior value in the probabilistic back-analysis (1: probabilistic back-analyses where laboratory tests are used as prior information. 2: probabilistic back-analysis where comparison of deterministic analysis and laboratory tests is used as prior information).

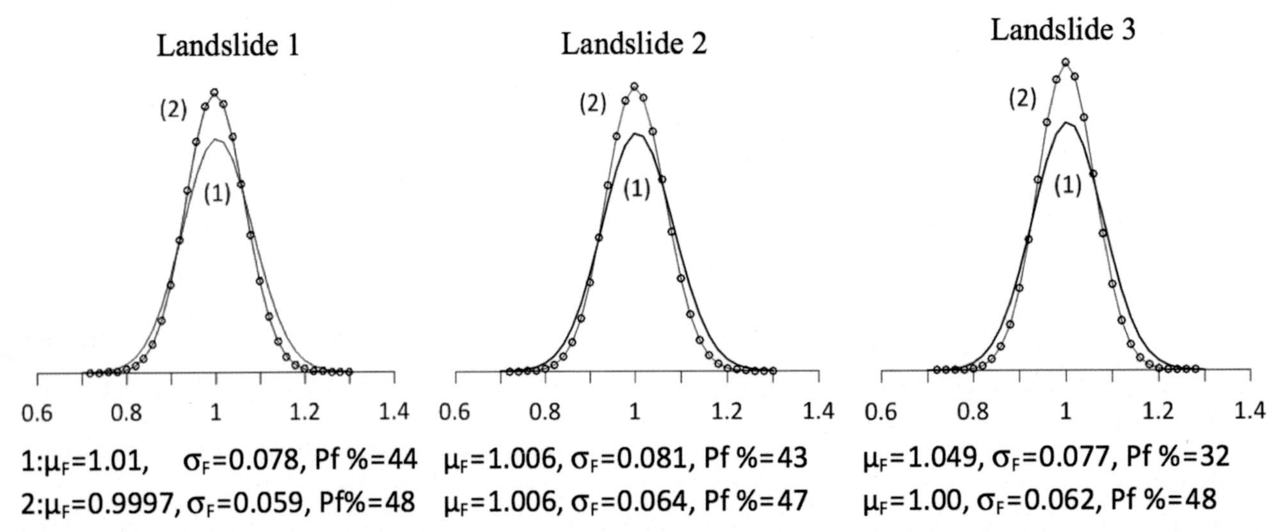

FIGURE 13.16 Results of the probabilistic stability analysis performed for the failure conditions obtained by the back-analysis (1: probabilistic back-analyses where laboratory tests are used as prior information. 2: probabilistic back-analysis where comparison of deterministic analysis and laboratory tests is used as prior information).

The deterministic back-analyses with pseudo-static method for three landslides resulted in c and ϕ values satisfying Fs = 1 (Table 13.5; Fig. 13.14). Then the multiple-intersection solution with a comparison with laboratory derived strength test results were performed by using the approach of Sancio (1981). For the clay layer, failure plane, the mean value of mobilized cohesion and friction angles were found to be 17.1° and 14.1 kPa, respectively (Fig. 13.14, Table 13.6). According to comparison, the results of the direct shear test and the deterministic back-analysis (Fig. 13.14), it was obtained that the c value is in the range 9–30 kPa, and friction angles vary between 12.4° and 16.2°, respectively. Assuming that these c and ϕ values display normal distribution, then the standard derivation values are determined as 6.19 kPa and 1.14° (prior information 2) by means of the Monte Carlo method using RocPlane program.

Given the variation coefficients, *prior information* 2 values of the shear strength parameters are lower relative to *prior information 1* (Fig. 13.15). In the back-analysis carried out using *prior information 1* and 2, the posterior mean value and standard derivation value of c were obtained as 14.47 and 5.61 kPa, and 17.45 and 5.56 kPa, respectively (Tables 13.7and 13.8; Fig. 13.15). According to the results of these analysis, the posterior mean value and standard deviation value of ϕ for prior information 1 are 14.57° and 1.024°. For the prior information 2 values, the values under interest are 14.20° and 0.77°, respectively. Given the COV, the posterior information of the shear strength parameters involves less uncertainty for the prior information 2 values (Table 13.6; Fig. 13.15).

In order to evaluate the performance of the probabilistic back-analysis, the probabilistic analysis was performed in failure condition obtained from the results of the said analysis given in Tables 13.7 and 13.8 (Fig. 13.16). The facts that

TABLE 13.9 The Classification of Expected Performance Level According to Reliability Index (After Yahiaoui et al., 2016)

Reliability index, β	Probability of failure, Pf = $\Phi(-\beta)$	Expected performance level
1.0	0.16	Hazardous
1.5	0.07	Unsatisfactory
2.0	0.023	Poor
2.5	0.006	Below average
3.0	0.01	Above average
4	0.00003	Good
5	0.0000003	High

FIGURE 13.17 The tension cracks developed behind the slopes formed after the landslides that occurred in 2005–2006.

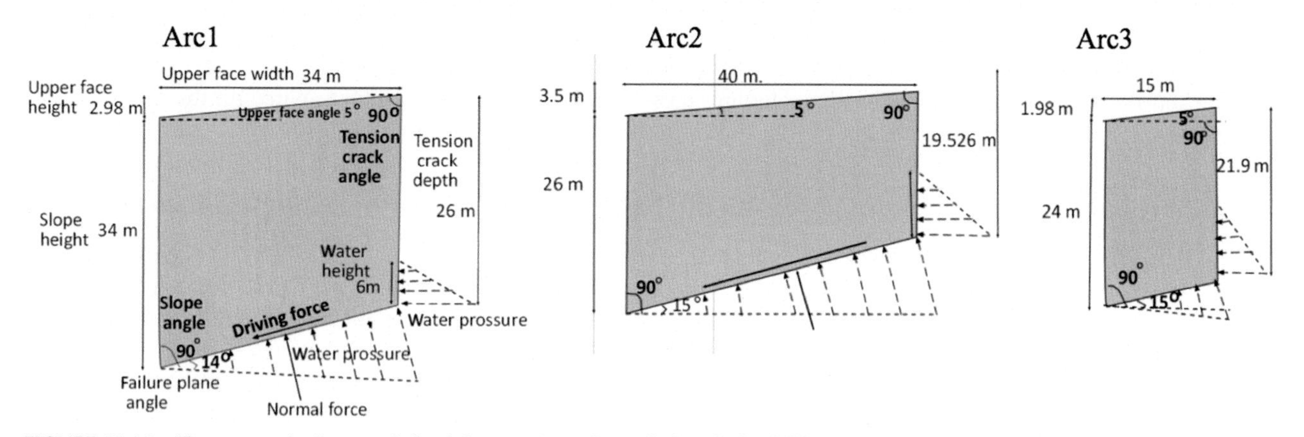

FIGURE 13.18 The geometric characteristic of the new slopes formed after the landslides that occurred in 2005–2006 and water condition in tension crack developed after the said slope.

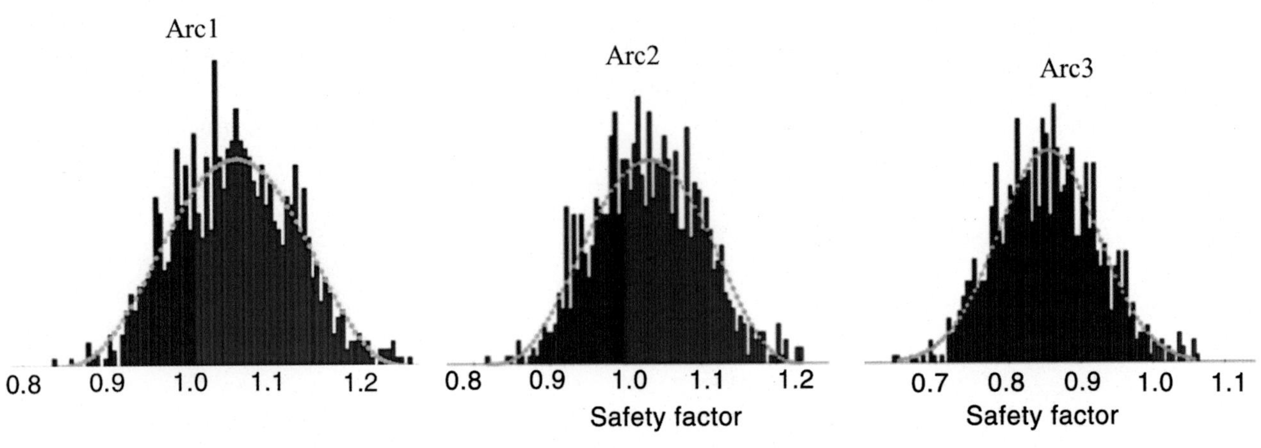

FIGURE 13.19 **Distribution of the safety factor obtained by the probabilistic stability analysis.**

the standard deviation of safety factor (σ_F) is small and probability of failure (Pf) is close to 50% show that there is small uncertainty in the limit equilibrium conditions. Although the difference between the two analysis results is not significant, the second analysis is closer to the limit equilibrium conditions (Fs = 1). Standard deviation of the safety factor obtained for the probabilistic analysis represented (2) in Fig. 13.16 is smaller than that for other analyses. In addition, probability of failures obtained from the probabilistic analysis represented (2) is equal to about 50.

The β value for the acceptable performance of geotechnical designs is given different by different researchers. According to Yahiaoui et al. (2016), geotechnical designs require a β value of at least 2 (i.e. Pf < 0.023) for an expected performance level better than "poor" (Table 13.9). If β value rise to 2.5, the expected performance level is "below average".

The uncertainty of the safety factor due to the uncertainty of the input parameters can be expressed by the standard deviation of the safety factor (or the COV of the safety factor). By taking advantage of this condition and the reliability index definition, the safety factor for a slope design with required robustness can be estimated using the following procedures.

Step 1. Select the reliability index value for the desired reliability attribute (e.g. $\beta = 2.5$). Let $\mathrm{Fs}_i = 1$ and $i = 1$)

Step 2. Find the mean and standard deviation of the input parameters (e.g. shear strength parameters) for the limit equilibrium using probabilistic back-analysis.

Step 3. Carry out probability stability analysis using these values. In the case the values of mean (μ_{Fs_i}), standard deviation (σ_{Fs_i}), and coefficient of variation ($\mathrm{CV_{Fs}}$) of the safety factor are obtained. During this process, $\mathrm{CV_{Fs}}$ will remain constant.

Step 4. Calculate $\mathrm{Fs}_{(i+1)} = (1 + \beta\sigma_{\mathrm{Fs}_i})$, $\sigma_{\mathrm{Fs}_{(i+1)}} = \mathrm{CV_{Fs}}\mathrm{Fs}_{(i+1)}$.

Step 5. $d = \mathrm{Fs}_{(i+1)} - \mathrm{Fs}_{(i)}$.

Step 6. If $d < 0.001$, then $\mathrm{Fs}_{(u)} = \mathrm{Fs}_{(i+1)}$; else ($i = i + 1$) and go to Step 4.

Step 7. Take the safety factor $\mathrm{Fs} > \mathrm{Fs}_{(u)}$ for the design of slopes under similar conditions.

The above method was applied to the planar failures investigated. And the safety factor required for the reliable design of Arc1, Arc2, and Arc 3 were calculated as 1.173, 1.188, and 1.182, respectively.

13.4.7 Probabilistic Stability Analyses for the New Rock Slope Formed the Said Landslides

The tension cracks developed behind and are parallel to these new slopes formed after the landslides that occurred in 2005–2006 (Fig. 13.17). The distance between the new rock slopes formed and tension cracks are 34 m for the first landslide, 40 m for the second landslide, and 15 m for the third landslide. The rock mass between the new rock slope surface formed by the three landslides and the tensile cracks limited to the well-developed fractures on the both sides. The said rock masses restricted were named Arc1, Arc2, and Arc3 (Fig. 13.9). Tensile cracks and the said new rock slopes are 90° inclined. The fractures well developed are inclined at 90° and their strike is parallel to the dip direction of the failure plane. There is a maximum difference of 5° between the strike of the clay layer forming the failure surface and the strike of the new rock slopes formed by the Arakli-Tasonu landslides. These geometric relations provide kinematically planar sliding conditions. In these tensile cracks, the movement in the vertical direction is 0.4 m and the movement in the horizontal direction is 0.6 m. The geometric characteristics of the rock masses Arc1, Arc2, and Arc3 are given in Fig. 13.18. Considering these information, it was thought that the new plane failure can be developed in these slopes. For this reason, no technical entry

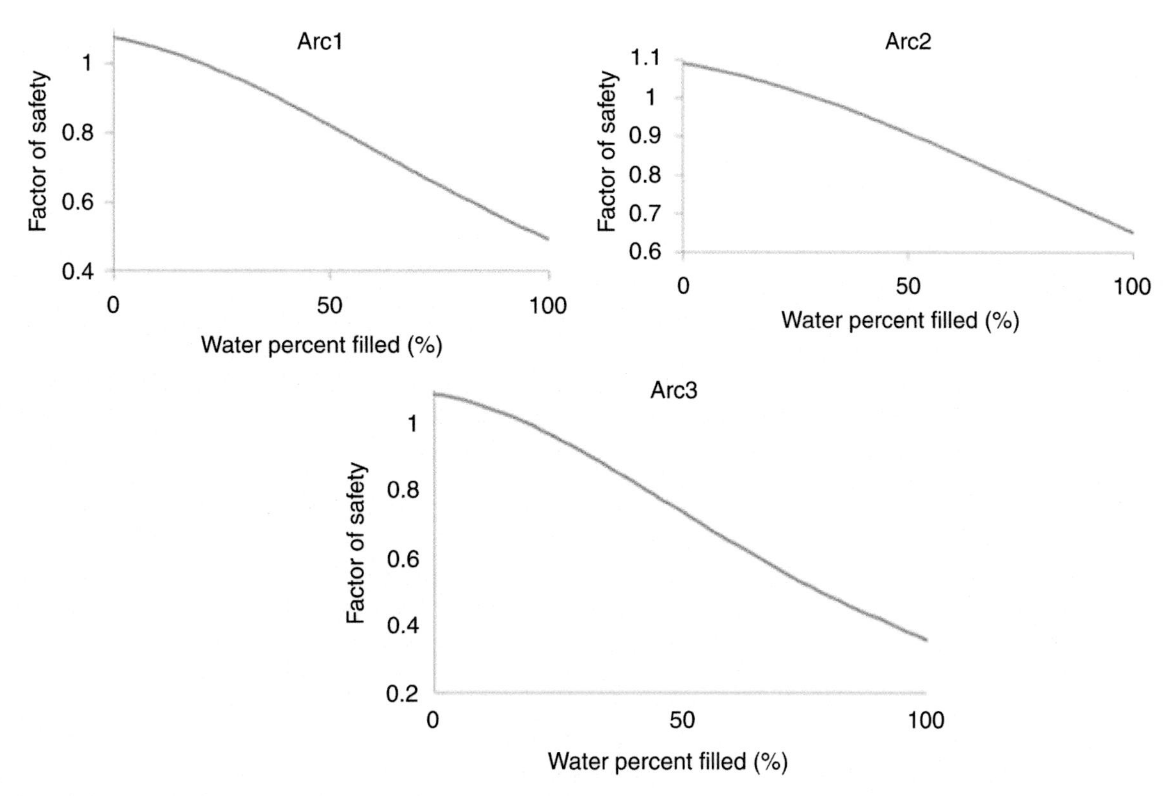

FIGURE 13.20 **The effect of the water in the cracks on the stability of the slopes.**

is allowed for this part of the quarry area. For this, probabilistic stability analysis of these new slopes was performed in this study. The values of the input parameters including shear strength parameters and the depth of the water in the tension crack obtained by probabilistic back-analyses were used in these analyses.

In the probabilistic stability analyses performed in this study, the Monte Carlo method and the reliability index were used. The Monte Carlo method is used when the PDF of each component variable is completely described. In the method, values of each component are generated randomly by its respective PDF and then these values are used to evaluate the factors of safety (Nguyen and Chowdhury, 1985; Park, 1999). By repeating this calculation, the probability of failure can be estimated as that proportion of the calculations in which the safety factor was less than 1. This estimation is reasonably accurate only if the number of simulations is very large and the probability distribution function of each variable is known (Nguyen and Chowdhury, 1985; Park, 1999). The reliability index (β) indicates safety based on the number of standard deviations separating the best estimate of Fs from its defined failure value of 1. When the shape of the probability distribution for the Fs is known, the reliability index can be related to the probability of failure (Christian et al., 1994).

In the probabilistic stability analyses for Arc1, Arc2, and Arc3, the value of the shear-strength parameters found from the results of the probabilistic back-analysis with prior information from comparison of direct shear test with deterministic back-analysis were used. The cohesion (c) and friction angles (ϕ) of the failure surface were obtained to vary according to normal distributions with the following means and standard deviations parameters: mean cohesion value = 17,45 kPa, standard deviation of cohesion = 5,60 kPa, mean friction angle = 14.20°, and standard deviation of friction angle = 0.77°. It is accepted that the depth of the water in the tension crack has normal distribution like value of ϕ and c of the failure plane. In these stability analyses, the mean value and standard deviation value of depth of the water in the tension cracks were used as the values in Table 13.8. In the probabilistic stability analysis for Arc1, the mean, standard deviation, minimum, and maximum values of the safety factor were found to be 1.05, 0.07038, 0.8388, and 1.22, respectively. In the analysis, probability of failure and reliability index were obtained as 25.6% and 0.71, respectively (Fig. 13.19).

According to the result of the analysis for Arc2, the mean, standard deviation, minimum, and maximum values of the safety factor are 1.031, 0.07158, 0.8131, and 1.238, respectively. Probability of failure is 35% and reliability index is 0.433 for Arc2. In the probabilistic stability analysis performed for Arc3, the mean, standard deviation, minimum, and maximum values of the safety factor obtained as 0.86, 0.0666, 0.6576, and 1.058, respectively. Probability of failure and reliability index were obtained as 97.9 and 2.10, respectively (Fig. 13.19).

Sensitivity analysis is performed to define the impact of effective parameters on Fs. The excavation in the landslides area was not allowed. For this reason, the geometry of the said slope cannot be changed. Only the condition of the water in the tensile cracks can be changed. For this reason, the effect of the water in the cracks on the stability of the slopes was evaluated (Fig. 13.20). As can be seen, in Fig. 13.20, even when water percent filled is zero, the safety factor does not exceed 1.1. So, to drainage the water in the cracks or to fill the cracks with clay does not provide stability for the slope.

13.5 CONCLUSION

In rock engineering applications, an engineer deals with a natural, inhomogeneous, and anisotropic rock mass that exhibits complex and in some cases, unpredictable behaviors. There are often many uncertainties in the analysis of rock slope stability owing to inadequate information for site characterization and inherent variability and measurement errors in geological and geotechnical parameters. Although uncertainty can be dealt with in different ways, statistical simulation is even more powerful. The aim of the probabilistic simulation is that it helps engineers to develop more robust and economic designs and solutions. In the probabilistic approach, in general, each combination of the strength parameters, which may be selected randomly or nonrandomly, creates a different realization of the slope, while the geometry remains the same. By analyzing each realization, a deterministic value for the Fs is obtained. Finally, a probabilistic distribution can be fitted to the values of Fs and accordingly, the probability of failure can be estimated. In this study, the methods of rock slope stability analysis, especially probability analysis methods, are emphasized. Some of the available probabilistic methods are MCS, LHM, FORM, FOSM, PEM, and RSM, and these probabilistic analysis methods were briefly discussed in this study. Although the probabilistic methods take into consideration the inherent variability and uncertainties in the analysis parameters, they include the significant drawbacks and limitations as given in this study. In addition, a case study from NE Turkey was given.

In this study, the probabilistic back-analysis was performed for the three landsides occurred on October 3, 2005 (Landslide 1), March 20, 2006 (Landslide 2), and October 19, 2006 (Landslide 3) at the Tasonu quarry in Trabzon, NE Turkey. The landslides developed as plane failures and the sliding plane of the landslides passed through the same clay layer. The prior information of the shear strength parameter in the back-analyses is determined in two different ways: (i) by direct shear tests (prior information 1), (ii) by comparison of direct shear test with deterministic back-analysis (prior information 2). Probabilistic stability analyses of the slopes under interest were performed using failure conditions obtained from probabilistic back-analysis. These analyses clearly reveal that results of the probabilistic back-analysis with prior information 2 contain less uncertainties and represents the failure conditions better. Conclusively, it can be said that prior to using the input parameters (such as cohesion and internal friction angle from the field or laboratory experiments) in the probabilistic back-analysis, they must be examined by deterministic back-analysis in terms of their reliability. In the back-analysis carried out, the posterior mean value and standard derivation value of c of the clay layer (plane of failure) were obtained as 17.45 and 5.56 kPa and the posterior mean value and standard deviation value of ϕ of the clay layer were obtained as 14.20° and 0.77°, respectively. In addition, the safety factor for a slope design with required robustness were estimated using a new approach based on the probabilistic back-analysis. According to the results of the estimation, it can be said that the said values for Arc1, Arc2, and Arc 3 should be bigger than 1.173, 1.188, and 1.182 respectively.

New tension cracks on the slopes occurred and the rock masses between the new rock slopes formed by the three landslides (main scarps of the landslide) and the tensile cracks and limited to fractures on the both sides. Because of this, the plane failure condition developed kinematically, the probabilistic stability analysis of these new slopes named Arc1, Arc2, and Arc3 were performed. In these analyses, the shear-strength parameter obtained from probabilistic back-analysis with prior information 2 and Monte Carlo method is used to obtain the PDF of safety factors. Moreover, the sensitivity analyses were carried out to define the impact of effective parameters on Fs obtained for Arc1, Arc2, and Arc3. The results of these analyses show that to drainage the water in the cracks or to fill the cracks with clay does not provide stability for the slope.

REFERENCES

Admassu, Y., Shakoor, A., 2013. DIPANALYST: a computer program for quantitative kinematic analysis of rock slope failures. Comput. Geosci. 54, 196–202.

Ahmadabadi, M., Poise, R., 2016. Probabilistic analysis of rock slopes involving correlated non-normal variables using point estimate methods. Rock Mech. Rock Eng. 49, 909–925.

Akgun, H., Kockar, M.K., 2004. Design of anchorage and assessment of a stability of openings in silty, sandy limestone: a case study in Turkey. Int. J. Rock Mech. Min. Sci. 41, 37–49.

Aksoy, H., Ercanoğlu, M., 2007. Fuzzified kinematic analysis of discontinuity-controlled rock slope instabilities. Eng. Geol. 89, 206–219.

Althuwaynee, O.F., Pradhan, B., Lee, S., 2012. Application of an evidential belief function model in landslide susceptibility mapping. Comput. Geosci. 44, 120–135.

Alzo'ubi, A.K., 2016. Rock slopes processes and recommended methods for analysis. Int. J. GEOMATE 11 (25), 2520–2527, Geotec., Const. Mat. & Env., ISSN: 2186-2982 (Print), 2186-2990 (Online), Japan.

Ang, A.H.S., Tang, W.H., 1984. Probability Concepts in Engineering Planning and Design, Vol. 2, Decision, Risk and Reliability. John and Sons, New York.

ASTM (American Society for Testing and Materials), 1994. Annual Book of ASTM Standard-Soil and Rock: Building Stone, Construction. ASTM Publication, Section 4, vol. 04.08.

Aydan, Ö., Ulusay, R., 2002. Back analysis of a seismically induced highway embankment failure during the 1999 Düzce earthquake. Environ. Geol. 42, 621–631.

Baecher, G.B., Christian, J.T., 2003. Reliability and Statistics in Geotechnical Engineering. John Wiley, Chichester.

Baker, J.W., Cornell, C.A., 2003. Uncertainty specification and propagation for loss estimation using FOSM methods. PEER Report 2003/07. Pacific Earthquake Engineering Research Center, University of California, Berkeley, CA.

Basarir, H., Saiang, D., 2013. Assessment of slope stability using fuzzy sets and systems. Int. J. Min. Reclam. Environ. 27 (5), 312–328.

Bieniawski, Z.T., 1976. Rock mass classifications in rock engineering. In: Bieniawski (Ed.), In: Proc. Symp. on Exploration for Rock Engineering, Johannesburg. Balkema, Rotterdam, pp. 97–106.

Bobet, A., 2010. Numerical methods in geomechanics. Arabian J. Sci. Eng. 35 (1B).

Böhme, M., Hermanns, R.L., Oppikofer, T., 2013. Analyzing complex rock slope deformation at Stampa, Western Norway, by integrating geomorphology, kinematics and numerical modeling. Eng. Geol. 154, 116–130.

Box, G.E.P., Wilson, K.B., 1951. On the experimental attainment of optimum conditions. J. R. Stat. Soc. Ser. B 13 (1), 1–38.

Brideau, M., Chauvin, S., Andrieux, P., Stead, D., 2012. Influence of 3D statistical discontinuity variability on slope stability conditions. In: Eberhardt et al., (Ed.), 11th International, 2nd North American Symposium on Landslides, Engineered Slopes: Protecting Society Through Improved Understanding. Banff, Canada.

Budetta, P., Santo, A., Vivenzio, F., 2008. Landslide hazard mapping along the coastline of the Cilento Region (Italy) by means of a GIS-based parameter rating approach. Geomorphology 94, 340–352.

Cai, M., Chen, X., 1987. Back-analysis of initial stress field in rocks by the simplex method. Proc. First Natl. Conf. Geomech. 1, 217–222.

Cai, M., Morioka, H., Kaiser, P.K., Tasaka, Y., Kurose, H., Minami, M., Maejima, T., 2007. Back-analysis of rock mass strength parameters using AE monitoring data. Int. J. Rock Mech. Min. Sci. 44, 538–549.

Calderon, A.R., 2000. The Application of Back-Analysis and Numerical Modeling to Design a Large Pushback in a Deep Open Pit Mine (MS thesis). Faculty and Board of Trustees of the Colarodo School of Mines.(Unpublished).

CANMET (Canada Centre for Mineral and Energy Technology), 1981. Pit slope manual: Chapter 2, Structural geology. CANMET Report 77-41, 123 pp.

Castaldini, D., Genevois, R., Panizza, M., Puccinelli, A., Berti, M., Simoni, A., 1998. An integrated approach for analyzing earthquake-induced surface effects: a case study from the Northern Apennins, Italy. J. Geodyn. 26 (2–4), 413–441.

Cervi, F., Berti, M., Borgatti, L., Ronchetti, F., Manenti, F., Corsini, A., 2010. Comparing predictive capability of statistical and deterministic methods for landslide susceptibility mapping: a case study in the Northern Apennines (Reggio Emilia Province, Italy). Landslides 7, 433–444.

Ceryan, N., 2009. Rock slope stability analyses by probability method and excavability in Tasonu Limestone Quarry (Trabzon). Ph.D. thesis, unpublished.

Ceryan, N., 2016. A review of soft computing methods application in rock mechanic engineering. In: Samui, P. (Ed.), Handbook of Research on Advanced Computational Techniques for Simulation-Based Engineering, Chapter 1, ISBN13: 9781466694798, 1–70.

Ceryan, N., Ceryan, S., 2008. An application of the interaction matrices method for slope failure susceptibility zoning: Dogankent Settlement Area (Giresun, NE Turkey). Bull. Eng. Geol. Environ. 67 (3), 375–385.

Chen, Z., 1995. Recent developments in slope stability analysis. In: Proc. 8th ISRM Congress, 25–29 September 1995. Tokyo, Japan. pp. 1041–1048.

Chiwaye, H.T., 2010. A comparison of the limit equilibrium and numerical modelling approaches to risk analysis for open pit mine slopes. Doctoral dissertation. Faculty of Engineering and the Built Environment, University of the Witwatersrand, Johannesburg.

Chiwaye, H.T., Stacey, T.R., 2010. A comparison of limit equilibrium and numerical modelling approaches to risk analysis for open pit mining. J. South Afr. Inst. Min. Metall. 110 (10), 571–580.

Chowdhury, R., Zhang, S., Flentje, P., 2004. Reliability updating and geotechnical back-analysis. In: Jardine, R.J., Potts, D.M., Higgins, K.G. (Eds.), Advances in Geotechnical Engineering: The Skempton Conference. Thomas Telford, London, pp. 815–821.

Christian, J.T., 2004. Geotechnical engineering reliability: how well do we know what we are doing? J. Geotech. Geoenviron. Eng. 130 (10), 985–1003.

Christian, J.T., Baecher, G.B., 1999. Point-estimate method as numerical quadrature. J. Geotech. Geoenviron. Eng. 125 (9), 779–786.

Christian, J.T., Baecher, G.B., 2002. The point-estimate method with large numbers of variables. Int. J. Num. Anal. Methods Geomech. 26 (15), 1515–1529.

Christian, J.T., Ladd, C.C., Beacher, G.B., 1992. Reliability and probability in stability analysis. In: Seed, R.B., Boulanger, R. (Eds.), Stability and Performance of Slopes and Embankments II. ASCE, Berkeley, Proceedings of a Speciality Conference, Berkeley, California, 29 June–1 July 1992.

Christian, J.T., Ladd, C.C., Beacher, G.B., 1994. Reliability applied to slope stability. J. Geotech. Eng. 120 (12), 2180–2207.

Chua, C.G., Goh, A.T.C., 2005. Estimating wall deflections in deep excavations using Bayesian neural networks. Tunnelling Underground Space Technol. 20, 400–409.

Cividini, A., Jurina, G., Gioda, G., 1981. Some aspects of characterization problems in geomechanics. Int. J. Rock Mech. Min. Sci. 18, 487–503.

Cividini, A., Maier, G., Nappi, A., 1983. Parameter estimation of a static geotechnical model using a Bayes' rule approach. Int. J. Rock Mech. Min. Sci. 20, 215–226.

Coggan, J.S., Stead, D., Eyre, J.M., 1998. Evaluation of techniques for quarry slope stability assessment. Trans. Inst. Min. Metall. Sect. B 107, B139–B147.

Corkum, A.G., Martin, C.D., 2004. Analysis of a rock slide stabilized with a toe-berm: a case study in British Columbia, Canada. Int. J. Rock Mech. Min. Sci. 41 (7), 1109–1121.

Cornell, C.A., 1971. First-order uncertainty analysis of soils deformation and stability. In: Proceedings of 1st International Conference on Application of Probability and Statistics in Soil and Structural Engineering (ICAPI). Hong Kong. pp. 129–144.

Cruden, D.M., 1978. A method of distinguishing between single and double plane sliding of tetrahedral wedges. Int. J. Rock Mech. Min. Sci. 15 (4), 217.

Cundall, P.A., 2002. The replacement of limit equilibrium methods in design with numerical solutions for factor of safety. Powerpoint presentation. Itasca Consulting Group, Inc.

Dadashzadeh, N., 2015. Reliability analysis of a rock slope in Sumela Monastery, Turkey, based on discrete element and response surface methods. MS thesis. Middle East Technical University, Turkey, 141 pp.

Deng, J.H., Lee, C.F., 2001. Displacement back analysis for a steep slope at the Three Gorges Project site. Int. J. Rock Mech. Min. Sci. 38, 259–268.

Deng, D., Nguyen Minh, D., 2003. Identification of rock mass properties in elasto-plasticity. Comput. Geotech. 30, 27–40.

Desai, C.S., Christian, J.T., 1977. Numerical Methods in Geomechanics. McGraw-Hill, New York.

Deschamps, R., Yankey, G., 2005. Limitations in the back-analysis of strength from failures. J. Geotech. Geoenviron. Eng. 4 (132), 532–536.

Donati, L., Turrini, M.C., 2002. An objective method to rank the importance of the factors predisposing to landslides with the GIS methodology: application to an area of the Apennines (Valnerina; Perugia, Italy). Eng. Geol. 63, 277–289.

Dong, J.J., Lee, W.R., Lin, M.L., Huang, A.B., Lee, Y.L., 2009. Effect of seismic anisotropy and geological characteristics on the kinematics of the neighboring Jiufengershan and Hungtsaiping landslides during Chi-Chi earthquake. Tectonophysics 466, 438–457.

Dong-ping, D., Liang, L., Jian-feng, W., Lian-heng, Z., 2016. Limit equilibrium method for rock slope stability analysis by using the Generalized Hoek-Brown criterion. Int. J. Rock Mech. Min. Sci. 89, 176–184.

Douglas, K.J., 2002. The shear strength of rock masses (Ph.D. thesis). The University of New South Wales, Australia.(unpublished).

Duncan, J.M., 2000. Factors of safety and reliability in geotechnical engineering. J. Geotech. Geoenviron. Eng. 126 (4), 307–316.

Duncan, J.M., Stark, T.D., 1992. Soil strengths from back analysis of slope failures. In: Proc. Stability and Performance of Slopes and Embankments II. ASCE, New York, pp. 890–904.

Duzgun, H., Bhasin, R., 2009. Probabilistic stability evaluation of Oppstadhornet Rock Slope, Norway. Rock Mech. Rock Eng. 42 (5), 729–749.

Eberhart, E., 2003. Rock slope stability analysis—utilization of advanced numerical techniques. Technical Report 41, UB C – Vancouver, Canada.

Einstein, H.H., Baecher, G.B., 1982. Probabilistic and statistical methods in engineering geology I. Problem statement and introduction to solution. In: Ingenieurgeologie und Geomechanik als Grundlagen des Felsbaues/Engineering Geology and Geomechanics as Fundamentals of Rock Engineering.

Eykhoff, P., 1974. System Identification—Parameter and State Estimation. John Wiley, UK, pp. 147–151, 519–526.

Fell, R., Mostyn, G.R., Maguire, P., O'Keeffe, L., 1988. Assessment of the probability of rain induced landsliding. Proc. Fifth Australia–New Zealand Conference on Geomechanics, 73–77.

Feng, X.T., An, H., 2004. Hybrid intelligent method optimization of a soft rock replacement scheme for a large cavern excavated in alternate hard and soft rock strata. Int. J. Rock Mech. Min. Sci. 41, 655–667.

Feng, X.T., Zhang, Z., Sheng, Q., 2000. Estimating mechanical rock mass parameters relating to the three gorges project permanent shiplock using an intelligent displacement back analysis method. Int. J. Rock Mech. 37, 1039–1054.

Ferentinou, M.D., Sakellariou, M.G., 2007. Computational intelligence tools for the prediction of slope performance. Comput. Geotech. 34 (5), 362–384.

Finno, R.J., Calvello, M., 2005. Supported excavations: the observational method and inverse modeling. J. Geotech. Geoenviron. Eng. 131 (7), 826–836.

Frayssines, M., Hantz, D., 2009. Modelling and back-analysing failures in steep limestone cliffs. Int. J. Rock Mech. Min. Sci. 46, 1115–1123.

Gao, W., 2015. Stability analysis of rock slope based on an abstraction ant colony clustering algorithm. Environ. Earth Sci. 73 (12), 7969–7982.

Gens, A., Ledesma, A., Alonso, E.E., 1996. Estimation of parameters in geotechnical backanalysis. II. Application to a tunnel excavation problem. Comput. Geotech. 18 (1), 29–46.

Gilbert, R.B., Wright, S.G., Liedtke, E., 1998. Uncertainty in back analysis of slopes: Kettleman Hills Case History. J. Geotech. Geoenviron. Eng. 124 (12), 1167–1176.

Gioda, G., 1985. Some remarks on back analysis and characterization problems. Proceedings, 5th International Conference on Numerical Methods in Geomechanics, Nagoya, Japan, 1–5 April, pp. 47–61.

Gioda, G., Maier, G., 1980. Direct search solution of an inverse problem in elastoplasticity: indentification of cohesion, friction angle and in situ stress by pressure tunnel tests. Int. J. Numer. Methods Eng. 15, 1823–1848.

Gioda, G., Sakurai, S., 1987. Back analysis procedures for the interpretation of field measurements in geomechanics. Int. J. Numer. Anal. Methods Geomech. 11, 555–583.

Gischig, V., Amann, F., Moore, J.R., Loew, S., Eisenbeiss, H., Stempfhuber, W., 2011. Composite rock slope kinematics at the current Randa instability, Switzerland, based on remote sensing and numerical modeling. Eng. Geol. 118, 37–53.

Gokceoglu, C., Sonmez, H., Ercanoglu, M., 2000. Discontinuity controlled probabilistic slope failure risk maps of the Altindag (settlement) Region in Turkey. Eng. Geol. 55, 277–296.

Goodman, R.E., 1976. Methods of Geological Engineering in Discontinuous Rocks. West Publishing, San Francisco, CA.

Goshtasbi, K., Ataei, M., Kalatehjary, R., 2008. Slope modification of open pit wall using a genetic algorithm-case study, Southern Wall of the 6th Golbini Jajarm Bauxite Mine. J. South Afr. Inst. Min. Metall. 108 (10), 651–656.

Griffiths, D.V., Fenton, G.A., 2004. Probabilistic slope stability analysis by finite elements. J. Geotech. Geoenviron. Eng. 130 (5), 507–518.

Griffiths, D.V., Huang, J., Fenton, G.A., 2009. Influence of spatial variability on slope reliability using 2-D random fields. J. Geotech. Geoenviron. Eng. 135 (10), 1367–1378.

Hack, H.R.G.K., 1996. Slope Stability Probability Classification. ITC publ. No. 43, Enschede, the Netherlands.

Hack, H.R.G.K., 1998. Slope Stability Probability Classification, second ed. SSPC. ITC, Enschede, The Netherlands, 258 pp, ISBN 90 6164 154 3.

Hack, H.R.G.K., 2002. An evaluation of slope stability classification. ISRM EUROCK'2002, Portugal, Madeira, Funchal, 25–28 November 2002.

Hack, R., Alkema, D., Kruse, G.A.M., Leenders, N., Luzi, L., 2007. Influence of earthquake on the stability of slopes. Eng. Geol. 91, 4–15.

Hack, R., Price, D., Rengers, N., 2003. A new approach to rock slope stability – a probability classification (SSPC). Bull. Eng. Geol. Environ. 62 (2), 167–184.

Hammah, R.E., Yaccoub, T.E., 2009. Probabilistic slope analysis with the finite element method. 43rd US Rock Mechanics Symposium and 4th U.S. Canada Rock Mechanics Symposium, Asheville, ARMA, 09-149.

Harr, M.E., 1987. Reliability Based on Design in Civil Engineering. McGraw-Hill, New York.

Harr, M.E., 1989. Probabilistic estimates for multivariate analyses. Appl. Math. Modell. 13 (5), 313–318.

Harris, S.J., Orense, R.P., Itoh, K., 2011. Back analyses of rainfall-induced slope failure in Northland Allochthon formation. Landslidesdoi: 10.1007/s10346-011-0309-1.

Hasofer, A.M., Lind, N.C., 1974. Exact and invariant second—moment code format. Journal of the Engineering Mechanics Division of American Society of Civil Engineers 100, 111–121.

Hatzor, Y.H., Levin, M., 1997. The shear strength of clay filled bedding planes in limestones-back analysis of a slope failure in a phosphate mine. Israel. Geotech. Geol. Eng. 15, 263–282.

Hisatake, M., Hieda, Y., 2008. Three-dimensional back analysis method for the mechanical parameters of the new ground ahead of a tunnel face. Tunnelling Underground Space Technol. 23, 373–380.

Hocking, G., 1976. A method for distinguishing between single and double plane sliding of tetrahedral wedges. Int. J. Rock Mech. Min. Sci. Geomech. Abstr. 13, 225–226.

Hoek, E., 2006. Practical rock engineering. Chapter 7: A slope stability problem in Hong Kong. http://www.rocscience.com/education/hoeks_corner.

Hoek, E., 2009. Fundamentals of slope design. In: Slope stability 2009, Santiago, Chile.

Hoek, E., Bray, J.W., 1981. Rock Slope Engineering. Institution of Mining and Metallurgy, London.

Hoek, E., Read, J., Karzulovic, A., Chen, Z.Y., 2000. Rock slopes in civil and mining engineering. In: Proceedings of the International Conference on Geotechnical and Geological Engineering, GeoEng2000, 19–24 November. Melbourne.

Hoek, E., Brown, E.T., 1980. Underground excavations in rock. The Institution of Mining and Metallurgy, London.

Hong, H.P., 1998. An efficient point estimate method for probabilistic analysis. Reliab. Eng. Syst. Saf. 59 (3), 261–267.

Honjo, Y., Liu, W.T., Soumitra, G., 1994. Inverse analysis of an embankment on soft clay by extended Bayesian method. Int. J. Numer. Anal. Methods Geomech. 18, 709–734.

Hoshiya, M., Yoshida, I., 1996. Identification of conditional stochastic field. ASCE, Eng. Mech. 122 (2), 101–108.

Hosseini, N., Gholinejad, M., 2014. Investigating the slope stability based on uncertainty by using fuzzy possibility theory. Arch. Min. Sci. 59 (1), 179–188.

Hudson, J.A., 1992. Rock Engineering Systems: Theory and Practice. Ellis Horwood, Chichester.

Huyse, L., 2001. Solving problems of optimization under uncertainty as statistical decision problems. In: AIAA-1519.

Iman, R.L., Davenport, J.M., Zeigler, D.K., 1980. Latin hypercube sampling. A program user's guide. Technical report SAND79-1473. Sandia Laboratories, Albuquerque, New Mexico.

Jaeger, J.C., 1971. Friction of rocks and stability of rock slopes. Geotechnique 21 (2), 97–134.

Jeon, Y.S., Yang, H.S., 2004. Development of a back analysis algorithm using FLAC. Int. J. Rock Mech. Min. Sci. 41 (3), 447–453.

Jing, L., 2003. A review of techniques, advances and outstanding issues in numerical modelling for rock mechanics and rock engineering. Int. J. Rock Mech. Min. Sci. 40, 283–353.

Jing, L.R., Hudson, J.A., 2002. Numerical methods in rock mechanics. Int. J. Rock Mech. Min. Sci. 39 (4), 409–427.

Kaiser, P.K., Zou, D., Lang, P.A., 1990. Stress determination by back-analysis of excavation induced stress changes—a case study. Rock Mech.Rock Eng. 23, 185–200.

Kamp, U., Owen, L.A., Growley, B.J., Khattak, G.A., 2010. Back analysis of landside susceptibility zonation mapping for the 2005 Kashmir earthquake: an assessment of the reliability of susceptibility zoning maps. Nat. Hazards 54, 1–25.

Kesimal, A., Ercikdi, B., Cihangir, F., 2005. Analysis of bench instability resulted from the planar failure at Araklı-Tasonu limestone quarry in Trabzon. Project Report 3, p 72 (in Turkish).

Kesimal, A., Ercikdi, B., Cihangir, F., 2008. Environmental impact of blast-induced acceleration on slope instability at a limestone quarry. Environ. Geol. 54, 381–389.

KhaloKakaie, R., Zare Naghadehi, M., 2012a. Ranking the rock slope instability potential using the interaction matrix (IM) technique: a case study in Iran. Arabian J. Geosci. 5, 263–273.

KhaloKakaie, R., Zare Naghadehi, M., 2012b. The assessment of rock slope instability along the Khosh-Yeylagh main road (Iran) using a systems approach. Environ. Earth Sci. 67, 665–682.

Kliche, C.A., 1999. Rock Slope Stability. SME, Littleton, CO.

Kola, N., 2014. Reliability based analysis of dam embankment, geocellreinforced foundation and embankment with stone columns using finite element method (e-Thesis). Master of Technology in Geotechnical Engineering, Department of Civil Engineering National Institute of Technology, Rourkale, June 2001. (https://www.noexperiencenecessarybook.com/YaJmY/reliability-based-analysis-of-dam-embankment-ethesis.html).

Kumar, M., Samui, P., Naıthanı, A.K., 2017. Determination of stability of epimetamorphic rock slope using minimax probability machine. Geomatics Nat. Hazards Risk 7 (1), 186–193.

Kutter, H.K., 1974. Analytical methods for rock slope analysis. In: Muller, L. (Ed.), Rock mechanics. Springer, New York, pp. 198–211.

Ledesma, A., Gens, A., Alonso, E.E., 1996. Parameter and variance estimation in geotechnical back analysis using prior information. Int. J. Numer. Anal. Methods Geomech. 20, 119–141.

Lee, J., Akutagawa, S., Yokota, Y., Kitagawa, T., Isogai, A., Matsunaga, T., 2006. Estimation of model parameters and ground movement in shallow NATM tunnel by means of neural network. Tunnelling Underground Space Technol. 21, 242.

Leroueil, S., Tavenas, F., 1981. Pitfalls of back-analysis. Proc. 10th Int. Conf. on Soil Mechanics and Foundation Engineering, Balkema, Rotterdam, Netherlands 1, 185–190.

Li, Y., 2016. A constitutive model of opened rock joints in the field. PhD thesis.

Li, W., Mei, S.H., 2004. Fuzzy system method for the design of a jointed rock slope. Int. J. Rock Mech. Min. Sci. 41, 569–574.

Li, D., Chen, Y., Lu, W., Zhou, C., 2011. Stochastic response surface method for reliability analysis of rock slopes involving correlated non-normal variables. Comput. Geotech. 38, 58–68.

Li, A.J., Cassidy, M.J., Wang, Y., Merifield, R.S., Lyamin, A.V., 2012. Parametric Monte Carlo studies of rock slopes based on the Hoek–Brown failure criterion. Comput. Geotech. 45, 11–18.

Li, G., Yang, M., Meng, Y., 2016. The assessment of correlation between rock drillability and mechanical properties in the laboratory and the field under different pressure conditions. J. Nat. Gas Sci. Eng. 30, 405–413.

Lindsay, P., Campbell, R.N., Fergusson, D.A., Gillard, G.R., Moore, T.A., 2001. Slope stability probability classification, Waikato Coal Measures, New Zealand. Int. J. Coal Geol. 45, 127–145.

Liu, Z., Shao, J., Xu, W., Xu, F., 2014. Comprehensive stability evaluation of rock slope using the cloud model-based approach. Rock Mech. Rock Eng. 47 (6), 2239–2252.

Loupasakis, C., Konstantopoulou, G., 2007. A failure mechanism of the fine neogene formations: an example from Thasos, Greece. Landslides 4, 351–355.

Low, B.K., 2007. Reliability analysis of rock slopes involving correlated nonnormals. Int. J. Rock Mech. Min. Sci. 44 (6), 922–935.

Low, B.K., Tang, W.H., 1997. Efficient reliability evaluation using spreadsheet. J. Eng. Mech. 123 (7), 749–752.

Lucas, J.M., 1980. A general stereographic method for determining possible mode of failure of any tetrahedral rock wedge. Int. J. Rock Mech. Min. Sci. Geomech. Abstr. 17, 57–61.

Luckman, P.G., Der Kiureghian, A., Sitar, N., 1987. Use of stochastic stability analysis for Bayesian back calculation of core pressures acting in a cut at failure. In: Lind, N. (Ed.), In: Proceedings of the 5th International Conference on Application of Statistics and Probability in Soil and Structural Engineering. University of Waterloo Press, Vancouver, Ontario, pp. 922–929.

Markland, J.T., 1972. A useful technique for estimating the stability of rock slopes when the rigid wedge sliding type of failure is expected. Imp. Coll. Rock Mech. Res. Rep. 19, p. 10.

Matherson, G.D., 1988. The collection and use of field discontinuity data in rock slope design. Q. J. Eng. Geol. Hydrogeol. 22, 19–30.

Mavrouli, O., Corominas, J., Wartman, J., 2009. Methodology to evaluate rock slope stability under seismic conditions at Sol De Santa Coloma, Andorra. Nat. Hazards Earth Syst. Sci. 9, 1763–1773.

Mazzoccola, D.F., Hudson, J.A., 1996. A comprehensive method of rock mass characterization for indicating natural slope instability. Q. J. Eng. Geol. Hydrogeol. 29, 37–56.

McMahon, B.K., 1971. A statistical method for the design of rock slopes. In: Proceedings of 1st Australia–New Zealand Geomechanics Conference. Melbourne. pp. 314–321.

Mello Franco, J.A., De Armelin, J.L., Santiago, J.A.F., Telles, J.C.F., Mansur, W.J., 2002. Determination of the natural stress state in a Brazilian rock mass by back analysing excavation measurements: a case study. Int. J. Rock Mech. Min. Sci. 39, 1005–1032.

Miller, S.M., Whyatt, J.K., McHugh, E.L., 2004. Applications of the Point Estimation Method for Stochastic Rock Slope Engineering, Gulf Rocks 2004, the 6th North America Rock Mechanics (NARMS), 5–9 June. American Rock Mechanics Association, Houston, Texas, Document ID ARMA-04-517.

Morgan, M.G., Henrion, M., 1990. Uncertainty: A Guide to Dealing with Uncertainty in Quantitative Risk and Policy Analysis. Cambridge University Press, New York.

Nadım, F., 2006. First order second moment (FOSM), first and second order reliability methods (FORM & SORM), Monte Carlo simulation, system reliability. In: Proceedings of the CISM Course "Probabilistic Methods in Geotechnical Engineering". Udine.

Nicholson, D.T., 2000. Deterioration of excavation rock slopes: mechanism, morphology and assessment. Ph.D. Dissertation. School of Earth Science, University of Leeds, Leeds.

Nicholson, D.T., 2003. Breakdown mechanisms and morphology for man-made rockslopes in North West England. North West Geogr. 3, 12–25.

Nicholson, D.T., 2004. Hazard assessment for progressive, weathering-related breakdown of excavated rock slopes. Q. J. Eng. Geol. Hydrogeol. 37, 327–346.

Nicholson, D.T., Hencher, S., 1997. Assessing the potential for deterioration of engineered rock slopes. Conference Proceedings of the IAEG Symposium, Athens.

Nguyen, V., Chowdhury, R., 1985. Simulation for risk analysis with correlated variables. Géotechnique 35 (1), 47–58.

Nilsen, B., 2000. New trends in rock slope stability analyses. Bull. Eng. Geol. Environ. 58, 173–178.

Nilsen, B., 2016. Rock slope stability analysis according to Eurocode 7: discussion of some dilemmas with particular focus on limit equilibrium analysis. Bull. Eng. Geol. Environ. 1–8 (online). doi: 10.1007/s10064-016-0928-9.

Oggeri, C., Oreste, P., 2012. Tunnel static behavior assessed by a probabilistic approach to the back-analysis. Am. J. Appl. Sci. 9, 1137–1144.

Oh, H.J., Pradhan, B., 2011. Application of a neuro-fuzzy model to landslide-susceptibility mapping for shallow landslides in a tropical hilly area. Comput. Geosci. 37, 1264–1276.

Okui, Y., Tokunaga, A., Shinji, M., Mori, S., 1997. New back analysis method of slope stability by using field measurements. Int. J. Rock Mech. Min. Sci. 34 (3–4), Paper no. 234.

Olsson, U., 2002. Generalized Linear Models: An Applied Approach. Studentlitteratur, Lund, 31–44.

Pantelidis, L., 2009. Rock slope stability assessment through rock mass classification systems. Int. J. Rock Mech. Min. Sci. 46, 315–325.

Pantelidis, L., 2010. An alternative rock mass classification system for rock slopes. Bull. Eng. Geol. Environ. 69 (1), 29–39.

Park, H., 1999. Risk Analysis of Rock Slope Stability and Stochastic Properties of Discontinuity parameters in Western North Carolina (Ph.D. thesis). Purdue University.

Park, H.J., Um, J.G., Woo, I., Kim, J.W., 2012. The evaluation of the probability of rock wedge failure using the point estimate method. Environ. Earth Sci. 65 (1), 353–361.

Park, H., West, T., 2001. Development of a probabilistic approach for rock wedge failure. Eng. Geol. 59 (3), 233–251.

Park, H.J., West, T.R., Woo, I., 2005. Probabilistic analysis of rock slope stability and random properties of discontinuity parameters. Interstate Highway 40. Eng. Geol. 79, 230–250.

Park, H.J., Um, J., Woo, I., 2008. The evaluation of failure probability for rock slope based on fuzzy set theory and Monte Carlo simulation. In: Chen et al., (Ed.), Landslides and Engineered Slopes—© 2008. Taylor & Francis Group, London, pp. 1943–1949, ISBN 978-0-415-41196-7.

Park, H.J., Lee, J.H., Kim, K.M., Um, J.G., 2016. Assessment of rock slope stability using GIS-based probabilistic kinematic analysis. Eng. Geol. 203, 56–69.

Phoon, K.K., 2008. Reliability-based design in geotechnical engineering. In: Phoon, K.-K. (Ed.), Computations and Applications. Routledge.

Pradhan, B., 2010a. Remote sensing and GIS-based landslide hazard analysis and cross-validation using multivariate logistic regression model on three test areas in Malaysia. Adv. Space Res. 45, 1244–1256.

Pradhan, B., 2010b. Application of an advanced fuzzy logic model for landslide susceptibility analysis. Int. J. Comput. Intell. Syst. 3 (3), 370–381.

Pradhan, B., 2010c. Landslide susceptibility mapping of a catchment area using frequencies ratio, fuzzy logic and multivariate logistic regression approaches. J. Indian Soc. Remote Sens. 38, 301–320.

Pradhan, B., Youssef, A.M., 2010. Manifestation of remote sensing data and GIS on landslide hazard analysis using spatial-based statistical models. Arabian J. Geosci. 3 (3), 319–326.

Qian, Q.H., 2012. Challenges faced by underground projects construction safety and countermeasures. Chin. J. Rock Mech. Eng. 31 (10), 1945–1956, (in Chinese).

Read, J., Stacey, P., 2009. Guidelines for Open Pit Slope Design. CSIRO Publishing, Australia.

Regmi, A.D., Yoshida, K., Pourghasemi, H.R., Dhital, M.R., Pradhan, B., 2014. Landslide susceptibility mapping along Bhalubang–Shiwapur area of mid-western Nepal using frequency ratio and conditional probability models. J. Mt. Sci. 11 (5), 1266–1285.

Robertson, A.M., 1988. Estimating Weak Rock Strength. AIME Annual General Meeting, 1–6 Jan 1988. Tucson, AZ, USA.

Romana, M., 1985. New adjustment ratings for application of Bieniawski classification to slopes. In: Proceedings of the International Symposium on the Role of Rock Mechanics in Excavations for Mining and Civil Works. International Society of Rock Mechanics, Zacatecas, pp. 49–53.

Romana, M., Serón, J.B., Montalar, E., 2003. SMR geomechanics classification: application, experience and validation. In: ISRM 2003—Technology Roadmap for Rock Mechanics, 8–12 September 2003, Gauteng, South Africa, pp. 1–4.

Rosenblueth, E., 1975. Point estimates for probability moments. Appl. Math. Modell. 72 (10), 3812–3814.

Rosenblueth, E., 1981. Two-point estimates in probabilities. Proc. Natl. Acad. Sci. 5 (2), 329–335.

Ross, T.J., 1995. Fuzzy Logic with Engineering Applications. McGraw-Hill, New York, 600 pp.

Rozos, D., Pyrgiotis, L., Skias, S., Tsagaratos, P., 2008. An implementation of rock engineering system for ranking the instability potential of natural slopes in Greek territory: an application in Karditsa County. Landslides 5 (3), 261–270.

Russelli, C., 2008. Probabilistic methods applied to the bearing capacity problem. PhD thesis. Universität Stuttgart.

Sakurai, S., Takeuchi, K., 1983. Back analysis of measured displacements of tunnels. Rock Mech. Rock Eng. 16 (3), 173–180.

Sakurai, S., Akutagawa, S., Takeuchi, K., Shinji, M., Shimizu, N., 2003. Back analysis for tunnel engineering as a modern observational method. Tunnelling Underground Space Technol. 18, 185–196.

Salmi, E.F., Hosseinzadeh, S., 2015. Slope stability assessment using both empirical and numerical methods: a case study. Bull. Eng. Geol. Environ. 74 (1), 1–13.

Sancio, R.T., 1981. The use of back-calculations to obtain the shear and tensile strength of weathered rocks. Proceedings of the International Symposium on Weak Rock, Tokyo. A.A. Balkema, Rotterdam, pp. 647–652.

Schweckendiek, T., 2006. Structural reliability applied to deep excavations coupling reliability methods with finite elements section. Hydraulic and Geotechnical Engineering, Delft University of Technology.

Selby, M.J., 1980. A rock mass strength classification for geomorphic purposes: with tests from Antarctica and New Zealand. Z. Geomorphol. 24, 31–51.

Shamekhi, S.E., 2014. Probabilistic assessment of rock slope stability using response surfaces determined from finite element models of geometric realizations. PhD thesis.

Shamekhi, E., Tannant, D.D., 2015. Probabilistic assessment of rock slope stability using response surfaces determined from finite element models of geometric realizations. Comput. Geotech. 69, 70–81.

Shang, Y., Park, H.D., Yang, Z., 2005. Engineering geological zonation using interaction matrix of geological factors: an example from one section of Sichuan–Tibet Highway. Geosci. J. 9 (4), 375–387.

Sharifzadeh, M., Sharifi, M., Delbari, S.M., 2010. Back analysis of an excavated slope failure in highly fractured rock mass: the case study of Kargar Slope Failure (Iran). Environ. Earth Sci. 60, 183–192.

Shen, H. Non-deterministic analysis of slope stability based on numerical simulation (Ph.D. thesis). Technische Universität Bergakademie Freiberg, 2012 (unpublished).

Sheorey, P.R., 1997. Emprical rock failure criteria. International Symposium on Weak Rock, Tokyo. A.A. Balkema, Rotterdam, Netherlands.

Singh, A., 2004. FRHI—a system to evaluate and mitigate rockfall hazard in stable rock excavations. J. Inst. Eng. India. Part CV. Civil Eng. Div. 85, 62–75.

Singh, R.N., Brown, D.J., Denby, B., Croghan, J.A., 1986. The development of a new approach to slope stability assessment in UK. Ground Movement and Control Related to Coal Mining Symp, Wollongong, Australia, pp. 59–63.

Singh, A., Connolly, M., 2003. VRFSR—an empirical method for determining volcanic rock excavation safety on construction sites. J. Inst. Eng. India. Part CV. Civil Eng. Div. 84, 176–191.

Smith, J.V., 2015. A new approach to kinematic analysis of stress-induced structural slope instability. Eng. Geol. 187, 56–59.

Sonmez, H., Ulusay, R., Gokceoglu, C., 1998. A practical procedure for the back analysis of slope failures in closely jointed rock masses. Int. J. Rock Mech. Min. Sci. 35 (2), 219–233.

Stanković, J.N., Filipović, S., Rajković, R., Obradović, L., Marinković, V., Kovaćević, R., 2013. Risk and reliability analysis of slope stability deterministic and probabilistic method. J. Trends Dev. Mach. Assoc. Technol. 17 (1), 97–100.

Stark, T.D., Eid, H.T., 1998. Performance of three-dimensional slope stability methods in practice. J. Geotech. Geoenviron. Eng. 124 (11), 1049–1060.

Startzman, R.A., Watterbarger, R.A., 1985. An improved computation procedure for risk analysis problems with unusual probability functions. In: Proc. Symp. Soc. Petroleum Engineers Hydrocarbon Economics and Evaluation. Dallas.

Stead, D., Coggan, J.S., Eberhardt, E., 2004. Realistic simulation of rock slope failure mechanisms: the need to incorporate principles of fracture mechanics. In SINOROCK2004, International Symposium on Rock Mechanics: Rock characterization, modelling engineering design methods. Int. J. Rock Mech. Min. Sci. 41, 460–466.

Stead, D., Eberhardt, E., Coggan, J.S., 2006. Developments in the characterization of complex rock slope deformation and failure using numerical modelling techniques. Eng. Geol. 83, 217–235.

Styles, T.D., Coggan, J.S., Pine, R.J., 2011. Back analysis of the Joss Bay chalk cliff failure using numerical modelling. Eng. Geol. 120, 81–90.

Subramanyan, K., Diwekar, U., Zitney, S.E., 2011. Stochastic modeling and multi-objective optimization for the APECS system. Comput. Chem. Eng. 35 (2011), 2667–2679.

Suikkanen, J.J., 2014. Modeling slope stability utilising fracture mechanics. MS thesis. Department of Civil and Environmental Engineering, Aalto University, 112 pp.

Tabba, M.M., 1984. Deterministic versus risk analysis of slope stability. 4th International Symposium on Landslides. Toronto, Canada. pp. 491–498.

Tan, X., et al., 2013. Response surface method of reliability analysis and its application in slope stability analysis. Geotech. Geol. Eng. 31, 1011–1025.

Tang, W.H., Stark, T.D., Angulo, M., 1999. Reliability in back analysis of slope failures. Soils Found Jap. Geotech. Soc. 39 (5), 73–80.

Tiwari, B., Brandon, T.L., Marui, H., Tuladhar, G.T., 2005. Comparison of residual shear strengths from back analysis and ring shear tests on undisturbed and remolded specimens. J. Geotech. Geoenviron. Eng. 131 (9).

Tokashiki, N., Aydan, Ö., 2011. Kita-Uebaru natural rock slope failure and its back analysis. Environ. Earth Sci. 62 (1), 25–31.

Tomas, R., 2008. Continuous Slope Mass Rating (SMR-C): Functions, Analysis and Spatial Application Methodology. In: The Young Laubscher DHA (1990) Geomechanical classification system for the rating of rock mass in mine design. J. South Afr. Inst. Min. Metall. 90 (10), 257–273.

Tomas, R., Delgado, J., Serón, J.B., 2007. Modification of slope mass rating (SMR) by continuous functions. Int. J. Rock Mech. Min. Sci. 44 (7), 1062–1069.

Tomas, R., Cuenca, A., Cano, M., Carcia, B.J., 2012. A graphical approach for slope mass rating (SMR). Eng. Geol. 124, 67–76.

Tonon, F., Amadei, B., Pan, E., Frangopol, D.M., 2001. Bayesian estimation of rock mass boundary conditions with applications to the AECL underground research laboratory. Int. J. Rock Mech. Min. Sci. 38 (7), 995–1027.

Tuncay, E., Ulusay, R., 2000. Probabilistic kinematical approach for the assessment of discontinuity controlled slope failures and an application. Bull. Earth Sci. Applic. Res. Center Hacettepe Univ. 22, 205–222.

Tuncay, E., Ulusay, R., 2001. Deterministic and probabilistic treatments of multiplanar pitfall failures at Himmetog˘lu (Turkey) coal mine. Can. Geotech. J. 38, 828–849.

Tutluoglu, L., Öge, I.F., Karpuz, C., 2011. Two and three dimensional analysis of a slope failure in a lignite mine. Comput. Geosci. 37 (2).

Valdivia, C., Lorig, L., 2000. Slope stability at Escondida mine. In: Hustrulid, McCarter, Van Zyl (Eds.), Slope Stability in Surface Mining. SME, Colorado, pp. 153–162.

Valley, B., Duff, D., 2011. Probabilistic analyses in Phase2 8.0. https://www.rocscience.com/documents/pdfs/rocnews/spring2011/Phase2stat-Probabilistic-Analys.

Valley, B., Kaiser, P.K., Duff, D., 2010. Consideration of uncertainty in modelling the behaviour of underground excavations. In: Van Sint Jan, M., Potvin, Y. (Eds.), In: Proceedings of the 5th International Seminar on Deep and High Stress Mining. Australian Centre for Geomechanics, Perth, pp. 423–435.

Vardakos, S., 2007. Back-analysis Methods for Optimal Tunnel Design (Dissertation). Faculty of the Virginia Polytechnic Institute and State University, Blacksburg, 179p.

Vardakos, S.S., Gutierrez, M.S., Barton, N.R., 2007. Back-analysis of Shimizu tunnel no. 3 by distinct element modeling. Tunnelling Underground Space Technol. 22, 401–413.

Vatanpour, N., Ghafoori, M., Talouki, H.H., 2014. Probabilistic and sensitivity analyses of effective geotechnical parameters on rock slope stability: a case study of an urban area in northeast Iran. Nat. Hazards 71, 1659–1678.

Vrıjlıng, J.K., 1997. Probabilities in civil engineering, Part 1: Probabilistic design in theory. CUR-Publicatie 190, Delft University of Technology, Stichting CUR, Gouda.

Wang, J.P., Huang, D., 2012. RosenPoint: a Microsoft excel-based program for the Rosenblueth point estimate method and an application in slope stability analysis. Comput. and Geosci. 48, 239–243.

Wang, L., Hwang, J.H., Luo, Z., Juang, C.H., Xiao, J.H., 2013. Probabilistic back analysis of slope failure –A case study in Taiwan. Comput. Geotech. 51, 12–23.

Wang, Y., Zhao, T., Cao, Z., 2015. Site-specific probability distribution of geotechnical properties. Comput. Geotech. 70, 159–168.

Wesley, L.D., Leelaratman, V., 2001. Shear strength parameters from back-analysis of single slips. Geotechnique 51 (4), 373–374.

Whu, J.H., Tsai, P.H., 2011. New dynamic procedure for back-calculating the shear strength parameters of large landslides. Eng. Geol. 123 (1–2), 29.

Wolff, T.F., 1996. Probabilistic slope stability in theory and practice. Conference on Uncertainty in the Geologic Environment. Madison, WI. pp. 419–433.

Wong, F.S., 1985. Slope reliability and response surface method. J. Geotech. Geoenviron. Eng.g 111 (1), 32–53.

Wyllie, D.C., Mah, C.W., 2001. Rock slope engineering, fourth ed. Civil and Mining. Spon Press, London and New York.

Wyllie, D., Mah, C., 2004. In: Hoek, E., Bray, J. (Eds.), Rock Slope Engineering: Civil and Mining. 4th edn. based on the 3rd edn. Spon Press, London and New York, p. 86.

Xu, C., Wang, L., Tien, Y.M., Chen, J.M., Juang, C.H., 2014. Robust design of rock slopes with multiple failure modes: modeling uncertainty of estimated parameter statistics with fuzzy number. Environ. Earth Sci. 72 (8), 2957–2969.

Yahiaoui, D., Kadid, A., Hakim, Z.A., 2016. Probability-based analysis of slope stability. Malaysian J. Civil Eng. 28 (2), 140–148.

Yang, Y., Zhang, Q., 1997. A hierarchical analysis for rock engineering using artificial neural networks. Rock Mech. Rock Eng. 30, 207–222.

Yoon, W.S., Jeong, U.J., Kim, J.H., 2002. Kinematic analysis for sliding failure of multi-faced rock slopes. Eng. Geol. 67, 51–61.

Yudhbir, R.K., Lemanza, W., Prinzl, F., 1983. An empirical failure criterion for rock masses. Proceedings of the 5th International Congress on Rock Mechanics, Melbourne 1983, Balkema, Rotterdam, 1, pp. B1–B8.

Zanbak, C., 1977. Statistical interpretation of discontinuity contour diagrams. Int. J. Rock Mech. Min. Sci.Geomech. Abstr. 14, 114–120.

Zang, T.A., Hemsch, M.J., Hilburger, M.W., Kenny, S.P., Luckring, J.M., Maghami, P., Padula, S.L., Stroud, W.J., 2002. Needs and Opportunities for Uncertainty-Based Multidisciplinary Design Methods for Aerospace Vehicle. Langley Research Center, Hampton, Virginia.

Zare Naghadehi, M., Jimenez, R., KhaloKakaie, R., Jalali, S.M.E., 2011. A probabilistic systems methodology to analyze the importance of factors affecting the stability of rock slopes. Eng. Geol. 118, 82–92.

Zare Naghadehi, M., Jimenez, R., KhaloKakaie, R., Jalali, S.M.E., 2013. A new open-pit mine slope instability index defined using the improved rock engineering systems approach. Int. J. Rock Mech. Min. Sci. 61, 1–14.

Zhang, L.Q., Yang, Z.F., Liao, Q.L., Chen, J., 2004. An application of the rock engineering system (RES) methodology to rockfall hazard assessment on the Chengdu-Lhasa Highway, China. Int. J. Rock Mech. Min. Sci. 41 (3), 833–838.

Zhang, L.Q., Yue, Z.Q., Yang, Z.F., Qi, J.X., Liu, F.C., 2006. A displacement-based back-analysis method for rock mass modulus and horizontal in situ stress in tunneling—illustrated with a case study. Tunelling Underground Space Technol. 21, 636–649.

Zhang, J., Tang, W.H., Asce, H.M., Zhang, L.M., Asce, M., 2010a. Efficient probabilistic back-analysis of slope stability model parameters. J. Geotech. Geoenviron. Eng. 136, 99.

Zhang, L.L., Zhang, J., Zhang, L.M., Tang, W.H., 2010b. Back analysis of slope failure with Markov chain Monte Carlo simulation. Comput. Geotech. 37, 905–912.

Zhang, J., Huang, H.W., Phoon, K.K., 2013. Application of the kriging-based response surface method to the system reliability of soil slopes. J. Geotech. Geoenviron. Eng. 139 (4), 651–655.

Zheng, J., Zhao, Y., Lu, Q., Deng, J., Pan, X., Li, Y., 2016. A discussion on the adjustment parameters of the slope mass rating (SMR) system for rock slopes. Eng. Geol. 206, 42–49.

Chapter 14

Civic Fire Control System for Historic District in Kiyomizu, Kyoto—Development Project and Its Techniques for "Environmental Water Supply System (EWSS) for Disaster Prevention" to Protect Traditional Wooden Cultural Heritage Zones from Postearthquake Fire

Takeyuki Okubo
Ritsumeikan University, Kyoto, Japan

14.1 CHARACTERISTICS OF WOODEN CULTURE AND ENVIRONMENTAL VALUE

As other Asian countries, Japan also experiences hot and humid monsoon climate. Its abundant rainfalls produce abundant water and greenery, and thus, our ancestors fostered unprecedented "wood culture" over a long period of history. The wooden architectures worldwide and the traditional townscapes consisting of such wooden architecture are irreplaceable cultural resources for the mankind at present.

It still cannot be denied that trees have risks such as combustibility, etc. However, wiping out the traditional wooden culture, which forms a part of the cultural heritage, in a hasty manner from the earth simply because they are susceptible to fire will be an immeasurable loss to the mankind from the perspective of cultural diversity and conservation of the global environment.

To preserve wooden cultural cities, it is important to make the buildings fireproof, and more than anything to realize safe environment which can extinguish the fire in no time just in case the building catches fire.

14.2 CONCEPT OF ENVIRONMENTAL WATER SUPPLY SYSTEM FOR FIRE DISASTER PREVENTION

One of the most important disasters from the aspect of conserving wooden cultural cities that are distributed worldwide is "seismic fire". This phenomenon involves dangerous fire accidents "occurring simultaneously" that wipe out any kind of cultural heritage.

Here, it should be noted that the major damage in the residential areas with wooden houses during the Great Hanshin Earthquake, which occurred on January 17, 1995, was because the water was not supplied from the modern urban infrastructure, which was otherwise excellent. This lack of water supply was due to the occurrence of severe seismic motion and scarcity of water to cope with "simultaneous fire accident" (Kobe City Fire Department, 1996).

Traditional wooden cultural cities need to be preserved in future along with the coexistence of fires after earthquake, which may also be caused from Japanese islands. Response to this proposition can be initially ensured by reviving "existing natural water resources in the region", such as rivers, water passages, ponds, seas, well water, rainwater harvesting, which can enable water supply even during earthquakes which make serious damage in modern firefighting hydrants using city water network. This proposition could also be ensured through firefighting by public firefighting services and maintaining

Integrating Disaster Science and Management. http://dx.doi.org/10.1016/B978-0-12-812056-9.00014-2

a safer environment focusing on first-aid fire extinguishing services on a voluntary basis by the local people who can immediately handle fire accidents.

Development of "Environmental Water Supply System (EWSS) for Fire Disaster Prevention" is a project that is aimed at fostering the climate and reviving natural water resources in an easy to use manner. In addition, this project also aims at preserving the wooden culture from seismic fires and realizing a beautiful and safer city environment with abundant water. Wooden culture and urban water resources have their similar origins in the water, which is the same natural blessing. It is also the duty of present generation to revive them from the perspective of environment and disaster prevention and pass on to the future generations.

14.3 COMPOSITION AND OVERVIEW OF THIS MANUSCRIPT

The strategy for the utilization of water resources, citizens, and facilities of the community in the wooden cultural cities and Kyoto City, Japan, has been formulated as the basic concept for Disaster Prevention Water Supply System (Research Committee on the Concept of Disaster Prevention Water Supply System, 2002) in 2002. This strategy has been formulated based on the experiences of the Great Hanshin Earthquake so as to complement the normal firefighting system for emergencies.

The following chapters also deal with a part of the existing cases which can be referred for Disaster Prevention Water Supply System using natural water resources specific to the region along with the improvement/development projects. Subsequently, we shall try out the recommendations of concrete plans for the case project regions employing these concepts. As for the target area, Kiyomizu area (inclusive of world heritage listed sites in the vicinity of Kiyomizu temple) with the heavy possibility of severe damages, especially from seismic fires, even in Kyoto area, which is considered as a precious social capital with a group of traditional wooden monuments and buildings, was selected.

Since 2006, Kyoto City became the main area among the relevant areas. Its primary maintenance was initiated as the Ministry of Internal Affairs and Communications and Ministry of Land, Infrastructure and Transport "Disaster Prevention Water Supply System maintenance service for the preservation of cultural heritage and its surrounding area." This maintenance was partially completed in 2011.

When it comes to the implementation of extensional maintenance project, there were many issues such as securing the budget, setting the operating body, and consensus building with the residents, and development of the project is not easy. Even in the case of project development techniques, which are important to manage the project, we will introduce policies toward subsequent additional maintenance by consolidating the requirements in the process of realizing the reference cases.

14.4 CASE RESEARCH OF DISASTER MITIGATION WATER SUPPLY SYSTEM USING NATURAL WATER SOURCES AND IMPLEMENTATION PROCESS OF MAINTENANCE SERVICES

First of all, we will deal with the system overview for typical cases and consolidate the project implementation process from the Disaster Mitigation Water Supply System maintenance cases using the investigated natural water sources. This manuscript overviews the maintenance of gravity water supply system in the area falling under Ogimachi, Shirakawa, Ono District in Gifu Prefecture (Shirakawa township) as typical examples for rural areas. Our research also overviews the maintenance of irrigation channels and maintenance of disaster prevention and snow removal in Kanazawa, Ishikawa Prefecture, as typical examples in urban areas.

14.4.1 District Overview

In December 1995, Shirakawa township, which at present inherits a group of traditional monuments listed in world cultural heritage as "Historic Villages of Shirakawa-go and Gokayama," is a village located in heavy snowfall areas in the northern part of Gifu Prefecture. It has a population of about 2000. The structures with rafter roofs, particularly thatched roofs, are weak against fire and disaster preparedness. However, in Ogimachi area, the disaster prevention maintenance is being done by utilizing the existing agricultural canals without damaging the landscape. This is distinctive as securing the budget and revision of laws for landscape preservation has been carried out simultaneously along with the disaster mitigation maintenance.

On the other hand, Kanazawa city in Ishikawa Prefecture which consists of urban zone stacked up on the structures such as narrow roads and artificial waterways, which were formed during Edo era from the 17th century. With a population of 460,000, the city also has remains of historical castle townscapes such as old samurai residences, red light districts, a group of sacred buildings, etc. It consists of a network of irrigation canals also used as castle's moat whose total length is 150 km and a large number of them turning into underground culverts because of modernization. Since it is the region with a lot of snow, a well-type sprinkler facility, which melts the snow on the road surface, has been introduced on the main routes, which are working from the beginning of December to the beginning of March next year.

14.4.2 Overview of Existing Disaster Mitigation Water Supply System

14.4.2.1 Fire Protection System Created by Utilizing Agricultural Water and the Height Difference in Shirakawa Township

In the Shirakawa-Ogimachi area, gravity-fed firefighting water supply system was established for sending the high-pressure water, which is produced due to height difference and gravity without any artificial power, to 62 fire hydrants and 59 water cannons that are arranged within the prefecture. The water is sent from 600 t water storage tank installed on the Mae-yama hill at 80 m higher place with the existing agricultural water supply route (Masami and Sunao, 1999) (Fig. 14.1).

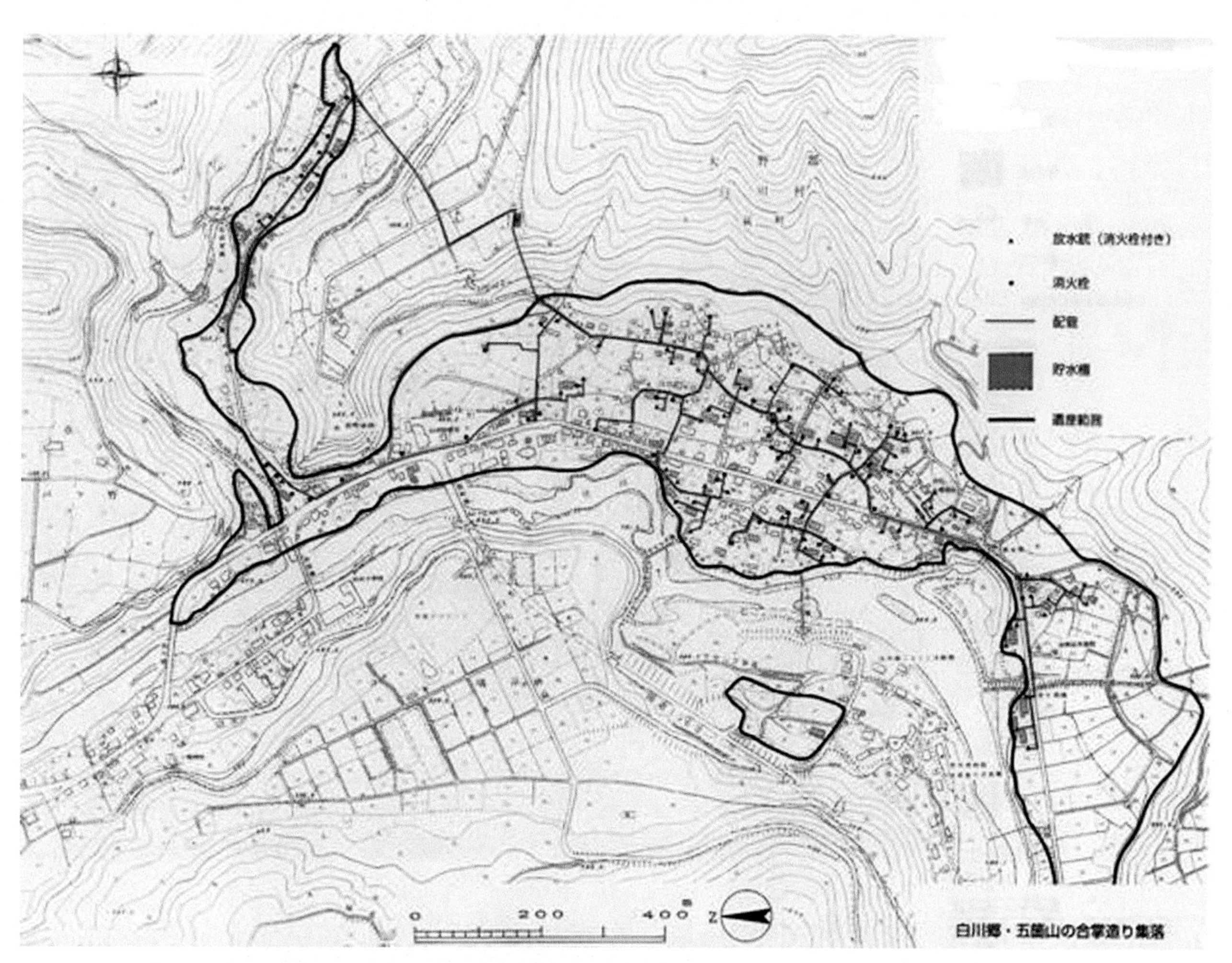

FIGURE 14.1 **Water distribution network from hilltop cistern to water cannons.**

The water flowing into the village is supplied to the eight fire-preventing cisterns/tanks installed along the waterway, which ensures multiple ways of securing water. Water cannons are installed diagonally in such a way that the houses with steep rafter roof are enclosed by water curtains at a height of 20 m to prevent fire spread by fire sparks (Photograph 14.1).

PHOTOGRAPH 14.1 **Water discharge exercise in Shirakawa township.**

14.4.2.2 Fire Protection System with Revival of Irrigation Canals and Utilization of Wells for Snow Removal in Kanazawa City

Taking advantage of "Landscape Regulations" of 1989, actions were initiated to preserve traditional environment and form beautiful landscape (Photograph 14.2).

PHOTOGRAPH 14.2 **Renovated Kuratsuki Channel from culvert.**

From 1996 onwards, apart from the landscape, activities such as maintenance of water canals (opening of culverts, ensuring access to the water surface) (Photograph 14.3) and installation of Kama-ba (water-absorbing pit for firefighting) were carried out with disaster prevention as one of the aims.

PHOTOGRAPH 14.3 **Examples for the maintenance of feed water in Kanazawa city.**

Improvements have been made so that the well systems for highway snow melting can be used as the disaster prevention water supply base and fire extinction base, and a pumping capacity of 1 t/min or more, which is the standard of firefighting water, can be secured (Photograph 14.4). Moreover, the electricity distribution boards have been improved so that power

PHOTOGRAPH 14.4 **Emergency water supply base using snow melting system.**

can be supplied from the outside, such as through lighting power supply cars, by considering the power failure due to disasters. Development has been completed at more than 60 locations.

14.4.3 Background Leading to the Development and Maintenance of Disaster Mitigation Water Supply System

14.4.3.1 Geographical Background

Irrigation canals exist in the area of Shirakawa and Ogimachi villages since the days when the rocks were hollowed out with chisel. A major renovation was carried out for agricultural water way (common for living and agriculture) at a total length of 2.5 km from 1973 to 1978. Water from these channels was transmitted through low mountains located behind the village, resulting in the implementation of the leading gravity system.

On the other hand, irrigation channels were arranged in the form of a network in Kanazawa town since Edo era (16th century). The Kanazawa city has a history of utilizing these water channels for many purposes such as agricultural irrigation, transporting supplies, and power source for polishing rice and refining botanical oil and cleaning dyes with Yuzen Nagashi (the process of removing glues and extra dyes in a clean river) prior to the installation of culverts in modern times. To cope with accumulated snow, 200 highway snow-removal devices were installed in the city roads after 1968, which resulted in the usage of ground water.

14.4.3.2 Causes Triggering Maintenance

The Shirakawa township became well known as a result of "Discover Japan" campaign by JNR in 1974. It was selected as an important preservation district for group of traditional buildings by the Central Government in 1976, which marks the beginning of the development of Disaster Mitigation Water Supply System to protect the villages from fire.

Decision on the creation of conduits for water channels followed by revival project of Korinbo in 1985 marks the beginning for the development in Kanazawa city. The addition of disaster mitigation function to water channels and addition of disaster mitigation function to snow removal wells system are being planned and executed by taking advantage of the Great Hanshin Earthquake of 1995.

14.4.4 Process for Development Project

14.4.4.1 Development Planning

Since the board of education was the in charge during the designation of historic villages of Shirakawa and Ogimachi as important preservation districts for a group of traditional buildings by the Central Government, entire operation was carried out mainly by the board of education till the implementation of Disaster Mitigation Water Supply System. Since there are no reference case examples for gravity utilization for Disaster Mitigation Water Supply System, the current unique system was formulated for the Shirakawa village.

Department for Urban Agriculture and Urban Forestry Infrastructure Development was engaged in the planning of irrigation channel maintenance project in Kanazawa city since 1994. However, the changes were made in the original plan due to the Great Hanshin Earthquake that occurred in the following year, and water scooping pits were established in 20 places to help firefighting. Kanazawa City General Affairs Division, General Affairs Section, and the Comprehensive Disaster Prevention Office formulated the disaster recovery plan for snow removal wells, aiming at multiple Disaster Mitigation Water Supply System.

14.4.4.2 Budget Securing and Utilization of the System

Regarding the development of gravity-fed firefighting water supply system in Shirakawa township, 50% of governmental subsidy was received from the Ministry of Culture in the name of "Introducing disaster prevention facilities in important preservation districts for group of traditional buildings" from 1977 to 1981 and again in 1989. The remaining 50% was granted as annual expenditure depending on the general finances of the village. Government subsidy for developing cisterns using cannel water was received for the titled purpose of aquaculture from the Ministry of Agriculture, Forestry and Fisheries of Japan in the name of "Comprehensive maintenance project to activate agricultural villages in hilly and mountainous areas" from 1994 to 1998. In addition to the maintenance of agricultural lands, agricultural roads and regional development, a part of the subsidy was also allotted to disaster mitigation development project.

The maintenance of irrigation channels in Kanazawa city was carried out using the subsidy from the Ministry of Agriculture, Forestry and Fisheries of Japan (water environment maintenance project). The budget burden rate was 50% for

Central Government, 20% for prefectures, and 30% for cities. Moreover, subsidy toward the maintenance of environment protection facilities (government subsidy rate 1/3) from the former Environment Ministry and loans/bonds from the former Ministry of Home Affairs were utilized. The city bears the entire expenditure for installation of bridges on water channels and maintenance of disaster prevention snow removal wells.

14.4.4.3 Implementation of Development Project

In the case of Shirakawa township, it is not the regular construction sector but a single section known as Board of Education that is responsible for continuing the project development. When it comes to consensus building with the community, there were no counter opinions since it was the project desired by the entire community. As for the right to water utilization, which is an issue, permission was granted by the Central Government for transmission of water from the First Grade Rivers on the condition of temporary usage of water only in emergencies.

Demolition and narrowing of about 30 privately owned bridges, revival of water banks by stone masonry, establishment of jetty pavements, and laying underground power lines were carried out in Kanazawa city, through the agreement between the owners of private bridges, which occupy irrigation channels, and the city authorities to create ease in getting around the town following water network. As for the departments responsible for the project till 1995, "Rivers Section of Civil Engineering Division" was responsible for negotiations for revival of open channels and maintenance, and management of irrigation channels in urbanization promoting areas. The Department for Urban Agriculture and Urban Forestry Infrastructure Development was responsible for the maintenance and management of irrigation channels in urbanization restricted areas. The Roads Improvement and Management Section of Civil Engineering Division was responsible for maintenance and management of city roads. However, the Irrigation Channels and Roads Maintenance Division, which consisted of three sections above, was established in 1996 to promote a broad range of development services to eliminate vertical structure in governmental sectionalism (Fig. 14.2).

When it comes to the right to irrigation water, permission was granted for the usage of water on the condition of temporary usage of water in emergencies. It has been scheduled that an advance payment be made by the Department for Urban Agriculture and Urban Forestry Infrastructure development to the Irrigation Water Management Association constituted by farmers toward water service charges for a period of 20 years and then to abandon its ownership rights to irrigation water.

Civil Engineering Division, Roads Construction Section is responsible for the installation of highway snow melting devices, and Civil Engineering Division, Livelihood Road Maintenance Section oversees the maintenance and administration. The Civic Life Division, Civic Safety Section, takes care of the maintenance and administration of the part of disaster prevention wells.

14.4.4.4 Disaster Mitigation Activities and Operations after Development

Fire broke out thrice, that is, in 1986, 1989, and 1996, in Shirakawa township after maintenance, but the fire was controlled in all these cases. The existence of mature community in Shirakawa village can be cited as a factor behind the strong disaster prevention system. The ongoing all-round inspection for firefighting equipment being conducted twice a year by the voluntary fire brigade virtually involves one person from every household. Since 2002, the policy is to let the community

FIGURE 14.2 **Built integrated section for development in Kanazawa City.**

people to conduct water discharge exercise (Photograph 14.5) and day-to-day maintenance such as snow removal in the surroundings of the equipment, etc.

PHOTOGRAPH 14.5 **Fire drill conducted by local community people.**

When viewed from the aspect of maintenance and operating status in Kanazawa city, efforts are being made to ensure water flow throughout the year including agriculture off-season not only from the perspective of scenery and tourism, but also from the perspective of disaster mitigation. The city is providing assistance to Irrigation Water Management Association. Water scooping pits have been installed in 27 areas in 2001, and firefighting activities are being carried out using these pits. When it comes to the power that drives the pumps in snow-removal wells, it is allowed to use it only for connecting the hose in winters since the contract is restricted only to winter season when the snow removal wells are operated. In the case of power cuts in seasons other than winter, it is not possible to operate them independently without power supply from outside. In this connection, it is proposed by voluntary disaster mitigation committee to arrange for a portable power generator in the future.

14.5 PROPOSAL FOR DEVELOPMENT ACTIVITIES IN CASE STUDY AREAS

Suggestions are provided on the development program for case study areas based on the information obtained from the investigation on the actual cases in six areas[1] across the country inclusive of the two examples mentioned above. This is aimed at practicing the concepts of EWSS for Disaster Prevention.

14.5.1 Performance that Needs to be Met by Disaster Mitigation Water Supply System

As for the basic requirements from the performance aspect of Disaster Mitigation Water Supply System, it is important to develop the plan while keeping in mind the following two points, that is, securing "Fail-safe measures (Substitutability & Safety) pertaining to firefighting activities." This is done in accordance with the firefighting activity scenario in 3 stages: (Opuscitatum, 2007) initial firefighting activity (quick response) by the civilians, normal scale firefighting by public firefighting services, and activities to control the fire spread by route/line protection inclusive of supportive fire organizations, and securing "Fail-safe measures pertaining to water supply" by utilizing various water sources (Fig. 14.3).

First of all, it is important to introduce development policies after understanding area's characteristics thoroughly to ensure various firefighting activities according to the firefighting stage and facilitate several water sources.

1. Apart from the examples of maintenance of irrigation water in Shirakawa township and Kuratsuki, it includes six examples of rain harvesting such as maintenance of Toga river in Hyogo Prefecture and other rivers to support disaster prevention, maintenance of Disaster Prevention Water Supply System inclusive of underground water in Osaka Prefecture, Matsubara City, maintenance of large scale fire-fighting system using sea water in Funabashi City, Chiba Prefecture, and maintenance of "Rojison" Rain Water Harvesting System at Sumida ward, Tokyo.

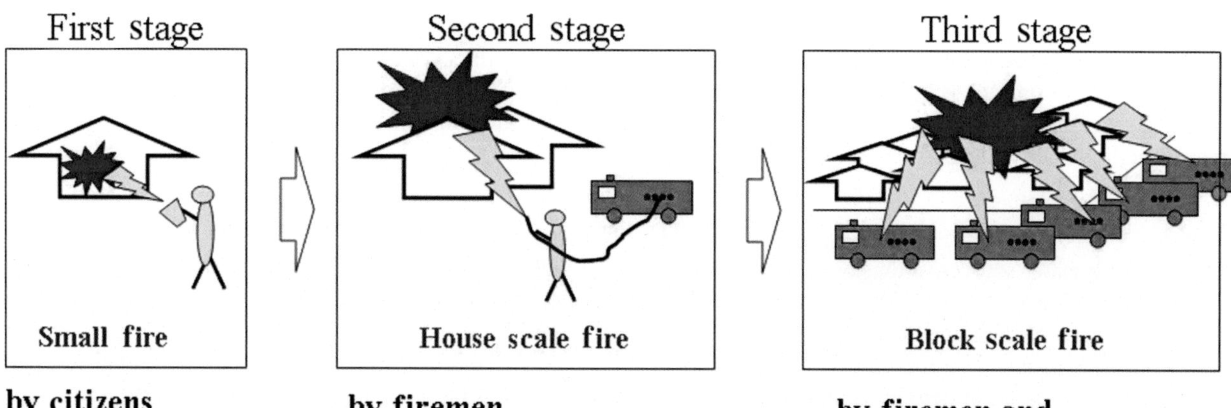

FIGURE 14.3 **Image showing activities of the three firefighting stages.**

14.5.2 Proposal for Basic Plan in Kiyomizu Area

14.5.2.1 Characteristics of Target Area

Kiyomizu area of Kyoto city is a valuable worldwide historical region. It has the Kiyomizu Temple that is registered in the world cultural heritage list. It also has various cultural assets such as government designated Sannenzaka Important Preservation District for Groups of Traditional Buildings (Photograph 14.6), etc.

PHOTOGRAPH 14.6 **Beautiful but vulnerable environment in Sannenzaka.**

It constitutes densely built traditional wooden buildings adjacent to Higashiyama and is characterized by narrow alley and bare wooden structures. In such an environment, the streets get blocked during earthquakes, making it difficult to proceed with firefighting activities. Since it becomes particularly difficult to control massive fire spread (Stage 3), it is considered an extremely important requirement to provide such an environment that makes it easy for civilians to carry out initial firefighting activities as soon as the fire breaks out.

14.5.2.2 Basic Plans Pertaining to the Development of EWSS for Disaster Mitigation

As for securing water sources, provision was made in the year 2006 for 1500-m^3 seismic-resistant rainwater storage tank underground in Kodaiji Park on the western side of Kodaiji temple. Provisions were also made for a pumping system to pump up the water of the aforementioned rainwater storage tank. A 1500-m^3 seismic resistant "gravity" pump-type water storage tank was installed in 2010 inside the premises of the Kiyomizu temple with required altitude in order to secure the substitutability of the system by complementing this feature. A natural pressurization system that does not use power is particularly effective in the event of an earthquake disaster in which a power system that relies on artificial energy can cause failures. In the case either of these two water sources is damaged, the project facilitates minimum pumping capacity in a mutually complementary manner (Fig. 14.4).

When it comes to water service pipe line inside the area, plastic polyethylene pipe was installed to minimize the damages due to earthquakes and tremors. It has been scheduled to form a network in future that will close the periphery of each city block into a loop in order to minimize the effects from ruptures.

FIGURE 14.4 CG image of operating WSS. Abbreviation: *WSS*, Water Shield System.

As for the water discharge facility, which uses this water, provision was made for fire hydrants for civilians that are used in the initial firefighting system (Stage 1) by civilians, and fire hydrants for public firefighting system (Stage 2) which is used by the firefighting team to control the fire.

In the future, it is being planned to provide Water Shield System (WSS), which is a street wall sprinkler system (currently in the planning stage), for controlling the fire crossing the city blocks (Stage 3). In the case of severe spreading fire, the system was designed in such a way that when the civilians opened the valve at the time of evacuation, water is automatically sprinkled on the wooden walls facing the street. The outer walls of the wooden structure being temporarily fire-resistant prevent the fire from spreading in the block (Fig. 14.5).

In particular, the fire hydrants for civilians can be used effectively in daily activities such as water sprinkling, which is a traditional custom for keeping down the dust on the streets in Kyoto, and watering the greenery, which ensures regular maintenance. Hence, there is no need for special emergency drills.

In order to ensure further fail-safe performance, it is currently being planned to provide a network of channels with fire cistern as a separate system, especially in the surroundings of Sannenzaka Preservation District for Important Groups of Traditional Buildings (Fig. 14.6). It is currently being planned to revive and open old river channels, such as Kikutani river, Todoroki river, etc., that have been mostly turned out into culverts to provide additional water storage tanks for circulating water regularly to complement the existing fire cisterns.

The development of intake pits, which draw shallow water in the case of open water channels, and the creation of new water spaces closer to the water surface in a safe manner ensure environmental maintenance and promote regular activities by civilians pertaining to water resources not only in emergencies, but also in daily basis. This becomes a major requirement to promote maintenance of water quality and maintenance and management of facilities at a high level.

14.5.3 Recommendations for Project Development Policies Pertaining to the Plan

Kiyomizu area of Kyoto city is a valuable worldwide historical region. It has the Kiyomizu Temple that is registered in the world cultural heritage list. It also has various cultural assets such as government designated Sannen.

14.5.3.1 Budget Securing and System Utilization Policy

As for budget securing and system utilization, efforts are being made to procure budget in the name of agricultural establishment maintenance, landscape maintenance, etc. and use a part of the amount towards the maintenance of Disaster Mitigation Water Supply System since budgetary mechanics have not been established especially for the development of EWSS for disaster prevention. Since the main usage of water resources in the project target area is not agricultural water resources, it may be important on the part of Ministry of Agriculture, Forestry and Fisheries to conduct a study on the implementation of the mechanics of water environment maintenance project. However, it is not required to consider water rights to utilize rainwater and ground water, which are originated mainly from rainwater.

Since the said area is the land of world cultural heritage, national treasure, and Important Preservation District for Groups of Traditional Buildings designated by the Nation, further aid can be expected from the cultural affairs agency in the same way as in the Shirakawa township.

Apart from this, it is necessary to procure the required budget in the name of recharge of rainwater and groundwater to utilize the subsidy system. This should be done toward the maintenance of environment protection facilities provided by the Ministry of Environment and utilization of the programs toward local governments offered by the Ministry of Internal Affairs and Communications, which are continuing till date as a result of decentralization of authority.

However, this kind of subsidy system requires private funds at a fixed rate, activities by civilians who become beneficiaries such as establishment of funds, establishment of civic associations prove indispensable not only from the aspect of funds, but also from the aspect of consensus building in civilians, and participation in planning and post maintenance operations. "Ordinance on the promotion of Kyoto civilians' participation," which was initially enforced in the year 2003 as government designated city and implementation of programs by partnership promotion office, which was established in 2001, are considered to prove effective in building a saucer of funds.

14.5.3.2 Maintenance Project Implementation Organization

The implementation of maintenance projects require cooperation from a large number of affiliated departments from the government side including fire prevention (Fire Department), rivers (Construction Bureau, etc.), groundwater (Water and Sewer Commission, Health and Welfare Department, etc.), roads (Construction Bureau, Transportation

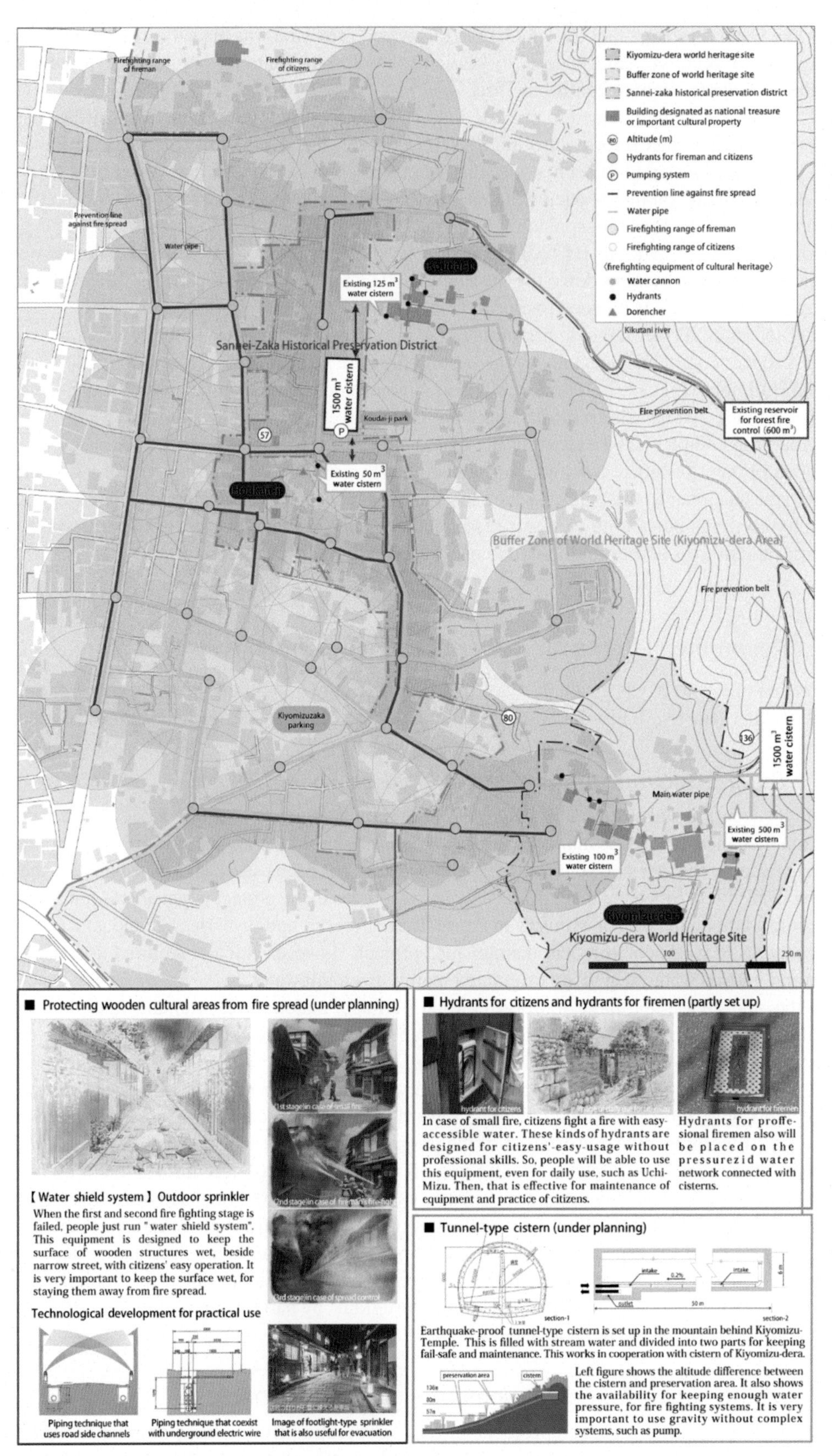

FIGURE 14.5 **Disaster mitigation water supply system development plan for Kiyomizu area.**

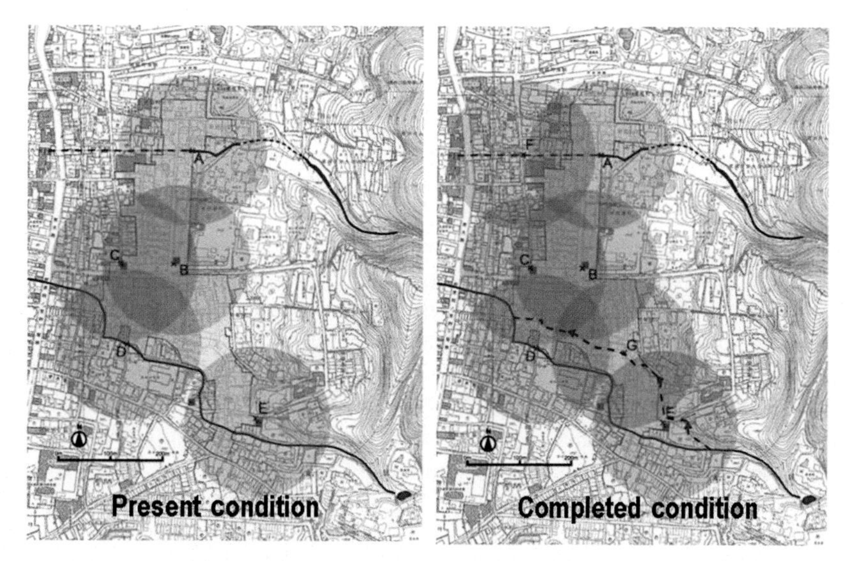

FIGURE 14.6 **Network of open water channels and fire cisterns (on the agenda).**

Authority, etc.), landscape (Urban Planning Department, etc.), environment (Environmental Bureau), and cultural assets (Culture and Civic Affairs Bureau, Board of Education, etc.). Thus, it is desirable to establish a single cross-sectional organization such as Comprehensive Planning Bureau, which plays a central role in handling the comprehensive project.

Implementation of the project in a rational and realistic manner under the cooperation of other affiliated bureaus by developing viable maintenance plans stage by stage becomes an extremely important requirement instead of completing the entire project in a broader range. Operations such as laying power lines underground, reconstruction of pavements, etc. have already been carried out in this system.

14.5.3.3 Postdevelopment Operation Policy

When it comes to posting development disaster prevention activities and operation, it becomes necessary to have a clear picture of the distribution of roles from the government side in connection with the maintenance and administration of facilities. Since the initial action taken by the civilians during disasters plays a crucial role, maintenance of vigorous community activities regularly by the civilians also becomes an important requirement for mitigating disasters just in case they occur. For this purpose, "Kiyomizu-Yasaka Disaster Prevention Water Supply System Network" was organized as the local community organization from 2012 onwards. However, instead of only using it as a self-defense organization that focuses solely on disaster mitigation activities in emergency situations, forming a local community for local activities such as preservation of water environment by the revival of local culture such as festivals becomes an important guideline for the success of the continuous maintenance project.

14.6 SUMMARY AND PROSPECTS

This document, which is quite different from the concept of earlier urban disaster mitigation aiming at making the entire city fireproof, deals with the study on the explanation of practical projects and relevant future plans on the basis of case studies and case projects. It aims to respect the culture of traditional wooden buildings and ensure "urban design that is centered around the cultural heritages" on the basis of disaster mitigation tactics to decide how to make preparations for "water environment for the people in the local area and easy to use techniques" in order to take action against multiple fire accidents during earthquakes. It further aims at contributing toward the revival of the safe and beautiful environment.

It is necessary to review the future plans and promote continuous research and activities toward the realization of "cultural city with safe and beautiful wooden buildings", but the challenge is in working toward the suggestions for comprehensive disaster mitigation project combined with the countermeasures for collapse, flooding, landslide disasters, and other disasters that occur from earthquakes.

ACKNOWLEDGMENT

The author, as a committee member and member of the Secretariat, has worked on this manuscript along with the Kyoto Fire Department, NPO for Protection of Cultural Heritages from Disaster, Institute of Disaster Mitigation for Urban Cultural Heritage and Historical Cities, Ritsumeikan University, and other organizations. This manuscript is based on the information obtained from the outcome of Research Committee for "Basic Concept of Disaster Prevention Water Resources in Kyoto" (Opuscitatum, 2017) in 2001, Report on "Research and development pertaining to the development of Environmental Water Supply System to protect the cities with wooden buildings from disasters like earthquakes" (Toki Kenzo et al., 2001) by the Ministry of Land, Infrastructure, Transport and Tourism—Construction Engineering Research Development Grant" in 2001, "Research Committee for Preservation of Cultural Heritage and Its Surrounding Area from Disasters" (Examination Committee, 2004) in 2001, "Research Committee for Planning Development of Environmental Water Supply System in Kyoto City" (Research Committee for Planning the Development of Environmental Water Supply System for Disaster Prevention in Kyoto City, 2004) in 2003, and Ministry of Land, Infrastructure, Transport and Tourism National Urban Revival Model Research Project "Revitalization of the region through the efforts to improve the disaster prevention capabilities of the region by focusing on its cultural heritage" (National Urban Revival Model Research Project Implemented, 2004) in 2004. Furthermore, all the civilians from each area, including Shirakawa village office, Ono District, Gifu Prefecture and Kanazawa City Office, and Ishikawa Prefecture, readily cooperated in the survey. I would like to express my gratitude to all the persons concerned.

REFERENCES

Examination Committee, 2004. Examination Committee (Toki Kenzo (Chairman), Masuda Kanefusa and 16 others) that protects the cultural heritage and its surrounding area from disasters: The whole concept of countermeasures to protect the cultural heritage and its surrounding area from disasters like earthquakes. Cabinet Office.

Kobe City Fire Department, 1996. Fire situation during Great Hanshin Earthquake, p. 8.

Masami, K., Sunao, N., 1999. Climate based water supply system for seismic fire. Earthquake 52, 199–212.

National Urban Revival Model Research Project Implemented, 2004. Revitalization of the region through the efforts to improve the disaster prevention capabilities of the region by focusing on cultural heritage. Ministry of Land, Infrastructure, Transport and Tourism.

Opuscitatum, 2007. Opuscitatum 2, p. 19.

Opuscitatum, 2017. Opuscitatum 2.

Research Committee for Planning the Development of Environmental Water Supply System for Disaster Prevention in Kyoto City, 2004. Report, Kyoto Fire Department, March 2004.

Research Committee on the Concept of Disaster Prevention Water Supply System, 2002. Toki Kenzo (Chairman), Kobayashi Masami (Vice Chairman), Tanaka Takeyoshi, et al. The Concept of Disaster Prevention Water Supply System in Kyoto City—Environmental Water Supply System for Disaster Prevention (The water of life) Countermeasures, pp. 12–15.

Toki Kenzo (Representative), Kobayashi Masami, Tanaka Takeyoshi and Okubo Takeyuki (Secretariat), et al., 2001. Research development pertaining to the development of environmental water supply system to protect the cities with wooden buildings from disasters like earthquakes. Report on Construction Engineering Research Development Grant in 2001. Ministry of Land, Infrastructure, Transport and Tourism, March 2001.

Chapter 15

Systematic Engineering Approaches for Ensuring Safe Roads

Ranja Bandyopadhyaya

National Institute of Technology, Patna, Bihar, India

"Let's manage safety recognizing how humans are and stop managing safety the way we wish humans were"

Alan Quilley

Development of a country lies in the development of its connectivity in the form of road infrastructure. The noble enterprise of road infrastructure development should ensure maintenance of safety standards. The road crash scenario worldwide is making it a significant manmade disaster, claiming 1 million lives and leaving 50 million grievously injured each year. The road crashes are outcomes of road user, vehicle and road environment and condition. The knowledge of safe behavior on roads for the road users and safe design of vehicles are also essential for reducing the chance of road crashes. But, the designers of the road transport system need to build and maintain a safe road infrastructure which will safe roads for the road users.

Section 15.1 discusses the worldwide scenario for road crash disaster and the global vision of reducing road crashes. Section 15.2 details the factors that make the roads unsafe. Section 15.3 provides details of the measures of safety and explains how to quantify and assess safety scenario of roads. Section 15.4 discusses the importance of systematic engineering approach to ensure safe roads. Section 15.5 summarizes the chapter and highlights the key concepts.

15.1 ROAD CRASH DISASTER—WORLDWIDE SCENARIO

Road traffic accidents constitute the highest percentage of all deaths due to unnatural causes and form a major public health concern. Worldwide, an estimated 1.2 million people are killed in road crashes each year and as many as 50 million are injured. In India, it is estimated that around 3 lakh people are killed and 80,000 injured annually. Also, fatality rates per 10,000 vehicles in India are 15–20 times higher than that in developed countries. The majority of road deaths occur among the vulnerable road users, viz. pedestrians, cyclists, and motorized two-wheeler users. The economic cost of road crashes and injuries is estimated to be 1% of the gross national product (GNP) in low-income countries, 1.5% in middle-income countries, and 2% in high-income countries. With the number of motor vehicles in use increasing globally, road accidents are projected to form the fifth biggest cause of death and injury to humanity by 2030. However, road fatality is a silent disaster attracting less mass media attention than other, less frequent types of tragedy.

Estimates of the annual number of road deaths may vary, due to problems of data collection and analysis, underreporting, and differences in interpretation. However, it is not an exaggeration to say that road crashes not only place a heavy burden on national and regional economies, but also on households.

Despite the large social and economic costs, there has been a relatively small amount of investment in road safety research and development, compared with other types of health loss. Road crashes and injuries or fatalities from them are predictable and preventable to a large extent. Thus, it is important to understand the causes of such crashes and focus on measures to systematically eliminate them so that the roads become safe for all types of users.

15.2 FACTORS THAT MAKE DRIVING ON ROADS UNSAFE

Road crashes are random events, which is the outcome of combination of many factors. Mainly the factors may be classified as human or driver factors, roadway factors, and vehicle factors. Human factors like psychological factors, visual acuity, etc. are taken into account in safe geometric design of roads. Driving is a human-driven process, and majority of road crashes are designated as an outcome of a human or driving error. The general practice for ensuring road safety is by enforcing safe driving measures for the drivers. But, the roadway factors play a vital role in road crashes, and a safe road infrastructure is to be provided to the road users to ensure safer roads.

Integrating Disaster Science and Management. http://dx.doi.org/10.1016/B978-0-12-812056-9.00015-4

The road factors are the inconsistencies in road geometric design, pavement surface conditions, presence of unwanted hazard element in roads, and presence of off shoulder hazards. The geometric design inconsistencies refer to improper design of various road elements. These include horizontal curves, vertical curves, intersection, and inadequate availability of sight distances. These also include the absence of proper road markings or misleading markings, improper signs, etc. The pavement surface conditions which affect safety performance of a road may be permanent defects like pavement distresses and cracks, or may be the pavement surface condition due to seasonal variation, for example, wet pavements which influence skidding action of vehicles. The presence of hazard elements on roads includes the presence of unwanted access points and the presence of some unwanted road furniture which block the visibility of drivers. The presence of off shoulder hazards are the presence of objects beyond road shoulders which may or may not directly affect the causing of road crashes but affect the severity of crashes. Along with these factors, another major element that affects the causing of road crashes is the visibility issues. Visibility issue may be due to street lighting or due to glare from opposing stream of vehicles. The presence of fog, smog, or heavy rains also affects visibility and in turn affect safety scenario of roads.

Combination of various factors discussed above may lead to crashes. Combinations of similar factors may or may not lead to crashes and also these crashes will vary in severity or intensity. To address the issue of road crash disaster, first it is important to quantify safety and then identify elements leading to crashes and the locations which are unsafe. Addressing problem elements and designing or adopting appropriate safety measures for problem road locations which are also known as hotspots or blackspots will lead to reduced chance of causing crashes.

15.3 QUANTIFYING AND ASSESSING SAFETY SCENARIO OF ROADS

Road safety refers to methods and measures that are used to reduce risks of injury, death, and harm to drivers, passengers, and pedestrians. Thus, the assessment of safety scenario of roads is important for designing safety improvement of the roads. A combination of various road features, viz. road geometric design, pavement conditions, road markings, speed limit restrictions, etc. along with the road user population and driving characteristics determine the safety scenario of a road. To address the issues of road safety hazards, it is important to measure the present safety scenario of roads and focus improvement efforts on these road locations. The measures of safety performance of roads may be classified as either proactive or reactive measures.

15.3.1 Proactive Road Safety Assessment

Proactive measure or road safety or road safety audit (RSA) for assessing safety performance of roads include identification of road elements and making focused efforts to improve these individual elements. The road elements include geometric design elements, road markings, pavement conditions, etc. which do not conform to the safety standards. However, this requires the list of standard features for which the roads are to be manually audited. Also, it does not consider the contribution of traffic in road crashes and the actual number of crashes the road locations are witnessing. The elements considered in RSA can broadly be classified into road geometric design elements, pavement conditions, road markings, posting of speed limits, and off-shoulder hazards.

The RSA has a checklist of features prepared by the experts for individual classes of roads. The checklist is exhaustive checks consistency for general features, alignment details, intersection details, special or vulnerable road user facility, lighting conditions, signs, pavement markings and delineation, presence of physical objects, and operation and traffic management and control facilities. The general feature includes checking of drainage facilities, access property and delineation, natural features, roadside hazards, and speed management facilities provided. Visibility of and at intersections is an important feature that is checked for in RSA as the intersections allow conflicting movements. This is more important when the intersection is uncontrolled or there is a channelization or roundabout at the intersection. The RSA calls for extensive audit of roads by experts.

15.3.2 Reactive Road Safety Assessment

The reactive road safety strategy attempts to act reactively addressing to safety problems in road segments which experience high number of crashes. The measures of safety for reactive road safety strategy take into account actual crash occurrence history. There are various safety measures available in literature. The road locations are ranked using these safety measures or metrics, and improvement strategy is designed for top ranked locations. The choice of appropriate safety metric is much debated.

The total number of crashes in a year or crash frequency (CF), or average annual crashes (averaged over a period of 3 years) is mostly used as a fundamental indicator of "safety" of a road location. CF at a particular road location or site *i* (CF_i) can be calculated using Eq. (15.1) and can be expressed as a number of crashes per year:

$$CF_i = \frac{\text{Number of crashes at site } i}{\text{Period of recording (duration in years)}} \tag{15.1}$$

To identify hotspots among a group of sites, CF for each site is calculated. The sites are then ranked in decreasing order of CF, that is, the site with highest CF gets rank 1 and the rank increases with decreasing CF. The top ranked locations are designated as hotspots or candidate locations for further investigation and engineering improvement. It is also used in the evaluation and estimation of effectiveness of safety improvement methods (Deacon et al., 1975; Hauer et al., 2002). This measure is easily quantifiable and cost effective (Hauer et al., 2002), but as the crashes are random events, CF has inherent bias of regressing to the mean value. Also, as Simon Washington et al. (2014) pointed out, both CF and severity dimensions are needed to fully assess the safety performance of roads and suggested the use of a combined frequency severity model for safety assessment of roads. Moreover, as crashes of lower severity levels are underreported, the use of CF as a measure of safety performance may lead to false indication of hazardous locations.

The CF and severity dimensions are made available by reliably weighting crash frequencies and converting all crashes to property damage-only crash equivalents (PDOEs) using comprehensive societal unit crash injury costs. The equivalent property damage-only (EPDO) considers crashes of all levels of severity occurring at a particular site for assessing safety condition of the site (see Table 15.1). The total accident cost for each site is calculated in terms of its EPDO cost. Thus, the $EPDO_i$ for a particular site *i* can be calculated using

$$EPDO_i = CF_{PDO} + \left(a \times CF_{Minor} + b \times CF_{Major} + c \times CF_{Fatal}\right) N_{inj} \tag{15.2}$$

The CF_{PDO}, CF_{Minor}, CF_{Major}, and CF_{Fatal} are the frequencies of property damage crash, minor injury, major injury, and fatal injury in a crash, respectively, and N_{inj} is the number of respective injuries for the total analysis period. The sites are then ranked in decreasing order of EPDO, that is, the site with highest EPDO gets rank 1. Moreover, the issue of under-reporting associated with minor injury and property damage-only crashes is addressed by weighing major injury crashes heavily (Blincoe et al., 2002; Sen et al., 2010; Simon Washington et al., 2014).

The traffic exposure of roads is an important parameter for comparing road safety scenario. While comparing number of crashes or economic burden of crashes across road locations, the traffic exposure of roads is not taken into account. Thus, some researchers suggested the use of crash rate (CR) or crash per unit exposure, measured in terms of traffic volume of the road (annual average daily traffic or AADT) as given in

$$CR_i = \frac{CF_i}{AADT_i} \tag{15.3}$$

But CR may be high even when both traffic volume and crashes are low for a site (Hauer et al., 2002) and is therefore not a preferred measure for assessing safety. Also, the effect of road and driving environment is not considered while comparing CR along road locations.

Moreover, crashes are rare random events influenced by vehicles and driving environment. The inherent safety of a location is influenced by the location characteristics and the traffic exposure of the location. This is taken into account for in the Bayesian technique of prioritization of unsafe road locations.

The Bayesian method uses expected crashes (obtained from crash prediction models or SPFs) of a site for hotspot identification and also help to eliminate the regression-to-mean bias associated with other methods discussed earlier (Elvik, 2008; Hauer and Persaud, 1984; Persaud and Lyon, 2007). Expected crashes are usually used for site ranking in Bayesian techniques—both empirical Bayes (EB) and full Bayes (FB) techniques. The Bayesian techniques combine prior

TABLE 15.1 Property Damage Equivalency Factor

	Calculated for India Scenario (Sen et al., 2010)	Calculated for Washington (Blincoe et al., 2002)
Property damage crashes	1	1
Minor injury	1.16	11
Incapacitating injury	15	949
Fatality	33	1330

information with the current information to calculate the expected safety of a site. The first step in using the Bayesian techniques is to develop a crash prediction model (CPM) or safety performance function (SPF) to estimate the expected number of crashes in the analysis period at locations with traffic volumes and other characteristics similar to the one being analyzed. The crash estimates are then combined with the count of crashes to obtain a better estimate of the expected number of crashes (Abbess et al., 1981). The EB method uses negative binomial (NB) or Poisson Gamma accident prediction models that give point estimates of expected mean and variance. Use of EB method for road hotspot detection started in the early 1980s (Persaud et al., 2010).

The EB procedure with SPF as a function of traffic exposure is illustrated with examples by Hauer et al. (2001). The detailed methodology of the development of CPMs or SPFs using negative binomial model and its application in the empirical Bayesian procedure of hotspot identification is detailed here. The probability distribution of number of accidents occurring in a road segment or intersection may be assumed to follow Poisson distribution. This assumption can be made as the number of accidents at a given time period has discrete nonnegative outcomes and each outcome does not depend on the previous outcomes. Probability of ith segment of experiencing n_i accidents per year can be written as

$$P(n_i) = \frac{e^{-\lambda_i}(\lambda_i)^{n_i}}{n_i!},\ (n \geq 0), \tag{15.4}$$

The Poisson parameter λ_i is the expected number of accidents per year at the ith segment which is obtained from the SPFs. The expected number of accidents per year is a function of the crash explanatory variables. The most common relationship between explanatory variables and the Poisson parameter can be given as a log-linear model as shown in

$$E[n_i] = \lambda_i = e([\beta][x]) \tag{15.5}$$

The Poisson assumption of crash data restricts mean and variance to be equal. This equality may not hold in actual practice, and the parameter estimate is biased if corrective measures are not taken. The data is overdispersed when variance is greater than the mean and underdispersed when variance is lesser than the mean. The dispersion in crash data can arise from a variety of reasons. The primary reason in many studies is that variables influencing the Poisson rate may have been omitted from the regression. This can be addressed by including a gamma-distributed error term in the Poisson parameter λ_i. The relationship between explanatory variables and the expected number of accidents per year λ_i, at the ith segment, can be written as

$$E[n_i] = \lambda_i = e([\beta][x] + s_i) \tag{15.6}$$

The gamma-distributed error term ε_i has a mean 1 and variance α^2, where α is called the overdispersion parameter. The addition of this term allows variance to differ from the mean and the variance can be written as

$$\mathrm{Var}[n_i] = E[n_i] + \alpha E[n_i]^2 \tag{15.7}$$

With this assumption, the probability of the ith segment experiencing n_i accidents per year can be said to follow negative binomial (NB) distribution. The Poisson regression model is a limiting model of NB model when α approaches 0. The NB distribution of probability of the ith segment experiencing n_i accidents per year can be written as

$$P(n_i) = \frac{\Gamma((1/\alpha)+n_i)}{\Gamma(1/\alpha)n_i!}\left(\frac{(1/\alpha)}{(1/\alpha)+\lambda_i}\right)^{1/\alpha}\left(\frac{\lambda_i}{(1/\alpha)+\lambda_i}\right)^{n_i} \tag{15.8}$$

The $\Gamma(\cdot)$ is the gamma function. The model is estimated using the maximum likelihood estimation (MLE) technique. The technique attempts to maximize the likelihood that the observed data comes from the hypothesized probability distribution by adjusting the model coefficients. The likelihood function is the joint probability of the function at the observation data points and can be written as

$$L(\beta, X, \alpha) = \prod_i \frac{\Gamma((1/\alpha)+n_i)}{\Gamma(1/\alpha)n_i!}\left(\frac{(1/\alpha)}{(1/\alpha)+e([\beta][x])+s_i}\right)^{1/\alpha}\left(\frac{\lambda_i}{(1/\alpha)+\lambda_i}\right)^{e([\beta][X]+s_i)} \tag{15.9}$$

Given the crash data and the explanatory variables, the parameters β and α can be obtained by maximizing the likelihood function given in Eq. (15.9). The parameters can be estimated with crash data and explanatory variable using any standard software.

Once the estimates of the parameters β and α are obtained, expected crashes λ_i at a site can be determined. The EB estimate of crashes with choice of negative binomially distributed accidents and gamma-distributed accident means can be written as

$$\mathrm{EB} = w \times E(n_i) + (1-w) \times \mathrm{crash}_{\mathrm{count}} \tag{15.10}$$

The weight factor w in the EB estimate of crashes of a site can be written as

$$w = \frac{1}{\left[1 + 1/\alpha \times E(n_i)\right]} \tag{15.11}$$

The t-test is useful to verify the inclusion of each explanatory variable in the model. The individual models are tested for their overall goodness-of-fit. The overall goodness-of-fit of the model is tested with the chi-square test. The test is used to assess whether the inclusion of explanatory variables has improved the performance of the model. The proposed model with explanatory variables is compared against a null model with no explanatory variables. The chi-square test statistic can be written as

$$\chi^2 = -2\left[\text{LL}(0) - \text{LL}(\beta)\right] \tag{15.12}$$

The LL(β) is the log-likelihood at convergence with the parameter vector β and LL(0) is the initial log-likelihood (with all parameters set to zero). The chi-square test statistic is chi-square distributed with degrees of freedom equal to the difference between the number of parameters in the restricted and unrestricted models. The null hypothesis is that there is no significant improvement in log-likelihood due to the inclusion of the explanatory variables.

The ρ^2 statistic is another measure to assess the goodness-of-fit and is specifically designed for the logistic regression models. It is used in a manner similar to the use of R^2 in linear regression analysis and can be calculated using

$$\rho^2 = 1 - \frac{\text{LL}(\beta)}{\text{LL}(0)} \tag{15.13}$$

The LL(β) is the log-likelihood at convergence with the parameter vector β and LL(0) is the initial log-likelihood (with all parameters set to zero).

The sites are then ranked in decreasing order of EB, that is, the site with highest EB gets rank 1. The top ranked locations are designated as hotspots. Many researchers suggested the use of improved variants of the Bayesian method for detecting high crash locations (Cheng and Wang, 2010a; Hauer et al., 2004; Miranda-Monero et al., 2007). Lord and Park (2008) examined the effect of traditional NB model with fixed dispersion parameter and generalized NB model with time-varying dispersion parameters on the performance of EB model. They found that the model with varying dispersion parameter provides better statistical performance than the generalized NB model. However, the use of generalized NB model affects the proper identification of hotspots and thus appropriate selection of functional form in the EB method is essential to avoid false identifications.

Researchers proposed that prioritization of unsafe locations on the basis of the location's potential for improvement. This measure ensures maximum benefits in improvement projects in terms of reduction of crash costs. The earliest such attempt is known as accident reduction potential (ARP) for hotspot identification and ranking. The ARP for a site is the potential for improvement for that site. The ARP for a site i, ARP_i, can be calculated using

$$\text{ARP}_i = \text{EB}_i - E(n_i) \tag{15.14}$$

The sites are then ranked in decreasing order of ARP, that is, the site with highest ARP gets rank 1. The top ranked locations are designated as hotspots.

Many high crash location prioritization matrices are discussed in this section, which in many cases are not very effective in identifying exhaustive list of high crash locations and also may have some false identification. Few researchers suggested the use of multiple criteria in combination to identify high crash locations (Cheng and Wang, 2010b; Cheng and Jia, 2014).

It is difficult to demarcate a location as truly safe or truly unsafe. It is reasonable to assume that all sites have a certain element of hazard and the extent of hazard determines the criticality of the site. Thus, Kononov and Allery (2003) suggested classifying sites based on level of service of safety (LOSS) for total crash count and LOSS for fatal and injury only crash counts for estimating the level of safety of a site. The concept is similar to the level of service of operation provided by any traffic facility and provides a frame of reference for deciding the level of safety of a site. The LOSS gives qualitative measures to characterize safety of a road segment with reference to its expected performance predicted by SPF.

Researchers also suggested grouping of road locations or sites into fuzzy groups in which each of the sites will belong to each group with certain belongingness which depends on the underlying safety of the site. A site is said to belong to a fuzzy group when the belongingness of that site to that particular group is much higher compared to the belongingness of that site to other groups (Bandyopadhyaya and Mitra, 2015). Also, in this method, multiple established Hotspot identification criteria may be used in combination for grouping sites into various safety levels.

This section details the various basic concepts of road safety measures. The next section discusses the importance of systematic approach toward road safety.

15.4 SYSTEMATICALLY ENGINEERING APPROACH TO ENSURE SAFE ROADS

In most countries, the traffic safety is ensured by the traffic police of the country, who lack the knowledge of benefits of systematic engineering approaches for road safety improvement. Traditionally, traffic safety improvement measures focus on law enforcement, such as speed management, reduced drunk-driving, seatbelt use, etc. These have positive impact in reducing injury severities, but these measures have been found to have a diminishing rate of return after a point. A safe systematic engineering approach is increasingly being recommended, which assumes that road users cannot be expected to be perfect and abide by rules and constraints of the system and crashes will occur. Thus, the designers of the road and road transport system need to accept and share the responsibility of safety of the system (Bandyopadhyaya and Mitra, 2015).

The primary systematic approach is extensive audit for roads for safety elements by road safety experts. The elements audited are then rectified to enhance safety performance of roads. However, this is a time-consuming process and requires involvement of experts.

The advanced systematic engineering approach tries to make roads safer, within budgetary constraints. It identifies locations which are inherently unsafe and focus improvement efforts toward them. To identify such unsafe road locations (also known as hotspots), it is practically impossible to travel and thoroughly examine the entire stretch. A more practical approach is scanning of the entire road network to identify a list of sites with inherent safety problems. Many standard hotspot identification methods are discussed in Section 15.3. After that, a *cutoff* is fixed, say top 2%–10% to mark them as hotspots or candidate sites for further investigation. This cutoff is decided based on the availability of resources and funds.

The hotspot identification methods or HSID methods should be very sound and rigorous, should be able to identify an exhaustive list of sites with safety problems, and minimize errors of false identifications. The choice of an appropriate method is still debatable but the use of multiple method or metric in combination is advisable.

15.5 SUMMARY AND HIGHLIGHTS

Road crashes are random events caused by many factors in combination. These may be road-related factors, driving error-related factors, or may be due to the fault of vehicles. As the process is human-driven, it is difficult to enforce and ensure complete safe driving. The driver safety rule enforcements reduce the risk of high severity crashes, but the road system should be designed to accommodate driving errors. Thus, to ensure road safety, systematic engineering approach is essential. The primary systematic approach for ensuring road safety is RSA. The advanced systematic engineering approach tries to make roads safer, within budgetary constraints by identifying crash hotspots and focusing on improvement in these road locations.

REFERENCES

Abbess, C., Jarrett, D., Wright, C.C., 1981. Accidents at blackspots: estimating the effectiveness of remedial treatment, with special reference to the regression-to-the-mean effect. Traffic Engineering and Control 22 (10), 1981.

Bandyopadhyaya, R., Mitra, S., 2015. Fuzzy cluster based method of hotspot detection with limited information. J. Transp. Saf. Secur., 307–323.

Blincoe, L., Seay, A., Zaloshnja, E., Miller, T., Romano, E., Luchter, S., et al., 2002. The Economic Impact of Motor Vehicle Crashes, 2000 (No. DOT HS 809 446). National Highway Traffic Safety Administration, Washington, DC.

Cheng, W., Jia, X., 2014. Exploring an alternative method of hazardous location identification: using accident count and accident reduction potential jointly. J. Transp. Saf. Secur. 7 (1), 40–55.

Cheng, W., Wang, X., 2010a. Evaluation of the joint use of accident count and accident reduction potential to identify hotspot. Presented at the 89th Annual Meeting of the Transportation Research Board. Washington, DC.

Cheng, Wang, X., 2010b. Evaluation of the joint use of accident count and accident reduction potential to identify hotspot. Presented at the 89th Annual Meeting of the Transportation Research Board. Washington, DC.

Deacon, J.A., Zegeer, V., Deen, R., 1975. Identification of hazardous rural highway locations. Transp. Res. Rec. 543, 16–33.

Elvik, R., 2008. The predictive validity of empirical Bayes estimates of road safety. Accid. Anal. Prev. 40 (6), 1964–1969.

Hauer, E., Persaud, B., 1984. Problem of identifying hazardous locations using accident data. Transp. Res. Rec. 975, 36–43.

Hauer, E., Harwood, D.W., Council, F.M., Griffith, M.S., 2001. Estimating safety by empirical Bayes method: a tutorial. Transportation Research Record 1784, 126–131.

Hauer, E., Kononov, J., Griffith, M.S., Allery, B., 2002. Screening the road network for sites with promise. Transportation Research Record 1784. TRB, National Research Council, Washington, DC, pp. 27–32.

Hauer, E., Allery, B., Kononov, J., Griffith, M., 2004. How best to rank sites with promise. Transportation Research Record 1897. TRB, National Research Council, Washington, DC, pp. 48–54.

Kononov, J., Allery, B., 2003. Level of service of safety: conceptual blueprint and analytical framework. Transportation Research Record 1840. TRB, National Research Council, Washington, DC, pp. 57–63.

Lord, D., Park, P.Y.J., 2008. Investigating the effects of the fixed and varying dispersion parameters of Poisson–gamma models on empirical Bayes estimates. Accid. Anal. Prev. 40, 1441–1457.

Miranda-Monero, L., Labbe, A., Fu, L., 2007. Bayesian multiple testing procedures for hotspot identification. Accid. Anal. Prev. 37, 1192–1201.

Persaud, B., Lyon, C., 2007. Empirical Bayes before–after safety studies: lessons learned from two decades of experience and future directions. Accid. Anal. Prev. 39 (3), 546–555.

Persaud, B., Lan, B., Lyon, C., Bhim, R., 2010. Comparison of empirical Bayes and full Bayes approaches for before–after road safety evaluations. Accid. Anal. Prev. 42, 38–43.

Sen, A.K., Tiwari, G., Upadhay, V., 2010. Estimating marginal external costs of transport in Delhi. Transp. Policy 17, 27–37.

Simon Washington, Md., Mazharul Haque, Oh, J., Lee, D., 2014. Applying quantile regression for modeling equivalent property damage only crashes to identify accident blackspots. Accid. Anal. Prev. 66, 136–146.

FURTHER READING

Transport Research Centre, OECD & International Transport Forum, 2008. Towards Zero: Ambitious Road Safety Target and Safe System Approach.

Part III

Analysis and Resilience

Chapter 16

Big Data Analytics and Social Media in Disaster Management

Joice K. Joseph**,†, Karunakaran Akhil Dev**,†, A.P. Pradeepkumar*,†, Mahesh Mohan**

*University of Kerala, Trivandrum, India

**Mahatma Gandhi University, Kottayam, Kerala, India

†CHAERT (Centre for Humanitarian Assistance and Emergency Response Training), Kottayam, Kerala, India

16.1 INTRODUCTION

Digital data is overwhelming the computing power of the world. But this data provides a plethora of opportunities too. This is especially true when it comes to the case of disaster management. In recent years, there has been a spurt of interest and much research into the role of social media (SM) in disaster management. Rather than as a means of communicating hazard, risk, and disaster perceptions and warnings, SM-generated data like Facebook posts and Twitter feeds are sought to be analyzed to arrive at the scale and spread analysis of disasters. This data is voluminous, different, and when used in new ways to monitor and manage a disaster, qualifies for the definition of Big Data and its five Vs (volume, velocity, variability, veracity, and variety). The data shadows on the internet of Facebook likes, Flipkart orders, Google searches, Research Index citations, Tumblr pictures, You Tube videos—all add up to billions of bytes of information, most of which are geotagged, and which, if tweaked properly can bring up hidden, but critical geographical patterns of crowd responses. Big Data snared in the geoweb can be of critical importance in big disaster events, if its analysis is mellowed by domain experts. Ushahidi, the Red Cross, the United States Geological Survey (USGS), all have had some success with SM analysis, for quite some time now. With mobile phone and 3G/4G services permeating to the lowest ranks of society in the developing countries of Asia and Africa, which are the most prone to disasters, SM analytics with Big Data tools need urgent attention as a disaster response, management, and mitigation tool. The information shared on disaster events in the SM, blog data on disasters, FB posts, and Twitter feed analysis, and the geographical real-time spread of SM messages as a monitor of disaster spread are techniques that could alter the way disaster information has so far been accessed and analyzed. This immense source of data, generated every microsecond, is the next great area of research and entrepreneurship, for the benefit of society through a process that is intimately interlinked with how the current population on Earth communicates and thinks. The Haiti earthquake could be thought of as the pathbreaking event which lead to the realization that crowd-sourced and social mode of data generation would deliver better results. Crowd-sourced data, using global positioning system (GPS)-enabled phones and the Internet, was widely used for the generation of maps to support emergency response efforts. But the lacuna was the inability to process the huge volumes of data and the United Nations realized that the potential of Big Data to be fully tapped would need technologically advanced inputs. This led to the project "Creative Technologies in support of Emergency and Humanitarian Response: Spacebased Information for Crowd-sourced Mapping" in 2011 (OOSA, 2011). Public behavior response analysis in disaster events utilizing visual analytics of microblog data is an innovative approach to crisis management spearheaded in part by the University of Stuttgart (Chae et al., 2014). SM is revolutionizing how people communicate not only in their daily lives, but also during perturbing events like natural disasters. In the May 2013, National Science Foundation (NSF) and Japan Science and Technology Agency (JST) Joint Workshop on Examining and Prioritizing Collaborative Research Opportunities in Big Data and Disaster Management Research (Arlington, VA, USA) discussions focused on the potential benefits of Big Data for disaster management as well as the Big Data research challenges arising from disaster management (Pu and Kitsuregawa, 2013). The five C characteristics of SM, viz. collectivity, connectedness, completeness, clarity, and collaboration make it attractive from a disaster management perspective. The three phases of disaster management—prevention, preparedness, and response & recovery—could each make use of the power of SM and associated Big Data. For instance, during the 2010 Haiti earthquake, Ushahidi Haiti was deployed to crowd source data. SM tools and mobile text messages were analyzed, and actionable information gleaned out to support the coordination efforts of the humanitarian relief services. But the incomplete, complex, and context-dependent

Integrating Disaster Science and Management. http://dx.doi.org/10.1016/B978-0-12-812056-9.00016-6

information certainly requires Big Data analysis capabilities to be made explicable. Often such an analysis will require a trained person, who could also suggest on the remediation efforts.

Every year, a number of natural disasters strike across the globe, causing large-scale loss of lives and infrastructure damage. The effective emergency management of these scenarios is a challenge for the modern world. Whether it is a natural or a manmade disaster, disaster managers always need critical data sets in order to respond quickly and efficiently (Alexander, 2014). These Big Data sets captured by mainly multipurpose sensor networks such as satellites, smart phones, Un-manned aerial vehicles (UAVs), and SM can help in all four phases of disaster management (National Research Council, 2006; Middleton et al., 2014). Although not much can be done to prevent natural disasters, comprehensive Big Data systems such as those developed by agencies like Palantir are being used to crack down on anthropogenic disasters such as those caused by terrorism. Companies such as Terra Seismic carry out near real-time monitoring of satellite data, and ecological factors which they say allow them to forecast earthquakes anywhere in the world with 90% accuracy (Marr, 2015). The usefulness of the analysis of Big Data and SM in the management of disasters is now discussed, and the major applications of Big Data analytics in SM in the four phases of disaster management are outlined here.

16.2 SOCIAL MEDIA

SM is the media for social interaction as a super-set beyond social communication (Wikipedia; https://en.wikipedia.org/wiki/Social_media). It includes web-based and mobile technologies used to turn communication into an interactive dialog. It has platforms that enable the interactive web by engaging users to participate in, comment on, and create content as a means of communicating with other users and the public. SM is defined as a group of Internet-based applications that build on the ideological and technological foundations of Web 2.0 and that allow the creation and exchange of user-generated content (Velve and Zlateva, 2012). SM technologies are characterized as being interactive and require users to generate, edit, or share information (Beneito-Montagut et al., 2013; Surowiecki, 2004). It includes a wide variety of content formats including texts, audio, video, photographs, portable document files, power points, and GPS coordinates. It lets interaction between one or many platforms through social sharing. It provides multistake engagement irrespective of gender, to create and comment on SM network. It swiftly facilitates information dissemination in an enhanced manner to many participants by extending engagement by creating real-time online events, extending online interactions offline, or augmenting live events online (Velve and Zlateva, 2012).

16.2.1 Types of SM

According to Kaplan and Haenlein (2010), there are six different types of SM: collaborative projects; blogs and microblogs; content communities; social-networking sites; virtual game worlds; and virtual social worlds. The applied technologies include blogs, picture sharing, vlogs, wall-postings, email, instant messaging, music sharing, crowdsourcing, and voice over IP (Manso and Manso, 2012). The top 10 SM platforms and their ownerships are as follows: Facebook and WhatsApp (Mark Zuckerberg), YouTube (Google), Twitter (Jack Dorsey), LinkedIn (Jeff Weiner), Pinterest (Ben Silbermann, Paul Sciarra, and Evan Sharp), Google Plus+ (Sundar Pichai, CEO), Tumblr, Inc. (Jerry Yang, David Filo), Instagram (Mark Zuckerberg, Dustin Moskovitz, Eduardo Saverin, Andrew McCollum, Chris Hughes), Reddit (Alexis Ohanian, Steve Huffman), VK (Doraview Limited), and Flickr (Jerry Yang, David Filo), (www.ebizmba.com/articles/social-networking-websites). Of these, Facebook and Twitter posts have prime importance as sources of Big Data in disaster situations.

16.3 NATURAL DISASTERS AND SM

The popularity, efficiency, and ease of use of SM have led to its pervasive use in disaster management (Denis et al., 2014; Hiltz et al., 2014; Hughes, 2014; Ngamassi et al., 2016; Yates and Paquette, 2011). SM services and networking stay active, while the conventional communication facilities may have been hindered. They help and enable individuals and organizations to collaborate in mutually beneficial ways in all stages of emergency management: mitigation, preparedness, and response and recovery (Velve and Zlateva, 2012). It is used as the primary news source at the time of natural disaster and emergency situations. It provides information about pre- and postdisaster scenarios through Internet or through messages. It also generates awareness amongst the affected communities, as well as helps in generating volunteers and donors for the emergency services. It becomes a connective link between displaced individuals, families, and friends. It provides information about aid and other resource centers available for the affected area. According to research reviews, there are four primary ways citizens use SM technologies: family and friend's communication; situation updates; situational/supplemental awareness; services access assistance. Before disaster, it helps communities to prepare, during disaster SM helps to communicate, and after disaster it helps communities come together again and enhance capabilities to build better recovery efforts and distribution of assistance.

16.4 BIG DATA AND BIG DATA ANALYTICS

Big Data is an evolving term that refers to data sets or mixture of data sets whose size, complexity, and rate of growth make them difficult to be analyzed by traditional technologies and tools (Kailser et al., 2013). Big Data is often characterized by the quantity of generated and stored data (volume), the type and nature of the data (variety), the speed at which the data is generated and processed (velocity), inconsistency of the data set (variability), and the quality of captured data (veracity). These 5 Vs aptly represent its true nature. Big Data is huge and is measured in petabytes (1024 terabytes) or exabytes (1024 petabytes). Our everyday online activities with digital devices (e.g. call data records, transaction data, access logs, banking records), user-generated data on the Internet activities (e.g. emails, SMS, blogs, comments, FB comments, Google+ posts, and Twitter tweets), various sensing technologies (e.g. remote sensing, network sensing, and participatory sensing), various small data sets (e.g. the Small Data Lab at the United Nations University), public-related, and crowd-sourced data sets are the major sources of Big Data. Big Data analytics involves making "sense" out of large volumes of varied data that in its raw form lacks a data model to define what each element means in the context of the others and is the process of examining large data sets to uncover hidden patterns and unknown correlations (Oracle White Paper, 2013). The four phases of a Big Data management system are data generation, data acquisition, data storage, and data analytics (Gallagher, 2013). Log files and Web Crawlers are the two important Big Data collection methods that are widely used today. The ultimate goal of Big Data analytics is to extract useful values, suggest conclusions, and support decision-making process (Fig. 16.1).

16.5 THE BIG DATA IN SM

The frequency of SM in sharing day-to-day information regarding all aspects of our life is ever increasing. According to the infographic Web site "Visual Capitalist" in every Internet minute there are 701,389 logins on Facebook, 1389 Uber rides, 527,760 photos shared on Snapchat, 120+ new LinkedIn accounts, 347,222 tweets on Twitter, 28,194 new posts to Instagram, 972,222 Tinder swipes, 1.04 million vine loops, 2.78 million video views on YouTube, and 20.8 million messages on WhatsApp (Visual Capitalist, 2016). The nature of SM itself is highly dynamic, multilingual, and geographically distributed (Mazumdar et al., 2012).

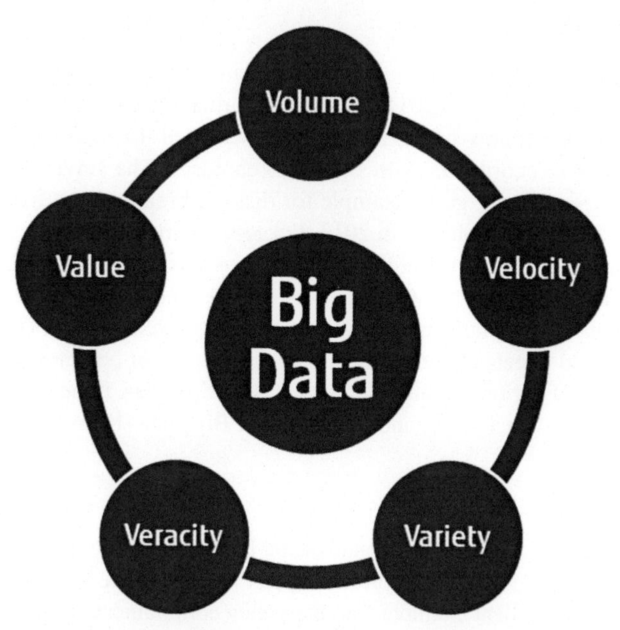

FIGURE 16.1 Five versus of Big Data. *Source: Martin (2015).*

16.5.1 Examples of Data Mining Software

Some examples of Data Mining Software are: Orange, Weka, rattle GUI, Apache mahout, SCaViS, RapidMiner, R, ML-Flex, Databionic, ESOM Tools, Natural Language Toolkit, SenticNet API, ELKI, UIMA, KNIME, Chemicalize.org, Vowpal Wabbit, GUI Octave, CMSR, Data miner, Mlpy, MALLET, Shogun, Scikit-Learn, LIBSVM, LIBLINEAR, Lattice Miner, Dlib, Jubatus, KEEL, Gnome-datamine-tools, Alteryx Project Edition, OpenNN, ADaM, ROSETTA, ADaMSoft, Anaconda, yooreeka, AstroML, StreamDM, jHepWork, TraMiner, ARMiner, arlues, CLUTO, and TANAGRA (Top 50 Data Mining Software's accessed from Predictive Analytics Today, 2016).

16.6 DATA MINING OF SM FOR DISASTER MANAGEMENT

The huge data sources of SM can be mined for enhancing communications before, during, and after a disaster (Houston et al., 2014). Much of the work on developing Big Data systems to help with disaster relief began in the wake of the 2010 Haiti earthquake and the 2011 Tohuku earthquake and tsunami. The United States and Japan started joint research initiatives in this area, especially, for earthquakes and tsunami. SM can also be used for peer-to-peer backchannel communications (users of SM can both receive and post messages) that increase the social capacity of information generation and dissemination (Xiao et al., 2015). Studies of SM data in disaster management have been reported only recently since real-time services such as Twitter have been introduced (Vieweg et al., 2010). According to the study of Lindsay (2010), SM was ranked as the fourth most popular source for retrieving emergency information. The efficient Big Data integration, aggregation, and visualization particularly from SM sources will assist emergency managers to optimize the situational awareness and in better decision-making procedure.

16.6.1 Mitigation Phase

Databases such as cyber GIS plays an important role in disaster mitigation efforts by producing various hazard maps to support different activities, operations, and decision-making processes. Social networks with the massive popularity are widely used as an intelligent "geosensor" network to detect and monitor extreme emergencies (Sutton et al., 2008). Multipurpose sensor networks and intelligent transportation systems (ITS) are now available for early warning and information dissemination, which basically works on the platforms of the Big Data analytics. Big Data from SM with larger coverage of people and target regions may lead to more accurate human response and prediction. These Big Data modeling are appropriate for preparing for the possibly cascading hazards to mitigate them. New sensor data technologies such as UAVs ("drones" can provide very high-resolution 2D and 3D imageries) and high-resolution satellite imagery can be paired with crowd sourcing to provide much accurate hazard zonation maps, risk modeling of natural disasters to the vulnerable communities.

16.6.2 Preparedness Phase

Studies show that during the immediate aftermath of the 2010 Haiti earthquake, information about the quake was first released through SM sources (Keim and Noji, 2011). Critical information can also be broadcast to the public, such as through SMS so that any preventable follow-up hazards are disenchanted. In addition, real-time analytics can assist aid organizations and first responders to coordinate with other stakeholders. The Environmental Extremes and Population (MDEEP) project in Bangladesh was a very good example for disaster preparedness from the mobile data early warning systems during the occurrence of cyclone Mahasen in 2013. Large-scale initiatives such as the UN Global Pulse Program was able to analyze tweets by using keywords about humanitarian crisis as this can contribute effective early warnings and the term "citizens as sensors" is now arising (UN Global Pulse, 2014). Another initiative by USGS has begun monitoring tweets for mentions of an earthquake on a platform called Twitter earthquake detector (TED) and this can filter place, time, and keywords to gather geo-located tweets about shaking—sometimes it can be detected faster by looking at SM (Earle et al., 2012). Disaster preparedness of various natural disasters such as droughts, floods, wild fires, etc. are very much possible with satellite image interpretation which is yet another source of Big Data (Kumar et al., 2014; Verbesselt et al., 2006).

16.6.3 Response Phase

During disaster response, timely action is a matter of life and death. In recent years, SM has emerged as an important source of multiway communication for information dissemination for disaster response. The Code Maroon system developed in Twitter by Texas A&M University is an excellent example for emergency communication that was used to disseminate warning messages to thousands of people shortly after the onset of emergencies (Villarreal and Sigman, 2010). Many organizations

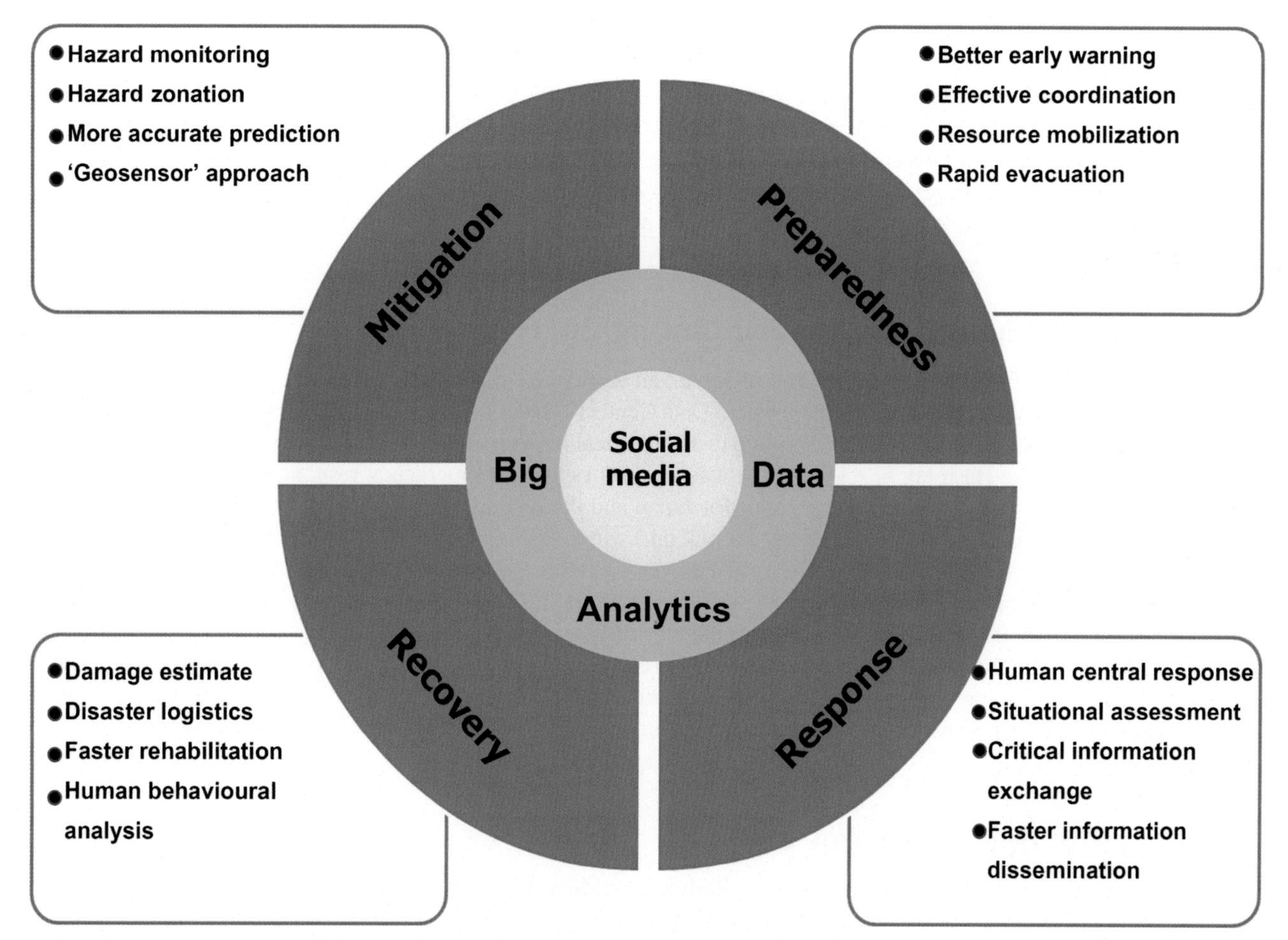

FIGURE 16.2 Schematic of Big Data analytics and social media in disaster management cycle.

working in the field of disaster relief have been encouraged to participate in the SM by broadcasting their own needs and perceptions of the disaster situation. SM analytics can also play a significant role in passive and active situational information generation, and situational awareness is a key factor in crisis response. Analysis of Big Data in real time from various SM can substantially enhance various disaster response aspects such as helping emergency response personnel to identify areas that need the most urgent attention, identification of critical resources, and in selecting most effective response methods.

16.6.4 Recovery Phase

Disaster recovery systems should support flexible communication processes, and it is important to explore how the most active participants in SM networks facilitate recovery processes in online communities. We can use SM to accelerate the damage estimate process by transmitting images of damaged structures such as dams, levees, bridges, and buildings taken from cell phones (White et al., 2009). During disaster recovery phase, we can estimate human behavior by Big Data analytics from various social network services and fast evacuation of victims trapped in isolated areas. The Big Data analysis can elucidate useful information for recovery procedures about volunteer coordination and logistics during the crisis. Technologies such as spatial video monitoring (SVM) and traditional satellite imagery analysis can be much quicker for damage assessments than deploying staff to the field (Lue et al., 2014) (Fig. 16.2).

16.7 CASE STUDIES IN BIG DATA IN EMERGENCY DISASTER MANAGEMENT

16.7.1 Case 1: Chennai Floods in India, 2015

In southern India during November and December 2015 heavy rainfall generated by the northeast monsoon lead to severe flooding in the cities of Chennai and Pondicherry. A small office desk at the National Disaster Response Force (NDRF) headquarters has picked up over 1000 strands of vital information on SM and the Internet and was posting responses to those requesting help. SOS calls were made online in the wake of the flood crisis. NDRF Chief OP Singh was getting hourly updates on his personal mobile phone about the SM response of the force and has also instructed the team to ensure that "each and every" tweet, email, WhatsApp message, SMS, or phone call is acknowledged. The 12,000-strong force had received and responded to 339 tweets on its official handle "@NDRFHQ" or through "#NDRF" along with 501 emails, 613 phone calls, and 13 messages on WhatsApp. A further 200 such messages were being replied to. The role of SM in this particular crisis was remarkable and the authorities are now considering to explore Big Data analytics from SM in its full extent.

16.7.2 Case 2: Tohoku Earthquake and Tsunami, 2011

On March 11, 2011 at Tohoku on the east coast of Japan, an earthquake of magnitude 9.0 was reported by USGS. Over 15,000 people died according to National Geophysical Data Center. The communication via cell phones broke down. Many of the affected took to SM to connect with others and to avail critical information from the authorities involved in the rescue and recovery efforts. The Hash tag #j _ j _help me was used on twitter for emergency responders to identify the people who were in need of rescue. The first letter j stands for Japan and the second one is for Jishin, which translates to "earthquake" in Japanese (White, 2014). Google tweeted a link on Twitter to its Google Person Finder tool which allows people to search for missing persons. This tweet was re-tweeted over 9000 times in the wake of the Tohoku disaster (Smith, 2011; White, 2014). After the earthquake/tsunami, people crowded into the Apple store in Tokyo. The free WiFi that was offered in the store allowed people to view critical information on the disaster over USTREAM, a video streaming application, and to contact loved ones on Twitter, Facebook, and other SM Web sites (White, 2014). The Mayor of Minami-Souma, Sakurai Katsunobu, posted a video on YouTube following the Tohoku disaster, pleadeding for help from the Japanese government for supply of food and fuel. The video drew international attention and was viewed almost half a million times on YouTube since it was posted.

16.7.3 Case 3: Typhoon Morakot, 2009

The typhoon caused damage to china, Japan, and the Philippines. Mostly Taiwan was affected. PPT, a popular online bulletin board system in Taiwan, was used by the people of Taiwan to post requests for volunteers and donations after Typhoon Morakot. The Association of Digital Culture Taiwan established an unofficial Morakot Online Disaster Report Center (Huang et al., 2010). They then asked people in the areas affected by the typhoon to monitor Twitter and other SM sites and post information on damage and people in need of assistance to the Online Disaster Report Center. The Taiwanese authorities set up a couple of Web sites for disaster relief based on Twitter and made them synchronize with other microblogs of individual bloggers and they established a canopy Web site incorporating other emergency management departments in the country (Mathew, 2005).

16.8 BIG DATA ANALYTICS CHALLENGES IN DISASTER MANAGEMENT

High speed and volume of data from social networking services pose a great challenge for the Big Data analytics for emergency management. Quality and services of Big Data is another challenge in addition to the continuous emerging technologies, privacy, security, and regulatory considerations and because of these challenges further research and international collaborations are very much required in this emerging field. Not all Big Data sources are free and public. While Facebook has an open Application Programming Interface (API) to access its data, access to twitter's data stream can be prohibitively costly for many and some other Big Data sources are free to view but not for downloading. Overestimation, noisy, false data are some other challenges in Big Data analytics for disaster management.

16.9 SUMMARY

Big Data from SM sources will become increasingly available for disaster management due to the advances of many kinds of capable sensors and their considerable growth. The combination of powerful computer systems and networks that include sensors, smart phones, and cyber-physical arrangements is creating massive data streams that can help decision-makers during disasters. Big Data analytics can play a role in monitoring of natural hazards, vulnerability assessment of human societies to the particular disaster risk, impact assessment, and in evaluating recovery and rehabilitation activities; and strengthening community coping capacity. Big Data analytics especially from SM sources can solve traditional disaster management problems in a better way and has the capability to come up with the right solutions to complex goals in disaster management. Hence, the Big Data analytics of SM in disaster management as a new technology has a number of pitfalls and perils that needs detailed scrutiny from international experts from related fields. For effective and efficient use of Big Data in disaster management local participation and guidance by emergency management professionals should be ensured.

ACKNOWLEDGMENTS

This work was partially supported by Paristhithi Poshini research fellowship from the Environment and Climatic Change Department, Government of Kerala, India to the first author and Rajiv Gandhi National Fellowship (RGNF) by University Grants Commission (UGC), India to the second author. The anonymous reviewers are thanked for comments that enhanced the quality of this chapter. The editors are thanked for the invitation to contribute this paper.

REFERENCES

Alexander, D.E., 2014. Social media in disaster risk reduction and crisis management. Sci. Eng. Ethics 20, 717–733.

Beneito-Montagut, R., Anson, S., Shaw, D., Brewster, C., 2013. Governmental social media use for emergency communication. In: Proceedings of the 10th International Conference on Information Systems for Crisis Response and Management. Baden-Baden, Germany.

Chae, J., Thom, D., Jang, Y., Kim, S.Y., Ertl, T., Ebert, D.S., 2014. Public behavior response analysis in disaster events utilizing visual analytics of microblog data. Comput. Graphics 38, 51–60.

Denis, L., Palen, L., Anderson, J., 2014. Mastering social media: an analysis of Jefferson County's communications during the 2013 Colorado floods. In: Proceedings of the 11th International Conference on Information Systems for Crisis Response and Management (ISCRAM). State College, PA, USA.

Earle, P.S., Bowden, D.C., Guy, M., 2012. Twitter earthquake detection: earthquake monitoring in a social world. Annals of Geophysics, http://www.ra.ethz.ch/CDStore/www2010/www/p851.pdf (accessed 18.12.16).

Gallagher, F., 2013. The Big Data value chain. http://fraysen.blogspot.sg/2012/06/big-data-value-chain.html (accessed 18.12.16).

Hiltz, S., Kushma, J., Plotnick, L., 2014. Use of social media by U.S. public sector emergency managers: barriers and wish lists. In: Proceedings of the 11th International Conference on Information Systems for Crisis Response and Management (ISCRAM). State College, PA, USA.

Houston, J.B., Hawthorne, J., Perreault, M.F., Park, E.H., Hode, M.G., Halliwell, M.R., McGowen, T., Davis, R., Vaid, S., McElderry, J.A., Griffith, S.A., 2014. Social media and disasters: a functional framework for social media use in disaster planning, response, and research. Disastersdoi: 10.1111/disa.12092.

Huang, C.M., Chan, E., Hyder, A.A., 2010. Web 2.0 and internet social networking: a new tool for disaster management? Lessons from Taiwan. BMC Medical Informatics and Decision Making 10, 57, http://www.biomedcentral.com/1472-6947/10/57.

Huffington, 2011. Post article "Google Launches Japan Earthquake Person Finder To Help Find Missing People" by Catharine Smith, March 11, 2011.

Hughes, A., 2014. Participatory design for the social media needs of emergency public information officers. In: Proceedings of the 11th International Conference on Information Systems for Crisis Response and Management (ISCRAM). State College, PA.

Kailser, S., Armour, F., Espinosa, A.J., Money, W., 2013. Big Data: issues and challenges moving forward. 46th Hawaii International Conference on System Sciencespp. 995–1004.

Kaplan, A.M., Haenlein, M., 2010. Users of the world, unite! The challenges and opportunities of social media. Business Horizons 53 (1), 59–68.

Keim, M.E., Noji, E., 2011. Emergent use of social media: a new age of opportunity for disaster resilience. Am. J. Disaster Med. 6 (1), 47–54.

Kumar, S.V., Peters-Lidard, C.D., Mocko, D., Reichle, R., Liu, Y., Arsenault, K.R., Xia, Y., Ek, M., Riggs, G., Livneh, B., Cosh, M., 2014. Assimilation of remotely sensed soil moisture and snow depth retrievals for drought estimation. J. Hydrometeorol. 15, 2446–2469.

Lindsay, B.R., 2010. Social media and disasters: current uses, future options and policy considerations. J. Curr. Issues Media Telecommun. 2 (4), 287–297.

Lue, E., Wilson, J.P., Curtis, A., 2014. Conducting disaster damage assessments with Spatial Video, experts, and citizens. Applied Geography 52, 46–54.

Manso, M., Manso, B., 2012. The role of social media in crisis: a European holistic approach to the adoption of online and mobile communications in crisis response and search and rescue efforts. 17th ICCRTS-Operationalizing C2 Agility. Fairfax, VA, June 19–21, 2012.

Marr, B., 2015. Using Big Data in a crisis: Nepal earthquake. Forbes. http://www.forbes.com/sites/bernardmarr/2015/04/28/nepal-earthquake-using-big-data-in-a-crisis/#1851d402532f (accessed 11.12.16).

Martin, T., 2015. The 5 Vs of Big Data and the Fujitsu M10, accessed 20 December 2016. http://blog.global.fujitsu.com/index.php/the-5-vs-of-big-data-and-the-fujitsu-m10/.

Mathew, D., 2005. Information technology and public health management of disasters—a model for South Asian countries. Prehospital Disaster Med. 20 (1), 54–60.

Mazumdar, S., Ciravegna, F, Gentile, A.L., Lanfranchi, V., 2012. Visualising context and hierarchy in social media. International Workshop on Intelligent Exploration of Semantic Data (IESD'2012) at EKAW 2012. Galway City, Ireland.

Middleton, S.E., Middleton, L., Modafferi, S., 2014. Real-time crisis mapping of natural disasters using social media. IEEE Intelligent Systems 29, 9–17.

National Research Council, 2006. The Indian Ocean Tsunami Disaster: Implications for U.S. and Global Disaster Reduction and Preparedness: Summary of the June 21, 2005 Workshop of the Disasters Roundtable. The National Academies Press, Washington, DC.

Ngamassi, L., Ramakrishnan, T., Rahman, S., 2016. Examining the Role of Social Media in Disaster Management from an Attribution Theory Perspective. In: Proc. of the ISCRAM. Conference – Rio de Janeiro, Brazil, May 2016.

Oracle White Paper, 2013. Big Data analytics—advanced analytics in oracle database. www.oracle.com/.../database/.../advanced-analytics/bigdataanalyticswpoaa-1930891.p (accessed 18.12.06).

OOSA 2011 Creative Technologies in support of Emergency and Humanitarian. Response - Space-based Information for Crowdsource Mapping. Technical cooperation concept note OOSA/FRT/CN/2006-7 (http://www.un-spider.org/sites/default/files/Crowdsource%20Mapping.pdf).

Pu C. and Kitsuregawa M. ST/NSF Joint Workshop Report on Big Data and Disaster Management, Technical Report No. GIT-CERCS-13-09; Georgia Institute of Technology, CERCS, 28p (https://grait-dm.gatech.edu/wp-content/uploads/2014/03/BigDataAndDisaster-v34.pdf).

Surowiecki, J., 2004. The Wisdom of Crowds. Why the Many are Smarter Than the Few and How Collective Wisdom Shapes Business, Economies, Societies and Nations. Doubleday, New York, NY.

Smith C. Google launches Japan Earthquake Person Finder to help find missing people. Huffington Post article, March 11, 2011.

Sutton, J., Palen, L., Shklovski, I., 2008. Backchannels on the front lines: emergent uses of social media in the 2007 Southern California wildfires. In: Proceedings of the 5th International ISCRAM Conference. Washington, DC.

Top 50 Data Mining Software's, 2016. http://www.predictiveanalyticstoday.com/top-free-data-mining-software/Social_media (accessed 20.12.16), http://en.wikipedia.org/wiki/Social_media (accessed 15.12.16).

UN Global Pulse, 2014. Mining Indonesian tweets to understand food price crises, methods paper. www.unglobalpulse.org/.../Global-Pulse-Mining-Indonesian-Tweets-Food-Price-Crise (accessed 18.12.16).

Velve, D., Zlateva, P., 2012. Use of social media in natural disaster management. International Proceedings of Economic Development and Research 39, 41–45.

Verbesselt, J., Jönsson, P., Lhermitte, S., Van Aardt, J., Coppin, P., 2006. Evaluating satellite and climate data-derived indices as fire risk indicators in savanna ecosystems. IEEE Transactions on Geoscience and Remote Sensing 44 (6), 1622–1632.

Vieweg, S., Hughes, A.L., Starbird, K., Palen, L., 2010. Microblogging during two natural hazards events: what twitter may contribute to situational awareness. In: Proceedings of the 28th International Conference on Human Factors in Computing Systems, CHI 2010. Atlanta, GA, USA, April 10–15, 2010. ACM, New York, NY, pp. 1079–1088.

Villarreal, S., Sigman, A., 2010. Explosion at Texas A&M Chemistry Annex Building. http://www.kbtx.com/home/headlines/93421299.html. Retrieved 1 March 2015.

Visual Capitalist, 2016. What happens in an internet minute in 2016? http://www.visualcapitalist.com/what-happens-internet-minute-2016/ (accessed 18.12.16).

White, E.T., 2014. The Application of Social Media in Disasters. International Institute of Global Resilience, Bethesda, MD, USA.

White, C., Plotnik, L., Kushma, J., 2009. An online social network for emergency management. International J.Emergency Manage. 6 (3–4), 369–382.

Xiao, Y., Huang, Q., Wu, K., 2015. Understanding social media data for disaster management. Natural Hazards 79, 1663–1679. doi: 10.1007/s11069-015-1918-0.

Yates, D., Paquette, S., 2011. Emergency knowledge management and social media technologies: a case study of the 2010 Haitian earthquake. Int. J. Inf. Manage. 31, 6–13.

FURTHER READING

Chennai Floods, 2015. National disaster response force uses social media to reach out to people. Tamil Nadu, Press Trust of India. Updated: December 04, 2015 18:26 IST. http://www.ndtv.com/tamil-nadu-news/chennai-floods-ndrf-uses-social-media-to-reach-out-to-people-1251104 (accessed 18.12.16).

Ebizmba, 2016. Top Big Data mining softwares. http://www.ebizmba.com/articles/social-networking-websites (accessed 20.12.16).

Lue, E., Wilson, J.P., Curtis, A., 2014. Conducting disaster damage assessments with Spatial Video, experts, and citizens. Applied Geography 52, 46–54. doi: 10.1016/j.apgeog.2014.04.014.

Chapter 17

Risk Assessment and Reduction Measures in Landslide and Flash Flood-Prone Areas: A Case of Southern Thailand (Nakhon Si Thammarat Province)

Indrajit Pal*, Pongpaiboon Tularug*, Sujoy Kumar Jana, Dilip Kumar Pal****

**Disaster Preparedness, Mitigation and Management (DPMM), Asian Institute of Technology, Thailand*

***Department of Surveying & Land Studies, The Papua New Guinea University of Technology, Papua New Guinea*

17.1 INTRODUCTION AND BACKGROUND OF STUDY AREA

Sichon District in Nakhon Si Thammarat Province is in the southwest of Thailand and surrounded by a high mountain on the west and plain area on the eastern side (Fig. 17.1). The mountain bedrock formation is granite and has high potential for landslide hazard. Furthermore, Sichon district is also prone to flash floods due to heavy rainfall or even for a moderate rainfall situation as well. The primary livelihoods in this area mostly come from agriculture activities such as rice farming and orchard plantation. Therefore, most of the people are trying to change the land-use pattern for better livelihoods without visualizing the repercussions (Pal, 2015). For example, some of the communities changed the forest area to agriculture land by cutting slopes of the mountains, which may trigger landslides during heavy and moderate rains due to unstable slopes (Pramojanee et al., 1997). The extensive floods in 2011 also affected the district and caused massive landslides. Apart from the private properties, lots of government facilities such as roads, bridges, small dams, and drainage system were also destroyed. The total cost of damages for the entire district due to 2011 floods was around 320 million baht (approximately $9.3 million).

Nakhon Si Thammarat is the second largest province in the Southern Thailand. The Nakhon Si Thammarat mountain range divides the area into three parts: mountain range in the middle, plain area at the eastern coast, and plain area in the west.

The study area is "Thepparat" Tambon (subdistrict) covering around 81.198 km^2 at 8°86′N and 99°77′E on the north of Nakhon Si Thammarat province and southwest of Sichon district (Fig. 17.2). Thepparat Tambon is mountainous with a maximum elevation of 1340 m above mean sea level and consists of Sam Thep hill, Teng hill, and Youn Thao hill, which create natural drainage canals, such as Tha Thon canal, Sam Thep canal, Phean canal, and Pean canal. The Tambon (subdistrict) consists of 15 villages with a total population of 7971 persons (3937males and 4034 females) and a total of 2567 households.

17.1.1 Climate and Overview of Rainfall

Northeast and southwest monsoons, especially during October–December, cast a significant influence on the climate of Nakhon Si Thammarat with an average annual rainfall of 2500 mL/year (Thai Meteorological Department, 2015). Rainy season starts from October to December in every year. However, in 2011, the monsoon came with a heavy rainfall that lasted 5–7 days, which led to landslides and flash floods on March 23–31, 2011.

Fig. 17.3 shows the rainfall pattern in Nakhon Si Thammarat province from 2002 to 2014. It also shows that in 2011, the rate of total rainfall was 4177.9 mm, which was much higher than the provincial average rainfall in the last 30 years, which was 2388.6 mm

Integrating Disaster Science and Management. http://dx.doi.org/10.1016/B978-0-12-812056-9.00017-8

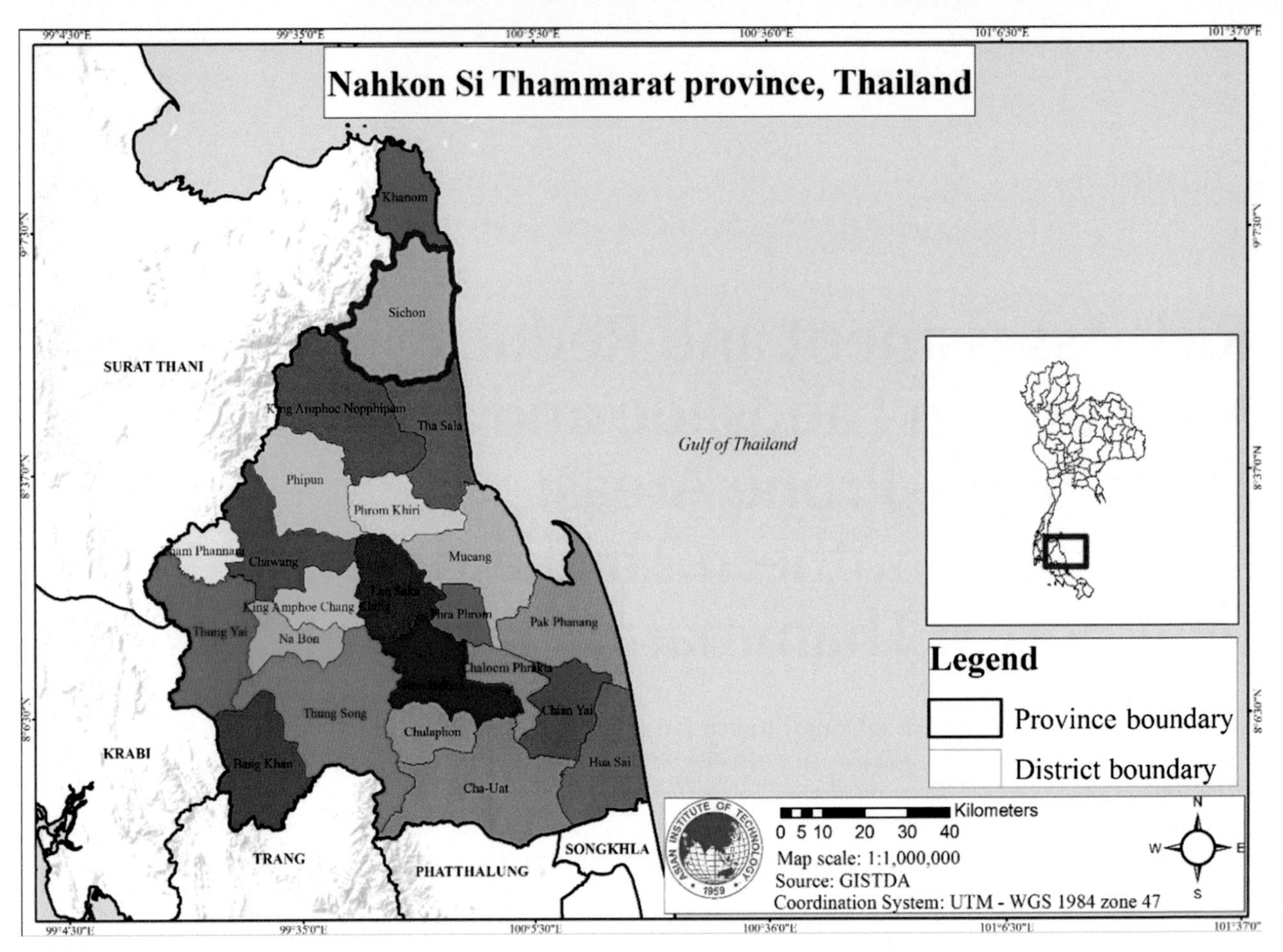

FIGURE 17.1 **Nahkon Si Thammarat province, Thailand.**

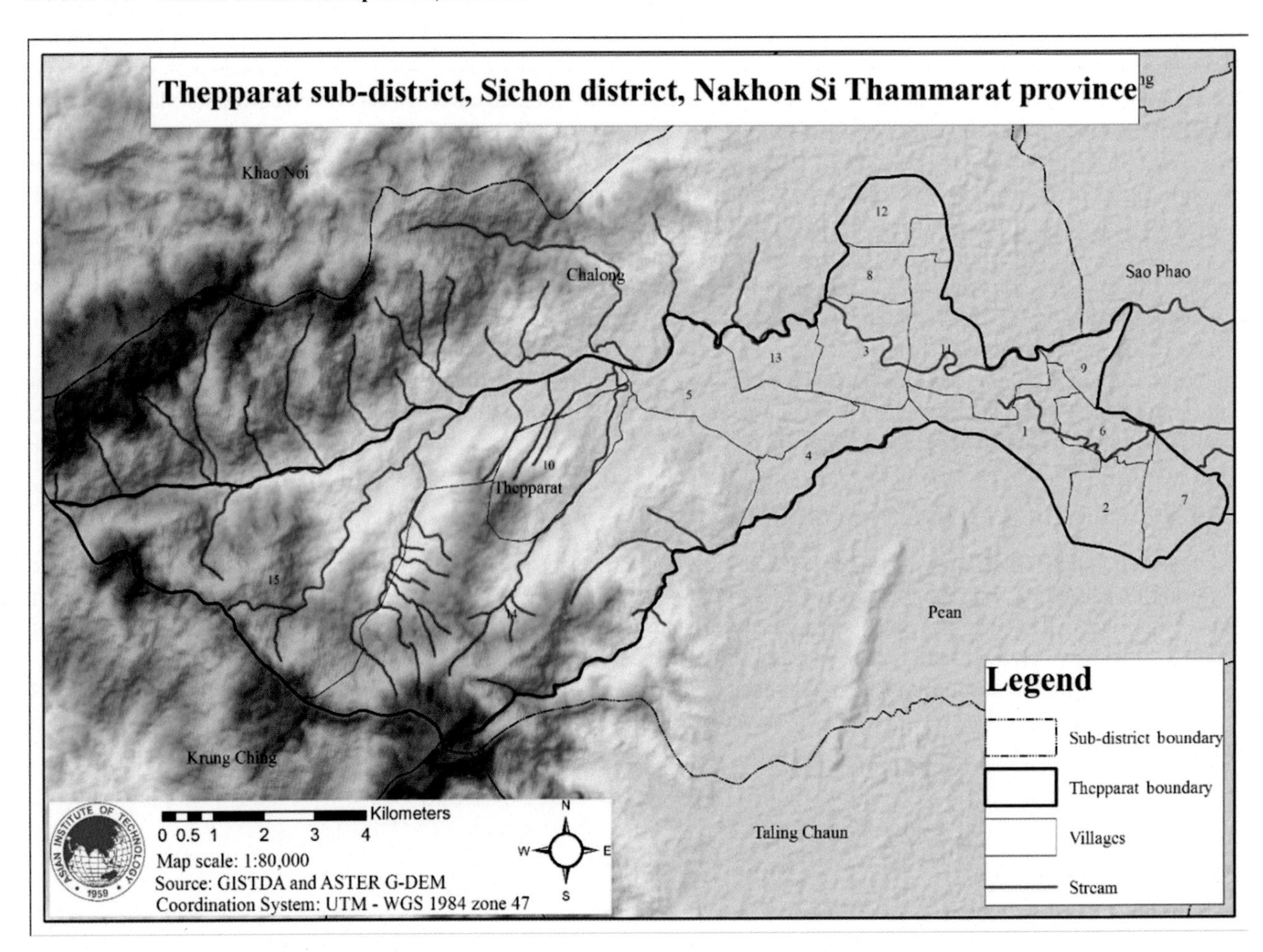

FIGURE 17.2 **Thepparat administration.**

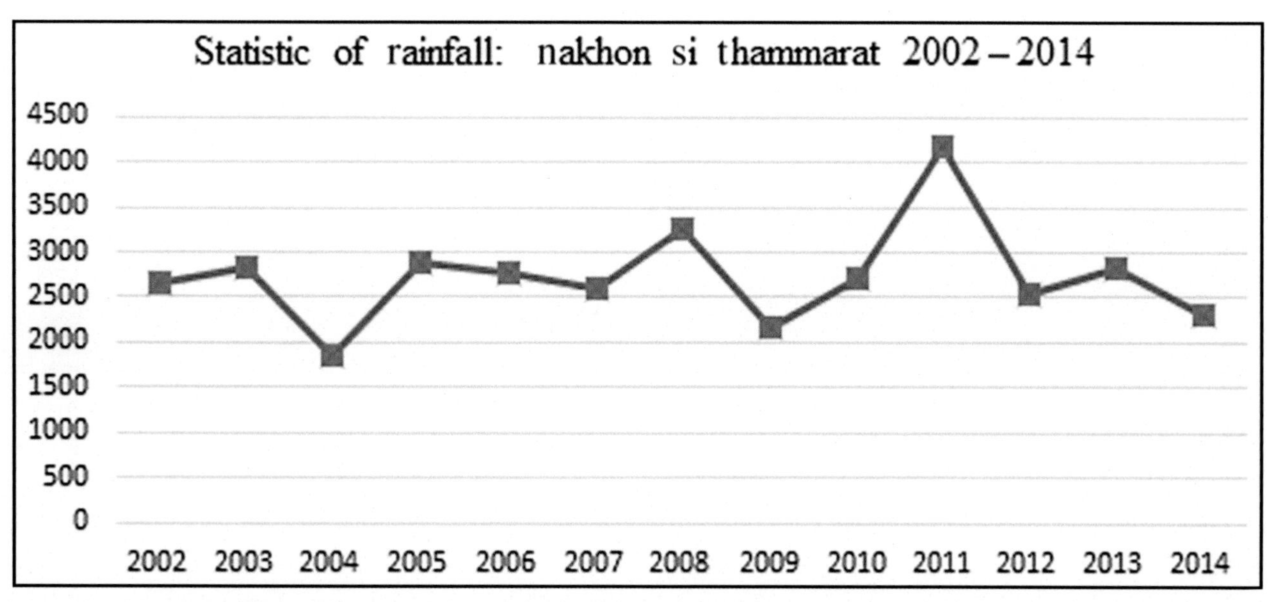

FIGURE 17.3 **Overall cumulative annual rainfall in Nakhon Si Thammarat from 2002 to 2014.** *Source*: Thai Meteorological Department, 2015. Annual Weather Summary over Thailand in 2015, Ministry of Information and Communication Technology.

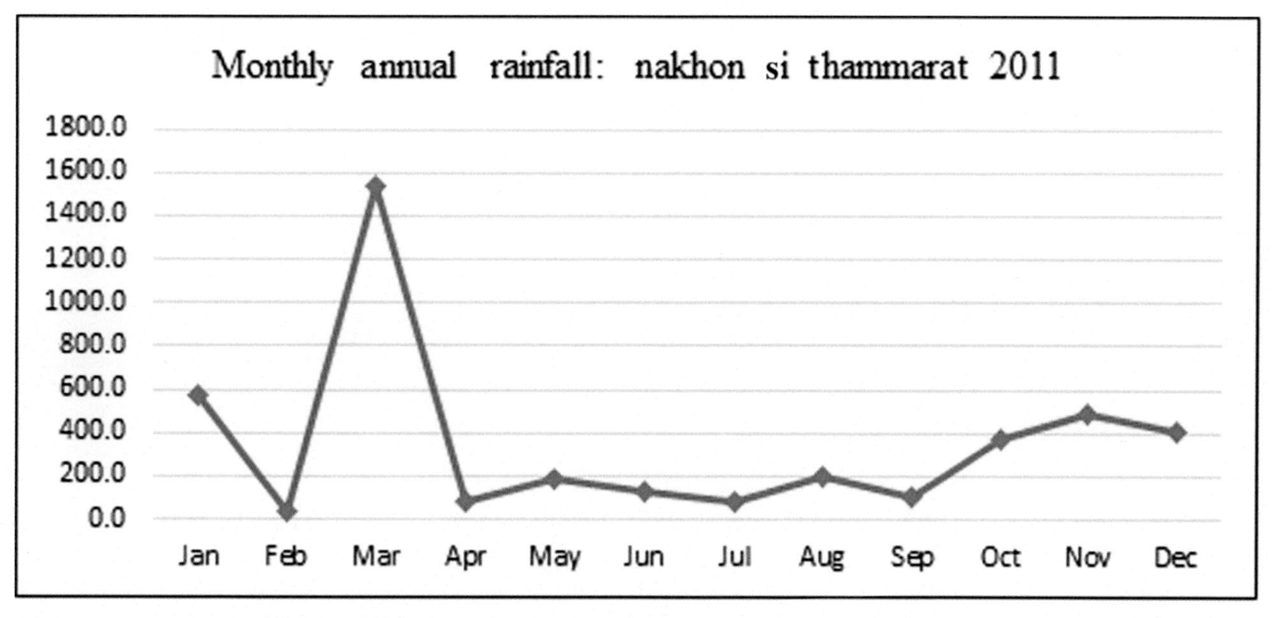

FIGURE 17.4 **Monthly annual rainfall in Nakhon Si Thammarat in 2011.** *Source*: Thai Meteorological Department, 2015. Annual Weather Summary over Thailand in 2015, Ministry of Information and Communication Technology.

(1976–2005). There was 1543.7 mm rainfall in March 2011 with the precipitation of 446 mL per day on March 31, 2011 (Fig. 17.4).

17.2 HAZARDS, VULNERABILITY, AND RISK OF NAKHON SI THAMMARAT

Nakhon Si Thammarat province located on the west coast of the Gulf of Thailand. The province experiences tropical monsoon climate with an annual rainfall of 1800–2200 mm. Landslides and flash flood hazards in Nakhon Si Thammarat are very closely associated with monsoon and typhoon where heavy rainfall can cause slope failure. Heavy rainfall combined with poor building construction practices and deforestations results into landslides and flash floods. Poor communities are generally affected most by the disasters because of the ineffective warning dissemination (Fig. 17.5).

FIGURE 17.5 **Before landslide and flash floods event in 2007 (left) and after event in 2011 (right).** *(Adapted from: Google Earth, Data source: CNES/Astrium SPOT 6 and SPOT 7 Resolution 50 cm.)*

TABLE 17.1 The Historical Landslide and Flash Floods in Nakhon Si Thammarat Province

Date	Hazard	Location	Losses and Damages
25–26 October 1962	Tropical storm Harriet can cause storm surge	Landfall at Laem Talumphuk, Pak Phanang District Nakhon Si Thammarat	900 people died and more than 10,000 people rendered homeless
22 November 1988	Landslides and debris flow	Ban Kathun Nuea, Phipun District and Ban Khiri Wong, Lan Saka District, Nakhon Si Thammarat	230 people died and damages worth 1 billion baht
29 November 1993	A tropical depression can cause flooding	Nakhon Si Thammarat	23 people died and damages 1.3 billion baht
3 November 1997	Tropical storm Linda	Nakhon Si Thammarat	164 people died
October–December 2010	Flooding	Manay provinces in Southern Thailand	80 people died and damages estimated up to 54 billion baht.
July 2011–January 2012	Flooding	65 provinces	815 people died and estimated economic loss of 1425 billion baht.
March 23–April 5, 2011	Landslide and flash flood	Thepparat, Cha Long and Khao Noi subdistrict	Five people dead from water wash out. 124 houses were totally destroyed and 3000 had some damages and estimated budget for recovery 320 million baht

(Modified after DDPM, 2014)

Sichon district administration has reported the flash floods and rainfall-induced landslides in March 23–31, 2011. The landslide and flash floods were triggered due to the heavy precipitation for several days (more than 100 mm per day). The community people in mountainous areas and community along the Tha Thon canal, especially, in village number 10 of Thepparat subdistrict incurred a lot of losses and damages: 4 people died, 48 houses damaged completely, 68 houses damaged moderately, and several government facilities, such as roads, bridges, small dams, and drainage system, were destroyed (Table 17.1) (Center for Excellence in Disaster Management & Humanitarian Assistance, 2015).

17.3 DATA COLLECTION METHOD AND TOOLS

The primary data have been acquired through semistructured interview, face-to-face interview, field observation, and focus group discussions. Secondary data have been collected from extensive document review, plans, and policies from DDPM provincial level office and Department of Mineral Resource (DMR).

17.3.1 Key Informant Selection

Key informant interview was one of the important methods for this study which has been used to acquire the critical primary data. The local community and government agency provided and shared their future plans and the way to enhance

TABLE 17.2 Key Information Selection

Agency	Administrative Level	Key Informant	Items Presented/Discussed/Observed
Department of disaster Prevention and Mitigation of Nakhon Si Thammarat Province and Sichon branch	Both provincial and district levels	Head of DDPM Office and staff	Role of DDPM for landslide/flash floods management for Nakhon Si Thammarat Province as well as the role of DDPM in Sichon branch incorporated with local level to take the disaster preparedness and mitigation measures, under the guidance of the national government-related document review
The Provincial Meteorological Station (MET)	Provincial level	Head of provincial MET office	Role of MET to provide the weather information to the landslide and flash flood-prone areas through information dissemination to local government and DPM as well
Provincial Irrigation Office	Provincial level	Head of provincial Irrigation office	Role of irrigation office for landslide/flash flood management Mitigation measures taken Related document review
Theppharat Tambon Administrative Organization	Theppharat Subdistrict (Tambon) level	Tambon Administration chief and officers, Community leader	Role of Theppharat Tambon Administrative Organization for landslide/flash flood management Field survey of landslide and flash flood structural mitigation measures Field observation along the river and mountain during landslide/flash flood events Related document review

the community resilience to find the solutions for risk reduction measures. Thus, the government organizations and local community have been selected as primary respondents for information and data collection (Table 17.2).

17.4 RISK ASSESSMENT AND RISK REDUCTION MEASURES

Risk assessment is an initial step to determine the risk by analyzing hazards and evaluating vulnerability conditions and exposure of the property, services, livelihoods, and the environment. Therefore, it is a mix of element of decision- and policy-making processes and requires coordination between various parts of society (UNDP, 2010).

17.4.1 The Elevation Profile of Thepparat Subdistrict

The cross-section has been generated to understand the elevation profile of Thepparat community. The profile was developed by the 3D Analysis function in ArcGIS software to display the elevation of area and categories of the landslide and flash flood group based on the topography and slope aspect. In addition, cross-sections of elevation in Thepparat subdistrict have been generated for three sections to understand the elevation in particular areas as illustrated in Fig. 17.6.

The first section A_1–A_2 represents the overall elevation of community from the west to east to represent the different land-use patterns from evergreen forest to settlement areas. The graph shows a sharp change in the gradient from 1200 to 600 m with a distance of 2500 m, which could be a significant insight for high landslide risk of the community around this area (Fig. 17.7).

The second cross-section between B_1 and B_2 from the north to south direction in the western part of the study area shows two high points, one at 920 m and another at 980 m. Between the hills, there is a lowest point, 200 m, which represents a stream channel. Second section B_1–B_2 represents the area of higher elevation which can represent the high-gradient zone to analyze the exposure. The cross-section in this zone has been generated to understand the topographic characteristics of hilly area or the high-gradient zone. In this zone, most of the areas are evergreen forest and sparsely populated. However, the result and analysis of this cross-section is not only to understand the topography of this community settlement but also to understand the exposure of the settlement to flash floods in high-gradient buffer (HGB) zones (Fig. 17.8).

The third section C_1–C_2 toward the eastern part of the community represents the settlement area of community and defined as low-gradient zone because of its moderate-to-low slope. Fig. 17.9 shows that the high number in the *Y*-axis varies between 30 and 45 m at a distance of around 10,000 m which represents almost a flat area. Most of the settlements are located in this zone, and population is also significant compared to the other part of the Tambon (Fig. 17.9).

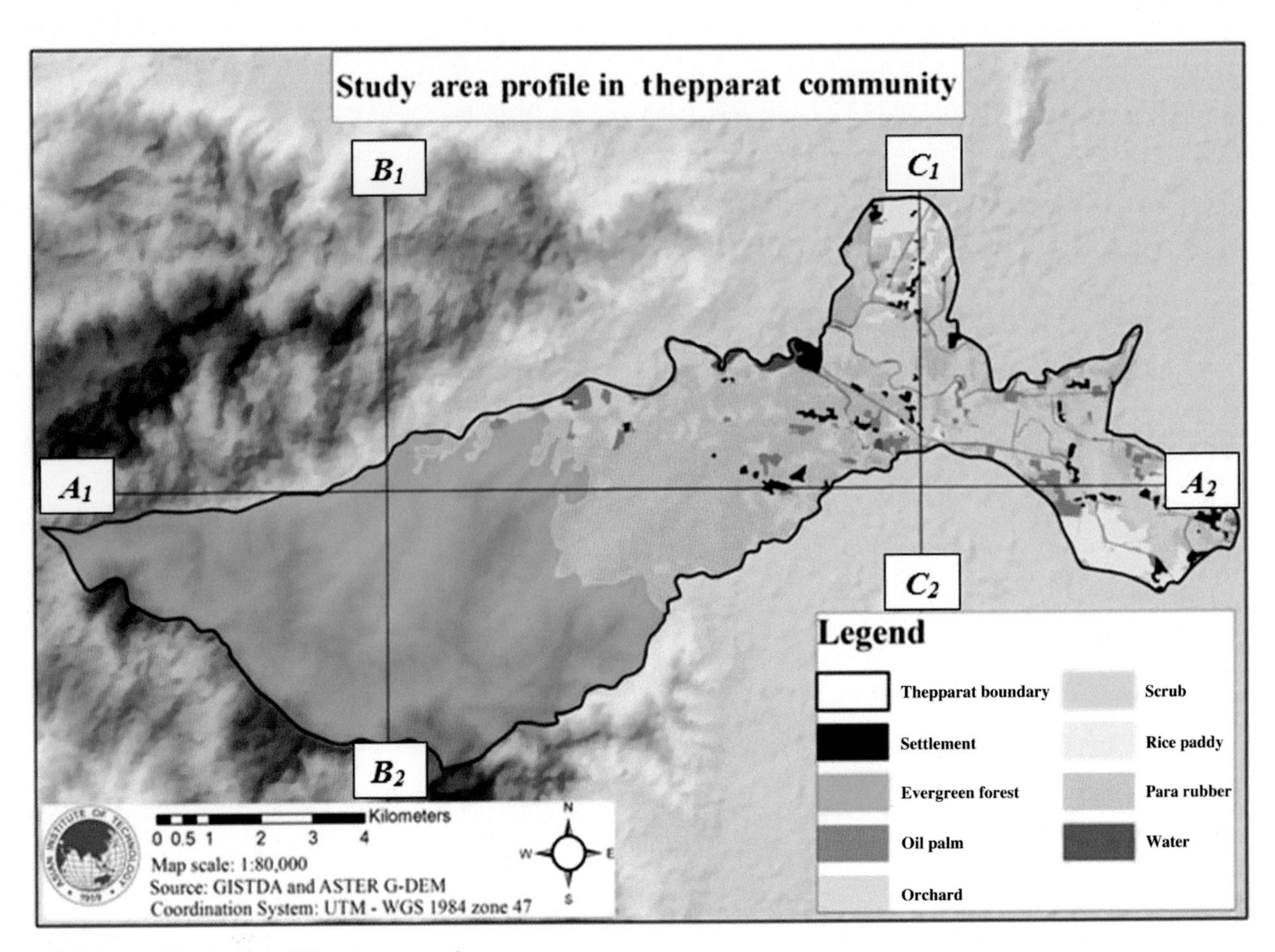

FIGURE 17.6 The elevation of Thepparat community.

In conclusion, the profile analysis of Thepparat subdistrict is the basis to understand the difference of elevation in the study area and classification of community from both landslide and flash flood hazards. Especially, for landslide hazard, the high elevation and high slope could be the primary influential factors. It can express the exposure of the community in the study area combining the elevation and distance from river and natural streams to determine the vulnerability of community in the risk analysis. According to the findings from profile analysis, the community groups in B_1–B_2 are located at the high elevation and high slope which are more prone to landslides than the community in C_1–C_2 which is located in the plane area. The location of point C is more vulnerable to widespread flooding (flash floods) due to the sharp difference in topography (Defra, 2011) (Fig. 17.10).

17.4.2 The Analysis of Settlement Areas for Flash Flood Exposure

The outcome and interpretation of the above-mentioned profile analysis provide the interrelations between exposure, slope, and hazards. Accordingly, the buffer zone area in Thepparat subdistrict has been generated using ArcGIS software through geo-processing tools. The site survey and questionnaire were followed through the generated buffer zones along the Tha Thon canal to understand the demographic distribution and related vulnerability of the community.

To analyze the exposure of the community, two sites have been identified which are located at high- and low-gradient zones. The buffer zone distance has also been created differently in two varied gradient sites. The two sites on different locations can have different impacts of flood, namely at the high-gradient zone, the stream channel is narrow and V-shaped, can lead the water with high velocity. On the other hand, at low-gradient zone, the flooding will spread to a wider area with a low velocity. So, the buffer scale in the high-gradient zones is expressed as 100, 300, and 500 m distance from the canal. Accordingly, the flood-prone area in low-gradient zone has been categorized into three scales as 200, 400, and 600 m. Because of the low slope aspect, flooding in low-gradient zones has slower water velocity with a wider coverage or the inundated area.

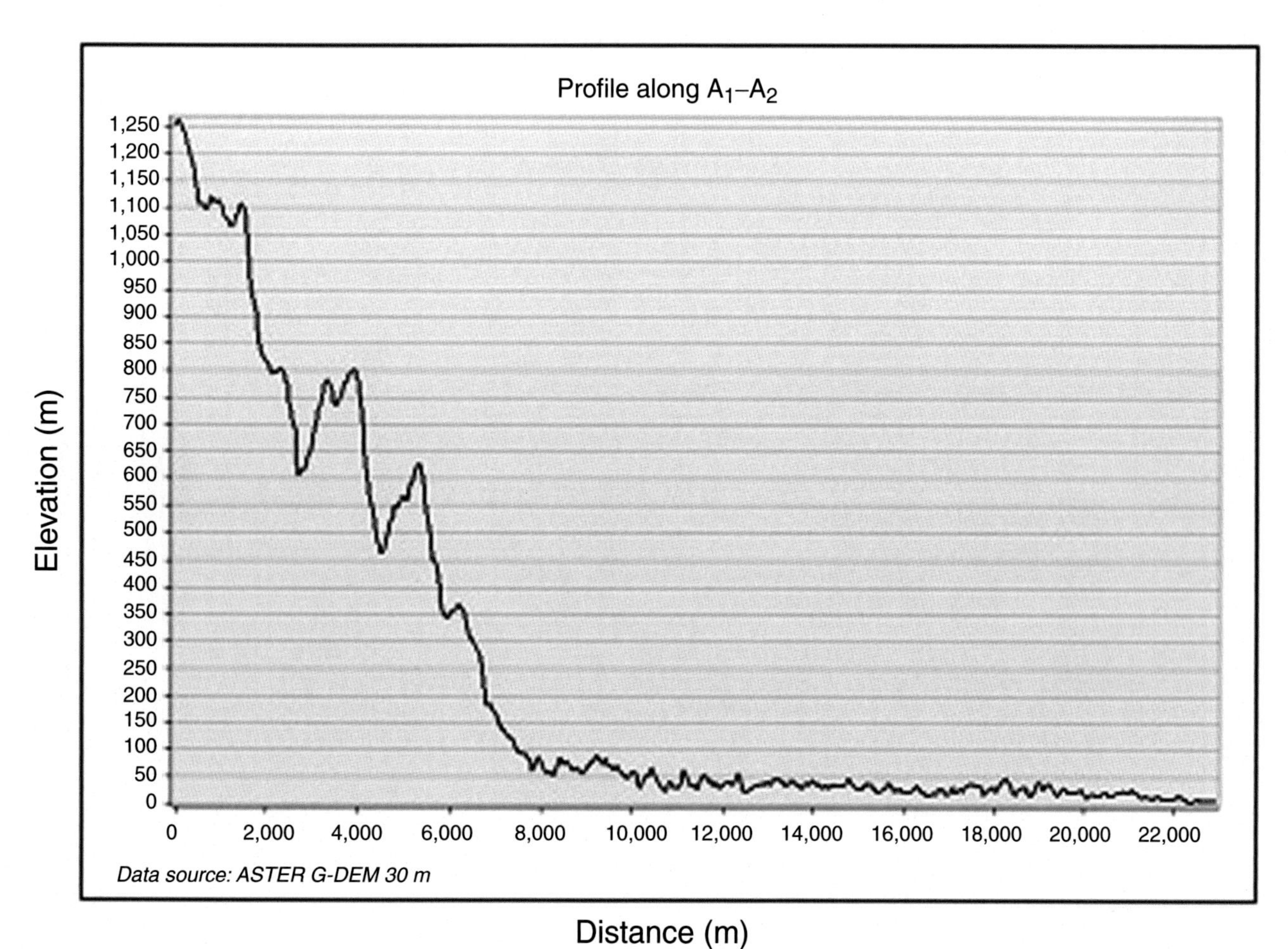

FIGURE 17.7 **Cross-section of elevation from point A_1 to A_2.**

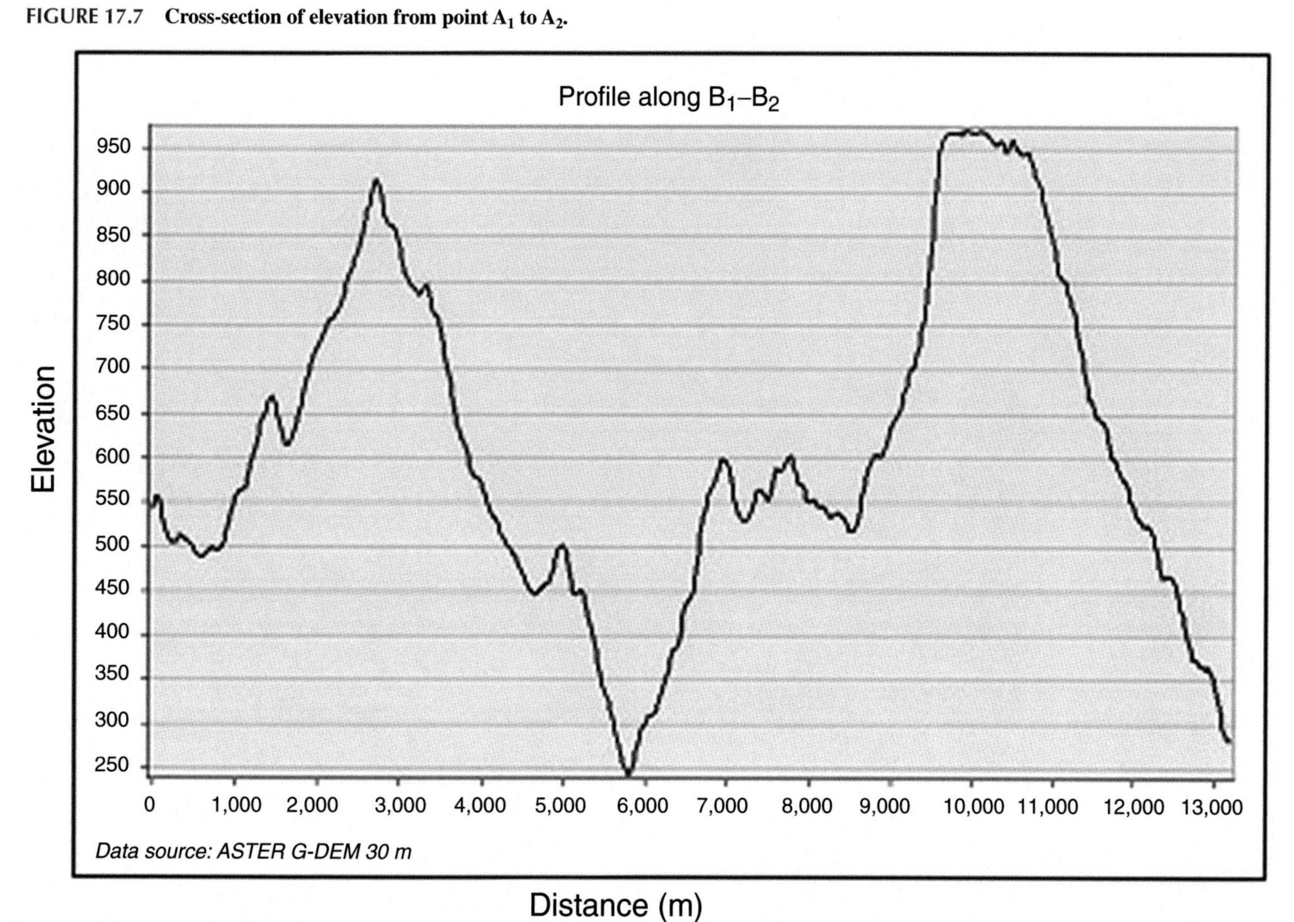

FIGURE 17.8 **Cross-section of elevation from point B_1 to B_2.**

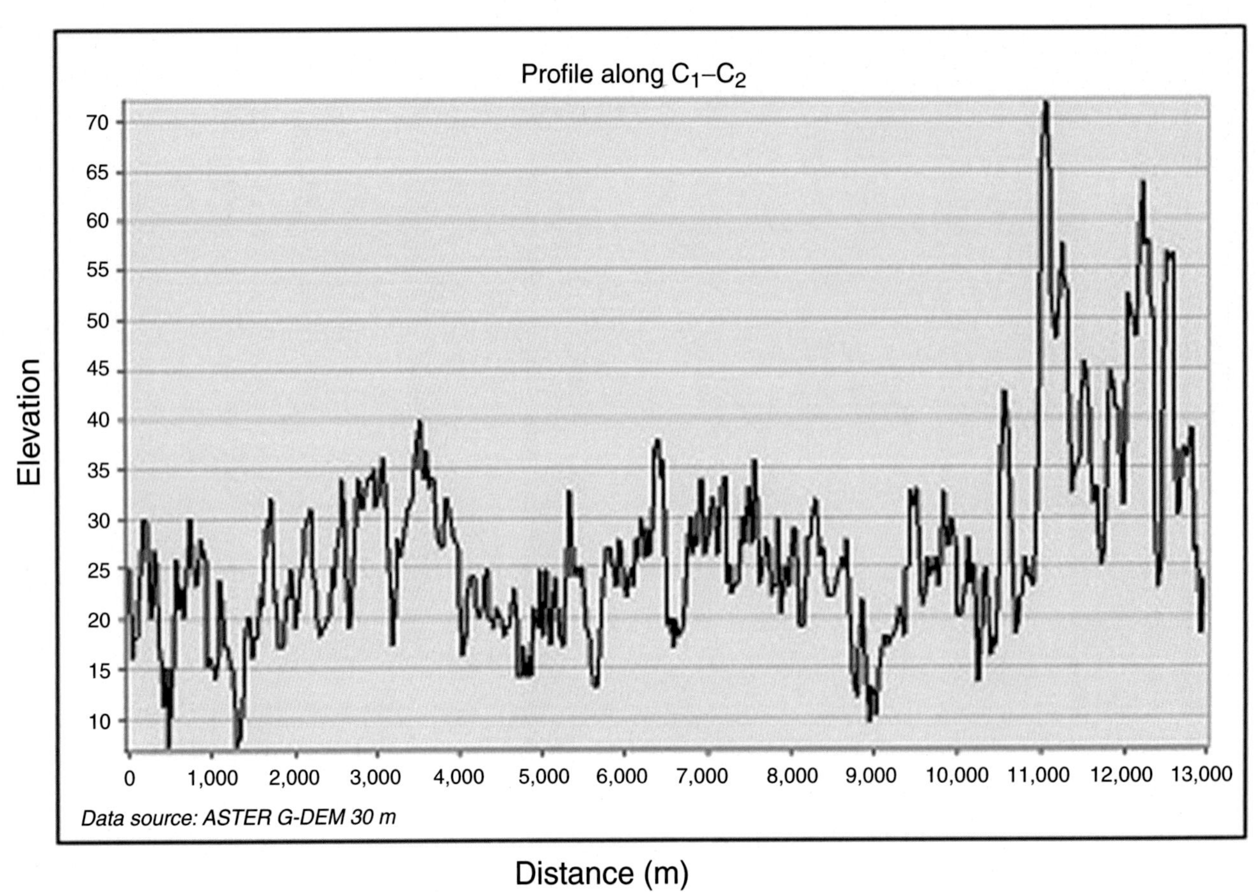

FIGURE 17.9 Cross-section of elevation from point C_1 to C_2.

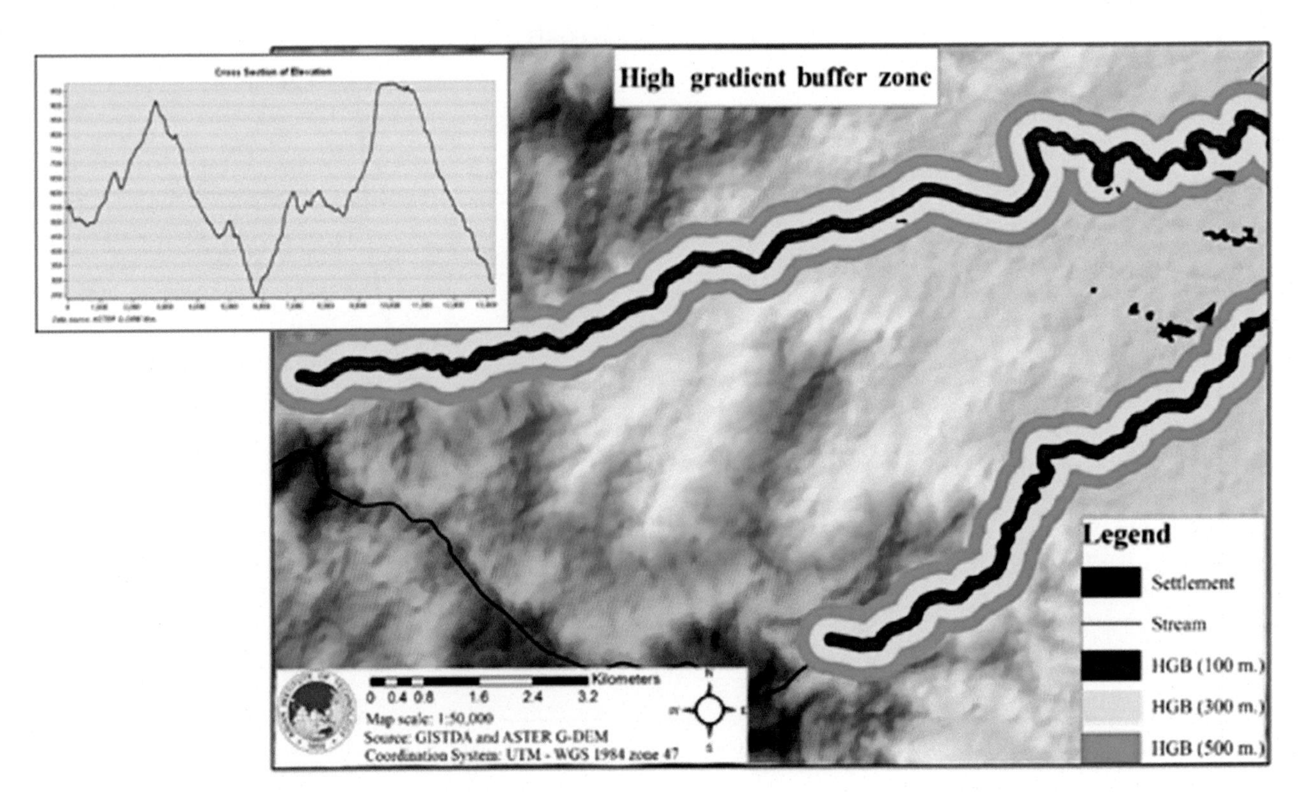

FIGURE 17.10 Settlement at high gradient buffer zone.

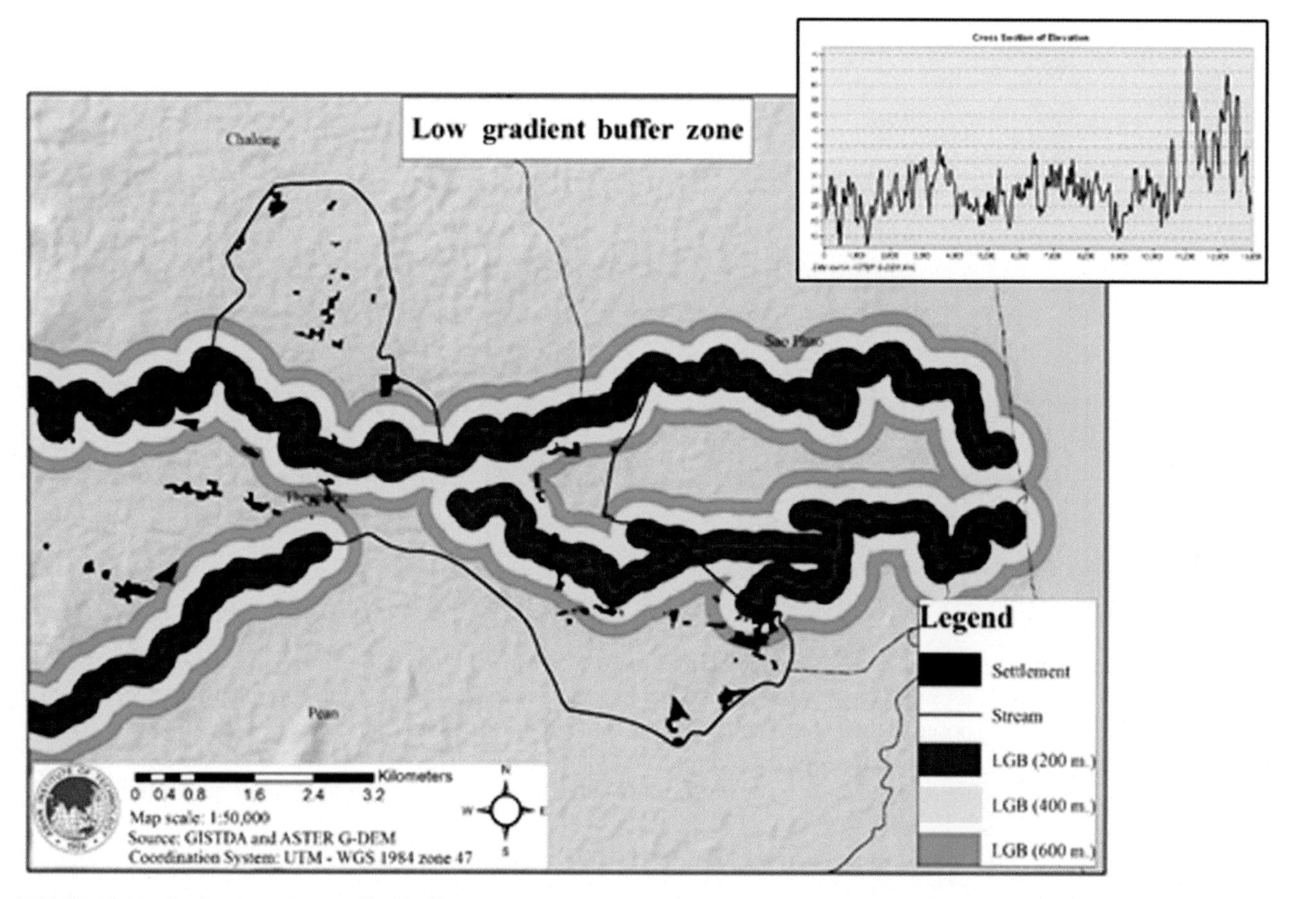

FIGURE 17.11 **Settlement at low gradient buffer zone.**

The exposure analysis of the community to flash flood hazard has been linked between the buffer zone areas with the vulnerability condition of housing. In this study, the overall percentage distribution of housing condition was categorized into three conditions. The community in Zone A has 55% of brick house with corrugated iron sheets; there are few residents located in the 100 and 300 m at HGB zones. In the context of the ground reality, the houses in Zone A are much better when compared to the wooden houses with corrugated iron sheets, which are 40% of this area, having high vulnerability to debris flow or flash floods (Fig. 17.12).

However, the community in Zone A also has 5% of concrete reinforcement houses with roof tiles which are stronger than the two types of house conditions mentioned above. It has been observed that the rate of monthly household income also has an influence on housing conditions. There is a positive correlation between the percentage of engineered reinforced concrete house and average monthly household income.

In addition to the residents exposed to flash floods in the high-gradient zone, this study also analyzed the residents at low-gradient buffer (LGB) zone vulnerable to flash floods. Most of the housing condition in this area or Zone B has 70% of engineered reinforced concrete houses with roof tiles which are considered to be a very good condition. The houses have been built from the reinforced concrete and are stronger than the houses built from wood or bricks. Moreover, the percentage of reinforced concrete houses in Zone B can indicate that most of the people in this zone have a higher monthly income than the community in Zone A. Fig. 17.11 depicts the residents exposed to flash floods with 200, 400, and 600 m of buffer zones. Thus, the community people in Zone B can build the houses more resistant to the flash floods in highly vulnerable areas which can decrease the risk of flash flood-induced impact (Pal et al., 2013).

Result from exposure analysis (Fig. 17.12) shows that the settlement in HGB zone has the percentage of settlement exposed to flash floods as 4.7% in 100 m buffer zone, 15.2% in 300 m buffer zone, and 23.5% in 500 m buffer zone.

However, in the LGB zone, the percentage of settlement exposed to flash floods is 6.7% in 200 m buffer zone, 26.7% in 400 m buffer zone, and 52.3% in 600 m buffer zone. The percentage of settlement vulnerable to flash floods in both the buffer zones indicates that the settlement in the HGB zone is less exposed to flash floods than that in the LGB zone. However, 56.6% settlement in HGB having negligible vulnerability or outside the estimated buffer zones and same evident for LGB is 14.3%.

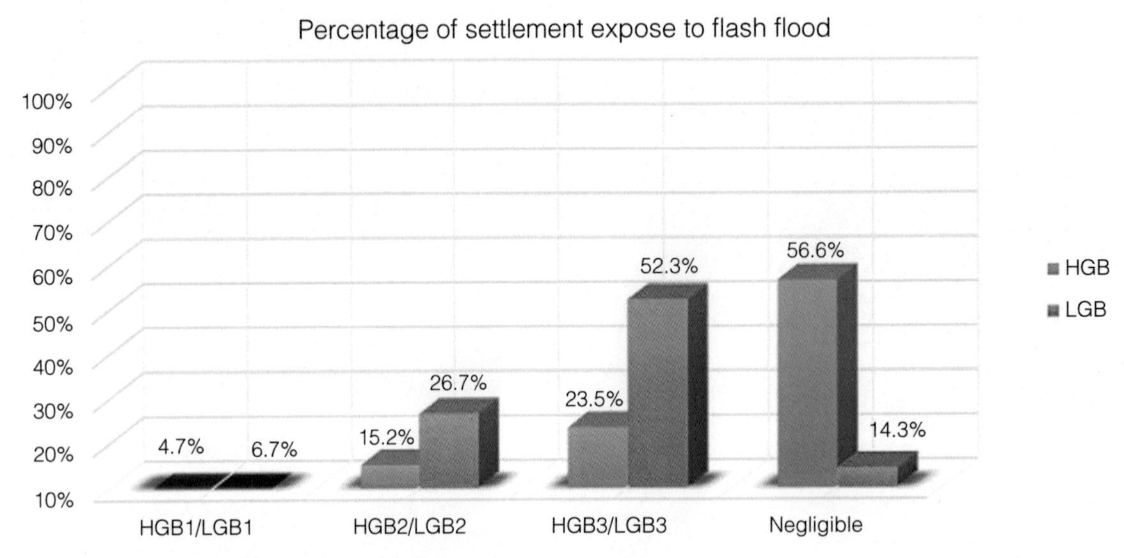

FIGURE 17.12 **Percentage of settlements for each buffer zone.**

In conclusion, the integration of land-use map or other maps with buffer zones can express the results from exposure analysis. Exposure analysis is one of the components of risk assessment which can help the community and local government to understand the risks posed to their community. The outcome of the exposure analysis can be useful for the decision-maker and planner in order to develop the future developmental plan for community which can reduce the risk in a specific area. However, for the existing settlements, the local government can implement the law or regulation to mitigate the risk persisting in that particular zone.

17.5 MITIGATION STRATEGIES FOR LANDSLIDE AND FLASH FLOODS IN COMMUNITY LEVEL

Landslide and flash floods are hard to predict and control but well-planned mitigation measures could reduce the risk to lives and property to a great extent. There are many approaches to reduce the potential impacts of landslide and flash flood hazards.

The combination of nonstructural measures and small-scale structural measures could be more effective in managing flash flood risk. Predicting magnitude and exact location of flash floods are still challenging and it is not necessary to take large-scale structural measures such as constructing embankment, dams, and levees. There are many nonstructural measures that can reduce the impact of floods such as land-use planning, building construction codes, soil management, acquisition policies, insurance and risk transfer, awareness raising, public information, emergency system, and recovery plans.

The nonstructural measures are suitable for local community because of its low cost and sustainability. Moreover, the small-scale structural measures are useful if community people use local materials such as check dams, small-scale levees, and sand bag embankments.

1. The implementation of plans and policies should include flash floods at national and regional levels.
2. Integration between flood management policies, water resource management, and disaster management plan needs to be taken as a priority.
3. A community flash flood risk management committee could be a good mechanism to involve community members and local authorities in order to carry out the local and indigenous knowledge.

17.6 DISASTER RISK REDUCTION MEASURES FOR LANDSLIDE-PRONE COMMUNITY

The community in Zone A is prone to landslide and mudflow due to the topographic characteristics. Risk analysis shows that Zone A is defined as low risk because of the low probability of occurrence of landslides. However, in terms of vulnerability, it is defined as moderate due to sparsely distributed settlement, fair resilient housing, and road conditions. However, in the capacity consideration, the community is found to be moderate because of some existing programs on early warning system. Thus, the recommendations for the community in Zone A will concern the following point.

17.6.1 Coordination for Emergency Response

Coordinated response becomes central for immediate aftermath of a huge landslide. The community and local government need to conduct the training and drill for community people and staff from the agencies to respond with prior defined coordination mechanism. The drinking water system and food might get affected severely due to the massive inflow of sediments, which might affect the water and agriculture systems adversely. Thus, the community needs to have a food and drinking water stock for their family to survive for at least 3–4 days before the local government and other agencies come for support. However, the livelihood of Thepparat subdistrict is agriculture and specifically to the steep slopes, which might increase the frequency of landslides. Communities need to be sensitized to avoid agricultural activities on high risk areas.

17.6.2 Increasing the Capacity of Volunteers

In general, Thepparat community organizes training and drill every year to increase the number of volunteers such as Mr. Warning to disseminate the warning information. However, in the case of landslide and flash floods, the volunteers for rapid response do play an important role in communicating with the rescue-affected communities. The national government provides some training on search and rescue to the selected volunteers to transfer the knowledge to the next generation.

17.7 DISASTER RISK REDUCTION MEASURES FOR FLASH FLOOD-PRONE COMMUNITY

The community in Zone B is defined as high risk due to the frequent impacts of flash floods. Apart from the frequent flash floods, the zone also lies in the plain area, where most of the settlements and government buildings are located, which eventually increase the vulnerability.

According to the exposure analysis on LGB zone, 85.7% of settlements are affected by flash floods. The outcome of the exposure analysis could be useful for the decision-maker and planners for future development and land-use planning which can reduce the risk of flash flood hazards. However, for the communities already settled, the local government can implement the law or regulation to mitigate the risk that could take place in a particular zone. Following are the specific recommendations for community in Zone B.

17.7.1 Early Warning System

In Thepparat subdistrict, the village head is well equipped with radio communication system and rain gauges. So, in the case of emergency, the staff and volunteers get alerts throughout the day to monitor the rainfall and the water volume in the stream channel. It has been observed in the past incident that the automatic rain gauge and installed siren were not functional during emergency because of the power shut off. However, the national warning center coordinated with the media during disaster.

17.7.2 Community-Based Disaster Risk Management (CBDRM)

The disaster risk management program will not be successful if the local community or people are not been involved. Therefore, community-based disaster risk management can create the connectivity and relationship between local community and government organizations. Since the community are well aware about the hazards in their place and safe areas through the experiences, the top-down risk governance mechanism is needed to integrate the local traditional wisdom with modern technologies to conduct efficient DRR activities in the community (ADPC, 2009, 2013).

17.7.3 Public Awareness Generation

It has also been observed that the local government systems such as the Head, Department of Disaster Prevention and Mitigation in Sichon district, and Head, Provincial Meteorological Office, interacted through local FM Radio channel for community awareness on weather situation and report on disaster news.

However, on March every year in Thepparat community, the Tambon Administrative Organisation (TAO) organize the exhibition on landslide and flash floods to recall the situation in 2011 to build the community awareness about the risk and knowledge for disaster risk management.

17.8 ANALYSIS AND DISCUSSION

17.8.1 An Analysis of Cluster Village Zone of Landslide and Flash Flood Risks

The result from risk analysis is being combined with the landslide/flash flood risk zones identified by the Department of Mineral and Resource in the cluster village zone of disaster-prone area (Fig. 17.13).

The community in Zone A faced major landslides for the first time in 2011, which has been triggered by the heavy rainfall along with the in situ characteristics of adverse geophysical conditions. The community people and local governments are not familiar with the large-scale landslide disasters due to the lack of risk information regarding hazards such as hazard vulnerability map and disaster preparedness program. Although majority of the population in Zone A is elderly and poor, because of the very low number of settlements and vast coverage of evergreen forest, Zone A poses low risk.

The community in Zone B is predominantly located on the plain area which is prone to flash floods. Thepparat community particularly in Zone B people are facing riverine and flash floods almost every year, and therefore, community people and local governments have taken flooding experience into account for risk management measures. Community people are well aware about the coping mechanism of their own along with the local government support through both structural and nonstructural mitigation measures. On the contrary to Zone A, most of the people in Zone B are in working age, with higher education, and high income which can indicate that community people in zone B are more resilient to cope up with the natural hazards. However, the high frequency of floods, densely populated settlements, location of government institutions, and establishments with high vulnerability make Zone B categorized in high risk.

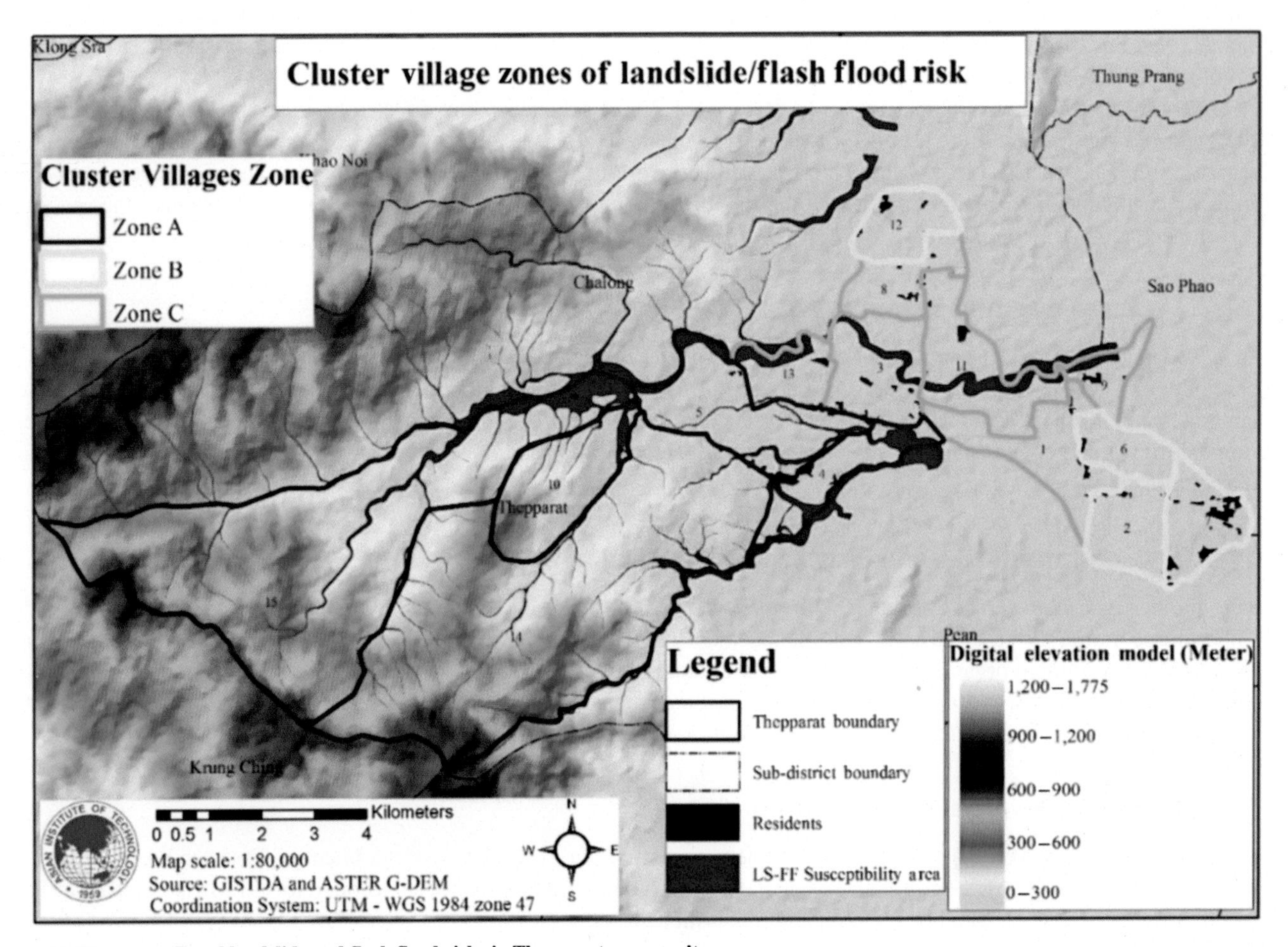

FIGURE 17.13 **Zonal landslide and flash flood risks in Thepparat community.**

TABLE 17.3 Disaster Risk with Village Cluster Zones

Cluster Zone	Zone A	Zone B	Zone C
Village No.	4, 5, 10, 14, and 15	2, 6, 7, and 12	1, 3, 8, 9, 11, and 13
Population	2013	2069	3889
Hazard identification	Landslide/mud flow	Flash floods	Landslide and flash floods
Hazard levels	Low	High	Moderate
Exposure	Low	High	Moderate
Vulnerability	Moderate	High	Moderate
Capacity	Moderate	Low	High
Risk result	Low	High	Moderate

The community in Zone C is located close to the Tha Thon canal and foothill of the nearby mountain. This zone faces both landslides and flash floods. Because of low occurrence of the events, the landslide hazard is not too high compared to the frequent Flash floods. Therefore, the community people are well aware to cope with flooding but not with the landslides. However, there are many villages located in Zone C and the number of population is also high but the hazard occurrence is not too high. So, Zone C is defined at moderate risk.

Table 17.3 shows the description and comparison of hazards, vulnerability, exposure, capacity, and risk of three zones with the village cluster. The community in Zone A is landslide-prone area which has low hazard and also low vulnerability, but because of the very low population density, the community is at low risk. In contrast, the community in Zone B which is in flash flood-prone area, which has high hazard potential and high vulnerability with high population density and infrastructures, puts the community at high risk. However, the community in Zone C is in landslide and flash flood-prone area, which has moderate hazard and moderate vulnerability with moderate population, resulting in moderate risk.

In brief, the different cluster village zone analysis can help decision maker or local governments and community people to understand the various types of disaster risk reduction measures. Each community zone may address the risks in different ways depending on the basic resources and capacities of community. Thus, the result from risk analysis and disaster cluster village zone analysis could be useful for disaster risk reduction strategies in particular village zones.

17.9 CONCLUSION

Due to the increase in frequency of hydro-meteorological hazards, both landslide and flash floods are increased in the study area, with time and local community being the first to be impacted. The disaster management and risk reduction activities may not be successful if done only at the national level; the local government and local community also need to be involved in disaster risk management initiatives for more effective implementation.

The study analyzes the degree of risk and vulnerability both quantitative (risk analysis) and qualitative (risk perception) way in landslide and flash flood-prone areas in order to recommend the disaster risk reduction measures (Peggion et al., 2008). The risk analysis method depicts the condition of risk in Thepparat subdistrict, the exposure of different communities at risk and vulnerability. The findings from risk analysis show that the three community groups in one Tambon (Thepparat subdistrict) have different risks due to the different topography, land-use, and settlements. The landslide susceptibility map has been used to find out the index to identify the factors influencing the landslide vulnerability areas. Even though the community in flash flood-prone areas is more resilient, the disaster risks are still high because of the ineffective coordination between government agencies. Most of the disaster preparedness and mitigation programs are addressed to the landslide groups.

However, the risk of community varies from the vulnerability components such as roads and house conditions as well as the community capacities such as time to receive early warning information and lead time to evacuation shelter. Vulnerability of the community grossly depends on the socioeconomic conditions, namely the community in landslide group is more vulnerable than the community in flash flood-prone areas because of the relatively low income, low rate of education, and less experience of disasters which can influence the perception of people about the social, physical, and environmental risks (Pal et al., 2017).

ACKNOWLEDGMENTS

Special thanks to Thailand Department of Mineral Resource, Nakhon Si Thammarat Provincial Disaster Prevention and Mitigation, Disaster Prevention and Mitigation Sichon Branch, Nakhon Si Thammarat Meteorological Station, Nakhon Si Thammarat Irrigation Office, and Thep-parat Tambon Administration Organization.

REFERENCES

ADPC, 2009. Using Risk Assessments to Reduce Landslide Risk. Program for Hydro-meteorological Disaster Mitigation in Secondary Cities in Asia, pp. 1–8.

ADPC, 2013. Assessment of disaster management planning, policies and response in Ayutthaya Province of Thailand. Master thesis. Asian Institute of Technology, Thailand.

Center for Excellence in Disaster Management & Humanitarian Assistance, 2015. Disaster Management Reference Handbook, Thailand.

Defra, 2011. Understanding the risk, empowering communities, building resilience.

Pal, I., 2015. Land use and land cover change analysis in Uttarakhand Himalaya and its impact on environmental risks. In: Shaw, R., Nibanupudi, H.K. (Eds.), Mountain Hazards and Disaster Risk Reduction. Springer, Japan, pp. 125–137, ISBN: 978-4-431-55241-3.

Pal, I., Singh, S., Walia, A., October 2013. Flood management in Assam, India: a review of Brahmaputra Floods, 2012. Int. J. Sci. Res. Publ. 3 (10), ISSN: 2250–3153.

Pal, I., Ghosh, T., Ghosh, C., 2017. Institutional framework and administrative systems for effective disaster risk governance—perspectives of 2013 cyclone Phailin in India. Int. J. Disaster Risk Reduct. 21, 350–359.

Peggion, M., Bernardini, A., Masera, M., 2008. Geographic Information System and risk. European Commission. Italy Center for Excellence in Disaster Management & Humanitarian Assistance (2015). Thailand Disaster Management Reference Handbook.

Pramojanee, P., Tanavud, C., Yongchalermchai, C., Navanugraha, C., 1997. An application of GIS for mapping of flood hazard and risk area in Nakorn Sri Thammarat Province, South of Thailand. In: Proceedings of the International Conference on Geo-Information for Sustainable Management, pp. 198–207.

Thai Meteorological Department, 2015. Annual Weather Summary over Thailand in 2015.

UNDP, 2010. Disaster risk assessment. Bureau for crisis prevention and recovery. One United Nations Plaza, New York.

FURTHER READING

Department of Disaster Prevention and Mitigation (n.d.). Department of Disaster Prevention and Mitigation. Available from: http://www.disaster.go.th/dpm/.

EM-DAT, 2015. The international disaster database. Environ. Sci. Policy Sustain. Dev. 21 (3), 14–39.

Grothmann, T., Reusswig, F., 2006. People at risk of flooding: why some residents take precautionary action while others do not. Nat. Hazards 38 (1–2), 101–120, http://www.emdat.be/database.

Paneetsin, K., 2014. Coping and Resilience of Ban Pho Flood Prone Community in Ayutthaya Province of Thailand (Master thesis), Asian Institute of Technology, Thailand.

Pramojanee, P., Tanavud, C., Yongchalermchai, C., Navanugraha, C., 2013. An application of GIS for mapping of flood hazard and risk area in Nakorn Sri Thammarat province south of Thailand. In: Proceedings of the International Conference on Geo-Information for Sustainable Management, pp. 198–207.

Slovic, P., Fischhoff, B., Lichtenstein, S., 1979. Rating the Risks. System and risk. European Commission, Italy.

Chapter 18

Advancements in Understanding the Radon Signal in Volcanic Areas: A Laboratory Approach Based on Rock Physicochemical Changes

Silvio Mollo*, Paola Tuccimei, Michele Soligo**, Gianfranco Galli†, Piergiorgio Scarlato†**

**Sapienza University of Rome, Rome, Italy;*

***Università "Roma Tre", Rome, Italy;*

†National Institute of Geophysics and Volcanology, Rome, Italy

Silvio Mollo Professor in petrology at Dipartimento di Scienze della Terra, Sapienza-Università di Roma in Rome, Italy. My researches concern about experimental petrology, magma dynamics related to fractional crystallization, assimilation and mixing, trace element partitioning, isotope geochemistry, equilibrium and disequilibrium cation exchanges between crystals and melts, thermobarometric and hygrometric modeling, and the physicochemical properties of rocks and fluids.

Paola Tuccimei Professor in geochemistry at Dipartimento di Scienze, Università "Roma Tre" in Rome, Italy. My research activity deals with radon as tracer of geological and environmental processes, natural radioactivity mapping, gas hazard assessment, isotope geochemistry applied to the study of carbonate precipitates (speleothems, travertinea, calcretes) with palaeoenvironmental and palaeoclimatic implications, hydro-geochemistry, and urban geology.

Michele Soligo Researcher in geochemistry at Dipartimento di Scienze, Università "Roma Tre", Italy. My research interests concern: (1) geochronology, with particular reference to U-series disequilibria of terrestrial carbonates such as travertines and speleothems, (2) palaeoenvironmental and palaeoclimatic reconstruction using oxygen and carbon isotopes on carbonates, and (3) environmental radioactivity, with special focus on radon problems.

Gianfranco Galli Head of Radionuclides Laboratory of the Istituto Nazionale di Geofisica e Vulcanologia (INGV) in Rome, Italy. My expertise is on radon, developing instruments for continuous and discrete measurements of radon concentration either in water or in soil/indoor and, in general, on design, realization, calibration, and maintenance of instruments used in research and surveillance geochemical networks.

Piergiorgio Scarlato Head of HPHT Laboratory of Experimental Volcanology and Geophysics of the Istituto Nazionale di Geofisica e Vulcanologia (INGV) in Rome, Italy. I investigate the most important processes controlling the eruptive dynamics of different volcanoes in the world, chemical and physical properties of silicate liquids, electrical conductivity and acoustic wave velocity of natural rocks and melts, rocks physical properties, and the dynamic mechanisms controlling earthquakes and volcanic eruptions.

18.1 RADON THEORY AND APPLICATIONS

Radon is a noble, rare, inert, invisible, colorless, odorless, and tasteless gas originated by radioactive decay of nuclides that naturally occur in groundwaters, rocks, and sediments of the Earth. Radon atoms are electrically neutral with an atomic radius of 1.2×10^{-10} m, and molecules are monatomic. Under standard pressures, gaseous radon converts into liquid at -65 °C, which then converts into solid at -113 °C. The melting and boiling points of radon are -71 and -61.8 °C, respectively. The density of radon is 9.73 kg/m^3, which is 7.5 times the air density. Despite the gas phase exhibiting preferential migration into the atmosphere, radon is also highly soluble in water and organic solvents (such as toluene, petroleum, and ethanol), and its solubility decreases with increasing temperature (Schubert, 2015). Radon is a natural gaseous radioactive tracer and has three isotopes that decay to daughter nuclides emitting alpha particles: ^{222}Rn (radon) is a short-lived decay product from the ^{238}U decay series (^{226}Ra is the direct parent nuclide) with a half-life of 3.823 days, ^{220}Rn (thoron) is a

Integrating Disaster Science and Management. http://dx.doi.org/10.1016/B978-0-12-812056-9.00018-X

decay product from the ^{232}Th decay series (^{224}Ra is the direct parent nuclide) with a relatively short half-life of 55 s, and ^{219}Rn (actinon) is a product of the ^{235}U decay series (^{223}Ra is the direct parent nuclide) with a very short half-life of 4 s. ^{222}Rn is the most widely monitored isotope because it is the longest lived of the three radon isotopes and hence can carry information over greater distances. ^{220}Rn and its daughters may also contribute in soil gas measurements or in small laboratory setups as a significant fraction of the decays when Th/U ratios are 3:1 or 4:1. For the case of higher Th/U ratios, the Th decay chain may be dominant. Despite heavy radon and thoron isotopes do not fractionate due to their low mass difference (0.01%), ^{220}Rn short half-life makes it useful in discriminating areas with very fast soil gas transport and/or Th-rich mineral outcrops. On the contrary, ^{219}Rn is not considered in geochemical exploration, partly because of its short lifetime, which strongly limits its migration, but primarily because of its low activity (4.6% relative to ^{222}Rn).

Alpha particles emitted by radon isotopes have a short penetration distance, being easily halted by the clothes and the thin epidermal layer of the skin. However, when ingested or inhaled, alpha particles can cause severe damage to stomach or bronchial tissue by releasing their entire penetration energy to a relatively small volume of tissue. Alpha particles are electrically charged and can attach themselves to tiny dust particles in indoor air. Finer dust particles (<1 μm) can easily be inhaled into the lungs where alpha radiation passes through a cell nucleus, damaging DNA molecules (Hauri et al., 2013). Long-term exposure to radon can cause lung cancer, the only cancer demonstrated to be associated with inhaling radon because alpha particles accumulated in the lungs travel only extremely short distances and cannot reach cells in any other organ. Cigarette smoking is the most common cause of lung cancer, but radon is the second leading cause of this disease. It is estimated that more than 10% of radon-related cancer deaths occur among nonsmokers. Indeed, radon can enter homes through cracks in floors, walls, and foundations, especially for buildings built on rock substrates that are rich in uranium and thorium (Tuccimei et al., 2009, 2011). Basement and first floors typically have the highest radon levels because of their proximity to the ground. Additionally, indoor radon can also be released from the building materials that act as a continuous source of radiation exposure because 80% of the population lifetime is spent in indoor air.

The transport behavior of radon in geological environments can be described on the basis of physicochemical processes such as fluid advection, diffusion, partition between liquid and gas phases, and radioactive decay. The most important mechanism controlling the release of radon from a rock is twofold (Fig. 18.1): (1) the formation of free-state radon and (2) the migration of radon gas (Zhang et al., 2016). The parent radium decays by emitting alpha particles to form radon daughters (Fig. 18.1). A fraction of the decay products may escape from the mineral grains and rock matrix to become free-state radon that resides and accumulates in the rock pore spaces (i.e. spaces between grains, fractures, vesicles, and voids). The nuclear recoil effect plays a key role in the formation of free-state radon atoms and is correlated with the composition

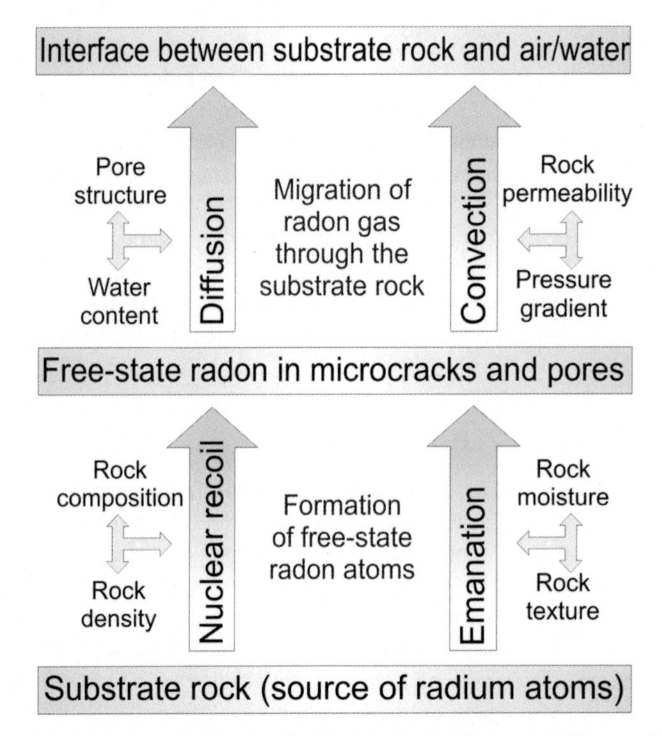

FIGURE 18.1 **Diagram illustrating the transport behavior of radon in geological environments.** The most important mechanism controlling the release of radon from a rock is twofold: (1) the formation of free-state radon and (2) the migration of radon gas.

and density of the decay medium determining the recoil distance of radon atoms (Fig. 18.1). Radon emanation is defined as the escape of radon atoms from Ra-bearing grains into pore spaces (Sakoda et al., 2011) by direct recoil effects due to the emission of the alpha particles and/or by diffusion across lattice imperfections (Andrews and Wood, 1972). The moisture content and grain size of the rock are the most important parameters controlling radon emanation (Fig. 18.1). In the presence of pressure/temperature gradients and/or fluxes of carrier gases, radon atoms migrate through the rock microcracks and pores to concentrate in soil air and/or groundwater (Fig. 18.1). The migration of radon through a medium is modeled as the effect of convection that, however, includes simultaneous diffusive and advective flows (Fig. 18.1). The molecular diffusive transport of radon is expressed by Fick's law of diffusion in terms of diffusion coefficient (i.e. the mobility of the radon gas toward the direction of its decreasing concentration). On the contrary, the advective transport of radon is described by Darcy's law in terms of permeability (i.e. the capacity of a medium for transmitting a fluid upon pressure gradients). Both diffusivity and permeability in porous rocks are also dependent on a number of parameters, such as the type of fluids present in the pores, fluid saturation, pore structure, adsorption phenomena in the solid matrix, and temperature.

The uranium and thorium contents of the most common volcanic rocks are 0.1–20 and 0.2–28 ppm from oceanic tholeiitic basalts to alkaline rhyolites, respectively (Barretto et al., 1974). The maximum radon emanation from effusive and intrusive igneous rocks is 9% and 16%, respectively (Barretto et al., 1974). The ratio of free-state radon generated by the nuclear recoil effect to overall radon content in the medium is termed emanation coefficient (Ferry et al., 2002). The emanation coefficient determines the quantity of free-state radon in the pore spaces of a rock, being a direct measurement of the number of radon atoms that diffuse from the inner to outer pores of the rock matrix. After the recoil motion, some radon atoms do not contribute to the overall emanation because retained within the U–Th-bearing minerals as atoms that remain stuck in the original grain lattice and/or enter another lattice. Radon has a short recoil length of typically 30–50 nm in solid materials, 95 nm in water, and 64,000 nm in air (Porstendörfer, 1994; Tanner, 1964, 1980). Thus, radon atoms display a poor intrinsic mobility with a diffusivity of 0.12 cm^2/s into air at 20 °C (Tanner, 1964). In contrast, radon is highly soluble in water (22.4 cm^3 per 100 cm^3 H_2O at 25 °C and atmospheric pressure); consequently, most of the gas is readily dissolved in groundwater rather than adsorbed onto the walls of the rock fractures. The upflow of gases through the subsurface waters can also form microbubbles carrying radon from deep sources toward the surface (Kristianson and Malmqvist, 1982). Consequently, radon nuclides may travel large distances because of their good inclination to concentrate in the water–gas surface zone of microbubbles (Baranyi et al., 1985; Somogyi and Lenart, 1985). The radon concentration in soil gas and groundwater is largely dependent on the surface area of rocks and only atoms produced at the surface of rock grains are released to the surrounding environment (Torgersen et al., 1990). Radon is a dense gas with a low diffusivity and a short half-life that limit the gas migration distance in the medium and make unlikely for deep radon to reach the surface. A deeper origin and the transport of radon over long distances toward the ground surface require a relatively fast-moving advective gas carrier that, alternatively, rapidly diffuse into the rock medium (Durrance and Gregory, 1990). This applies also to natural magma because the diffusivity of radon is very slow and, at magmatic temperatures, approaches that of other heavy noble gases ($\sim 10^{-12}$ m^2/s at $\sim$1225 °C; Gauthier et al., 1999). Neither diffusion processes (which are too slow and balanced by radioactive decay of radon) nor exsolution of pure radon bubbles (because of its exceedingly low abundance in the magma) can be responsible for the complete radon depletion observed in erupted lavas (Gauthier et al., 1999). The exsolution of water from magma (and other volatiles acting as carrier gases) are the main cause for radon degassing. The most important parameters controlling radon activity concentrations at the surface are the emanating power of the host rock, the permeability of the substrate, and the flow rate of the carrying gas (Ball et al., 1991). Usually, radon activity increases with increasing flow rate because gas velocity increases, generating both less time for decay and more radon extraction from the medium. However, at higher flow rates, dilution of radon by the carrying fluid may also take place leading to an overall low radon signal. Underground water, carbon dioxide, nitrogen, and methane are these main fluids (Shapiro et al., 1982; Toutain et al., 1992). Importantly, increasing radon emissions highlight not only the presence of fractures which undergo fluid circulation, but also fractured rocks containing U–Th-bearing minerals. Therefore, radon can be used for the mapping of hydrothermal/geothermal systems and active faults in volcanic and tectonic environments. Moreover, radon signal is often used as a tool for exploration of the hidden underground uranium deposits and for groundwater flow classification (Toutain and Baubron, 1999), where the radon concentration in groundwater is proportional to the uranium concentration in rocks of the aquifer. Groundwater plays a focal role in the transport mechanism of radon because the gas dissolved in water can be transported very far away from its original place in a short time. Radon concentration in groundwater is highly variable as a function of several parameters, such as the contents of uranium and thorium as parent nuclides in the original rocks, the grain size of the rock, the type of crystal lattice, the dilution with rain water, the temperature, and the pH of the water itself (Milvy and Cothern, 1991). More complex phenomena such as mixing, contamination, chemical reactions, and differential solubilities may alter the original gas concentration, resulting in broad radon changes within a confined area (Toutain and Baubron, 1999). On the other hand, measurements of soil gas radon are remarkably influenced by a number of meteorological factors related to rainfall, barometric pressure, temperature, moisture, and wind, causing fluctuations in the radon signal and potential misinterpretations of acquired data (Kozlova and Yurkov, 2005). Under such circumstances, radon monitoring conducted over long timescales (e.g. several months) appears to be more appropriate to overcome the weather-related fluctuation effects that, conversely, are encountered over short timescales (e.g. a few days). Uneven weathering of the exposed rocks may also produce inhomogeneities within the monitored formations and extremely low radon emissions (Washington and Rose, 1992). Notably, the effects related to Earth's tides are also known to cause most of distortion to radon data (Lenzen and Neugebauer, 1999). The increase in vertical extension and the decrease in gravity induced by Earth's tides in the rocks may open supplementary pathways for the flow of radon-bearing carrier gases, enhancing radon transport through the rocks (Kies et al., 1999).

18.2 RADON MONITORING IN TECTONIC AND VOLCANIC ENVIRONMENTS

Owing to its importance, radon monitoring has become a continuously growing study area in the search of premonitory signals in tectonic and volcanic areas, and a great number of models are reported in the literature for interpreting radon anomalies in groundwater, wells, and soils before earthquakes and volcanic eruptions (Finkelstein et al., 1998; Planinic et al., 2000; Plastino et al., 2002; Singh et al., 1999; Wakita, 1998; Zmazek et al., 2000). According to the early dilatancy-diffusion model (Scholz et al., 1973), radon anomalies are related to an increasing elastic strain associated with opening of small cracks and inelastic increase in volume of the material (i.e. dilatancy) surrounding the fault. Influx of water into open fractures increases fluid pressure, lowering the rock strength, and prompting rupture. However, the limit of the dilatancy model lies in the requirement of unreasonably stress–strain conditions, much higher than those typically observed in nature. Most of the dilatant cracks are also order of micrometers smaller than the rock grain size, so that water cannot penetrate into microcracks resulting in a low pore fluid diffusivity. It is therefore proposed that radon anomalies depend on a crack growth mechanism at relative low strain rates upon the effect of water-induced stress corrosion (Anderson and Grew, 1977). Furthermore, the reactive surface area model (Thomas, 1988) points out that microfracturing prior to major seismic events is responsible for precursory increase in radon gas concentrations due to the exposure of fresh rock surfaces, escape of trapped gas from the rock matrix, and development of alteration reactions with groundwater. Conversely, in the pore collapse model (Schery and Gaeddert, 1982; Thomas, 1988; Toutain and Baubron, 1999), the elastic deformation involving the rock substrate is responsible firstly for the pore pressure changes and secondarily of its rupture (Bernard, 1992). Only after pore ruptures, the contained gas and fluids escape towards the surface, determining the increase in the radon concentration. If microcrack propagation with increasing stress causes an increase in the surface area of the rock substrate and anomalously high radon concentrations (Cicerone et al., 2009; Igarashi et al., 1995; Teng, 1980), strain changes in the rock substrate are accompanied by compressive mechanisms responsible for an anomalous decrease in the radon signal (Kuo et al., 2006; Roeloffs, 1999; Silver and Wakita, 1996). At the early deformative stage, the closure of cracks by small increase in compressive stress decreases the radon emanation, providing explanation for the rare but existing cases of negative radon anomalies recorded prior to a major event (Sultankhodzhayev et al., 1976; Fleischer and Mogrocampero, 1985). When the creeping rate of active faults decreases over time, the radon background level is documented to substantially decrease by 50% several days before a mainshock (Kuo et al., 2006). Additionally, rock microfracturing may produce, at its onset, a negative radon anomaly because of dilution of the gas by an increased atmospheric circulation in the soil (Yasuoka et al., 2009; Wang et al., 2014).

The intense seismicity recorded before earthquakes and volcanic eruptions induces rock fracturing and opening of new voids, thus enhancing the rock discontinuities that facilitate degassing flux. A fault zone is expected to change both before and during seismic and eruptive processes due to a number of factors, such as the formation of new cracks, changes from ineffective (or isolated) to effective (or connected) porosity, grain size comminution, and gouge evolution. The characteristic high concentrations of radon over faults in seismic and volcanic areas reflect gas migration dominated by deformations both at macro- and/or microscale. Active faults are weakened zones composed of highly fractured materials with a high porosity and permeability compared to the surrounding country rock. This property can turn fault zones into preferential pathways for advective gas-carrying fluid transport (King, 1986). Degassing occurs by pressure drop during ascent of fluids to the surface that allows the gases to escape from the fluids into soil gas and eventually into the atmosphere (Toutain and Baubron, 1999). Variations of radon concentrations in volcanic environments account for an increased heat flow or dry steam discharge, pushing up the available underground radon (Cox et al., 1980). In areas of high subsurface temperature and structural permeability, the carrier for shallow radon is both steam and heated ground air. In zones where permeability and porosity are high, the upward flux of the gas mixture brings up radon from a relatively important depth, such as several hundred meters. H_2O vapor prevails in fumarolic fields, whereas CO_2 is normally the most abundant volatile specie driving radon emissions in volcanic and geothermal areas (Allard et al., 1991; Farrar et al., 1995; Alparone et al., 2005; Baubron et al., 2002; Chiodini and Marini, 1998; Chiodini et al., 2008; Giammanco et al., 1998; Lewicki et al., 2005; Neri et al., 2006; Notsu et al., 2006; Sorey et al., 1998). Along the main fault plane, CO_2 advection is expected to be high enough to bring radon atoms to the ground surface, but not so high as to dilute the radon concentration with excess CO_2 (Giammanco et al., 2009). Low ^{222}Rn and ^{220}Rn concentrations measured in correspondence to a CO_2 anomaly suggest that the ascending deep gas dilutes both radon isotopes (Siniscalchi et al., 2010). This inverse correlation is addressed to substantial substrate fracturing, so that the CO_2 flux is high enough to overwhelm the source of radon radionuclides (Giammanco et al., 2007). Since only the long-lived isotope ^{222}Rn can reach the surface through the faulted material, $^{222}Rn/^{220}Rn$ ratios higher than those measured in other parts of an active fault indicate local conditions of relatively low soil permeability (Bonforte et al., 2013). On several volcanic edifices, faults do not necessarily reach the surface, being buried by more recent eruptions. Therefore, radon monitoring is also a suitable tool for identifying, locating, and mapping buried faults and hidden fracture patterns (Burton et al., 2004; De Gregorio et al., 2002; Giammanco et al., 2007; King et al., 1996; Neri et al., 2011; Siniscalchi et al., 2010; Tanner, 1980). However, fresh, compact, impermeable, and less fractured lava flows covering the active faults may result in comparatively low radon concentrations with respect to deformed subsurface rocks in the far field (Heiligamann et al., 1997). It is also possible that hydrothermal fluids circulating inside the volcanic edifice produce chemical alteration phenomena that significantly reduce the rock permeability by formation of crack-sealing minerals and vein deposits (Gratier et al., 1994). The radon signal is also observed to decrease when the ground is sealed by the development of an impermeable clay layer associated with an acid reflux zone that acts as a barrier to magmatic gases (Heiligamann et al., 1997). Surface manifestations of hydrothermal activity, such as vapors emitted from fumaroles, frequently produce chemical alteration of the surface rocks, and the development of soil layers with elevated radon emanating radium concentrations that provide a secondary contribution to the radon signal (Nishimura and Katsura, 1990). Mapping of ground gas radon concentrations is therefore used in the identification of subvolcanic geothermal systems (Cox et al., 1980; Thomas et al., 1986).

Volcanic eruptions are commonly preceded from days to months by an increasing volcanic tremor and substantial variations in near-surface radon concentrations at distances up to tens of kilometers from the events, especially when the summit part of the volcano and/or its flanks are interested by magmatic intrusions, deformations, and earthquakes (Cox et al., 1980; Cox, 1983). In permanently active volcanoes, the main conduct is constantly filled with magma, and the eruptive events are produced either by sharp increases of gas pressure or by the opening of new fissures. The expulsion of a large amount of volcanic gas carriers for radon may occur by the reactivation of fractures and faults, immediately before the volcanic explosion (La Delfa et al., 2007). A small dyke injection generates stress and minor deformation along faults and fissures in the volcanic edifice. Compression and contraction of the material adjacent to the dyke cause that most of the trapped gases are pushed to the surface, so that more radon atoms are detected (Heiligamann et al., 1997). In contrast, large volumes of intrusive magma result in substantial changes in the stress field within the volcanic edifice and stress propagation through the subvolcanic substrate. Consequently, radon anomalies are measured away from the location of the magmatic intrusion by aseismic deformation and microfracturing of the subsurface rocks over distances of several tens to hundreds of kilometers long distances (Dobrovolsky et al., 1979; Fleischer, 1981; Thomas et al., 1986). Positive radon anomalies are also induced by faulting resulting from seismically induced ground shaking and by thermal effects due to heat flow. Increasing volcanic temperatures for months or even years are frequently accompanied by high radon emissions. Stream-driven ground gas movement and direct volatile release from magma are effective mechanisms for the removal of radon along the rock surfaces bounding intensively fractured volcanic rocks (Gasparini and Mantovani, 1978). Increased radon concentrations are due to the fast advective gas transport near the surface and the great availability of radon at shallow levels along faults and fissures. In this view, measurements of radon concentrations are frequently used to estimate the volume of magma chambers feeding eruptions, as well as radon temporal variations are correlated with the onset of volcanic eruptions (Connor et al., 1993; Monnin and Seidel, 1997; Alparone et al., 2004; Segovia et al., 2003; Immè et al., 2006; Morelli et al., 2006). As ascending magmas approach the surface, radon atoms are removed from shallower and shallower fluids filling the overlying rocks, accounting for the higher solubility of radon in gases than in liquid water (Gasparini and Mantovani, 1978). This effect is enhanced by the increase in temperature and/or abundant gas flushing that extracts radon from water, so that the increase in the radon signal is proportional to the rate of magma uprising and/or the rate of gas release through the fractured rocks. However, the stress distribution on a volcano is far from uniform, and local stress accumulation may deeply control the radon behavior (Chirkov, 1976). Prolonged inflation and deflation of the magma chamber, as well as frequent intrusions of magma, generate deformation of the volcanic edifice and microfracturing in near surface rocks. During the inflation of the magma chamber by input of new magma from depth (or increased magmatic volatile pressure), the caldera floor rises and cracks and/or fissures open, allowing radon to be released from rocks (Chirkov, 1976). Conversely, during deflation stage, the caldera floor subsides by removal of overpressure with eruption and cracks and/or fissures close, reducing the number of pathways for radon release.

Summarizing, the radon signal measured on active volcanoes can be addressed to: (1) the concentration of parent radionuclides in the different layers of rocks from the subvolcanic substrate, (2) the preferential escape of radon from the rock matrix with increasing surface area to volume ratio, (3) the average bulk permeability of the rock substrate and its primary (interconnected pores and vesicles of the rock) or secondary (rock fracture and foliation) characteristics, (4) changes in the advection-driven transport due to variations in the deep gas flux and in the gas carrier velocity (Neri et al., 2016). Although there is a great number of different approaches that attempt to explain radon changes in active tectonic and volcanic areas, the position of the monitoring station and the physicochemical properties of the rock substrate remain undoubtedly the most important factors controlling the measured radon signal. This is also testified by simultaneous observations of positive and negative radon anomalies prior to earthquakes and volcanic eruptions, accounting for the variable stress–strain conditions and the different local rock lithologies (King, 1980; Liu et al., 1985). If positive radon anomalies are observed at distances of hundreds of kilometers from an earthquake epicenter and/or a volcanic eruptive vent (Cox et al., 1980; Cox, 1983; Wakita, 1996), preseismic sharp drops of radon are also measured from a few days to a few months prior to major ruptures (Choubey et al., 1999; Igarashi et al., 1995; Wakita et al., 1980; Kuo et al., 2006, 2013), as well as no relationships are found between the radon signal and magmatic activity (Giammanco et al., 2007). Because local geology has different effects on radon emissions, contrasting radon anomalies can be measured by monitoring stations located on adjacent rock substrates with different lithologies. Under such circumstances, the precursor time, distance, and magnitude of major events do not exhibit a universal relationship valid for the interpretation of geochemical anomalies. Radon monitoring alone is not yet successful in correctly predicting earthquakes or volcanic eruptions because many problems still remain unresolved due to the influence of rock physicochemical properties on radon emissions measured in tectonic and volcanic environments.

18.3 RADON SIGNAL AND DEFORMATION EXPERIMENTS

In the early laboratory study by Giardini et al. (1976), a variety of rock samples of igneous, metamorphic, and sedimentary origin are uniaxially stressed up to failure at room temperature, with the aim to investigate the release of occluded H_2, CH_4, H_2O, N_2 CO, O_2, and CO_2 gases. The novelty of this study consists of the early experimental observation that stress-induced emission of highly mobile gases serve as potential indicator of a critical stress build-up in crustal rocks. Similarly, Honda et al. (1982) analyze the degassing behavior of He and Ar from granite, basalt, and volcanic tuff samples subjected to uniaxial compression. The authors conclude that degassing of rare gases from the compressed samples depends primarily on the generation of new surface areas by dilatancy. The frontier research conducted by Holub and Brady (1981) documents for the first time the behavior of radon from rock placed under uniaxial stress conditions. The testing material is a granite impregnated with secondary (radioactive) mineralization. Uniaxial compressive tests are performed under deformation control to perform loading and unloading cycles. At the end of the cycles, the granite sample is loaded up to failure, when the unconfined compressive strength (UCS) is ~130 MPa. The most important feature of this experimental setup is that radon (i.e. ^{222}Rn) emanation is recorded continuously during the test (Fig. 18.2). A rotary pump and a tubing system connect the individual parts of the setup, so that the radon-containing atmosphere flows into a chamber whose walls are covered with a scintillator compound made of zinc sulfide (ZnS). At the inlet, a filter is placed to remove all radon daughters and to ensure that only radon is allowed into the chamber (Fig. 18.2). The chamber is attached to a photomultiplier supplied with a high-voltage source, preamplifier, and amplifier followed by a counting system (Fig. 18.2). The alpha particles impinging on the sensitive ZnS layer from the inner volume of the chamber or from a molecule deposited on the chamber walls produce a scintillation which is converted in an electric pulse, and then counted. This feature has, however, two types of limitations: (1) the radon-detecting efficiency depends on whether the measured species is in the gas phase or is deposited on the chamber walls, but a proper calibration can easily account for that and (2) radon trends are affected by the delayed decay of ^{214}Po. At the initial loading condition (Fig. 18.3), a decrease in radon emanation is observed due to the closure of preexisting microcracks and pores within the granite (i.e. rock inelastic deformation or rock inelasticity; Scholz, 1968). When the uniaxial stress is approximately one-half of the ultimate strength (Fig. 18.3), a temporary increase corresponding to 50% of

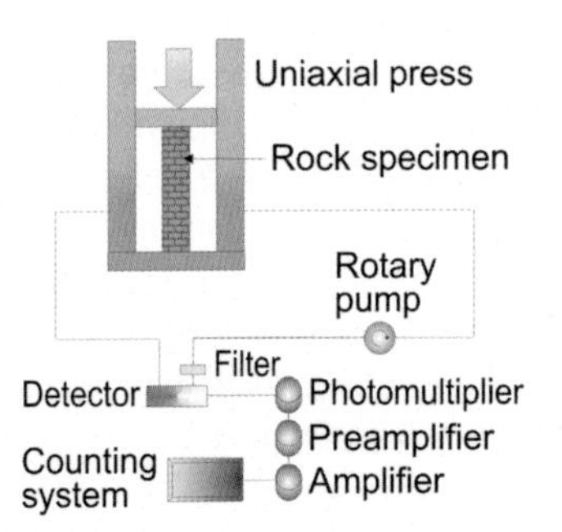

FIGURE 18.2 Schematic representation of the experimental setup designed by Holub and Brady (1981) to measure radon emissions from rock samples placed under uniaxial stress conditions.

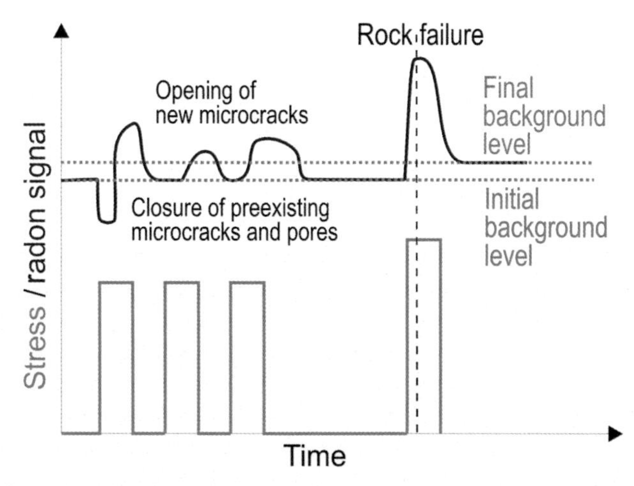

FIGURE 18.3 **Simplified sketch illustrating changes in stress condition and radon signal as a function of time.** The uniaxial experiments are conducted on a granite rock.

the initial radon emanation is observed, in concert with an increase in microcracking within the rock (i.e. rock dilatancy or rock dilatant region; Scholz, 1968). When the stress is removed for several hours, the original radon emanation is restored (Fig. 18.3). But, once the granite is failed, a temporary steep increase up to 120% in radon emanation occurs, attesting that the time duration of the prefailure stress was enough to store radon nuclides in the newly created microfractures before rock failure (Fig. 18.3). Subsequently, the radon emanation decreases down to 5%, remaining invariably higher than the pre-failure background level (Fig. 18.3). In this view, most of the rocks forming geological substrates and/or volcanic edifices consist of a matrix with randomly distributed open cracks and closed pores. Radon emissions from the matrix take place partly in the space of the closed pores, partly in the cracks, and some is adsorbed by the free inner surface (Utkin, 2000). Thus, the measured radon concentration at the initial stress corresponds to the background level of the rock. Under compressive conditions, the stress in the rock substrate increases and the radon first increases as the volume of cracks contracts releasing interstitial gas, but then decreases after the cracks close. Anomalous high radon concentrations are also addressed to compressive mechanisms in the crust, leading to substantial outgassing at the ground surface by rock pore collapse and upflow of deeper radon-rich gases (King, 1978). As the stress grows, rocks become subject to failure breaking down the pore walls to develop macroscopic fractures and radon activity increases remarkably. In contrast, under stress release, radon first decreases as the volume of cracks expands, but then increases after that cracks and pores open (Utkin and Yurkov, 2010).

In the same years, Katoh et al. (1984, 1985) develop an experimental system to measure radon (i.e. both ^{220}Rn and ^{222}Rn) emissions from granite specimens stored in a Tedlar bag that is directly uniaxially compressed by the uniaxial press. The air in the bag containing radon and thoron emitted from the specimens is circulated by a pump in a counting system where the daughters generated by radon and thoron decays are collected electrostatically and the alpha particles emitted by them are counted by a scintillation detector. The deformation tests are performed by increasing the stress stepwise and by applying constant stresses intermittently, in order to investigate the relationship between radon emanation and stress condition. The radon emanation is observed to increase only when the axial stress corresponds to 70% of UCS (Fig. 18.4). Then, the emanation increases proportionally to the increasing stress condition, as the diffusion of the radon gas through the channels of the stress-induced microcracks is more favored (Fig. 18.4). A remarkable increase of radon emanation appears only after the ultimate fracture, when the granite is broken into blocks exposing an increased surface area ~150% greater than the original one (Fig. 18.4). This result is consistent with radon data collected by Hishinuma et al. (1999) using an almost identical experimental setup in which a granite sample is gradually compressed over several days. With respect to the background level, a weak radon increase of ~15% is measured after ~8 days, whereas the signal increases by ~400% when the granite is crushed after ~9 days. If the fractured granite sample is gradually recompressed using the same steps, the progressive closure of microcracks and pores causes an early radon increase of ~300% after ~4 days, testifying the expulsion of residual gas entrapped within the rock. As the compression further increases, the radon signal decreases down to the background level after ~5 days, due to complete closure of microcracks and pores in the granite.

In a more comprehensive study, Koike et al. (2015) conduct uniaxial compression tests on a set of granite specimens showing different grain size, mineral phases, radioelement contents, UCS, and failure types. The experimental system is composed of a material testing machine, a self-made cell unit for uniaxial compression of specimens, a pump to generate gas flow in the cell unit, and a portable alpha-scintillation detector for continuous measurement of the total number of alpha particles per minute emitted from the granite, resulting from the decays of ^{222}Rn and ^{220}Rn and their daughter nuclides,

^{218}Po (186 s half-life) and ^{216}Po (0.158 s half-life), respectively. The disadvantages of this experimental setup lie in the detector that does not differentiate the separate contribution of each nuclide. The alpha particles are simply associated with a general radon count by assuming that this count is mostly proportional to the ^{222}Rn. The rock samples are loaded linearly up to the failure, corresponding to UCS values from 146 to 260 MPa. Common to all specimens, radon emissions remain at or slightly decline from the background level after loading, responding to the closure of original microcracks. Then, radon emissions begin to increase at 46%–57% stress level and continue to increase even after the granite failure. Interestingly, the rock failure pattern occurs as both conjugate shear failure type (i.e. development of cross-fractures in the failed sample) and axial splitting failure type (i.e. development of vertical fractures in the failed sample) at high and low values of UCS, respectively. Apparently, axial splitting failure generates a larger surface area in the granite specimens due to the development of longitudinal fractures that act as continuous, open conduits for the migration of radon (Fig. 18.5). The opposite occurs during conjugate shear failure, suggesting that granites with higher UCS are less prone to release radon. In rocks characterized by brittle deformations, opening of tensile cracks and fissures are produced in proximity of the maximum state of stress (Heap et al., 2010). Since the measured radon emanation is proportional to the exhaling surface area, rock samples with smaller grain sizes tend to expose higher grain boundary surfaces, thus enhancing remarkably the radon signal (Fig. 18.5). A more rapid development of microcracks (i.e. fast failure velocity) is also correlated with a higher radon emission per unit time (i.e. transient concentration peak of radon). This points out that radon emission may be a function not only of the rate of gas accumulation within the pore spaces of the rock, but also of the rate at which a new surface area develops by crack growth (Fig. 18.5).

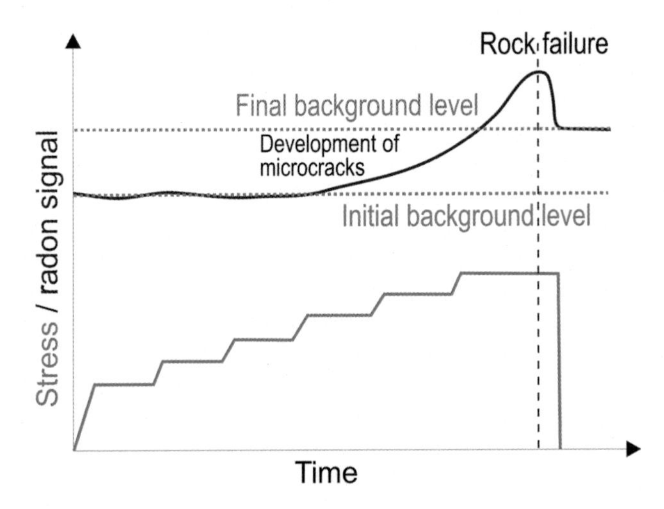

FIGURE 18.4 Simplified sketch illustrating changes in stress condition and radon signal as a function of time. The uniaxial experiments are conducted on a granite rock.

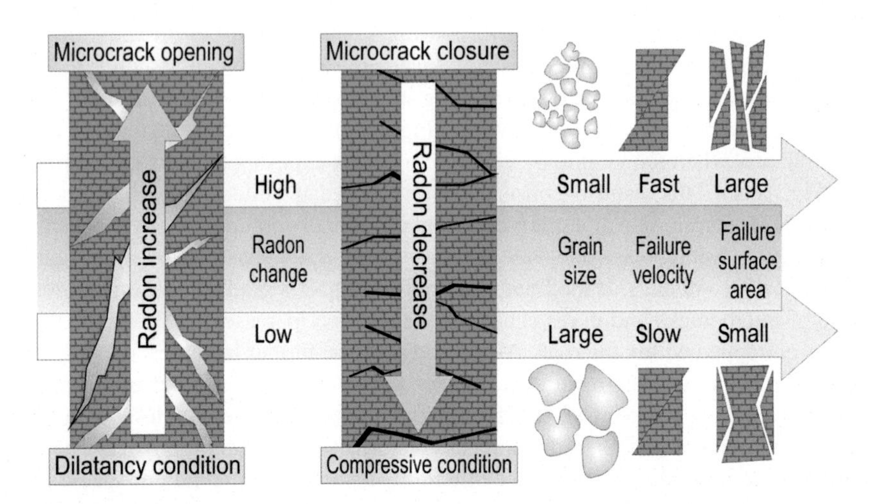

FIGURE 18.5 Diagram illustrating the most important parameters controlling radon changes during rock deformation. The uniaxial experiments are conducted on a granite rock.

In order to elucidate the seemingly confusing array of radon anomalies observed prior to earthquakes and volcanic eruptions in nature, Tuccimei et al. (2010) and Mollo et al. (2011a) present uniaxial experiments in which radon emissions are measured on intact, deformed, and failed volcanic samples characterized by very different petrophysical characteristics: (1) a loosely consolidated, vesicular tuff formed by explosive eruptions and (2) a highly consolidated, massive phonolite lava typical of effusive volcanic activity. The tuff has low cohesion and high porosity (~47%) due to the occurrence of centimetric to micrometric vesicles dispersed in a matrix of poorly aggregated volcanic ashes, mineral fragments, and scoria clasts. In contrast, the lava has high cohesion and low porosity (~4%) resulting from a dense mosaic of millimetric minerals surrounded by an intricate groundmass of micrometric crystals. The deformative mechanism of the tuff is characteristically ductile showing a short duration of the elastic phase and a large strain region before failure (Fig. 18.6). This poorly welded tuff exhibits also a low strength corresponding to ~2 MPa. By increasing the deviatoric loading, the onset of pervasive pore collapse process causes increased densification and compaction (Hudyma et al., 2004; Sonmez et al., 2004, 2006; Avar and Hudyma, 2007), causing that the radon emission decreases (Fig. 18.6). This well correlates with the inelastic pore collapse typically observed by applying hydrostatic pressure on high porosity tuffs (Vinciguerra et al., 2006, 2009). Under loading conditions, an increase in radon emission is measured only after rock failure (Fig. 18.6). Despite the compaction of the material continues to reduce the rock porosity, the newly formed macroscopic faults influence the exhalation rate so massively that an overall increase in emission is recorded (Fig. 18.6). The concentration of radionuclides in a rock is dependent on the microstructural fabric, attesting that microfractures (and, obviously, macroscopic faults) act as preferential pathways for migration of radionuclides (Sengupta et al., 2005). In contrast, the deformative mechanism of the lava is characteristically brittle, with minor deviation from linear elastic deformation behavior until immediately prior to failure, where a small degree of strain hardening is observed (Fig. 18.7). At the early stage of applied stress condition, the uniaxial strain is very low accounting for minimum compaction of preexisting microcracks and pores in the intact lava. As a consequence, the closure of a negligible number of internal pore spaces does not produce any variation in radon concentration at the beginning of deformation (Fig. 18.7). In the elastic region, the stress–strain curve is linear and no microcracks develop, resulting in a constant radon concentration (Fig. 18.7). Despite the rock does not macroscopically fail at the end

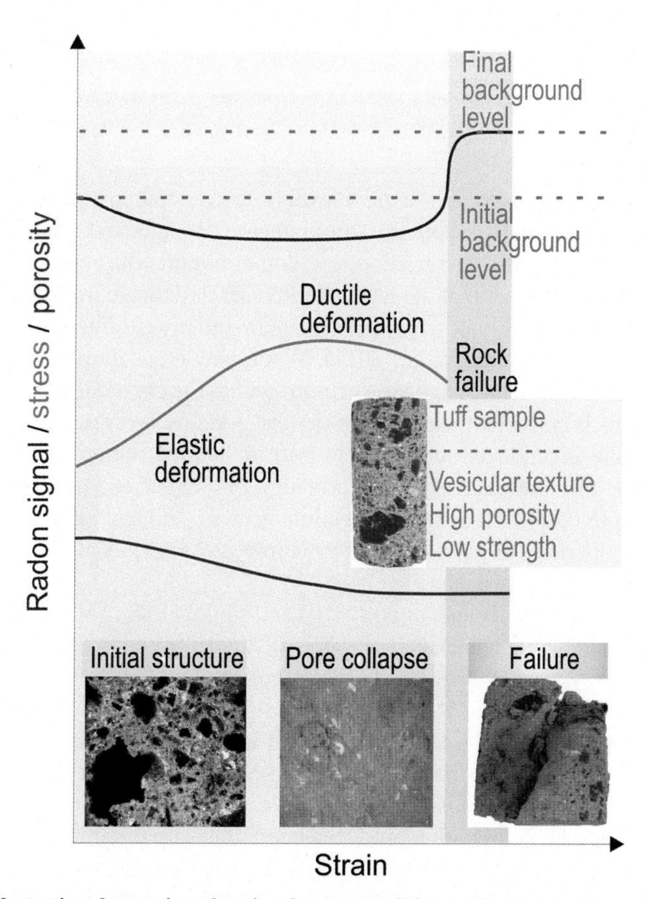

FIGURE 18.6 Simplified sketch illustrating changes in radon signal, stress condition, and rock porosity as a function of strain level. The uniaxial experiments are conducted on a loosely consolidated, vesicular tuff.

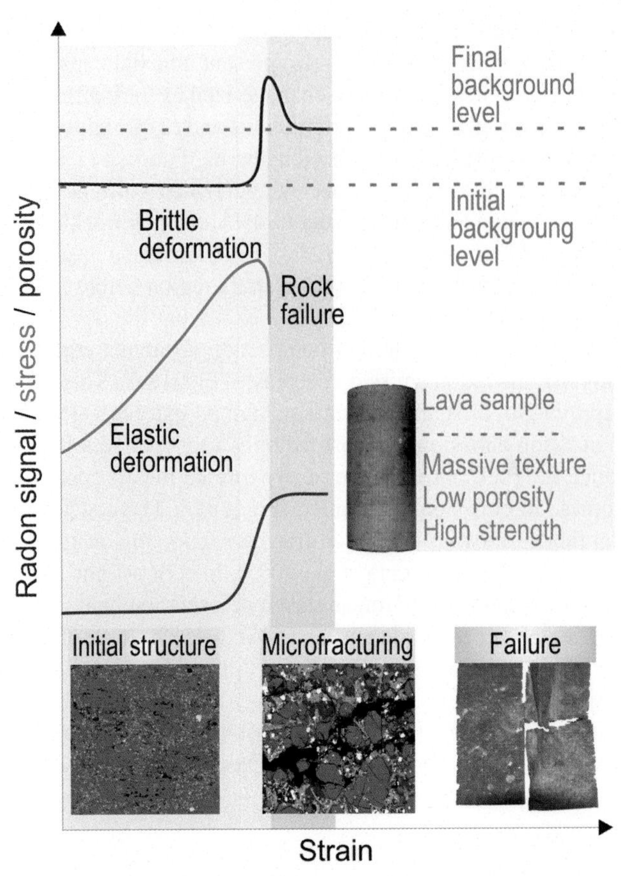

FIGURE 18.7 **Simplified sketch illustrating changes in radon signal, stress condition, and rock porosity as a function of strain level.** The uniaxial experiments are conducted on a highly consolidated, massive phonolite lava.

of the elastic deformation, the lava sample contains a great level of irreversible microcrack damage. The dilatant nature of the deformation is highlighted by an increase in porosity from intact to damaged to failed sample (Fig. 18.7). However, the damaged sample does not show any significant variation in radon emission with respect to the background signal of the intact rock (Fig. 18.7). This is due to the fact that the level of microcrack damage imparted on the sample is not sufficient to drive changes in radon emanation. Notably, low-porosity rocks invariably exhibit a reduced radon flux, in spite of their enhanced radioactive source content (Banerjee et al., 2011). When new emanation surfaces are created in the damaged sample, most of them occur as isolated fractures that are enclosed within the crystalline matrix and the internal radon atoms cannot escape from the rock. This is consistent with studies on low-porosity lavas demonstrating that a significant portion of the damage required for failure is produced immediately prior to sample rupture (Heap et al., 2010) and that significant changes in permeability are only observed after the formation of a large macroscopic fault (Fortin et al., 2011). The lava failure occurs at high strength (~240 MPa) by propagation, growth, and aggregation of microcracks. A macroscopic fracture develops parallel to the direction of the maximum principal stress (i.e. axial splitting) and it acts as new exhaling surface for substantial increase in radon emission (Fig. 18.7).

In the comparative data presented by Tuccimei et al. (2010) and Mollo et al. (2011a), the measured radon signal is collected by a stepped experimental strategy in which the sample is firstly deformed in the uniaxial machine and then its radon emission is analyzed in a counting system. In other words, the authors present "snapshot experiments" that can only provide radon emissions from a rock sample deformed at predetermined stress conditions. Possible interpretations about changes in radon emission rest on the assumption that a rock produces an identical radon emission over time irrespective of load paths and stress trajectories. To overcome this restriction, Tuccimei et al. (2015) develop a novel real-time experimental setup for measuring in continuum the radon emissions from uniaxially deformed rocks. With respect to previous systems in which radon fluctuations during rock deformation produce uncertainties up to ~35% (cf. Holub and Brady, 1981; Katoh et al., 1984, 1985; Hishinuma et al., 1999; Koike et al., 2015), this novel experimental setup allows to analyze the alpha-emitting daughters of radon with uncertainty of ~10%. The real-time setup includes a radon accumulation chamber obtained from a DryPack material made up of extremely tough and durable layers (Fig. 18.8). The rock specimen is inserted and

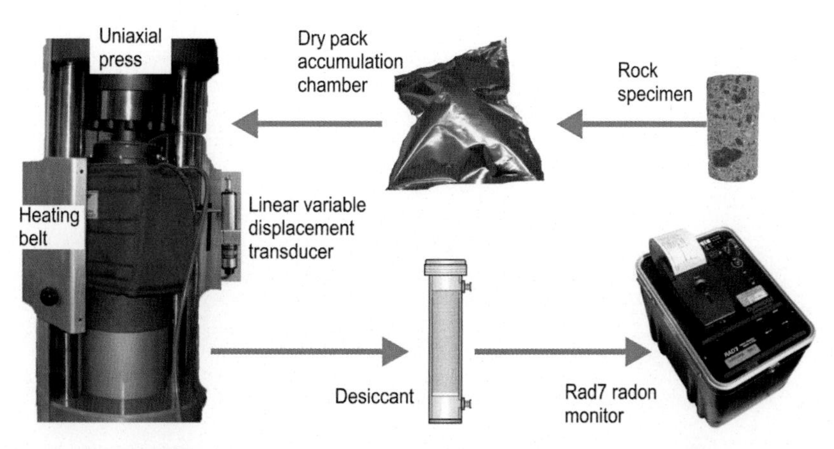

FIGURE 18.8 **Schematic representation of the experimental setup designed by Tuccimei et al. (2015) to measure radon emissions from rock samples placed under uniaxial stress conditions.**

sealed into the accumulation chamber that, in turn, is placed into the uniaxial testing apparatus (Fig. 18.8). The axial deformation is measured throughout a linear variable displacement transducer (Fig. 18.8), whereas a dedicated software allows us to conduct a variety of stress–strain experiments from fast-to-slow deformation rates. A heating belt surrounds the accumulation chamber (Fig. 18.8) connected to a desiccant and to a RAD7 radon monitor (Durridge Company, Inc.). Since the radioactive decay is a stochastic process, statistical fluctuations are the main source of measurement errors. This is particularly true under deformation conditions in which rock mechanical and physical properties continuously change over time. The heating belt enhances the number of effective collision of radon atoms with other molecules and grain walls, favoring the diffusion of radon gas through the pore spaces of the rock (Tuccimei et al., 2009, 2011). The end result is that the radon signal increases and the analytical uncertainty decreases. A recirculating pump in the RAD7 moves the enhanced gas from the accumulation chamber to the detection volume where alpha-emitting radon and thoron progenies are electrostatically collected onto the solid-state detector for counting. Using this experimental setup, it is found that the signal of radon emitted from the tuff rock can drastically change when the same stress condition is applied over different time durations. In a short-term experiment, the tuff rock is deformed by applying two different stress conditions, corresponding to 40% and 80% of UCS. The stress is maintained constant for a few days during each step and then it is increased up to sample failure (Fig. 18.9). Due to the overall effect of pore collapse, the radon concentration is observed to continuously decrease over time, especially at the high stress level (Fig. 18.9). After sample rupture, the radon signal drastically increases as the result of the formation of new large exhaling surfaces (Fig. 18.9). The long-term experiment is conducted at the same stress levels used for the short-term one, but the time duration of each step is about twice as much (Fig. 18.9). At low stress level, the concentration of radon decreases over time, accounting for the compaction of the tuff material and the reduction of the pore space (Fig. 18.9). On the contrary, at high stress level, pervasive microfracturing prevails over tuff densification, new exhaling surfaces develop, and the radon signal progressively increases until sample failure (Fig. 18.9). The contrasting radon behavior observed in long- and short-term experiments evidences that, in volcanic areas characterized by large volumes of ignimbrite deposits (i.e. tuff rocks) subjected to a constant stress level, both negative and positive radon anomalies may be recorded over short and long timescales. Constant magmatic pressures caused by dike intrusions and hydrothermal fluid injections may cause pervasive pore collapse, in concert with significant radon decrease over short timescales. Conversely, after compaction and densification of the material, microfracturing emerges as the most effective deformation mechanism over long timescales. Thus, for high-porosity volcanic rocks, the radon signal may first decrease and subsequently increase over a considerable period of time (days and perhaps weeks and months), with paramount repercussions for the interpretation of field monitoring data.

It is found in the literature that variable radon concentrations can also result from different types of deformation of the volcanic edifice, including cycles of inflation and deflation triggered by new magma supply at subvolcanic levels (Neri et al., 2006). Repeated stress cycles over time are investigated in laboratory by Tarquini (2012) and Scarlato et al. (2013) using the real-time experimental setup described in Tuccimei et al. (2015). Four types of experiments are conducted with the aim to highlight the deformation mechanism of the tuff rock prior to macroscopic rupture (Fig. 18.10). Experiments 1 and 2 consist of a low number of stress cycles where the tuff sample is loaded and unloaded over time. At the end of the cycles, the stress condition is released for several days, and then increased to low stress for Experiment 1 and to high stress for Experiment 2 (Fig. 18.10). Experiments 3 and 4 are performed by adopting the same conditions of Experiments 1 and 2, but the tuff sample is subjected to a high number of stress cycles (Fig. 18.10). The most striking result is that repeated stress cycles invariably leads to the development of a diffusive pattern of microcracks in the tuff rock, causing a substantial increase in the radon signal

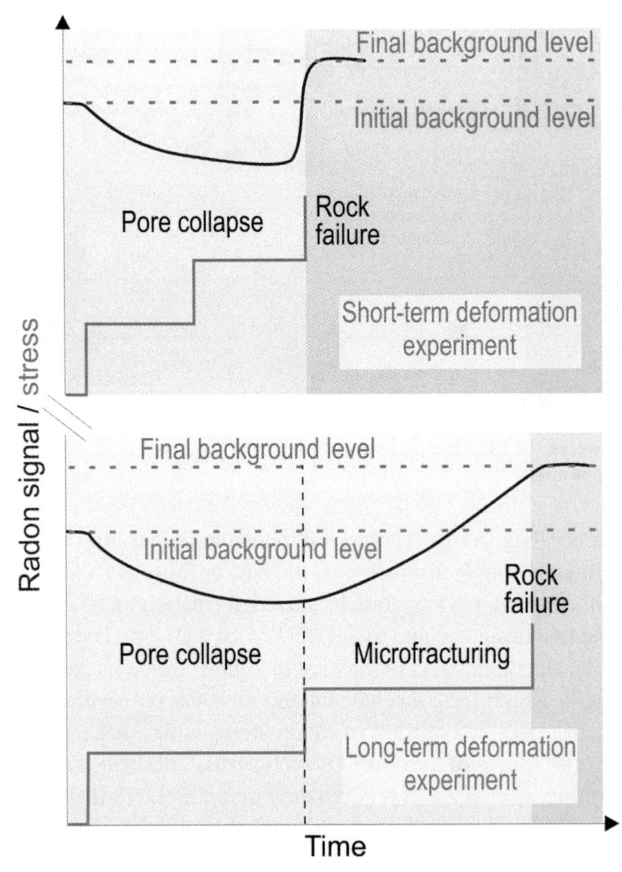

FIGURE 18.9 **Simplified sketch illustrating changes in stress condition and radon signal as a function of time.** Short- and long-term uniaxial deformation experiments are conducted on a loosely consolidated, vesicular tuff.

over time (Fig. 18.10). The damage imparted to the tuff sample by microfracturing produces new emanation surfaces and migration channels for gas release. As a consequence, the radon signal is observed to irremediably increase either when the stress condition is released or maintained a constant (Fig. 18.10). However, microfracturing and radon escape are more effective when the number of stress cycles is low and the applied stress condition is high (i.e. Experiment 2; Fig. 18.10). Otherwise, an increased compaction and densification of the material partly reduce the number of pathways for gas release, so that the radon signal weakly increases (i.e. Experiment 3; Fig. 18.10). This demonstrates that the imposed stress conditions have contrasting repercussions on the radon signal produced by voluminous and highly porous volcanic deposits. Repeated cycles of stress, such as those resulting from magma recharge events and inflation/deflation deformations of the volcanic edifice, are mostly accompanied by microfracturing rather than pore collapse phenomena. These opposite mechanisms are consistent with the notion that the rock emanation surface area is the most important factor controlling radon exhalation from rocks (Banerjee et al., 2011; Sengupta et al., 2005). Therefore, in the case of the same rock type in the ground, variable volcanic stress regimes may drastically increase the emanating power of the substrate and, consequently, the background level of radon.

Different from uniaxial experiments measuring radon under unconfined pressure conditions, Sammis et al. (1981) analyze the radon nuclides dissolved in water percolating through the microfractures of a granitic rock under variable triaxial stress conditions. The granite samples are loaded from ~30 to ~270 MPa in a triaxial cell which is placed between the anvils of a hydraulic press. Confining pressure and flow pressure are generated by air-driven hydraulic pumps, while radon-free distilled water is flowed through the sample. An outlet at the base of the triaxial pressure vessel allows to sample the pore water in which radon nuclides (adsorbed to the walls of the fracture network) are dissolved. Radon is extracted from the water using recirculating helium as a carrier, then the two gases are separated by a liquid nitrogen cold trapping. A drying column is used to capture CO_2 and H_2O, whereas radon is transferred to a scintillation cell where ^{222}Rn nuclides are measured by counting scintillations associated with their decays. In the first series of experiments, dry granite samples are stressed uniaxially under confining pressure and a single water sample is collected. Despite new microfractures develop at high stress conditions, no correlation is found between radon released and stress applied to the dry samples. Microfractures carry a remarkable fraction of the percolating water but they do not contain appreciable radon. This unexpected phenomenon

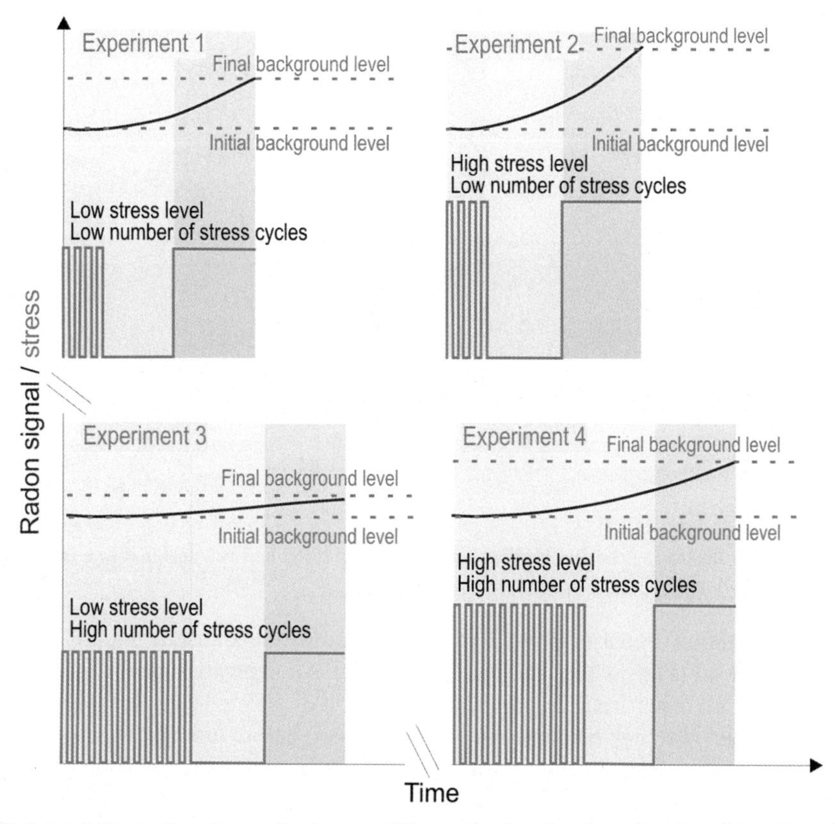

FIGURE 18.10 **Simplified sketch illustrating changes in stress condition and radon signal as a function of time.** Four different uniaxial deformation experiments are presented. Experiment 1 and Experiment 2 consist of a low number of stress cycles where the specimen is loaded and unloaded over time. At the end of the cycles, the stress condition is released for several days and then increased to low stress for Experiment 1 and to high stress for Experiment 2. Experiments 3 and 4 are performed by adopting the same conditions of Experiments 1 and 2, but the specimen is subjected to a high number of stress cycles. The experiments are conducted on a loosely consolidated, vesicular tuff.

is addressed either to the fact that only preexisting cracks had enough time to concentrate and trap significant radon nuclides or to the occurrence of radium atoms as secondary surface coatings on preexisting cracks and virtually responsible of all the observed emanation. In the second series of experiments, the granite samples are initially saturated with water, equilibrated for 1 month, and then stressed while successive water samples are collected. For a constant value of permeability, saturated samples produce between 2 and 10 times more radon than initially dry samples, delineating that most of radon is dissolved in water rather than adsorbed to the walls of the rock fracture network. However, granite permeability and radon release are inversely correlated. This surprising result appears to indicate that water flows through the rock so quickly that radon does not have time to diffuse from side channels or, alternatively, that most of the water flows through a few, large channels (rather than many small channels), thus removing a smaller fraction of the accessible radon atoms in the rock. In the more recent study by Nicolas et al. (2014), a long-term radon experiment is conducted on a granite sample deformed under triaxial conditions. The sample is prefractured through a heat treatment in order to increase the crack surface area, the rock permeability, and the efficiency of water–rock interaction. In the experiment, the sample is kept for 21 days under constant isotropic conditions (no differential stress and confining pressure). Then, cycles of differential stress are performed by loading the sample to a given differential stress and unloading to the initial isotropic condition at a constant rate. Pore fluid sampling is conducted at the end of each cycle. The differential stress is increased each day up to rock failure. The constant pore pressure is controlled by argon flow that permeates through the sample and carries radon with it. At the time of the sampling, argon is released in a gas sampling device and then analyzed with total uncertainty on radon concentration of 9%. Results of the continuous radon measurements show that the signal is constant under isotropic conditions (Fig. 18.11). During the cycles of differential stress, the radon concentration progressively increases up to ~130%. Repeated differential cycles yield a large release of radon, which is maximum near the sample rupture (Fig. 18.11). Differential stress modifies the porous network first by the closing of microcracks and micropores and then by the opening and coalescence of new fractures. The fractures serve as channels to connect the isolated micropores to the permeable rock network. Radon liberated from occluded pores to interconnected fractures is thus flushed producing a transient gas anomaly (Fig. 18.11). When rup-

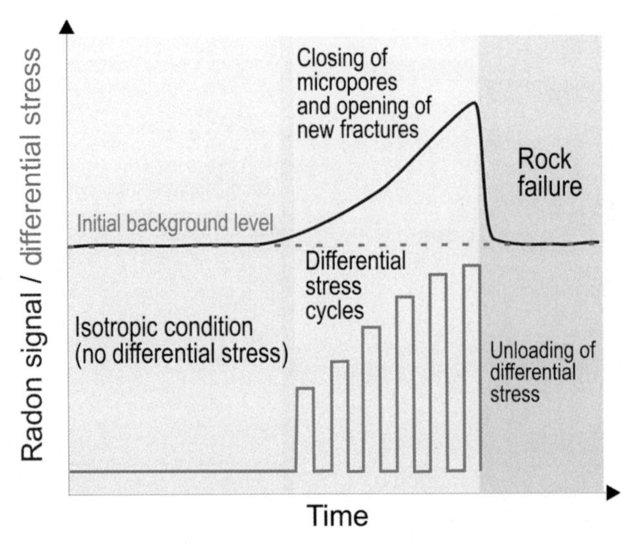

FIGURE 18.11 **Simplified sketch illustrating changes in differential stress condition and radon signal as a function of time.** A granite rock is deformed under triaxial conditions during a long-term experiment.

ture occurs, the radon concentration decreases by ~170% and, after unloading of the differential stress, radon signal returns to values of the background level (Fig. 18.11). The main conclusion is that positive radon anomalies from volcanic rocks with a purely brittle deformation result from the transient connection of isolated pore spaces in which radon nuclides are stored at radioactive equilibrium (i.e. high radon activity concentration) before rupture. This is particularly true when most of the pore spaces between grains serve as traps for radon gas whose activity concentration progressively increases over time. The rate at which the migration channels develop in the rock structure and connect the isolated pores to the external rock surface plays a major role in controlling the release of entrapped gas and the radon emission.

18.4 RADON SIGNAL AND THERMAL EXPERIMENTS

Most of the monitored and active volcanoes around the world are characterized by continuous injections of magma that stall at very shallow levels or feed complex dike networks, even at a few meters from the ground surface. Intense heat is provided to the rock substrate by large, long-lived magmatic bodies (Bonaccorso et al., 2010; Wohletz et al., 1999) and circulating hot fluids (Merle et al., 2010; Siniscalchi et al., 2010). The conductive heat flow produces thermal gradients in the host rocks from 1100 °C at the contact to 200 °C at a distance of thousands of meters, incorporating several cubic kilometers of the substrate (Mollo et al., 2012a). Subvolcanic substrates, including lava flows, pyroclastic deposits, and sedimentary materials, are frequently accompanied by intense hydrothermal alteration reactions controlled by temperature changes (Merle et al., 2010). There is a great number of volcanic districts and/or volcanic edifices overlying sedimentary and pyroclastic successions containing in their paragenesis H_2O- and CO_2-bearing minerals, such as clays, zeolites, calcites, and dolomites (Conte et al., 2009; Mollo et al., 2011b; Passaglia et al., 1990). Thermally induced mineral devolatilization reactions may contribute significantly in changing the chemical, physical, and mechanical properties of subvolcanic substrates. After mineral thermal decomposition, the rock porosity is observed to increase up to five times by reducing the rock strength up to 90% and possibly triggering instability processes without any additional tectonic or magmatic forces (Heap et al., 2013; Mollo et al., 2013). For the most common clay and zeolite minerals, thermal dehydration mostly occurs from 140 to 550 °C, in concert with H_2O liberation up to ~30 wt.% (Mollo et al., 2011c; Passaglia et al., 1990). Conversely, calcite and dolomite are thermally decomposed from 600 to 910 °C by releasing CO_2 up to ~44 wt.% (Di Rocco et al., 2012; Mollo et al., 2012b) that may also contribute to the anomalously high emissions measured before volcanic eruptions and/or from magma chambers emplaced in thick carbonate substrates (Allard et al., 2006; Freda et al., 2008; Heap et al., 2013; Mollo et al., 2010). Undoubtedly, CO_2 of magmatic origin remains the most important carrier gas for radon, however, increasing radon emissions are not necessarily related to carrier gases of magmatic origin. Decoupled spatial and temporal behaviors of monitored gases are observed during eruptive and seismic activities (Heiligamann et al., 1997), as well as the recorded radon signal appears spatially heterogeneous and nonstationary in time (Kotsarenko et al., 2016; Neri et al., 2016).

If radon emissions from rock substrates exposed to subvolcanic thermal gradients are expected to be controlled by the dependency of the gas diffusion coefficient on temperature (Beckman and Balek, 2002; Scarlato et al., 2013; Voltaggio et al., 2006), thermally induced dehydration and decarbonation reactions may affect significantly the emanating power of the substrate and the background level of radon (Xue et al., 2010). Such a mechanism can be threefold: (1) formation of a disturbed material, with a vesicular and highly permeable microstructure favorable for gas transport, (2) production of abundant carrier gases by thermal decomposition of H_2O- and CO_2-bearing minerals, and (3) development of pressure gradients that contribute to increase the pore pressure, the gas ascent velocity, and the rate of outgassing. From this prospective, Manniello (2015) and Mattia (2017) investigate in laboratory the complex interplay between temperature and radon signal using as comparative materials the tuff rock and the phonolite lava studied by Tuccimei et al. (2010) and Mollo et al. (2011a), respectively. The rock samples are heated in a high-temperature chamber furnace where the temperature is constantly monitored by a factory-calibrated thermocouple with an uncertainty of ±3 °C (Fig. 18.12). The inside walls of the chamber are insulated by a gas-impermeable alumina cement, so that the radon gas in the chamber can only pass through a Mullite tube extending outside the chamber. A Teflon tube connects the Mullite tube to the desiccant and to the RAD7 radon monitor (Fig. 18.12). A recirculating pump in the RAD7 moves the gas from the chamber to the detection volume where alpha emitting radon and thoron progenies are electrostatically collected onto the solid-state detector for counting. The experimental strategy consists of heating the rock samples in eight steps of 100, 200, 300, 400, 500, 600, 700, and 800 °C. The time duration of each step is of several hours to ensure thermal homogenization of the sample, complete release of volatile at each specific experimental temperature, and low analytical uncertainty of ±20%. For the phonolite lava, radon measurements evidence that the signal monotonically increases with increasing temperature (Fig. 18.13). From a mineralogical point of view, the lava sample preserves its original phase assemblage at the end of the thermal experiment. This is due to the presence of anhydrous minerals that, in the applied temperature range, undergo negligible structural changes. Hence, the number of effective collision of radon atoms with other molecules and grain boundaries is enhanced only by the effect of temperature, favoring the diffusion of radon through the pores of the material in which radon nuclides are temporarily stored (Tuccimei et al., 2015). Therefore, radon diffusion process is prevalently controlled by temperature showing a typical nonlinear Arrhenius behavior (Fig. 18.13), where radon emissions result in a new value of equilibrium activity concentration for each temperature step (Beckman and Balek, 2002; Voltaggio et al., 2006; Molo et al., 2017). In contrast, for the case of the tuff rock, the radon emission rate cannot be interpreted and modeled in the frame of a temperature-dependent diffusion model (Fig. 18.13). The tuff rock contains abundant zeolite minerals (~35 wt.%) originated by hydrothermal alteration processes. The onset of zeolite dehydration reactions produces two positive radon peaks at 200 and 700 °C (Fig. 18.13), corresponding to thermal regions in which the release of H_2O reaches maximum values. H_2O vapor increases the pore pressure and serves as a carrier for radon atoms entrapped in the rock structure. At the highest temperature of 800 °C, the radon emission abruptly decreases due to the total thermal dissolution of zeolite minerals accompanied by the collapse of their crystal structure (Fig. 18.13). In this scenario, the radon signal mirrors the transient state of thermally induced devolatilization reactions that act as an important perturbation mechanism, causing radon fluctuations as a function of mineral dehydration.

It is important to know that many densely populated towns around the world rest on piles of pyroclastic flow deposits that extend over tens to hundreds to thousands of square kilometers (Forni et al., 2016). Tuff rocks are the most common products of pyroclastic density current flows and, since ascent times, zeolitic tuffs have been considered as sustainable building materials. A survey of indoor radon has evidenced that houses built with volcanic materials, such tuff and pozzolana, have radon activity concentrations 60% higher than houses built with typical industrial materials (Sabbarese et al., 1993). It is apparent that the high radon activity concentrations of tuff rocks from the subvolcanic substrates can be further increased by the onset of devolatilization reactions triggered by magmatic and/or geothermal processes. Extemporaneous radon fluctuations may represent a serious danger to public health, being potential source of extremely high potential radon indoor accumulations.

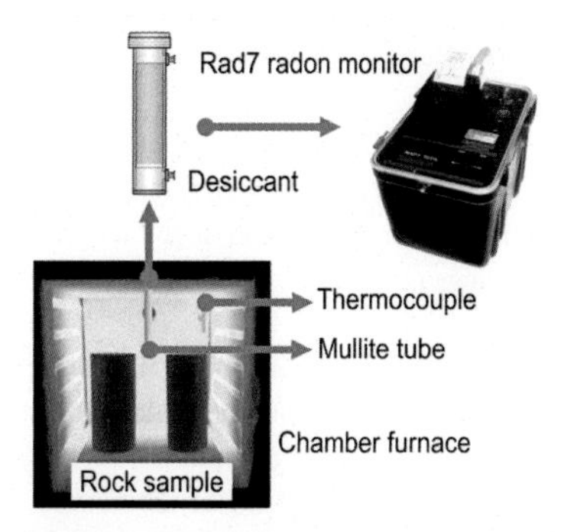

FIGURE 18.12 **Schematic representation of the experimental setup designed by Manniello (2015) and Mattia (2017) to measure radon emissions from rock samples exposed to high-temperature conditions.**

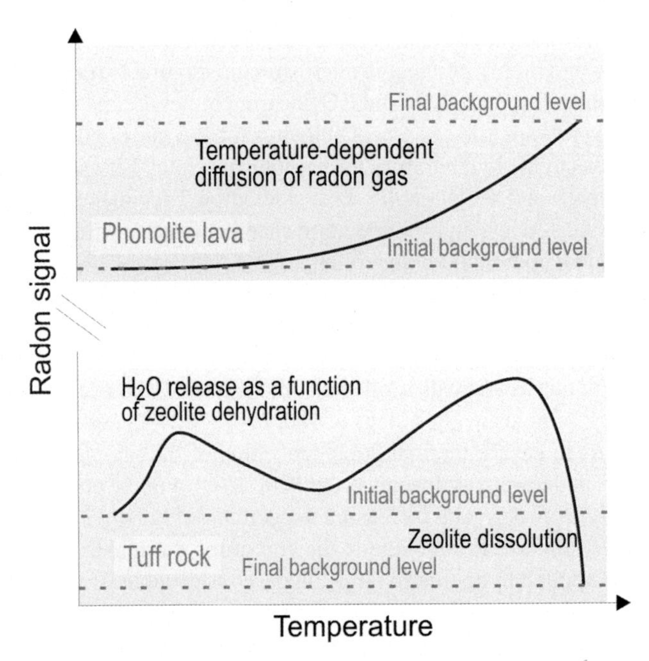

FIGURE 18.13 **Simplified sketch illustrating changes in radon signal as a function of temperature.** The experiments are conducted on a tuff rock containing hydrous minerals and on phonolite lava containing anhydrous minerals.

ACKNOWLEDGMENTS

The authors acknowledge Emanuele Tarquini, Maria Antonietta Manniello, and Martina Mattia for the valuable work done in the framework of their master thesis. Parts of this research were supported by MIUR, Premiale project—NoRth: New hORizons of the Technology applied to experimental researches and geophysical and volcanological monitoring, and by INGV, FISR 2016 Grant Number 0865.025.

REFERENCES

Allard, P., Behncke, B., D'Amico, S., Neri, M., Cambino, S., 2006. Mount Etna 1993–2005: Anatomy of an evolving eruptive cycle, Earth-Science Reviews 78, 85–114.

Allard, P., Carbonelle, J., Dajlevic, D., Le Bronec, J., Morel, P., Robe, M.C., Maurenas, J.M., Faivre-Pierret, R., Martin, D., Sabroux, J.C., Zettwoog, P., 1991. Eruptive and diffuse emissions of CO_2 from Mount Etna. Nature 351, 387–391.

Alparone, S., Andronico, D., Giammanco, S., Lodato, L., 2004. A multidisciplinary approach to detect active pathways for magma migration and eruption at Mt. Etna (Sicily, Italy) before the 2001 and 2002–2003 eruptions. J. Volcanol. Geotherm. Res. 136, 121–140.

Alparone, S., Behncke, B., Giammanco, S., Neri, M., Privitera, E., 2005. Paroxysmal summit activity at Mt. Etna (Italy) monitored through continuous soil radon measurements. Geophys. Res. Lett. 32, L16307.

Anderson, O.L., Grew, P.C., 1977. Stress corrosion theory of crack propagation with applications to geophysics. Rev. Geophys. 15, 77–104.

Andrews, J.N., Wood, D.F., 1972. Mechanism of radon release in rock matrices and entry into groundwater. Transactions of the Institution of Mining and Metallurgy B 81, 198–209.

Avar, B.B., Hudyma, N.W., 2007. Observations on the influence of lithophysae on elastic (Young's) modulus and uniaxial compressive strength of Topopah Spring Tuff at Yucca Mountain, Nevada, USA. International Journal of Rock Mechanics and Mining Sciences 44, 266–270.

Ball, T.K., Cameron, D.G., Colman, T.B., Roberts, P.D., 1991. Behaviour of radon in the geological environment: a review. Quarterly Journal of Engineering Geology and Hydrogeology 24, 169–182.

Banerjee, K.S., Basu, A., Guin, R., Sengupta, D., 2011. Radon (^{222}Rn) level variations on a regional scale from the Singhbhum Shear Zone, India: a comparative evaluation between influence of basement U-activity and porosity. Radiat. Phys. Chem. 80, 614–619.

Baranyi, I., Gerzon, I., Varhegyi, A., 1985. A new hypothesis of radon migration and its practical application in the emanation exploration method of uranium occurrence. In: Karus, E.V. (Ed.), In: Proceedings of the 30th International Geophysics Symposium, 23–27 September. Min. Geol. USSR Publ., Moscow, pp. 16–25.

Barretto, P.M.C., Clark, R.B., Adams, J.A.S., 1974. Physical characteristics of Rn-222 emanation from rocks, soils and minerals: its relation to temperature and alpha dose. In: Adams, J.A.S., Gesell, T.F., Lowder, W.M. (Eds.), The Natural Radiation Environment, II. USERDA, Houston, USA, pp. 731–740.

Baubron, J.-C., Rigo, A., Toutain, J.-P., 2002. Soil–gas profiles as a tool to characterise active tectonic areas: the Jaut Pass example (Pyrenees, France). Earth Planet. Sci. Lett. 196, 69–81.

Beckman, I.N., Balek, V., 2002. Theory of emanation thermal analysis. XI. Radon diffusion as the probe of microstructure changes in solids. J. Therm. Anal. Calorim. 67, 49–61.

Bernard, P., 1992. Plausibility of long distance electrotelluric precursor to earthquakes. J. Geophys. Res. 97, 17531–17546.

Bonaccorso, A., Currenti, G., Del Negro, C., Boschi, E., 2010. Dike deflection modelling for inferring magma pressure and withdrawal, with application to Etna 2001 case. Earth Planet. Sci. Lett. 293, 121–129.

Bonforte, A., Federico, C., Giammanco, S., Guglielmino, F., Liuzzo, M., Neri, M., 2013. Soil gases and SAR data reveal hidden faults on the sliding flank of Mt. Etna (Italy). J. Volcanol. Geotherm. Res. 251, 27–40.

Burton, M., Neri, M., Condarelli, D., 2004. High spatial resolution radon measurements reveal hidden active faults on Mt. Etna. Geophys. Res. Lett. 31, L07618.

Chiodini, G., Marini, L., 1998. Hydrothermal gas equilibria: the H_2O–H_2–CO_2–CO–CH_4 system. Geochimica et Cosmochimica Acta 62, 2673–2687.

Chiodini, G., Caliro, S., Cardellini, C., Avino, R., Granieri, D., Schmidt, A., 2008. Carbon isotopic composition of soil CO_2 efflux, a powerful method to discriminate different sources feeding soil CO_2 degassing in volcanic–hydrothermal areas. Earth Planet. Sci. Lett. 274, 372–379.

Chirkov, A.M., 1976. Radon as a possible criterion for predicting eruption as observed at Karymsky volcano. Bulletin of Volcanology 37, 126–131.

Choubey, V.M., Bist, K.S., Saini, N.K., Ramola, R.C., 1999. Relation between soil–gas radon variation and different lithotectonic units, Garhwal Himalaya, India. Applied Radiation and Isotopes 51, 487–592.

Cicerone, R.D., Ebel, J.E., Britton, J., 2009. A systematic compilation of earthquake precursors. Tectonophysics 476, 371–396.

Connor, C.B., Clement, B.M., Song, X.D., Lane, S.B., West-Thomas, J., 1993. Continuous monitoring of high-temperature fumaroles on an active lava dome, Volcano Colima, Mexico: evidence of mass low variation in response of atmospheric forcing. J. Geophys. Res. 98 (B11), 19713–19722.

Conte, A.M., Dolfi, D., Gaeta, M., Misiti, V., Mollo, S., Perinelli, C., 2009. Experimental constraints on evolution of leucite–basanite magma at 1 and 10^{-4} GPa: implications for parental compositions of Roman high-potassium magmas. European Journal of Mineralogy 21, 763–782.

Cox, M.E., 1983. Summit outgassing as indicated by radon, mercury and PH mapping, Kilauea volcano, Hawai. J. Volcanol. Geotherm. Res. 16, 131–151.

Cox, E.M., Cuff, E.K., Thomas, M.D., 1980. Variations of ground radon concentrations with activity of Kilauea volcano, Hawaii. Nature 288, 74–76.

De Gregorio, S., Diliberto, I.S., Giammanco, S., Gurrieri, S., Valenza, M., 2002. Tectonic control over large-scale diffuse degassing in eastern Sicily (Italy). Geofluids 2, 273–284.

Di Rocco, T., Gaeta, M., Freda, C., Mollo, S., Dallai, L., 2012. Magma chambers emplaced in carbonate substrate: petrogenesis of skarn and cumulate rocks and implication on CO_2-degassing in volcanic areas. J.Petrology 53, 2307–2332.

Dobrovolsky, I.P., Zubkov, S.I., Miachkin, V.I., 1979. Estimation of the size of earthquake preparation zones. Pure and Applied Geophysics 117, 1025–1044.

Durrance, E.M., Gregory, R.G., 1990. Helium and radon transport mechanism in hydrothermal circulation systems of Southwest England. In: Durrance, E.M., Galimov, E.M., Hinckle, M.E., Reimer, G.M., Sugisaki, R., Autustithis, S.S. (Eds.), Geochemistry of Gaseous Elements and Compounds. Theophrastus Pupl., Athens, pp. 337–352.

Farrar, C.D., Sorey, M.L., Evans, W.C., Howle, J.F., Kerr, B.D., Mack Kennedy, B., King, C.-Y., Southon, J.R., 1995. Forest-killing diffuse CO_2 emission at mammoth mountain as a sign of magmatic unrest. Nature 376, 675–678.

Ferry, C., Richon, P., Beneito, A., Cabrera, J., Sabroux, J.-C., 2002. An experimental method for measuring the radon-222 emanation factor in rocks. Radiation Measurements 35, 579–583.

Finkelstein, M., Brenner, S., Eppelbaum, L., Ne'eman, E., 1998. Identification of anomalous radon concentrations due to geodynamics processes by elimination of Rn variations caused by other factors. Geophys. J. Int. 133, 407–412.

Fleischer, R.L., 1981. Dislocation model for radon response to distant earthquakes. Geophys. Res. Lett. 8, 477–480.

Fleischer, R.L., Mogrocampero, A., 1985. Association of subsurface radon changes in Alaska and the northeastern United States with earthquakes. Geochimica et Cosmochimica Acta 49, 1061–1071.

Forni, F., Bachmann, O., Mollo, S., De Astis, G., Gelman, S.E., Ellis, B.S., 2016. The origin of a zoned ignimbrite: insights into the Campanian Ignimbrite magma chamber (Campi Flegrei, Italy). Earth Planet. Sci. Lett. 449, 259–271.

Fortin, J., Stanchits, S., Vinciguerra, S., Guéguen, Y., 2011. Influence of thermal and mechanical cracks on permeability and elastic wave velocities in a basalt from Mt. Etna volcano subjected to elevated pressure, Tectonophysics 503, 60–72.

Freda, C., Gaeta, M., Misiti, V., Mollo, S., Dolfi, D., Scarlato, P., 2008. Magma–carbonate interaction: an experimental study on ultrapotassic rocks from Alban Hills (Central Italy). Lithos 101, 397–415.

Gasparini, P., Mantovani, M.S.M., 1978. Radon anomalies and volcanic eruptions. J. Volcanol. Geotherm. Res. 3, 325–341.

Gauthier, P.-J., Condomines, M., Hammouda, T., 1999. An experimental investigation of radon diffusion in an anhydrous andesitic melt at atmospheric pressure: implications for radon degassing from erupting magmas. Geochimica et Cosmochimica Acta 63, 645–656.

Giammanco, S., Gurrieri, S., Valenza, M., 1998. Anomalous soil CO_2 degassing in relation to faults and eruptive fissures on Mount Etna (Sicily, Italy). Bulletin of Volcanology 60, 252–259.

Giammanco, S., Sims, K.W.W., Neri, M., 2007. Measurements of ^{220}Rn and ^{222}Rn and CO_2 emissions in soil and fumarole gases on Mt. Etna volcano (Italy): implications for gas transport and shallow ground fracture. Geochemistry, Geophysics, Geosystems 8, Q10001.

Giammanco, S., Immè, G., Mangano, G., Morelli, D., Neri, M., 2009. Comparison between different methodologies for detecting radon in soil along an active fault: the case of the Pernicana fault system, Mt. Etna. Applied Radiation and Isotopes 67, 178–185.

Giardini, A.A., Subbarayudu, G.V., Melton, C.E., 1976. The emission of occluded gas from rocks as function of stress: its possible use as a tool for predicting earthquakes. Geophys. Res. Lett. 13, 355–358.

Gratier, J.P., Chen, T., Hellmann, R., 1994. Pressure solution as a mechanism for crack sealing around faults. In: Proceedings of Workshop LXIII on the Mechanical Involvement of Fluids in Faulting. USGS Open-file Report 94-228. Menlo Park, CA, pp. 279–300.

Hauri, D., Spycher, B., Huss, A., Zimmermann, F., Grotzer, M., von der Weid, N., Weber, D., Spoerri, A., Kuehni, C.E., Röösli, M., 2013. Domestic radon exposure and risk of childhood cancer: a prospective census-based cohort study. Environmental Health Perspectives 121, 1239–1244.

Heap, M.J., Faulkner, D.R., Meredith, P.G., Vinciguerra, S., 2010. Elastic moduli evolution and accompanying stress changes with increasing crack damage: implications for stress changes around fault zones and volcanoes during deformation. Geophys. J. Int. 183, 225–236.

Heap, M.J., Mollo, S., Vinciguerra, S., Lavallée, Y., Hess, K.-U., Dingwell, D.B., Baud, P., Iezzi, G., 2013. Thermal weakening of the carbonate basement under Mt. Etna volcano (Italy): implications for volcano instability. J. Volcanol. Geotherm. Res. 250, 42–60.

Heiligamann, M., Stix, J., Williams-Jones, G., Sherwood Lollar, B., Garzon, G.V., 1997. Distal degassing of radon and carbon dioxide on Galeras volcano, Colombia. J. Volcanol. Geotherm. Res. 77, 267–283.

Hishinuma, T., Nishikawa, T., Shimoyama, T., Miyajima, M., Tamagawa, Y., Okabe, S., 1999. Emission of radon and thoron due to the fracture of rock. Il Nuovo Cimento 22C, 523–527.

Holub, R.F., Brady, B.T., 1981. The effect of stress on radon emanation from rock. J. Geophys. Res. 86, 1776–1784.

Honda, M., Kurita, K., Hamano, Y., Ozima, M., 1982. Experimental studies of He and Ar degassing during rock fracturing. Earth Planet. Sci. Lett. 59, 429–436.

Hudyma, N., Burcin Avar, B., Karakouzian, M., 2004. Compressive strength and failure modes of lithophysae-rich Topopah Spring Tuff specimens and analog models containing cavities. Eng. Geol. 73, 179–190.

Igarashi, G., Saeki, S., Takahata, N., Sumikawa, K., Tasaka, S., Sasaki, Y., Takahashi, M., Sano, Y., 1995. Ground-water radon anomaly before the Kobe earthquake in Japan. Science 269, 60–61.

Immè, G., La Delfa, S., Lo Nigro, S., Morelli, D., Patanè, G., 2006. Soil radon concentration and volcanic activity of Mt. Etna before and after the 2002 eruption. Radiation Measurements 41, 241–245.

Katoh, K., Ikeda, K., Kusunose, K., Nishizawa, O., 1984. An experimental study of radon (^{222}Rn and ^{220}Rn) emanated from granite specimens under uniaxial compression. Bulletin of the Geological Survey of Japan 35, 1–11, (in Japanese with English abstract).

Katoh, K., Nishizawa, O., Kusunose, K., Ikeda, K., 1985. An experimental study on variation of radon emanation from Westerly granite under uniaxial compression Part 1. J. Seismological Soc. Jpn. 38, 173–182, (in Japanese with English abstract).

Kies, A., Majerus, J., D'Oreye de Lantremange, N., 1999. Underground radon gas concentrations related to earth tides. Il Nuovo Cimento 22C, 287–293.

King, C.Y., 1978. Radon emanation on San Andreas fault. Nature 271, 516–519.

King, C.-Y., 1980. Episodic radon changes in subsurface soil gas along active faults and possible relation to earthquakes. J. Geophys. Res. 85, 3065–3078.

King, C.-Y., 1986. Gas geochemistry applied to earthquake prediction: an overview. J. Geophys. Res. 91, 12269–12281.

King, C.-Y., King, B.-S., Evans, W.C., Zhang, W., 1996. Spatial radon anomalies on active faults in California. Applied Geochemistry 11, 497–510.

Koike, K., Yoshinaga, T., Suetsugu, K., Kashiwaya, K., Asaue, H., 2015. Controls on radon emission from granite as evidenced by compression testing to failure. Geophys. J. Int. 203, 428–436.

Kotsarenko, A., Yutsis, V., Grimalsky, V., Koshevaya, S., 2016. Detailed study of radon spatial anomaly in Tlamacas Mountain area, Volcano Popocatepetl, Mexico. Open Journal of Geology 6, 158–164.

Kozlova, I.A., Yurkov, A.K., 2005. Soil-gas radon-222 monitoring: methodological issues. Uralskii Geofizicheskii Vestnik 7, 31–34.

Kristianson, K., Malmqvist, L., 1982. Evidence for nondiffusive transport of Rn-222 in the ground and a new physical model for the transport. Geophysics 47, 1444–1452.

Kuo, T., Fan, K., Kuochen, H., Han, Y., Chu, H., Lee, Y., 2006. Anomalous decrease in groundwater radon before the Taiwan M 6.8 Chengkung earthquake. J. Environ. Radioact. 88, 101–106.

Kuo, T., Liu, C., Su, C., Chang, C., Chen, W., Chen, C., Lin, C., Kuochen, H., Hsu, Y., Lin, Y., Huang, Y., Lin, H., 2013. Concurrent concentration declines in groundwater-dissolved radon, methane and ethane precursory to 2011 Mw 5.0 Chimei earthquake. Radiation Measurements 58, 121–127.

La Delfa, S., Immè, G., Lo Nigro, S., Morelli, D., Patanè, G., Vizzini, F., 2007. Radon measurements in the SE and NE flank of Mt. Etna (Italy). Radiation Measurements 42, 1404–1408.

Lenzen, M., Neugebauer, H.J., 1999. Measurements of radon concentration and the role of earth tides in a gypsum mine in Walferdange, Luxembourg. Solar Health Physics 77, 154–162.

Lewicki, J.L., Bergfeld, D., Cardellini, C., Chiodini, G., Granieri, D., Varley, N., Werner, C., 2005. Comparative soil CO_2 flux measurements and geostatistical estimation methods on Masaya volcano. Bulletin of Volcanology 68, 76–90.

Liu, K.K., Yui, T.F., Yeh, Y.H., Tsai, Y.B., Teng, T.L., 1985. Variations of radon content in ground waters and possible correlation with seismic activities in Northern Taiwan. Pure and Applied Geophysics 122, 231–244.

Manniello, M.A., 2015. Studio sperimentale dell'emissione radon da materiali geologici a temperature subvulcaniche. Unpublished Master thesis. Università "Roma Tre", Rome, Italy.

Mattia, M., 2017. Studio sperimentale dell'emissione radon da una lava fonolitica dei Colli Albani (Roma, Italia) a temperature magmatiche. Unpublished Master thesis. Università "Roma Tre", Rome, Italy.

Merle, O., Barde-Cabusson, S., van Wyk de Vries, B., 2010. Hydrothermal calderas. Bulletin of Volcanology 72, 131–147.

Milvy, P., Cothern, C.R., 1991. Scientific background for the development of regulations for radionuclides in drinking water. In: Cathern, C.R., Rebers, P.A. (Eds.), Radon, Radium and Uranium in Drinking Water. Lewis Publishers, Michigan, pp. 1–16.

Mollo, S., Gaeta, M., Freda, C., Di Rocco, T., Misiti, V., Scarlato, P., 2010. Carbonate assimilation in magmas: a reappraisal based on experimental petrology. Lithos 114, 503–514.

Mollo, S., Tuccimei, P., Heap, M.J., Vinciguerra, S., Soligo, M., Castelluccio, M., Scarlato, P., Dingwell, D.B., 2011a. Increase in radon emission due to rock failure: an experimental study. Geophys. Res. Lett. 38, L14304.

Mollo, S., Scarlato, P., Freda, C., Gaeta, M., 2011b. Basalt–crust interaction processes: insights from experimental petrology. In: West, J.P. (Ed.), Basalt: Types, Petrology and Uses. Nova Science Publishers, New York, NY, pp. 33–61.

Mollo, S., Vinciguerra, S., Iezzi, G., Iarocci, A., Scarlato, P., Heap, M.J., Dingwell, D.B., 2011c. Volcanic edifice weakening via devolatilization reactions. Geophys. J. Int. 186, 1073–1077.

Mollo, S., Misiti, V., Scarlato, P., Soligo, M., 2012a. The role of cooling rate in the origin of high temperature phases at the chilled margin of magmatic intrusions. Chemical Geology 322–323, 28–46.

Mollo, S., Heap, M.J., Iezzi, G., Hess, K.-U., Scarlato, P., Dingwell, D.B., 2012b. Volcanic edifice weakening via decarbonation: a self-limiting process? Geophys. Res. Lett. 39, L15307.

Mollo, S., Heap, M.J., Dingwell, D.B., Hess, K.-U., Iezzi, G., Masotta, M., Scarlato, P., Vinciguerra, S., 2013. Decarbonation and thermal microcracking under magmatic P-T-fCO_2 conditions: the role of skarn substrata in promoting volcanic instability. Geophys. J. Int. 195, 369–380.

Mollo, S., Tuccimei, P., Galli, G., Iezzi, G., Scarlato, P., 2017. The imprint of thermally-induced devolatilization phenomena on radon signal: Implications for the geochemical survey and public health in volcanic areas. Geophys. J. Int. 211, 558–571.

Monnin, M.M., Seidel, J.L., 1997. Physical models related to radon emission in connection with dynamic manifestations in the upper terrestrial crust: a review. Radiation Measurements 28, 703–712.

Morelli, D., Immè, G., La Delfa, S., Lo Nigro, S., Patanè, G., 2006. Evidence of soil radon as tracer of magma uprising at Mt. Etna. Radiation Measurements 41, 721–725.

Neri, M., Behncke, B., Burton, M., Giammanco, S., Pecora, E., Privitera, E., Reitano, D., 2006. Continuous soil radon monitoring during the July 2006 Etna eruption. Geophys. Res. Lett. 3, L24316.

Neri, M., Giammanco, S., Ferrera, E., Patanè, G., Zanon, V., 2011. Spatial distribution of soil radon as a tool to recognize active faulting on an active volcano: the example of Mt. Etna (Italy). J. Environ. Radioact. 102, 863–870.

Neri, M., Ferrera, E., Giammanco, S., Currenti, G., Cirrincione, R., Patanè, G., Zanon, V., 2016. Soil radon measurements as a potential tracer of tectonic and volcanic activity. Scientific Reports 6, 24581.

Nicolas, A., Girault, F., Schubnel, A., Pili, E., Passelegue, F., Fortin, J., Deldicque, D., 2014. Radon emanation from brittle fracturing in granites under upper crustal conditions. Geophys. Res. Lett. 41, 5436–5443.

Nishimura, S., Katsura, I., 1990. Radon in soil gas: applications inexploration and earthquake prediction. In: Durrance, E.M. et al., (Ed.), Geochemistry of Gaseous Elements and Compounds. Theophrastus Pupl., Athens, pp. 497–533.

Notsu, K., Mori, T., Chanchah Do Vale, S., Kagi, H., Ito, T., 2006. Monitoring quiescent volcanoes by diffuse CO_2 degassing: case study of Mt. Fuji, Japan. Pure and Applied Geophysics 163, 825–835.

Passaglia, E., Vezzalini, G., Carnevali, R., 1990. Diagenetic chabazites and phillipsites in Italy: crystal chemistry and genesis. European Journal of Mineralogy 2, 827–839.

Planinic, J., Radolic, V., Culo, D., 2000. Searching for an earthquake precursor: temporal variations of radon in soil and water. FIZIKA B (Zagreb) 9, 75–82.

Plastino, W., Bella, F., Catalano, P., Di Giovambattista, R., 2002. Radon groundwater anomalies related to the Umbria-Marche, September 26, 1997, earthquakes. Geofísica Internacional 41, 369–375.

Porstendörfer, J., 1994. Properties and behaviour of radon and thoron and their decay products in the air, tutorial/review. J. Aerosol Sci. 25, 219–263.

Roeloffs, E., 1999. Radon and rock deformation. Nature 399, 104–105.

Sabbarese, C., De Martino, S., Signorini, C., Gialanella, G., Roca, V., Baldassini, P.G., Cotellessa, G., Sciocchetti, G., 1993. A survey of ^{222}Rn in Campania region. Radiation Protection Dosimetry 48, 257–263.

Sakoda, A., Ishimori, Y., Yamaoka, K., 2011. A comprehensive review of radon emanation measurements for mineral, rock, soil, mill tailing and fly ash. Applied Radiation and Isotopes 69, 1422–1435.

Sammis, C.G., Banerdt, M., Hammond, D.E., 1981. Stress induced release of Rn-222 and CH_4 to percolating water in granitic rock. In: Proceedings of the Seventh Workshop on Geothermal Reservoir Engineering, Stanford University Press, SGP-TR-55pp. 133–138.

Scarlato, P., Tuccimei, P., Mollo, S., Soligo, M., Castelluccio, M., 2013. Contrasting radon background levels in volcanic settings: clues from ^{220}Rn activity concentrations measured during long-term deformation experiments. Bulletin of Volcanology 75, 751.

Schery, S.D., Gaeddert, D.H., 1982. Measurements of the effect of cyclic atmospheric pressure variation on the flux of 222-Rn from the soil. Geophys. Res. Lett. 9, 835–838.

Scholz, C.H., 1968. Mechanism of creep in brittle rock. J. Geophys. Res. 73, 3295–3302.

Scholz, C.H., Sykes, L.R., Aggarawal, Y.P., 1973. Earthquake prediction: a physical basis. Science 181, 803–810.

Schubert, M., 2015. Using radon as environmental tracer for the assessment of subsurface non-aqueous phase liquid (NAPL) contamination—a review. The European Physical Journal Special Topics 224, 717–730.

Segovia, N., Armienta, M.A., Valdes, C., Mena, M., Seidel, J.L., Monnin, M., Pena, P., Lopez, M.B.E., Reyes, A.V., 2003. Volcanic monitoring for radon and chemical species in the soil and in spring water samples. Radiation Measurements 36, 379–383.

Sengupta, D., Ghosh, A., Mamtani, M.A., 2005. Radioactivity studies along fracture zones in areas around Galudih, East Singhbhum, Jharkhand, India. Applied Radiation and Isotopes 63, 409–414.

Shapiro, M.H., Melvin, J.D., Tombrello, T.A., Fong-Liang, J., Gui-Ru, L., Mendenhall, M.H., 1982. Correlated radon and CO_2 variations near the San-Andreas fault. Geophys. Res. Lett. 9, 503–506.

Silver, P.G., Wakita, H., 1996. A search for earthquake precursors. Science 273, 77–78.

Singh, A.K., Sengupta, D., Prasad, R., 1999. Radon exhalation rate and uranium estimation in rock samples from Bihar uranium and copper mines using the SSNTD technique. Applied Radiation and Isotopes 51, 107–113.

Siniscalchi, A., Tripaldi, S., Neri, M., Giammanco, S., Piscitelli, S., Balasco, M., Behncke, B., Magrì, C., Naudet, V., Rizzo, E., 2010. Insights into fluid circulation across the Pernicana Fault (Mt. Etna. Italy) and implications for flank instability. J. Volcanol. Geotherm. Res. 193, 137–142.

Somogyi, G., Lenart, L., 1985. Time integrated radon measurements in spring and well waters by track technique. Nuclear Tracks 12, 731–734.

Sonmez, H., Tuncay, E., Gokceoglu, C., 2004. Models to predict the uniaxial compressive strength and the modulus of elasticity for Ankara Agglomerate. International Journal of Rock Mechanics and Mining Sciences 41, 717–729.

Sonmez, H., Gokceoglua, C., Medley, E.W., Tuncay, E., Nefeslioglu, H.A., 2006. Estimating the uniaxial compressive strength of a volcanic bimrock. International Journal of Rock Mechanics and Mining Sciences 43, 554–561.

Sorey, M., Evans, B., Kennedy, M., Farrar, C.D., Hainsworth, L.J., Hausback, B., 1998. Carbon dioxide and helium emissions from a reservoir of magmatic gas beneath Mammoth Mountain, California. J. Geophys. Res. 103, 15303–15323.

Sultankhodzhayev, A.N., Chernov, I.G., Zakirov, T., 1976. Hydroseismic precursors to the Gazli earthquake. Academy of Sciences of the Republic of Uzbekistan 7, 51–53.

Tanner, A.B., 1964. Radon migration in the ground: a review. Paper Presented at Symposium the Natural Radiation Environment, 10–13 April. University of Chicago Press, Houston, TX.

Tanner, A.B., 1980. Radon migration in the ground A: supplementary review. In: Gedsell, T.F., Lowder, W.M. (Eds.), The Natural Radiation Environment III. University of Chicago Press, University of Chicago Press, Chicago, USA , pp. 5–56.

Tarquini, E., 2012. Variazione delle emissioni di radon da parte di campioni di tufo sottoposti a stress uniassiale con cicli rapidi di carico e di scarico. Relazione tra meccanismi deformativi e variazioni dell'emanazione. Unpublished Master thesis. Università "Roma Tre", Rome, Italy.

Teng, T.L., 1980. Some recent studies on groundwater radon content as an earthquake precursor. J. Geophys. Res. 85, 3089.

Thomas, D., 1988. Geochemical precursors to seismic activity. Pure and Applied Geophysics 126, 241–266.

Thomas, D.M., Cox, M.E., Cuff, K.E., 1986. The association between ground gas radon variations and geologic activity in Hawaii. J. Geophys. Res. 91, 2186–12198.

Torgersen, T., Benoit, J., Mackie, D., 1990. Controls on groundwater 222Rn concentrations in fractured rock. Geophys. Res. Lett. 17, 845.

Toutain, J.-P., Baubron, J.-C., 1999. Gas geochemistry and seismotectonics: a review. Tectonophysics 304, 1–27.

Toutain, J.-P., Baubron, J.-C., Le Bronec, J., Allard, P., Briole, P., Marty, B., Miele, G., Tedesco, D., Luongo, G., 1992. Continuous monitoring of distal gas emanations at volcano, Southern Italy. Bulletin of Volcanology 54, 147–155.

Tuccimei, P., Castelluccio, M., Soligo, M., Moroni, M., 2009. Radon exhalation rates of building materials: experimental, analytical protocol and classification criteria. In: Cornejo, D.N., Haro, J.L. (Eds.), Building Materials: Properties, Performance, Applications. Nova Science Publishers, Hauppauge, NY, pp. 259–273, ISBN: 978-1-60741-082-9.

Tuccimei, P., Mollo, S., Vinciguerra, S., Castelluccio, M., Soligo, M., 2010. Radon and thoron emission from lithophysae-rich tuff under increasing deformation: an experimental study. Geophys. Res. Lett. 37, L05406.

Tuccimei, P., Castelluccio, M., Moretti, S., Mollo, S., Vinciguerra, S., Scarlato, P., 2011. Thermal enhancement of radon emission from rocks. Implications for laboratory experiments under increasing deformation. Veress, B., Szigethy, J. (Eds.), Horizons in Earth Science Research, 4, Nova Science Publishers, New York, USA, pp. 247–256, ISBN 978-1-61122-763-5, Chapter 9.

Tuccimei, P., Mollo, S., Soligo, M., Scarlato, P., Castelluccio, M., 2015. Real-time setup to measure radon emission during rock deformation: implications for geochemical surveillance. Geoscientific Instrumentation, Methods and Data Systems 4, 111–119.

Utkin, V.I., 2000. Space–time radon monitoring as a basis for medium-term earthquake prediction. Uralskii Geofizicheskii Vestnik 1, 101–106.

Utkin, V.I., Yurkov, A.K., 2010. Radon as a tracer of tectonic movements. Russian Geology and Geophysics 51, 220–227.

Vinciguerra, S., Trovato, C., Meredith, P.G., Benson, P.M., Troise, C., De Natale, G., 2006. Understanding the seismic velocity structure of Campi Flegrei Caldera (Italy): from the laboratory to the field scale. Pure and Applied Geophysics 163, 2205–2221.

Vinciguerra, S., Del Gaudio, P., Mariucci, M.T., Marra, F., Meredith, P.G., Montone, P., Pierdominici, S., Scarlato, P., 2009. Physical properties of tuffs from a scientific borehole at Alban hills volcanic district (Central Italy). Tectonophysics 471, 161–169.

Voltaggio, A., Masi, U., Spadoni, M., Zampetti, G., 2006. A methodology for assessing the maximum expected radon flux from soils in Northern Latium (central Italy). Environmental Geochemistry and Health 28, 541–551.

Wakita, H., 1996. Geochemical challenge to earthquake prediction. Proceedings of the National Academy of Sciences of the United States of America 93, 3781–3786.

Wakita, H., 1998. Radon observation for earthquake prediction. In: Katakase, A., Shimo, M. (Eds.), In: Proceedings of the Seventh Tohma University International Symposium Radon and Thoron in the Human Environment. World Scientific Publishing Co. Pte. Ltd., Singapore, pp. 124–130.

Wakita, H., Nakamura, Y., Notsu, K., Noguchi, M., Asada, T., 1980. Radon anomaly: a possible precursor of the 1978 Izu-Oshima-kinkai earthquake. Science 207, 882–883.

Wang, X., Li, Y., Du, J., Zhou, X., Asada, T., 2014. Correlations between radon in soil gas and the activity of seismogenic faults in the Tangshan area, North China, Radiation Measurements 60, 8–14.

Washington, J.W., Rose, A.W., 1992. Temporal variability of radon concentration in the interstitial gas of soils in Pennsylvania. J. Geophys. Res. 97, 9145–9159.

Wohletz, K., Civetta, L., Orsi, G., 1999. Thermal evolution of the Phlegraean magmatic system. J. Volcanol. Geotherm. Res. 91, 381–414.

Xue, S., Wang, J., Xie, J., Wu, J., 2010. A laboratory study on the temperature dependence of the radon concentration in coal. Int. J. Coal Geol. 83, 82–84.

Yasuoka, Y., Kawada, Y., Nagahama, H., Omori, Y., Ishikawa, T., Tokonami, S., Shinogi, M., 2009. Pre-seismic changes in atmospheric radon concentration and crustal strain. Physics and Chemistry of the Earth 34, 431–434.

Zhang, W., Zhang, D., Wu, L., Li, J., Cheng, J., 2016. Radon release from underground strata to the surface and uniaxial compressive test of rock samples. Acta Geodynamica et Geomaterialia 13, 407–416.

Zmazek, B., Vaupotic, J., Zivicic, M., Premru, U., Kobal, I., 2000. Radon measurements for earthquake prediction in Slovenia. Fizika B 9, 111–118.

FURTHER READING

Chyi, L.L., Chou, C.Y., Yang, F.T., Chen, C.H., 2001. Continuous radon measurements in faults and earthquake precursor pattern recognition. Western Pacific Earth Sciences 1, 227–243.

Yang, X., Wang, F., Liu, C., Zhang, B., 1989. Relationship between rock failure and gas emission. Physics and Chemistry of the Earth 17, 85–90.

Chapter 19

GIS Based Macrolevel Landslide Hazard Zonation Using , Newmark's Methodology

T.G. Sitharam*, Naveen James**

*Indian Institute of Science, Bangalore, India; **Indian Institute of Technology Ropar, Punjab, India

19.1 INTRODUCTION

The disaster management authority defines a landslide as the sudden downward/outward movement of earthen materials along a slope under gravity. From the geotechnical engineering perspective, a landslide is the eventuality of slope instability. The most common causes for the slope instability are intense and prolonged rainfall, earthquakes, snow melting, etc. Apart from these, the stability of the earthen slope also depends on the geotechnical properties of slope material, drainage pattern, land cover, slope gradient, etc. In India, the main landslide-prone areas are the Himalayan region, northeast region, and the Western Ghats. The statistics provided by disaster management authority of India show that about 15% of the total land area (0.49 million km^2) are affected by landslide hazard (NDMA, 2009). Major rain-induced landslides occurred in various parts of India during the last decade itself have claimed more than 300 lives. Some serious and fatal landslides in India from 1948 to 2016 are, Guwahati Landslide, Darjeeling Landslide, Malpa Landslide, Mumbai Landslide, Amboori Landslide, Kedarnath Landslide, Malin Landslide, and Tawang Landslide. All these landslides locations are either part of the Himalayas or the Western Ghats.

One of the earliest and most catastrophic rainfall-induced landslides in India was the 1948 Guwahati landslide in the state of Assam, which resulted in the death of about 500 people. A similar rainfall-induced landslide was also reported in Darjeeling, West Bengal, in the year 1968, resulting in the death of thousands of people. The 1998 landslide in Malpa village of Uttarakhand resulted in 380 death toll along with the destruction of an entire village. The rainfall-induced landslide of Mumbai in 2000 resulted in the death of 67 people.

The Amboori landslide of 2001 is one of the disastrous landslides in Kerala, killing around 40 people. A series of landslides were reported along the Western Ghat regions of Karnataka near Karwar–Mangalore belt, disrupting the train services. The 2013 Kedarnath landslide in Uttarakhand is the most catastrophic of all landslides, with over 5700 were reportedly missing/dead and several villages had been affected (CBS, 2013). In 2014, a rain-triggered landslide occurred in the village of Malin in Maharashtra (Fig. 19.1), India, has claimed 134 lives. Sixteen construction workers were killed when a landslide buried them alive in Tawang, Arunachal Pradesh, on April 22, 2016.

In addition to rainfall-induced landslides, numerous seismically induced landslides were also reported in various parts of India. A series of landslides were reported along Uttarkashi–Harsil road due to 1991 Uttarkashi earthquake. Studies by Ravindran and Philip (1999) and Shrikhande et al. (2000) have showed that several old and stabilized landslides were also reactivated during 1999 Chamoli earthquake. Series of prominent earthquake landslides occurred in the state of Sikkim due to 6.9 magnitude earthquake of 2011 along the India–Nepal border (Fig. 19.2). Similarly, the 7.8 magnitude Nepal earthquake of 2015 has triggered many landslides along the Nepal–China border and also in Sikkim. With such high casualties and huge economic losses caused during landslide, it is high time to develop a landslide zonation map at microlevel.

Integrating Disaster Science and Management. http://dx.doi.org/10.1016/B978-0-12-812056-9.00019-1

FIGURE 19.1 **Debris flow during 2014 rainfall-induced landslide at Malin, Maharashtra.** Source: *BBC (2014).*

FIGURE 19.2 **Series of landslides in Sikkim after 2011 earthquake.** Source: *Chakraborty et al. (2011).*

19.2 LANDSLIDE HAZARD ANALYSIS AND MAPPING

Landslides can be considered as local events and their impact is restricted to a smaller region. Hence, for the analysis purpose, the landslide is often treated as a slope stability problem. Landslide hazard assessment and mapping is an important step toward the evaluation of hazard and risk management. Evaluating the stability of slopes is one of the challenging problems in civil engineering which requires the knowledge of soil mechanics. The slope instability can occur under both static and dynamic conditions. Most landslide cases in India are mainly rainfall-induced ones, where the slope instability was caused by a rise in pore water pressure under static condition. The stability of a slope is predominantly governed by the shear strength parameters, cohesion, and friction angle of the slope material. As the slope material (mainly soil) gets saturated, its shear strength reduces, resulting in the slope instability. Hence, the analysis to determine whether the given slope is susceptible to landslide requires the knowledge of shear strength parameters of the slope materials. Apart from reducing the shear strength of the slope material, the flowing pore water also induces seepage force, which adds up to the destabilizing/driving forces. The analysis procedure involves limit equilibrium method and numerical finite element method. The limit equilibrium method involves quantification of the driving and the resisting forces along a predefined failure surface. The bureau of Indian standards (BIS) has given guidelines for macrolevel landslide hazard zonation in India (IS:14496-Part-2, 1998). Landslide hazard mapping has been carried out in different parts of the world by different investigators using different approaches such as inventory-based mapping, heuristic approach, probabilistic approach, deterministic approach, statistical analyses, and multicriteria decision-making approach. The multivariate statistical approach considers relative contributions of each thematic data layer to the total landslide susceptibility and classification of hazard for the given area. Some of the notable quantitative macrolevel landslide hazard mapping was carried out by GIS and NDMA (2009) for entire India.

Another scenario where the possibility of landslides is when the slope is subjected to dynamic loading such as an earthquake. The slope failure mechanism under dynamic loading is different from that under static conditions. Progressive stiffness degradation of slope materials (soil) due to stress reversal under dynamic loading is the predominant reason for the slope failure. The analysis of such a dynamic slope stability requires knowledge of cyclic stress–strain behavior of soil materials. However, the earliest of analysis approximates the cyclic transient load by the equivalent static horizontal and vertical inertial forces. This methodology is popularly known as pseudostatic method (Terzaghi, 1950). Once the equivalent inertial forces are determined, the limit equilibrium method procedure is followed to determine the factor of safety (FS) along a predetermined failure surface. An improved limit equilibrium method for analyzing the dynamic slope stability problems was introduced by Newmark (1965) known as the "sliding block approach". This approach considers the slope at the verge of failure where the driving force exceeds shear strength of slope material thus the FS falling below 1.

The soil mass tends to move along the failure surface just like rigid block slides along and inclined surface (Fig. 19.3). The other methodology for analyzing slope stability problem is dynamic finite element analysis. The analysis using finite element method is an approximate numerical analysis, in which the entire domain is discretized into finite element meshes. The strain or stress compatibility equations are solved numerically at the mesh nodes and the deformation of entire slope is obtained by integrating elemental level deformations.

19.3 SEISMIC LANDSLIDE HAZARD ANALYSIS AT MACROLEVEL

The above listed methodologies are well suited to analyze the stability of a particular slope. However, carrying out a quantitative landslide hazard analysis for a larger region is a challenge to scientists and engineers. A macrolevel landslide hazard analysis is an important step toward macrolevel landslide hazard zonation. The landslide hazard zonation map is an important tool for designers and planners to identify the vulnerable sites and plan the infrastructural development accordingly. BIS (IS:14496-Part-2, 1998) landslide hazard zonation procedure is based on landslide hazard evaluation factor rating scheme for different causative factors such as lithology, structure, slope morphometry, relative relief, land-use- land-cover, and hydrological condition. The challenge of carrying out macrolevel landslide hazard analysis can be addressed effectively with use of geographic information systems (GIS) and availability of high-speed computers.

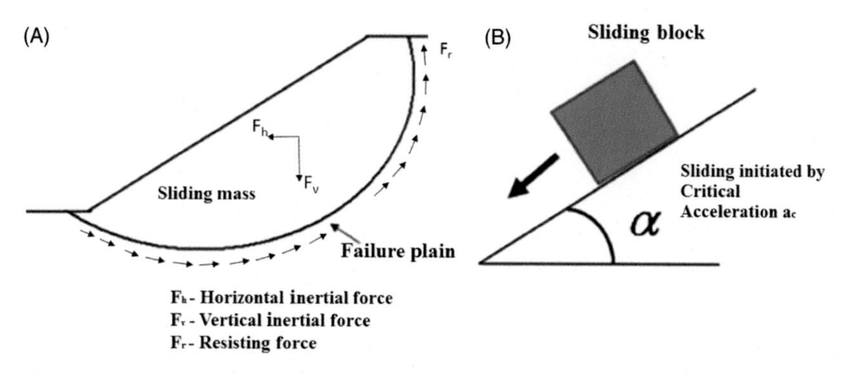

FIGURE 19.3 Schematic representation of Newmark's model: (A) Conventional slope failure. (B) Sliding mass is modeled as a rigid block which is on the verge of failure due to the threshold acceleration a_c. Source: *James and Sitharam (2014).*

Many researchers (Jibson, 1993; Wieczorek et al., 1985; Wilson and Keefer, 1983, 1985) have employed the Newmark's method for predicting earthquake-induced landslides. James and Sitharam (2014, 2015) demonstrated the use of Newmark's methodology for the quantitative macrolevel seismic landslide hazard assessment for the states of Karnataka and Sikkim. The first and foremost step toward the assessment of seismically induced landslide hazard is the quantification of seismic hazard of the study area.

19.3.1 Seismic Hazard Analysis

Seismic hazard analysis is generally carried out to evaluate the intensity of ground shaking at a given location. There are two methodologies available for carrying out seismic hazard analysis at a given location: probabilistic seismic hazard analysis (PSHA) and deterministic seismic hazard analysis (DSHA). DSHA only considers the critical scenario by assuming the occurrence of the maximum credible earthquake (MCE) at the closest possible distance to the site. Hence, DSHA often gives an upper bound value for the seismic hazard at the site. However, the PSHA considers and quantifies all major uncertainties in the earthquake process for the calculation of seismic hazard at the given site (Kramer, 1996). Thus, it provides different values for seismic hazard for different return periods. For the design of small structures, the peak ground acceleration and spectral acceleration for lower return period can be used. Krinitzsky (1995) recommends the use of DSHA for determining peak ground acceleration and spectral acceleration for the design of critical structures. James and Sitharam (2014, 2015) have used DSHA to evaluate the seismic hazard for the states of Karnataka and Sikkim. Fig. 19.4 presents a schematic of methodology that they adopted for DSHA. As a first step, they have prepared the earthquake event map for the two states as shown in Figs. 19.5 and 19.6. Smoothened point source model along with region-specific attenuation relationships were used in DSHA.

For the state of Karnataka, three major attenuation relations (Atkinson and Boore, 2006; Campbell and Bozorgnia, 2003; Kanth and Iyengar, 2007) combined with a logic tree methodology (Bommer et al., 2005) were used. However, for the state of Sikkim, the attenuation relation by Sharma et al. (2009) was used to predict PHA. Figs. 19.7 and 19.8 present the spatial variation of surface PHA values throughout the states of Karnataka and Sikkim. Further, the PHA values at the surface level were predicted using nonlinear site amplification technique proposed by Kanth and Iyengar (2007).

19.3.2 Development of Slope Map

The information about the slope map was derived from the digital elevation model (DEM) of the two study areas. DEM provides the terrain elevation details for a given region in a raster form. James and Sitharam (2014, 2015) have used ASTER

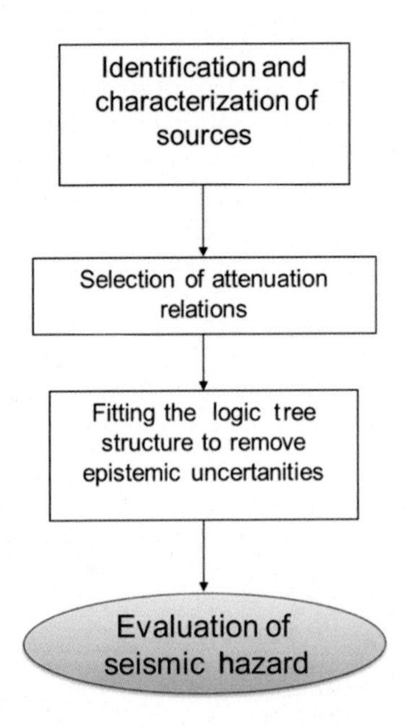

FIGURE 19.4 **Steps involved in DSHA.**

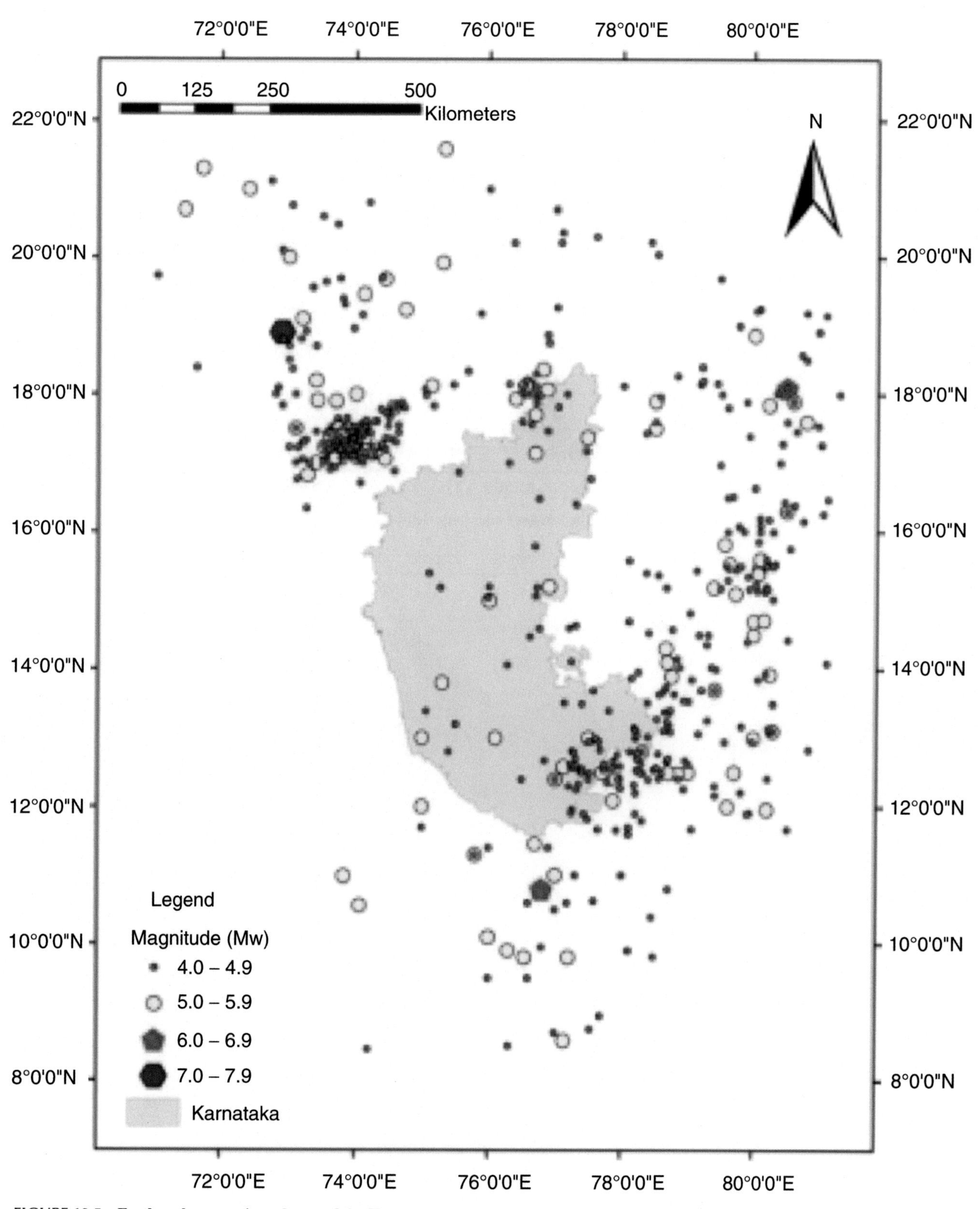

FIGURE 19.5 **Earthquake events in and around the Karnataka state.** Source*: James and Sitharam (2014).*

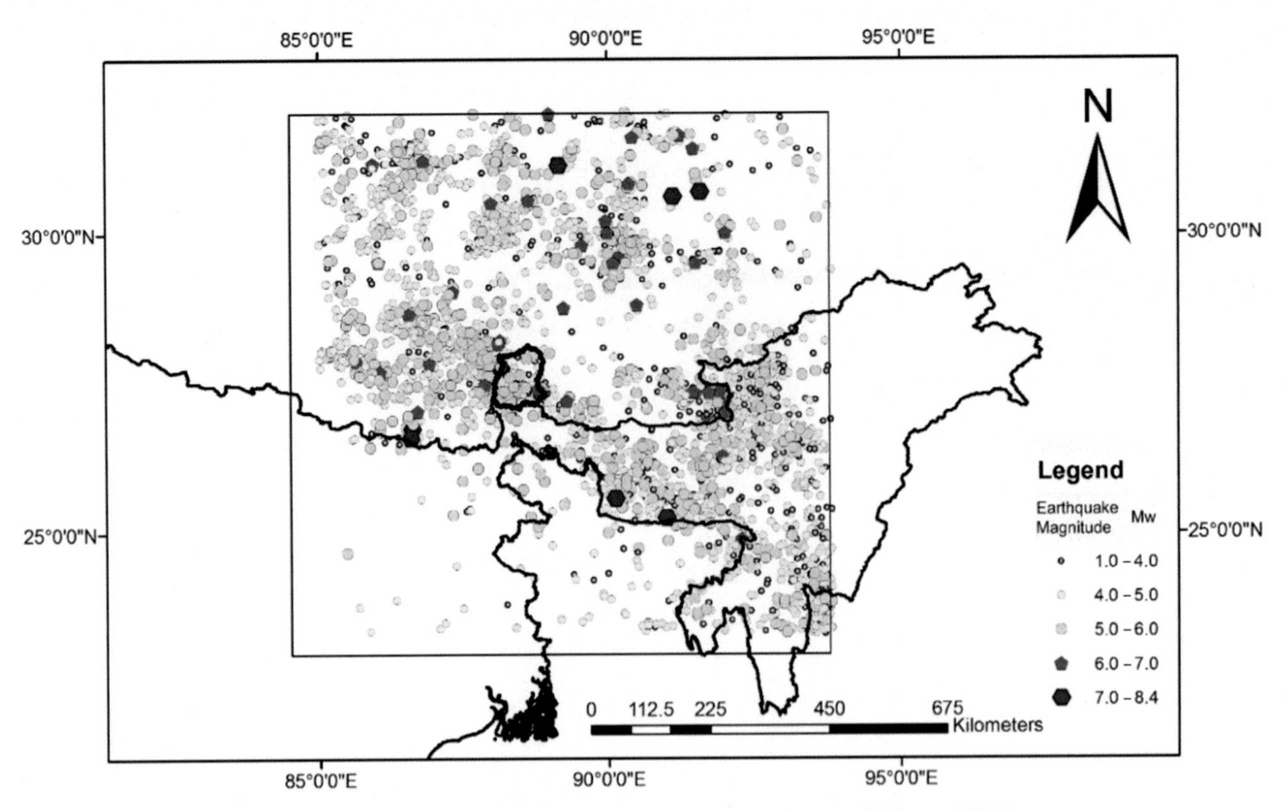

FIGURE 19.6 **Earthquake events within 500 km from the Sikkim's boundary.** Source: *James and Sitharam (2015).*

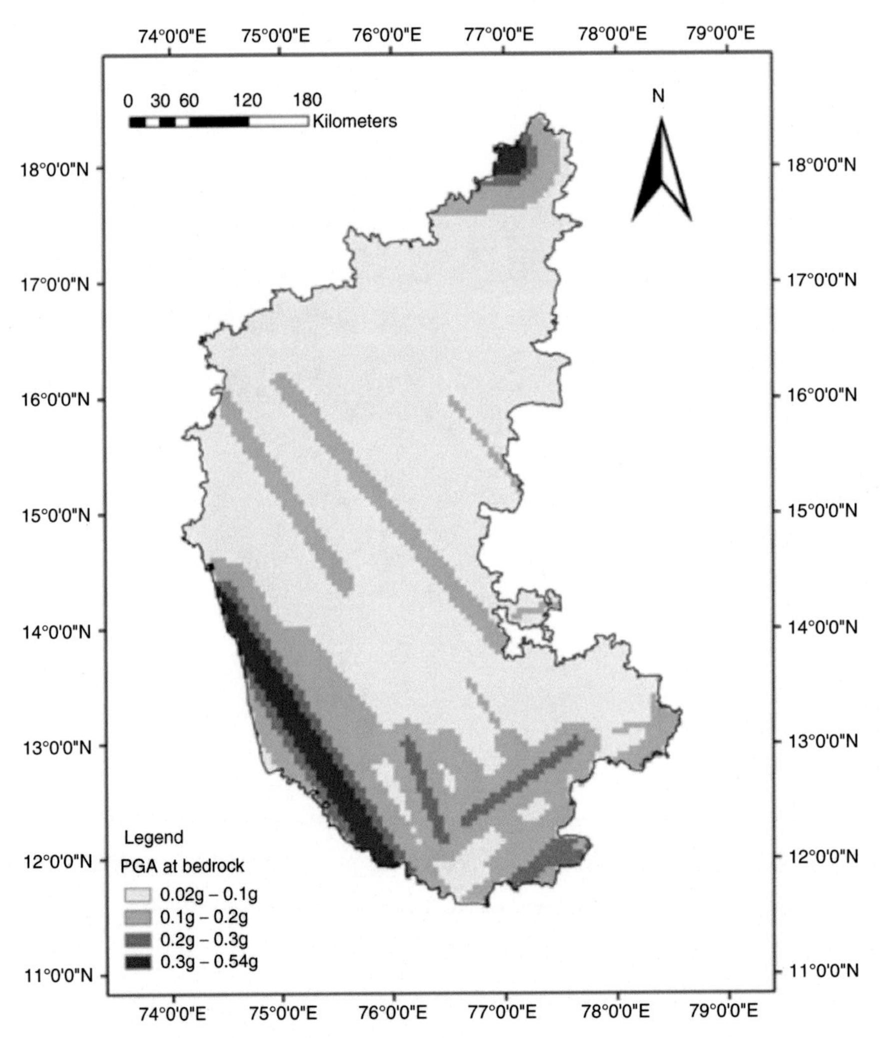

FIGURE 19.7 **Spatial variation of PHA values throughout Karnataka using DSHA.** Source: *James and Sitharam (2014).*

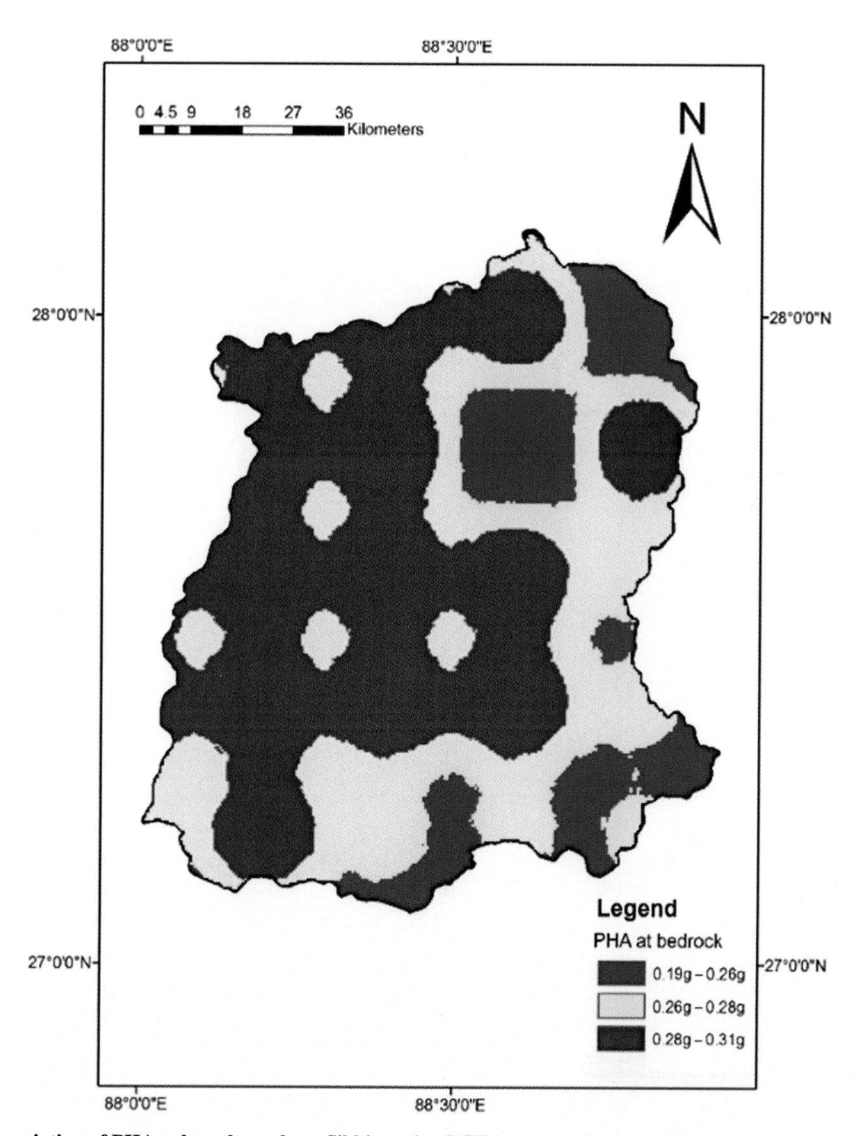

FIGURE 19.8 **Spatial variation of PHA values throughout Sikkim using DSHA.** Source: *James and Sitharam (2015).*

Global Digital Elevation Model (ASTER GDEM) having 30 m resolution, to develop slope for Karnataka and Sikkim. The ASTER GDEM was imported in ArcGIS software to develop the slope map (Figs. 19.9 and 19.10).

19.3.3 Landslide Hazard Map

In the Newmark's method, it is assumed that the sliding block requires a threshold acceleration termed as critical acceleration (a_c) that depends upon the static FS and slope angle (α), as in Eq. (19.1)

$$a_c = (\mathrm{FS} - 1) g \sin\alpha \tag{19.1}$$

James and Sitharam (2014) evaluated the landslide hazard in terms of critical static FS_c required to resist the initiation of seismic landslides. The surface-level peak horizontal acceleration (PHA) evaluated using deterministic seismic hazard analysis was considered to produce the inertial driving force, and the slope angle at a given location in the study area was derived from the slope map. For a given value of PHA and terrain, the critical value, FS_c, required to resist landslide was evaluated as per Eq. (19.2). Figs. 19.11 and 19.12 present the spatial variation of critical value, FS_c, required to resist landslide throughout Karnataka and Sikkim.

$$\mathrm{FS}_c = \frac{\mathrm{PHA}}{\sin\alpha} + 1 \tag{19.2}$$

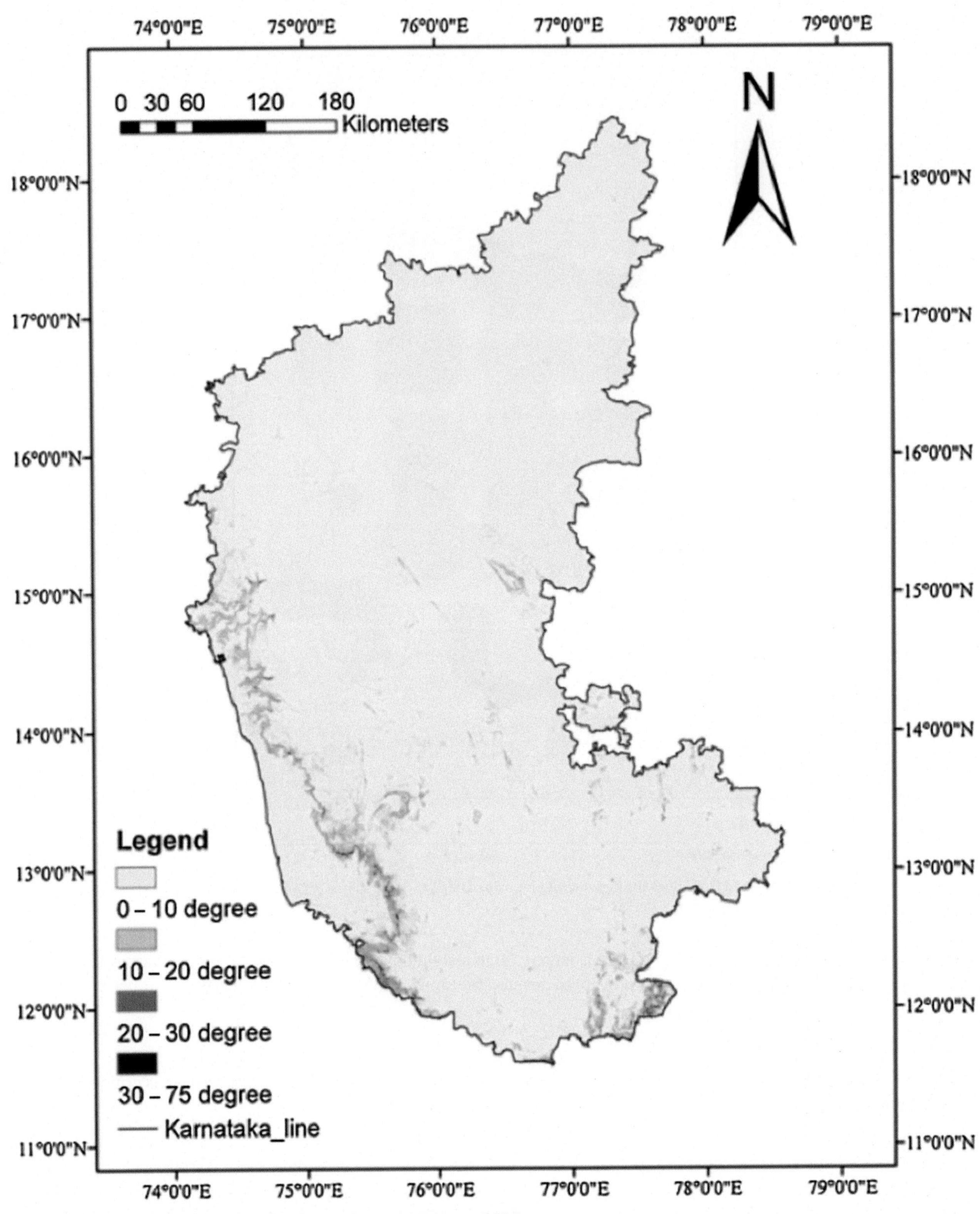

FIGURE 19.9 **Slope map of Karnataka.** Source: *James and Sitharam (2014).*

19.4 INTEGRATED LANDSLIDE HAZARD ANALYSIS

The above section primarily deals with the assessment of seismically induced landslide hazard at the macrolevel. However, earthquake is one of the causes for the landslide. Hence, to get a complete picture of the landslide hazard, one must need to quantify the earthquake-induced as well as rainfall-induced landslide hazard at a given location. However, carrying out dynamic slope stability analysis using in situ geotechnical properties at every location is an impossible task. Hence, there is a need to develop

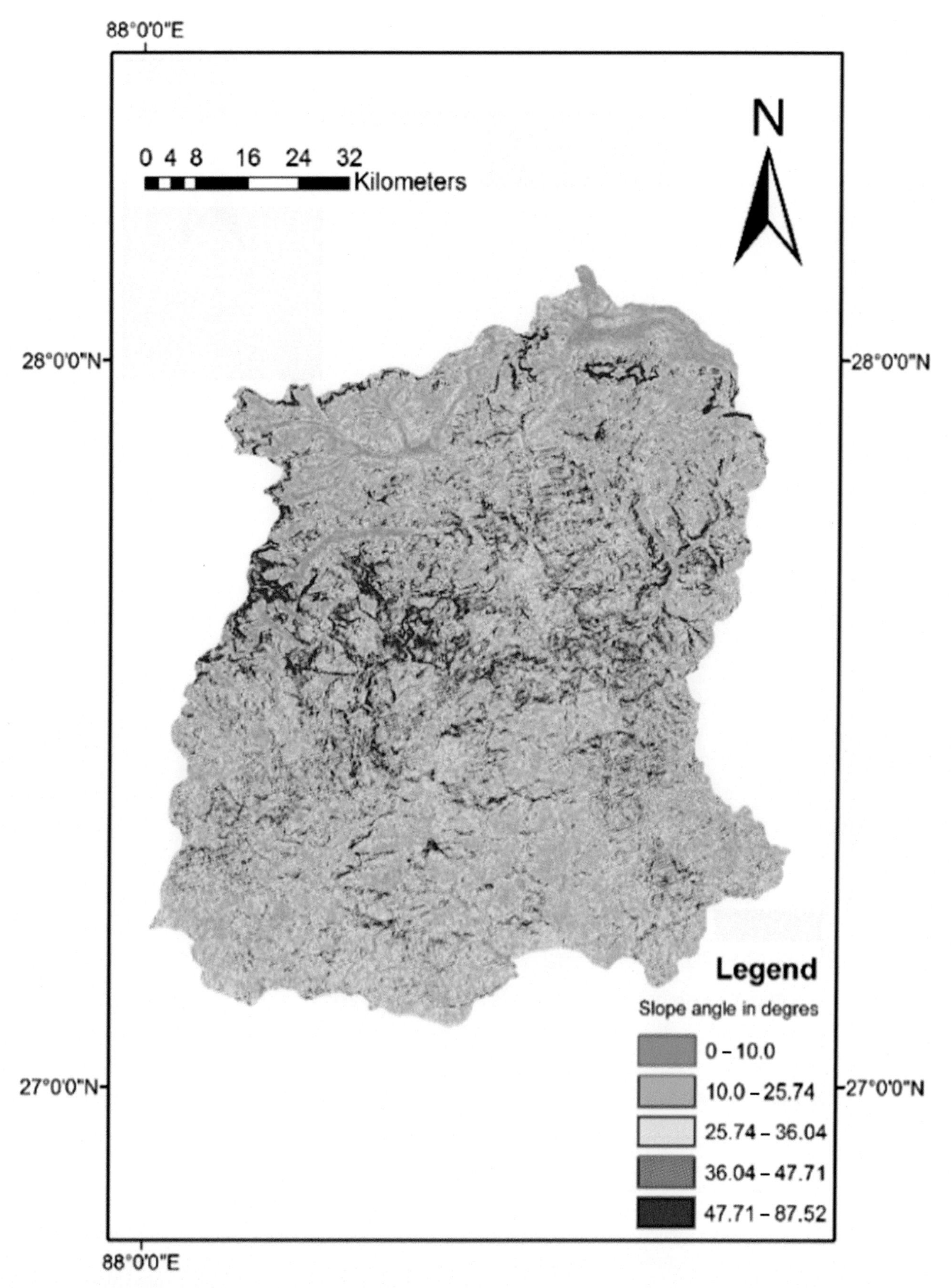

FIGURE 19.10 Slope map of Sikkim. Source: *James and Sitharam (2015).*

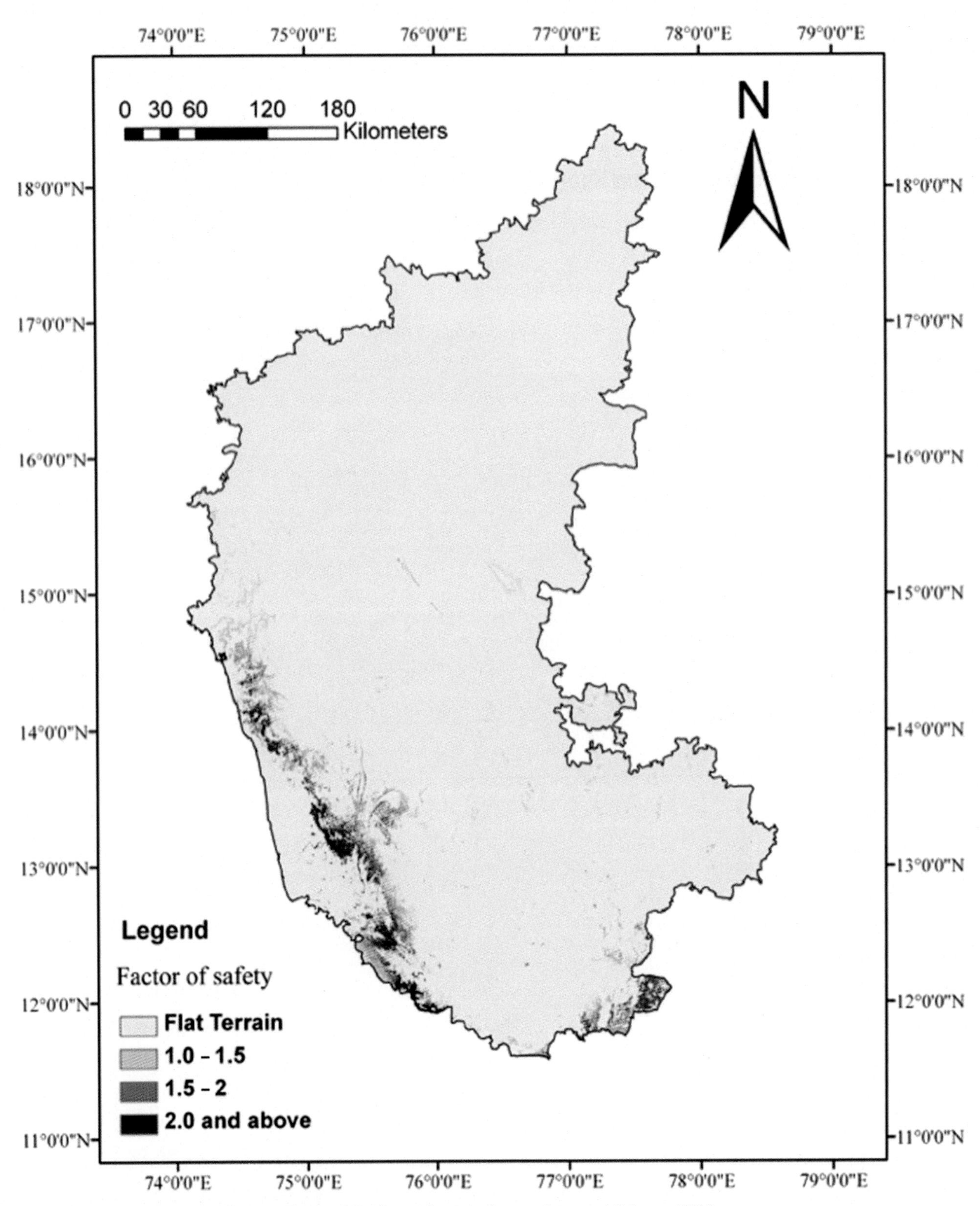

FIGURE 19.11 **Seismic landslide hazard map of the Karnataka state.** Source*: James and Sitharam (2014).*

an integrated landslide hazard map using analytical hierarchical approach (AHP). James and Sitharam (2017) have employed multicriteria AHP to evaluate integrated landslide hazard index. As the landslide hazard at a given location depends on many factors such as rainfall intensity, earthquake shaking, and slope of the terrain, AHP (Saaty, 1980) is best suited for such a multicriteria problem. The first step in AHP involves identification and prioritization of various hazard influencing parameters. James and Sitharam (2017) identified the landslide hazard parameter and arranged according to priority as described in Table 19.1.

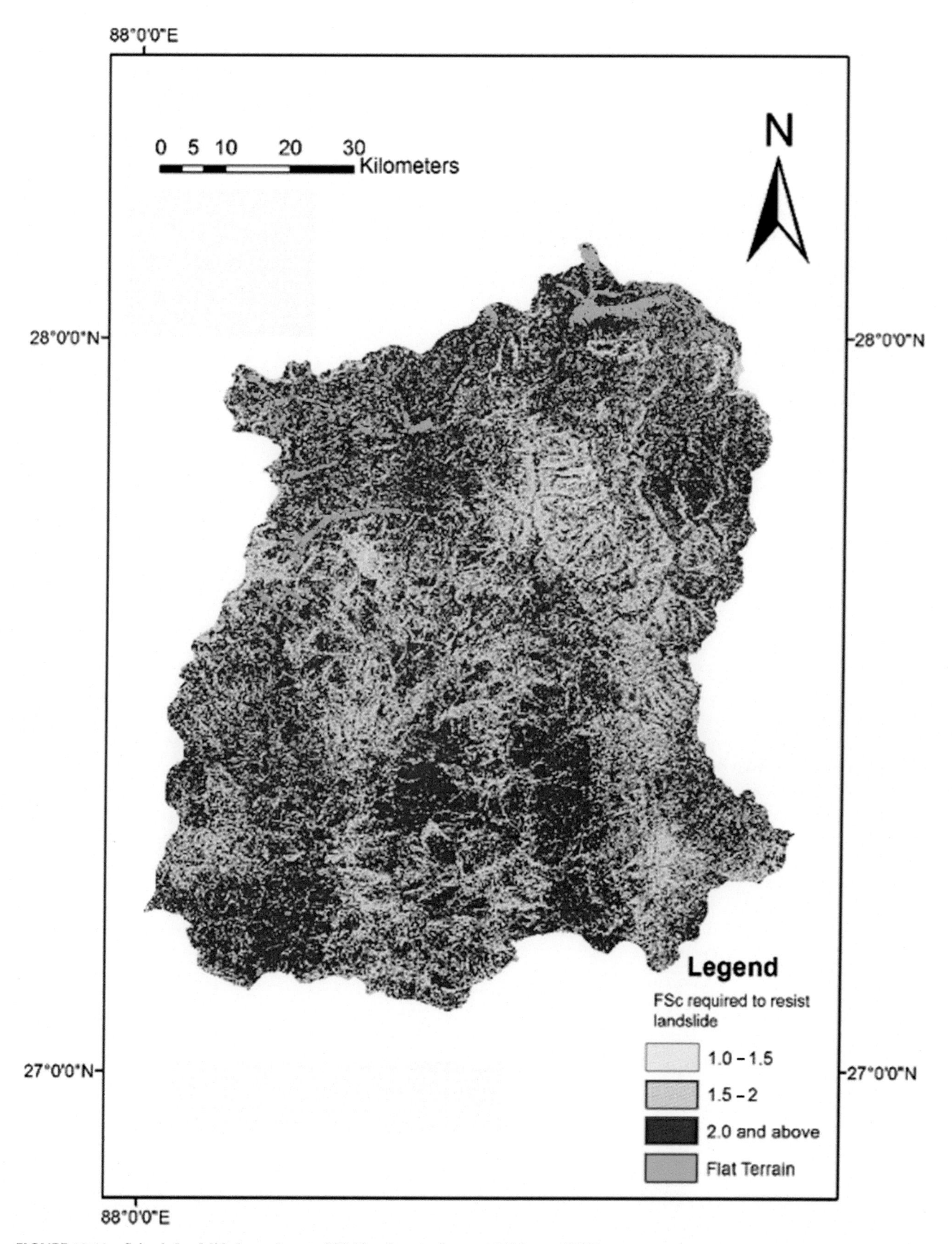

FIGURE 19.12 **Seismic landslide hazard map of Sikkim.** Source: *James and Sitharam (2015).*

Furthermore, they built a pairwise comparison matrix, and normalized weights of each parameters were established (Table 19.2). Each landslide hazard parameter is then categorized into various ranges (as in Table 19.3) and a rating was assigned to each category. The normalized rating for each category was then determined using the following equation:

$$X_i = \frac{R_i - R_{\min}}{R_{\max} - R_{\min}} \tag{19.3}$$

TABLE 19.1 Hierarchical Arrangement of Landslide Hazard Parameters

Landslide hazard parameter (LHP)	Weights
Terrain slope (TS)	4
Peak ground acceleration at bedrock (PGA)	3
Rainfall intensity (RF)	2
Amplification factor (AF)	1

TABLE 19.2 Pair-Wise Comparison of Each Landslide Hazard Parameters

LHP	TS	PGA	RF	AF	Normalized weights
TS	4/4	4/3	4/2	4/1	0.4
PGA	3/4	3/3	3/2	3/1	0.3
RF	2/4	2/3	2/2	2/1	0.2
AF	1/4	1/3	1/2	1/1	0.1

TABLE 19.3 Normalized Rating for Each EHP Category for Microlevel Hazard Integration

LHP	Value range	Weight	Rank	Normalized rank
TS	<10°	0.4	1	0
	10°–30°		2	0.33
	30°–50°		3	0.67
	50°–90°		4	1
PGA	<0.1 g	0.3	1	0
	0.1–0.15 g		2	0.25
	0.15–0.2 g		3	0.5
	0.2–0.25 g		4	0.75
	>0.25 g		5	1
RF	<100 cm	0.2	1	0
	100–200 cm		2	0.33
	200–300 cm		3	0.67
	>300 cm		4	1
AF	<1	0.1	1	0
	1–2		2	0.5
	2>		3	1

Here, R_i is the rank assigned to each category of a single theme and $R_{\min}$ and $R_{\max}$ are the minimum and the maximum rating values of that theme, respectively. The landslide hazard index values were then estimated based on normalized weights and ranks by integration of all themes using the following equation:

$$\mathrm{LHI} = \frac{\mathrm{TS}_w \times \mathrm{TS}_r + \mathrm{PGA}_w \times \mathrm{PGA}_r + \mathrm{RF}_w \times \mathrm{RF}_r + \mathrm{AF}_w \times \mathrm{AF}_r}{\sum w} \quad (19.4)$$

where LHI is the landslide hazard index, w is the normalized weight of the landslide hazard parameter (EHP) and r is the normalized rank of a category in the EHP. The integrated landslide hazard map of Sikkim (as in Fig. 19.13) was developed using AHP which considers the combined effect of seismic- and rainfall-induced landslide hazard.

19.5 CONCLUSIONS

This chapter discusses about quantification of landslides at macrolevel for two states in India. The study focuses on a Himalayan region and Western Ghats area in Karnataka, which are one of the most landslide-prone and Himalayan seismically active regions. Hence, there is a need to quantify landslide hazards in these regions. This chapter discusses the quantitative seismic landslide hazard assessment by Newmark's methodology using GIS tool. Landslide hazard zonation was performed for two states based on the spatial variation of static FS required to prevent landslides. Furthermore, the chapter also describes the application of microzonation principles in carrying out integrated landslide hazard zonation, considering the combined effect of rainfall- and earthquake-induced landslide. The integrated landslide hazard assessment can be treated as a multicriteria problem, and AHP is one of the best methodologies to address these kinds of problems. The integrated landslide hazard assessment map should be combined with population and land-use map to identify high landslide risk zones. Rigorous analysis involving extensive geotechnical and geophysical investigations can be carried out for predicting accurate landside hazard and selecting suitable mitigation works. Such sensitive regions are recommended to have advanced instrumentation to monitor movement slope material and early warning systems. Additionally, these locations also require most suitable landslide mitigation measures such as good drainage, surface water diversions away from landslide-prone

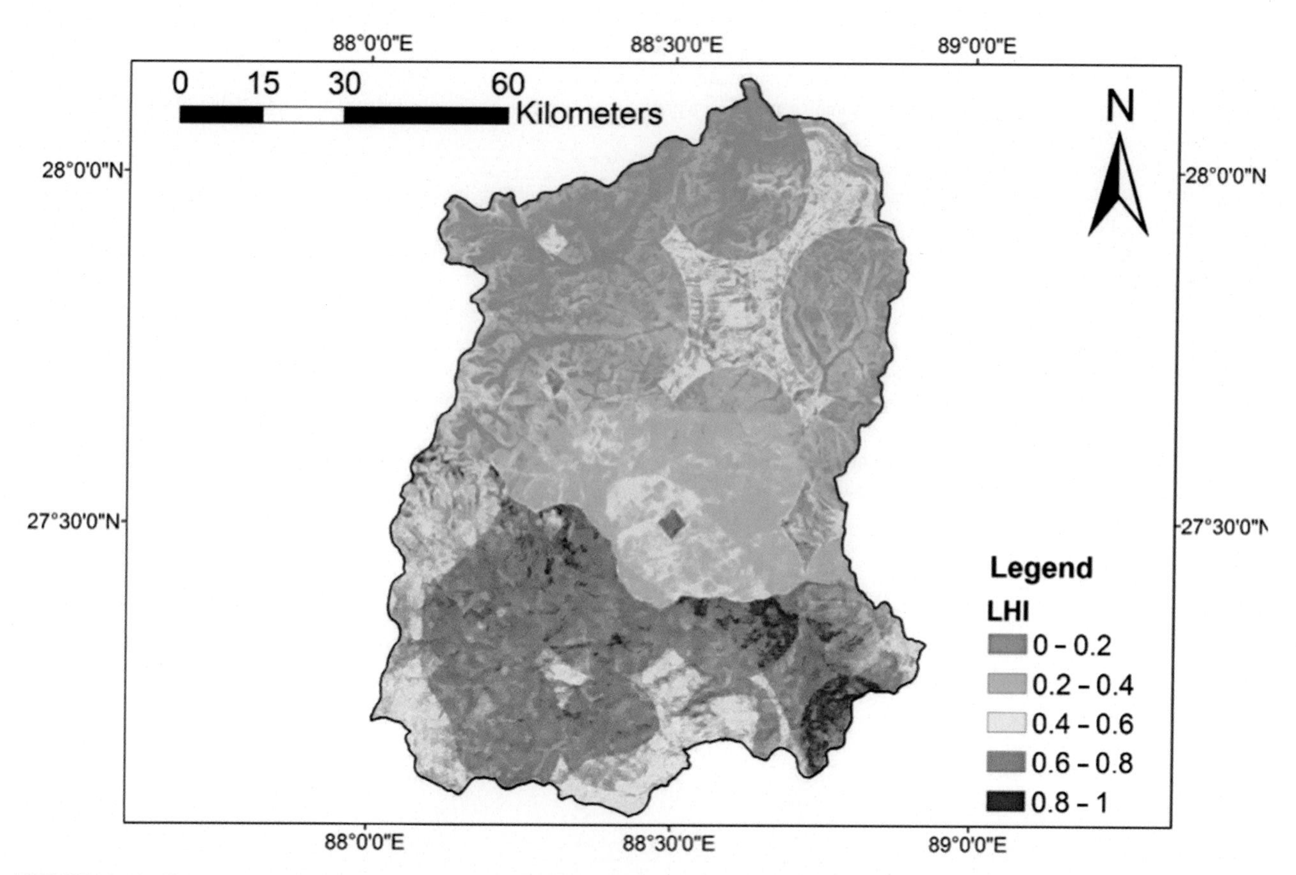

FIGURE 19.13 **Integrated landslide hazard index map of Sikkim.** Source: *James and Sitharam (2017).*

areas, construction of toe retaining walls, rock bolting with wire mesh and shotcreting or soil nailing, or control of surface water infiltration by providing appropriate drainage and/or planting of vegetation (like vetiver grass).

REFERENCES

Atkinson, G.M., Boore, D.M., 2006. Earthquake ground-motion prediction equations for eastern North America. Bulletin of the Seismological Society of America 96 (6), 2181–2205.

BBC, 2014. Indian landslide: dozens trapped in Pune village of Malin. BBC online: http://www.bbc.com/news/world-asia-28559617.

Bommer, J.J., Scherbaum, F., Bungum, H., Cotton, F., Sabetta, F., Abrahamson, N.A., 2005. On the use of logic trees for ground-motion prediction equations in seismic-hazard analysis. Bulletin of the Seismological Society of America 95 (2), 377–389.

Campbell, K.W., Bozorgnia, Y., 2003. Updated near-source ground-motion (attenuation) relations for the horizontal and vertical components of peak ground acceleration and acceleration response spectra. Bulletin of the Seismological Society of America 93 (1), 314–331.

CBS, 2013. India raises flood death toll reaches 5700 as all missing persons now presumed dead. http://www.cbsnews.com/news/india-raises-flood-death-toll-reaches-5700-as-all-missing-persons-now-presumed-dead/.

Chakraborty, I., Ghosh, S., Bhattacharya, D., Bora, A., 2011. Earthquake induced landslides in the Sikkim-Darjeeling Himalayas—an aftermath of the 18th September 2011 Sikkim earthquake. Geological survey of India, Kolkata.

IS:14496-Part-2, 1998. Preparation of landslide hazard zonation maps in mountainous terrains-guidelines, Part 2: Macro-zonation. Bureau of Indian Standards (BIS), New Delhi.

James, N., Sitharam, T.G., 2014. Assessment of seismically induced landslide hazard for the State of Karnataka using GIS technique. J. Indian Soc. Remote Sens. 42 (1), 73–89.

James, N., Sitharam, T., 2015. Macro-level assessment of seismically induced landslide hazard for the state of Sikkim, India based on GIS Technique. Paper read at IOP Conference Series: Earth and Environmental Sciencep. 012027.

James, N., Sitharam, T., 2017. Landslide hazard assessment and vulnerability studies for the state of Sikkim, India based on multicriteria analytical hierarchical approach. Paper Read at 16th World Conference on Earthquake. Santiago, Chile. .

Jibson, R.W., 1993. Predicting earthquake-induced landslide displacements using Newmark's sliding block analysis. Transp. Res. Rec. 1411, 9–17.

Kanth, S.R., Iyengar, R., 2007. Estimation of seismic spectral acceleration in Peninsular India. J. Earth Syst. Sci. 116 (3), 199–214.

Kramer, S.L., 1996. Geotechnical Earthquake Engineering. Pearson Education India, New Delhi.

Krinitzsky, E.L., 1995. Deterministic versus probabilistic seismic hazard analysis for critical structures. Engineering Geology 40 (1–2), 1–7.

NDMA, 2009. National Disaster Management Guidelines—Management of Landslides and Snow Avalanches. National Disaster Management Authority, Government of India, New Delhi.

Newmark, N.M., 1965. Effects of earthquakes on dams and embankments. Géotechnique 15 (2), 139–160.

Ravindran, K., Philip, G., 1999. 29 March 1999 Chamoli earthquake: a preliminary report on earthquake-induced landslides using IRS-1C/1D data. Current Science 77 (1), 21–25.

Saaty, T.L., 1980. The Analytic Hierarchy Process. McGraw-Hill, New York, NY.

Sharma, M.L., Douglas, J., Bungum, H., Kotadia, J., 2009. Ground-motion prediction equations based on data from the Himalayan and Zagros regions. J. Earthquake Eng. 13 (8), 1191–1210.

Shrikhande, M., Rai, D.C., Narayan, J., Das, J., 2000. The March 29, 1999 earthquake at Chamoli, India. 12th World Conference on Earthquake Engineering. Auckland, New Zealand. .

Terzaghi, K., 1950. Mechanism of landslides. In: Paige, S. (Ed.), Application of Geology to Engineering Practice. Geological Society of America, New York, pp. 83–123.

Wieczorek, G.F., Wilson, R.C., Harp, E.L., 1985. Map showing slope stability during earthquakes in San Mateo County, California.

Wilson, R.C., Keefer, D.K., 1983. Dynamic analysis of a slope failure from the 6 August 1979 Coyote lake, California, earthquake. Bulletin of the Seismological Society of America 73 (3), 863–877.

Wilson, R., Keefer, D., 1985. Predicting areal limits of earthquake-induced landslide earthquake hazards in the Los Angeles region—an earth-science perspective. US Geological Survey Professional Paper 1360, 317–345.

Chapter 20

What Behaviors We Think We Do When a Disaster Strikes: Misconceptions and Realities of Human Disaster Behavior

Tatsuya Nogami
Japan Fire and Crisis Management Association, Tokyo, Japan

Tatsuya Nogami is a research fellow in Japan Fire and Crisis Management Association, Tokyo, Japan. He has been conducting psychological research on how humans behave in anonymous situations, in cyberspace, and in postdisaster situations.

20.1 INTRODUCTION

When something unexpected or unusual happens, we behave differently from the way we do under normal circumstances. For example, a calm person may lose her/his composure and scream loud, when s/he finds a big spider on the wall of her/his room. Likewise, our behavioral patterns do not stay the same across different situations, but they can vary depending on what situation we are in. A layperson who generally abides by the laws and rules could occasionally engage in rule-breaking behavior (e.g. ignoring a traffic light) in a situation where nobody is around, for instance. Disasters can occur unexpectedly and suddenly, putting us in unusual situations. Therefore, it is no surprise that when a disaster strikes without warning, we ourselves could behave differently from the way we do in everyday life, as well as we may see others showing behavioral patterns that we rarely encounter under normal circumstances.

Whenever disasters, in particular, large ones, occur, we tend to bring to mind certain types of human behavior in postdisaster situations. One of such "assumed" postdisaster behaviors is *panic* among the general public. It is not rare to find a newspaper headline or a media report on a news program using the term panic after a major disaster has occurred. Also, we occasionally see pictures of disaster victims fleeing from a danger on TV, the Internet, and/or in a movie. Thus, nowadays even people who have never been affected by any kind of major disasters may not find it difficult to imagine a crowd of people desperately fleeing from a danger when a disaster suddenly strikes.

Another common belief that many of us tend to associate with postdisaster situations is looting and other kinds of criminal acts flourishing in disaster-affected areas. Disasters can alter our everyday lives dramatically, possibly leaving houses and stores more vulnerable to break-in through disaster damage to buildings, power blackouts, and/or evacuations of local residents to safer places. It may not be difficult for anybody to imagine that some people take advantage of a disaster and pursue their own self-interest in the confusion by committing a criminal act, resulting in increases in the crime rate of the disaster-affected area.

Moreover, we generally think that it is a good thing to help others in trouble. This sense of ethics is likely to exist in humans almost universally. Therefore, when it comes to disasters, it is not surprising that many of us are willing to alleviate the distress suffered by disaster victims as much as we can. Interestingly enough, in many cases, we prefer to send supplies of food, clothing, and necessities to disaster-affected areas rather than donate money, believing with confidence that such relief materials will be more helpful to disaster victims than monetary donations.

I assume that many of you reading this book think that some or all of these three behaviors described in the above three paragraphs are quite likely. Some may even insist that you have actually witnessed others getting panicky, looting a store, and/or relief materials sent by individuals comforting disaster victims in past disasters. You are not completely wrong. However, more than half a century of research on human disaster behavior indicates that such assumed disaster behaviors do not give us true pictures of postdisaster situations. This chapter is all about such persistent images of inappropriate disaster

Integrating Disaster Science and Management. http://dx.doi.org/10.1016/B978-0-12-812056-9.00020-8

behavior, which are also called *disaster myths*. Despite the long-standing denial by disaster researchers (Barton, 1969; Goltz, 1984; Quarantelli and Dynes, 1972), disaster myths are so widespread and firmly rooted in our minds. Some myths have been taken as a given in postdisaster situations even by disaster-response professionals and policy makers, as well as by the general public and media personnel.

The main purpose of this chapter is to disconfirm some of the commonest disaster myths by scrutinizing how people actually responded to and what actually happened in past disasters, with particular attention paid to three of the most popular disaster myths: panic, criminal acts, and donating behavior in postdisaster situations. I have been feeling that these three myths need to be more publicly shared and discussed not only by laypeople, but by disaster-response professionals, policy makers, media personnel, and scholars, because these myths could possibly lead to disorganized and inadequate disaster response and management that may put further burden on disaster victims and affected areas. In reference to human behavior observed in past natural and human-inflicted disasters, this chapter also outlines psychological research on our innate tendencies related to these three disaster myths.

20.2 MISCONCEPTIONS AND REALITIES OF HUMAN BEHAVIOR IN DISASTERS

Disasters, regardless of whether nature- or human-inflicted, can occur in any place at any time. It is very much the case in Japan, which has been a disaster-prone country, particularly, when it comes to natural disasters. Japan has been vulnerable to natural disasters, such as typhoons, floods, landslides, earthquakes, tsunamis, and volcanic eruptions. In fact, 10.1% of the world's major natural disasters with more than 1000 casualties have actually occurred in Japan since 1900 (16 out of 159; Cabinet Office, 2015). Considering the land area occupied by the country, accounting only for 0.28% of the globe's total land area (Japan Institute of Country-ology and Engineering, 2015), Japan is no doubt one of the most disaster-prone countries in the world.

Japan has been particularly vulnerable to large earthquakes and tsunamis since 1990s. In the 1993 Hokkaido Earthquake (Mw, 7.7), 230 people were killed or missing in devastating tsunami waves. The death toll from the 1995 Great Hanshin-Awaji Earthquake (Mw, 6.9) reached more than 6000. During the 2004 Niigata Chuetsu Earthquake (Mw 6.6), 68 were crushed to death. The 2011 Great East Japan Earthquake (Mw, 9.0) claimed lives of nearly 20,000 locals on the pacific coast of the Tohoku region. The year 2016 was no exception: In April, two big earthquakes (Mw, 6.2 on 14th; Mw, 7.0 on 16th) struck in Kumamoto, in the southern part of Japan, leaving 50 people dead and more than 2000 injured (National Police Agency, 2016). Even before these earthquakes, Japan was devastated by the Great Kanto Earthquake in September 1923, in which more than 100,000 people were victimized (Muroi and Takemura, 2006).

Other than earthquakes and tsunamis, there have also been many unforeseen contingency situations observed in Japan since 1990s. In particular, typhoons and heavy rain have frequently caused disastrous consequences. From August to October 2004, a series of typhoons left 160 casualties in the Japanese islands, while huge landslides triggered by heavy rain killed 36 people in Oshima, Tokyo, in October 2013. Torrential rain also triggered landslides in Hiroshima in August 2014, killing 74 local residents.

Volcanic activity can occasionally bring many deaths in Japan, too. The 1991 eruption of Mount Unzen in Nagasaki generated a pyroclastic flow that took the lives of 43 people. Mount Ontake, located in between Nagano and Gifu, erupted in September 2015, and 58 bodies were later found dead on the slopes and near the summit of the mountain with five still missing. Needless to mention, there have been far more disasters in Japan than these that had no/fewer casualties but serious impacts on local residents and communities.

As mentioned above, Japan, sitting atop four tectonic plates with one of the commonest typhoon paths, has been meteorologically and geophysically very prone to natural disasters. In other words, Japanese, including disaster researchers, have ample opportunity to see how humans actually react to sudden, unexpected events like disasters. Considering Japan's high vulnerability to disasters, it sounds plausible that the Japanese are more insightful into disaster behavior than people from other areas of the world. Unfortunately, such an inference is probably not right. Japanese researchers looked into disaster myths among Japanese people after the 2011 Great East Japan Earthquake and found that the Japanese still did give credit to some of the commonest disaster myths (Nogami and Yoshida, 2014). Thus, it seems that the Japanese, living in one of the most disaster-prone countries, are at least no less susceptible to disaster myths than those from less disaster-prone countries.

Since its inception in the middle of the 20th century, research on human behavior in postdisaster situations has been active mainly in North America (Fritz and Mathewson, 1957; Quarantelli, 1954). At least by the 1970s, numerous researchers disconfirmed disaster myths relating to panic, looting, and donating behavior in postdisaster situations (Barton, 1969; Fritz and Mathewson, 1957; Quarantelli, 1954, 1960; Quarantelli and Dynes, 1972; Wenger et al., 1975). However, such myths still seem to firmly stick to the minds of people in North America (Auf der Heide, 2004; Clarke,

2002; Clarke and Chess, 2008; Tierney, 2003; Tierney et al., 2006) as well as in Europe (Alexander, 2007; Drury et al., 2013a) and in Japan (Nogami and Yoshida, 2014). In this section, disaster myths relating to panic, criminal acts, and donating behavior in postdisaster situations are scrutinized in detail in reference to past natural and human-inflicted disasters. With respect to each of the myths, I will describe why they are assumed to be untrue based on findings of previous research.

20.2.1 Panic in a Disaster

People use the term panic in everyday life to express a variety of emotions and behavior. The term may be used to describe fear, anxiety, anger, and/or surprise about something, or to explain some sort of unexpected behavior that people are forced to do under unusual circumstances. Cambridge Dictionaries Online defines the term as "a sudden strong feeling of fear that prevents reasonable thought and action" (Panic, 2015a). Likewise, panic is defined as "sudden uncontrollable fear or anxiety, often causing wildly unthinking behaviour" in Oxford Dictionaries (Panic, 2015b). In both definitions, the term panic basically refers to sudden strong feelings of fear and/or anxiety that can provoke some form of *irrational* emotional and behavioral responses.

Panic has also been strongly associated with disasters. For more than half a century, panic has been one of the commonest disaster behaviors that the general public assumes (Clarke, 2002; Quarantelli, 1954, 2001; Wenger et al., 1975). People tend to think that major disasters can trigger panic among disaster-affected individuals, who flee hysterically in the face of great threats at the expense of others (Quarantelli, 1960; Quarantelli and Dynes, 1972). Not only among the general public, but whenever and wherever a major disaster occurs, the term panic is frequently used in news headlines by the mass media. For instance, as soon as a 7.8 magnitude earthquake hit Nepal at the end of April 2015, the term panic was quickly transmitted worldwide through the Internet, television, and other media (BBC News, 2015, April 26; Cullinane and Park, 2015, April 27; Mainichi Newspapers, 2015, April 26). Not surprisingly, Emergency management personnel and policy makers also tend to assume that panic is likely to occur among the general public during a disaster (Alexander, 2007; Clarke, 2002; Clarke and Chess, 2008; Drury et al., 2013a,b; Tierney, 2003; Tierney et al., 2006), although these professionals may be aware, more or less, of the collective resilience that the public and the community can show in emergency situations (Drury et al., 2013a, 2015).

Despite this common assumption about panic among the general public, media personnel, and professionals related to disaster response, empirical evidence has indicated repeatedly that panic rarely occurs in emergency situations (Quarantelli, 1954, 1960, 2001). Quarantelli (1954) already pointed out in the 1950s that "the frequency of panic has been exaggerated ... Compared to other reactions panic is a relatively uncommon phenomenon" (p. 275). He further claimed that even when observers see disaster victims fleeing away from danger as panic flight, it may be functional and reasonable behavior from a standpoint of those in danger. Mass flight, even when it occurs, is thought to rarely result in a massive stampede trampling other evacuees to death (Quarantelli, 1954). In addition to that, panic is rarely on a large scale, with only a few persons locally engaging in flight behavior for a short duration of time (Quarantelli, 1960, 2001).

Research on disaster behavior has been mainly performed in Western societies. However, cultural differences in the lack of disaster panic do not seem significant, at least in Japan. Hiroi (2003), who was one of the pioneer disaster researchers in Japan, looked back at his 25-year research career and stated that "I have never seen or heard of any single known case in which a disaster triggered panic" (p. 716). The same can be said for the misconception about panic, as the Japanese also seem to hold the assumption that people are likely to panic in disaster situations (Nogami and Yoshida, 2014). As many disaster researchers have been claiming over the last decades, the individual's initial response to an emergency situation is generally prosocial and altruistic rather than selfish and/or irrational flight behavior (Auf der Heide, 2004; Sandin and Wester, 2009; Tierney, 2003; Tierney et al., 2006; Quarantelli, 2001). Contrary to this common misconception about panic, which is also known as the panic myth, one of the major problems in emergency situations is to get people to evacuate, not flight behavior (Quarantelli and Dynes, 1972).

According to Quarantelli (2001), there seems to be a huge discrepancy between laypeople's interest in disaster panic and its actual frequency. Although it is still unknown what is actually creating this discrepancy, one possible source of the discrepancy has been thought to be the mass media and popular culture. As pointed out by numerous disaster researchers, media coverage of a disaster and disaster movies are assumed to be responsible for spreading inconsistent, distorted images of disasters (Ali, 2013; Clarke and Chess, 2008; Fahy et al., 2012; McEntire, 2006; Mitchell et al., 2000; Quarantelli, 1960, 1980, 2001; Scanlon, 2007; Tierney, 2003; Tierney et al., 2006; Wenger, 1985; Wenger et al., 1975; Wenger and Friedman, 1986), which will be discussed later in this chapter. Another possibility is that there may be some definitional difference between what laypeople consider as panic and what disaster researchers construe as panic.

20.2.1.1 *Definition of Disaster Panic*

As mentioned earlier, the general definition of panic is sudden strong feelings of fear and/or anxiety that can provoke some form of irrational emotional and behavioral responses. Laypeople also use the term panic to indicate mass flight in emergencies (Nogami and Yoshida, 2014; Wenger et al., 1975). The meanings of the term used in the mass media and popular culture seem the same, but collective emotional and behavioral responses to emergency incidents (e.g. mass flight) are more often described as panic by the media (e.g. BBC News, 2015, April 26; Cullinane and Park, 2015, April 27; Mainichi Newspapers, 2015, April 26). In either use of the term, emotional and behavioral responses triggered by feelings of fear and/or anxiety are not strictly defined. That is, any emotion and behavior responding to incidents that can elicit fear and/or anxiety could be described as panic from laypeople's perspectives.

Indeed, recent research found that laypeople use the term panic to indicate multiple different behavioral and emotional responses to emergency situations (Nogami, 2016). For some people, the term may mean losing composure and shouting, while some others could construe it as trembling and fleeing. Such responses can also vary between emergency situations. The term panic can indicate fleeing in an earthquake, whereas it may not imply such behavior in an aviation accident where fleeing is not a possible behavioral choice. Considering these uses of the term panic among laypeople, it can be said that panic is quite elusive in meaning. Clarke and Chess (2008) also noted that "the word is sometimes used to convey the seriousness of an incident or to exaggerate the danger so as to impress an audience" (p. 998). This arbitrary use of the term may be one reason why the term panic is so widespread among laypeople, especially when describing one's and/or others' emotions and behavior in unexpected incidents like disasters.

In the field of disaster research, Quarantelli (1954) initially defined panic as "an acute fear reaction marked by a loss of self-control which is followed by nonsocial and nonrational flight behavior" (p. 272). However, the definition of panic employed in disaster research also carries a variety of different meanings and there is little consensus on the definition even among disaster researchers (Gantt and Gantt, 2012; Mawson, 2005, 2007; Quarantelli, 1954, 2001). Despite this diversity in its definition, Quarantelli (2001) argued that the nature of panic can be grouped into two categories: extreme, groundless fear, and flight behavior. He also noted that these two are not necessarily related to one another, and groundless, extreme fear does not inevitably trigger flight behavior (Quarantelli, 2001). On the other hand, Mawson (2007) stated that two classes of behavior are consistently associated with definitions of panic found in the literature: wild running and immobility reactions.

In short, even in disaster research, the term panic can be used to express a variety of different emotional and behavioral responses to emergency situations. Considering that the meanings of the term can also vary among laypeople, it seems inevitable that what laypeople consider as panic is not consistent with what disaster researchers construe as panic. Indeed, researchers occasionally found that disaster survivors and emergency response professionals frequently use the term panic in their postdisaster accounts, but the meanings of the term are often different from what disaster researchers typically define as panic (e.g. irrationality, selfishness, and flight behavior; Clarke, 2002; Cocking and Drury, 2014; Drury et al., 2009b, 2015; Fahy et al., 2012). Supposedly, this definitional difference may let what laypeople call panic occur more frequently in a disaster than disaster researchers think.

Summarizing the present discussion on the definition of disaster panic, there seem to be two types of panic that are assumed to occur in emergency situations: individual panic and mass panic. The former is characterized by the individual's behavioral and emotional responses to an unexpected event that are normally elicited by strong feelings of fear and/or anxiety (e.g. shouting, crying). The precise meaning of this panic may differ depending on the person and the situation, but it is probably used, in particular, among laypeople, in reference to a fear- and/or anxiety-eliciting event regardless of whether such an event is a disaster. The latter usage of panic puts more focus on collective behavior in disasters, particularly mass also seems common among laypeople, the mass media, and emergency management personnel (Alexander, 2007; BBC News, 2015, April 26; Clarke, 2002; Clarke and Chess, 2008; Cullinane and Park, 2015, April 27; Drury et al., 2013b; Mainichi Newspapers, 2015, April 26; Nogami and Yoshida, 2014; Tierney, 2003; Tierney et al., 2006; Quarantelli and Dynes, 1972; Wenger et al., 1975). This usage of the term is in line with what disaster researchers define as panic (Quarantelli, 2001; Mawson, 2007), but as discussed so far, the idea of mass panic has been generally discredited by disaster researchers (Cocking and Drury, 2014; Cornwell et al., 2001; Drury et al., 2009a–c, 2015; Johnson, 1987, 1988; Quarantelli, 1954, 1960, 2001; Rosengren et al., 1975).

20.2.1.2 *Human Behavior in Past Emergency Situations*

It is now evident that the meanings of the same term panic can differ between laypeople and disaster researchers, as well as between laypeople themselves. On the one hand, it is no surprise that when something unexpected happens, we may go into an individual panic (the individual's behavioral and emotional responses to a fear- and/or anxiety-eliciting event). However, such behavioral and emotional patterns seem to be quite normal human reactions to fear- and/or anxiety-eliciting events.

On the other hand, mass panic sounds far more detrimental than individual panic since it may make disaster situations even worse (e.g. causing more casualties due to a stampede for an exit). Therefore, it is worth figuring out in what situations mass panic is assumed to occur despite its rare occurrence. Although the occurrence of mass panic is thought to be subject to certain conditions, just like its definition, conditions required to cause mass panic also vary from researcher to researcher. Quarantelli (1954) suggested that conditions for mass panic are the person's perceptions and/or feelings of possible entrapment, collective powerlessness, and individual isolation in an emergency situation. Mawson (2007) sorted out two key factors leading to mass panic based on past findings: "a belief that physical danger is present from which escape must be sought, and the belief that escape routes are either limited or closing" (p. 155). Hence, although mass panic is not unique to disasters and its occurrence is supposed to be rare, at least this phenomenon is assumed to occur in emergency situations where people *perceive* that they are facing physical danger.

Obviously, the conditions required to cause mass panic should have been met in some, if not all, past disasters. Has mass panic actually occurred in past disasters? Quarantelli (2001) pointed out that, even in disaster situations where these panic-eliciting conditions are apparently met, people do not necessarily engage in flight behavior. In fact, no mass flight was explicitly observed in some of the past major disasters, such as the 1977 Beverly Hills Supper Club fire (Johnson, 1988), the 1979 Who concert crush tragedy (Johnson, 1987), the 1989 Hillsborough stadium disaster (Cocking and Drury, 2014), and the 2005 London bombings (Drury et al., 2009b). The occurrence of panic flight was also not confirmed under some unique circumstances where mass panic was reported to have occurred in the past (radio broadcasts of a fictitious alien invasion from Mars, Quarantelli, 2001, and of a fictitious nuclear accident, Rosengren et al., 1975).

Human behavior observed in past maritime accidents could also provide evidence for whether mass flight and other irrational behavior actually occurred under the apparently panic-eliciting conditions. According to Frey et al. (2011), mass flight did not seem to have occurred in the 1912 *Titanic* disaster, in which a social norm that women (and children) were to be saved first was thought to have existed. In this particular disaster, the survival rate of the females was 72.0%, whereas only 20.6% of the males managed to survive. Surprisingly, females with children had the highest survival rate (94.7%) in this tragic accident.

Unlike the *Titanic* disaster, there seemed to be no such social norm in the 1915 *Lusitania* disaster. The survival rate of the females in this disaster was 28.0%, while 34.3% of the males managed to survive (Frey et al., 2011). The main difference between these two maritime disasters could be attributed to the sinking time. The *Titanic* sank in the North Atlantic in 2 h and 40 min, but it took the *Lusitania* only 18 min to sink into the Celtic Sea. Frey et al. (2011) concluded that self-interested flight behavior predominates when a time frame for evacuation is narrow in a life-threatening situation.

Cornwell et al. (2001), however, drew a different conclusion from another maritime accident with similar circumstances to the *Lusitania*. In the 1994 sinking of the *M/V Estonia*, which sank in the Baltic Sea within 50 min since the initial damage to the ship, 22.0% of the males were rescued, while the survival rate of the females was only 5.4%. Based on quantitative and qualitative analyses on available information, Cornwell et al. argued that, rather than selfish flight behavior, the extremely physically challenging situation (e.g. severe weather conditions, violent ship movements, flying objects on the listing ship), and the narrow time frame for evacuation might have prevented the crew members and passengers from effective rescue activities, resulting in the survival rate of the females being significantly lower than that of the males.

Even in other emergency situations observed in Japan, mass panic of affected individuals did not seem to be predominant during a disaster. For instance, Japan Airlines Flight 123 suffered a serious accident on its way from Tokyo to Osaka on August 12, 1985. The aircraft crashed into the top of a mountain in Gunma, located in the center of the main island, 32 min after the accident occurred, taking the lives of 505 passengers and 15 crew members. During this 32-min period, the aircraft lost control and became extremely unstable with violent shaking in the air. However, after reviewing available evidence, it was thought that neither passengers nor crew members had showed selfish or irrational flight behavior. Rather, the majority of these victims were reported to have managed to behave with composure in such an extreme situation (Yoshioka, 1986).

Japanese researchers from the University of Tokyo also looked into how local residents had responded to the 1986 volcanic eruption in Izu Oshima. In this volcanic disaster, all the local residents (more than 10,000) had to evacuate from the island within a day. The results of the study indicated that more than 80% of the 807 respondents had not thought that mass panic would occur during the evacuation, while 18.3% of them had actually concerned about mass panic (Mikami, 1988). Moreover, 95.3% of them stated that the local residents had evacuated in compliance with instructions from the municipal organizations, with only 0.8% witnessing others behaving in a selfish way during the evacuation.

After the 2011 Great East Japan Earthquake, Central Disaster Prevention Council (2011) conducted a survey on the disaster survivors living in the three most affected prefectures, asking them what they had done right after the initial earthquake (multiple answers allowed). Of 870 respondents, 57% answered that they had evacuated to safer places in anticipation of possible tsunami waves coming. The rest of the respondents stated that they had not evacuated immediately because they had returned home (22%), tried to find/pick up one's family members (21%), or because they thought that past earthquakes had not triggered tsunami waves (11%). Even in Japan's largest earthquake, mass panic did not seem to have occurred.

It is still overreaching to conclude that mass panic never occurred in the above-mentioned emergency situations and that it will never occur in future disasters. However, at least such mass flight, as well as other kinds of selfish and irrational behavior, did not appear to be predominant even under the extreme conditions described above. On the contrary, the findings of previous research on the above-mentioned disasters and other available information apparently failed to provide any clear evidence that mass panic had actually occurred in these disasters. Thus, in line with what disaster researchers have been claiming (Hiroi, 2003; Quarantelli, 1954, 1960, 2001), it is safe to assume at present that mass panic (mass flight) is not a common reaction to disasters. To say the least, it is a much rarer phenomenon than is commonly supposed.

20.2.2 Increased Crime in Disaster-Affected Areas

Crime occurs every day in human society, regardless of the occurrence of a disaster. We may occasionally break existing rules, with or without intention. Ranging from relatively minor misdeeds to capital felonies, it can be regarded as a criminal act when someone has violated an existing law, even if s/he has an apparently justifiable reason. Also, there is no doubt that certain people are more likely to commit crime than most of us. Individuals with criminal lifestyles can commit crime in unusual circumstances as well as in normal circumstances.

Needless to mention, certain places and certain periods of time are more vulnerable to crime. For example, some city/country has a higher crime rate than others, and crime generally rises when social issues, such as unemployment and poverty, are widespread in a society. Therefore, when it comes to the discussion of criminal acts in postdisaster situations, we should carefully pay attention to such background factors before making a definite conclusion. Also, the important point here is whether crime actually occurs *more* in postdisaster situations than in predisaster situations, not just whether crime *merely* occurs in postdisaster situations.

Along with the panic myth, the assumption that crime is on the rise in disaster-affected areas is another common misconception about postdisaster behavior, which is labeled as the crime myth in this chapter. Although looting can easily be imagined in postdisaster situations, such a misdeed has been found very rare in disaster-affected areas (Quarantelli, 2007; Quarantelli and Dynes, 1970, 1972; Rodríguez et al., 2006; Tierney et al., 2006). Furthermore, the crime rate normally decreases in affected areas after a disaster (Jacob et al., 2008; Quarantelli, 2007; Quarantelli and Dynes, 1972) or at least is not significantly affected by a disaster (Varano et al., 2010). Still, many people anticipate that looting and other types of criminal acts will be widespread in disaster-affected areas (Alexander, 2007; Nogami and Yoshida, 2014; Wenger et al., 1975). As with the panic myth, the crime myth is held not only by the general public, but professionals related to disaster response and the mass media seem to believe it (Alexander, 2007; Ali, 2013; Rodríguez et al., 2006; Sandin and Wester, 2009). In the following paragraphs, I will scrutinize looting and other kinds of crime in past disasters separately, with emphasis on postdisaster crime in Japan.

20.2.2.1 Looting After a Disaster

Oftentimes, the mass media tend to sensationally report looting in disaster-affected areas, whenever a major disaster takes place. For example, when Hurricane Katrina hit New Orleans in August 2005, sensational accounts of looting in the largest city of Louisiana were frequently broadcast on TV and the Internet, many of which were later found seriously distorted or incorrect (Rodríguez et al., 2006; Tierney et al., 2006). Intriguingly enough, when the Great East Japan Earthquake occurred in March 2011, the foreign media expressed surprise by reporting that "why is there no looting in Japan?" (BBC News Magazine, 2011, March 18; Cafferty, 2011, March 14; Picht, 2011, March 14; West, 2011, March 14). Such media accounts were clearly premised on that looting should take place whenever and wherever a major disaster occurs.

Looting is generally defined as "the activity of stealing from shops during a violent event" (Looting, 2016) or "both grand and petty larceny of personal property during and after disaster impact" (Gray and Wilson, 1984). In many cases looting is not counted as it stands in the official crime statistics, but is normally categorized as burglary or property crime. Therefore, I will refer to the number of burglary or property crime, when citing the frequency of the misdeed in postdisaster situations.

Looting, as already mentioned, is not generally considered pervasive in postdisaster situations by disaster researchers (Quarantelli and Dynes, 1970, 1972; Rodríguez et al., 2006; Tierney et al., 2006). Quarantelli (2007) stated that "In natural disasters looting was very rare, covertly undertaken in opportunistic settings, done by isolated individuals or very small groups, and socially condemned" (p. 2). Nevertheless, as Quarantelli (2007) noted, "very rare" means that looting does occur after a disaster. In fact, looting actually took place in some past disasters, including the 1985 Hurricane Hugo

(LeBeau, 2002; Quarantelli, 2007; Rodríguez et al., 2006; Tierney et al., 2006), the 2005 Hurricane Rita (Leitner and Helbich, 2010), and the 2005 Hurricane Katrina (Frailing, 2007; Rodríguez et al., 2006). Quarantelli (2007) also pointed out that "there are occasional atypical instances of mass lootings that only emerge if a complex set of prior social conditions exist" (p. 3). In line with this statement, Frailing (2007) argued that looting in New Orleans after the 2005 Hurricane Katrina was no myth but ubiquitous due to the lack of social control and poverty deeply rooted in the affected community. Therefore, we need to consider preexisting social issues surrounding a disaster-affected community, when looking at postdisaster looting.

In the example of Hurricane Katrina, there is no doubt that looting occurred, more or less, in New Orleans after the hurricane. However, as mentioned at the beginning of the section, certain places are more vulnerable to crime than others regardless of the occurrence of a disaster. In reality, New Orleans had 5162 property crimes per 100,000 residents in 2004 (City of New Orleans, 2016), while the national average of the same crime was 3517.1 per 100,000 inhabitants for the same year period (Department of Justice, 2006a). That is, New Orleans' property crime rate was 1.47 times higher than the national average. New Orleans was vulnerable to property crimes even before the hurricane. Looting-like behavior did occur there, but some researchers pointed out that compared to a nondisaster time period, looting or similar offenses actually decreased in New Orleans following Hurricane Katrina (Barsky et al., 2006).

Moreover, many of the affected in New Orleans were disadvantaged who had been in difficult conditions even before the hurricane struck in the city and could not have afforded to escape beforehand (Simo, 2008). If people in such conditions cannot expect timely support from government institutions in a postdisaster situation, some might well decide to help themselves by taking necessities from closed stores and homes to survive the miserable situation (Barsky et al., 2006; Sandin and Wester, 2009). Such behavior certainly falls into the definition of looting and definitely breaks existing laws in many countries. However, at the same time such a deed should also significantly differ from what many of us generally associate with the word looting: stealing others' property from stores and homes with force. A difficult question is posed here: Is it still sensible to call disaster victims looters, who take necessities for emergency survival from a closed store and home in a desperate situation?

Thus, some, if not all, of the looting-like behavior observed in New Orleans (thefts by individuals with criminal lifestyles, inevitable stealing of necessities by the poor) was what had already existed in the community before the disaster (a high property crime rate, poverty). With these preexisting social issues and available evidence in mind, can we still claim for sure that looting or similar offenses actually occurred more in post-Katrina New Orleans or these misdeeds were solely due to the disaster? Obviously, it is not a simple yes–no question. Nonetheless, at least the hurricane did not seem to intensify the antisocial tendency of the majority of the local residents. As Rodríguez et al. (2006) argued, prosocial behavior was by far the primary response to the disaster in New Orleans at the time of the disaster.

Now let us turn our attention to Japanese disasters. Over the last 50 years, Japan has been a relatively stable country in terms of economic and social equality among the general public. Japan has also been quite safe in terms of crime victimization even when compared to other developed countries (OECD, 2005). If preexisting social issues play a significant role in the occurrence of looting during a disaster, Japan might have been relatively free from this misdeed in recent natural disasters (e.g. media reports on "no looting in Japan" after the Great East Japan Earthquake). However, is it really true that there has been no looting in Japan after disasters? From here, I will focus on looting in some of the worst natural disasters that have occurred in Japan.

Right after the 1995 Great Hanshin-Awaji Earthquake, the number of property crime in January went down by 58.5% in comparison with the same month in the previous year, with the number of burglary being decreased by 50.2% (Adachi, 2012; Saito, 2001). These figures also showed decreases when compared in a six-month period (from January to June of 1994 and 1995): Property crime decreased by 16.4% in 1995 and burglary dropped by 42.9%. While other property crimes showed decreases, the number of motorcycle theft actually increased by 33.3% in a 6-month period. This increase in motorcycle theft might have been due to the fact that the motorcycle was the most convenient means of transportation in the affected areas in which roadways had been severely damaged by the earthquake (Saito, 2001). Presumably, those who needed to move a long distance right after the disaster for some reason (e.g. evacuation, rescue activities) might have illegally taken it.

The number of burglary temporarily increased in the three affected areas (Iwate, Miyagi, and Fukushima) after the 2011 Great East Japan Earthquake. Burglary increased by 19.7% in the first two months after the disaster (March and April) when compared to the same period in the previous year (2010 and 2011), although the total number of property crime decreased by 12.6% (National Police Agency, 2011a). National Police Agency (2012) later issued a more detailed report on postdisaster crime relating to the Great East Japan Earthquake. From March to December 2011, the number of burglary actually decreased in Iwate (−19.6%) and Miyagi (−12.4%) in comparison with the same period in the previous year. Only in Fukushima, however, the number increased by 35.0%. The National Police Agency (2012) attributed this increase to the designated no-go areas around the Fukushima nuclear power plant, where burglars had virtually been free to break into abandoned homes and stores. It is also worth noting that JPY 684 million (approximately USD 5.8 million) was stolen from ATMs in convenience stores and local banks in the three affected areas after the disaster (JPY 27 million in Iwate, JPY 180 million in Miyagi, and JPY 477 million in Fukushima; National Police Agency, 2012).

As for the 2016 Kumamoto Earthquake, in which 8329 houses completely collapsed and 31,692 half collapsed (Cabinet Office, 2016), National Police Agency (2016) announced that there were 20 cases of theft in the affected areas from April 14 to August 15 following the earthquake. These 20 criminal acts (16 burglaries and four nonburglaries) were committed by 14 individuals. Their targets ranged from emergency goods, such as a helmet, batteries, and a pot, to disaster-unrelated valuable items, such as money, a necklace, and television sets (National Police Agency, 2016). Available statistics show that property crime in Kumamoto from January to August 2016, dropped by 18.1% on a year-to-year comparison (Kumamoto Prefectural Police, 2016).

Here, I bring one last case from Japanese disasters, which is way older than the above-mentioned disasters. After the Great Kanto Earthquake occurred on September 1, 1923, property crime in Tokyo, the most disaster-affected area in this disaster, decreased by 36.5% in September when compared to the same month in the previous year (Adachi, 2012). At the time of the disaster, which occurred approximately five years after World War I, Japan was not as economically stable and socially mature as it had been for the past 50 years: unemployment, poverty, socioeconomic disparities, and anxiety about public safety among the general public should have been much more visible in society than they are now. In reality, some criminal acts, such as murder and arson, significantly increased after this disaster, which will be discussed in the next section. However, at least the available crime statistics do not support the claim that property crime was widespread in the disaster-affected Tokyo (Adachi, 2012).

To sum up the present discussion on looting in postdisaster situations, it is obvious that looting-like behavior did occur after the disasters, and it even seemed to have occurred during some of the past natural disasters in Japan, where preexisting social issues have not been so striking. That being said, the majority of such behavior might also differ from the typical mass looting that many of us imagine. A small portion of such behavior might have met our stereotypical images of looting, which were most likely to have committed by individuals with criminal lifestyles (e.g. property crime in New Orleans, burglaries in the no-go areas in Fukushima). However, the majority seems to be "justifiable" looting by disaster-affected locals (e.g. the taking of necessities for emergency survival after the Hurricane Katrina, motorcycle theft in the Great Hanshin-Awaji Earthquake). On the one hand, the illegal taking of someone's property can never be justifiable in any situation. On the other hand, at least I myself cannot promise from my heart that I would never do the same misdeed as those seen in New Orleans, if I were in the same difficult condition with no immediate rescue activities expected from government institutions.

Of course, I am not saying that "ordinary" people never loot items unnecessary for emergency survival in postdisaster situations. As Rodríguez et al. (2006) claimed, people who do not engage in everyday criminal behavior could steal more than necessities in the confusion of a disaster. While we need to consider this possibility likely, at the same time we should neither unnecessarily overestimate nor exaggerate it. Up to the present, available evidence does not seem to support the notion that looting is pervasive in postdisaster situations, as disaster researchers have been claiming over the past decades (Quarantelli, 2007; Quarantelli and Dynes, 1970, 1972; Rodríguez et al., 2006; Tierney et al., 2006). Without preexisting "push" factors that favor such antisocial behavior, disasters themselves are unlikely trigger to increase looting or other similar offenses.

20.2.2.2 Other Criminal Acts in Postdisaster Situations

As with looting, many people also anticipate other types of crime to increase in disaster-affected areas when a disaster strikes (Nogami and Yoshida, 2014; Wenger et al., 1975). However, as described earlier, the crime rate generally drops after a disaster (Jacob et al., 2008; Quarantelli, 2007; Quarantelli and Dynes, 1972), or stays at the same level as a nondisaster time period (Varano et al., 2010). Again, preexisting social issues, such as poverty, unemployment, and/or socioeconomic disparities in a disaster-affected community, play their roles in the rate of postdisaster crime (Barsky et al., 2006; Genevie et al., 1987; Varano et al., 2010; Weems et al., 2007). In other words, when these issues exist before a disaster strikes, some types of criminal acts may increase.

First, I will look into how Hurricane Katrina affected the crime rate in New Orleans, which had had several preexisting push factors influencing the crime rate. As noted in the looting section, New Orleans was facing poverty, socioeconomic disparities (Simo, 2008), and a high property crime rate (City of New Orleans, 2016; Department of Justice, 2006a) even before the hurricane hit the city. Violent crime, including murder, rape, and robbery, was also widespread in the city. In the pre-Katrina year (2004), the average of violent crime in the United States was 465.5 offenses per 100,000 inhabitants

(Department of Justice, 2006b), whereas New Orleans' rate of the same crime was 948 per 100,000 residents (City of New Orleans, 2016). New Orleans' violent crime rate was more than twice as high as the national average.

Did violent crime in New Orleans increase in the post-Katrina year? Contrary to expectations, New Orleans actually saw decreases in the crime rate: The city's overall crime rate was 3348 cases per 100,000 residents in 2006, which was 6110 per 100,000 in 2004 (City of New Orleans, 2016). The violent crime rate decreased to 523 per 100,000 residents in 2006, and the murder rate also dropped to 37.6 per 100,000 residents in 2006, which was 56 per 100,000 residents in 2004. However, things are not as straight as these figures indicate. VanLandingham (2007) argued that according to his analysis, the murder rate in New Orleans during the post-Katrina year (2006) showed a 49% increase over 2004 and a 30% increase over 2005, even when the most optimistic figures were used. Furthermore, based on the city's official statistics (City of New Orleans, 2016), the total crime rate, the violent crime rate, and the murder rate all indicated significant increases in 2007 (8628 crimes per 100,000 people, 1564 violent crimes per 100,000, 94.7 murders per 100,000 respectively).

The changes in New Orleans' postdisaster crime rates were somehow complex, but the preexisting social issues were likely to have contributed to these changes (in particular, the increases seen in 2007). In Japan, the crime rate in postdisaster years generally indicates decreases. The overall crime rate in January, including motorcycle theft, decreased by 58.4% in the affected areas after the 1995 Great Hanshin-Awaji Earthquake, when compared to the same month in the previous year (Adachi, 2012; Saito, 2001). In a 6-month period, the overall rate slightly went up, but it was still well below the previous year's overall rate (a 16.2% decrease), although motorcycle theft increased by 33.3%.

Approximately three weeks after the 2011 Great East Japan Earthquake, the National Police Agency (2011b, April 1) issued an official statement that homicide, robbery, and rape had not increased in the three most disaster-affected areas. Six weeks after the statement, they reported that felonious offenses (homicide, robbery, arson, and rape) in March and April dropped by 21.7% in the three affected areas on a year-to-year comparison (National Police Agency, 2011a). This figure showed a 29.8% decrease in a 10-month period from March to December 2011 (National Police Agency, 2012). The overall crime rate also decreased in Iwate (−15.4%), Miyagi (−17.7%), and Fukushima (−20.0%), although Fukushima saw a 35.0% increase in burglary.

In the 1923 Great Kanto Earthquake, property crime (−36.5%), sexual assault (−57.1%), bodily injury (−36.6%), fraud (−74.3%), and embezzlement (−53.6%) all decreased in Tokyo in a month-period (September) when compared to the previous year (Adachi, 2012). The overall crime rate in the same period also showed a 39.9% decrease. However, the number of homicide in Tokyo rose from 5 in September 1922, to 110 in September 1923 (+2100%), and arson also increased from 5 to 25 (+400%). The number of the homicide victims in all the affected areas after the disaster ranged from 578 (Suzuki, 2008a) to more than 6000 (Suzuki, 2008b), depending on available statistical sources. Either way, homicide and arson significantly increased in the aftermath of the disaster.

So, what triggered these misdeeds there? At that time, there were vicious rumors spreading in the affected areas that Koreans who had been forced to work in Japan took advantage of the disaster: setting fire to collapsed houses and buildings and killing the Japanese (Sato, 2008; Suzuki, 2008a). These rumors were untrue, but some Japanese took them seriously, forming vigilante groups, and struck back against the innocent Koreans. Chinese and even Japanese people were mistakenly victimized by these vigilantes (Suzuki, 2008a). The so-called "Korean massacre" during this disaster is assumed to have been triggered by growing fear of and discrimination against the Koreans among the Japanese at the time of the disaster. This incident has also been thought to be the only disaster-related massacre ever recorded in Japanese disaster history.

Based on these statistics, some types of crime occasionally increased in the aftermath of the disasters. Nonetheless, taking a close look at each of them reveals that the increases in these crimes seem to have depended on specific circumstances caused by the disasters (motorcycle theft in the Great Hanshin-Awaji Earthquake, burglary in Fukushima after the Great East Japan Earthquake) or on the preexisting social issues (violent crime after the Hurricane Katrina, homicide and arson in the aftermath of the Great Kanto Earthquake). As shown above, it is important to note that the majority of criminal acts did not increase after these disasters as well as the overall crime rates also showed decreases. Having seen how the criminal rates changed after the disasters, it does not seem reasonable to assume that the crime rates of the disaster-affected areas increased in postdisaster years.

What conclusion can we draw from the present discussions on looting and postdisaster crime? While some exceptions can occasionally be found, both looting and other criminal acts do not seem to occur as frequently as many of us generally imagine. Of course, we need to keep in mind that crime record keeping is not as easy during postdisaster periods as during nondisaster periods. Public safety agencies may be busy performing rescue activities, while disaster victims may not be able to recognize whether their property has been lost due to the impact of a disaster or has actually been stolen by someone else in the confusion of a disaster. Thus, it is still sensible that when affected by a disaster we should be a little cagier about our own security against criminal acts than under normal circumstances.

At the same time, however, we should also not unnecessarily overestimate the probability of criminal acts in postdisaster situations. Police agencies generally strengthen security in disaster-affected areas in order to prevent postdisaster crime after major disasters (National Police Agency, 1996, 2012), making postdisaster situations not ideal for criminals to do their business. Moreover, contrary to the crime myth, many disaster researchers have been claiming for decades that people generally tend to engage in prosocial behavior in postdisaster situations rather than antisocial behavior (Auf der Heide, 2004; Drury et al., 2013a, 2015; Rodríguez et al., 2006; Sandin and Wester, 2009). Crime occurs anywhere at any time. Although the causal relationship between criminal acts and disasters still needs to be examined, we should not easily ascribe every single wrongdoing observed in a postdisaster situation to a disaster itself.

20.2.3 Donating Behavior in Postdisaster Situations

When we see others in trouble, we generally try to help them. This tendency seems to stay unchanged after a disaster strikes. In the immediate aftermath of a disaster, disaster victims try to help each other even when still in the face of a great danger (Auf der Heide, 2004; Drury et al., 2013a, 2015; Rodríguez et al., 2006; Sandin and Wester, 2009). Such altruistic behavior in postdisaster situations has also been frequently observed in Japan. When the 1995 Great Hanshin-Awaji Earthquake occurred in the early morning, for instance, many people were sleeping in bed but suddenly got trapped in collapsed houses. One survey revealed that 45.6% of 840 respondents saw victims trapped in collapsed houses in their neighborhood, and 76.5 % of these witnesses also saw the trapped victims being rescued. Surprisingly, 60.5% of such life-saving activities were performed by neighbors (Kuraoka, 1995), rather than by emergency response professionals.

Not only disaster victims but nonvictims are also willing to mitigate the agony of disaster victims. After the 1995 Great Hanshin-Awaji Earthquake, approximately 1.4 million people volunteered in the affected areas in a year period (Hyogo-Ken, 1996). More than a million volunteers went to Iwate, Miyagi, and Fukushima in the space of a year after the Great East Japan Earthquake and dedicated themselves to helping the affected areas and victims (Cabinet Office, 2013).

Moreover, people will donate whatever they think is helpful to others in trouble. According to previous research, both internal and external factors will influence the way we donate to those in need of help. As for internal influencing factors, past donation experience, self-efficacy, and moral obligation are thought to significantly affect our intention to donate (Cheung and Chan, 2000; Oosterhof et al., 2009). Other internal factors, such as the perception of emotional intensity caused by humanitarian crises (Huber et al., 2011), subjective well-being (Aaker and Akutsu, 2009), and unambiguous responsibility for a single recipient (Cryder and Loewenstein, 2012) are also assumed to enhance our altruistic behavior.

External factors that affect our donating behavior include the type of a disaster (Zagefka et al., 2011), the number of fatalities (Evangelidis and Van den Bergh, 2013), and identifiable information of victims (Kogut and Ritov, 2007; Small and Loewenstein, 2003). The mass media also play a role in our donating behavior. We are more likely to donate to disaster relief charities, as the media coverage of the disaster increases (Brown and Minty, 2008; Eisensee and Strömberg, 2007).

Interestingly enough, after a major disaster occurs, people tend to help disaster victims by sending supplies of food, clothing, and necessities rather than by giving money. However, such relief materials generally far exceed actual needs, creating unnecessary chaos in disaster-affected areas (Fritz and Mathewson, 1975; Fukumoto et al., 2007; Wenger et al., 1975). In many cases, monetary donations are the most effective aid for disaster victims (Wenger et al., 1975). Still, many of us persistently believe that sending relief materials will be more helpful to disaster victims than monetary donation (Nogami and Yoshida, 2014), which is called the donation myth in this chapter.

How does sending relief materials create unnecessary chaos in disaster-affected areas? Why do many of us choose to send relief materials to disaster victims rather than to donate money? In the following sections, I will pick up two issues surrounding the donation myth: chaos created by relief materials and our reluctance to donate money. Again, I will occasionally refer to how the Japanese reacted to the past disasters in order to help disaster victims.

20.2.3.1 Chaos Created by Relief Materials

When a massive earthquake struck in Nepal on April 25, 2015, which killed more than 8000 people, I was teaching disaster psychology at a technical college in Tokyo. One day when I went into the classroom for the first time since the earthquake, a few students came to me and said that they really wanted to help those affected by the earthquake. They suggested sending "traditional" relief materials, including emergency food, some of their clothing, and origami cranes (we, Japanese people, have the custom of sending a thousand origami cranes to those in a difficult situation for encouragement). When I replied that sending materials to disaster-affected areas can often cause trouble and donating money is in many cases the most helpful, they all put confused looks on their faces.

These of my students are not the only ones who wish to send relief materials to disaster-affected areas. Whenever and wherever a major disaster occurs, people start sending whatever they think is helpful to disaster victims to disaster-affected areas. These "helpful" materials often include emergency food, clothing, and blankets. Some people even send opened food, perishable food, used clothing that they themselves will never wear again, and encouraging letters. If such materials are sent to disaster-affected areas from all over the country or the world in a short period of time, the affected municipal organization will have to consume its limited resources looking after and distributing massive amounts of the relief materials (Fukumoto et al., 2007). As a result, what we think is helpful to disaster-affected victims often fails to reach the victims in time due to its enormous amount (e.g. some food found moldy when distributed to victims, winter clothing arriving at evacuation center in spring or summer).

Such material convergence has been a major issue for disaster-affected areas in Japan. In the 1993 Hokkaido Earthquake, Okushiri Town in Okushiri Island, the most affected municipality in this disaster, had to handle at least 300,000 parcels coming from all over the country, which consumed a great amount of manpower of the organization and volunteers (Nagasaki, 2012). If such amounts of parcels had not been sent to the island by nonvictims, manpower of the municipal organization, and volunteers would have been more effectively used to reduce the toll of the disaster. Furthermore, this small town with the population of approximately 4000 inhabitants at the time of the disaster was forced to spend JPY 120 million (approximately USD 1 million) to build a storehouse for the relief materials (Koshimori, 1995). Thus, what people thought would help the disaster-affected area and victims actually ended up bothering them.

The former mayor of Ojiya City, one of the most affected cities in the 2004 Niigata Chuetsu Earthquake, stated that huge amounts of relief materials sent from all over the country had given the city several problems (Seki, 2007). The incoming relief materials caused a traffic congestion in the city and the city office became full of the materials due to a lack of storage space. The massive amounts of the materials also exhausted the city officials, who could otherwise have performed other emergency tasks. Material convergence repeatedly made trouble in past disasters, including the 1995 Great Hanshin-Awaji Earthquake and the 2011 Great East Japan Earthquake. In the 2011 Great East Japan Earthquake, material convergence occurred not only in the disaster-affected areas, but also in nonaffected areas. Some nonaffected areas started receiving relief materials sent by their residents after the disaster, but the affected areas refused to accept unneeded relief materials from the nonaffected areas in order to avoid material convergence (Iwate Nippo, 2011, October 25; Yamagata Shimbun, 2011, April 15; Yomiuri Shimbun, 2011, April 21).

Of course, there have been actual cases in which relief materials sent by individuals were profitably used. For example, three days after the 2016 Kumamoto Earthquake occurred, Fukuoka City, located in the north part of Kyushu Island, informed its citizens what relief materials would be needed for disaster victims in Kumamoto (Takashima, 2016). The city only accepted a limited range of relief materials which were most likely to be wanted by the victims (e.g. bottles of water, rolls of toilet paper). Such a systematic approach can make individually sent relief materials useful. However, although the central and municipal governments have been taking countermeasures against material convergence occurring after disasters, piles of materials are repeatedly sent to affected areas in disorganized fashions by individuals every time a major disaster occurs.

As we have seen so far, many of us tend to send what we think is useful to disaster victims to disaster-affected areas. Unfortunately, however, in many cases such materials are not as useful to victims as we initially think they would be. On the contrary, these materials often give further physical and financial burdens to the disaster-affected area. When a major disaster occurs in Japan, the central government, other municipal organizations, and private companies, in particular, retailers, take quick action and provide disaster-affected areas with emergency essentials, which can oftentimes meet immediate demands by disaster victims. Thus, it is worth remembering that without any instructions from government institutions and/or local public authorities in a disaster-affected area, we should refrain from sending any of what we think will help disaster victims to the affected area.

20.2.3.2 Reluctance to Donate Money

If someone happens to ask me the way to the nearest station in Tokyo, I will of course tell him/her how to get there, without doubting whether the person has really got lost on the way. When one of my friends looks down, I will be more than happy to treat and encourage him/her regardless of the details and the seriousness of his/her trouble. However, if someone asks me for money in the street, I may be less willing to help that person. Even if the person asks for a tiny amount of money for a charitable purpose, I will usually be very reluctant to respond to that kind of a request (I honestly admit that). In the same vein, if one of my friends asks me for money, I will ask him/her back why in great detail.

It does not necessarily mean that I am too greedy for money, but many people have a similar tendency in terms of helping others. We are willing to give a helping hand to others. We can dedicate considerable amounts of time and effort to supporting and helping others in trouble. However, things are slightly different, when money gets involved.

Why do we become reluctant to give money for helping others? One key factor is uncertainty surrounding whether our money will actually be used for the initial purpose. A good example comes from our concerns about charities' transparency in handling collected donations. Of 1772 Japanese people who had not donated money for the victims of the Great East Japan Earthquake, 53.3% were skeptical about whether their donations would actually go to the victims, how their donations would be spent, and whether donations would really help the victims (Japan NPO Research Association, 2013). Previous research revealed that nondonors' skepticism about the effectiveness of donating money is stronger than donors' (Nogami, 2014; Sargeant and Lee, 2002). Donors are also skeptical about how donated money is used by charitable organizations, as 92% of Canadian donors think that charitable organizations should disclose how donated money has been spent (Muttart Foundation, 2013). Thus, donors and nondonors are both dubious about whether their money will actually be used for charitable purposes.

Our preference to sending relief materials over giving money may be largely due to the fact that many of us are not sure about whether our donations will be actually used to help disaster victims. Nevertheless, people do donate money for disaster victims, when a major disaster strikes. For instance, the total amount of donations made by the general public for the 2011 Great East Japan Earthquake reached more than JPY 440 billion (approximately USD 3.8 billion). In reality, three in four Japanese people have actually made some form of donations for the disaster victims through government institutions, disaster relief organizations, and nonprofit organizations (Nikkei, 2012).

There are two types of donations for disaster victims in Japan: the monetary donation ("gien-kin" in Japanese) and the charitable donation ("sien-kin"). The monetary donation goes directly to disaster victims without any commission charge taken by government institutions and/or charitable organizations from the donation, while the charitable donation is given to nonprofit and/or disaster relief organizations to support their disaster relief activities (Kifusuru.com, 2011). Interestingly enough, the monetary donation made up more than 93% of the total donated money for the Great East Japan Earthquake (JPY 411 billion out of JPY 440 billion), whereas the charitable donation only accounted for approximately 7% (JPY 29 billion) of the total amount (Nikkei, 2012). The difference in the amount of the donated money for the disaster victims between the two types of donations may imply that we tend to choose ways of donation that have more transparency in handling our money.

Now we know that we tend to choose sending relief materials over donating money, when it comes to helping disaster victims. Even when donating money, we prefer the monetary donation over the charitable donation, as the former is generally more accountable than the latter for our donated money. Having said, it does not mean that the charitable donation is less effective than the monetary donation. The charitable donation does have several advantages over the monetary donation (Kifusuru.com, 2011). For example, nonprofit and/or disaster relief organizations funded with the charitable donation can perform some disaster relief work in disaster-affected areas faster than government institutions (e.g. supplying victims with necessities, cleaning up in affected areas). Furthermore, these organizations can provide disaster victims with additional services for particular needs of children, women, the disabled, and the elderly that government institutions often fail to provide (e.g. providing house cleanup work for elderly victims, casual consultation for female victims).

To summarize our donating behavior in postdisaster situations, contrary to our stereotypes about disaster relief donations, sending disaster relief materials can often cause additional problems in disaster-affected areas. Also, our persistent belief that sending relief materials is more helpful than donating money may be ascribed to our uncertainty about charities' handling of donated money. If we know these facts, the most effective and the only way of helping disaster victims for us is to donate money to well-known charitable organizations. Otherwise, we might as well not do anything for disaster-affected areas and victims, so that at least we will not unnecessarily bother them. When we help others in trouble, we should also pay attention to the fact that our helping will not bring any more trouble to them.

20.3 IMPACTS OF THE DISASTER MYTHS ON DISASTER RESPONSE AND MANAGEMENT

So far, we have seen the panic myth, the crime myth, and the donation myth in detail. Despite the lack of empirical evidence and the long-standing denial by disaster researchers, these disaster myths still seem to be widespread and firmly rooted in modern society. At first glance, the disaster myths do not seem very harmful, provided that people just assume that these phenomena are likely in postdisaster situations. However, these three myths can affect the way we behave before, during, and after a disaster. Here, I will briefly explain how these three disaster myths can affect disaster response and management both at an organizational level and an individual level.

The panic myth held by certain types of professionals, in particular, those related to disaster response, can be problematic. For instance, it has been predicted in Japan that a massive earthquake is likely to occur around the Kanto region including Tokyo in the next three decades (Headquarters for Earthquake Research Promotion, 2014). To make Tokyo residents prepare for this potential disaster, the Tokyo Metropolitan Government published an emergency booklet for its residents (Tokyo Metropolitan Government, 2015). In this booklet, there is a paragraph entitled "Be careful about panic in crowds" (p. 48), implying that panic is likely to occur during a disaster. Previous research also found that in several UK emergency guidance documents, panic was represented as an unreasonable, dysfunctional, emotional, and behavioral reaction to which crowds of people are susceptible in an emergency (Drury et al., 2013b).

Because emergency management personnel and policy makers, as well as media personnel and anybody in positions of authority, can exercise more powerful influence on organizational action and society than the general public, their fear of public panic could result in ineffective disaster responses in emergency situations (e.g. withholding urgent disaster information and warnings to the public, spreading wrong information; Clarke and Chess, 2008; Gantt and Gantt, 2012; Quarantelli, 1960; Tierney, 2003; Tierney et al., 2006; Wester, 2011). This tendency is known as *elite panic* among disaster researchers (Clarke and Chess, 2008). If elites, those in positions of authority, panic about the possibility of panic during a disaster, their decisions and actions could expose the public to more realistic dangers than what panic would be assumed to cause (see also Scanlon, 2007).

Elite panic was actually observed during the 2011 Great East Japan Earthquake. After a massive explosion took place at the Fukushima Daiichi Nuclear Power Plant on March 11, 2011, the central government of that time decided to conceal some of the crucial information about the consequences of the explosion from the general public (e.g. forecasts of the diffusion of radiation from the plant). In May 2011, the government officially disclosed the information and apologized to the public by stating that they had refrained from disclosing the information in order to avoid causing panic in society (Shikoku News, 2011). Clearly, the government thought that panic among the general public would be more dangerous to the nation than the diffusion of radiation would be to the public (see also Tierney, 2003).

The crime myth, like the panic myth, can also have some potential negative side effects on myth believers. For example, while the fear of crime may make people more vigilant against potential crime (Jackson and Gray, 2010), it has also been found to be associated with poorer mental health and physical functioning of the person (Jackson and Gray, 2010; Jackson and Stafford, 2009; Skogan, 1986; Stafford et al., 2007), lower life satisfaction (Hanslmaier, 2013), as well as with deterioration of a community (Skogan, 1986). Assuming that a perceived high frequency of crime in a postdisaster situation accompanies the fear of crime, it could further impose extra burdens on disaster victims who have already been physically and psychologically devastated by a disaster.

Moreover, if people think that their property becomes vulnerable to looting or other similar acts in a disaster, they may become reluctant to evacuate their homes and/or stores even in the face of a great danger. In the same vein as elite panic, the crime myth could also make emergency response professionals take inappropriate responses to a disaster, if they firmly believe that looting and other kinds of crime increase in postdisaster situations. After the 2005 Hurricane Katrina hit New Orleans, numerous media reports on looting spread widely throughout the nation. As a result, Louisiana Governor and the mayor of New Orleans prioritized public safety activities over life-saving activities (Coates and Eggen, 2005; Tierney et al., 2006). Needless to mention, the latter activities should generally be taken on top priority in government disaster response in the aftermath of a disaster.

I have already touched on how the donation myth can affect governmental responses as well as on our individual responses to a disaster in the donating behavior section. On the one hand, the disaster-affected municipal organization will face mountains of emergency tasks and services for victims in a short period of time as soon as a disaster occurs. On the other hand, we tend to think that sending relief materials to affected areas will help disaster victims. Contrary to our good intentions, if tens of thousands of us follow that belief in a disorganized way, we will end up imposing additional troublesome work on the affected area, which can negatively affect the municipality's tasks and services for disaster victims. In Japan, material convergence in postdisaster situations has been thought to be a "secondary disaster" to the disaster-affected area (Hino, 2010).

To wrap up the present discussion on how the disaster myths affect disaster response and management, it is obvious that these myths can have, more or less, negative impacts on official disaster responses and management, as well as on our individual responses to a disaster. The myths are not true and they have been negatively affecting the way we respond to a disaster both at an organizational and an individual level. If these misconceptions only do harm to us, why do many of us still stick to them? In order to prevent the potential negative consequences to disaster responses caused by the myths, we have to look at the root of the misconceptions and why we persistently believe them.

20.4 WHAT GIVES RISE TO DISASTER MYTHS?

Now, we need to find out where these disaster myths come from and why they have been commonly held by many of us for more than half a century. Numerous disaster researchers have been arguing that the mass media is a major source of disaster myths. Thus, first I will briefly look at the effects of the mass media and popular culture, such as movies and fiction, on the creation of disaster myths based on findings of previous research. Secondly, as a disaster researcher with a sociopsychological background, I will also try to analyze the sources of the disaster myths in reference to some of the well-established concepts from the field of psychology.

20.4.1 Effects of Mass Media and Popular Culture on Disaster Myths

One survey revealed that after the 2011 Great East Japan Earthquake occurred, 80% of Japanese people sought disaster-related information through the national TV channel, 56.9% through commercial TV channels (Nomura Research Institute, 2011). Many people also looked at internet portal sites (43.2%) and newspaper accounts (36.3%) for following what was going on after the disaster. On the one hand, the mass media are no doubt principal sources of information in postdisaster situations. On the other hand, sensational news reports and exaggerated images of disaster behavior in movies and fiction have been thought to contribute to creating and spreading disaster myths (Ali, 2013; Barsky et al., 2006; Koshiro, 1997; McEntire, 2006; Mitchell et al., 2000; Quarantelli, 1980; Rodríguez et al., 2006; Tierney et al., 2006; Tsuganesawa, 1999; Wenger et al., 1975).

However, the actual effects of the mass media and popular culture on the creation of disaster myths have been somehow inconclusive. While media reports on disaster behavior were found to contain mythical elements (Wenger and Friedman, 1986), only a minority of news stories were found to have dealt with such elements (Goltz, 1984). Also, one's preferences for the type of media (Nogami and Yoshida, 2014) and media coverage (Nogami, 2015), and the amount of media consumption (Nogami, 2015) were not found to have as significant effects on disaster myths as initially predicted. Another possible factor that can contribute to creating disaster myths is the degree to which the media exaggerate and sensationalize disaster behavior (see also Wenger and Friedman, 1986). This possibility is yet to be examined, but even a brief glance at exaggerated, sensationalized disaster behavior in a news program and/or a movie could be sufficient to make people accept unreal, rare aspects of human behavior as real and widespread in postdisaster situations. Although it is still not clear that how much contribution the mass media actually make to creating disaster myths, they surely enhance, more or less, the degree to which we believe disaster myths (Nogami, 2015; Nogami and Yoshida, 2014).

Considering their negative influence on our images of disaster behavior, do the mass media do no good to us? This is not correct. There is no doubt that the media can play an important role before, during, and after disasters (Ali, 2013; Scanlon, 2007). Media coverage can enhance public awareness in local communities before a disaster occurs, as well as yield quick relief activities from government institutions and the general public during a disaster (Ali, 2013). The media can also increase the amount of donation we do for those in need of help after a disaster (Brown and Minty, 2008; Eisensee and Strömberg, 2007).

It is true that due to their economic priorities the media occasionally spread half-baked information about disasters that can propagate erroneous images among the public (Ali, 2013). However, exaggeration and sensationalism in the media are not limited to disasters. These can be found in media reports on politics, economics, daily crime, and celebrity gossips. Therefore, the important thing is that every one of us needs to deepen our understanding of disaster behavior a little bit more and take what the media say about how people respond to a disaster with a grain of salt.

20.4.2 Psychological Mechanisms Behind Disaster Myths

Other than the mass media and popular culture, what can possibly contribute to the creation of disaster myths? So far, no specific provider of disaster myths has been suggested in the disaster literature. Nonetheless, here I propose some human innate tendencies found in the field of psychology as potential candidates for the creator of disaster myths.

First I pick up the negativity bias. Humans tend to pay more attention to negative information than to positive one (Baumeister et al., 2001). This tendency has also been found in our preference of news content, as we react more strongly to negative news content than to positive one (Soroka and McAdams, 2015). With this tendency, we might overreact to media accounts of disaster victims getting confused, looting-like behavior occurring, and/or victims suffering a shortage of food and necessities in a disaster-affected area, even when such negative accounts are only briefly broadcast and/or we just have a glance at them.

We also tend to estimate the frequency of an event based on how easily similar events come to our minds. This tendency, known as the availability heuristic (Tversky and Kahneman, 1973), may help us overestimate the actual frequency and the probability of panic and postdisaster crime due to the striking impact of a one-shot rare event and/or media sensationalism. Along with the negativity bias, it may be our innate tendency to assume that negative postdisaster events, such as panic and looting, are typical human reactions to disasters, although it could be media exaggeration and sensationalism that trigger such assumptions of ours.

When we see others' behavior, we often draw inferences about others' dispositions from their observable behavior without considering situational factors. This tendency is known as the correspondence bias (Gilbert and Malone, 1995). For example, when the 2015 Nepal earthquake occurred, approximately three-quarters of casualties were due to building collapse (Cross, 2015, April 30). In this particular disaster, running away from homes and buildings was actually logical behavior for the disaster victims. Nevertheless, such flight behavior was reported as panic flight by the mass media (BBC News, 2015, April 26; Cullinane and Park, 2015, April 27; Mainichi Newspapers, 2015, April 26).

In many cases it is difficult to judge what behavior is appropriate in an emergency situation through objective observation (Quarantelli, 1954). As a result, people may possibly end up making incorrect assumptions from others' behavior during a disaster (e.g. victims merely running away from collapsing buildings as panic flight, taking their own belongings from their collapsed home as looting). The correspondence bias seems common among disaster survivors, witnesses, disaster-response professionals, and media personnel, as they often refer to the term panic in their postdisaster accounts despite no clear evidence for such behavior at a disaster site (BBC News, 2015, April 26; Cocking and Drury, 2014; Cullinane and Park, 2015, April 27; Drury et al., 2009b, 2015; Fahy et al., 2012; Mainichi Newspapers, 2015, April 26). This tendency may also contribute to why we persistently believe that panic and looting frequently occur in postdisaster situations.

Finally, in relation to the crime myth and the donation myth, I would like to note our uncertainty about how we behave when we become unaccountable for our behavior. We tend to think that looting and other criminal acts increase in postdisaster situations, probably because looters and criminals could go unpunished for such acts in the confusion of a disaster. We are a little reluctant to donate money, as we are suspicious about whether charitable organizations properly spend our money. These concerns are all related to how we behave in an anonymous situation in which we are held unaccountable for whatever we do. Previous research found that even ordinary people may occasionally engage in rule-breaking behavior when they are held unaccountable for their misdeed in an anonymous situation, particularly in order to pursue material self-interest (e.g. money; Nogami, 2009; Nogami and Takai, 2008; Nogami and Yoshida, 2013a, b). That is, our persistent beliefs in the crime myth and the donation myth may be partly due to our concerns over the possibility that some could do bad things in an unmonitored situation.

20.5 CONCLUSION

In this chapter, three disaster myths and their impacts on disaster response and management were explained. There are more disaster myths than the three described here (Alexander, 2007), but I believe that these three seem to be more widespread and influential on our society before, during, and after a disaster than other myths. In the meantime, I need to remind every reader of this chapter that disaster researchers, including me, are not saying that panic and looting never occur in a disaster or that sending relief materials never does good to disaster victims. We have been arguing that many people tend to overestimate the frequencies and/or the impacts of these behaviors. I also briefly discussed what gives rise to disaster myths. Although nothing conclusive has been suggested in this chapter regarding the root of the myths, our innate tendencies, along with media exaggeration and sensationalism, may also contribute to creating disaster myths and letting them stay alive in modern society.

More than 60 years ago, Quarantelli (1960) explicitly denied the popular image of panic in a disaster. More than 60 years later from that time, I still found that many people believe that panic is likely in a disaster (Nogami, 2016). Misconceptions relating to disaster behavior are that persistent. While more work clearly needs to be done to demolish the long-standing disaster myths by disaster researchers, every one of us also needs to deepen understanding of our own behavior in an emergency situation.

REFERENCES

Aaker, J.L., Akutsu, S., 2009. Why do people give? The role of identity in giving. J. Consum. Psychol. 19 (3), 267–270. doi: 10.1016/j.jcps.2009.05.010.

Adachi, M., 2012. Daishinsai chokugo no hanzai nit suite [On the crime just after the great earthquake disaster outbreak]. Kantogakuin Hougaku 23 (4), 49–70.

Alexander, D.E., 2007. Misconception as a barrier to teaching about disasters. Prehosp. Disaster. Med. 22 (2), 95–103. doi: 10.1017/S1049023X00004441.

Ali, Z.S., 2013. Media myths and realities in natural disasters. Eur. J. Bus. Soc. Sci. 2 (1), 125–133, Retrieved from http://www.ejbss.com/Data/Sites/1/vol2no1april2013/ejbss-1238-13-mediamythsandrealitiesinnaturaldisasters.pdf.

Auf der Heide, E., 2004. Common misconceptions about disasters: panic, the "disaster syndrome," and looting. In: O'Leary, M. (Ed.), The First 72 Hours: A Community Approach to Disaster Preparedness. iUniverse Publishing, Lincoln, NB, pp. 340–380, Retrieved from https://www.atsdr.cdc.gov/emergency_response/common_misconceptions.pdf.

Barsky, L., Trainor, J., Torres, M., 2006. Disasters realities in the aftermath of Hurricane Katrina: revisiting the looting myth. Quick Response Rep. 184, 1–6, Retrieved from http://udspace.udel.edu/handle/19716/2367.

Barton, A.H., 1969. Communities in Disaster: A Sociological Analysis of Collective Stress Situations. Doubleday, Garden City, NY.

Baumeister, R.F., Bratslavsky, E., Finkenauer, C., Vohs, K.D., 2001. Bad is stronger than good. Rev. Gen. Psychol. 5 (4), 323–370. doi: 10.1037/1089-2680.5.4.323.

BBC News, 2015, April 26. Nepal earthquake: panic as strong aftershock hits Kathmandu. Retrieved from http://www.bbc.com/news/world-asia-32472157.

BBC News Magazine, 2011, March 18. Why is there no looting in Japan after the earthquake? Retrieved from http://www.bbc.co.uk/news/magazine-12785802.

Brown, P.H., Minty, J.H., 2008. Media coverage & charitable giving after the 2004 tsunami. South. Econ. J. 75 (1), 9–25, Retrieved from http://www.jstor.org/stable/20112025.

Cabinet Office, 2013. Borantelia katsudoushasuu no suii [The number of volunteers registered at disaster volunteer centers]. Retrieved from http://www.bousai.go.jp/kaigirep/kentokai/hisaishashien2/pdf/dai2kai/siryo4_7.pdf.

Cabinet Office, 2015. White paper: disaster management in Japan 2015. Retrieved from http://www.bousai.go.jp/kaigirep/hakusho/pdf/WP2015_DM_Full_Version.pdf.

Cabinet Office, 2016. Kumamoto ken kumamoto chihou wo shingen tosuru jishin ni kakawaru higai joukoyu tou nitsuite (18:00 on 14th November) [Disaster damage relating to the earthquake in the region of Kumamoto in Kumamoto Prefecture (18:00 on 14th November)]. Retrieved from http://www.bousai.go.jp/updates/h280414jishin/pdf/h280414jishin_36.pdf.

Cafferty, J., 2011, March 14. Why is there no looting in Japan? CNN Cafferty File. Retrieved from http://caffertyfile.blogs.cnn.com/2011/03/15/why-is-there-no-looting-in-japan/.

Central Disaster Prevention Council, 2011. Tohoku chihou taiheiyouoki jishin wo kyoukun toshita jishin tsunami taisaku ni kansuru senmon chousa iinkai houkoku [Report: expert panel on measures against earthquakes and tsunamis based on the Great East Japan Earthquake experience]. Retrived from http://www.bousai.go.jp/kaigirep/chousakai/tohokukyokun/pdf/sankou.pdf.

Cheung, C.K., Chan, C.M., 2000. Social-cognitive factors of donating money to charity, with special attention to an international relief organization. Eval. Program Plann. 23 (2), 241–253. doi: 10.1016/S0149-7189(00)00003-3.

City of New Orleans, 2016. Historical crime data 1990–2004. Retrieved from http://www.nola.gov/getattachment/NOPD/Crime-Data/Crime-Stats/Historic-crime-data-1990-2014.pdf/.

Clarke, L., 2002. Panic: myth or reality? Contexts 1 (3), 21–26. doi: 10.1525/ctx.2002.1.3.21.

Clarke, L., Chess, C., 2008. Elites and panic: more to fear than fear itself. Soc. Forces 87 (2), 993–1014. doi: 10.1353/sof.0.0155.

Coates, S., Eggen, D., 2005, September 2. A city of despair and lawlessness. The Washington Post. Retrieved from http://www.washingtonpost.com/wp-dyn/content/article/2005/09/01/AR2005090100533.html.

Cocking, C., Drury, J., 2014. Talking about Hillsborough: "Panic" as discourse in survivors' accounts of the 1989 football stadium disaster. J. Commun. Appl. Soc. Psychol. 24 (2), 86–99. doi: 10.1002/casp.2153.

Cornwell, B., Harmon, W., Mason, M., Merz, B., Lampe, M., 2001. Panic or situational constraints? The case of the *M/V Estonia*. Int. J. Mass Emerg. Disasters 19 (1), 5–25.

Cross, R., 2015, April 30. Nepal earthquake: a disaster that shows quakes don't kill people, buildings do. The Guardian. Retrieved from http://www.theguardian.com/cities/2015/apr/30/nepal-earthquake-disaster-building-collapse-resilience-kathmandu.

Cryder, C.E., Loewenstein, G., 2012. Responsibility: the tie that binds. J. Exp. Soc. Psychol. 48 (1), 441–445. doi: 10.1016/j.jesp.2011.09.009.

Cullinane, S., Park, M., 2015, April 27. Nepal earthquake: "People are panicked, running down to streets". Cable News Network. Retrieved from http://edition.cnn.com/2015/04/25/asia/nepal-earthquake-witnesses/.

Department of Justice, 2006a. Property crime. Retrieved from https://www2.fbi.gov/ucr/cius_04/offenses_reported/property_crime/index.html.

Department of Justice, 2006b. Violent crime. Retrieved from https://www2.fbi.gov/ucr/cius_04/offenses_reported/violent_crime/index.html.

Drury, J., Cocking, C., Reicher, S., 2009a. Everyone for themselves? A comparative study of crowd solidarity among emergency survivors. Br. J. Soc. Psychol. 48 (3), 487–506. doi: 10.1348/014466608X357893.

Drury, J., Cocking, C., Reicher, S., 2009b. The nature of collective resilience: survivor reactions to the 2005 London bombings. Int. J. Mass Emerg. Disasters 27 (1), 66–95.

Drury, J., Cocking, C., Reicher, S., Burton, A., Schofield, D., Hardwick, A., Graham, D., Langston, P., 2009c. Cooperation versus competition in a mass emergency evacuation: a new laboratory simulation and a new theoretical model. Behav. Res. Methods 41 (3), 957–970. doi: 10.3758/BRM.41.3.957.

Drury, J., Novelli, D., Stott, C., 2013a. Psychological disaster myths in the perception and management of mass emergencies. J. Appl. Soc. Psychol. 43 (11), 2259–2270. doi: 10.1111/jasp.12176.

Drury, J., Novelli, D., Stott, C., 2013b. Representing crowd behaviour in emergency planning guidance: "Mass panic" or collective resilience? Resilience 1 (1), 18–37. doi: 10.1080/21693293.2013.765740.

Drury, J., Novelli, D., Stott, C., 2015. Managing to avert disaster: explaining collective resilience at an outdoor music event. Eur. J. Soc. Psychol. 45 (4), 533–547. doi: 10.1002/ejsp.2108.

Eisensee, T., Strömberg, D., 2007. News droughts, news floods, and US disaster relief. Q. J. Econ. 122 (2), 693–728. doi: 10.1162/qjec.122.2.693.

Evangelidis, I., Van den Bergh, B., 2013. The number of fatalities drives disaster aid: increasing sensitivity to people in need. Psychol. Sci. 24 (11), 2226–2234. doi: 10.1177/0956797613490748.

Fahy, R.F., Proulx, G., Aiman, L., 2012. Panic or not in fire: clarifying the misconception. Fire Mater. 36 (5–6), 328–338. doi: 10.1002/fam.1083.

Frailing, K., 2007. The myth of a disaster myth: potential looting should be part of disaster plans. Nat. Hazard Observation 31 (4), 3–4.

Frey, B.S., Savage, D.A., Torgler, B., 2011. Behavior under extreme conditions: the Titanic disaster. J. Econ. Perspect. 25 (1), 209–222. doi: 10.1257/jep.25.1.209.

Fritz, C., Mathewson, J.H., 1957. Convergence behavior in disasters: a problem in social control. National Research Council Disaster Study Number 9. National Academy of Sciences, Washington, DC.

Fukumoto, J., Inoue, R., Okubo, K., 2007. Higashi nihon daishinsai ni okeru kinkyuu shien bultushi no ryuudou ziltutai no teiryou teki haaku [Quatitative grasp of distribution of relief materials in the Great East Japan Earthquake]. *Heisei 23 nendo kokudo seisaku kankei kenkyuu shien zigyou kenkyuu seika houkokusho* [research report for year 2011 research support program relating to grant national land policy]. Ministry of land, infrastructure, transport and tourism. Retrieved from http://www.mlit.go.jp/common/000999574.pdf.

Gantt, P., Gantt, R., 2012. Disaster psychology: dispelling the myths of panic. Prof. Saf. 57 (8), 42–49.

Genevie, L., Kaplan, S.R., Peck, H., Struening, E.L., Kallos, J.E., Muhlin, G.L., Richardson, A., 1987. Predictors of looting in selected neighborhoods of New York city during the blackout of 1977. Sociol. Soc. Res. 71 (3), 228–231.

Gilbert, D.T., Malone, P.S., 1995. The correspondence bias. Psychol. Bull. 117 (1), 21–38. doi: 10.1037/0033-2909.117.1.21.

Goltz, J.D., 1984. Are the news media responsible for the disaster myths? A content analysis of emergency response imagery. Int. J. Mass Emerg. Disasters 2 (3), 343–366.

Gray, J., Wilson, E.A., 1984. Looting in disaster: a general profile of victimization. DRC Working Paper #7. Disaster Research Center, The Ohio State University, Columbus, OH.

Hanslmaier, M., 2013. Crime, fear and subjective well-being: how victimization and street crime affect fear and life satisfaction. Eur. J. Criminol. 10 (5), 515–533. doi: 10.1177/1477370812474545.

Headquarters for Earthquake Research Promotion, 2014. Sagami torafu zoi no zishin katsudo no chouki hyouka (dai 2 ban) nit suite [Long-Term Estimates of Earthquake Activities Along Sagami Trough, second ed.]. Headquarters for Earthquake Research Promotion. Retrieved from http://www.jishin.go.jp/main/chousa/kaikou_pdf/sagami_2.pdf.

Hino, M., 2010. Chiiki bousai ziltusen nouhau: Kyuuen bulltushi ha hisaichi wo osou daini no saigai dearu [Know-how of local disaster risk reduction practice No. 63: Sending relief materials could strike disaster-affected areas as a secondary disaster]. Kikan shoubou kagaku to jouhou, No. 100 (Haru gou) [Quarterly Issue Fire-Fighting Science and Information, No. 100 (Spring)]. Institute of Approaches for Fire & Disaster. Retrieved from http://www.isad.or.jp/cgi-bin/hp/index.cgi?ac1=IB17&ac2=knowhow63&ac3=5902&Page=hpd_view.

Hiroi, O., 2003. Saigai zi no ningen koudou [Human behavior in a disaster]. Jpn. Sci. Mon. 56 (7), 716–720.

Huber, M., Van Boven, L., McGraw, A.P., Johnson-Graham, L., 2011. Whom to help? Immediacy bias in judgments and decisions about humanitarian aid. Organ. Behav. Human Decis. Processes 115 (2), 283–293. doi: 10.1016/j.obhdp.2011.03.003.

Hyogo-Ken, 1996. Hanshin awaji daishinsai: Hyogo ken no 1 nenkan no kiroku [The great Hanshin-Awaji Earthquake: a year-long record of Hyogo ken]. Hyogo-Ken. Retrived from http://www.lib.kobe-u.ac.jp/directory/eqb/book/4-367/.

Iwate Nippo, 2011, October 25. *Kyuuen bultushi wo shoukyaku shobun he Ken, kabi ya kigengire de* [Iwate has decided to dispose of moldy and out-of-date relief materials]. Retrieved from http://www.iwate-np.co.jp/311shinsai/sh201110/sh1110252.html.

Jackson, J., Gray, E., 2010. Functional fear and public insecurities about crime. Br. J. Criminol. 50 (1), 1–22. doi: 10.1093/bjc/azp059.

Jackson, J., Stafford, M., 2009. Public health and fear of crime: a prospective cohort study. Br. J. Criminol. 49 (6), 832–847. doi: 10.1093/bjc/azp033.

Jacob, B., Mawson, A.R., Payton, M., Guignard, J.C., 2008. Disaster mythology and fact: hurricane Katrina and social attachment. Public Health Rep. 123 (5), 555–566.

Japan Institute of Country-ology and Engineering, 2015. Kokudo wo shiru/Igaito shiranai nihon no kokudo [Information on Japan's national land/Not many people know about Japan's national land]. Retrieved from http://www.jice.or.jp/knowledge/japan/commentary02.

Japan NPO Research Association, 2013. Shinsai go no kifu, borantelia tou ni kansuru ishiki chousa houkokusho [A report on donations and volunteer activities after the Great East Japan Earthquake]. Retrieved from http://www.osipp.osaka-u.ac.jp/janpora/shinsaitokubetsuproject/seika/seika1208.pdf.

Johnson, N.R., 1987. Panic at "The Who concert stampede": an empirical assessment. Soc. Prob. 34 (4), 362–373. doi: 10.2307/800813.

Johnson, N.R., 1988. Fire in a crowded theater: a descriptive investigation of the emergence of panic. Int. J. Mass Emerg. Disasters 6 (1), 7–26.

Kifusuru.com, 2011. Kifukin ya gienkin ha dou tsukawarerunoka [How are your monetary and charitable donations spent?]. Retrieved from http://kifusuru.com/explanation.html.

Kogut, T., Ritov, I., 2007. "One of us": outstanding willingness to help save a single identified compatriot. Organ. Behav. Human Decis. Processes 104 (2), 150–157. doi: 10.1016/j.obhdp.2007.04.006.

Koshimori, Y., 1995. Chihou bunken ni kansuru tokubetsu iinkaigiroku dai 3gou [The 3rd Proceedings of Special committees on decentralization of authority in the Lower House]. Dai 132 kai koltukai shuugiin [The 132nd session in the Diet]. National Diet Library, Tokyo.

Koshiro, E., 1997. Hanshin daishinsai to masukomi houdou no kouzai: Kisha tachi no mita daishinsai [The Great Hanshin-Awaji Earthquake and Contributions of the Mass Media: A Correspondent's Viewpoint on the Disaster]. Akashi Publishing, Tokyo, p. 6.

Kumamoto Prefectural Police, 2016. Kumamoto kennai no hanzai jousei (heisei 28 nen 8 gatsu matsu genzai) [Crime trend in Kumamoto (as of the end of August 2016)]. Retrieved from https://www.pref.kumamoto.jp/police/page830.html.

Kuraoka, K., 1995. Zishin haltusei! Sonotoki shimin ha...: Hanshin-Awaji daishinsai ni okeru shimin koudou chousa no keltuka [The Great Hanshin-Awaji Earthquake and responses of the locals to the disaster: analyses of post-disaster behavior after the Great Hanshin-Awaji Earthquake]. Kobe City. Retrieved from http://www.city.kobe.lg.jp/safety/fire/hanshinawaji/syukihensyuubu4.html.

LeBeau, J.L., 2002. The impact of a hurricane on routine activities and on calls for police service: Charlotte, North Carolina, and Hurricane Hugo. Crime Prev. Commun. Saf. 4 (1), 53–64. doi: 10.1057/palgrave.cpcs.8140114.

Leitner, M., Helbich, M., 2010. The impact of hurricanes on crime in the city of Houston, TX: a spatio-temporal analysis. Paper Presented at a Special Joint Symposium of ISPRS Technical Commission IV & AutoCarto in Conjunction with ASPRS/CaGIS. Orlando, FL.

Looting, 2016. In Cambridge dictionaries online. Retrieved from http://dictionary.cambridge.org/dictionary/english/looting.

Mainichi Newspapers, 2015, April 26. Magnitude 6.7 aftershock hits Nepal, causes panic. Retrieved from http://mainichi.jp/english/english/newsselect/news/20150426p2g00m0in096000c.html.

Mawson, A.R., 2005. Understanding mass panic and other collective responses to threat and disaster. Psychiatry 68 (2), 95–113. doi: 10.1521/psyc.2005.68.2.95.

Mawson, A.R., 2007. Mass Panic and Social Attachment: The Dynamics of Human Behavior. Ashgate Publishing Limited, Hampshire, England.

McEntire, D.A., 2006. Disaster Response and Recovery. Wiley, New York, NY.

Mikami, S., 1988. 2 bu izu oshima funka ni okeru juumin no taiou [Chapter 2: responses of the locals in Izu Oshima]. 1986 nen izu oshima funka ni okeru saigai jouhou no dentasu to juumin no taiou [The transmission of disaster information and responses of the locals in the 1986 volcanic eruption in Izu Oshima]. Institute of Journalism and Communication Studies at the University of Tokyo, Tokyo, pp. 81–91.

Mitchell, J.T., Thomas, D.S.K., Hill, A.A., Cutter, S.L., 2000. Catastrophe in reel life versus real life: perpetuating disaster myth through Hollywood films. Int. J. Mass Emerg. Disasters 18 (3), 383–402.

Muroi, T., Takemura, M., 2006. Dai 1 shou: Higai no zentaizou [Chapter 1: the whole picture of the disaster damage]. Saigai kyoukun no keishou ni kansuru senmon chousakai houkokusho: 1923 Kanto Daishinsai [Report of the expert panel on inheritance of lessons from past disasters: 1923 Great Kanto Earthquake Issue 1 the disaster and its mechanism]. Cabinet Office, Tokyo, pp. 1–25. Retrieved from http://www.bousai.go.jp/kyoiku/kyokun/kyoukunnokeishou/rep/1923--kantoDAISHINSAI/pdf/1923--kantoDAISHINSAI-1_04_chap1.pdf.

Muttart Foundation, 2013. Talking about charities 2013: Canadians' opinions on charities and issues affecting charities. Retrieved from http://www.muttart.org/sites/default/files/survey/3.Talking%20About%20Charities%202013.pdf.

Nagasaki, T., 2012. Houkoku 3: Hokkaido nansei oki zisihin saigai to fultukou, machi zaisei heno eikyou ni tsuite [Report 3: Hokkaido Earthquake and its effects on the town's recovery and finances]. Kako no daikibo saigai to kaigai zirei kara miru higashi nihon daishinsai to toshi zaisei [The Great East Japan Earthquake and municipal finance from the perspectives of the past major disasters and foreign cases]. Japan Center for Cities, Tokyo, pp. 28–39.

National Police Agency, 1996. Dai 7 shou: Saigai, ziko to Keisatsu katsudou [Chapter 7: police activities on disasters and accidents]. Heisei 7 nen Keisatsu Hakusho [Year 1996 Police White Paper]. Retrieved from https://www.npa.go.jp/hakusyo/h08/h080700.html.

National Police Agency, 2011a. Hisaichi tou ni okeru hanzai jousei [Crime trend in the disaster-affected areas]. Retrieved from http://www.kantei.go.jp/jp/singi/hanzai/dai17/siryou1.pdf.

National Police Agency, 2011, April 1. Futashikana jouhou ni madowasarezu, ochitsuitekoudouwo! [Act calmly and do not get swayed by unreliable information!]. Retrieved from http://warp.da.ndl.go.jp/info:ndljp/pid/3487619/www.npa.go.jp/archive/keibi/biki/cyber/futashika.pdf.

National Police Agency, 2012. Heisei 23 nen no hanzai jousei [Crime trends in year 2011–2012]. Retrieved from http://www.npa.go.jp/toukei/seianki/h23hanzaizyousei.pdf.

National Police Agency, 2016. Heisei 28nen Kumamoto zishin ni tomonau higai joukyou to keisatsu shochi [Damage caused by the 2016 Kumamoto earthquake and the response of the police to the disaster]. Retrieved from https://www.npa.go.jp/kumamotoearthquake/pdf/zyoukyou.pdf.

Nikkei, 2012. Shinsai shien kokumin no 4nin ni 3nin kifu sougaku 4400 okuen [Three in four donated for the disaster areas and victims worth 440 billion yens, according to private survey]. http://www.nikkei.com/article/DGXNASDG08029_T10C12A2CR8000/.

Nogami, T., 2009. Reexamination of the association between anonymity and self-interested unethical behavior in adults. Psychol. Rec. 59 (2), 259–272.

Nogami, T., 2014. What makes disaster donors different from non-donors. Disaster Prev. Manag. 23 (4), 484–492. doi: 10.1108/DPM-04-2014-0080.

Nogami, T., 2015. The myth of increased crime in Japan: a false perception of crime frequency in post-disaster situations. Int. J. Disaster Risk Reduct. 13, 301–306. doi: 10.1016/j.ijdrr.2015.07.007.

Nogami, T., 2016. Who panics and when: a commonly accepted image of disaster panic in Japan. Int. Perspect. Psychol. Res. Pract. Consult. 5 (4), 245–255. doi: 10.1037/ipp0000050.

Nogami, T., Takai, J., 2008. Effects of anonymity on antisocial behavior committed by individuals. Psychol. Rep. 102 (1), 119–130. doi: 10.2466/pr0.102.1.119-130.

Nogami, T., Yoshida, F., 2013a. Rule-breaking in an anonymous situation: when people decide to deviate from existing rules. Int. J. Psychol. 48 (6), 1284–1290. doi: 10.1080/00207594.2012.736024.

Nogami, T., Yoshida, F., 2013b. The pursuit of self-interest and rule breaking in an anonymous situation. J. Appl. Soc. Psychol. 43 (4), 909–916. doi: 10.1111/jasp.12056.

Nogami, T., Yoshida, F., 2014. Disaster myths after the Great East Japan disaster and the effects of information sources on belief in such myths. Disasters 38 (s2), s190–s205. doi: 10.1111/disa.12073.

Nomura Research Institute, 2011. Shinsai ni tomonau medelia seltushoku doukou ni kansuru chousa wo ziltushi [A survey on information-seeking behavior after the Great East Japan Earthquake]. Retrieved from http://www.nri.co.jp/news/2011/110329.html.

OECD, 2005. OECD Factbook 2005: economic, environmental and social statistics. Retrieved from http://www.oecd-ilibrary.org/docserver/download/3005041e.pdf?expires=1482044791&id=id&accname=guest&checksum=F468C41541C83FE020711D51A14D6F04.

Oosterhof, L., Heuvelman, A., Peters, O., 2009. Donation to disaster relief campaigns: underlying social cognitive factors exposed. Eval. Program Plann. 32 (2), 148–157. doi: 10.1016/j.evalprogplan.2008.10.006.

Panic, 2015a. In Cambridge Dictionaries online. Retrieved from http://dictionary.cambridge.org/dictionary/british/panic.

Panic, 2015b. In Oxford Dictionaries. Retrieved from http://www.oxforddictionaries.com/definition/english/panic.

Picht, J., 2011, March 14. Where are the Japanese looters? The Washington Times Communities. Retrieved from http://communities.washingtontimes.com/neighborhood/stimulus/2011/mar/14/where-are-japanese-looters/.

Quarantelli, E.L., 1954. The nature and conditions of panic. Am. J. Sociol. 60 (3), 267–275.

Quarantelli, E.L., 1960. Images of withdrawal behavior in disasters: some basic misconceptions. Soc. Prob. 8 (1), 68–79. doi: 10.2307/798631.

Quarantelli, E.L., 1980. The study of disaster movies: Research problems, findings and implications. DRC Preliminary paper #64. Disaster Research Center, The University of Delaware, Newark, DE.

Quarantelli, E.L., 2001. The sociology of panic. In: Smelser, N.J., Baltes, P.B. (Eds.), International Encyclopedia of the Social and Behavioral Sciences. Elsevier Science, Oxford, UK, pp. 11020–11023.

Quarantelli, E.L., 2007. The myth and the realities: keeping the "looting" myth in perspective. Nat. Hazard Observ. 31 (4), 2–3.

Quarantelli, E.L., Dynes, R.R., 1970. Property norms and looting: their patterns in community crises. Phylon (1960-) 31 (2), 168–182. doi: 10.2307/273722.

Quarantelli, E.L., Dynes, R.R., February 1972. When disaster strikes: it isn't much like what you've heard and read about. Psychol. Today, 67–70.

Rodríguez, H., Trainor, J., Quarantelli, E.L., 2006. Rising to the challenges of a catastrophe: the emergent and prosocial behavior following Hurricane Katrina. Ann. Am. Acad. Polit. Soc. Sci. 604 (1), 82–101. doi: 10.1177/0002716205284677.

Rosengren, K.E., Arvidson, P., Sturesson, D., 1975. The Barsebäck "panic": a radio programme as a negative summary incident. Acta Sociol. 18 (4), 303–321. doi: 10.1177/000169937501800403.

Saito, T., 2001. Hanshin daishinsai go no hanzai mondai [The crime issue in the aftermath of the Great Hanshin-Awaji earthquake]. Library No. 63: Research Institute at Konan University. Konan University, Hyogo, Japan.

Sandin, P., Wester, M., 2009. The moral black hole. Ethical Theor. Moral Pract. 12 (3), 291–301. doi: 10.1007/s10677-009-9152-z.

Sargeant, A., Lee, S., 2002. Improving public trust in the voluntary sector: an empirical analysis. Int. J. Nonprofit Voluntary Sect. Mark. 7 (1), 68–83. doi: 10.1002/nvsm.168.

Sato, K., 2008. Dai 1 setsu: Ryuugen higo to toshi [Rumors and the city]. Saigai kyoukun no keishou ni kansuru senmon chousakai houkokusho: 1923 Kanto Daishinsai dai 2 hen [Report of the expert panel on inheritance of lessons from past disasters: 1923 Great Kanto Earthquake Issue 2 the disaster and its mechanism]. Cabinet Office, Tokyo, pp. 179–205. Retrieved from http://www.bousai.go.jp/kyoiku/kyokun/kyoukunnokeishou/rep/1923-kantoDAISHINSAI_2/pdf/18_chap4-1.pdf.

Scanlon, J., 2007. Research about the mass media and disaster: never (well hardly ever) the twain shall meet. In: McEntire, D.A. (Ed.), Disciplines, Disasters and Emergency Management: the Convergence and Divergence of Concepts, Issues and Trends from the Research Literature. Charles C. Thomas Publisher, Springfield, IL, pp. 75–95.

Seki, H., 2007. Chuuetsu daishinsai jichitai no sakebi [The Great Chuetsu Earthquake: The Voice of the Local Government]. Gyosei, Tokyo.

Shikoku News, 2011, May 2. *Houshanou kakusan sisan zu 5 sen mai wo koukai he/Hosono shushou hosakan ga chinsha* [Five thousand pieces of image information on diffusion of radiation are to be disclosed/Assistant to the prime minister Mr. Hosono apologizes to the public]. Retrieved from http://www.shikoku-np.co.jp/national/science_environmental/20110502000634.

Simo, G., 2008. Poverty in New Orleans: before and after Katrina. Vincentian Heritage J. 28 (2), 309–320.

Skogan, W., 1986. Fear of crime and neighborhood change. Crime Justice 8, 203–229. doi: 10.1086/449123.

Small, D.A., Loewenstein, G., 2003. Helping a victim or helping the victim: altruism and identifiability. J. Risk Uncertainty 26 (1), 5–16. doi: 10.1023/A:1022299422219.

Soroka, S., McAdams, S., 2015. News, politics, and negativity. Polit. Commun. 32 (1), 1–22. doi: 10.1080/10584609.2014.881942.

Stafford, M., Chandola, T., Marmot, M., 2007. Association between fear of crime and mental health and physical functioning. Am. J. Public Health 97 (11), 2076–2081. doi: 10.2105/AJPH.2006.097154.

Suzuki, J., 2008a. Dai 2 setsu: Saltushou ziken no haltusei [Section 2: The occurrence of the massacre]. Saigai kyoukun no keishou ni kansuru senmon chousakai houkokusho: 1923 Kanto Daishinsai dai 2 hen [Report of the expert panel on inheritance of lessons from past disasters: 1923 Great Kanto Earthquake Issue 2 the disaster and its mechanism]. Cabinet Office, Tokyo, pp. 206–213. Retrieved from http://www.bousai.go.jp/kyoiku/kyokun/kyoukunnokeishou/rep/1923-kantoDAISHINSAI_2/pdf/19_chap4-2.pdf.

Suzuki, J., 2008b. Column 8: Saltushou ziken no kenshou [Column 8: Examination of the massacre]. Saigai kyoukun no keishou ni kansuru senmon chousakai houkokusho: 1923 Kanto Daishinsai dai 2 hen [Report of the expert panel on inheritance of lessons from past disasters: 1923 Great Kanto Earthquake Issue 2 the disaster and its mechanism]. Cabinet Office, Tokyo, pp. 218–219. Retrieved from http://www.bousai.go.jp/kyoiku/kyokun/kyoukunnokeishou/rep/1923-kantoDAISHINSAI_2/pdf/22_column8.pdf.

Takashima, S., 2016. Heisei 28 nen Kumamoto zishin: Fukuoka shi hisaichi sien katsudou repouto [2016 Kumamoto Earthquake: a report on support activities of Fukuoka City]. Fukuoka City. Retrieved from http://www.city.fukuoka.lg.jp/data/open/cnt/3/53301/1/report_280512.pdf.

Tierney, K., 2003. Disaster beliefs and institutional interests: recycling disaster myths in the aftermath of 9–11. Res. Soc. Prob. Public Policy 11, 33–51. doi: 10.1016/S0196-1152(03)11004-6.

Tierney, K., Bevc, C., Kuligowski, E., 2006. Metaphors matter: disaster myths, media frames, and their consequences in Hurricane Katrina. Ann. Am. Acad. Polit. Soc. Sci. 604 (1), 57–81. doi: 10.1177/0002716205285589.

Tokyo Metropolitan Government, 2015. Disaster Preparedness Tokyo. Tokyo Metropolitan Government, Tokyo.

Tsuganesawa, T., 1999. Ryuugen higo to media [rumors and media]. In: Kuroda, N., Tsuganesawa, T. (Eds.), Shinsai no shakaigaku: Hanshin Awaji daishinsai to minshuu ishiki [Sociology of Disaster: The Great Hanshin-Awaji Earthquake and Public Awareness]. Sekaishisosha, Kyoto, pp. 159–191.

Tversky, A., Kahneman, D., 1973. Availability: a heuristic for judging frequency and probability. Cogn. Psychol. 5 (2), 207–232. doi: 10.1016/0010-0285(73)90033-9.

VanLandingham, M.J., 2007. Murder rates in New Orleans, 2004–2006. Am. J. Public Health 97 (9), 1614–1616. doi: 10.2105/AJPH.2007.110445.

Varano, S.P., Schafer, J.A., Cancino, J.M., Decker, S.H., Greene, J.R., 2010. A tale of three cities: crime and displacement after Hurricane Katrina. J. Crim. Justice 38 (1), 42–50. doi: 10.1016/j.jcrimjus.2009.11.006.

Weems, C.F., Watts, S.E., Marsee, M.A., Taylor, L.K., Costa, N.M., Cannon, M.F., Carrion, V.G., Pina, A.A., 2007. The psychosocial impact of Hurricane Katrina: contextual differences in psychological symptoms, social support, and discrimination. Behav. Res. Ther. 45 (10), 2295–2306. doi: 10.1016/j.brat.2007.04.013.

Wenger, D., 1985. Mass media and disasters. DRC Preliminary paper #98. Disaster Research Center, The University of Delaware, Newark, DE.

Wenger, D., Friedman, B., 1986. Local and national media coverage of disaster: a content analysis of the print media's treatment of disaster myths. Int. J. Mass. Emerg. Disasters 4 (3), 27–50.

Wenger, D.E., Dykes, J.D., Sebok, T.D., Neff, J.L., 1975. It's a matter of myths: an empirical examination of individual insight into disaster response. Mass Emerg. 1 (1), 33–46.

West, E., 2011, March 14. Why is there no looting in Japan? The Telegraph. Retrieved from http://blogs.telegraph.co.uk/news/edwest/100079703/why-is-there-no-looting-in-japan.

Wester, M., 2011. Fight, flight or freeze: assumed reactions of the public during a crisis. J. Contingencies Crisis Manag. 19 (4), 207–214. doi: 10.1111/j.1468-5973.2011.00646.x.

Yamagata Shimbun, 2011, April 15. Kyuuen bultushi no "fuyumono iryou" atsukai ni konwaku Yonezawa no hokanjo ni yamazumi [Winter clothing accumulated in the warehouse in Yonezawa]. Retrieved from http://yamagata-np.jp/feature/shinsai/kj_2011041500618.php.

Yomiuri Shimbun, 2011, April 21. Ikiba nai kyuuen bultushi hinanchi ni todokerarezu yamazumi [Large amounts of relief materials have nowhere to go]. Retrieved from http://www.yomiuri.co.jp/national/news/20110421-OYT1T00630.htm.

Yoshioka, S., 1986. Tsuiraku no natsu [The Summer of the Crash]. Shinchosha, Tokyo.

Zagefka, H., Noor, M., Brown, R., de Moura, G.R., Hopthrow, T., 2011. Donating to disaster victims: responses to natural and humanly caused events. Eur. J. Soc. Psychol. 41 (3), 353–363. doi: 10.1002/ejsp.781.

Chapter 21

A Quantitative Study of Social Capital in the Tertiary Sector of Kobe—Has Social Capital Promoted Economic Reconstruction Since the Great Hanshin Awaji Earthquake?

Go Shimada*,**,†

*Meiji University, Tokyo, Japan; **Columbia University, NY, USA; †JICA Research Institute, Tokyo, Japan

21.1 INTRODUCTION

In 1995, a powerful earthquake (magnitude 7.3) occurred in the Kobe region of Japan, killing 6434 people and destroying more than 200,000 homes.[1] It is estimated that the cost of the damage to the area's industry was around 5 trillion yen, of which direct damage to business property and equipment accounted for half, while indirect damage, such as business closures, accounted for the rest (Kuramochi, 1997).

More than 20 years have passed since the earthquake. Has Kobe fully recovered and rebuilt its economy? How should we assess its reconstruction? If it has been reconstructed, what factors contributed to the process? The objective of this paper is to revisit the experience of the Great Hanshin Awaji Earthquake.

21.2 HAS KOBE RECOVERED AND BEEN RECONSTRUCTED?

Has Kobe fully recovered from the 1995 earthquake? This is a question that is difficult to answer in one word.[2] As we will see in the case of Nagata Ward in Kobe, some areas of Kobe languished after the disaster.

As indices for reconstruction, this paper focuses on population growth and employment because these are the most important factors for recovery and reconstruction, respectively. As we will see, a disaster has an immediate impact on population. People move out of the affected area fall into two categories: The first group of people are those who evacuate to temporary housing or to a relative's house as it takes time to reconstruct their house and workplace and some of them are forced to start a new life, getting a job in a new place. In many cases, these people then decide to stay in the new place rather than going back to their place of origin as they have found their new job and their children go to a new school nearby with new friends. As time goes on, the chance of returning decreases. This is what happened in Kobe. The second group of people are those who move out of the affected area to avoid possible future disasters. Sometimes people simply move to a new home and stay in the same job, but some people find another job in another location.

Here, job is the key to people's movement during the mid- and long-term reconstruction phase. Population growth and jobs are inseparable. Jobs provide an income for people to spend and local stores can then start to sell products. These two cogs are especially important in the phase of reconstruction beyond recovery; without this mechanism, it is not possible for an economy to recover and be reconstructed. Since this paper will focus on the reconstruction phase, we will concentrate on employment in Kobe.

The population of Kobe declined drastically the year the disaster occurred. Compared with the previous year, the population declined by around 95,000 inhabitants (Figs. 21.1 and 21.2). In this, 6434 people were killed in the earthquake and the

1. This was the first major earthquake to hit an urban area in Japan since the Tokyo Earthquake (Great Kanto Earthquake) of 1923.

2. For a detailed discussion on the definition of the terms "resilience," "recovery," and "reconstruction," see Shimada (2015a).

Integrating Disaster Science and Management. http://dx.doi.org/10.1016/B978-0-12-812056-9.00021-X

rest, around 85,000 people, moved outside of the Kobe region. With each passing year, the population gradually returned. In 2004, almost a decade after the earthquake, the population of Kobe had returned to its predisaster level (Fig. 21.3). However, ward-by-ward data give a different picture (Fig. 21.4). As the map of Kobe shows, there are nine wards in Kobe. Among the nine wards, the eastern parts of the coastal wards were severely damaged (Higashi-nada, Nada, Chuo, Hyogo, and Nagata), while the western parts of the coastal wards and the mountainous wards were relatively less affected (Suma, Tarumi Nishi, and Kita). There was a stark difference in the death toll ratio between the two groups (Table 21.1).

So, does the magnitude of damage affect the population growth trends after a disaster? As Fig. 21.4 shows, in terms of the population trend, the wards of Kobe can be categorized into four types: (1) wards where the population declined after the earthquake but recovered well (Higashi-nada, Nada, Chuo); (2) wards where the population declined after the earthquake and then continued to decline (Nagata); (3) wards where the population decline was small (in other words, damage was small) at the time of the earthquake, but the population continued to decline (Suma, Tarumi); and (4) wards where there was almost no impact (Kita, Nishi). Hence, the population growth trend after the disaster has nothing to do with the impact of the disaster. The Nishi and Kita wards were less affected, so it is natural for people to move to these wards. However, even though Tarumi was affected less than other wards, its population has continued to decline. While the wards in the first category were hit harder, population growth after the earthquake was much faster than in the other wards. In other words, the recovery situation is mixed. Because of this mixed picture, it is not easy to say in one word whether Kobe has recovered or not.

So, what did economic recovery in the Kobe region look like? Fig. 21.5 shows that after the earthquake its economy quickly improved mostly because of the investment in reconstruction. However, the economic trend soon reversed and declined in terms of gross output. Further, the gap between Kobe and the rest of Japan widened up until 2003.[3] After 2004, the economic trend in Kobe equaled that of the rest of Japan but still has not totally "filled the gap." By 2001, the gross output had also recovered to its predisaster level. This overall picture, however, needs to be looked at industry by industry, in more detail.[4]

Figs. 21.6 and 21.7 show the share of the working population in the secondary and tertiary sectors, respectively. It is very clear that after the disaster, there was a working population shift from the secondary sector to the tertiary sector. The secondary sector has not recovered to the predisaster level. The drop is especially steep in Nagata ward, where a huge number of small factories producing artificial leather shoes were traditionally located, forming an industrial cluster. According to Yamaguchi (2013), before the earthquake, there were around 450 shoe manufacturers and around 1680 related companies employing 15,000 people. The earthquake completely or partially destroyed 90% of those companies. The loss from the earthquake was estimated at 300 billion yen (Seki and Otsuka, 2001). So the impact was huge. Part of the reason for the loss was the fire after the earthquake. The artificial leather shoe factories use chemicals which are highly flammable, such as paint thinner. According to the Japan Chemical Shoes Industry Association, it is reported that by 2010 sales of the associated companies had dropped from 70 billion yen to around 45 billion yen and employment had dropped from 6500 people to below 3000.

On the other hand, the tertiary sector recovered well and in most wards the number of people working in this sector has increased beyond predisaster levels. According to the statistics of the city of Kobe (2006 and 2009), the medical and welfare industries, and the education support and service sectors have grown rapidly in terms of both the number of offices and employment.[5] In terms of employment, the tertiary industry accounted for 83.5% of total employment in 2006.[6] As of 2012, the medical and welfare industry alone employed 13% (93,618 people) of those employed in Kobe.

This shift from an industrial structure to a tertiary industry is significant compared with the overall figure for Japan (Fig. 21.8). As the figure shows, in comparison to the rate of expansion for Japan overall, the tertiary industry in Kobe has expanded rapidly since 1996, moving from 70.8% to 78% in 2009. One important aspect is that in the case of Kobe, the expansion occurred largely within the first 8 years (1996–2003) after the 1995 earthquake.

We will now compare Kobe's third tertiary sector with other major urban cities in Japan. Fig. 21.9 shows the ratio of tertiary sector offices among all industries. The ratio within Kobe had been high compared with other major cities, but it declined slightly before the earthquake. This trend suddenly changed after the earthquake, rapidly becoming an upward trend. The same is true for the ratio of employees working in the tertiary sector (Fig. 6.10). The trends in major cities have been the same. In the case of Kobe, the trend was slightly different from other cities. The ratio is higher compared with other major cities, excluding the Tokyo metropolitan area, but before the earthquake it became flat. After the earthquake, the trend has regained its momentum and became steeper than other cities. As we have discussed, it would be reasonable to say that in Kobe the tertiary sector was the driver of economic recovery.

3. The data shows that the trend in Hyogo Prefecture, to which Kobe belongs, is similar to that of Kobe. This is probably because most of the prefecture's economic activity is concentrated in Kobe.

4. Beniya et al. (2007) analyzed local industries such as the artificial leather shoes industry cluster and concluded that industry in Kobe had not yet fully recovered.

5. The number of those employed in the manufacturing industry declined.

6. Primary industry accounted for 0.1% of the employed population and secondary industry accounted for the remaining 16.4%.

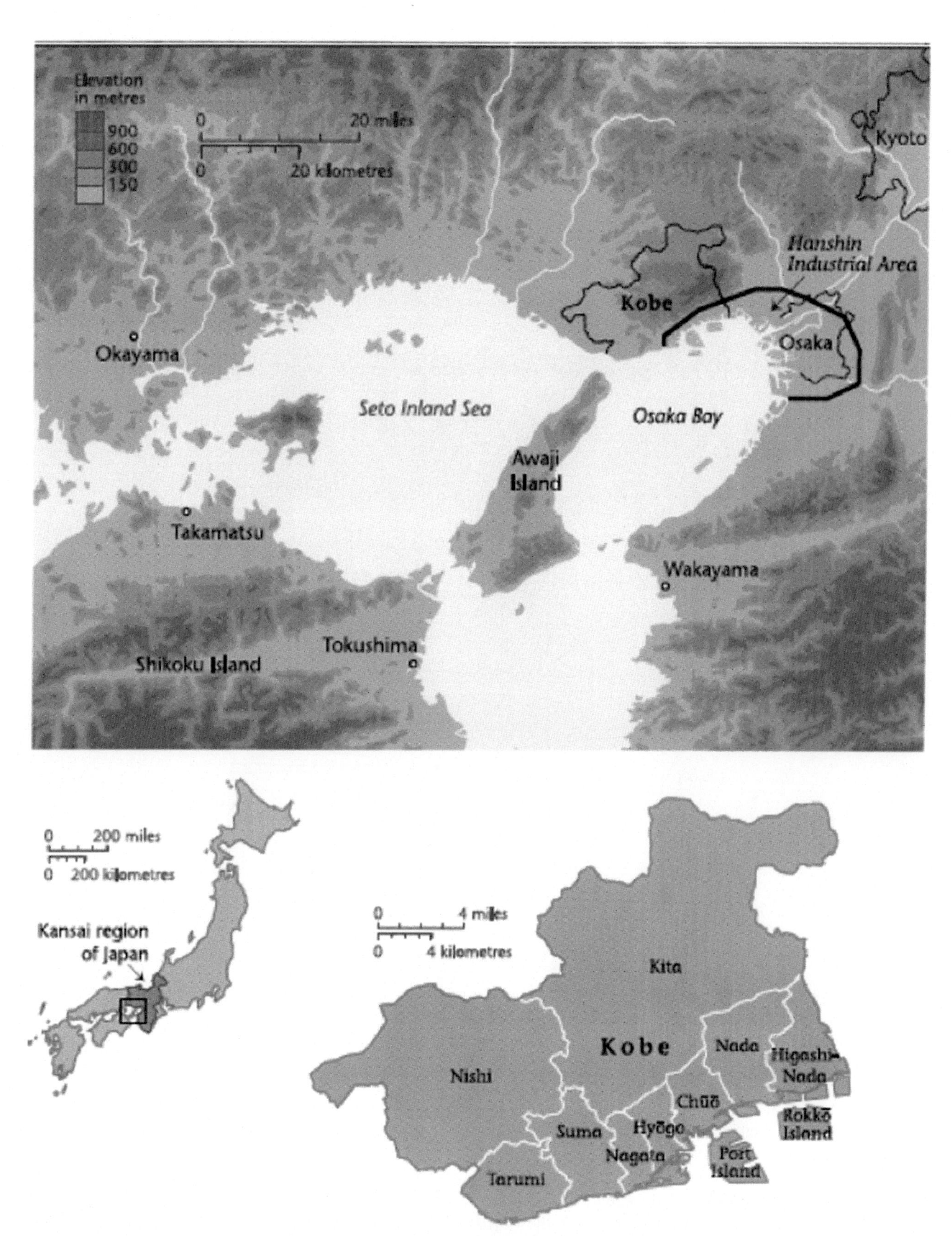

FIGURE 21.1 Map of Kobe. *Source: Adopted from Edington (2010), originally from Fujimori (1980).*

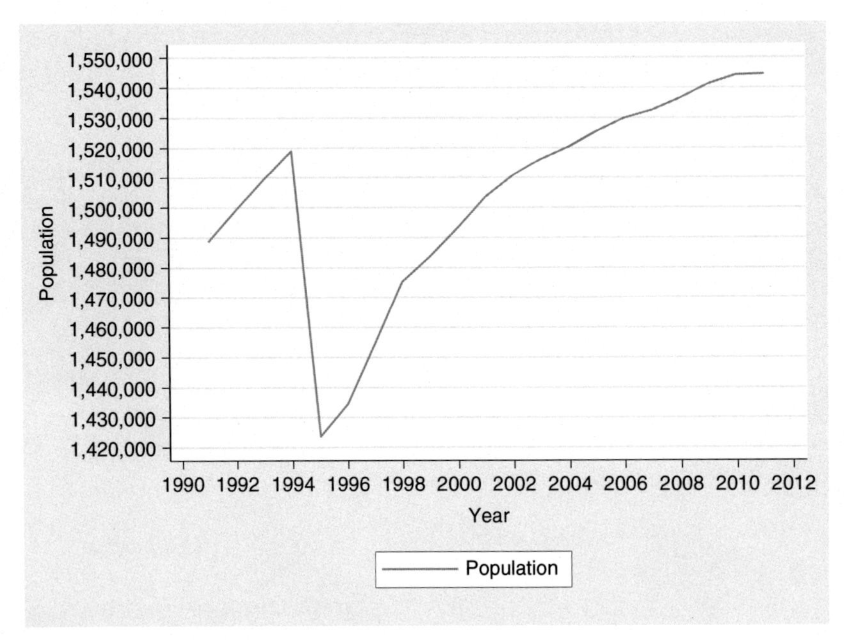

FIGURE 21.2 Population of Kobe. Created by this author based on statistics from The City of Kobe (2013).

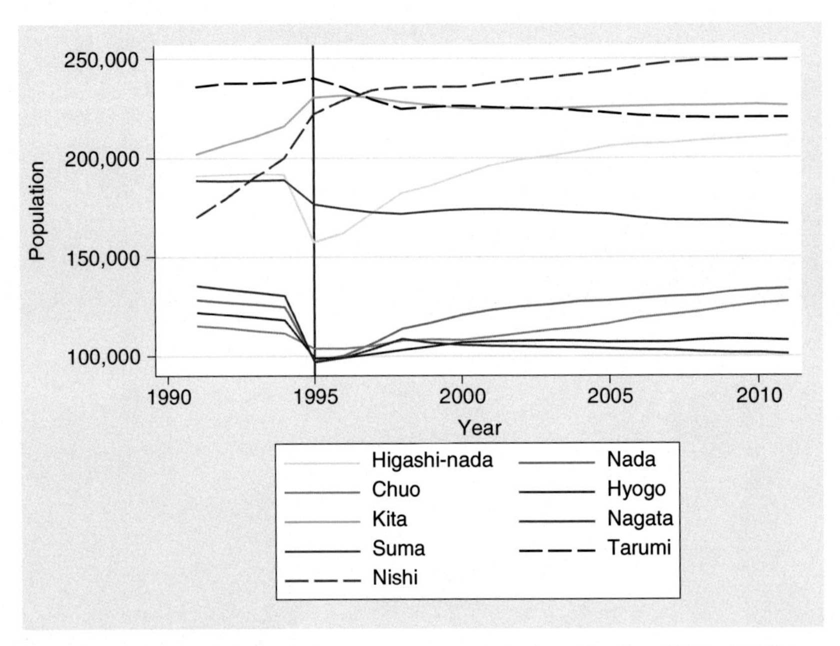

FIGURE 21.3 Population of Kobe by ward. Created by this author based on statistics from The City of Kobe (2013).

One of the reasons for the development of the tertiary sector is the Kobe Bio-medical Innovation Cluster (KBIC) in Port Island. So far, more than 220 companies have invested in KBIC. KBIC was initiated by the city of Kobe in 1998, soon after the earthquake, as part of the recovery plan. Once the cluster developed, the economic effects spilled over into the related industries, leading to higher employment in those industries. Then, as the population grew, the opportunities for small businesses such as retail shops and restaurants increased. This dynamic process of development will have a significant economic impact through multiplier effects (Shimada, 2016).

Another aspect which the city of Kobe focused upon was community business by providing public support to those who introduce initiatives. With the help of public intervention, community business has become active (Ozawa, 2000). Many successful cases have been reported, for example, the TOR-Road Town Planning Corporation (which utilizes vacant stores for glassware sales as well as town planning consultation) and Hyogo Transfer Service Network (which helps disabled and senior people to move).

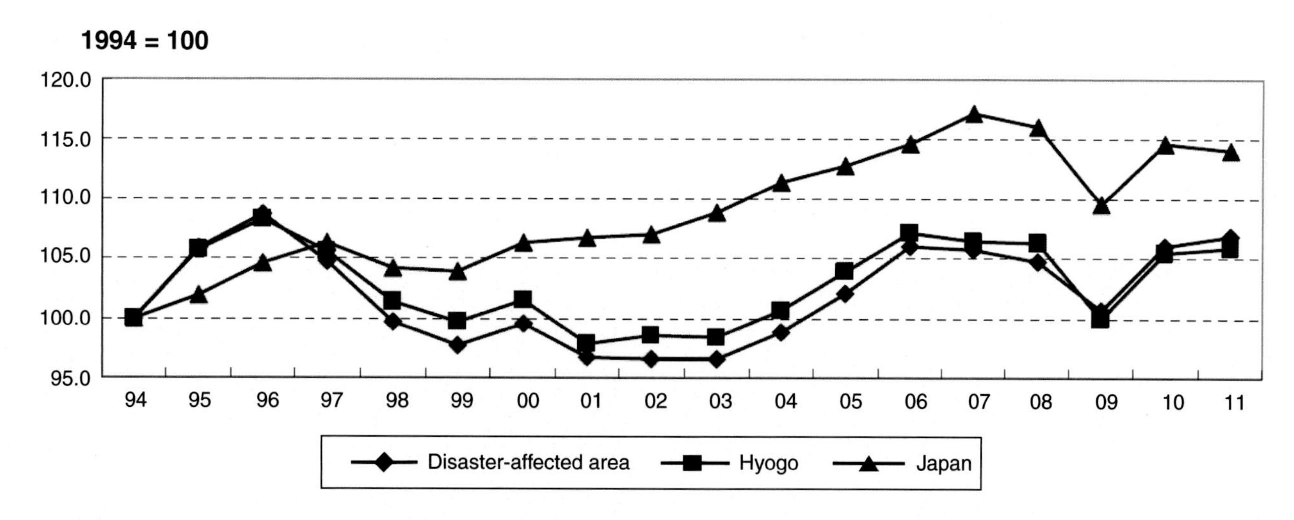

FIGURE 21.4 **Time-series data on gross output.** *Source: Hyogo Prefecture (2013).*

TABLE 21.1 Death Toll by Wards (Made by this Author Based on Statistics from The City of Kobe (2013))

Wards	Death toll	Population	Death toll ratio (death toll/population)
Higashi-nada	1,471	157,599	0.933
Nada	933	97,473	0.957
Chuo	244	103,711	0.235
Hyogo	555	98,856	0.561
Nagata	919	96,807	0.949
Suma	401	176,507	0.227
Tarumi	25	240,203	0.010
Nishi	11	222,163	0.005
Kita	12	230,473	0.005

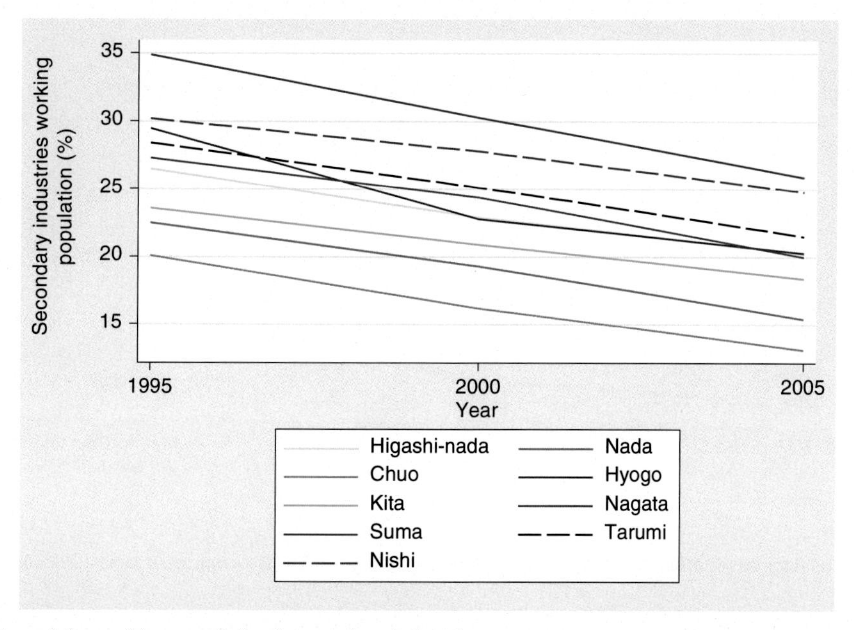

FIGURE 21.5 **Percentage of the working population in secondary industries.**

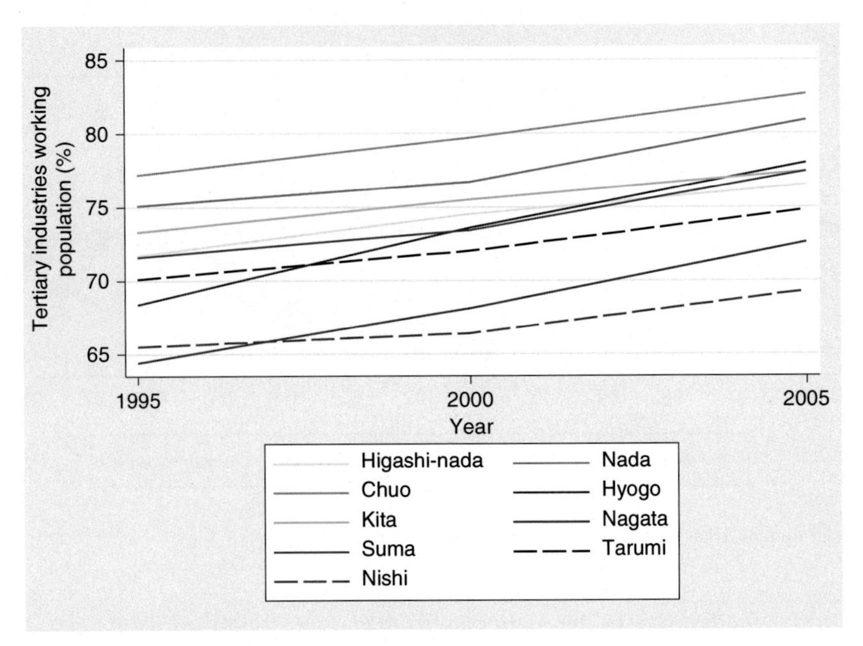

FIGURE 21.6 **Percentage of the working population in tertiary industries.**

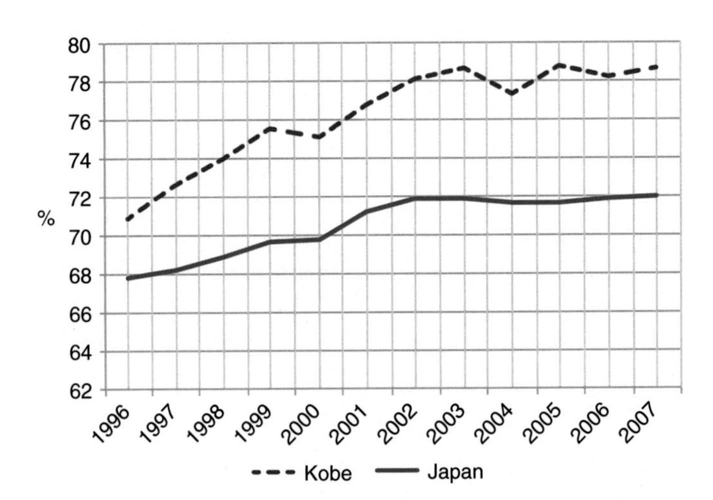

FIGURE 21.7 **Industrial structure of Kobe.** By this author based on the database from The City of Kobe (2013).

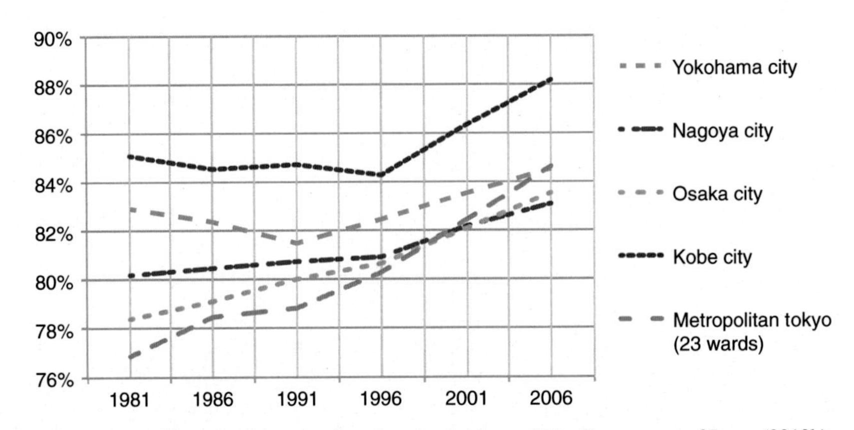

FIGURE 21.8 **Ratio of tertiary sector offices.** By this author based on the database of The Government of Japan (2013b).

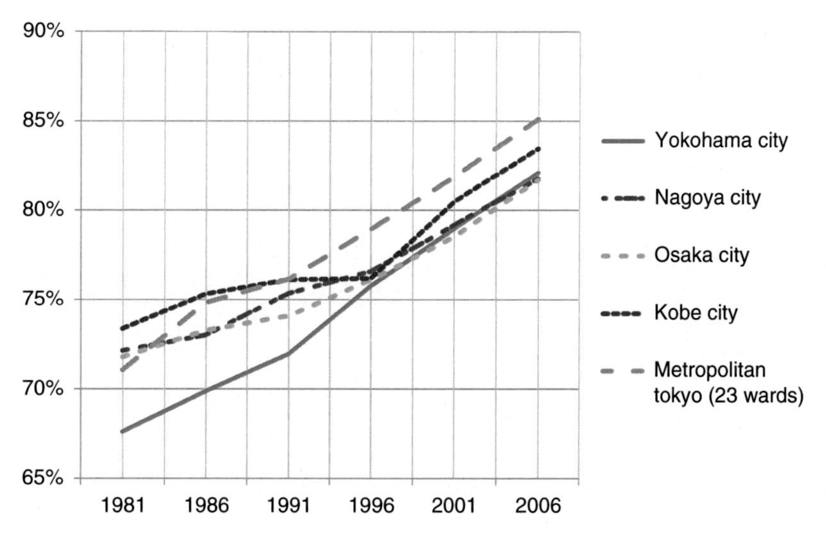

FIGURE 21.9 **Ratio of tertiary sector employees.** By this author based on the database of The Government of Japan (2013b).

Aside from these subsectors, attention needs to be paid to the fact that according to the 2009 economic census, more than 90% of offices in Kobe employ less than 20 workers. In other words, small medium enterprises (SMEs) are the driver of economic reconstruction, providing jobs through multiplier effects.

As we have seen, the data shows a mixed picture of recovery and reconstruction among sectors. This is why it is difficult to say in one word whether Kobe has recovered or not; the situation in each ward and sector is different. After the earthquake, a structural change in industry occurred, whereby it shifted from being a secondary to a tertiary industry. This paper focuses on the tertiary sector and the ways that social capital plays a role in promoting the sector.

21.3 LITERATURE REVIEW

There is a vast amount of literature dedicated to post-earthquake Kobe (Aldrich, 2011; Edington, 2010; Hayashi, 2011a, b; Horwich, 2000; Sawada and Shimizutani, 2008; Seki and Otsuka, 2001; Shibanai, 2007).

One of the characteristics of the discourse following the Great Hansin Awaji Earthquake was the emphasis on social capital. The city of Kobe formed the Social Capital Study Group in 2006 and invited social scientists to act as advisors; they published a report just before the Great East Japan Earthquake. The Study Group organized a workshop among stakeholders to study the Mano area, a downtown section with both a residential area and an artificial leather shoes industrial cluster, and the northern part of Noda in Kobe's Nagata Ward. The report found that the community was a catalyst between the city administration and residents, which is critical in the process of recovery. It also concluded that the community functioned well even before the earthquake and that people actively participated in the reconstruction process.

The Study Group published a number of articles on social capital in the city of Kobe (Shibanai, 2007; Tatsuki et al., 2005; Tatsuki, 2007). The former uses elementary school areas as the unit for social capital, which seems to be a useful alternative to disaggregate prefectural data as in Japan much of the community effort centers on elementary schools. The latter proposed the Seven Elements Model of life recovery after the Kobe earthquake. These seven elements are: housing, social ties, townscape, physical/mental health, preparedness, economic/financial situation, and relation to government. Tatsuki et al. (2005) and Tatsuki (2007) found that these seven critical elements accounted for nearly 60% of the life recovery variance. Nakagawa and Shaw (2004) also studied the Mano area and found that a community with social capital shows the highest satisfaction rates for new town planning and has the speediest recovery rate. Aldrich (2011) employed econometric analysis to study the impact of social capital on population growth and found out that the amount of social capital (measured by the number of nonprofit organizations, created per capita) most strongly determines recovery rates.

A great number of studies have been carried out on social capital and recovery (Chamlee-Wright and Storr, 2009; Chamlee-Wright, 2010; Aldrich, 2012; Shimada, 2015b). What seems to be lacking, however, is analysis of the mid- and long-term reconstruction phases. Social capital is considered to promote the start-up of business (Nam et al., 2010; Todo et al., 2013).[7] There are four causal relations (Table 21.2), namely: (1) job matching; (2) business information and technology transfer; (3)

7. The benefit of accumulation is not confined to the manufacturing sector but can be applied to the service sector as well. Shopping streets are one particular case. After the Great East Japan Earthquake, to the location for re-opening stores in the tsunami-affected area became an issue. There is no point in opening a store that is isolated from other stores. They cannot return to their original location. However, it has taken a long time to decide where communities should be moved to and where offices should be established. So, it is difficult to decide where shopping streets should be located.

TABLE 21.2 Social Capital in Postdisaster Application (in the Reconstruction Phase)

Broad mechanism	Postdisaster application
Strong social capital provides information, knowledge, and access to members of the network (decreases asymmetry of information)	Social capital promotes job matching between the employer and the employee, reducing asymmetry of information
	Social capital promotes knowledge transfer among networks (e.g. technology and business information) to make industrial clusters more competitive
	Social capital provides access to distant markets
Strong ties create trust among network members (decreases transaction costs)	Strong social capital reduces transaction costs among neighbors and private sector activities

Source: This author.

the provision of access to distant markets; and (4) the reduction of transaction costs. The issues in the reconstruction phase are chronic problems the community faced even before disasters; however, these problems were amplified by the disaster rather than the acute external shock itself. As previously discussed, jobs are an important factor in reconstruction.

In the four causal relationships, two common factors are crucial. The first is the need to decrease asymmetry of information. Where there is asymmetry of information, it is known that the market fails, and investment drops to less than the desirable level (underinvestment) (Dasgupta and Stiglitz, 1988; Stiglitz, 2010, 2012). Social capital complements this market failure. As an entrepreneur gets more information from a social network, asymmetry of information decreases, which in turn promotes investment.

Asymmetry of information is also common in labor markets. In this situation, it is difficult to match actual jobs with the labor available. For the employer, it is not easy to find somebody suitable using terms of reference since it is difficult to get accurate information on the capacity or human capital of job applicants. Studying the US labor market, Granovetter (1974) found that social networks raised the efficiency of the job matching process and sped up the job search for workers. Put more simply, information in the form of personal recommendations addresses the asymmetry of information and catalyzes job matching.

The second important factor is the promotion of knowledge transfer among networks so as to make industrial clusters more competitive (Inkpen and Tsang, 2005; Urata and Ito, 1994;). The study by Otsuka and Sonobe (2011) and various related empirical studies by Sonobe et al. (2011) and Kuchiki and Tsuji (2008) have shown that without introducing new ideas and knowledge, industrial clusters never grow in a sustainable way.

21.4 TESTABLE HYPOTHESIS

To translate these arguments into a testable hypothesis, this paper postulates the following hypothesis:

Social capital promotes reconstruction, through contributing to business and the growth of employment. Regarding social capital, it is important to note that there are negative and positive aspects of social capital. The same social capital that gives the members privileged access to certain resources may exclude nonmembers access from those same resources (Arrow, 2000; Portes, 1998).[8] As discussed in the last section, social capital promotes knowledge transfer, thus reducing the transaction cost to the market. On the other hand, it may exclude nonmembers. It is well known that the FDNY (the Fire Department of the City of New York) is dominated by Italian Americans and the diamond trade in New York is dominated by Jewish dealers. The mafia and the caste system in certain parts of south Asia are extreme examples. It is possible to have high-bonding social capital (where members help each other) but a lack of bridging social capital (the exclusion of members of other social groups). This is particularly true for bonding social capital, which may result in nepotism and crony capitalism; this in turn causes market failure and hampers the healthy development of the private sector and employment growth. Thus, social capital does not always benefit the society.

21.5 METHODOLOGY

To test the above hypotheses, this paper employs the following equation. The dependent variable is the employment growth rate in the tertiary industry ($Emp_{i,t}$), where i and t denote ward and time, respectively. This variable is chosen because employment is the most suitable index for the mid- and long-term reconstruction phases.

8. Arrow (2000, p. 3) stated that "... social interactions can have negative as well as positive effects.... Good behavior spreads; so does bad."

$$\Delta \text{Emp}_{i,t} = \alpha_i + \beta\, \Delta \text{Emp}_{i,\,t-1} + \gamma_0\, \text{SC}_{i,t} + \gamma_1\, \text{HC}_{i,t} + \gamma_2\, \Delta \text{population\_growth}_{i,t} + \varepsilon_{i,t}$$

Following the literature on the New Keynesian Phillips Curve (NKPC), wages, prices, and employment levels in the labor market are assumed to be volatile and the adjustment to market equilibrium is gradual. This assumption is appropriate for Japan's labor market, where lifetime employment is common. This assumption is different from that of the neoclassical Phillips curve. Therefore, the equation contains lagged $\text{Emp}_{i,t}$. The model with a lagged *Y* variable is known as an autoregressive model (Beck and Katz, 2004).

$\text{SC}_{i,t}$ is the social capital variable and $\text{HC}_{i,t}$ is the human capital. For the social capital variable, the following three proxy variables will be used. The possible proxies to be used from past literature are as follows: PTA (Coleman, 1988; Putnam, 2001); living arrangements with parents and intensity of interactions with parents (Teachman et al., 1997), crime rate (Putnam, 2001), and newspaper reading (Putnam and Robert, 1996). Among these, this chapter uses the following proxies: crime rate and number of households, where three generations live together.

The crime rate is selected because communities with high social capital are considered to have a lower crime rate (Akcomak and Weel, 2008; Aldrich, 2012; Buonanno et al., 2009; Deller, 2010; Putnam, 2001). The study conducted by the Cabinet Office of the Government of Japan (2003) also used crime rates as a proxy for social capital. In a community with high social capital, members feel they have a responsibility for the security of the neighborhood as a way of protecting their families. They organize community meetings, walking patrols, and inform the police if they spot any suspicious individuals (Aldrich, 2012). Coleman (1988, p. S104) stated that "effective norms that inhibit crime make it possible to walk freely outside at night." Without tight community control, it would be difficult for parents to send their children to play outside.

The number of households with three generations living together was selected because those households are considered to have strong family ties and provide a social safety net. Recently, Abe (2014) conducted a comprehensive study that looked at Japan's poor in 2007 and 2010 and found that household structure is a very important factor behind poverty. According to her study, among all households, the poverty rate is the highest in households with a single parent and children, followed by households made up of a single elderly person. On the contrary, households where three generations live together have the lowest poverty rate. This is because in households where three generations live together, household members help each other. In other words, the social safety net is rich in these households.

The rate of population growth is shown as *population growth*$_{i,t}$. Population growth and employment are considered to be closely associated with one another. The causality is not one way but is probably two way. People will come back to an area where there are employment opportunities. At the same time, if people move into an area, the need for various consumer products and goods within that area increases. This creates good business opportunities for SMEs, which in turn increases demand for labor. The variable $\text{disaster}_{i,t}$ is related to damage caused by the disaster. Here, we will use the death toll rate (= death toll number/population).

For the purposes of this paper, the standard panel estimation (random effects (REs), fixed effects (Fes), pooling cross-section across time), Prais–Winsten estimation, and system generalized method of moments (GMMs) have been used. Since this model is a dynamic model containing a lagged dependent variable on the right-hand side of the equation, this paper uses a Prais–Winsten estimation to ensure the findings. Prais–Winsten is a method of multiple linear regression with AR(1) and exogenous explanatory variables. The Prais–Winsten standard errors account for serial correlation, which the RE, FE, and pooling estimations do not.

The system GMM is used to tackle other possible biases by endogeneity and omitted variables in addition to the bias caused by the lagged dependent variable. In our system of GMM estimation, all regressors are considered to be endogenous. Arellano and Bond (1991) first established the "difference-GMM" estimator for dynamic panels (Roodman, 2003). Arellano and Bond's estimation starts by transforming all regressors via differencing and uses the GMM. This method regards lagged dependent variables as not exogenous but predetermined. A problem with the original Arellano–Bond difference-GMM estimator is that if there is an issue of a random walk of endogenous variables, the estimation becomes a biased coefficient estimation.

To tackle the above problem, Blundell and Bond (1998) articulated an improvement of the augmented difference GMM put forward by Arellano and Bover (1995), adding more assumptions that first differences of instrument variables are uncorrelated with the FEs, allowing more instruments to be introduced and making them exogenous to the FEs. The augmented estimator is called "system GMM." The STATA command *xtabond2* implements both estimations.

The major advantage of the system-GMM estimation, compared with the difference-GMM estimation, is that it effectively controls autocorrelation and heteroskedasticity. This chapter uses one-step estimation and implements the Hansen test and the Sargan test for joint validity of the instruments; it also implements the AR test for autocorrelation.

21.6 DATA

For the empirical study, this paper uses the variables listed in Table 21.3. The database contains unbalanced panel data, covering all nine of the wards in Kobe from 1995 to 2010, with some gaps. This data is used because ward-by-ward data is only available for the years after 1995, not before the earthquake. This is as far as this paper can go back. The data used in this paper is from the existing data from The City of Kobe (2006, 2012, 2013) and census data from The Government of Japan (2013a, b).

The standard deviation of "death toll" is large; this is due to the earthquake. Considering the standard deviation, this paper takes the natural logarithm for this variable to analyze. As people commute from one ward to the other or commute into or out of the city, the data include: outbound commuter ratio, inbound commuter ratio, and day population ratio. Chuo-ward and Hyogo-ward are business districts, so people commute to these two wards from other wards and from outside of the city. With regard to crime rates, the data on crime refers to offences such as murder, robbery, and rape. It does not include a number of minor offences and traffic accidents. The number of observations on the crime rate is small because ward-by-ward data on crime rates is available only after 2000, not from 1995 as with other data.

21.7 ESTIMATION RESULTS

Tables 21.4 and 21.5 show estimation results. As discussed, the dependent variable is the employment growth rate in the tertiary industry. Models 1–3 of Table 21.4 show the results from the standard panel estimations, FE is a better method than RE and pooling, judging from the results of the Hausman test, *F*-test, and Breusch and Pagan test. It shows that population growth has a positive impact on employment growth in the tertiary sector. The tertiary sector in Kobe is mainly wholesale, retail, restaurant, medical, and education. More than 90% of those businesses employ less than 20 workers, so they are basically very small enterprises. A higher population means they have more consumers for their services. However, the direction of causality may go in both directions with employment growth driving population growth.

The social capital proxy, share of members of households where three generations live together becomes significantly positive. The coefficient is also large for the FE. In other words, the stronger the social capital, the more jobs have been created. Judging from the results of the three models, it would be safe to say that the results are robust.

As for the death toll, this became insignificant for the FE. This is in line with the earlier discussion based on Fig. 21.4. The magnitude of damage by the earthquake does not predict the recovery. Higasni-nada and Nagata were equally hit by earthquake but the recovery afterwards was very different. One of the possible reasons for the insignificance may be that if one ward is hit harder than another ward, the recovery budget for that ward is larger. As the ward-by-ward data on the budget is not available, this point is difficult to examine further.

TABLE 21.3 Descriptive Statistics

Variable	Obs.	Mean	Std. Dev.	Min	Max
Death toll	36	106.750	285.920	0	1292
Employment growth rate in tertiary industry	36	1.023	14.426	−27.881	49.234
Population growth rate	36	1.305	12.753	−29.280	40.100
Share of households with three generations living together	36	7.319	2.660	3.577	13.677
Crime rate	27	0.023	0.015	0.009	0.073
Outbound commuter ratio	36	57.914	10.398	34.800	74.800
Inbound commuter ratio	36	79.864	97.925	12.000	369.400
Day time population ratio	36	112.964	52.763	68.100	275.300

TABLE 21.4 Estimation Results (Dependent Variable: Employment Growth Rate in Tertiary Industry)

	Model 1	Model 2	Model 3	Model 4	Model 5	Model 6	Model 7	Model 8	Model 9
	Pooling	FE	RE	Pooling	FE	RE	Pooling	FE	RE
Population growth	1.147	1.106	1.147	1.134	1.110	1.134	1.159	1.111	1.159
	[21.62]***	[20.30]***	[21.62]***	[26.39]***	[23.10]***	[26.39]***	[22.03]***	[19.37]***	[22.03]***
Share of households with three generations living together	0.944	2.269	0.944	0.752	0.678	0.752	0.631	2.250	0.631
	[3.76]***	[4.53]***	[3.76]***	[4.29]***	[1.06]	[4.29]***	[2.91]***	[4.38]***	[2.91]***
Death toll by the earthquake (log)	0.817	0.360	0.817	0.702	0.332	0.702	0.959	0.333	0.959
	[2.95]***	[1.16]	[2.95]***	[3.02]***	[1.24]	[3.02]***	[3.42]***	[1.02]	[3.42]***
Ratio of outbound commuters				0.362	0.596	0.362			
				[4.27]***	[3.09]***	[4.27]***			
Ratio of inbound commuters				0.026	0.100	0.026			
				[2.72]***	[2.32]**	[2.72]***			
Ratio of daytime population							−0.020	0.027	−0.020
							[−1.91]*	[0.34]	[−1.91]*
_cons	−8.350	−17.458	−8.350	−29.806	−48.264	−29.806	−3.978	−20.288	−3.978
	[−4.46]***	[−5.05]***	[−4.46]***	[−5.10]***	[−4.89]***	[−5.10]***	[−1.74]*	[−2.27]**	[−1.74]*
N	36	36	36	36	36	36	36	36	36
R^2		0.972			0.981			0.972	
Adj-R^2		0.959			0.971			0.958	
F-test	Prob > *F* = 0.031			Prob > *F* = 0.164			Prob > *F* = 0.094		
Breusch and Pagan Lagrangian multiplier test	Prob > chibar2 = 0.433			Prob > chibar2 = 1.0000			Prob > chibar2 = 1.0000		
Hausman test	Prob > chi2 = 0.042			Prob > chi2 = 0.021			Prob > chi2 = 0.017		

*$P < 0.1$.
**$P < 0.05$.
***$P < 0.01$.

TABLE 21.5 Estimation Results 2 (Dependent Variable: Employment Growth Rate in Tertiary Industry)

	Model 10	Model 11	Model 12	Model 13
	RE	Pooling	Prais–Winsten	System GMM
Employment growth rate in the tertiary industry (lagged)	−0.0394	−0.0394	−0.072	−0.0276
	[−0.89]	[−0.89]	[−1.71]	[−1.03]
Population growth	1.1478	1.1478	1.2018	1.1193
	[11.20]***	[11.20]***	[15.89]***	[11.93]***
Crime rate	−1.4083	−1.4083	−60.7822	−89.5034
	[−0.03]	[−0.03]	[−2.02]*	[−4.94]***
Share of members of households with three generations living together	0.8926	0.8926	0.7153	0.1401
	[3.26]***	[3.26]***	[3.98]***	[2.09]**
_cons	−7.8562	−7.8562	−4.834	
	[−3.17]***	[−3.17]***	[−2.95]***	
N	27	27	27	27
R^2			0.9174	
Adj-R^2			0.9023	
Hansen test				0.999
Sargan test				0.386
Arellano-Bond statistic				0.415

*$P < 0.1$.
**$P < 0.05$.
***$P < 0.01$.

However, as we saw in the description data section, people commute from one ward to another. In particular, Hyogo and Chuo are two big business districts in the city. It is necessary to control people's movement. The models from 4 to 6 control inbound and outbound commuters. Additionally, so as to check for robustness, the models from 7 to 9 used the ratio of the daytime population. Among the standard panel estimations, according to the Hausman test, the *F*-test, and the Breusch and Pagan test, the most appropriate estimation is: pooling for models 4–6 and FE for models 7–9.

Controlling the commuter population (model 4) and the daytime population (model 8), both population growth and three generations living together became robustly significant. Even if coefficients are small, both inbound and outbound commuter ratios became positive in model 4. The interpretation of inbound commuter ratios is straightforward because they will become consumers within the ward that they commute to, thus creating more demand. Regarding the outbound commuter ratios, their family members stay in the ward and their main place of consumption still remains the ward they live in. That is probably the reason why the outbound commuter ratios become positive. Comparing the coefficients, outbound commuter ratios are larger than the inbound commuter ratios. This is probably because for the tertiary industry, final consumption mainly occurs within their residential area, rather than within their working area. Model 8 is in line with this argument: daytime population ratios are not significant. In controlling the movement of people, the most important conclusion is that the social capital variable still remains robustly positive. The result of death toll is different; it became positive. This is probably because of the recovery effort and budget after the earthquake, as we discussed earlier.

In Table 21.5, this paper examined another social capital proxy, namely crime rates. As discussed in the section on descriptive statistics, the crime rate data is available only after 2000, not before. In Table 21.4, crime rate was not included so as not to lose observations and to be able to see the impacts of disaster damage. Further, models 10–13 included lagged dependent variables, as discussed earlier.

According to models 10 and 11, the families where three generations live together remained positive, but the crime rate was insignificant for both RE and pooling estimations. Then, as discussed in the methodology section, this paper estimated the Prais–Winsten, in which standard errors account for serial correlation, unlike the RE, FE, and pooling estimations. With the Prais–Winstein, in model 12 the crime rate became negatively significant. This means that if the crime rate is lower thanks to high social capital, then it has positive impacts on employment. This model also confirms the positive impact of another social capital (three generations living together).

Finally, considering the possible endogeneity and omitted variable biases, model 13 checked the results with the system-GMM (one step). The results also confirmed that both households with three generations living together and crime rates became significant. In the same way as in model 12, the three generations living together became positive, and the crime rate became negative. Hence, with regard to the social capital variable, we would be able to say that these are robust results. Therefore, the results are concordant with the hypotheses.

21.8 CONCLUSIONS

Even if the recovery experienced by the city of Kobe after the 1995 earthquake was positive overall, the ward-by-ward analysis shows a different picture. Even if damage levels are the same, some wards recover quickly and some wards do not. The empirical analysis of this paper shows that with more social capital (three generations living together and less crime rates), more employment in the tertiary sector was generated. In the case of Kobe, most offices and work places in Kobe employ less than 20 workers. Under this economic structure, this paper found that social capital is the factor that causes different patterns of recovery among wards. This has important policy implications for making recovery after possible future natural disasters faster, and for building resilient societies. This is because the importance of infrastructure still tends to be over emphasized in the process of recovery.

This paper has several limitations. First, even if it is impossible to construct panel data afterwards, it is desirable to get more detailed data at the community level rather than at the ward level. For future studies, it would be advisable to start collecting more detailed data soon after a natural disaster in order to construct the panel data. Second, related to the first issue, due to data limitations, this paper used existing data as social capital proxies. Ideally, it is better to collect social capital data using methods such as dictator games and other methods. The selected social capital variables for this paper—crime rates and households where three generations live together—might be affected by other factors such as police efforts and family structures. These are the challenges for future studies. Even if there are some limitations, the findings of this paper provide hints for building resilient societies.

REFERENCES

Abe, A., 2014. Poverty of children in Japan II. Iwanami Press, Tokyo.

Akcomak, I.S., Weel, B.T., 2008. The impact of social capital on crime: evidence from the Netherlands. IZA Discussion Paper 3603.

Aldrich, D.P., 2011. The power of people: social capital's role in recovery from the 1995 Kobe earthquake. Natural Hazards 56 (3), 595–611.

Aldrich, D.P., 2012. Building Resilience—Social Capital in Post-Disaster Recovery. University of Chicago Press, Chicago.

Arellano, M., Bond., S., 1991. Some tests of specification for panel data: Monte Carlo evidence and an application to employment equations. Rev. Econ. Stud. 58, 277–297.

Arellano, M., Bover, O., 1995. Another look at the instrumental variables estimation of error components models. J. Econometrics 68, 29–51.

Arrow, K.J.A., 2000. Observations on social capital. In: Dasgupta, P., Serageldin, I. (Eds.), Social Capital: A Multifaceted Perspective. World Bank, Washington, DC, pp. 3–5.

Beck, N., Katz, J.N., 2014. Time-series-cross-section issues: dynamics, 2004. Working paper. The Society for Political Methodology. http://polmeth.wustl.edu/media/Paper/beckkatz.pdf (accessed 18.01.04).

Beniya, S., Hokugo, A., Murosaki, Y., 2007. Time-series analysis on industrial recovery indexes after a disaster and outline of public support programs for small-sized business. Departmental Bulletin Paper 11, 149–158.

Blundell, R., Bond, S., 1998. Initial conditions and moment restrictions in dynamic panel data models. J. Econometrics 87, 11–143.

Buonanno, P., Montolio, D., Vanin, P., 2009. Does social capital reduce crime? J. Law Econ. 52, 145–170.

Cabinet Office of the Government of Japan, 2003. Social Capital: Looking for a Good Cycle of Rich Human Relationships and Civic Activities. Government Printing Office, Tokyo.

Chamlee-Wright, E., 2010. The Cultural and Political Economy of Recovery: Social Learning in a Post-Disaster Environment. Routledge, New York.

Chamlee-wright, E., Storr, V.H., 2009. There's no place like New Orleans: sense of place and community recovery in the ninth ward after hurricane Katrina. J. Urban Affairs 31 (5), 615–634.

Coleman, J.S., 1988. Social capital in the creation of human capital. Am. J. Sociology 94, S95.

Dasgupta, P., Stiglitz, J., 1988. Learning-by-doing, market structure and industrial and trade policies. Oxford Economic Papers New Series 40 (2), 246–268.

Deller, S., Deller, M., 2010. Rural crime and social capital. Growth and Change 41 (2), 221–275.

Edington, D.W., 2010. Reconstructing Kobe: The Geography of Crisis and Opportunity. UBC (University of British Columbia) Press, Vancouver.

Fujimori, T., 1980. Hanshin region. In: Murata, K., Ota, I. (Eds.), An Industrial Geography of Japan. St Martin's, New York, NY, pp. 81–90.

Granovetter, M., 1974. Getting a Job: A Study of Contacts and Careers, second ed. Chicago University Press, Chicago, 1995.

Hayashi, T., 2011a. Economics of Disaster (Daisaigai no keizaigaku). PHP Press, Tokyo.

Hayashi, T., 2011b. Handbook of Disaster Management (Saigai taisaku zensyo). Gyosei, Tokyo.

Horwich, G., 2000. Economic lessons of the Kobe earthquake. Econ. Dev. Cult. Change 48 (3), 521–542.

Hyogo Prefecture, 2013. Present situation of recovery and reactivation after the Great Hanshin Awaji Earthquake. Hyogo Prefecture.

Kuchiki, A., Tsuji, M., 2008. The Flowchart Approach To Industrial Cluster Policy. Palgrave Macmillan, New York.

Inkpen, A.C., Tsang, E.W.K., 2005. Social capital, networks, and knowledge transfer. Acad. Manag. Rev. 30 (1), 146–165.

Kuramochi, H., 1997. Industrial reconstruction after the Great Hanshin-Awaji earthquake. Earthquakes and People's Health. WHO, Kobe, http://helid.digicollection.org/en/d/Jh0220e/6.4.html (accessed 03.10.16).

Nakagawa, Y., Shaw, R., 2004. Social capital: a missing link to disaster recovery. Int. J. Mass Emergencies and Disasters 22 (1), 5–34.

Nam, V.H., Sonobe, T., Otsuka, K., 2010. An inquiry into the development process of village industries: the case of a knitwear cluster in northern Vietnam. J. Dev. Stud. 46 (2), 312–330.

Otsuka, K., Sonobe, T., 2011. A cluster-based industrial development policy for low-income countries. Policy Research Working Paper 5703. World Bank, Washington, DC.

Ozawa, Y., 2000. The economy of Kobe through statistics (Data ni miru Kobe keizai no genjo to kadai). Toshi Seisaku 98, 3–16.

Portes, A., 1998. Social capital: its origins and applications in modern sociology. Annu. Rev. Sociology 24, 1–24.

Putnam, R., 2001. Social capital: measurement and consequences. Isuma: Can. J. Policy Res. 2, 41–51, Spring 2001.

Putnam, R., Robert, D., 1996. The strange disappearance of civic America. The American Prospect 7 (24), 34–49.

Roodman, D., 2003. XTABOND2: Stata module to extend xtabond dynamic panel data estimator. Statistical software components S435901. Boston College Department of Economics (revised 17 Jan, 2012).

Sawada, Y., Shimizutani, S., 2008. How do people cope with natural disasters? Evidence from the Great Hanshin-Awaji (Kobe) Earthquake in 1995. J. Money, Credit and Banking 40, 463–488.

Seki, M., Otsuka, Y., 2001. The Reactivation of the Hanshin Area and Local Industry (Hanshin fukko to chiiki sangyo). Shin-Hyoron, Tokyo.

Shibanai, Y., 2007. Kobe shinai no social capital ni kansuru jissyo bunseki (An empirical study of social capital in Kobe City). Toshi Seisaku 127.

Shimada, G., 2015a. Towards community resilience—the role of social capital after disasters. In: Chandy, L., Kato, H., Kharas, H. (Eds.), The Last Mile in Ending Extreme Poverty. Brookings Institutions, Washington, DC.

Shimada, G., 2012b. The role of social capital after disasters: an empirical study of Japan based on time-series-cross-section (TSCS) data from 1981 to 2012. Int. J. Disaster Risk Reduct. 14, 388–394.

Shimada, G., 2016. Inside the Black Box of Japan's Institution for Industrial Policy—an institutional analysis of development bank, private sector and labour. In: Noman, A., Stiglitz, J. (Eds.), Efficiency, Finance and Varieties of Industrial Policy. Columbia University Press, New York.

Sonobe, T., Suzuki, A., Otsuka, K., 2011. Kaizen for managerial skills improvement in small and medium enterprises: an impact evaluation study (Background paper). In: Dinh et al., (Ed.), Light Manufacturing in Africa. World Bank, Washington, D.C.

Stiglitz, J., 2010. Learning, growth, and development: a lecture in honor of Sir Partha Dasgupta. Paper Presented at the World Bank's Annual Bank Conference of Development Economics. Stockholm.

Stiglitz J., Creating a learning society, In: *Paper Presented at the Amartya Sen Lecture*, The London School of Economics and Political Science, London, 2012.

Tatsuki, S., 2007. Social capital to chiiki zukuri. Toshi Seisaku 127, 4–19, (in Japanese).

Tatsuki, S., et al., 2005. Long-term life recovery processes of the survivors of the 1995 Kobe earthquake: causal modeling analysis of the Hyogo Prefecture life recovery panel survey data. A Paper Presented at the First International Conference on Urban Disaster Reduction. Kobe.

Teachman, J., Paasch, K., Carver, K., 1997. Social capital and the generation of human capital. Social Forces 75 (4), 1–17.

The City of Kobe, 2006. Offices of Kobe (Kobe no jigyosyo). http://www.city.kobe.lg.jp/information/data/statistics/toukei/ (accessed 26.06.13).

The City of Kobe, 2012. Offices of Kobe (Kobe no jigyosyo). http://www.city.kobe.lg.jp/information/data/statistics/toukei/ (accessed 26.06.13)

The City of Kobe, 2013. Statistics of Kobe (Kobe no tokei). http://www.city.kobe.lg.jp/information/data/statistics/toukei/ (accessed 24.06.13).

The Government of Japan, 2013a. Census data (Tokei de miru sicho-son no sugata). http://www.stat.go.jp/data/ssds/5b.htm (accessed 24.06.13).

The Government of Japan, 2013b. National economic statistics (Kokumin keizai keisan). http://www.stat.go.jp/data/ssds/5b.htm (accessed 06.12.13).

Todo, Y., Nakajima, K., Matous, P., 2013. How do supply chain networks affect the resilience of firms to natural disasters? Evidence from the Great East Japan Earthquake. RIETI Discussion Paper 13-E-028.

Urata, S., Ito, M., 1994. Small and medium-size enterprise support policies in Japan. Policy Research Paper 1403. The World Bank, Washington, DC.

Yamaguchi, J., 2013. The issues of local industry and SMEs based on the lessons from the Great Hanshin Awaji Earthquake (in Japanese). http://www.jepa-hq.com/Report (accessed 18.02.13).

FURTHER READING

Anthony, E.J., 1987. Risk, vulnerability, and resilience. In: Anthony, E.J., Bertram, C.J. (Eds.), The Invulnerable Child. Guilford Press, New York, pp. 3–48.

Beck, N., 2004. Longitudinal (panel and time series cross-section) data. http://weber.ucsd.edu/~tkousser/Beck%20Notes/longitude20041short.pdf (accessed 25.10.12).

Carter, M.R., Maluccio, J.A., 2003. Social capital and coping with economic shocks: an analysis of stunting of South African children. World Development 31 (7), 1147–1163.

Dacy, D., Kunreuther, H., 1969. The Economics of Natural Disasters: Implications for Federal Policy. Free Press, New York.

Dinh, H.T., et al., 2012. Light Manufacturing in Africa. World Bank, Washington, DC.

Fujii, S., 2011. Resilient Japan (Retto Kyojinka-ron). Bungei-Syunju, Tokyo, (in Japanese).

Government of Japan, 2013a. National resilient committee meeting memorandum. Government of Japan, Tokyo. http://www.cas.go.jp/jp/seisaku/resilience/dai1/1sidai.html (accessed 01.08.13).

Government of Japan, 2013b. Database of regional statistics (Chiki tokei database). http://www.stat.go.jp/data (accessed 20.12.13).

Guillaumont, P., 2009. An economic vulnerability index: its design and use for international development policy. Oxford Development Studies 37 (3), 193–228.

Hayashi, H., 2012. Resilience—power of recovery from disasters (Saigai kara tachinaoru chikara). Kyoiku to Igagu (in Japanese).

Kates, R.W., Pijawka, D., 1977. From rubble to monument: the pace of reconstruction. In: Haas, E.J., Kates, R.W., Bowden, M.J. (Eds.), Disaster and Reconstruction. MIT Press, Cambridge, MA, pp. 1–23.

Longstaff, P.H., et al., 2010. Community resilience: a function of resource and adaptability. White Paper. Institute for National Security and Counterterrorism, Syracuse University.

McCreight, R., 2010. Resilience as a goal and standard in emergency management. J. Homeland Secur. and Emergency Manage. 7 (1), 1–7.

Narayan, D., 1999. Bonds and bridges: social capital and poverty groups, PREM. World Bank, Washington, DC.

Norman, G., 1971. Vulnerability research and the issue of primary prevention. Am. J. Orthopsychiatry 41 (1), 101–116.

Norris, F., Stevens, S., Pfefferbaum, B., Wyche, K., Pfefferbaum, R., 2008. Community resilience as a metaphor, theory, set of capacities and strategies for disaster readiness. Am J. Community Psychology 41, 127–150.

Okada, K., 2005. Methodologies of disaster risk management and economic analysis (Sigai risk management no houhou-ron to keizai bunseki no kosa, in Japanese). In: Takagi, A., Tatano, H. (Eds.), Disaster Economics (Saigai no Keizai Gaku in Japanese). Keiso Press, Tokyo.

Shimada, G., 2015. The economic implications of comprehensive approach to learning on industrial development (policy and managerial capability learning): a case of Ethiopia. In: Noman, A., Stiglitz, J. (Eds.), Industrial Policy and Economic Transformation in Africa. Columbia University Press, New York.

Tatsuki, S., Hayashi, H., 2005. Seven critical elements model of life recovery: general linear model analyses of the 2001 Kobe panel survey data. Proceedings of the 2nd Workshop for Comparative Study on Urban Earthquake Disaster Management, February 2002. pp. 14–15.

UNISDR, 2005. Hyogo framework of action 2005–2015: building the resilience of nations and communities to disasters. UNISDR. Geneva.

UNISDR, 2009. Terminology on disaster risk reduction. http://www.unisdr.org/we/inform/terminology (accessed 18.02.13).

Chapter 22

Resilience and Vulnerability: Older Adults and the Brisbane Floods

Evonne Miller*, Lauren Brockie**

*QUT Design Lab, Brisbane, QLD, Australia; **School of Design, Queensland University of Technology, Brisbane, QLD, Australia

Evonne Miller is an environmental psychologist at the Creative Industries, Queensland University of Technology in Brisbane, Australia. Her recent research has explored the disaster experience of older Australians during the 2011/2013 Brisbane floods and 2011 Cyclone Yasi in Far North Queensland.

Lauren Brockie is a registered psychologist who works in private clinical practice in Brisbane, Australia. Her research interests center around trauma, posttraumatic stress, and posttraumatic growth and her PhD focused on exploring the unique experiences of older adults who experienced the 2011/2013 Brisbane floods.

Mother Earth can seem like an uncaring planet. The impact of geohazards on our lives and economy is very great, and will never go away. Every year floods, tsunamis, severe storms, drought, wildfires, volcanoes, earthquakes, landslides and subsidence claim thousands of lives, injure thousands more, devastate homes and destroy livelihoods (International Union of Geological Science, 2007, cited in Australian Bureau of Statistics, 2008).

As the above quote highlights, those who live through a natural disaster never forget it. Wherever disaster strikes, contemporary media technologies connect us directly to this uniquely catastrophic experience and challenge our trust in the world. From the recent earthquakes in Nepal, New Zealand, and Haiti, tsunamis in Japan, floods in Australia, and hurricanes in the United States, natural disasters continually remind individuals, communities, policymakers, and researchers of the significant economic, environmental, social, and personal impacts. Critically, the severity of impacts is due not solely to extreme weather, but to the interaction between individual characteristics, community systems, the physical environment, and these events—the "convergence of hazards with vulnerabilities" (Jha et al., 2010, p. 339).

When thinking about who might be the most vulnerable during a disaster, emergency management research, policy, and practice have identified the aged, the very young, the poor, the socially and physically isolated, people with disabilities, and some ethnic groups as particularly vulnerable (Buckle, 1998). In this chapter, we focus particularly on one of these vulnerable groups—older adults. We review older adults' vulnerability and resilience, the disaster lifecycle, and introduce the novel creative methodology of poetic inquiry as a tool to better understand and communicate the lived experience of disasters. Finally, drawing on our own in-depth qualitative research with older adults evacuated during the 2011 and 2013 Brisbane floods, we map their experiences on the disaster lifecycle and share two poems created from the interview data. We argue that the unique poetic presentation might help reach and engage policymakers, practitioners, planners, the wider community, and older people themselves in a discussion about the disaster experience, vulnerability, and resilience, while also acting as a reminder of the importance of planning, preparation, and communication.

22.1 OLDER ADULTS' DISASTER EXPERIENCE

When disaster strikes, older adults are often uniquely vulnerable and disproportionality negatively affected. When Hurricane Katrina hit New Orleans in 2005, 71% of people killed were aged 60 years and over, yet older people only accounted for 15% of the local population at the time (Adams et al., 2011). Even if older residents survived the initial flood event, their health declined at four times the rate of a national sample of older adults unaffected by the disaster (Burton et al., 2009). While researchers might argue that these statistics from Hurricane Katrina illustrate issues of poverty or social class (rather than simply aging), statistics from the World Health Organization (WHO, 2008) show disproportionately poorer outcomes for older adults who experience a natural disaster. Over half of those killed by the 1995 Kobe Earthquake were older adults; during the 2003 Paris heat wave, the highest death rate was among those aged over 70 years, and after the 2004 Aceh tsunami, the highest death rate was adults aged 60 years and above (WHO, 2008).

Integrating Disaster Science and Management. http://dx.doi.org/10.1016/B978-0-12-812056-9.00022-1

In explaining older adults' vulnerability, researchers point to age-related physical and cognitive factors such as decreased physical health, chronic health conditions, mobility issues, and cognitive decline, as well as social characteristics including smaller social networks and socioeconomic changes (Fernandez et al., 2002). Combined, these factors mean that it is often much harder for older adults to effectively prepare for and move away from an impending disaster (Astill and Miller, 2017; Pekovic et al., 2007), with many also frequently drawing on past disaster experiences to justify a decision to delay leaving the familiar settings of home. At an evacuation shelter, an older person may present as independent and functional but can quickly deteriorate if chronic health conditions (e.g. diabetes and hypertension) are not well managed and there is limited access to proper medications or medical records. Thus, while older adults "aging in place" in their own homes can typically remain independent and cope in everyday situations, a disaster may push them over their coping threshold (Astill and Miller, 2017; Tuohy and Stephens, 2011, 2012).

22.2 THE DISASTER LIFECYCLE, VULNERABILITY, AND RESILIENCE

When policymakers, practitioners, educators, and researchers discuss disasters, the disaster lifecycle is typically grouped into four different phases/stages: preparedness, response, recovery, and mitigation (Neal, 1997). Mitigation is a proactive action designed to prevent and/or lessen the impact of a disaster, for example, through improved urban design, stricter building codes, hazard mapping and control, as well as taking out insurance. Preparation is advanced planning and organization, including both community and individual emergency response plans and actions (e.g. stocking food, water, torches). Response is putting preparedness plans into action (including evacuation, search, and rescue), while recovery is the final phase of the disaster lifecycle and is the action taken to "return to normal" as quickly as possible afterwards (e.g. clean-up, insurance claims, rebuild). These four phases are not necessarily linear, are often cyclical, and may overlap, with more vulnerable populations potentially moving between these disaster phases at different times.

In exploring the disaster experience, the focus is increasingly on understanding and fostering "disaster resilience"—the "ability to prepare and plan for, absorb, recover from, and more successfully adapt to adverse events" (National Research Council, 2012, p. 1). The word resilience originates from the Latin root resi-lire, which means to spring back; this focus on "bouncing back" and returning "to normal" as quickly as possible underpins contemporary disaster resilience discourse, alongside the mantra "building back better" (Roberts and Pelling, 2016). Of course, people are changed by traumatic experiences and often it is the most vulnerable who are least able to cope before, during, and after the disaster, as they are most disadvantaged in terms of both economic and social resources (e.g. reduced bridging and bonding social capital; Hawkins and Maurer, 2010; Brockie and Miller, 2017a). Conceptually, disaster resilience (the capacity to respond) is the opposite of vulnerability (a risk factor). While older adults have traditionally been viewed as more vulnerable during disasters, several recent studies have challenged this narrative and demonstrated their resilience (Astill and Miller, 2017; Brockie and Miller, 2017a,b; Henderson et al., 2009). For example, New Zealand research on a flood experience concluded that a lifetime experience provides "resources for psychological resilience and strength rather than vulnerability in the face of disaster" (Tuohy and Stephens, 2012, p. 33). Similarly, Cornell (2015) described how older people do not see themselves as vulnerable; they acknowledge and accept their limitations, but feel mentally prepared and confident in their ability to cope during an emergency.

To date, relatively little empirical work has focused explicitly on the unique needs, experiences, and contributions of older people during natural disasters. This knowledge gap is particularly worrying, given the intersection of two key global trends: climate change and population aging. Most researchers agree that climate change means that the frequency and intensity of natural disasters posing a threat to communities will increase (Roberts and Pelling, 2016). At the same time, the world population is rapidly aging with predictions that by 2050, the number of people aged 65 and over will double—from 650 million (11% of the global population) to 2 billion people, representing 22% of humanity (WHO, 2012). This will create a turning point in history—for the first time, older people will outnumber children aged 14 and under. Clearly, in addition to affecting the physical infrastructure and social functioning of exposed communities, disasters will increasingly impact the individual psychological and physical health of a growing older adult population (Doherty and Clayton, 2011).

22.3 APPLYING A LIFECYCLE, TEMPORAL AND POETIC LENS TO THE DISASTER EXPERIENCE

In this chapter, we document the vulnerability and resilience of older residents evacuated from their homes during the 2011 and 2103 floods in Queensland in two ways. First, we use the disaster lifecycle as an analytical framework to explore older residents flood experience, capturing key moments in each of the four stages: preparation, response, recovery, and mitigation. As many of these older adults have experienced repeated flood disasters in their lifetime (floods in 1955, 1974, 2011, and 2013), a unique temporal lens is applied to capture not only the impact of advancing age on vulnerability and resilience, but also the changing sociocultural context (e.g. characteristics of the local community and the impact of information communication technologies, such as social media).

Second, we use an unusual arts-based qualitative data analysis technique, variously termed as poetic transcription, poetic inquiry, or found poems (Faulkner, 2007; Richardson, 2002). Emerging approximately two decades ago, poetry-based research is where researchers "merge the tenets of qualitative research with the craft and rules of traditional poetry" (Leavy, 2009, p. 64). Drawing on interview transcripts, researchers rearrange the words of participants to create poems (or poem-like prose). In a world of text, poetry is a special language encouraging slower, multisensory, and visceral engagement that forces us "to pay attention, to listen, to be awake. Poetry cuts deep to the bone, makes vivid the flesh and sounds of the world and the pilgrimages of the mind and heart" (Glenn, 2016, p. 99). Moreover, during times of change (such as disasters) poetry can act as both "an anchor and an inspirational kite" (Borhani, 2013, p. 1). As a form of arts-based research, poetic inquiry provides a unique opportunity for public engagement, learning, and education, whilst fostering empathic understanding as "research participants' stories and lives become audible, visible, felt by them, in visceral and potentially lasting ways" (Sinding et al., 2008, p. 465). Although a growing number of disaster researchers are engaging with alternative creative arts-based approaches to engage the community, to our knowledge, to date we are the only researchers who have used poetic inquiry to convey the disaster experience (see Miller and Brockie, 2015).

22.4 THE CASE STUDY: THE 2011 AND 2013 BRISBANE FLOODS

This chapter explores the experience of older adults during the Brisbane floods. Located on the Brisbane River estuary and natural floodplain, Brisbane is the state capital of Queensland in Australia. It is home to over 1 million people, with the broader metropolitan area of South East Queensland (SEQ) housing approximately 2.5 million people. Our participants lived in the smaller city of Ipswich, population 190,000, located approximately 45 min west of Brisbane.

In the summer of January 2011, torrential rain led to the second highest flood in the last 100 years. The climate in SEQ is subtropical and the unexpected volume of rain combined with a conservative water catchment policy after several years of severe drought meant that the Wivenhoe Dam (purposely built for water storage and as part of a flood mitigation strategy after the 1974 floods) was too full to operate effectively (van den Honert and McAneney, 2011). Three-quarters of the state of Queensland were declared a disaster zone—for context, this is an area of approximately 500,000 km^2 (larger than France and Germany combined). Thousands of people were evacuated from over 70 towns and cities, with major flooding through most of the Brisbane River catchment area when the river peaked at 19.25 m. Approximately 22,000 homes and 7600 businesses across 94 city suburbs were affected with the damage estimated at A$ 2.38 billion (Apelt, 2011; Brisbane City Council, 2012) and the official death toll reached 35 people (Apelt, 2011).

As a slow-onset disaster limited to riverside suburbs, many residents could safely evacuate homes and businesses. An estimated 200,000 anonymous "mud-army" volunteers (Brisbane residents who were experiencing only constant rain and short-term power outages) assisted in this evacuation and cleanup process. Working alongside emergency services, police, and the "real army", these "mud-army" volunteers took on "legendary status as the grassroots heroes of the floods" (Besley and Were, 2014, p. 45), helping ensure more socially isolated and vulnerable residents without strong local support networks were not forgotten. The quote below provides a vivid description not only of the scale of the flood, but the goodwill and support from the broader community.

My South Brisbane community was engulfed by natural disaster in January 2011. At that time, the floodwaters of the swollen Brisbane River came up through drains, spilled over parks and roads, and invaded homes. For three tense days, we watched from a distance as evacuees trying to gauge how high the water level was inside our house. When the flood waters receded, floods of people took their place. Friends, relatives, work colleagues, and more distant acquaintances arrived on our doorstep armed with mops, brooms, shovels, and a determination to assist our clean-up. Over the following days and weeks, this burden-sharing continued and helped us manage the disorientation, despair, and emotional exhaustion that accompanied our efforts to find and furnish a new home and reestablish family life. Although we had lost our home and most of our material goods, we felt greatly supported by the many people who had shown us generosity (George, 2013, p. 41).

After the 2011 floods, a large free exhibition entitled *Floodlines: A living memory* was held at the State Library of Queensland in Brisbane and then toured the state as part of the process of "recovering, remembering, reflecting, and rebuilding". This participatory exhibition was a "digitally dense and sensory experience" and included a soundscape, wall of digital stories, and an app for tracking the rise and fall of floodwaters (Besley and Were, 2014, p. 44). The exhibition encouraged a strong interactive dimension as community members contributed their flood images, items, and ideas on how to better plan for and respond to future floods. Donated exhibition items included mud-covered mobile phone, a dirty shovel, and gumboots used during the cleanup effort, and a marker cone used to block a flood-hit street. The exhibition encouraged residents to view themselves as agents of change, not victims, with a strong emphasis on the importance of building community and social connections with neighbors before a disaster—for example, one installation of stickers said "*don't wait for a flood to say g'day*" (George, 2013). Unfortunately, only 2 years after the 2011 floods, rainfall associated with ex-tropical Cyclone Oswald resulted in another major flood in 2013. While there were fortunately no casualties, these floods were a mere 2 years from the 2011 experience (in contrast, the 2011 floods were nearly 40 years after the last major floods in 1974). After two major floods in as many years, floods and the possibility of future floods were very much on local resident's minds. Below, we outline our qualitative research methodology to better understand the lived experience of 10 older Brisbane residents during the 2011 and 2013 floods.

22.5 PROJECT OVERVIEW

Theoretically, our overall frame of reference was a qualitative phenomenological approach that examines how people make sense of their major life experiences. Phenomenology focuses on the "lived experience" and is particularly appropriate for investigating unstudied populations (Smith et al., 2009). We also drew on a sociological framework, Elder's (1998) life course perspective that emphasizes the intersection of personal biography with social, cultural, and historical factors. As just over half the sample ($n = 6$) had experienced the 1974 floods (and two of the 1955 floods), our study offered a unique sociohistorical perspective on the disaster experience and how interactions among time, place, age, and lives intertwined to affect each individuals' flood experiences. In-depth narrative interviews were conducted with 10 older adults (six women and four men) evacuated from their homes during the 2011 and 2013 floods in Brisbane, Australia. The average age was 73 years (ranging from 67 to 83 years), with two married couples (interviewed as couple) and the remaining widowed. All had adult children (average of 3.5, range 2–7 children), had lived in their current detached home for more than 15 years (all but 1 were home owners), and all were retired. After the 2011 flood, four were out of their homes for a week, with six displaced for between 7 and 12 months.

After obtaining formal ethical clearance, residents were recruited through a combination of purposive and snowball sampling including personal contacts, word of mouth, industry contacts, radio interviews, and a targeted mail-out to the 356 houses listed in the 2011 Ipswich flood map. Interested older residents were instructed to contact the researcher to participate in a face-to-face interview exploring their flood experiences. Those meeting the inclusion criteria (aged over 65 years, lived independently in own home, and evacuated in both 2011 and 2013 Queensland floods) were mailed an information package and interviewed in their own home by the second author (a registered psychologist). Before the interviews were conducted, during initial phone conversations with the participants, they were encouraged to collect any photographs or memorabilia they might like to share and add to their stories, during the interview. The presence of these visual images and/or memorabilia provide "resource tool" to access deeper information about the floods, as did holding the interviews at residents' homes where the flood events occurred. During the interviews, residents would frequently gesture at walls to indicate flood heights, show their emergency preparation kit' and gesture at valued possessions they had been able to save from the rising flood-waters. These audio-recorded interviews ranged in length from 60 to 190 min. All started with a single open-ended question: "so, tell me about your experience with the Ipswich floods?". The over-arching focus was to understand how these older adults had "made it" through the floods, with follow up prompts exploring the disaster lifecycle as well as their assessment of the impact of the flood (immediate and ongoing), their emotional and physical health, social and broader community support. Critically, the data from one interview were compared with data from another, resulting in constant comparison enabling the researcher to ask increasingly targeted questions. All interviews were fully transcribed by the second author.

Data was analyzed using a combination of thematic and constructivist grounded theory (CGT), an iterative process that involves ongoing coding, constant comparison, and analysis to identify similarities and differences in participants' lived experiences. For clarity, we grouped the concepts into the phases of the disaster lifecycle and, through the process of poetic inquiry, created a poem for each phase. In many ways, poetic inquiry can be viewed as an extension of more traditional qualitative data analysis. The first step is data immersion, with transcripts read repeatedly and key "nugget" words and phrases identified. The second step is to create more rhythmic and engaging sentences, taking interesting phrases from throughout the transcript, altering word and sentence order, and arranging and rearranging each line of the poem for maximum impact. If words or phrases are repeated, it is because participants repeated them. Thus, in creating the poems, we followed approach and only used the exact words of our participants. While crafting this poem, we focused on conveying the raw emotions of the flood experience, and ending in a way which echoes the major theme of the participants' story. Our method and analysis has been described in detail in several recent publications (Brockie and Miller, 2017a,b; Miller and Brockie, 2015).

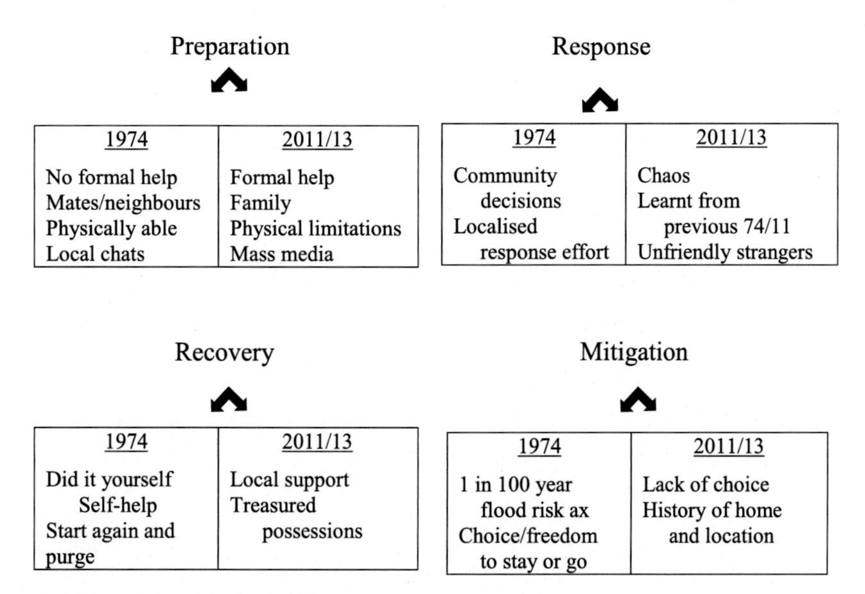

FIGURE 22.1 Visual representation of changing disaster lifecycle, 1974–2011/2013.

22.6 A POETIC APPROACH TO THE CHANGING DISASTER LIFECYCLE (1974–2011/2013)

This research highlighted how these older adults' experiences of floods had changed markedly over the years, particularly, the different ways the risk of a flood is communicated in modern society and how their increasing age-related frailty was impeding both disaster preparation and recovery. Below we show how the four phases of the disaster lifecycle have changed over time, with a particular focus on the rebuild process and the challenges these residents had in accessing support when multiple natural disasters hit their state. For each phase, we present one poem created through poetic inquiry to more emotively convey these older peoples vulnerability and resilience during the flood disaster.

Fig. 22.1 visually illustrates the four phases of the disaster lifecycle and how these older adults' experience of floods has changed markedly over the years, as their increasing age-related frailty and ill health has impeded both preparation and recovery. Of course, in addition to changes in their own physical and coping abilities, there have been significant societal changes in the nearly 40 years between the 1974 and 2011 floods. Our participants described how, in 1974, there was very little formal support (from governments, nongovernment organizations, insurance, etc.), with informal support from friends and family critical. As one explained, you simply "relied on your workmates to come and give you a hand to rebuild" (P4). In contrast, in 2011 and 2013, there was a heavy reliance on "friendly strangers"—people who drove and walked by, asking if they could help. These changes are particularly reflected in the preparation and response categories in Fig. 22.1 which highlight how 2011/2013 felt more chaotic and some older residents felt their planning, preparation, and evacuation process was complicated by "unfriendly strangers" who were simply "gawking" and taking photographs of them and the flood waters.

22.6.1 Phase 1: Preparation

What am I going to do?

I'm 74,
1939, I was born.
And, when you get older
you can't lift things, by yourself
you're not as strong
as you used to be.
And, when it starts to rain
for a long time
And, when they predict a storm -
say strong winds,
could be flash flooding

its hair raising.
you think—is it on again?
you think—what am I going to do?
you think—what if it is just the Mrs and I here?
you think—oh, what am I going to do?
got to sink
or
swim.
Charles, Age 74.

Poem 1 "*What am I going to do*?" captures the unique challenge of a slow-onset disaster, like a flood. While there is often time for planning and preparation, there is also time for anxiety and anticipation, and the reality of aging was making the preparation process much harder for these older residents. All described difficulties of preparing, packing, and moving furniture explaining that declining health and mobility quite simply meant they could not physically accomplish the task. As one older man explained: "When you get older you're not as strong as you used to be... You run out of puff... You think, what am I going to do... what do I have to do, what have I got done... it's a bit hair raining".

At the same time, although there was a sense of anxiety around the floods and concern about how severe the impact might be, the slow-onset and past disaster experience brought a degree of comfort and calmness. Practically, many had a plan regarding when to leave. For example, one older man described his own preparation system: "after 74 I measured the height of that gully to the top of the road here, so I know how high it is to our front yard. So when you see the water coming up the gully and coming across the road you know it's time to move". Similarly, another described how "it was all sitting back in my mind, where all the flood waters came too". One couple described how their daughter "booked a truck" for them to remove all they could from their house: "we took everything, even down to the plants, we nearly killed ourselves, but we got everything into the truck, it was an effort you know, we had to get out quick". Emotionally, although most described feeling calm, prepared and stoic, there was an undercurrent of apprehension, powerlessness, stress, and worry about the size and impact of the flood, and their ability to respond.

Second, communication and warnings changed. No longer were police walking the street instructing people to leave, now this information was primarily disseminated through social media and mobile phones. Most described getting a text or voice message on their phone alerting them to the impending disaster and advising them to leave; however, this was often of limited utility. When asked by the interviewer what time of day the message came through, one replied: "I don't know, it came through as an SMS on the mobile phone and um ... mine is always turned off". Such generational differences were further reflected in how this cohort received information, with a strong preference for the radio (or as they referred to it, the wireless). The national broadcaster was praised, as was how they would put "specific music beforehand and you know to listen then". Social media was not a key factor, with no one using Twitter and Facebook managed by younger family members. For example, one older woman recalled how her step-daughter posted on Facebook to recruit friends to help pack her house and move furniture. About half felt there was little support beforehand, in that they were "doing their own thing" to prepare—"when you get a phone call to say get out, you don't think of anything, just you've got to go. We sort of threw clothes in and kept throwing things in till we filled up the cars and had to go". As the quote below illustrates, one older woman felt very alone in light of the impending disaster, given her physical limitations and lack of social networks.

I can't lift furniture you know, impossible. I did what I could, I loaded my car until I couldn't get any more in it and I just had to say goodbye to the rest of it. That's all you can do, you know you put it up in hope. So no, nobody came to help me with that I didn't have anyone. "I don't see anyone around here. I couldn't tell you the name of anyone apart from next door and people in the house next door to her. I have been here nine years and I wouldn't know them if I fell over them".

22.6.2 Phase 2: Response

On my own

I am on my own.
Been here nine years
But, I don't see anyone -
I couldn't tell you,
the name of anyone –
wouldn't know them,
if I fell over them.

I am on my own here.
There were a lot of people,
the police and emergency crews,
going around
but they never stopped.
I did get a text message,
to say get out.
I did get that.
Carolyn, 67 years

Although Carolyn's poem "On my own" poignantly captures her sense of feeling alone and unsupported during the flood, most residents described the presence of both officials and technology to support and warn them. One described how there were "neighborhood meetings at 3 am" to look at and discuss the water and when to leave, while others drew on their experience of past floods to decide when they might leave their homes. Several older men discussed how they knew where all the flood waters came to, and were ready to leave when the water hit a specific marker. In very practical terms, all residents reported power loss, challenges moving furniture and belongings, difficulties leaving and accessing their homes, and just a general feeling of chaos as they left and later returned to heavily flooded homes. With no work or social commitments linking them with the neighborhood, some felt very isolated and forgotten by the broader community describing their area as "a forgotten island".

Residents also recalled how they had easy access to major local buildings (golf clubs, school halls, etc.) in 1974. In contrast, those local evacuation points were not available in 2011/2013 and were missed by residents. As P4 commented: "they wouldn't open it. They said security, to a certain degree I think is fair enough… but you've got 1000 houses this side, and then 2000 homes the other side… they (police officers) should have keys to major buildings, in case something happens". Instead there were formal evacuation centers, which were described very positively by those who used them. The kindness and generosity of strangers was memorable, with a woman sharing the story of a man "in a blue shirt" who made her a cup of tea and talked with her. The memory of his kindness has stuck with her, and is her enduring, defining memory of the floods.

When I got there I was number 19, but I think at the end they had about 280 or something like that, but a lot of them were travelers that couldn't get through, they had nowhere to go as the motels were full. They did an excellent job up there, it is the Red Cross that run it. They fed us and looked after us really well. They had counselors there if you needed them, and they had a St Johns Ambulance in a room for people with medical problems and there was always somebody there, they rotated the shifts. They were very, very good.

22.6.3 Phase 3: Recovery

Help

we had a lot of help
they came from everywhere
relations and friends
and strangers
in buses
from Brisbane.
they would walk the streets
with brooms and buckets
saying -
what will we do,
what will we do?
they came down
and helped.
community spirit
in action
Mary, 83 years

There were significant differences in the recovery process; four participants were evacuated for less than a week, while others were out of their home for over 12 months. In terms of who helped in the first few days and weeks afterwards, most described a change: from family and friends in the 1955 and 1974 floods to more "friendly strangers" in 2011 and 2013.

This is captured in Mary's poem above, where she describes help with the cleanup process, not only from friends and family but also from strangers. Although a few expressed concerns about the trustworthiness of strangers, most were greatly appreciative of those in the so-called "mud-army", who came up in buses from Brisbane to help.

Many praised the practical and emotional social support they received from family and friends, noting "it's not what you know, it's who you know. If you've got friends who are willing, then that's all you need". Indeed, one couple described how two sets of distant friends gave their daughter significant amounts of cash (A$ 500 and 1000) to pass on to them, to help them out in this challenging time. However, approximately one-third of our older participants described very limited support from their existing social networks and a sense of truly being alone and unsupported during this challenging time. This sense of isolation is reflected in the poem "On my own", as well as the words of one older woman who poignantly explained that a disaster shows who your real friends are: "as you go through life you have a heap of friends, you get lots of friends and when something goes wrong, you find out who your friends are. You can count them on one hand basically." She felt alone, frightened, and unable to cope with the magnitude of the recovery task in front of her and her experience was not of "friendly strangers", but of "unfriendly strangers" who watched her struggle and did not help.

The rubbernecking. That's what annoyed me in the first place, people coming to look at the water and because they couldn't get through they would stop. They stood there watching my neighbour and myself, we are both on our own, and they stood there watching us lug out furniture and god knows what and never got out of their cars to help. I felt like hitting them. Now I think I should have been out there with a coffee trolley - I could have made a fortune.

In terms of broader recovery resilience, the majority described how they made a choice to persevere rather than give into despair, viewing themselves as survivors rather than "victims" of the floods. As one explained, "my philosophy is you just carry on anyway… [take the] ups and downs". The presence of valued keepsakes facilitated resilience and the recovery process, with one widow explaining how the only thing that was saved was an antique dining room suite from her late husband's aunt. It gave her great pleasure to have saved that, while other older residents also emphasized the personal and symbolic importance of saving photographs and other valued mementos.

22.6.4 Phase 4: Mitigation

Water

When I left
it was just
little ripples
sitting there
not doing much,
grey water.
Yet, it really come in
tipped over
the fridge
the cupboards.
you wonder
how the hell it does that?
to pick up beds, mattresses, clothes,
the sewing machine?
all picked up and swung them around
How does it do that?
Unless you have seen it
it is hard to comprehend
the amount
the strength
the power
of the
water

Carolyn, 67 years.

Poem 4 "Water" captures the quiet yet devastating power of the flood waters, highlighting how if you have not lived through it, it is hard to comprehend the impact and strength of the water. In looking to the future, these older residents all feared a time when they would not be able to effectively look after themselves during a disaster. As many of these homes are relatively modest properties of older construction, they explained that they could not afford to move elsewhere and were resigned to the risk that came with living in low-lying properties near a river. As one explained, while he has been here since 1949 and loves the place, it's in "a position where nobody would want to buy it. They would say oh no you've been flooded twice. Two and a half times." Residents also described mixed experiences with insurance and the rebuild process, explaining how times had really changed. As the quote below illustrates, there was no insurance in 1974 and you simply did it yourself, with a little help from your friends and workmates.

> *In 74, it wasn't half the trouble and there was no insurance in those days, well there was but not many had it and they wouldn't insure anybody for a couple of years after that. '74, there was nothing, you just relied on your workmates in 74 to come and give you a hand to rebuild... I came in here with a big hammer... this was all fibro, the asbestos (P4).*

In 2011, a lucky few (although they had to leave their home for three days) were not significantly affected "physically, mentally, property wise" and thus did not have to engage in the formal insurance claim process (although they did access the A$ 1000 from the federal government for those affected by the power blackout and lost food). Approximately, one-third reported a very positive insurance experience, with "no complaints whatsoever", "we had $74,000 contents and there was a cheque in the mail, you know it was put into our bank, I shouldn't say cheque in the mail". The other third described an endless process of "running-around", of phone calls, missed meetings, and just lack of clarity about what to do, who to talk, and the final outcomes. Participant 9 felt the entire insurance business process was just "lack-luster" and described how, as she did not know the "game", they missed out: "the assessor, he said list everything you lost, age and cost. He gave me one sheet of paper and I needed about 20. He didn't tell me it was new for old, I was insured for about $68,000 and I got about $26000... at what I paid, not at replacement value, but I will know next time" (P9).

One older man was quite upset about how insurance companies defined "flood" and the specific conditions under which they would or would not pay; "they were arguing whether this one was riverine water coming up the gully here, and then when it gets up into your house they say no it was a flash flood. Well the water running down the gully here, it can't get away because of the river" (P5). He explained how, in dealing with the annoying onlookers, he would walk out and say "is anybody here a hydrologist, study water?" He wanted to have a debate with them and have them prove some of the "stupid things you stooges, working for the insurance companies, are talking about." He described an unsatisfactory town meeting about the issue, attended by state and federal politicians, and how "it was a bloody disgrace it was, and people were yelling out and crying. It was a bit rugged". Others were frustrated at the transparency of the work that the insurance company approved and the pace at which they were asked to complete and sign paperwork to authorize the repairs. Complicating matters was that Queensland experienced multiple large natural disaster events around the same time. Thus, many of these older residents were assigned a builder but then had to wait six months (forced to live in the homes of friends/family) as the building workforce was reassigned to far north Queensland to deal with the aftermath of Cyclone Yasi.

> *There was a big hiccup, I was assigned to one early, and then Yasi happened so a lot of the workforce went up there, so then it was 5 months down the track before I even met a builder. He had 11 houses here in Goodna at the same time. He was very good and very easy to deal with, the only thing I would say in the future, is you really have to be careful of what is on the scope of work, as they will only do what is on the scope of work, and you have about a minute to go through the paperwork and sign it. You know my landing at the back is a mess and my front landing is a mess, because it wasn't on the scope of work.*

In terms of recovery and mitigation, most were philosophical about accepting what "Mother Nature" throws at them. P1, a loner at heart and not connected to her local community, explained her mitigation plan: if it happens again, she would leave her rented home: "I think I would have to accept it and have to move on and find something else. I've got my tent and stove and sleeping bag…I just need a bed, and I might just get the car and travel round".

22.7 LEARNING FROM OLDER AUSTRALIANS' FLOOD EXPERIENCE

At the outset, it is important to acknowledge that we explored older adults' vulnerability and resilience within the context of relatively affluent Australia, which has an effective, well-trained, and well-resourced emergency response team combined with generally high levels of community and institutional trust. That said, Australians currently have an estimated lifetime exposure to natural disasters of one in six (McFarlane, 2005) and the forecast for climate change suggests that this exposure will only increase. At a policy level, Australia's *National Strategy for Disaster Resilience* explicitly describes disaster management to be the responsibility of the individual. This self-help approach is a challenge for older adults who are especially reliant on external support during an emergency (Astill, 2017), with researchers raising concerns about this shift from government's responsibility for safety and welfare to one of personal responsibility and accountability. Like

Douglas (1992) and Lupton (2013), we question the efficacy of disaster management policies that seem to focus on resilience so wholeheartedly that they fail to acknowledge or address the natural increased vulnerability of older adults due to the breakdown of more traditional certainties (e.g. support structures, finances, social capital, physical, and mental health).

Reflecting on our findings, there are three core messages. First, a critical finding was that the experience of the same natural disasters—the 2011 and 2013 floods—varied greatly between older participants. For many older adults, age-related losses (physical, psychological, financial, and social) can create greater vulnerability during the disaster lifecycle and increase the need for more social support avenues. The effect of trauma can make it difficult for any age cohort to absorb information and make decisions, while the effort involved in flood cleanup and recovery is exhausting. Although the participants in this study did not overtly discuss these vulnerabilities (perhaps a reflection of their resilience and generally good physical health), they did offer the occasional joke or passing remark about starting to notice physical frailty, the consequent limitations and the increased need for assistance from their networks.

It is important to note that, overall, our cohort of older adults coped well; this is consistent with research conceptualizing age as a predictive factor, suggesting that a "lifetime of experience provides resources for psychological resilience and strength rather than vulnerability" (Tuohy and Stephens, 2012, p. 33). Indeed, previous disaster experience and the presence (or absence) of community networks and support from family, friends, and the community were central features of individual stories of resilience. Their narratives demonstrate that resilience was not merely a trait or behavior, but a multidimensional and ongoing process. Indeed, the findings suggest that an older adult may struggle with one area—such as physical or financial wellbeing—but be highly socially resilient and psychologically resilient. Stories of past traumas were described matter-of-factly by residents and the recent floods were filed away as "just another experience"; indeed, the tone of these respondents' resilience was nicely encapsulated in one participant's statement: "what does not kill us only makes us stronger". Of course, some described being extremely isolated and struggling through alone. Two poems in particular ("What am I going to do" and "On my own") palpably capture the growing sense of fear and apprehension as these older adults realize that the challenge of preparing for and living through the disaster flood event is ultimately up to them. In a very real way, these poems convey the challenge and lived experience of Australia's self-help disaster management policy.

Overall, despite decreasing personal and social resources, these generally active and healthy community-dwelling older adults adapted and managed to maintain independence during the disaster lifecycle. However, rapid global population aging means that the number and proportion of older people needing help during a disaster is increasing significantly. In Australia, 15% of the population (3.5 million people) is currently aged 65 and over; by 2054 this will increase to 21% (8.4 million people), with those aged 85 years and over doubling by 2032 (Australian Bureau of Statistics, 2013). It is likely that many of those older people will not cope as well as our sample during a disaster, thus putting more strain on local communities and formal emergency response systems. With research typically showing "an almost universal lack of consultation from older community members in developing emergency response plans" (Toner and Alami, 2010), we reiterate the importance of genuinely participatory, inclusive, and reflective engagement with older and potentially more vulnerable members of the community. Older adults contribute to their communities through decades of accumulated experience, knowledge, and understanding. Taking into consideration the generational experiences of this age group, it is reasonable to suggest that by surviving traumas such as World War II, economic depression, and numerous natural disasters, this age group possesses a wealth of strategies for coping with the challenges associated with natural disasters. This insight makes them an essential resource and potential partner in developing emergency preparedness and response programs, yet they described a lack of consultation and inability to contribute input into decision-making about local or state-wide disaster management policies relating to older adults.

Second, this research captures a unique temporal and historical perspective as some participants had lived through four floods (1955, 1974, 2011, and 2013) over five decades. In reflecting on their unique lived experience of disaster, residents highlighted changes in communication, technology, and sources of social support. Communication methods have evolved from a simple chat with the locals and/or people with megaphones walking the streets to the use of text messages, Facebook, and mass media. However, many older adults simply do not engage with these new forms of technology, not on Facebook and describing how they often left their cellphone switched off (and thus missing official text messages directing them to leave). They made individual phone calls for assistance, whereas younger family members turned to mass social media for information and to recruit help. As these older adults preferred to use much more traditional methods of communication (landline, telephone, radio, and word of mouth), there is a critical need for further research into how rapidly changing technology can be better integrated into older adults' worlds for the purposes of disaster management and communication.

These older adults reported that big decisions (when to evacuate, recovery processes, etc.) were mainly influenced by their own past experiences and members of their social support network, not the mass media. Nearly 40 years ago, Wenger (1978) described how some communities and individuals with repeated exposure and experience with disasters learn to read and respond to the environment, developing "disaster subcultures" best viewed as "a blueprint for residents' behaviour before during and after impact" (p. 41). At its best, the tangible and intangible beliefs, practices, routines, and folklore that comprise a disaster subculture may facilitate preparation, preparedness, and a disaster-resilient culture; at its worst, complacency, risk-taking, and unsafe patterns of behavior. The disaster subculture amongst our sample of older adults was resilient, generally responsible, and conservative, although our research identified many localized beliefs and practices that were potentially risky (e.g. decision to evacuate based on when water reaches a specific local landmark, desire to access shared community spaces rather than formal evacuation centers, etc.). Additionally, these long-term residents generally reported a strong place attachment and emotional connection to their community, verbalizing a desire to stay in their homes long term (despite living in a flood-prone zone; notably, economic considerations were a factor as housing in this area is affordable for the region).

Consistent with a large body of research, this study demonstrated the importance of community networks during and after a disaster, highlighting the value of community bonds, support, and social capital (George, 2013; Hawkins and Maurer, 2010; Meyer, 2013). These older residents identified a temporal change in the sense of community (feelings of belonging and attachment) and social support, explaining that the 1974 and 2011/2013 flood experiences were very different. In 1974, the majority of support was highly dependent on friends, family, and neighbors, whereas in 2011/2013 there was a great deal of reliance on "friendly strangers". Notably, while there are obvious issues with relying on memories of the past and reinterpretation of events, it would be fair to say that the community and world these older adults live in has significantly changed over the past four decades, with a general decrease in community interactions and social capital. Indeed, reflecting on the 2011 Brisbane floods, also concluded that stories of local support were rarer than those of strangers (the mud-army) and institutional actors (state emergency services and the army). As disaster recovery is frequently dependent on people's social networks (their social capital), this research reinforces the importance of predisaster mitigation activities designed to build a sense of community and social connections amongst neighbors.

Third, the unique application of poetic inquiry allowed for these data-rich stories to be reported in novel, engaging, and emotive ways, highlighting the vulnerable but resilient voices of these older adults. As the first researchers to utilize poetic inquiry as an analysis tool in the context of aging and the disaster experience, our hope is that our work inspires others to experiment with this alternative but engaging way of presenting qualitative data (as a starting reference, see Leavy, 2009). Compared to more traditional quantitative and qualitative reporting, we would argue that the descriptive, evocative, and authentic poetic approach may better reach and resonate with the broader community, motivating both disaster preparedness and recovery. We hope our description of the poetic inquiry process might encourage others to integrate creative arts-based approaches (e.g. poetry, creative writing, videos, music, drama, etc.) into their research process and better engage the wider community in a conversation on disaster preparation, response, recovery, and mitigation.

In conclusion, let us remind readers that a disaster occurs when a physical hazard meets a vulnerable population. Although our older adults described significant resilience, there is no doubt that they struggled both physically and emotionally, were spatially and often technologically isolated, with limited participation or power in local political decision-making. By sharing their personal narratives, especially in poetic form, we have highlighted the unique experiences, vulnerability, and resilience of older adults during disasters. Next to the critical "hard" infrastructure of regulations, control systems, technology, and physical risk-mitigation assets, the "soft" infrastructure of disaster subcultures (values, beliefs, evolving social practices, social networks, connections) contributes significantly to vulnerability, resilience, and risk reduction during the disaster lifecycle. While multiple environmental, social, cultural, political, economic, and local characteristics combine to uniquely determine vulnerability, risk, and resilience during disasters, as Bendix (1990) explains below, our unique personal narrative of the disaster experience is also critically important.

> *Personal narratives are…the primary means at an individual's disposal to regain order out of chaos. While fire trucks, bulldozers, construction crews and money allow for removal of rubble and rebuilding of physical structures, personal narratives accomplish the same work with our heads and hearts (Bendix, 1990, p. 333).*

REFERENCES

Adams, V., Kaufman, S., Van Hattum, T., Moody, S., 2011. Aging disaster: mortality, vulnerability, and long-term recovery among Katrina survivors. Med. Anthropol. 30 (3), 247–270.

Apelt, C., 2011. January 2011 flood event—report on the operation of Somerset Dam and Wivenhoe Dam. St Lucia, Australia: Uniquest Pty Ltd., Brisbane.

Astill, S., 2017. Ageing in remote and cyclone-prone communities: geography, policy, and disaster relief. Geogr. Res. doi: 10.1111/1745-5871.12228.

Astill, S., Miller, E., 2017. "The trauma of the cyclone has changed us forever": self-reliance, vulnerability and resilience among older Australians in cyclone prone areas. Ageing Soc., https://doi.org/10.1017/S0144686X1600115X.

Australian Bureau of Statistics, 2008. Natural Disasters in Australia. 1301.0—Yearbook Complete, 2008. ABS, Canberra.

Australian Bureau of Statistics, 2013. Population Projections, Australia, 2012 (base) to 2101. ABS, Canberra, ABS cat. no. 3222.0.
Bendix, R., 1990. Reflections on earthquake narratives. West. Folklore 49 (4), 331–347.
Besley, J., Were, G., 2014. Remembering the Queensland floods: community collecting in the wake of natural disaster. In: Convery, I., Corsane, G., Davis, P. (Eds.), Displaced Heritage: Responses to Disaster, Trauma, and Loss. Boydell Press, Woodbridge, Suffolk, UK, pp. 41–49.
Borhani, M., 2013. Riding the bus, writing on the bus: a self in transition. UNESCO Observatory Multi-Disciplinary J. Arts 3 (2), 1–21.
Brisbane City Council, 2012. Fact sheet: Brisbane City Council 12-month flood recovery report. Brisbane City Council, Brisbane.
Brockie, L., Miller, E., 2017a. Older adults' disaster lifecycle experience of the 2011 and 2013 Queensland floods. Int. J. Disast. Risk Reduct., DOI: http://dx.doi.org/10.1016/j.ijdrr.2016.08.001.
Brockie, L., Miller, E., 2017b. Understanding older adults' resilience during the Brisbane floods: social capital, life experience and optimism. Disast. Med. Public Health Prep., DOI: https://doi.org/10.1017/dmp.2016.161.
Buckle, M., 1998. Disaster Resilience: An Integrated Approach. Social Science, New York, NY.
Burton, L., Skinner, E., Uscher-Pines, L., Lieberman, R., Leff, B., Clark, R., et al., 2009. Health of medicare advantage plan enrollees at 1 year after Hurricane Katrina. Am. J. Manage. Care 15, 13–22.
Cornell, V., 2015. What do older people's life experiences tell us about emergency preparedness? Aust. J. Emerg. Manage. 30 (1), 27–30.
Doherty, T.J., Clayton, S.D., 2011. The psychological impacts of global climate change. Am. Psychol. 66, 265–276.
Douglas, M., 1992. Risk and Blame. Psychology Press, New York, NY.
Elder, G.H., 1998. The life course as developmental theory. Child Dev. 69, 1–12.
Faulkner, S.L., 2007. Concern with craft: using Ars Poetica as criteria for reading research poetry. Qual. Inq. 13 (2), 218–234.
Fernandez, L.S., Byard, D., Chien-Chih, L., Benson, S., Barbera, J., 2002. Frail elderly as disaster victims: emergency management strategies. Prehospital Disast. Med. 17 (2), 67–74.
George, N., 2013. "It was a town of friendship and mud": "flood talk", community, and resilience. Aust. J. Commun. 40 (1), 41–56.
Glenn, L., 2016. Resonance and aesthetics: no place that does not see you. In: Galvin, K., Prendergast, M. (Eds.), Poetic Inquiry II: Seeing, Understanding, Caring: Using Poetry as and for Inquiry. Sense Publishers, Rotterdam, The Netherlands, pp. 99–105.
Hawkins, R., Maurer, K., 2010. Bonding, bridging and linking: how social capital operated in New Orleans following Hurricane Katrina. Br. J. Soc. Work 40 (6), 1777–1793.
Henderson, T., Roberto, K., Kamo, Y., 2009. Older adults' responses to Hurricane Katrina: daily hassles and coping strategies. J. Appl. Gerontol. 29, 48–64.
Jha, A.K., Barenstein, J.D., Phelps, P.M., Pittet, D., Sena, S., 2010. Safer Homes, Stronger Communities: A Handbook for Reconstructing After Natural Disasters. World Bank, Washington, DC.
Leavy, P., 2009. Method Meets Art: Arts-based Research Practice. Guilford Press, New York, NY.
Lupton, D., 2013. Risk, second ed. Routledge, London, UK.
McFarlane, A., 2005. Psychiatric morbidity following disasters: epidemiology, risk and protective factors. In: López-Ibor, J.J., Christodolou, G., Maj, M., Sartorious, N., Okasha, A. (Eds.), Disasters and Mental Health. Wiley, West Sussex, pp. 37–63.
Meyer, M., 2013. Social capital and collective efficacy for disaster resilience: connecting individuals with communities and vulnerability with resilience in hurricane-prone communities in Florida. Unpublished doctoral dissertation. Colorado State University, Fort Collins.
Miller, E., Brockie, L., 2015. The disaster flood experience: older people's poetic voices of resilience. J. Aging Stud. 34, 103–112.
National Research Council, 2012. Disaster Resilience: A National Imperative. The National Academies Press, Washington, DC.
Neal, D.M., 1997. Reconsidering the phases of disaster. Int. J. Mass Emerg. Disast. 15 (2), 239–264.
Pekovic, V., Seth, L., Rothman, M., 2007. Planning for and responding to special needs of elders in natural disasters. Generations 31 (4), 37–41.
Richardson, L., 2002. Poetic representations of interviews. Health Care Women Int. 31 (11), 981–996.
Roberts, E., Pelling, M., 2016. Climate change-related loss and damage: translating the global policy agenda for national policy processes. Clim. Dev. doi: 10.1080/17565529.2016.1184608.
Sinding, C., Gray, R., Nisker, J., 2008. Ethical issues and issues of ethics. In: Knowles, J.G., Cole, A.L. (Eds.), Handbook of the Arts in Qualitative Research: Perspectives, Methodologies, Examples, and Issues. Sage, Los Angeles, CA, pp. 459–467.
Smith, J.A., Flowers, P., Larkin, M., 2009. Interpretative Phenomenological Analysis: Theory, Method and Research. Sage, London, UK.
Toner, J., Alami, O., 2010. A primer for disaster and emergency preparedness and evidence-based care practices in geriatric mental health. In: Toner, J.A., Mierswa, T.M., Howe, J.L. (Eds.), Geriatric Mental Health and Emergency Preparedness. Springer, New York, NY, pp. 3–30.
Tuohy, R., Stephens, C., 2011. Exploring older adults' personal and social vulnerability in a disaster. Int. J. Emerg. Manage. 8 (3), 60–74.
Tuohy, R., Stephens, C., 2012. Older adults' narratives about a flood disaster: resilience, coherence, and personal identity. J. Aging Stud. 26, 26–34.
van den Honert, R.C., McAneney, J., 2011. The 2011 Brisbane floods: causes, impacts and implications. Water 3 (4), 1149–1173.
Wenger, D.E., 1978. Community response to disaster: functional and structural alterations. In: Quarantelli, E.L. (Ed.), Disasters: Theory and Research. Sage Publications, London, UK, pp. 17–47.
WHO, 2008. Older Persons in Emergencies: An Active Ageing Perspective. World Health Organization, Geneva, Switzerland.
WHO, 2012. World Health Day 2012: Ageing and Health. World Health Organization, Geneva, Switzerland.

FURTHER READING

Bonanno, G., Brewin, C., Kaniasty, K., La Greca, A., 2010. Weighing the costs of disaster: consequences, risks and resilience in individuals, families, and communities. Psychol. Sci. Public Interest 11 (1), 1–49.

Carroll, B., Morbey, H., Balogh, R., Araoz, G., 2009. Flooded homes, broken bonds, the meaning of home, psychological processes and their impact on psychological health in a disaster. Health Place 15 (2), 540–547.

Miller, E., Buys, L., 2008. The impact of social capital on residential water-affecting behaviours in a drought-prone Australian community. Soc. Nat. Resour. J. 21 (3), 244–257.

Norris, F.H., Stevens, S.P., Pfefferbaum, B., Wyche, K.F., Pfefferbaum, R.L., 2008. Community resilience as a metaphor, theory, set of capacities, and strategy for disaster readiness. Am. J. Commun. Psychol. 41 (1-2), 127–150.

Chapter 23

Postdisaster Relief Distribution Network Design Under Disruption Risk: A Tour Covering Location-Routing Approach

Zohreh Raziei*, Reza Tavakkoli-Moghaddam*,, Mohammad Rezaei-Malek*,**, Ali Bozorgi-Amiri*, Fariborz Jolai***

**School of Industrial Engineering, College of Engineering, University of Tehran, Tehran, Iran;*

***LCFC, Arts et Métiers ParisTech, Metz, France*

23.1 INTRODUCTION

Nowadays, a significant increase in the number of disasters in all around the world and their serious damages result in paying the head of many researchers in the field of disaster management. Indeed, aid and rescue operations are regarded as one of the most important pillars of disaster management (Rezaei-Malek et al., 2014; Rezaei-Malek and Tavakkoli-Moghaddam, 2014). In the last two decades, the relief operations were taken into consideration by a considerable number of researchers. Today, operations research knowledge provides the mathematical models and methods for solving various relief operation problems. Based on the definition provided by the International Organization of Red Cross and Red Crescent, "disaster is a precipitate incident, which leads to serious disruptions in the performance of a society and causes human, environmental and economic damage, which is beyond the capability of a society to deal with" (Guha-Sapir et al., 2014). Hence, it is a fundamental need to carry out research studies on appropriate activities, which are generally required to manage these disasters efficiently.

In the first step of disaster management, it is of paramount importance to identify clearly different phases of the management practice. Despite some disagreements, most of the scholars have classified it into four programmatic phases, namely mitigation, preparedness, response, and recovery (Altay and Green, 2006; Dorasamy et al., 2012; Rezaei-Malek and Tavakkoli-Moghaddam, 2012). According to Galindo and Batta (2013), most of the previous research papers have focused mainly on the first phase of disaster management, whereas response and recovery are the most critical phases in disaster management; therefore, the number of studies on these phases is rapidly increasing.

A comprehensive review on the studies about disaster management indicates that most of the presented mathematical models are separately concentrated on location and routing problems, while it is required definitely to combine these two problems due to the rapid development of integrated logistics system designs. Additionally, considering these problems separately leads to a high increase in supply chain costs (Salhi and Rand, 1989).

Ukkusuri and Yushimito (2008) presented a methodology for a reliable path location-inventory problem (IRP) to solve the inventory-prepositioning problem for the humanitarian supply chain under disruption risk of the selected locations. The model was formulated based on a reliable path and integer programming for a location-routing problem (LRP). Mingang et al. (2009) proposed a two-echelon LRP for a relief distribution system and solved it by using a clustering method for an emergency facility location problem and resource routing subproblem. Rennemo et al. (2014) developed a three-stage stochastic programming model for the response phase. The availability of vehicles, state of infrastructure, and demand were supposed as stochastic elements. Rath and Gutjahr (2014) proposed a multiobjective LRP to provide relief goods for affected people after a disaster. Wang et al. (2014) proposed a multiobjective and reliable LRP model for relief operations in a postearthquake phase. They considered Open Vehicle Routing Problem (OVRP) with split delivery and investigated the Sichuan earthquake in China as a case study.

Integrating Disaster Science and Management. http://dx.doi.org/10.1016/B978-0-12-812056-9.00023-3

Ahmadi et al. (2015) presented an LRP with network failure and the standard relief time for last mile distribution after an earthquake. In addition, they extended their model into a two-stage stochastic programming model with random travels and considered the San Francisco district as a case study. Bozorgi-Amiri and Khorsi (2016) developed a multiobjective dynamic LRP with uncertainty in demand and travel time parameters for the response phase. The objectives were the minimization of the maximum shortage of the damage area, travel time, and total cost of pre- and postdisaster phases. Xu et al. (2016) proposed a bi-level LRP under a random fuzzy condition for the initial 72-h postearthquake. The rescue control centers were considered as upper-level decision makers and logistic companies as lower level ones. They solved this problem by using their proposed random fuzzy simulation-based interactive genetic algorithm (GA) and investigated the Lushan earthquake as a case study.

Several studies about LRPs in disaster management have focused on investigating disruptions. Xie et al. (2015) proposed a reliable LRP with probabilistic disruptions for facilities. Backup plans were considered where locations were disrupted and a set covering approach was applied for finding appropriate facility locations. They solved the problem with a meta-heuristic algorithm based on a maximum-likelihood sampling method, route-reallocation improvement, two-stage neighborhood search, and simulated annealing. Zhang et al. (2015) presented a scenario-based LRP with depots disruption assumption. One backup emergency depot was considered to satisfy the unmet emergency demand of the victims.

In the network disruptions, most of the relief chains are risk-averse, so the risk measures can be applied in order to deal with these disruptions in relief networks. Conditional value-at-risk (CVaR) is a popular risk management tool that measures a percentile of loss distribution such as value-at-risk (VaR). It can be applied for the convex and linear programming (Sarykalin et al., 2008). Noyan (2012) proposed a risk-averse two-stage stochastic programming model for disaster management. Noyan (2012) determined the facility location and inventory level for each commodity while considering uncertainty in demand, capacity of transportation, and damage level of supplies. Hu et al. (2016) developed a two-stage stochastic programming model for the predisaster phase with three preventive measures in disaster management. These measures principally aim at reducing risks posed by an earthquake.

To the best of our knowledge, there is no study developing a two-echelon LRP model to determine optimal locations of temporary distribution centers (TDCs) and optimal routes for the shipment of relief items (RIs) to victim temporary residences (VTRs) by vehicles in the initial 72-h postdisaster period, while considering the most important three types of disruptions, namely TDC capacity, number of vehicle, and route capacity. In this chapter, a new model for investigating the disruptions of routes and TDC capacities is proposed based on the CVaR risk measurement to evaluate diverse decisions with respect to the degree of risk aversion. Moreover, the fairness level constraint is taken into account to distribute fairly relief goods among victims.

The rest of this chapter is organized as follows. Section 23.2 presents risk-measurement policies. Section 23.3 provides the problem definition and deterministic formulation of the model, linearization procedure, and model formulation with CVaR. Section 23.4 explains the proposed meta-heuristic algorithm, and Section 23.5 belongs to computational results and sensitivity analysis. Finally, Section 23.6 concludes this study.

23.2 RISK MEASURE

The risk function ρ assigns a real value $\rho(x)$ to the random variable x or risk position. Then, it selects a reasonable outcome from set X of allowable uncertain outcomes. The general form of the optimization problem is shown as

$$\underset{x \in X}{\text{Min}}\, \rho(x) \tag{23.1}$$

Let (Ω, F, P) be a probability space, where Ω is the measurable space, F is the σ-algebra subsets of Ω, P is a probability measure of F, and X is a linear space of Borel measurement function $(X: \Omega \rightarrow R)$. Also, risk function ρ can be defined as $\overline{R} = R \cup \{+\infty\} \cup \{-\infty\}$, where $\overline{R}$ is the extended real line. In the literature review, several properties have been described in the extended risk measure $(\rho: X \rightarrow \overline{R})$ (Ruszczynski and Shapiro, 2006). The most important properties are shown below:

(i) *Convexity:* $\forall x, y \in X$ and $\alpha \in [0,1], \rho(\alpha x + (1-\alpha)y) \leq \alpha\rho(x) + (1-\alpha)\rho(y)$.

(ii) *Monotonicity:* $\forall x, y \in X$ and $x \geq y, \rho(x) \geq \rho(y)$.

(iii) *Translation invariance:* $\forall x \in X$ and $c \in R, \rho(x+c) = \rho(x) + c$.

(iv) *Positive homogeneity:* $\forall x \in X$ and $\lambda \geq 0, \rho(\lambda x) = \lambda\rho(x)$.

Risk measure ρ that satisfies properties (i)–(iv) is called coherent (Ahmadi-Javid and Seddighi, 2013; Ruszczynski and Shapiro, 2006).

In most practical situations, one may use composite random variables $X = H(Z;\Theta)$, where Z is the real decision vector of subset ς from w-dimensional ($\varsigma \subseteq R^w$), Θ is a random vector ($\Theta \in R^n$) with known distribution, and $H(Z;.): R^n \rightarrow R, \forall Z \in \varsigma$ is a Borel measurable function. So, function (23.1) transforms into the following function:

$$\underset{Z\in\varsigma}{\text{Min}}\ \rho(H(Z;\Theta)) \tag{23.2}$$

The model presented in this chapter is a special case of function (23.2), where Z includes the U, x, R, r, and y vectors, and Θ includes the random vectors Γ and Ψ.

It is necessary to describe the behavior of the Borel measurement function ($X:\Omega\rightarrow R$) for each policy. The set of all Borel measurement functions is shown with L on a probability space (Ω, F, P). The subset of L is shown with L_p for each function X, the $\int_{\Omega}|X(\omega)|^p\,\mathrm{d}P(\omega)$ of which is finite. The moment generating ($M_x(z) = E(e^{zx})$) function exists for each L_p and $z \in R$.

CVaR in $1-\alpha$ confidence level has been defined as a cautious risk measure policy for quantifying risk $X \in L_p$. According to the properties (i)–(iv), CVaR is a coherent risk measurement that is equal to the conditional average of VaR to a continuous value of X (see the following equation):

$$\text{CVaR}_{1-\alpha}(X) := E(X \mid X \geq \text{VaR}_{1-\alpha}(X)) \tag{23.3}$$

Generally, the following equation can be used for all random variables:

$$\text{CVaR}_{1-\alpha}(X) = \frac{1}{\alpha}\int_0^{\alpha} \text{VaR}_{1-t}(X)\,\mathrm{d}t \tag{23.4}$$

Rockafellar and Uryasev (2002) proposed a convex formulation based on the minimization rule as follows (see the following equation):

$$\text{CVaR}_{1-\alpha}(X) = \inf_{t\in R}\left\{t + \frac{1}{\alpha}E(\max\{0, X-t\})\right\} \tag{23.5}$$

23.3 PROBLEM DESCRIPTION AND MATHEMATICAL FORMULATION

23.3.1 Modeling Framework

This section presents a mixed-integer linear programming (MILP) model for relief distribution in initial 72 h of the post-earthquake phase considering disruption risks. According to the schematic view of the network in Fig. 23.1, humanitarian

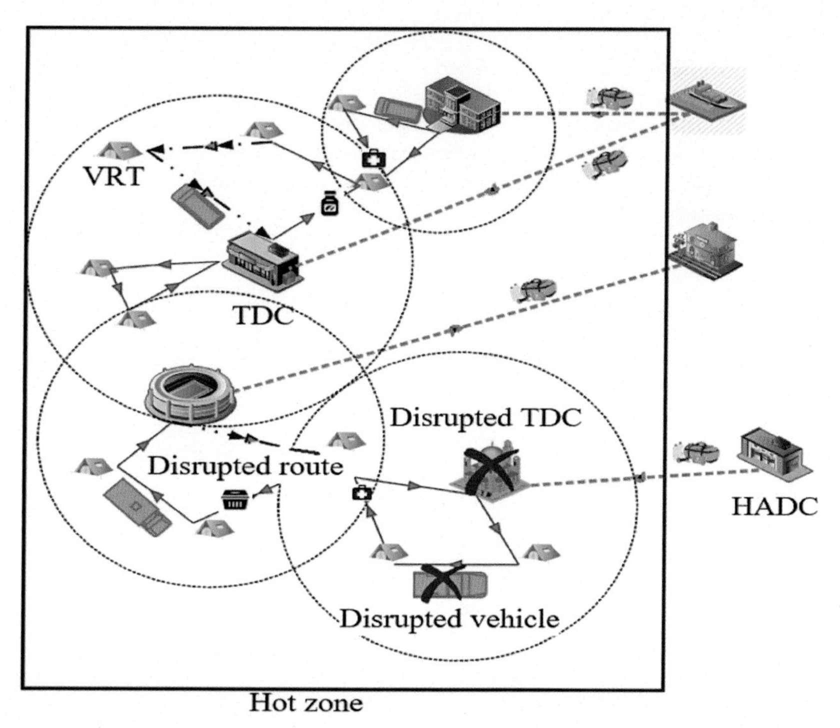

FIGURE 23.1 **Schematic view of the relief network.**

relief are sent from humanitarian aid distribution centers (HADCs) from a cold zone to TDCs in a hot zone and then RIs are sent from TDCs to VTRs. To get closer to a real situation, it is assumed that the demands of each VTR can be satisfied by several TDCs and VRTs are located in a coverage radius of TDCs for each type of RI. The main assumptions that distinguish this model from the other ones are as follows:

1. Considering the possibility of disruption on the number of vehicles.
2. Developing new approaches to incorporate the disruption on distribution capacity of TDCs and the route flow capacity within the golden 72 h.
3. Considering a multicommodity case with a coverage radius for each type of RIs and heterogeneous fleet assumptions.
4. Developing a new approach for fairness level in distribution of humanitarian RIs based on a combination approach of Rezaei-Malek et al. (2016a, b) and Vitoriano et al. (2011).
5. Using a CVaR measure to cope with the facility and route disruptions.
6. Considering the standard relief time for visiting each VTR as a soft time window.

23.3.2 Mathematical Modeling

23.3.2.1 Sets

i index of HADCs nodes ($i \in I$)
j index of TDCs nodes ($j \in J$)
k index of VTRs nodes ($k \in K$)
n $\in N = \{1, \ldots, J \cup K\}$
p index of RIs ($p \in P$)
v index of vehicles of ($v \in V$)
s index of scenarios ($s \in S$)

23.3.2.2 Parameters

cf_j fixed cost for opening TDC j
cht_{ij} fixed transportation cost between HADC i and TDC j
$\mathrm{cv}_{nn'}$ fixed transportation cost between n and n'
cah_{ip} capacity of HADC i for RI p
cav_v capacity of vehicle v
cw_k penalty cost for time-window violation
d_{kp} demand of VTR k for RI p
$t_{nn'}$ travel time between n and n'
RA_{jp} coverage radius of TDC j for RI p
$\mathrm{dis}_{nn'}$ distance between n and n'
δ_j distribution of TDC j for initial 72 h, which has a discrete random variable from support set, $j \in J$
$\xi_{nn'}$ flow capacity of arc (n, n') for initial 72 h, which has a discrete random variable from support set, $j \in J$
ω_v number of vehicle v for initial 72h, which has discrete random variable from support set, $v \in V$
$\mathrm{rc}_{n'n}$ disruption cost of arc capacity (n, n')
dc_j disruption cost of distribution of TDC j
β additional percentage of routing cost for each vehicle disrupted
LTW_k lower bound of time-window for each VTR
UTW_k upper bound of time-window for each VTR
FL maximum satisfaction rates between two demand points
Q penalty cost for unmet demand

23.3.2.3 Decision Variables

$x_{jnn'v}$ 1; if arc(n,n') is allocated to TDC j and used by vehicle v; and 0, otherwise
U_j 1; if TDC jth is selected; and 0, otherwise
s_{jv} 1; if vehicle vth is allocated to TDC jth; and 0, otherwise
y_{jkpv} 1; if RI p delivered to VTR k from TDC j with vehicle v; and 0, otherwise
h_{ij} 1; if TDC jth is allocated to HADC ith; and 0, otherwise
at_{vk} arrival time to VTR k for vehicle v
EL_k time-window violation for VTR k (earliness)
LL_k time-window violation for VTR k (lateness)
UD_{kp} unmet demand of VTR k for RI p
del_{vkp} amount of delivered RI p to TDC j with vehicle v

23.3.2.4 Auxiliary Variables

$\varphi_{js} = \text{Max}\left\{\sum_{v\in V}\sum_{k\in K}\sum_{p\in P} del_{vkp} \times s_{vj} - \delta_{js}, 0\right\}$ **(TDC disruption)**

$\psi_{nn's} = \text{Max}\left\{\sum_{j\in J}\sum_{v\in V} x_{jnn'v} - \xi_{nn's}, 0\right\}$ **(Route disruption)**

θ_s **= auxiliary variable of CVaR**

23.3.2.5 Linearization Variables

$\text{d}s_{jvkp} = \text{del}_{vkp} \times s_{vj},$

$\text{d}sh_{ijvkp} = \text{del}_{vkp} \times s_{vj} \times h_{ij},$

$Ix_{jvn'n} = I_{jvs} \times x_{jn'nv}.$

23.3.2.6 Disruption Terms of the Objective Function

The objective function includes the following disruption cost components for the initial 72 h.

1. The distribution disruption cost for the initial 72 h is given by $\sum_{v\in V}\beta(\sup\omega_v - \omega_v)\sum_{j\in J}\sum_{n'\in N}\sum_{n\in N} cv_{nn'} \times \text{dis}_{nn'} \times x_{jnn'v}$
2. The TDC capacity disruption cost for the initial 72 h is given by $\sum_{j\in J}\sum_{p\in P} \text{d}c_j \times \max\left\{\sum_{k\in K}\text{del}_{vkp} \times s_{vj} - \delta_j \cdot 0\right\}$
3. The route capacity disruption cost for the initial 72 h is given by $\sum_{n'\in N}\sum_{n\in N}\text{rc}_{n'n} \times \max\left\{\sum_{j\in J}\sum_{v\in V} x_{jnn'v} - \xi_{nn's} \cdot 0\right\}$

23.3.2.7 Model Formulation

The proposed model is provided as follows:

$$\begin{aligned}
\text{Min Z} = \rho\Bigg[&\sum_{j\in J}\text{cf}_j \times U_j + \sum_{i\in I}\sum_{j\in I}\text{cht}_{ij} \times h_{ij} + \sum_{k\in K}(\text{EL}_k + \text{LL}_k)\times \text{cw}_k + \sum_{k\in K}\sum_{p\in P}\text{UD}_{kp} \times Q \\
&+\sum_{v\in V}\sup\omega_v \times \sum_{j\in J}\sum_{n'\in N}\sum_{n\in N}\text{cv}_{nn'} \times \text{dis}_{nn'} \times x_{jnn'v} + \sum_{j\in J}\sum_{p\in P}\text{dc}_j \times \max\left\{\sum_{k\in K}\text{del}_{vkp} \times s_{vj} - \delta_{j}.0\right\} \\
&+\sum_{v\in V}\beta(\sup\omega_v - \omega_v)\sum_{j\in J}\sum_{n'\in N}\sum_{n\in N}\text{cv}_{nn'} \times \text{dis}_{nn'} \times x_{jnn'v} + \sum_{n'\in N}\sum_{n\in N}\text{rc}_{n'n} \times \max\left\{\sum_{j\in J}\sum_{v\in V} x_{jnn'v} - \xi_{nn's}.0\right\}\Bigg]
\end{aligned} \tag{23.6}$$

s.t.

$$\sum_{i\in I} h_{ij} \geq U_j \quad \forall j \in J \tag{23.7}$$

$$\sum_{i\in I} h_{ij} \geq 1 \quad \forall j \in J \tag{23.8}$$

$$\sum_{j\in J}\sum_{v\in V} y_{jkpv} = 1 \quad \forall k \in K, p \in P \tag{23.9}$$

$$\sum_{j\in J} s_{vj} \leq 1 \quad \forall v \in V \tag{23.10}$$

$$\sum_{k\in K}\sum_{p\in P} y_{jkpv} \leq M \times s_{vj} \quad \forall v \in V, j \in J \tag{23.11}$$

$$y_{jkpv} \times \text{dis}_{jk} \leq \text{RA}_{jp} \quad \forall j \in J, k \in K, p \in P \tag{23.12}$$

$$\sum_{v\in V}\sum_{p\in P}\sum_{k\in K}\text{del}_{vkp} \times s_{jv} \leq \sup(\delta_{j'})\times U_j \quad \forall j \in J \tag{23.13}$$

$$\sum_{j\in J} h_{ij} \times \sum_{k\in K}\sum_{v\in V}\text{del}_{vkp} \times s_{jv} \leq \text{cah}_{ip} \quad \forall i \in I, p \in P \tag{23.14}$$

$$\sum_{j\in J}\sum_{v\in V} x_{jnn'v} \leq \sup(\xi_{nn's}) \quad \forall n, n' \in N \tag{23.15}$$

$$\sum_{k\in K}\sum_{p\in P}\text{del}_{vkp} \le \sup(\omega_V)\times \text{cav}_v \quad \forall v\in V \tag{23.16}$$

$$\sum_{k'\in K} x_{jjk'v} \le s_{vj} \quad \forall v\in V, j\in J \tag{23.17}$$

$$\sum_{n\in N} x_{jnk'v} = \sum_{n\in N} x_{jk'nv} \quad \forall j\in J, k'\in K, v\in V \tag{23.18}$$

$$\sum_{k'\in K} x_{jjk'v} = \sum_{k'\in K} x_{jk'jv} \quad \forall j\in J, v\in V \tag{23.19}$$

$$\sum_{n\in N} x_{jnkv} \ge y_{jkpv} \quad \forall j\in J, k\in K, p\in P, v\in V \tag{23.20}$$

$$\sum_{k'\in K} x_{jk'kv} \le \sum_{p\in P} y_{jkpv} \quad \forall j\in J, k\in K, v\in V \tag{23.21}$$

$$\sum_{k'\in K} x_{jkk'v} \le \sum_{p\in P} y_{jkpv} \quad \forall j\in J, k\in K, v\in V \tag{23.22}$$

$$\sum_{j\in J} x_{jnnv} = 0 \quad \forall n\in N, v\in V \tag{23.23}$$

$$\sum_{j'\in J, j'\ne J}\sum_{n\in N} x_{jnj'v} = 0 \quad \forall j\in J, v\in V \tag{23.24}$$

$$\text{at}_{vk} \ge \text{at}_{vk'} + t_{k'k} - M\times\left(1-\sum_{j\in J} x_{jk'kv}\right) \quad \forall j\in J, k\in K, k'\in K, v\in V \tag{23.25}$$

$$\text{at}_{vk} \ge t_{jk} - M\times\left(1-\sum_{j'\in J} x_{j'jkv}\right) \quad \forall j\in J, k\in K, v\in V \tag{23.26}$$

$$\text{at}_{vk} \le M\times\left(1-\sum_{j\in J}\sum_{n'\in N} x_{jn'kv}\right) \quad \forall k\in K, v\in V \tag{23.27}$$

$$\sum_{v\in V}\text{at}_{vk} + \text{EL}_K \ge \text{LTW}_k \quad \forall k\in K \tag{23.28}$$

$$\sum_{v\in V}\text{at}_{vk} - \text{LL}_K \le \text{UTW}_k \quad \forall k\in K \tag{23.29}$$

$$\text{del}_{vkp} \le d_{kp}\times\sum_{j\in J} y_{jkpv} \quad \forall k\in K, v\in V, p\in P \tag{23.30}$$

$$\sum_{p\in P}\text{del}_{vkp} \le M\times\left(\sum_{j\in J}\sum_{n\in N} x_{jnkv}\right) \quad \forall k\in K, v\in V \tag{23.31}$$

$$\sum_{v\in V}\text{del}_{vkp} + \text{UD}_{kp} = d_{kp} \quad \forall k\in K, p\in P \tag{23.32}$$

$$\sum_{v\in V}\frac{\text{del}_{vkp}}{d_{kp}} - \sum_{v\in V}\frac{\text{del}_{vk'p}}{d_{k'p}} \le \text{FL} \quad \forall k.k'\in K, p\in P \tag{23.33}$$

$$\sum_{v\in V}\frac{\text{del}_{vk'p}}{d_{k'p}} - \sum_{v\in V}\frac{\text{del}_{vkp}}{d_{kp}} \le \text{FL} \quad \forall k.k'\in K, p\in P \tag{23.34}$$

$$x_{jnn'v}, y_{jk}, U_j, s_{jv}, y_{jkpv}, h_{ij} \in \{0.1\} \quad \text{and} \quad I_{jvn}, at_{vk}, \text{EL}_k, \text{LL}_k, \text{UD}_{kp}, \text{del}_{vkp} \ge 0 \tag{23.35}$$

Objective function (23.6) minimizes the total cost including the fixed opening cost of TDCs, transportation cost between HADCs and TDCs, penalty cost for time-window violation, unmet demand penalty, cost of routing in a case of no disruption, and disruption costs including the TDC's capacity, the number of vehicles visiting, and capacity of route disruptions in the initial 72 h. Constraints (23.7) and (23.8) make sure that each TDC is allocated exactly to one HADC. Constraint (23.9) demands satisfaction of VTRs for different commodities which receive exactly each of them from one TDC. Constraints (23.10) assure that each vehicle assign to exactly on TDC. Constraint (23.11) is allocation equations. Constraint (23.12) imposes tour selection of TDCs with regard to the radius of convergence of them for each RI. Constraint (23.13) imposes the limitation of capacity of TDCs. Constraint (23.14) imposes the limitation of capacity of HADCs. Constraint (23.15) determines the arc flow capacity based on the number of vehicles passing through each route, and Constraint (23.16) enforces vehicle capacities. Constraint (23.17) guarantees the construction of a route for each open TDC. Constraints (23.18) and (23.19) are flow conservation equations. Constraint (23.20) links the allocation and routing parts of the model. Constraints (23.21) and (23.22) ensure that the commodities allocate to active nodes. Constraints (23.23) and (23.24) confirm tour construction for TDCs. Constraints (23.25)–(23.27) determine the arrival time for each VTR and confirm as the subtour elimination equations. Constraints (23.28) and (23.29) assure that the time-window violation does not exceed the determined earliness and lateness. Constraints (23.30)–(23.32) assure that the delivered RIs do not exceed the VTRs demand and determine the amount of unmet VTRs demand. Constraints (23.33) and (23.34) guarantee that the fairness is established based on the maximum satisfaction rates between two VTRs. Constraint (23.35) defines the variable types.

23.3.3 Linearization Procedure

Due to the nonlinear terms appeared in the model, a linearization procedure for Constraints (23.13) and (23.14) is implemented as follows (Azadeh et al., 2015, 2017; Tan and Khoshnevis, 2004):

$$\sum_{v\in V}\sum_{p\in P}\sum_{k\in K} \mathrm{d}s_{jvkp} \le \sup(\delta_{j'})\times U_j \quad \forall j\in J \tag{23.36}$$

$$\mathrm{d}s_{jvkp} \le M\times s_{jv} \quad \forall j\in J, k\in K, v\in V, p\in P \tag{23.37}$$

$$\mathrm{d}s_{jvkp} \le \mathrm{del}_{vkp} \quad \forall j\in J, k\in K, v\in V, p\in P \tag{23.38}$$

$$\mathrm{d}s_{jvkp} \ge \mathrm{del}_{vkp} - M\times(1-s_{jv}) \quad \forall j\in J, k\in K, v\in V, p\in P \tag{23.39}$$

Constraint (23.14) is linearized as follows:

$$\sum_{j\in J}\sum_{k\in K}\sum_{v\in V} \mathrm{d}sh_{ijvkp} \le \mathrm{cah}_{ip} \quad \forall i\in I, p\in P \tag{23.40}$$

$$\mathrm{d}sh_{ijvkp} \le M\times h_{ij} \quad \forall i\in I, j\in J, k\in K, v\in V, p\in P \tag{23.41}$$

$$\mathrm{d}sh_{ijvkp} \le \mathrm{d}s_{jvkp} \quad \forall i\in I, j\in J, k\in K, v\in V, p\in P \tag{23.42}$$

$$\mathrm{d}sh_{ijvkp} \ge \mathrm{d}s_{jvkp} - M\times(1-h_{ij}) \quad \forall i\in I, j\in J, k\in K, v\in V, p\in P \tag{23.43}$$

23.3.4 Conditional Value-at-Risk

According to the existing random variables in the disruption terms and incorporating them in a numerical optimization procedure, it needs a risk measure to scalarize them. There are several risk measures that the VaR and CVaR are the mostly used in the engineering applications. VaR risk constraints are equivalent to the so-called chance constraints on probabilities of losses. There is a close relationship between CVaR and VaR since VaR is a lower bound for CVaR for the same confidence level and CVaR is a so-called "coherent risk measure." Several reasons affect the choice between VaR and CVaR which are based on the differences in optimization procedures clearness, mathematical properties, statistical estimation stableness, etc. (Sarykalin et al., 2008). Rockafellar and Uryasev (2000) demonstrated that CVaR is conspicuously premiere to VaR in optimization applications. CVaR can be used to optimize and constrain with convex and linear programming methods, while VaR is comparatively difficult to optimize. CVaR prepares a sufficient illustration of risks reflected in the extreme tails since it is a greatly important property if the extreme tail losses are accurately estimated. So, CVaR is used as a risk-cautions policy toward the risk that considers the mean and the worst value of the risk in the disaster management and applies for evaluating and analyzing of a decision in an emergency response. According to Eq. (23.5), it is required to calculate $\rho(x)=E(X-t)$ as the following mixed-integer program:

$$\begin{aligned}\text{Min } Z = &\sum_{j\in J} u_j \times \text{cf}_j + \sum_{i\in I}\sum_{j\in I} cht_{ij}\times h_{ij} + \sum_{i\in I}(\text{EL}_k+\text{LL}_k)\times cw_k \\ &+\sum_{k\in K}\sum_{p\in P}\text{UD}_{kp}\times Q+(1+\beta)\sum_{v\in V}\sup\omega_v\times\sum_{j\in J}\sum_{n\in N}\sum_{n'\in N}\sum_{v\in V} cv_{nn'}\times\text{dis}_{nn'}\times x_{jnn'v}+\sum_{s\in S}pr(s)\\ &\times\left[\sum_{j\in J}\text{dc}_j\times\varphi_{js}-\beta\sum_{v\in V}\omega_{vs}\sum_{n'\in N}\sum_{n\in N}cv_{nn'}\times\text{dis}_{nn'}\times x_{jnn'v}+\sum_{n'\in N}\sum_{n\in N}rc_{nn'}\times\psi_{nn's}\right]\end{aligned} \quad (23.44)$$

$$\sum_{v\in V}\sum_{k\in K}\sum_{p\in P}\text{ds}_{jvkp}-\delta_{js}\le\varphi_{js}\quad \forall\ s\in S\cdot j\in J \quad (23.45)$$

$$\sum_{j\in J}\sum_{v\in V}x_{jnn'v}-\xi_{nn's}\le\psi_{nn's}\quad \forall\ s\in S.\ n.n'\in N \quad (23.46)$$

$$\text{Constraints (23.7)–(23.12) and (23.15)–(23.43)} \quad (23.47)$$

Finally, the proposed model is rewritten for cautions risk measurement policy by considering $\rho(x)=\text{CVaR}_{1-\alpha}(X)$ as follows:

$$\text{Min } Z=\sum_{j\in J}U_j\times cf_j+\sum_{i\in I}\sum_{j\in I}cht_{ij}\times h_{ij}+\sum_{k\in K}(\text{EL}_k+\text{LL}_k)\times cw_k \quad (23.48)$$

$$+\sum_{k\in K}\sum_{p\in P}\text{UD}_{kp}\times Q+(1+\beta)\sum_{v\in V}\sup\omega_v\times\sum_{j\in J}\sum_{n\in N}\sum_{n'\in N}\sum_{v\in V}cv_{nn'}\times\text{dis}_{nn'}\times x_{jnn'v}+t+\frac{1}{\alpha}\sum_{s\in S}Pr(s)\times\theta_s$$

$$\sum_{j\in J}\text{dc}_j\times\varphi_{js}-\beta\sum_{\omega_{vs}}\omega_{vs}\sum_{v\in V}\sum_{n'\in N}\sum_{n\in N}cv_{nn'}\times\text{dis}_{nn'}\times x_{jnn'v}+\sum_{n'\in K}\sum_{n\in N}rc_{nn'}\times\psi_{nn's}-t\le\theta_s \quad (23.49)$$

Constraints (23.7)–(23.12), (23.15)–(23.43), (23.45), and (23.46).

23.4 META-HEURISTIC ALGORITHM

A GA is a mechanism based on natural genetics and natural selection that was first articulated by Holland (1975) (Rezaei-Malek et al., 2017). It consists of a family of randomized and parallel search optimization heuristics. According to the survey of the GA, it refers to the reproduction, mutation, crossover, and selection mechanism. In this study, the selected mechanisms are one-point crossover operator as discrete crossover operator and linear crossover as a continuous one. For a mutation mechanism, swap and inversion as discrete operators and randomized mutation as a continuous operator are used. Also, roulette wheel selection is chosen as a mechanism for selecting individuals in the proposed GA.

23.4.1 Solution Representation

The arrays represent the chromosome structures and are used to represent the solution structure. The number of genes in the chromosome is equal to the number of decision variables in the model. In this chapter, six structures of chromosomes are presented. One array for representing tours solution regarding the assumption that each demand point can be allocated to more than one TDC. First, it is applied an array to the length of $V+K-1$ for producing vehicle tours. Furthermore, a matrix with V columns and $V+K-1$ rows (named Tour matrix) is produced which represents V structures of vehicle tours since each VTR may be met more than one time. Finally, for further diversifying in a solution representation, a cell array is produced, which includes V number of Tour matrices. The procedure for selected tour is shown in Fig. 23.2.

Three arrays are created in a matrix structure that include the solution representations for allocating TDCs to HADCs for each RI with a random variable between [0, $|I|$] (with dimensions ($|P|\times|J|$)).

The solution representation for the delivered RIs is shown by the multiplication of two matrices. The former one is the matrix of the amount of delivered RIs for each VTR ($|M|\times|L|$). The latter one is the matrix with dimension ($|P|\times|K|$) of a random variable between [0, 1] which is created for representation of the percentage of delivering RIs. The representation matrix is shown in Fig. 23.3. The RI in the first column of this matrix shows the delivered items for the first VRTs in a tour of the first vehicle.

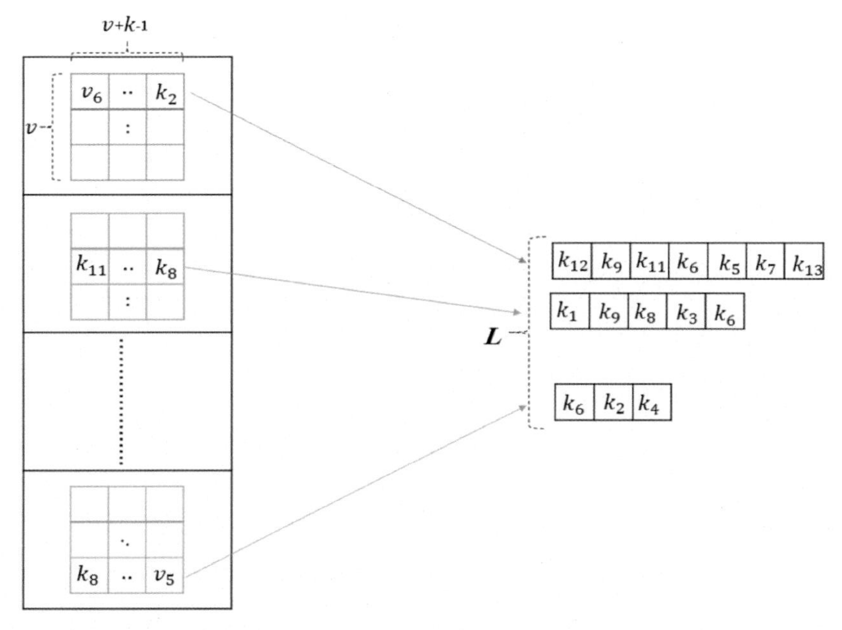

FIGURE 23.2 **Tour representation chromosome.**

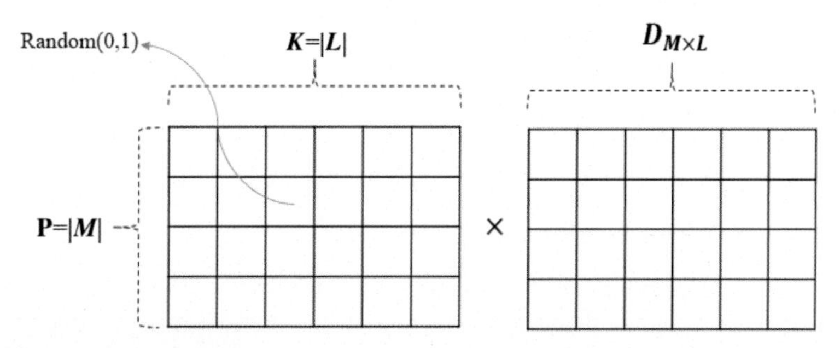

FIGURE 23.3 **Delivery representation chromosome.**

23.5 COMPUTATIONAL RESULTS

In this section, the performance of the proposed model is evaluated over a number of randomly generated test problems with regard to the cautious risk-measurement policy (i.e. CVaR). The test problems are defined in three different sizes with the same distributions as shown in Table 23.1. Seven test problems are generated as shown in Table 23.2. The test problems are solved using GAMS 24.1 using CPLEX solver to obtain optimal solutions. All the computations are performed on Intel Core i5 computer with 2.6 GHs CPU and 6 GB RAM. Additionally, the VaR measures are provided in Table 23.2. As indicated by the obtained results, CVaR is more sensitive than VaR to form the distribution in the right tail, which is a percentile of a loss distribution with the same confidence level (the value of α is considered 0.1).

In order to solve the model for large-sized problems and compare its results with the exact solution, all the algorithm parameters should be tuned. In order to tune parameters for different problem sizes, several experiments should be designed and then the parameters are optimized based on those experiments. The Taguchi method is used for the design of experiments (DOEs). The average output of three run times is considered for the test and then we use the Minitab 16 statistical software for implementing the Taguchi test. The results of parameter tuning for small, medium, and large sizes are shown in Table 23.3. The meta-heuristic algorithm has a stochastic behavior, so it should be run several times for each test problem then the objective function and the CPU time of the GA and GAMS are compared with the percentage of a relative gap between them, which is calculated by $[(G_{GA} - G_{GAMS})/G_{GAMS}] \times 100$ (see Table 23.4). The termination condition of the proposed GA is considered the strong convergence rate in order to make a more realistic comparison with the GAMS results. According to the results, the obtained relative gap for the different test problems seems logical. Because of the high possibility of falling into a local optimum for the GA in the large-sized problems, the relative gap between the GA and GAMS solution is increasing while the size of test problems is growing.

TABLE 23.1 Parameter Value

Parameter	Value	Parameter	Value
cah_{ip}	$\sim \text{Uniform}(2000, 4000)$	$rc_{nn'}$	$\sim \text{Uniform}(20, 100)$
cav_v	$\sim \text{Uniform}(100, 200) \times \left\lceil \frac{2 \times \lvert K \rvert}{\lvert V \rvert} \right\rceil$	dc_j	$\sim \text{Uniform}(100, 300)$
cf_j	$\sim \text{Uniform}(1000, 1200)$	β	$\sim \text{Uniform}(0.1, 0.5)$
cht_{ij}	$\sim \text{Uniform}(300, 700)$	LTW_k	$\sim \text{Uniform}(1, 3)$
$cv_{nn'}$	$\sim \text{Uniform}(10, 50)$	UTW_k	$\sim LTW_k + 72$
cw_k	$\sim \text{Uniform}(1, 5)$	FL	$\sim \text{Uniform}(0.1, 1)$
d_{kp}	$\sim \text{Uniform}(100, 150)$	Q	$\sim \text{Uniform}(500, 600)$
$t_{nn'}$	$\sim \text{velocity} \times 0 / 0036 \times dis_{nn'}$	ω_v	$\sim \text{Binomial}(100, 0.1)$
RA_{ip}	$\sim \text{Uniform}(50, 150)$	p_s	$\sim \text{Uniform}(0,1), \sum_{s \in S} p_{s=1}$
$dis_{nn'}$	$\sim \text{Uniform}(0, 100)$		
δ_j	$\sim \text{Uniform}\left(0, r\,\text{and}(1.5, 2) \times \sum_{k \in K} \sum_{p \in P} d_{kp} \cdot \lvert J \rvert^{-1}\right)$		
$\xi_{nn'}$	$\sim \text{Uniform}\left(1, (\max\{d_{kp}\} \times \lvert V \rvert) / \sum_{v \in V} cav_v\right)$		

The sensitivity analysis of the demand-increasing scenarios is shown in Fig. 23.4. The results show that the cost objective is increased because the cost of unmet demand has increased. In addition, it is possible that the unmet demand was provided by the newly opened TDCs, and therefore, the routing and the disruption cost increased.

In the disaster situation, the balance between the standard relief time, demand satisfaction, and fairness level is a crucial factor for the relief distribution, so the sensitivity analysis is implemented for an investigation about the effect of increasing the penalty cost when the network exceeds the standard relief time. In Table 23.5, it can be seen that the cost objective function is increased because the cost of unsatisfied demand, routing, and opening TDCs may be increased. Also, the effect of increasing in standard relief time violation cost on the unmet demand is shown in Fig. 23.5. It is clear that the objective function value increases in 10 illustrated scenarios of increasing penalty, which results in decreasing the demand satisfaction at the minimization problem.

TABLE 23.2 Comparison of Test Problems Under CVaR and VaR Measures

Test Problem	$\lvert I \rvert \times \lvert J \rvert \times \lvert K \rvert \times \lvert V \rvert \times \lvert P \rvert \times \lvert S \rvert$	Obj.	CPU Time (s)	CVaR	VaR
T01	$2 \times 2 \times 5 \times 3 \times 2 \times 2$	167,217.3	12.8	278.2	278.2
T02	$2 \times 2 \times 6 \times 4 \times 2 \times 3$	239,153.7	13.6	1320.2	414.4
T03	$2 \times 2 \times 7 \times 5 \times 2 \times 4$	152,662.4	31.6	1353.2	1142.9
T04	$2 \times 2 \times 8 \times 6 \times 2 \times 4$	206,431.8	37.9	19,798.8	14,380.8
T05	$2 \times 3 \times 8 \times 6 \times 2 \times 4$	227,742.1	1515.7	21,546.8	21,518.5
T06	$2 \times 3 \times 10 \times 7 \times 3 \times 6$	763,616.6	4980.8	76,714.4	64,884
T07	$3 \times 4 \times 12 \times 8 \times 3 \times 6$	987,382.3	9103.2	87,323.5	73,453.2

TABLE 23.3 Parameter Tuning

	Value		
Parameters	Large	Medium	Small
Mutation	0.5	0.4	0.4
Crossover	0.6	0.7	0.8
Npop	90	100	120

TABLE 23.4 Comparison Between Results of GA and GAMS

Test Problem	GA		GAMS		Gap
$\|I\| \times \|J\| \times \|K\| \times \|V\| \times \|P\| \times \|S\|$	CPU Time (s)	Obj.	CPU Time (s)	Obj.	
2 × 3 × 7 × 3 × 2 × 4	20.3	138,982.3	36.4	138,456.2	0.3
2 × 3 × 10 × 3 × 2 × 4	48.4	329,456.3	113.2	327,619.4	0.5
3 × 3 × 10 × 3 × 3 × 5	138.2	518,321.4	467.3	498,878.6	3.8
3 × 4 × 5 × 3 × 3 × 5	108.2	310,432.1	347.2	287,631.3	7.9
3 × 4 × 7 × 4 × 3 × 5	776.3	672,001.2	1289.1	631,432.5	6.4
3 × 4 × 10 × 4 × 3 × 5	1266.1	3,094,728.4	6347.3	2,800,819.1	6.6
4 × 5 × 12 × 5 × 4 × 5	3067.1	7,531,446.3	17321.2	6,934,761.3	8.6
4 × 5 × 15 × 6 × 4 × 5	4286.3	10,500,592.1	–	–	–
4 × 5 × 20 × 7 × 4 × 5	5766.2	20,780,183.1	–	–	–
5 × 6 × 25 × 8 × 5 × 5	7892.2	58,148,828.4	–	–	–

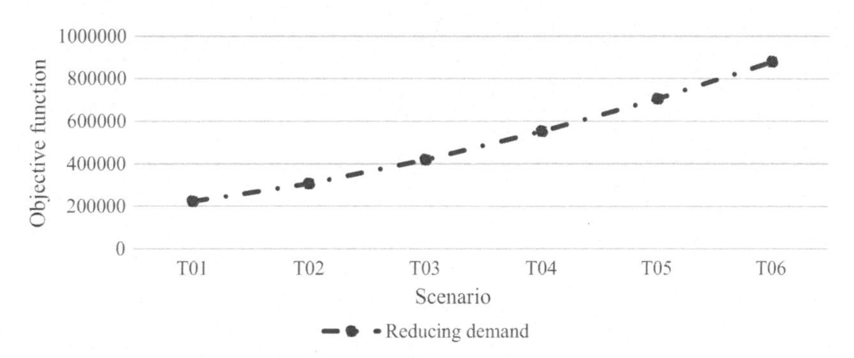

FIGURE 23.4 **Sensitivity analysis of the objective function on demand scenarios.**

TABLE 23.5 Sensitivity Analysis of the Cost Objective Function on the Penalty Cost of Time-Window Violation

	Scenario	Obj.	Scenario	Obj.
T04	cw_k	206,431.81	$25 \times cw_k$	586,426.35
	$5 \times cw_k$	273,424.81	$30 \times cw_k$	574,868.61
	$10 \times cw_k$	353,913.73	$35 \times cw_k$	661,238.93
	$15 \times cw_k$	419,982.21	$40 \times cw_k$	784,956.76
	$20 \times cw_k$	485,334.16	$45 \times cw_k$	779,361.54

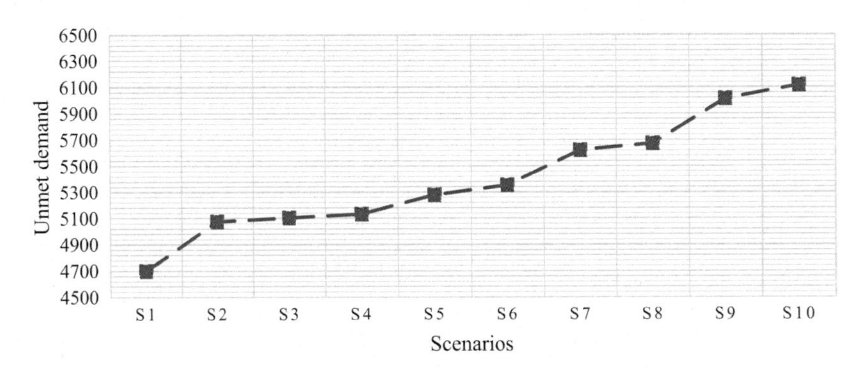

FIGURE 23.5 Amount of unmet demand when the penalty cost of standard relief time violation is increasing.

TABLE 23.6 Scenario Analysis of Random Parameters

			Changed Capacity of TDC			Changed Number of Vehicle Visited			Changed Route Capacity	
		Scenario	Obj.	CVaR	Scenario	Obj.	CVaR	Scenario	Obj.	CVaR
T04	1.	$0.95 \times \delta_j$	207,348.2	19,843.4	$0.8 \times \omega_v$	206,746.6	19,901.6	$0.8 \times \eta_{nn'}$	209,092.3	19,928.8
	2.	$0.85 \times \delta_j$	224,927.4	41,334.4	$0.7 \times \omega_v$	206,769.9	19,924.9	$0.7 \times \eta_{nn'}$	249,236.8	20,064.9
	3.	$0.70 \times \delta_j$	337,479.9	45,217.7	$0.6 \times \omega_v$	206,849.4	20,004.4	$0.6 \times \eta_{nn'}$	279,300.2	22,110.4
	4.	$0.55 \times \delta_j$	417,825.5	49,576.2	$0.5 \times \omega_v$	206,929.8	20,084.8	$0.5 \times \eta_{nn'}$	319,325.1	26,180.3
	5.	$0.40 \times \delta_j$	487,259.6	54,997.2	$0.4 \times \omega_v$	207,004.2	20,159.5	$0.4 \times \eta_{nn'}$	349,671.6	27,500.1
	6.	$0.25 \times \delta_j$	547,538.1	61,275.9	$0.3 \times \omega_v$	212,683.5	20,811.7	$0.3 \times \eta_{nn'}$	369,830.6	30,658.8
	7.	$0.10 \times \delta_j$	650,334.9	64,072.7	$0.2 \times \omega_v$	223,080.2	22,320.8	$0.2 \times \eta_{nn'}$	569,574.7	454,513.2
	8.	$0.05 \times \delta_j$	690,452.8	70,022.2	$0.1 \times \omega_v$	330,703.9	22,874.5	$0.1 \times \eta_{nn'}$	569,574.7	454,513.2

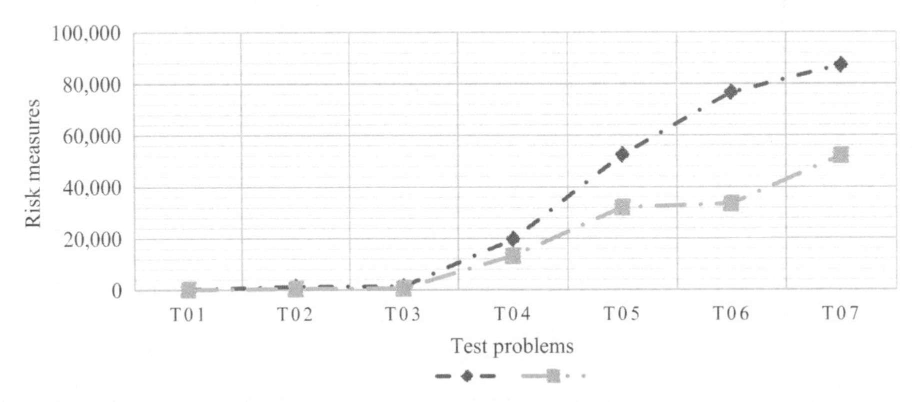

FIGURE 23.6 Comparison between CVaR and expected values.

The results of the sensitivity analysis on random parameters (e.g. TDC capacities, the route capacities, and the number of vehicles visited in the initial 72 h) are shown in Table 23.6. Decreasing these parameters results in increasing the cost objective function. The cost of disruptions, unmet demand, and standard relief time violation increase in this situation. In addition, decreasing the capacity of TDCs results in increasing in the cost of a number of vehicles visited and route capacities disruption since it is not possible to visit all VTRs by increasing the number of allocating VTRs to TDCs. Similarly, the cost of disruption in the TDC capacities may be increased by decreasing the route capacity and the number of vehicle visit.

The results of comparing the cost objective function of applying CVaR risk measure with the cost objective function of expectation risk measures (i.e. $E(X)$) as a risk-neutral condition (i.e. confidence level $\alpha = 1$) is shown in Fig. 23.6. This risk-neutral condition paves the way to achieve the lowest expected cost objective function because it considers all conditions. Actually, it would attain a situation that considers all good and bad scenarios.

23.6 CONCLUSION

This chapter proposed a two-echelon, multicommodity, and capacitated tour covering location-routing problem with disruption risk to the distribution capacity (i.e. the capacity of TDCs), the number of vehicles, and the route capacity for the initial 72 h in a relief distribution plan. The disruption in the route capacity was expressed by decreasing the number of vehicles passing through each arc and the disruption in the capacity of TDCs was taken into account by the amount of RIs delivered to VTRs. In order to get closer to a real situation, a covering tour was considered for each type of RIs in TDCs. In addition, the RIs were distributed in standard relief time regarding a fairness level. The aim was to minimize the total cost of location, routing, and disruption in the initial 72 h. Among the risk measures, CVaR was used for the risk-averse decision maker in disaster management. Because of the NP-hard nature of the LRP, a GA was implemented for solving large-sized problems. A new solution representation for the GAs chromosomes was presented to improve the result of the algorithm. The computational results showed the confidence level of CVaR effects on the decisions and the value of objective function. Furthermore, a sensitivity analysis on the comparison of CVaR with the expectation risk measure illustrated that CVaR is coherently more reliable to cope with a different risk aversion levels.

REFERENCES

Ahmadi, M., Seifi, A., Tootooni, B., 2015. A humanitarian logistics model for disaster relief operation considering network failure and standard relief time: a case study on San Francisco district. Transp. Res. Part E: Logist. Transp. Rev. 75, 145–163.

Ahmadi-Javid, A., Seddighi, A.H., 2013. A location-routing problem with disruption risk. Transp. Res. Part E: Logist. Transp. Rev. 53, 63–82.

Altay, N., Green, W.G., 2006. OR/MS research in disaster operations management. Eur. J. Oper. Res. 175 (1), 475–493.

Azadeh, A., Rezaei-Malek, M., Evazabadian, F., Sheikhalishahi, M., 2015. Improve design of CMS considering operators decision-making style. Int. J. Prod. Res. 53 (11), 3276–3287.

Azadeh, A., Ravanbakhsh, M., Rezaei-Malek, M., Sheikhalishahi, M., Taheri-Moghaddam, A., 2017. Unique NSGA-II and MOPSO algorithms for improved dynamic CMS by considering human factors. Appl. Math. Modell. doi: 10.1016/j.apm.2017.02.026.

Bozorgi-Amiri, A., Khorsi, M., 2016. A dynamic multi-objective location–routing model for relief logistic planning under uncertainty on demand, travel time, and cost parameters. Int. J. Adv. Manufact. Technol. 855, 1633–1648.

Dorasamy, M., Raman, M., Kaliannan, M., Muthaiyah, S., 2012. Knowledge management systems for emergency management: a situational approach. Int. J. Bus. Continuity Risk Manage. 3 (4), 359–372.

Galindo, G., Batta, R., 2013. Review of recent developments in OR/MS research in disaster operations management. Eur. J. Oper. Res. 230 (2), 201–211.

Guha-Sapir, D., Hoyois, P., Below, R., 2014. Annual Disaster Statistical Review 2013: The Numbers and Trends. Available from: http://cred.be/sites/default/files/ADSR_2013.pdf. (Accessed 12 May 2015).

Holland, J.H., 1975. Adaptation in Natural and Artificial Systems: An Introductory Analysis With Applications to Biology, Control and Artificial Intelligence. University of Michigan Press, London, England.

Hu, S.L., Han, C.F., Meng, L.P., 2016. Stochastic optimization for investment in facilities in emergency prevention. Transp. Res. Part E: Logist. Transp. Rev. 89, 14–31.

Mingang, Z., Zeng, M., Xiaoyan, W., 2009. Research on location-routing problem of relief system based on emergency logistics. In: Proceedings of the 16th IEEE International Conference on Industrial Engineering and Engineering Management (IE&EM'09), 228 232.

Noyan, N., 2012. Risk-averse two-stage stochastic programming with an application to disaster management. Comput. Oper. Res. 39 (3), 541–559.

Rath, S., Gutjahr, W.J., 2014. A math-heuristic for the warehouse location-routing problem in disaster relief. Comput. Oper. Res. 42, 25–39.

Rennemo, S.J., Rø, K.F., Hvattum, L.M., Tirado, G., 2014. A three-stage stochastic facility routing model for disaster response planning. Transp. Res. Part E: Logist. Transp. Rev. 62, 116–135.

Rezaei-Malek, M., Tavakkoli-Moghaddam, R., 2012. A new multi-objective mathematical model for relief logistic network under uncertainty. In: Proceedings of the 16th IEEE International Conference on Industrial Engineering and Engineering Management (IE&EM'09). Hong Kong, December 12–14. pp. 1878–1882.

Rezaei-Malek, M., Tavakkoli-Moghaddam, R., 2014. Robust humanitarian relief logistics network planning. Uncertain Supply Chain Manage 2 (2), 73–96.

Rezaei-Malek, M., Tavakkoli-Moghaddam, R., Salehi, N., 2014. Robust planning of medical supplies with time windows in a humanitarian relief logistic network. In: Proceedings of the 44th International Conference on Computers and Industrial Engineering (CIE44). Istanbul, Turkey, October 14–16. 134 147.

Rezaei-Malek, M., Tavakkoli-Moghaddam, R., Zahiri, B., Amiri-Bozorgi, A., 2016a. An interactive approach for designing a robust disaster relief logistics network with perishable commodities. Comput. Ind. Eng. 94, 201–215.

Rezaei-Malek, M., Tavakkoli-Moghaddam, R., Cheikhrouhou, N., Taheri-Moghaddam, A., 2016b. An approximation approach to a trade-off among efficiency, efficacy, and balance for relief pre-positioning in disaster management. Transp. Res. Part E: Logist. Transp. Rev. 93, 485–509.

Rezaei-Malek, M., Razmi, J., Tavakkoli-Moghaddam, R., Taheri-Moghaddam, A., 2017. Towards a psychologically consistent cellular manufacturing system. Int. J. Prod. Res. 55 (2), 492–518.

Rockafellar, R.T., Uryasev, S., 2000. Optimization of conditional value-at-risk. J. Risk 2, 21–42.

Rockafellar, R.T., Uryasev, S., 2002. Conditional value-at-risk for general loss distributions. J. Banking Finance 26 (7), 1443–1471.

Ruszczynski, A., Shapiro, A., 2006. Optimization of convex risk functions. Math. Oper. Res. 31 (3), 433–452.

Salhi, S., Rand, G.K., 1989. The effect of ignoring routes when locating depots. Eur. J. Oper. Res. 39 (2), 150–156.

Sarykalin, S., Serraino, G., Uryasev, S., 2008. Value-at-risk vs. conditional value-at-risk in risk management and optimization. State-of-the-Art Decision-Making Tools in the Information-Intensive Age, 270 294, Informs.

Tan, W., Khoshnevis, B., 2004. A linearized polynomial mixed integer programming model for the integration of process planning and scheduling. J. Intell. Manufact. 15 (5), 593–605.

Ukkusuri, S., Yushimito, W., 2008. Location-routing approach for the humanitarian prepositioning problem. Transp. Res. Rec.: J. Transp. Res. Board 2008, 18–25.

Vitoriano, B., Ortuño, M.T., Tirado, G., Montero, J., 2011. A multi-criteria optimization model for humanitarian aid distribution. J. Global Optim. 51 (2), 189–208.

Wang, H., Du, L., Ma, S., 2014. Multi-objective open location-routing model with split delivery for optimized relief distribution in post-earthquake. Transp. Res. Part E: Logist. Transp. Rev. 69, 160–179.

Xie, W., Ouyang, Y., Wong, S.C., 2015. Reliable location-routing design under probabilistic facility disruptions. Transp. Sci. 50 (3), 1128–1138.

Xu, J., Wang, Z., Zhang, M., Tu, Y., 2016. A new model for a 72-h post-earthquake emergency logistics location-routing problem under a random fuzzy environment. Transp. Lett. 8 (5), 270–285.

Zhang, Y., Qi, M., Lin, W.H., Miao, L., 2015. A metaheuristic approach to the reliable location routing problem under disruptions. Transp. Res. Part E: Logist. Transp. Rev. 83, 90–110.

FURTHER READING

Adams, W.P., Sherali, H.D., 1990. Linearization strategies for a class of zero-one mixed integer programming problems. Oper. Res. 38 (2), 217–226.

Vos, F., Rodríguez, J., Below, R., Guha-Sapir, D., 2010. Annual Disaster Statistical Review 2009. Centre for Research on the Epidemiology of Disasters, University Catholique de Louvain, Brussels, Belgiumpp. 8–38.

Chapter 24

Climate Change and Typhoons in the Philippines: Extreme Weather Events in the Anthropocene

William N. Holden, Shawn J. Marshall
University of Calgary, Calgary, Alberta, Canada

A very solid scientific consensus indicates that we are presently witnessing a disturbing warming of the climatic system. In recent decades this warming has been accompanied by a constant rise in the sea level and, it would appear, by an increase of extreme weather events.

Pope Francis (2015, p. 18)

24.1 INTRODUCTION: SUPER TYPHOON HAIYAN 8 NOVEMBER 2013

In the early morning hours of November 8, 2013, Super Typhoon Haiyan (referred to in the Philippines as Super Typhoon Yolanda) ravaged the Eastern and Central Visayan Islands of the Philippines. Haiyan was an exceptional storm; it brought precipitation in some places of up to 615 mm between the 3rd and the 12th of November, had an air pressure at its center of only 895 mbar, generated wind gusts of up to 375 km/h, and was one of the strongest typhoons to ever make a landfall in the entire Western North Pacific (Primavera et al., 2016; Takagi and Esteban, 2016; Takagi et al., 2015). When Haiyan reached Concepcion, Iloilo, on the island of Panay, it still had sustained wind speeds of 215 km/h, with gusts up to 250 km/h, and this was its *fifth landfall* (National Disaster Risk Reduction Management Council, 2014)! Dr. Wei Mei, a climate scientist at the Scripps Institution of Oceanography, in La Jolla, CA, stated that he was "shocked" by the strength of Haiyan (Personal Communication, November 4, 2015). Perhaps the most remarkable aspect of Haiyan was its storm surge of 7.4 m, which inundated 98 km^2 of the island of Leyte and 93 km^2 of the island of Samar (Cardenas et al., 2015). Lander et al. (2014, p. S114) described the storm surge as having "nearly the same force and rapidity as a destructive tsunami." Not only was Haiyan a powerful storm, it was also believed to be the fastest storm on record and it was traveling westward at a speed of 41 km/h when it made a landfall (Takagi and Esteban, 2016). While meteorologists may have wondered in amazement at the sheer power of Haiyan, its consequences for the people of the Philippines were anything but wonderful. Eight of the 17 regions of the Philippines were affected by Haiyan and it generated 6245 deaths, 28,626 injuries, and caused 1039 people to be reportedly missing, with much of this damage occurring when the City of Tacloban (Box 24.1) was essentially destroyed (Takagi and Esteban, 2016). The storm caused between US $12 and US$ 15 billion worth of damages and resulted in the destruction of 1 million homes (Primavera et al., 2016). Six months after Haiyan, 2 million people remained without secure shelter (Rodgers, 2016). In some locations on the island of Samar, up to 90% of all water wells were inundated with seawater and rendered undrinkable (Cardenas et al., 2015). This chapter discusses the relationship between climate change and typhoons in the Philippines and addresses whether such powerful storms are merely outliers, extreme events attracting attention, or are they, instead, a consequence of climate change and something one can expect more of as we progress further into that portion of Earth's natural history which Syvitski and Kettner (2011, p. 957) characterized as "the Anthropocene?"

Integrating Disaster Science and Management. http://dx.doi.org/10.1016/B978-0-12-812056-9.00024-5

BOX 24.1 The Tacloban Disaster

Although Super Typhoon Haiyan made its first landfall in Guiuan, Eastern Samar, it was in the City of Tacloban, a city of approximately 220,000, on the island of Leyte where the consequences of Super Typhoon Haiyan where most acutely felt. Tacloban bore the brunt of the winds and storm surge, which is estimated to have killed approximately 2678 people (National Disaster Risk Reduction Management Council, 2014). Many people sought refuge in the Tacloban Convention Center, a large concrete sports complex, where roof sections collapsed due to wind and lower sections flooded with storm surge. Although many residents of Tacloban had experienced similar storms before they were unprepared for its storm surge. In the days after the storm, Tacloban became a picture of chaos and desperation as widespread looting and lawlessness broke out.

The Tacloban Convention Center (photo credit, the authors).

24.2 TYPHOONS: EXTREME TROPICAL STORMS

"Typhoon," originating from the Chinese *tai* (strong) and *fung* (wind), is the term used to describe a tropical cyclone in the Western Pacific Ocean (Holden and Jacobson, 2012). Tropical cyclones develop in the northern hemisphere during the months of July–November in an area just north of the equator (Fig. 24.1) in a large area ranging from 130° to 180° East and 5° to 15° North (Mei et al., 2015). To develop their rotation, tropical cyclones need to be located at a latitude where the relative speed of the earth's rotation differs sufficiently between their northern and southern sides; consequently, they usually do not develop or strike within 10° latitude of the equator (Sheppard et al., 2009). Tropical cyclones generally occur over the oceans in regions where sea surface temperatures exceed 26 °C (Trenberth, 2005). Such sea surface temperatures are found in the tropics because solar energy per unit area of the ocean surface is greatest there and declines substantially as one moves away from the equator towards the poles (Sheppard et al., 2009). Tropical cyclones develop when strong clusters of thunderstorms drift over warm ocean waters, and warm air from these thunderstorms combines with warm air and water vapor from the ocean's surface and begins rising; as this air rises, there is a reduction in air pressure on the ocean surface, along with latent heat release that deepens the low and fuels the instability. As these clusters of thunderstorms consolidate into one large storm, convergent winds blowing toward the surface low, along with rotation due to the Coriolis effect, cause the storm to begin spinning (counterclockwise in the northern hemisphere), while rising warm air creates divergence aloft;

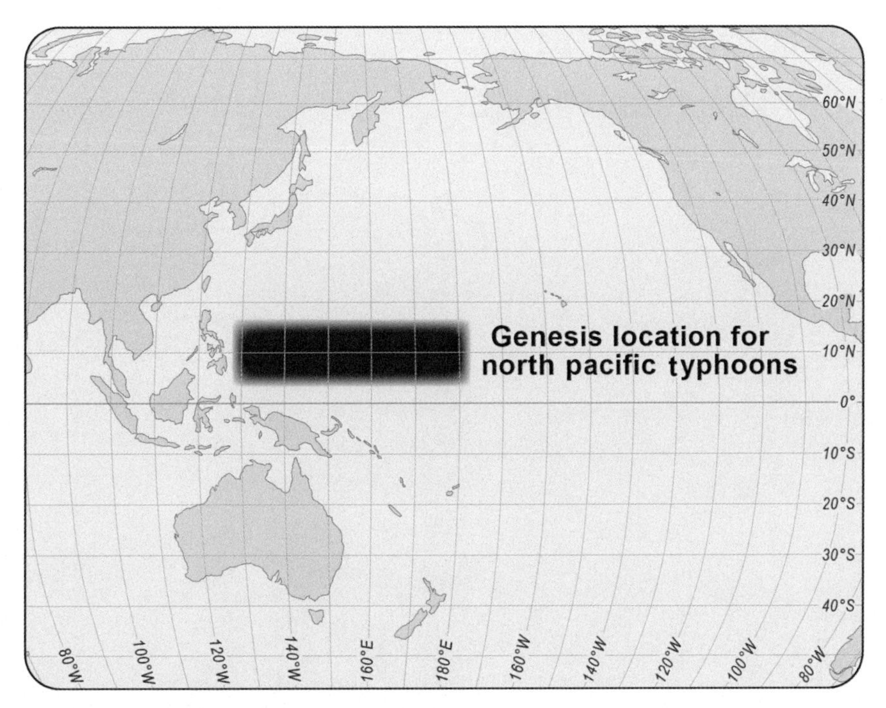

FIG. 24.1 Genesis location for north Pacific Typhoons. Source: *The Authors.*

eventually, the storm will have a low pressure center (the eye) with no clouds and calm winds, while winds in the eyewall and the outer part of the storm can be extremely strong. All typhoons have five characteristics: low air pressure, strong winds, cyclonic rotation, heavy rains, and storm surge. The air pressure reduction associated with a typhoon can cause the sea level to rise by up to 1 cm for every 1 mbar reduction in air pressure and there have been documented instances where sea levels have risen by 1.5 m due to air pressure reductions alone (Wang et al., 2005). This means that when a typhoon strikes land, with its heavy rains and strong winds, the local sea level will be higher due to the reduction of the atmospheric pressure. Onshore winds add to the strength of the storm surge and the greatest storm surges, in excess of 5 m, occur when a typhoon makes a landfall during a high spring tide. Storm surges are one of the most destructive aspects of a tropical typhoon and are feared by people living in coastal regions (Loy et al., 2014).

Although typhoons rapidly lose power as they move inland, they are capable of causing massive amounts of damage to coastal areas, with a fully developed typhoon releasing the energy equivalent to an atomic bomb and they move in an unpredictable manner that can be difficult to track. According to Mei et al. (2015, p. 1), "Tropical cyclones are among the most devastating and destructive natural hazards on earth." Since 2009, tropical cyclones have been divided into six categories, which are presented in Table 24.1, and the Western North Pacific basin experiences, on average, 26 named tropical cyclones each year, accounting for about 33% of the global total (Wu and Wang, 2004). Most typhoons develop between July and November, but the season runs year round and typhoons can develop in any month.

TABLE 24.1 The Six Categories of Tropical Storms

Type of storm	Wind speeds
Tropical depression	63 km/h or lower
Tropical storm	Between 63 and 89 km/h
Severe tropical storm	Between 90 and 119 km/h
Typhoon	Between 120 and 149 km/h
Severe typhoon	Between 150 and 190 km/h
Super typhoon	Greater than 190 km/h

Source: Abdullah et al. (2015).

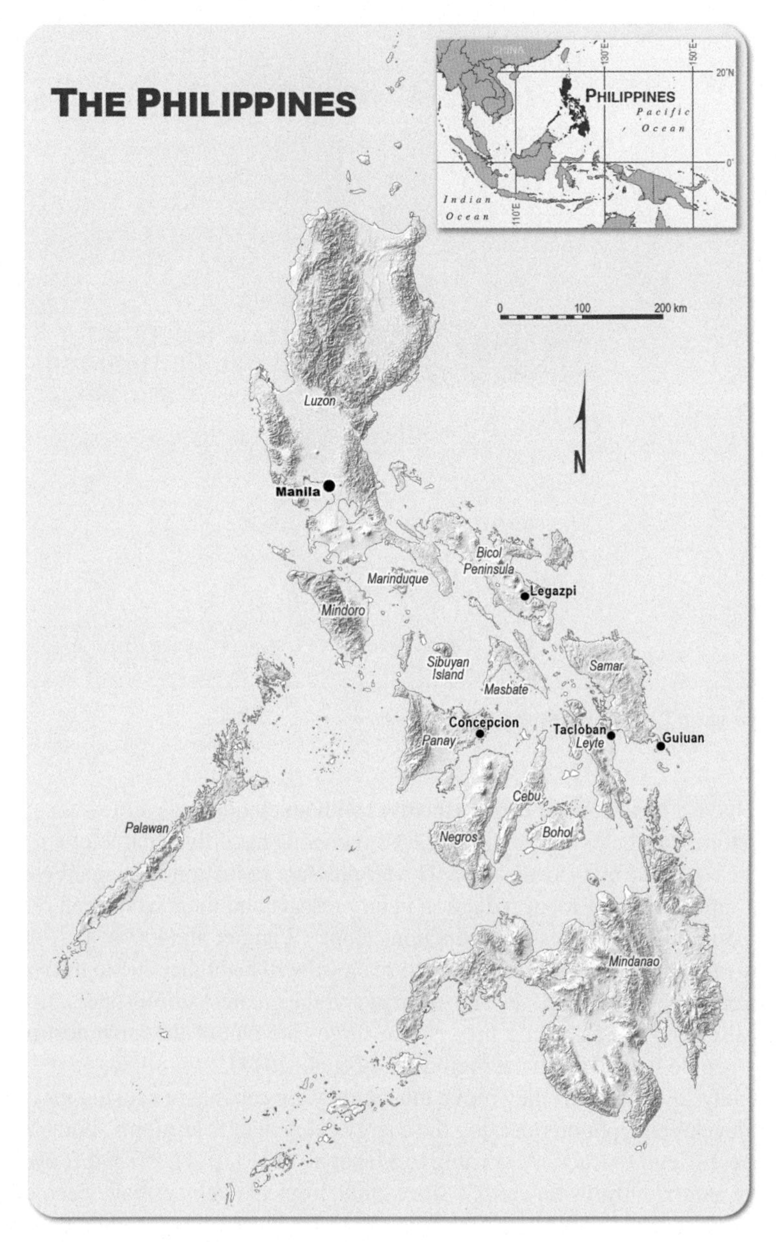

FIG. 24.2 **The Philippines, an archipelago of 7100 islands in southeast Asia.** Source: *The Authors.*

24.3 TYPHOONS AND THE PHILIPPINES

The Philippines (Fig. 24.2) are an archipelago of 7100 islands located in Southeast Asia lying between 5° and 21° North, and between 117° and 126° East. In 2017, the population of the archipelago was approximately 105 million people spread over roughly 300,000 km^2 of land area generating a population density of 352 people/km^2 (Worldometers, 2017). The seas, and life in close proximity to it, are integral components of life in the Philippines; the archipelago has approximately 36,289 km of coastline and 25,000 km^2 of coral reefs (Sheppard et al., 2009). More than 80% of its population live within 50 km of the coast (Magdaong et al., 2014) and the majority of all rice and foodstuffs are grown on land that is marginally above sea level (Broad and Cavanagh, 2011).

Much of the Philippines is at risk from typhoons and each year about 20 of them, equivalent to 25% of the total number of such events in the world, occur in the Philippines' coastal waters (Holden and Jacobson, 2012). From 1970 to 2013, 856 tropical cyclones entered Philippine waters and 322 of these were destructive (National Disaster Risk Reduction Management Council, 2014). Approximately 95% of these typhoons originate in the Pacific Ocean, south and east of the

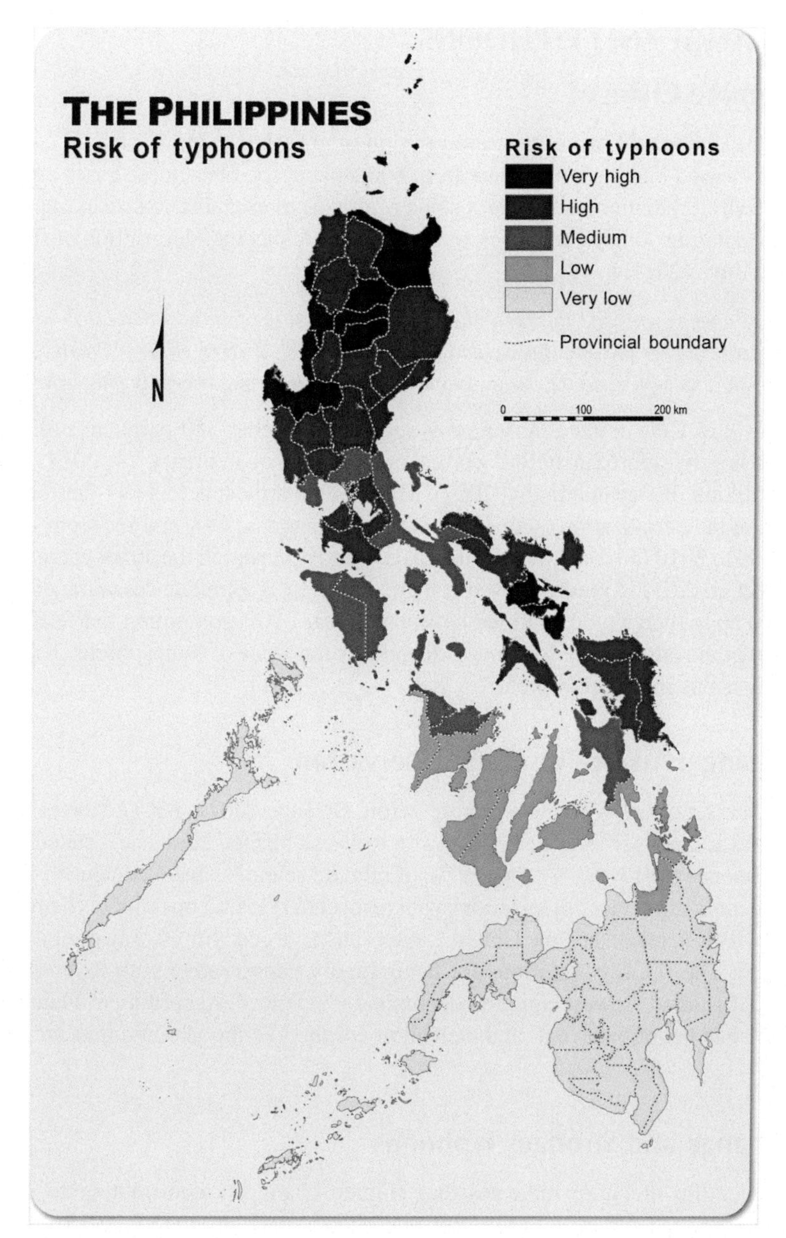

FIG. 24.3 **The Philippines, risk of typhoons.** Source: *The Authors, based on Holden and Jacobson (2012), used with the permission of Anthem Press.*

archipelago, between the months of July and November, and they travel in a northwesterly direction. They mainly affect the eastern half of the country, with the risk of typhoons on a provincial basis shown in Fig. 24.3. Mindanao and Palawan are islands with a very low risk, Mindoro, Panay, and Leyte are at moderate risk and the most heavily affected portions of the Philippines are Northern Luzon, the Bicol Peninsula, and Samar. The province of Eastern Samar, on the east side of the island of Samar, experienced 25 tropical depressions, 51 tropical storms, and 58 typhoons between 1948 and 2009 (National Disaster Risk Reduction Management Council, 2014). Although the moisture provided by these storms has somewhat of a positive effect (providing between 38% and 47% of the archipelago's average annual rainfall), overall their effects are profoundly negative because they set off landslides, cause severe and recurrent flooding of lowland areas, and are responsible for more loss of life and property than any other natural hazard. "Straddling the western edge of the Pacific Ocean," stated Primavera et al. (2016, p. 744), "the Philippines has the dubious distinction of being visited by the greatest number of storms that also are the most intense." Duncan et al. (2016, p. 773) described the archipelago as "among the most typhoon-ravaged countries in the world."

24.4 CLIMATE CHANGE AND TYPHOONS

24.4.1 What is Climate Change?

Climate change is occurring due to the increasing concentration of greenhouse gases, such as carbon dioxide (CO_2), methane, and nitrogen oxides, released into the atmosphere from human activity. Once these gases are concentrated in the atmosphere, they intercept terrestrial radiation and prevent some of this from escaping to space, trapping the energy within the lower atmosphere and re-radiating some of this back to the surface. A succinct description of the basic physics of climate change is provided by De Buys (2011, p. 10):

> *The basic physics of climate change work like this: greenhouse gases trap more of the heat that Earth would otherwise radiate back into space. The retained heat charges the atmosphere and oceans—the main drivers of the planetary climate system—with more energy, loading them with more oomph to do the things they already do, but more powerfully than before.*

The maximum safe level of CO_2 in the atmosphere is regarded as being 350 parts per million (ppm) and, in 2016, it topped 400 ppm and is rising by approximately 2 ppm every year; as of February 14, 2017, atmospheric CO_2 stood at 406.21 ppm (Scripps Institution of Oceanography, 2017). Much of the atmosphere's CO_2 has been emitted in recent years as a consequence of modern industrialization, with 75% of all anthropogenic CO_2 emitted from 1950 to 2010 and 50% having been emitted from 1980 to 2010 (Nixon, 2011). The CO_2 currently present in the atmosphere will have a warming effect for years to come; Gillett et al. (2011) found that even if there was to be a complete cessation of all CO_2 emissions in 2100 the impact of CO_2 emitted up to then would continue beyond the year 3000. According to Pfeiffer et al. (2016, p. 2), "CO_2 emissions remain resident in the atmosphere for centuries and it is the *stock* of atmospheric CO_2 that affects temperatures, rather than the flow of emissions in any given year."

24.4.2 Climate Change: An Undisputed Observation

"Although anomalies and uncertainties will always exist," wrote De Buys (2011, p. 61), "the case for a warming climate is about as solid as any scientific case will ever be." "Based on well-established evidence," stated the American Association for the Advancement of Science (2013, p. 1), "about 97% of climate scientists have concluded that human-caused climate change is real." One of the most important organizations with respect to research on climate change is the Intergovernmental Panel on Climate Change (IPCC), created by the United Nations and charged with issuing periodic assessments of the status of climate change research. The IPCC is notorious for being highly conservative with its predictions of climate change, what Flannery (2005, p. 246) called "lowest common denominator science." According to Flannery (2005, p. 246), "If the IPCC says something, you had better believe it- and then allow for the likelihood that things are far worse than it says they are."

24.4.3 Climate Change and Stronger Typhoons

A substantial body of scientific literature indicates that climate change is contributing to stronger tropical cyclones (Combest-Friedman et al., 2012; Elsner et al., 2008; Emanuel, 2005, 2013; Mei et al., 2015; Mei and Xie, 2016; Peduzzi et al., 2012; Rozynski et al., 2009; Trenberth, 2005; Takagi and Esteban, 2016; Webster et al., 2005), and this trend can be expected to continue as a consequence of the thermodynamics that drive tropical cyclones. The principal mechanism by which climate change generates stronger typhoons is the higher temperature of the world's oceans. As the surface of the oceans warms, the oceans provide more energy to convert into tropical cyclones (Elsner et al., 2008). The higher sea surface temperatures, and increased water vapor, act to increase the energy available for thunderstorm, and tropical cyclone formation (Trenberth, 2005). During 2013, for example, sea surface temperatures in an extensive area of the Western North Pacific exceeded 29 °C, providing ample thermodynamic energy for the formation of storms such as Super Typhoon Haiyan (Takagi and Esteban, 2016). The area of the world's oceans with ocean temperatures that favor cyclone development is also expanding.

An important component of how climate change leads to stronger tropical cyclones is the increase in *subsurface* sea temperatures occurring over the last 30 years. Normally, during a tropical cyclone, the violent disturbance of the ocean's surface causes an upwelling of cold water from below the ocean's surface and, as this happens, sea surface temperatures decline and this acts to serve as a natural break on the strength of tropical cyclones. Sheppard et al. (2009) report that such upwelling of cold water can reduce surface temperatures by as much as 8 °C, which is enough to reduce surface water temperatures below that needed for tropical cyclone maintenance. However, research conducted by Mei et al. (2015) shows that over the 1985–2015 period there has been a 0.75 °C rise in the temperature of the world's oceans at a depth of 75 m. These higher subsurface sea temperatures remove a natural buffer on the strength of tropical cyclones, favor rapid tropical cyclone intensification, and go a long way toward explaining why typhoon intensity from 2005 to 2015 has been, on average, the strongest over the time period from 1955 to 2015 (Mei et al., 2015). According to Mei et al. (2015), by the end of the 21st century, the average tropical storm will increase from being a category 3 (severe tropical storm) to a category 4 (typhoon) and even typhoons of moderate intensity will increase by 14%. The intensity of tropical cyclones making landfall in Southeast Asia increased by 12% from 1977 to 2014, with a doubling of the number of category 4 and 5 typhoons over this time (Mei and Xie, 2016). Takagi and Esteban (2016) predict an increase in the mean maximum tropical cyclone wind speed of between 2% and 11% by the end of the century, in association with deeper low pressures in the core of these systems. "The strengthened typhoon intensity," wrote Mei et al. (2015, p. 4), "poses heightened threats to human society."

24.4.4 Climate Change and Wetter Typhoons

Not only will climate change lead to stronger typhoons, it also will lead to *wetter* typhoons as they will carry more moisture and generate heavier rainfall. Warmer air holds more moisture and drives increased evaporation, so more heavy rainfall events are expected as the Earth warms (Milly et al., 2002; Trenberth, 2011; Westra et al., 2013). For every 1 °C increase in temperature, the air holds 7% more water (Trenberth, 2011). This means that warmer tropical storms are supplied with more moisture and, thus, will produce more intense precipitation events. Trenberth (2011, p. 128) wrote that there is "a distinct link between higher rainfall extremes and temperatures." Even tropical cyclones of weaker intensity now have very intense associated rains and these heavier rainfall events pose serious challenges for those affected by them (Thomas et al., 2013). Preparing for, and predicting runoff from, such heavy rainfall events is very difficult (Clutario and David, 2014; Milly et al., 2008).

In the Philippines, tropical cyclones approaching the archipelago may not make any landfall but may still adversely affect the islands, either as a result of their wide cloud bands, resulting in rainfall, or by interacting with other weather systems, such as the southwest monsoon and causing precipitation anomalies (Yumul et al., 2012). Such precipitation anomalies can lead to flash floods, higher sediment influx, and landslides. Consider, for example, Typhoon Ketsana (referred to in the Philippines as Typhoon Ondoy), which impacted Metro Manila on 26 September 2009, depositing 341 mm of rain in just six hours. This shattered the previous Philippine precipitation record set 42 years earlier in 1967, when 334 mm fell over a 24-h period (Holden and Jacobson, 2012). Amalie Obusan, a climate and energy campaigner for Greenpeace Southeast Asia, recalled how after Typhoon Ketsana, the Philippine Red Cross asked Greenpeace Southeast Asia to use its inflatable boats (usually used for campaign purposes) to rescue people from the flooded streets of Marikina City in Metro Manila (Personal Communication, November 4, 2009).

24.4.5 How Typhoons Track Differently and Move Faster

There are some who maintain that typhoons in the Western North Pacific have become more unpredictable, tracking (as Fig. 24.4 shows) in an *east* to *west* trajectory instead of their normal *southeast* to *northwest* trajectory (Thomas et al., 2013; Wu and Wang, 2004; Yumul et al., 2012). In December 2011, Tropical Storm Washi (known in the Philippines as Sendong) tracked from east to west over Mindanao, causing severe flooding in the city of Cagayan de Oro and killing one thousand people. In December 2012, Typhoon Bopha (known in the Philippines as Pablo) tracked from east to west over Mindanao killing a similar number of people and forcing one million people to evacuate their homes (Internal Displacement Monitoring Center, 2013). Super Typhoon Haiyan also tracked in more of an east to west trajectory instead of a southeast to northwest trajectory. Takagi and Esteban (2016, p. 218) found that typhoon landfalls between latitudes 10° and 12° North have increased by around 0.02 times per year over the time period from 1945 to 2013.

As indicated at the outset of this chapter, Super Typhoon Haiyan was believed to be the fastest storm on record and was traveling at a speed of 41 km/h when it made a landfall; this propagation speed is nearly twice as fast as the average tropical cyclone (Takagi and Esteban, 2016). Faster typhoons pose two very serious challenges for the inhabitants of the archipelago. First, faster-tracking typhoons reduce the time available for preparation and, if necessary, evacuation; second, unlike, slower moving storms, where the peak storm surge occurs before landfall, the storm surge associated with a rapidly moving storm occurs *after* landfall. During Super Typhoon Haiyan (Box 24.1), these caused problems for people who were hesitant to leave their shelters while the strong winds and heavy rains raged outside or were forced outside after their homes were destroyed by the wind and rain. The rapid speed of Haiyan, and its post-landfall storm surge, may explain why so many people were killed by the storm surge even though people in this part of the Philippines were experienced with typhoons.

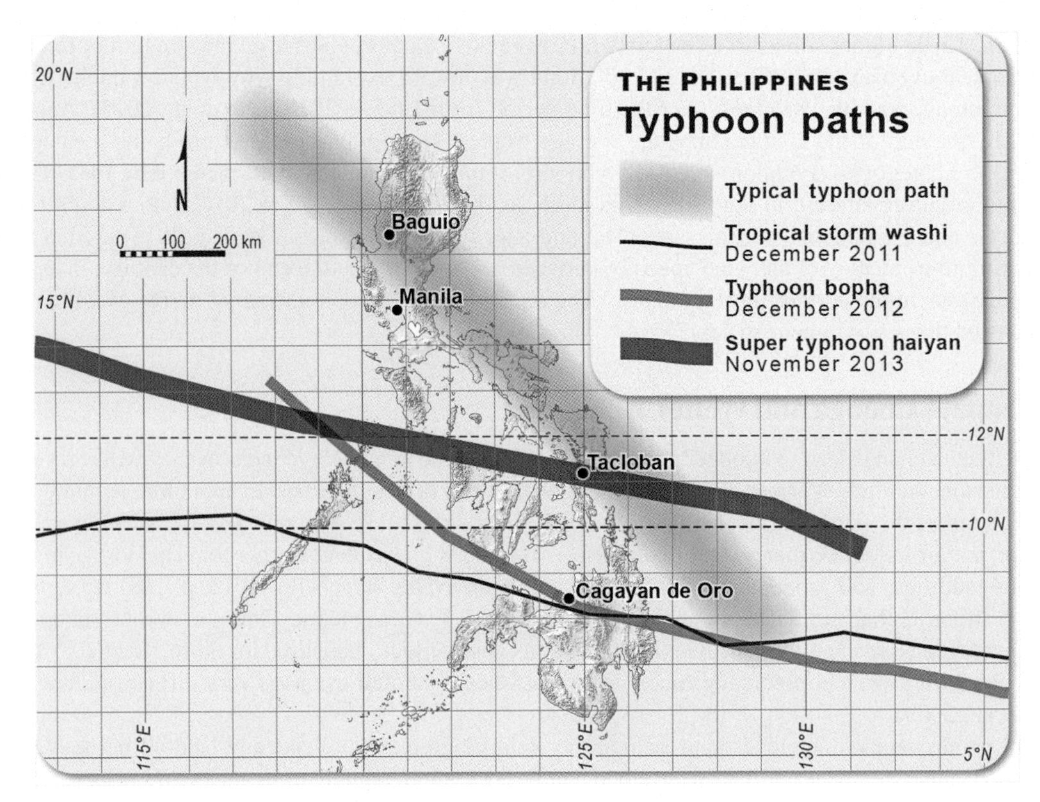

FIG. 24.4 **The Philippines, typhoon paths.** Source: *The Authors.*

24.4.6 Typhoons and Sea Level Rise

An important interface between climate change and typhoons is the occurrence of stronger typhoons concomitant with rising sea levels. More powerful typhoons, on their own, are very serious, but when taken in conjunction with higher sea levels, their severity increases even more as a higher sea level generates an even higher storm surge. The global rise in sea levels is referred to by Bellard et al. (2014, p. 203) as "one of the most certain consequences of global warming." Climate change leads to sea level rise through two mechanisms. First, as water warms, it expands; research conducted by Brauch (2012) found that an increase in global mean temperature between 2 and 4 °C will lead to a thermal expansion of the world's oceans ranging between 0.4 and 2.4 m above preindustrial levels, with much of this expansion occurring beyond the present century. Second, as the world's climate warms terrestrial glaciers will melt, particularly those in Greenland and the Arctic. The melting of mountain glaciers and the polar ice could cause sea levels to rise by an additional 0.5 m or more by the year 2100 (Church et al., 2013).

If all of the currently attainable carbon resources were burned, much of the Antarctic Ice Sheet could be lost over the next 1000 years and this would lead to a sea-level rise in the tens of meters (Winkelmann et al., 2015). Indeed, one cannot be certain that multimeter sea level rise will not occur even if we allow global warming of only 2 °C above preindustrial levels (Gillett et al., 2011; Hansen et al., 2016).

It is also important to stress that sea-level changes will not be uniform or linear across the world, due to gravitational effects, local tectonic processes, differential thermosteric changes, and potential shifts in ocean circulation (Milne et al., 2009; Rhein and Rintoul, 2013). The western tropical Pacific experienced above-average increases in the sea-level over the last two decades (Mass and Carius, 2012; Rhein and Rintoul, 2013). The Philippines themselves have experienced above-average increases in sea-level; since 1970, mean sea-level readings taken at Legazpi, in the Bicol Peninsula of Luzon, indicate an increase of 0.2 m per year (Lander et al., 2014). From 1960 to 2012, sea level in Manila Bay rose by 15 mm per year, which is approximately nine times the average rate of global sea level rise (1.7 mm per year) attributed to climate change (Morin et al., 2016). Bellard et al. (2014) estimate that sea level rise could cause the Philippines to lose up to 17% of its land area. The Philippines, along with the Caribbean and Sundaland, is one of the three places in the world most vulnerable to land loss due to sea-level rise.

TABLE 24.2 Vulnerability of Countries to Natural Hazards and Climate Change

Vulnerability ranking	Country
1	Vanuatu
2	**Philippines**
3	Tonga
4	Guatemala
5	Bangladesh
6	Solomon Islands
7	Costa Rica
8	El Salvador
9	Cambodia
10	Papua New Guinea
–	–
143	Canada

Source: Alliance Development Works (2014).

24.5 THE PHILIPPINES AND STRONGER TYPHOONS

24.5.1 The Vulnerability of the Filipino People to Stronger Typhoons

The Philippines is a country highly at risk to the capricious vicissitudes of climate change. According to Alliance Development Works, in its *World Risk Report 2014*, the archipelago is ranked second only to Vanuatu among the countries of the world most at risk to natural hazards and climate change (Table 24.2). Much of this vulnerability emanates from so many Filipinos living in close proximity to the sea. In terms of coastal population exposure to climate risks, the Philippines has the second largest number of people exposed, behind only China (a country with over 13 times as many people) and the highest percentage of people exposed (Busby et al., 2012). The combination of stronger typhoons and rising sea levels make the coastal population of the Philippines highly vulnerable to stronger typhoons; at the same time that climate change generates stronger typhoons; it also causes sea-level rise and contributes to higher storm surges such as the one associated with Super Typhoon Haiyan. "Considering the location of the Philippines within the typhoon belt," wrote Villanoy et al. (2012, p. 494), "the synergistic effects of sea level rise and waves that strong typhoons will bring, can generate wave run-up invading farther into the coast."

An important contributor to the vulnerability of the Philippines to stronger typhoons is the poverty experienced by many Filipinos. "Resilience" is defined by Smith and Vivekanda (2012, p. 77) as "the capacity of a system to withstand shocks and to rebuild and respond to change." According to the Philippine Statistical Authority, in 2015, 26% of all Filipinos were living in poverty; these people, lacking resilience are highly vulnerable to typhoons when they impact. Earlier, it was indicated that it is expected that more typhoons will impact the archipelago between latitudes of 10° and 12° North (Fig. 24.4); these are the provinces of Eastern Samar, Leyte, and Southern Leyte and in 2015 these provinces had an average poverty rate of 44.43% (Philippine Statistical Authority, 2015). This means that over time some of the archipelago's residents who are least capable of coping with stronger typhoons will become more exposed to them. The large cities of the Philippines, especially Metro Manila, are inhabited by millions of urban poor who similarly lack resilience to climate change related effects such as floods (Porio, 2011). The urban poor tend to live on the most marginal lands, such as riverbanks and coastal lowlands. In Metro Manila, a vast drainage basin, there are approximately 540,000 people living along river banks where they are vulnerable to flooding (Alliance Development Works, 2014).

Another important contributor to the vulnerability of the Philippines to stronger typhoons is the high, and rising, population of the archipelago. As mentioned earlier, the 2016 population of the Philippines was approximately 104 million people with a population density of 352 people per km^2 people per square km. This population is growing at a rate of 1.55% a year, and is projected to double in approximately 46 years (Worldometers, 2017). While "the scientific evidence linking population growth, family planning, and climate change," is what Sasser (2014, p. 103) called "complex and contradictory," such high levels of population growth make adaptation to climate change, and the development of better resilience to its effects, substantially more difficult (Engelman, 2010). By the year 2063, barring a dramatic decrease in population growth, the Philippines will have a population of roughly 210 million people, with a population density of 700 people per km^2. Such a large, and densely concentrated, population can only increase the likelihood, and consequence, of disasters, such as the one befalling Tacloban City (Box 24.1).

24.5.2 Synergies Between Stronger Typhoons and Other Types of Environmental Degradation

Something with great potential to aggravate stronger typhoons are the perverse synergies between stronger typhoons and other types of environmental degradation prevalent in the Philippines. Over the years, the archipelago has experienced environmental degradation from a number of anthropogenic sources, both local and global, and these, when taken into conjunction with stronger typhoons, make the Philippines even more vulnerable to their harmful effects. In the words of Rees (2016, p. 268), "Continuing environmental deterioration and unsustainable development practices will aggravate the country's vulnerability to climate change."

24.5.2.1 Coral Reef Loss

Coral reefs provide protection to coastal areas from the strong waves generated by typhoons (Sheppard et al., 2009; Villanoy et al., 2012). In the Philippines, however, only 5% of all reefs are estimated to be in excellent condition (Villanoy et al., 2012). The archipelago's reefs have been damaged by localized environmental degradation, which Broad and Cavanagh (1993, p. 37) described as "siltation from denuded mountains, tailings from mines, and harmful fishing techniques." While this local environmental degradation has been occurring, Philippine reefs have also been damaged by global environmental degradation, namely ocean acidification and the bleaching of reefs (Sheppard et al., 2009). The world's oceans have absorbed almost half of the CO_2 emissions since the beginning of the industrial revolution and are currently absorbing about one metric ton of CO_2 per year for every person on earth. As the oceans absorb more CO_2 they become more acidic and this reduces the ability of corals to secrete limestone and build reefs. According to the American Association for the Advancement of Science (2013), the current rate of ocean acidification is the fastest in the last 300 million years. At the same time that higher CO_2 levels prevent reef formation, the higher temperatures caused by their presence in the atmosphere causes ocean warming and this, in turn, causes a bleaching of coral reefs, wherein large tracts of coral have died. In the words of Sheppard et al. (2009, p. 278), "It is likely that reefs will be the first major ecosystem in the modern era to become ecologically extinct." At the very time that climate change produces stronger typhoons, it gravely weakens coral reefs, which are a natural protection against the high-amplitude waves generated by these stronger typhoons.

24.5.2.2 Mangrove Loss

Mangroves, the trees capable of growing in brackish water in estuaries, also provide protection from typhoons and a 100-m band of mangroves can cause a reduction of wave energy between 13% and 60% (Primavera et al., 2016). The National Disaster Risk Reduction Management Council (2014) found that after Super Typhoon Haiyan, those coastal communities with thick mangrove forests had fewer deaths than those lacking such forests. Unfortunately, the Philippines lost approximately 50% of all mangroves during the last 100 years, with much of this being attributable to mangrove removal for aquaculture, namely prawn farming (Duncan et al., 2016). Mangroves are also threatened by sea-level rise and tropical storms (Long et al., 2014). "The occurrence of high intensity typhoons in the Philippines," wrote Salmo et al. (2014, p. 86), "threaten the growth, development, and regeneration of both planted and natural mangroves." Considering that more typhoons are expected to impact the archipelago between latitude 10° and 12° North (Fig. 24.4), it is noteworthy to observe that the province of Eastern Samar has experienced a decline in mangrove forest area and from 1990 to 2010 mangrove forest area in Eastern Samar, where more typhoons are expected to make landfall, declined by 10% (Long et al., 2014). This means that a diminution of a natural defense against typhoons has occurred precisely in the area where it is most needed.

24.5.2.3 Land Subsidence Due to Groundwater Withdrawal

Many parts of the Philippines are highly vulnerable to El Niño-induced drought (Holden and Jacobson, 2012). During these droughts, which can become quite severe, extensive use is made of groundwater resources. As groundwater is withdrawn, aquifers are reduced in size, and this causes land subsidence (Rodolfo and Siringan, 2006). Land subsidence is a particularly serious concern in the area around Metro Manila where sea level has risen very quickly. In Metro Manila, the entirety of sea level rise cannot be attributed to a rise in the world's oceans and some of it must be attributed to land subsidence; as Bankoff (2003, p. 230) stated, "Such an increase cannot be explained as solely a consequence of global warming and bears

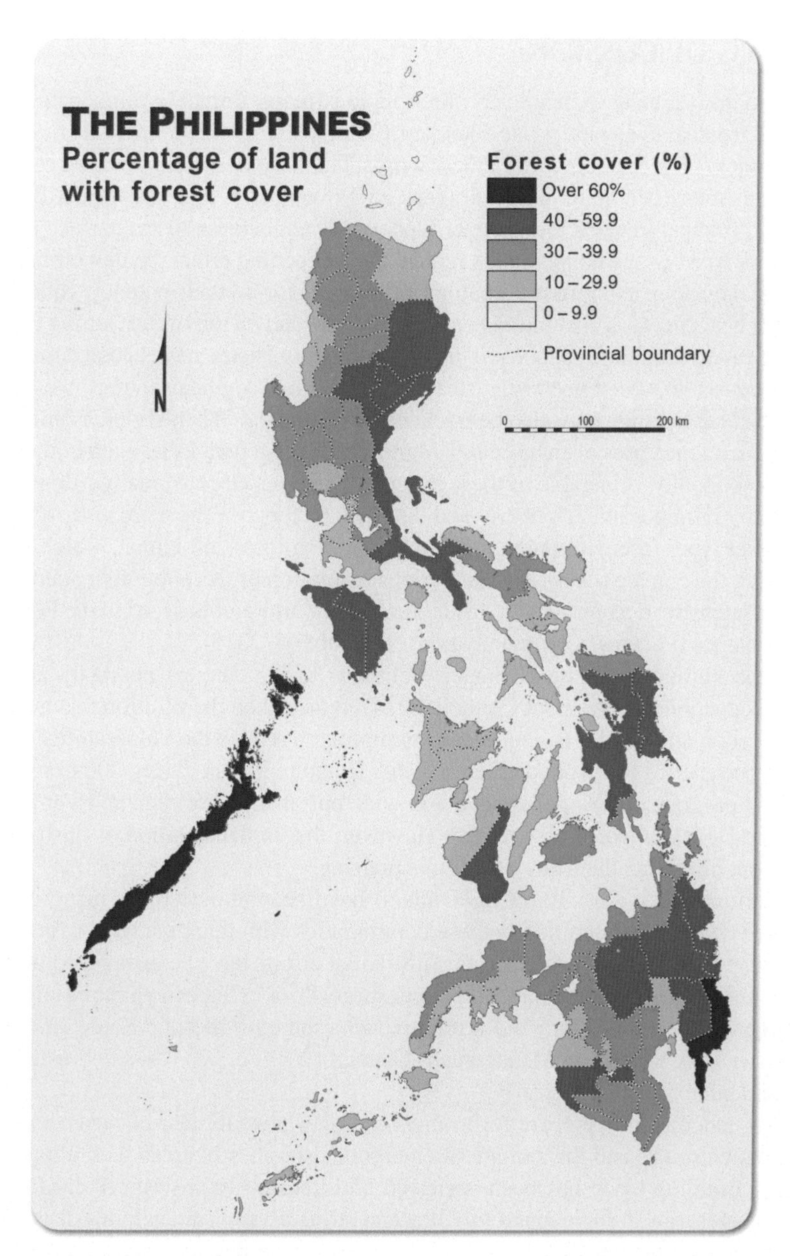

FIG. 24.5 **The Philippines, percentage of land with forest cover.** Source: *The Authors, based on Holden and Jacobson (2012), used with the permission of Anthem Press.*

a marked correlation to the rise in both groundwater extraction and population growth." Land subsidence, concomitant with sea level rise and stronger typhoons makes the archipelago even more vulnerable to stronger typhoons. At the very time, stronger typhoons send sea water ashore as storm surge, the land onto which the storm surge is flowing is subsiding.

24.5.2.4 Deforestation

Another form of localized environmental degradation with a perverse synergy with stronger typhoons is deforestation (Fig. 24.5). In 1521, when the Spanish first arrived in the Philippines, 95% of the land area of the islands was covered by forests. By 1934, 57% of the land area had forest cover, by 1990 only 20% of the archipelago's land area had forest cover, and in 2010 it was estimated that only 8% of the original forest cover remained (Rees, 2016). Should there be a typhoon (with its associated heavy rains) landslides will occur much more quickly on deforested hillsides. Deforestation also reduces the mitigating effect availed to inland locations by their remoteness from the ocean since the deforestation of coastal areas allows typhoons to penetrate further inland and inflict damage over wider areas (Myers, 1988).

24.6 CONCLUDING DISCUSSION

This chapter has discussed how climate change is contributing to stronger tropical cyclones in the Western North Pacific. Climate change enhanced tropical cyclones, wrote Flannery (2005, p. 314), "have the potential to kill many more people than the largest terrorist attack"; 8 years after writing these words, Flannery was (sadly) proved correct when Super Typhoon Haiyan killed over twice as many people in the Philippines on November 8, 2013 as were killed in the United States on September 11, 2001. The principal impetus for stronger typhoons is an increase in sea surface temperatures. While this is clearly occurring, higher *subsurface* sea temperatures reduce the dampening effect the upwelling of cold water usually has on the strength of a tropical cyclone and remove a natural limiting factor on their potency. Since warmer air carries more moisture, additional latent heat energy is also fueling stronger storms, and on top of this, strong typhoons are carrying more moisture and generating heavier rainfall. Typhoons in the Western North Pacific have become more unpredictable and seem to be tracking in more of an east to west trajectory, instead of their normal southeast to northwest trajectory, and (at least in the case of Super Typhoon Haiyan) they may also be tracking more quickly. When stronger typhoons are superimposed on higher sea levels, their severity increases even more as a higher sea level generates an even higher storm surge. The islands of the Philippines are exceptionally vulnerable to these stronger typhoons since so many Filipinos live in close proximity to the sea and since so many Filipinos live in poverty and thus lack resilience. There are also substantial synergies between stronger typhoons and other types of environmental degradation, both local and global, which impact the archipelago. In the opinion of Dr. Wei Mei, the climate scientist at the Scripps Institution of Oceanography, people in the Philippines have to be concerned about the intensity of tropical cyclones in the coming future; if he lived in the Philippines he would be very worried about tropical cyclones (Personal Communication, November 4, 2015).

The government of the Philippines is by no means oblivious to the dangers posed by amplified tropical cyclones (National Disaster Risk Reduction Management Council, 2014). It has taken the position that a disaster is a natural hazard that impacts a vulnerable population and it has set about attempting to reduce the vulnerability of the archipelago's population; after the flooding occasioned by Typhoon Ketsana and Typhoon Parma during 2009 the government implemented an ambitious flood control program ensuring there is a 3-m wide buffer zone between all river banks and residential areas in Metro Manila (Alliance Development Works, 2014). However, the implementation of this policy will displace at least 500,000 urban poor without providing them any alternative housing.

One billion Pesos (approximately 26 million US Dollars) have been allocated for mangrove forest rehabilitation to reduce the vulnerability of coastal communities as those communities with thick mangrove forests had fewer deaths after Super Typhoon Haiyan than those lacking such forests. Substantial efforts have been engaged in to ensure better warnings are given about approaching tropical cyclones and their storm surge. Prior to Super Typhoon Haiyan, many people had been warned about storm surge but they stated they did not know what the term meant. "Some emergency warning officers," wrote the National Disaster Risk Reduction Management Council (2014, p. 32), "avoided using the term storm surge as they themselves were hard put explaining it clearly."

Efforts have also been undertaken to prepare for tropical cyclones prior to their occurrence; this involves the preparation of emergency evacuation routes and the storage of emergency supplies in areas with a high vulnerability to tropical cyclones. The Philippine Building Code has been reviewed and updated to ensure critical facilities such as hospitals, schools, and evacuation centers are strong enough to withstand stronger tropical cyclones. It is, however, difficult to rely upon technology as being something capable of preventing a large-scale disaster; perhaps, the best example of this was Marikina City, the area of Metro Manila severely flooded by Typhoon Ketsana in September 2009. In October 2008, the Asian Development Bank issued a report touting the flood control measures in Marikina City, declaring it "Flood Ready Marikina City" (Asian Development Bank, 2008). Nevertheless, less than 1 year later, Marikina City was severely flooded and images of its residents attempting to cope with the floodwaters were displayed worldwide by the global media (British Broadcasting Corporation, 2009).

An interesting dimension of climate change amplified typhoons in the Philippines is how they are a departure from the slow violence inherent in climate change. In many ways, climate change is an example of what Nixon (2011, p. 2) calls "a violence that occurs gradually and out of sight, a violence of delayed destruction that is dispensed across time and space, an attritional violence that is typically not viewed as violence at all." Climate change *usually* has consequences, such as gradually receding glaciers, occurring *slowly* in what Nixon (2011, p. 6) calls "unspectacular time." This goes a long way toward explaining why politicians are so reluctant to engage in policies mandating the deep emission cuts necessary to address climate change; in the words of Nixon (2011, p. 9), "Because preventative or remedial environmental legislation targets slow violence, it cannot deliver dependable electoral cycle results, even though those results may ultimately be life saving." Spectacularly violent tropical storms, however, transcend the slow violence inherent in climate change. Unlike a gradually retreating glacier, Super Typhoon Haiyan did not need a montage of images taken over several decades to reveal the effects of climate change- pictures of the City of Tacloban taken on November 7–8, 2013 would be more than enough evidence. Climate change amplified typhoons take the discussion of climate change and move it from unspectacular time and place it in what Nixon (2011, p. 6) would call "spectacular time." As Flannery (2005, p. 314) wrote, tropical cyclones "focus attention on climate change in a way that few other natural phenomena do."

Ultimately, the discussion of how climate change contributes to stronger tropical cyclones must involve a discussion of climate injustice, the situation where some people enjoy the benefits of energy use, and other emission-generating activities, while those activities cause other people to suffer the burdens of climate change (Bell, 2013). At the 19th yearly session of the UNFCCC Conference of the Parties (COP 19), coincidentally held in Warsaw, Poland, in the days immediately after Super Typhoon Haiyan, Yeb Sano, the lead negotiator of the Philippine delegation (and a Tacloban City resident), gave an impassioned speech calling for urgent action on climate change; in his speech, Sano announced he would commence a fast during COP 19 in order to be in solidarity with the people of Tacloban (Vidal, 2014). Sano's speech and fast were heart-wrenching examples of the need for climate justice. The Philippines (as indicated in Table 24.2) is second only to Vanuatu in terms of its vulnerability to natural hazards and climate change.

While the archipelago is highly vulnerable to the effects of climate change, most notably stronger typhoons, it bears a low responsibility for causing climate change. In 2013, the Philippines emitted only 1.007 metric tons of CO_2 for every person in the archipelago (World Bank, 2017). In contrast, Canada (the country of residence of this Chapter's authors), ranked 143rd in terms of its vulnerability to the effects of climate change (Table 24.2) while emitting 20.65 metric tons of CO_2 for every Canadian—over 20 times as much CO_2 on a per capita basis as the Philippines (Conference Board of Canada, 2017). This means that countries such as Canada, which experience a substantially lower risk to climate change while also doing disproportionately more to cause it, have a responsibility to reduce their emissions and to assist those countries most adversely impacted by climate change. Canada, unfortunately, has been a country reluctant to make the transition to a low carbon economy (Rodgers, 2016). Canada is among the few countries that might (at least in the short term) benefit from a warming climate as crops may be grown further north and the loss of Arctic sea ice opens up new fuel and mineral resources for exploitation. Canada also has large supplies of oil in its tar sands, which themselves require large amounts of energy to extract and thus generate large CO_2 emissions; Alberta (the province of residence of this Chapter's authors) hosts the tar sands and each Albertan was responsible for emitting 66.67 metric tons of CO_2 in 2013 (Conference Board of Canada, 2017). Nevertheless, as humans we only have one planet and all of humanity must share this planet. The countries causing the problem of climate change and standing to benefit from causing it must assist those countries that do not cause the problem and stand only to be hurt by it. In the words of Pope Francis (2015, p. 125), "Reducing greenhouse gases requires honesty, courage, and responsibility, above all on the part of those countries which are more powerful and pollute the most."

REFERENCES

Abdullah, K., Anukklarmphai, A., Kawasaki, T., Neopmuceno, D., 2015. A tale of three cities: water disaster policy responses in Bangkok, Kuala Lumpur, and Metro Manila. Water Policy 17 (S1), 89–113.

Alliance Development Works, 2014. World risk report 2014. Alliance Development Works, Berlin.

American Association for the Advancement of Science, 2013. What we know: the reality, risk, and response to climate change. https://www.google.ca/search?q=American(Association(for(the(Advancement(of(Science((2013).(What(We(Know:(The(Reality,(Risk,(and(Response(to(Climate(Change.(&ie=utf-8&oe=utf-8&gws_rd=cr&ei=iIqkWKSPNsyKjwSU7Z7YBw (accessed 15.02.17).

Asian Development Bank, 2008. Country water action: flood-ready Marikina city. https://www.adb.org/results/country-water-action-flood-ready-marikina-city (accessed 15.02.17).

Bankoff, G., 2003. Cultures of Disaster: Society and Natural Hazard in the Philippines. Routledge, London.

Bell, D., 2013. How should we think about climate justice? Environ. Ethics 35 (2), 189–208.

Bellard, C., Leclerc, C., Courchamp, F., 2014. Impact of sea level rise on the 10 insular biodiversity hotspots. Global Ecol. Biogeogr. 23 (2), 203–212.

Brauch, H.G., 2012. Policy responses to climate change in the Mediterranean and MENA region during the Anthropocene. In: Scheffran et al., (Ed.), Climate Change, Human Security and Violent Conflict. Springer, Berlin, Germany, pp. 719–794.

British Broadcasting Corporation, 2009. In pictures: Philippines floods. Online: http://news.bbc.co.uk/2/hi/in_pictures/8276374.stm (accessed 15.02.17).

Broad, R., Cavanagh, J., 1993. Plundering Paradise: The Struggle for the Environment in the Philippines. University of California Press, Berkley, CA.

Broad, R., Cavanagh, J., 2011. Reframing development in the age of vulnerability: from case studies of the Philippines and Trinidad to new measures of rootedness. Third World Q. 32 (6), 1127–1145.

Busby, J.W., Smith, T.G., White, K.L., Strange, S.M., 2012. Locating climate insecurity: where are the most vulnerable places in Africa. In: Scheffran, J. et al., (Ed.), Climate Change, Human Security and Violent Conflict. Springer, Berlin, Germany, pp. 463–511.

Cardenas, M.B., Bennett, P.C., Zamora, P.B., Befus, K.M., Rodolfo, R.S., Cabria, H.B., Lapus, M.R., 2015. Devastation of aquifers from Tsunami-like storm surge by super Typhoon Haiyan. Geophys. Res. Lett. 42 (8), 2844–2851.

Church, J.A., et al., 2013. Sea level change. In: Stocker, T.F. et al., (Ed.), Climate Change 2013: The Physical Science Basis. Cambridge University Press, Cambridge, pp. 1137–1216, Contribution of Working Group I to the Fifth Assessment Report of the Intergovernmental Panel on Climate Change.

Clutario, M.V.A., David, C.P.C., 2014. Event-based soil erosion estimation in a tropical watershed. Int. J. For. Soil Erosion 4 (2), 51–57.

Combest-Friedman, C., Christie, P., Miles, E., 2012. Household perceptions of coastal hazards and climate change in the central Philippines. J. Environ. Manage. 112 (1), 137–148.

Conference Board of Canada, 2017. Greenhouse gas emissions. http://www.conferenceboard.ca/hcp/provincial/environment/ghg-emissions.aspx (accessed 15.02.17).

De Buys, W., 2011. A Great Aridness: Climate Change and the Future of the American Southwest. Oxford University Press, New York, NY.

Duncan, C., Primavera, J.H., Pettorelli, N., Thompson, J.R., Loma, R.J., Koldewey, H.J., 2016. Rehabilitating mangrove ecosystem services: a case study on the relative benefits of abandoned pond reversion from Panay island, Philippines. Mar. Pollut. Bull. 109 (2), 772–782.

Elsner, J.B., Kossin, J.P., Jagger, T.H., 2008. The increasing intensity of the strongest tropical cyclones. Nature 455 (7209), 92–95.

Emanuel, K.A., 2005. Increasing destructiveness of tropical cyclones over the past 30 years. Nature 436 (4), 686–688.

Emanuel, K.A., 2013. Downscaling CMIP5 climate models shows increased tropical cyclone activity over the 21st century. Proc. Natl. Acad. Sci. U.S.A. 110 (30), 12219–12224.

Engelman, R., 2010. Population, climate change, and women's live. Worldwatch report 183. Worldwatch Institute, Washington, DC.

Flannery, T., 2005. The Weathermakers: How We are Changing the Climate and What It Means for Life on Earth. Harper Collins, Toronto, Canada.

Gillett, N.P., Arora, V.K., Zickfeld, K., Marshall, S.J., Merryfield, W.J., 2011. Ongoing climate change following a complete cessation of carbon dioxide emissions. Nat. Geosci. 4 (1), 83–87.

Hansen, J., Sato, M., Hearty, P., Ruedy, R., Kelley, M., Masson-Delmotte, V., Russell, G., Tselioudis, G., Cao, J., Rignot, E., Velicogna, I., Tormey, B., Donovan, B., Kandiano, E., von Schuckmann, K., Kharecha, P., Legrande, A.N., Bauer, M., Lo, K.W., 2016. Ice melt, sea level rise and superstorms: evidence from paleoclimate data, climate modeling, and modern observations that 2 °C global warming could be dangerous. Atmos. Chem. Phys. 16 (6), 3761–3812.

Holden, W.N., Jacobson, R.D., 2012. Mining and Natural Hazard Vulnerability in the Philippines: Digging to Development or Digging to Disaster? Anthem Press, London, UK.

Internal Displacement Monitoring Center, 2013. Living in the Shadows: Displaced Lumads Locked in a Cycle of Poverty. Internal Displacement Monitoring Center, Geneva, Switzerland.

Lander, M., Guard, C., Camargo, S.J., 2014. Tropical cyclones, Super-typhoon Haiyan, in "state of the climate in 2013", J. Blunden and D.S. Arndt (editors). Bull. Am. Meteorol. Soc. 95 (7), S112–S114.

Long, J., Napton, D., Giri, C., Graesser, 2014. A mapping and monitoring assessment of the Philippines' mangrove forests from 1990 to 2010. J. Coastal Res. 30 (2), 260–271.

Loy, K.C., Sinha, P.C., Liew, J., Tangang, F., Husain, M.L., 2014. Modeling storm surge associated with super Typhoon Durian in South China Sea. Nat. Hazards 70 (1), 23–37.

Magdaong, E.T., Fujii, M., Yamano, H., Licuanan, Y., Maypa, A., Campos, W.L., Alcala, A.C., White, A.T., Apistar, D., Martinez, R., 2014. Long-term change in coral cover and the effectiveness of marine protected areas in the Philippines: a meta-analysis. Hydrobiologia 733 (1), 5–17.

Mass, A., Carius, A., 2012. Territorial integrity and sovereignty: climate change and security in the Pacific and beyond. In: Scheffran, J. et al., (Ed.), Climate Change, Human Security and Violent Conflict. Springer, Berlin, Germany, pp. 651–665.

Mei, W., Xie, S.P., 2016. Intensification of landfalling Typhoons over the northwest Pacific since the late 1970s. Nat. Geosci. 9 (10), 753–757.

Mei, W., Xie, S.P., Premeau, F., McWilliams, J.C., Pasquero, C., 2015. Northwestern Pacific Typhoon intensity controlled by changes in ocean temperatures. Sci. Adv. 4 (1), 1–8.

Milly, P.C.D., Wetherald, R.T., Dunne, K.A., Delworth, T.L., 2002. Increasing risk of great floods in a changing climate. Nature 415 (6871), 514–517.

Milly, P.C.D., Betancourt, J., Falkenmark, M., Hirsch, R.M., Kundzewicz, Z.W., Lettenmaier, D.P., Stouffer, R.J., 2008. Stationarity is dead: whither water management? Science 319 (5863), 573–574.

Milne, G.A., Gehrels, W.R., Hughes, C.W., Tamisiea, M.E., 2009. Identifying the causes of sea-level change. Nat. Geosci. 2 (7), 471–478.

Morin, V.M., Warnitchai, P., Weesakul, S., 2016. Storm surge hazard in Manila bay: Typhoon (Pedring) and the SW monsoon. Nat. Hazards 81 (3), 1569–1588.

Myers, N., 1988. Environmental degradation and some economic consequences in the Philippines. Environ. Conserv. 15 (3), 303–311.

National Disaster Risk Reduction Management Council, 2014. It Happened: Learning from Typhoon Yolanda. National Disaster Risk Reduction Management Council, Quezon City.

Nixon, R., 2011. Slow Violence and the Environmentalism of the Poor. Harvard University Press, London, UK.

Peduzzi, P., Chatenoux, B., Dao, H., De Bono, A., Herold, C., Kossin, J., Mouton, F., Nordbeck, O., 2012. Global trends in tropical cyclone risk. Nat. Clim. Change 2, 289–294.

Pfeiffer, A., Millar, R., Hepburn, C., Beinhocker, E., 2016. The "2 °C Capital Stock" for electricity generation: committed cumulative carbon emissions from the electricity generation sector and the transition to a green economy. Appl. Energy 179, 1395–1408.

Philippine Statistical Authority, 2015. Official Poverty Statistics of the Philippines. Philippine Statistical Authority, Quezon City.

Pope Francis, 2015. Laudato Si: On Care for Our Common Home. Vatican, Vatican City, http://w2.vatican.va/content/francesco/en/encyclicals/documents/papa-francesco_20150524_enciclica-laudato-si.html (accessed 16.06.16).

Porio, E., 2011. Vulnerability, adaptation, and resilience to floods and climate change-related risks among marginal, riverine communities in Metro-Manila. Asian J. Social Sci. 39 (4), 425–445.

Primavera, J.H., de la Cruz, M., Montilijao, C., Consunji, H., de la Paz, M., Rollon, R.N., Maranan, K., Samson, M.S., Blanco, A., 2016. Preliminary assessment of post-Haiyan mangrove damage and short-term recovery in Eastern Samar, Central Philippines. Mar. Pollut. Bull. 109 (2), 744–750.

Rees, C., 2016. The Philippines: A Natural History. Ateneo de Manila University Press, Quezon City.

Rhein, M., Rintoul, S.R., 2013. Observations: ocean. In: Stocker, T.F. et al., (Ed.), Climate Change 2013: The Physical Science Basis. Cambridge University Press, Cambridge, pp. 255–316, Contribution of Working Group I to the Fifth Assessment Report of the Intergovernmental Panel on Climate Change.

Rodgers, P., 2016. Irregular War: ISIS and the New Threat from the Margins. IB Tauris, London, UK.

Rodolfo, K.S., Siringan, F.S., 2006. Global sea-level rise is recognized, but flooding from anthropogenic land subsidence is ignored around Northern Manila bay, Philippines. Disasters 30 (1), 118–139.

Rozynski, G., Hung, N.M., Ostrowski, R., 2009. Climate change related rise of extreme Typhoon power and duration over South-East Asia seas. Coast. Eng. J. 51 (3), 205–222.

Salmo, S.G., Lovelock, C.E., Duke, N.C., 2014. Assessment of vegetation and soil conditions in restored mangroves interrupted by severe tropical Typhoon Chan-hom in the Philippines. Hydrobiologia 733 (1), 85–102.

Sasser, J.S., 2014. The wave of the future? Youth advocacy at the Nexus of population and climate change. Geogr. J. 180 (2), 102–110.

Scripps Institution of Oceanography, 2017. https://scripps.ucsd.edu/programs/keelingcurve/ (accessed 15.02.17).

Sheppard, C.R., Davy, S.K., Pilling, G.M., 2009. The Biology of Coral Reefs. Oxford University Press, Oxford, UK.

Smith, D., Vivekananda, J., 2012. Climate change, conflict and fragility: getting the institutions right. In: Scheffran, J. et al., (Ed.), Climate Change, Human Security and Violent Conflict. Springer, Berlin, Germany, pp. 77–90.

Syvitski, J.P., Kettner, A., 2011. Sediment flux and the Anthropocene. Philos. Trans. R. Soc. A 369 (1938), 957–975.

Takagi, H., Esteban, M., 2016. Statistics of tropical cyclone landfalls in the Philippines: unusual characteristics of 2013 Typhoon Haiyan. Nat. Hazards 80 (1), 211–222.

Takagi, H., Esteban, M., Shibayama, T., Mikami, T., Matsumaru, R., De Leon, M., Thao, N.D., Oyama, T., Nakamura, R., 2015. Track analysis, simulation, and field survey of the 2013 Typhoon Haiyan storm surge. J. Flood Risk Manage. 8 (4), 1–11.

Thomas, V., Albert, J.R.G., Perez, R.T., 2013. Climate-related disasters in Asia and the Pacific. Asian Development Bank Economics Working Paper Series No. 358. Asian Development Bank, Manila.

Trenberth, K., 2005. Uncertainty in hurricanes and global warming. Science 308 (5729), 1753–1754.

Trenberth, K.E., 2011. Changes in precipitation with climate change. Clim. Res. 47 (1), 123–138.

Vidal, J., 2014. Yeb Sano: unlikely climate justice star. https://www.theguardian.com/environment/2014/apr/01/yeb-sano-typhoon-haiyan-un-climate-talks (accessed 21.10.16).

Villanoy, C., David, L., Cabrera, O., Atrigenio, M., Siringan, F., Alino, P., Villaluz, M., 2012. Coral reef ecosystems protect shore from high-energy waves under climate change scenarios. Clim. Change 112 (2), 493–505.

Wang, Y.H., Lee, I.H., Wang, D.P., 2005. Typhoon induced extreme coastal surge: a case study at Northeast Taiwan in 1994. J. Coast. Res. 21 (3), 548–552.

Webster, P.J., Holland, G.J., Curry, J.A., Chang, H.R., 2005. Changes in tropical cyclone number, duration, and intensity in a warming environment. Science 309 (5742), 1844–1846.

Westra, S., Alexander, L.V., Zwiers, F.W., 2013. Global increasing trends in annual maximum daily precipitation. J. Clim. 26 (3), 3904–3918.

Winkelmann, R., Levermann, A., Ridgwell, A., Caldeira, K., 2015. Combustion of available fossil fuel resources sufficient to eliminate the Antarctic Ice Sheet. Sci. Adv. 1, 1–5.

World Bank, 2017. Per capita CO_2 emissions 2013. http://data.worldbank.org/indicator/EN.ATM. CO2E.KT?view=chart (accessed 13.02.17).

Worldometers, 2017. Population of the Philippines. http://www.worldometers.info/world-population/population-by-country (accessed 13.02.17).

Wu, L., Wang, B., 2004. Assessing impacts of global warming on tropical cyclone tracks. J. Clim. 17 (8), 1686–1698.

Yumul, G.P., Cruz, N.A., Servando, N.T., Dimalanta, C.B., 2012. Extreme weather events and related disasters in the Philippines, 2004–08: a sign of what climate change will mean? Disasters 35 (2), 362–382.

Chapter 25

The Role of Disaster Medicine in Disaster Management and Preparedness

Hüseyin KoÇak*,**
*Çanakkale Onsekiz Mart University, School of Health, Department of Emergency and Disaster Management, Turkey
**Bezmialem Vakif University, Institute of Health Science, Disaster Medicine Doctorate Program, Turkey

25.1 INTRODUCTION

The health of individuals and the society are physically, socially, psychologically, and economically affected in an adverse manner by natural and technological disasters (James et al., 2010). Disasters are complex events leading to rapid and multidimensional emergency health services in the short term and public health and psychosocial problems in the long term (Hogan and Burstein, 2010). Annually, a total of 270 million people are affected by natural disasters as of 1990 (Leaning and Guha-Sapir, 2013). Disasters have started to pose a greater risk to people with each passing day. The factors such as the rapid growth of population in the last century, the risks resulting from technological development (nuclear technology, chemical risks, etc.), climate change, increasing imbalances between countries, and the reduction of world resources pose an even greater risk to people. A professional approach is essential in managing such risks. The main purpose of disaster management is to prevent harm to any living being and requires a multidisciplinary approach (Altıntaş, 2007b). Today, "the Integrated Disaster Management System," which takes into consideration all and any dangers in a comprehensive manner, is commonly used in disaster management. The system consists of the phases of operation, risk reduction, preparedness, response, and recovery (Maya and Çalışkan, 2015). The most important element in disasters is people. For that reason, health is one of the most critical sectors during predisaster, disaster, and postdisaster UN, 2014.

This chapter explains the studies on disaster management and the discipline of disaster medicine (DM) in preparedness activities based on case studies.

25.2 DISASTER RISK MANAGEMENT

The fundamental strategy of risk reduction in disasters is the activities of risk evaluation, risk reduction, and risk insurance (Ardalan et al., 2015). Risk in disasters in terms of medical services is associated with healthcare system during and after any event causing a disaster (Jaiswal et al., 2016). Emergency managers in the healthcare system and hospital management should provide healthcare facilities and perform a comprehensive risk management as part of disaster management. It is essential that such management follows these criteria: (1) The physical structure of the health system and the risk of damage to the infrastructure; (2) patients, visitors, and staff risks within anxiety and panic-related hazards; (3) the risks arising out of the failure of other facilities in selecting patients and procedures so as to provide future services due to the lack of protection, damage, and contamination of the facility; (4) legal and economic problems resulting from the drawbacks in the actions of the hospital staff during or after an incident.

Many disasters and major accidents have negative impacts on human health in the short, medium, and long term. These impacts place considerably a higher burden on the healthcare system (Boyd et al., 2014). Disaster plans and management aiming to meet such demands have a sophisticated nature. In general, emergency and disaster plans seek the effective and efficient use of resources on a local, regional, and national scale. Strengthening response capacity at local level, planning, practicing, and learning from mistakes would increase resistance and ensure a better response. Depending on the nature and severity of a danger, the mobilization of professional health equipment and resources or the provision of a long-term healthcare service in the disaster area may be necessary (Boyd et al., 2014).

Integrating Disaster Science and Management. http://dx.doi.org/10.1016/B978-0-12-812056-9.00025-7

Capacity development: The process by which people, organizations, and society systematically stimulate and develop their capacities over time to achieve social and economic goals by improvement of knowledge, skills, systems, and institutions.
Critical facilities: The primary physical structures, technical facilities, and systems which are socially, economically, or operationally essential to the functioning of a society or community, both in routine circumstances and in the extreme circumstances of an emergency.
Disaster health: The main elements of the health system are public health, prehospital emergency medical services and the preparation of hospitals for risk and crisis management; protection, prevention, preparedness principles, and the society is all the activities to increase the resistance against disaster.
Disaster medicine (DM): It is a science that studies and researches to be able to intervene in all the health problems caused by disasters, with many specialties of medical sciences, outside the hospital (in field, wreckage, field hospitals), in cooperation with other disciplines in disaster management.
Disaster risk management: The systematic process of using administrative directives, organizations, and operational skills and capacities to implement strategies, policies, and improved coping capacities in order to lessen the adverse impacts of hazards and the possibilities of a disaster.
Disaster risk reduction (DRR): The concept and practice of reducing disaster risks through systematic efforts to analyze and manage the causal factors of disasters, by reduced exposure to hazards, lessened vulnerability of people and property, wise management of land and the environment, and improved preparedness for adverse events.
Emergency medical services (EMS): All of the medical services provided by specially trained teams in the field, at the scene of the medical device support, at the time of transfer to the hospital and at the hospital.
Emergency medicine (EM): The discipline of the first care and treatment of the patients brought to the hospitals and urgently needed to be looked after (after triage classification).
Hazard: A dangerous phenomenon, substance, human activity, or condition that may cause loss of life, injury, or other health impacts, property damage, loss of livelihoods and services, social and economic disruption, or environmental damage.
Healthcare facilities: Health-care facilities are hospitals, primary health-care centres, isolation camps, burn patient units, feeding centres and others. In emergency situations, health-care facilities are often faced with an exceptionally high number of patients, some of whom may require specific medical care (e.g. treatment of chemical poisonings).
Preparedness: The knowledge and capacities developed by governments, professional response and recovery organizations, communities and individuals to effectively anticipate, respond to, and recover from, the impacts of likely, imminent, or current hazard events or conditions.
Prevention: The outright avoidance of adverse impacts of hazards and related disasters.
Resilience: The ability of a system, community, or society exposed to hazards to resist, absorb, accommodate to, and recover from the effects of a hazard in a timely and efficient manner by the preservation and restoration of its essential basic structures and functions.
Response: The provision of emergency services and public assistance during or immediately after a disaster in order to save lives, reduce health impacts, ensure public safety, and meet the basic subsistence needs of the people affected.
Risk: The combination of the probability of an event and its negative consequences.
Vulnerability: The characteristics and circumstances of a community, system, or asset that make it susceptible to the damaging effects of a hazard.

Mitigation is defined as comprehensive measures to be taken prior to disasters so as to prevent diseases, injuries, and death and also to limit the structural losses (Gougelet, 2016). The efforts taken to mitigate potential hazard (in particular, international DRR and disaster risk management) have been in favor of the disaster preparedness cycle. Following the principles and practices of damage reduction for disasters completely has become a must due to the deaths, diseases, injuries, psychological effects, those who are forced to emigrate from their home or from the community, and social and financial results in relation to disasters. The Third United Nations World Conference on Disaster Risk Reduction was held by the United Nations in the city of Sendai in Japan in 2015. As the final declaration of the conference, the member states determined the framework of action toward DRR between the years 2015 and 2030. The "Sendai Framework for Disaster Risk Reduction 2015–2030" features important goals on DM. These are the following global targets: (i) to substantially reduce global disaster mortality by 2030, aiming to lower the average per 100,000 global mortality rate in the decade 2020–2030 compared to the period 2005–2015 (18/a); (ii) to substantially reduce the number of affected people globally by 2030, aiming to lower the average global figure per 100,000 in the decade 2020–2030 compared to the period 2005–2015 (18/b); and (iii) to substantially reduce disaster damage to critical infrastructure and disruption of basic services, among them health and educational facilities, by developing their resilience by 2030 (18/d) (UNISDR, 2015b). DM, as a term, has been first mentioned in Sendai Framework, among DRR Frameworks. The UN framework action plans on risk reduction strongly emphasize the strengthening and increasing of the resistance of health services in disasters (UNISDR, 2015a).

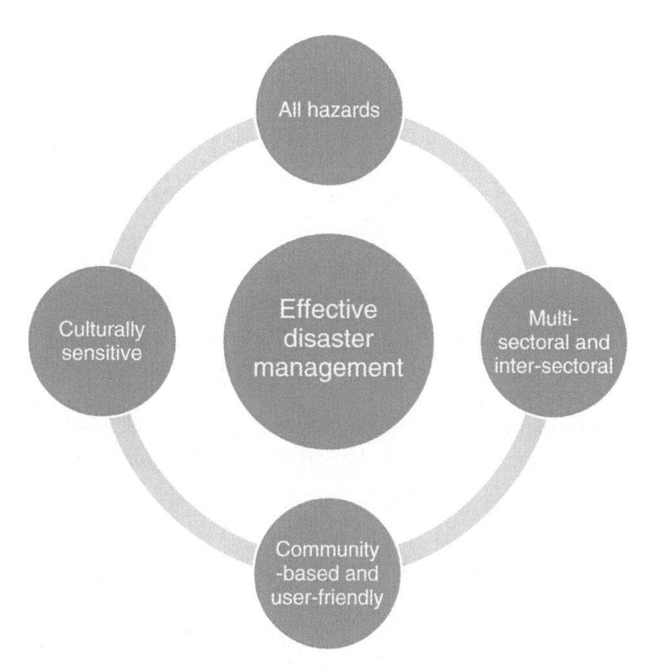

FIG. 25.1 **Essentials for effective disaster management.**

There are certain fundamental principles to initiate the preparation phase of disasters in an efficient way (Fig. 25.1) (Keim, 2016). The all hazards approach is essential in developing and practicing emergency management strategies for both natural as well as human- and technology-induced emergencies and disasters.

25.3 HEALTH AND DISASTERS

Traditionally, the role of health sector in disaster management has been considered to respond to emergencies (Dar et al., 2014). This is very critical in decreasing injuries and death in postdisaster period. However, multisector DRR strategies enable the health sector to have a comprehensive and proactive role so as to gain a better resilience against disasters (Dar et al., 2014). In other words, the activities of *prevention, protection, and mitigation* in predisaster period are of importance in making use of the capacities and skills of health services. These activities can be considered as part of maintaining and improving health.

25.3.1 Objectives of Health Services in Disaster Preparedness

The objective of preparedness for health services in emergencies are to: (1) prevent morbidity and mortality, (2) provide care for the injured, (3) manage poor climate and environmental conditions, (4) ensure the improvement of normal health, (5) restore health services, (6) protect the staff, and (7) maintain public health and medical resources (Keim, 2016).

In Sendai Framework, health sector is regarded as a key component in order to reinforce the activities of DRR. It is required for communities and countries to strengthen their capacity and skills so as to manage all types of hazards and health risks in relation to emergencies and disasters. These activities include the provision of safe drinking water, safe human waste management, energy requirements, first aid, and immediate aid in prehospital period, safe patient transport system, safe hospitals, trauma care centers, psychosocial support services, emergency immunization programs, monitoring, and early warning systems for diseases (Merlin, 2009).

The Marmara earthquake on August 17, 1999, in Turkey caused a total of 17,127 deaths and 43,953 injuries. In the earthquake region, 9 state hospitals were partially damaged; 48 healthcare centers were damaged, 20 of them completely destroyed. Twelve doctors, 18 nurses, and 4 health officers lost their life (Republic of Turkey Ministry of Health, 2001). The significant medical and public health needs of the people affected during the Haiyan Typhoon in the Philippines were not associated with any injury. Rather, these needs stemmed from the lack of sufficient measures to prevent infectious and noncommunicable diseases due to the inadequate access to water, food, accommodation, and health services (Aitsi-Selmi et al., 2015). Following the earthquake in Nepal in 2015, the destruction of healthcare centers in the rural region, where health services were provided under difficult conditions as a result of the earthquake, led to the unavailability of basic health services (Hall et al., 2017). Many hospitals were damaged by the earthquake. For that reason, the healthcare services were provided outside the facilities. Coordination has a key role in order to make use of resources in a more efficient and effective way (Hall et al., 2017).

DRR strategies and health sector are profoundly associated with each other (Aitsi-Selmi et al., 2015). The building stone of health sector is people. In the same way, the most important factor in disasters is people (Altıntaş, 2012). Therefore, almost all the DRR activities seek to protect human health. The item number 30(i) in Sendai DRR Framework is as follows: "Enhance the resilience of national health systems, including by integrating disaster risk management into primary, secondary and tertiary healthcare, especially at the local level; developing the capacity of health workers in understanding disaster risk and applying and implementing DRR approaches in health work; and promoting and enhancing the training capacities in the field of disaster medicine; and supporting and training community health groups in DRR approaches in health programs, in collaboration with other sectors, as well as in the implementation of the 2005 International Health Regulations of the World Health Organization"(UNISDR, 2015b).

A global evaluation in 2008 on the emergency preparedness and response of national health sector in 62 countries indicates that about 70% of the countries have a unit related to emergency preparedness and response within the Ministry of Health (UN, 2014).

25.4 ROLE OF DM IN DISASTER MANAGEMENT AND PREPAREDNESS

25.4.1 History of DM

The concept of DM, similar to the concept of disaster management, dates back to military institutions (Suner, 2015). However, this study only focuses on civil institutions in a historical context. DM has become an important part of human medicine today (Stehrenberger and Goltermann, 2014). In the 20th century, efforts were made to respond to the large explosions in mines, railway tunnels, and factories, which are a part of civil life (Stehrenberger and Goltermann, 2014). Although the need of practicing scientific principles in the field of DM was identified by the pioneers of the field nearly 40 years ago, today, the science of disaster health and medicine slowly continues to improve (Stratton, 2014). Among the most prominent reasons of that, there is a lack of control on variables in the field due to the nature of disasters, the high risks that these studies involve, and the lack of staff specialized in the field of DM. DM was institutionalized as a discipline in the early 1980s (Stehrenberger and Goltermann, 2014). It is necessary to reveal the time to which the concept of disaster dates back in order to analyze the studies on disaster in a historical context. For that reason, the difference between the concepts such as disaster, accident, crisis, and incident is required to be explained in detail, prior to any study. The World Association for Disaster and Emergency Medicine was founded on October 2, 1976, as a multidisciplinary professional association, aiming to globally improve the activities of prehospital and emergency healthcare services, public health, disaster health, and preparedness (WADEM, 2017). WADEM is the earliest organization with 55 member countries involved in governmental or nongovernmental medicine, nursing, emergency management, academy, military, veterinary medicine, psychology, and sociology (WADEM, 2017). One of the important efforts by WADEM is the formation of a Master's Program on European DM. The program was initiated in the year 2000–2001. The purpose of the program is to graduate students, who have learnt the concepts of and developments on the preparedness and management of DM and are able to perform scientific studies on the aspects of DM (Altıntaş, 2007a). The first PhD program in the field of DM has been initiated in 2014 by the Institute of Health Sciences in Bezmialem Vakıf University.

25.4.2 What is DM?

The purpose and treatment environment of DM is different from that of emergency services at hospitals. There is a need for rescue staff to treat the injured in a disaster area even under poor environmental conditions (Hou et al., 2016). Emergency medical rescue and disaster relief activities are among the central tasks of DM, depending on the development of the practice and research activities of the related health services (Hou et al., 2016).

The objective of DM is to restore health facilities in predisaster period in order to prevent and reduce the poor health conditions of the affected population from disasters and to ensure the availability of healthcare services and facilities in postdisaster period (Hubloue and Debacker, 2010). DM is a discipline with a very broad application and research area, including the field of disaster management as well as nonmedical fields such as communication, logistics, and transportation in addition to medical fields such as EM, Traumatology, Public Health (Altıntaş, 2005; Manni and Magalini, 2017). DM is an area of expertise where many different institutions are specialized in disasters and interact with each other (Ciottone, 2015). Thus, the medical response to disasters is provided in different ways (Ciottone, 2015). In particular, it is considered a highly important discipline with the increase in mass injuries, terrorism, and public health emergencies in the recent years (Ragazzoni et al., 2014). Although the need for practicing scientific principles in the studies in the field of disaster has been identified by the pioneers of the field of DM, the progress of the studies in the field is considerably slow (Stratton, 2014).

Chemical, biological, radiological, and nuclear events and accidents are relatively rare. Yet, managing the environmental and health effects of these events and accidents is considerably challenging (Boyd et al., 2014). There are specially trained and equipped teams for such events in states. The directive published by the Ministry of Health, Turkey, in 2014 explains the content of the training on the intervention in CBRN, the way to intervene in incidents, the coordination of all health-related institutions, and other related institutions in a legal context (Republic of Turkey Ministry of Health, 2014). Public transportation accidents (e.g. train accident, ship accident, airplane crash), which are of importance in terms of DM, are among the accidents that commonly take place. An emergency plan and management beyond local facilities and resources is necessary for such accidents (Boyd et al., 2014).

25.4.3 Characteristics of DM

The working principle of DM is considerably different. The characteristics of DM are: (1) *Temporary organization*: In general, when an incident occurs, the medical team is dispatched to the scene. A comprehensive emergency medical rescue is provided within 12 h or less (Wang et al., 2015). (2) *Difficult working conditions*: A temporary medical team is generally formed when assigned. However, in the recent times, there has been a tendency to form medical teams which are only assigned with disasters (Wang et al., 2015). (3) *Mass losses*: Thousands of people can lose their lives within minutes after disasters such as earthquakes, floods, volcanic eruptions, and tsunamis (Wang et al., 2015). (4) *Mass injuries leading to multiple injuries*: Certain incidents lead to very serious physical injuries in the affected area. These injuries require a special clinical treatment and a long-term care in trauma centers. For instance, Crush syndrome and acute renal failure are common after an earthquake. When a fire breaks out after an earthquake, many people can suffer from burns (Wang et al., 2015). (5) *First aid*: Mass injuries occur after disasters. According to the postearthquake data in Armenia, 90% of the injured can be rescued in the first 3 h and 50% in the first 6 h. The first 72 h after the earthquake is defined as gold hours for rescue operations (Wang et al., 2015). (6) *It involves nonmedical aspects:* DM requires the discipline of medicine as well as other command, control, coordination practices, and skills (Altıntaş, 2013). It involves many aspects of management knowledge such as logistics management, human resources management, and financial resource management (Altıntaş, 2013). (7) *Triage*: DM mainly aims to rescue as many patients and injured people as possible with scarce resources. For that reason, the concept of triage can be considered the key word in the field of DM. The word triage is a word of French origin and means selection–distinction (Jaiswal et al., 2016). It is the first and foremost step of medical response (Jaiswal et al., 2016). The purpose of triage is to rescue the maximum number of the injured through scarce sources. (8) *DM and planning:* Emergency help and disaster managers are required to plan emergency situations as well as to enable these plans to ensure business continuity and management in health organizations (Boyd et al., 2014).

Healthcare services are of vital importance in the stages of prevention, preparation, and response in disasters (Boyd et al., 2014). Disasters, with a negative impact on the substructures of the health facilities in the affected regions, reduce the delivery of healthcare services. Therefore, emergency planning should include business continuity planning and management to ensure an uninterrupted delivery of healthcare services. There are certain efforts to reinforce and increase the capacity of healthcare systems prior to disasters. These efforts include the structural reinforcement of hospitals, the special trainings, plannings, and practices of healthcare staff in relation to disasters and the reinforcement of infrastructures. The most important service is healthcare services after disasters in the short- and long-term. In order to respond to disasters accurately and effectively, it is essential that all healthcare staff have an understanding of the principles of DM. Healthcare staff are particularly required to comprehend their roles so that they can respond to different types of incidents (Ciottone, 2015). DM comprises the integration of public health, emergency health services, patient care services (hospitals), definite treatment centers (special centers, advanced hospitals, university hospitals), and disaster management areas (Fig. 25.2). Thus, the working principles and practices in all of these fields should be well understood in the pre-, during, and postdisaster periods.

25.5 EMS IN DISASTERS

EMS is defined by the World Health Organization as part of an effective and functional healthcare system (Rifino and Mahon, 2016). It is a sophisticated system involving the processes of EMS, triage, rapid cynic evaluation, critical therapeutic interventions, medical communication, management of a large number of injuries, and the transportation capacity of victims

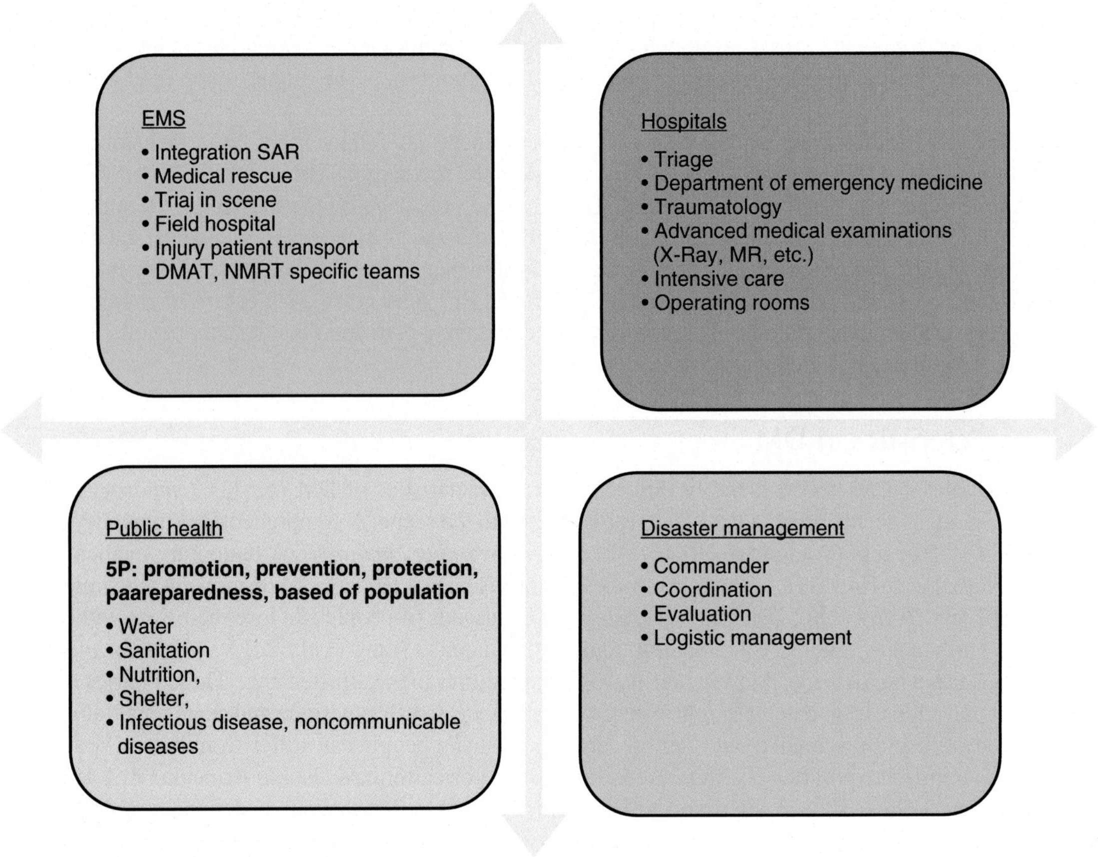

FIG. 25.2 **Disaster medicine components and functions.**

(Miller, 2010). The most important factors in performing these activities are the rate of medical intervention, time, the number of injuries, and the characteristics of incident. Particularly, in fire, hazardous material accidents, and earthquake rescue operations, different institutions are required to cooperate and coordinate with each other (Miller, 2010). EMS is the first point of contact toward the injured/patient in the healthcare system in the event of an emergency or any incident leading to injuries. For that reason, the damage reduction and preparation activities of EMS system in relation to disasters are of considerable importance. One of the first activities is to take structural and nonstructural measures for the house where the family of the EMS staff resides. It is not possible that a staff with a family being the victim of a disaster can work properly during or after the incident. EMS standby points, training areas, and administrative buildings are required to be reinforced against the available risks. The provision of EMS means that the injured and response teams are protected against the risks in the rescue area. An efficient EMS area management would positively contribute to the success of damage reduction and response activities. One of the fundamental objectives of Sendai Framework is to decrease the rate of mortality (UNISDR, 2015b). Therefore, fast and effective search and medical rescue during and after disasters is vital in decreasing mortality.

The delivery of EMS in a typical situation is very different from that in disasters. EMS is frequently faced with a number of injured/patients that force or exceed the local capacity in multiple traffic accidents, fire and explosions, train accidents, ship accidents, etc. However, these are the incidents that can be managed with local or regional resources (Nagele and Hüpfl, 2015). The disasters that affect a very broad area and population, such as earthquakes, hurricanes, and tsunamis, can be only responded with national and international capacity (Nagele and Hüpfl, 2015). An efficient response with local resources is vital in the first 72 h until the specialized medical rescue teams such as disaster medical assistant team (DMAT) and NMRT from other regions arrive at the scene. It is necessary that a single coordinator in the area organizes all the relevant institutions (fire departments, hospitals, police, etc.) (Rifino and Mahon, 2016). EMS staff are needed to strengthen the physical and psychological aspects in the recovery phase following the incident. Special efforts should be made in regard to the loss of their family, economic loss, physical, and psychological exhaustion.

EMS staff are required to participate in special training for disasters. In particular, the volunteer staff receive special trainings such as hazardous materials, approach to chemical incidents, and rescue in water. The National Medical Rescue Team (NMRT) was established in 2003, in Turkey. The NMRT provides special training on medical rescue in disasters for healthcare staff, who are volunteers in EMS, hospitals, and other health institutions. Every community, region, and nation has a different system around the world (Rifino and Mahon, 2016). Efforts were made to invest to increase the capacity of UMKE, which was among the first responding teams to disasters within the Istanbul Seismic Risk Mitigation Project (IPKB, 2014). Accordingly, these efforts for UMKE involved the procurement of 4 × 4 heavy-duty health recovery, health recovery tools, and operation management tools: vaccine, blood, and blood products transport vehicle; 3-axle semitrailers with sliding tilting; 3-axle refrigerated semitrailers; various medical first aid equipment (stretchers, splints sets, rescue vests); warehouse containers; mobile generators and towers; cordless and diesel forklifts; communication equipment (analog wireless zone transmitters, analog wireless center units, analog wireless mobile rollovers), which increased and strengthened the capacity (IPKB, 2014).

The National Disaster Medical System was established in 1984, in the United States. Following that, in 1985, the DMAT was founded (Suner, 2015). They provided medical assistance in many disasters in the United States.

25.6 HOSPITAL IN DISASTERS

The services in hospitals are routinely provided for 24 h a day, 7 days a week. There is a need for substantial plans and resources so that they can provide uninterrupted services in disasters. Hospitals should be prepared for disasters in three different categories: structural, nonstructural, and DM planning (functional) (Andress, 2010). The buildings that provide structural services are required to be reinforced against the risks in the area. The risks created by nonstructural materials should be eliminated in the second step. Among them, the monitors and mobile medical equipment in operating rooms, intensive care, and emergency services; imaging equipment, laboratory equipment, cabinets, items, TVs in patient rooms, etc. should be fixed in a secure way. It is necessary to ensure that the healthcare staff and their family are not affected by the incident and they are taken to a secure zone. The third and the most important preparation involves a hospital disaster plan for the provision of healthcare services. Many countries have specific guidelines on this issue. The Hospital Disaster and Emergency Plan/Hastane Afet ve Acil Durum Planı (HAP) Preparation Guide was published by the General Directorate of Emergency Health Services of the Ministry of Health in Turkey in December 2015 (Ministry of Health of the Republic of Turkey, 2017). The HAP Preparation Guide includes general introduction information, disaster and emergency management system, hospital incident management system, hospital intervention steps in the case of mass injury incidents, incident reporting, hospital medical capacity, hospital disaster triage area, emergency room and in-hospital patient traffic flow, logistics and equipment, psychosocial support, and operations related to deaths and losses. Furthermore, it features specific planning examples such as chemical events and fire.

The formation of healthcare facilities and the teaching and education of healthcare professionals require a serious capital investment in developing and developed countries (WHO, 2007).

After the 2003 Algerian earthquake, 50% of the health facilities in the affected region were no longer functional due to damage. In the region of Pakistan worst affected by the 2005 South Asia Earthquake, 49% of health facilities, from sophisticated hospitals to rural primary care clinics and drug dispensaries, were completely destroyed (UN, 2009). The 2011 earthquake in Japan severely damaged some of the hospitals in the area. The surrounding area including Saka General Hospital in the nearby region suffered from the lack of water, electricity, gas and diesel for several weeks (Wang et al., 2015). The 1999 Marmara earthquake in Turkey led to approximately 17,127 deaths and affected 1,358,953 (EMDAT, 1999). Following the earthquake, many efforts were made to increase the capacity. Among them, there was the Istanbul Seismic Risk Mitigation and Emergency Preparedness Project, costing about 2 billion euros and initiated in 2006 (IPKB, 2014). The project was funded by World Bank, European Investment Bank, Council of Europe Development Bank, Islamic Development Bank, German Development Bank. Within the Istanbul Seismic Risk Mitigation and Emergency Preparedness Project:

1. The structural reinforcement or reconstruction of hospitals (Tables, 25.1 and 25.2)
2. The structural reinforcement or reconstruction of outpatient clinics
3. The reinforcement of the equipment of UMKE teams.

25.7 PUBLIC HEALTH IN DISASTERS

According to the World Health Organization, public health "refers to all organized measures—whether public or private—to prevent disease, promote health, and prolong life among the population as a whole" (Ardalan et al., 2015).

Millions of people suffer from disasters in developed and underdeveloped countries around the world every year. Disasters have direct and indirect impacts on healthcare services. The direct impacts include the destruction of facilities and the interruption of basic health needs such as clean water, food, and sanitation. The infrastructure problems and the difficulties in meeting the fundamental needs cause serious health problems in the society. The efforts on public health should be considered an essential part of disaster management in order to minimize the public health problems of disasters. It is necessary to obtain the fundamental information, such as the population structure, infrastructure facilities, immunization, developmental level of the region, and transportation facilities for the risk evaluation in relation to public health (Ardalan et al., 2015).

TABLE 25.1 The Structural Reinforcement or Reconstruction of Healthcare Facilities in Istanbul as Part of ISMEP Project (ISMEP, 2014)

	Reinforced hospital buildings	Reconstructed hospital buildings	Reinforced or reconstructed clinics/family health centers
Completed	29	2	47
Ongoing	7	(Planned) 6	14
In tender phase	6	3	–
Other	2	6	–
Total number of hospitals	44	17	61
Total bed availability	2,154	3,839	–
Number of patients treated daily	20,855	20,412	22,523
Total construction area (m^2)	217,685	884,598	108,584
Total cost (TL)	143,588,498	1,300,000,000	46,356,621

TABLE 25.2 Prioritization Criteria for Hospitals (ISMEP, 2014)

No	Criteria	Points
1	Access during disasters (×0.20)	20
	Airborne access (×0.05) (yes) 100, (no) 0	
2	Technical features of the building (×0.20)	20
	Construction year after 1980 (40)	
	Construction year before 1980 (100)	
3	Distance to center radius (×0.10)	10
	Distance to fault line >20 km (40)	
	Distance to fault line <20 km (100)	
4	Importance in disaster management plan (strategic location) (×0.40)	40
5	Capacity	10
	0–100 beds (30)	
	100–500 beds (60)	
	500 beds and above (100)	
	Total	100

The most serious problem for those living in rural areas after the 2015 earthquake in Nepal was the access to clean water (Hall et al., 2017). Following the first few days, relief organizations delivered water to the region by means of water reservoirs. Yet, the quality and testability of water was a major problem (Hall et al., 2017).

25.8 Future and DM

The studies on DM have been increasing every day. In future, it will become one of the most important areas of expertise studying on disasters. Sendai DRR emphasizes that as follows: "promotion and enhancing the training capacities in the field of DM." This discipline, globally developing more and more, will soon complete its academic structure. There are some studies performed by certain countries to strengthen the healthcare system and increase the capacity. All these studies contribute to damage reduction and recovery efforts in disasters.

REFERENCES

Aitsi-Selmi, A., Egawa, S., Sasaki, H., Wannous, C., Murray, V., 2015. The Sendai framework for disaster risk reduction: renewing the global commitment to people's resilience, health, and well-being. Int. J. Disaster Risk Sci. 6 (2), 164–176, http://doi.org/10.1007/s13753-015-0050-9.

Altıntas¸, K.H., 2005. Teaching and education in disaster medicine. Hacettepe Med. J. 36, 139–146.

Altıntas¸, K.H., 2007a. European disaster medicine master program. In: Eryılmaz, M., Dizer, U. (Eds.), Disaster Medicine. second ed. Ünsal Press, Ankara, pp. 147–152.

Altıntas¸, K.H., 2007b. Health management master's program in disasters. In: Eryılmaz, M., Dizer, U. (Eds.), Disaster Medicine. second ed. Ünsal Press, Ankara, pp. 152–153.

Altıntas¸, K.H., 2012. Disasters and disaster medicine. In: Güler, Ç., Akın, L. (Eds.), Public Health Basic Information. second ed. Hacettepe University Press, Ankara, pp. 1106–1107.

Altıntas¸, K.H., 2013. Basic concepts about disaster and disaster medicine. In: Altıntas¸, K.H., Bayraktar, N., Erden, Z., Koçer, B., Demiröz, F. (Eds.), HAMER Health Management in Emergency and Disaster Situations. Hacettepe University Press, Ankara, pp. 16–25.

Andress, K., 2010. Healthcare facility preparedness. In: Powers, R., Daily, E. (Eds.), International Disaster Nursing. University Press, Cambridge, pp. 13–28.

Ardalan, A., Ordun, C.Y., Riley, J.M., 2015. Public health and disasters, second ed. Ciottone's Disaster Medicine. Elsevier, Philadelphia, http://doi.org/10.1016/B978-0-323-28665-7.00002-9.

Boyd, A., Chambers, N., French, S., Shaw, D., King, R., Whitehead, A., 2014. Emergency planning and management in health care: priority research topics. Health Syst. (Basingstoke, England) 3 (2), 83–92, http://doi.org/10.1057/hs.2013.15.

Ciottone, G.R., 2015. Introduction to disaster medicine. Ciottone's Disaster Medicine, second ed. Elsevier, Philadelphia, http://doi.org/10.1016/B978-0-323-28665-7.00001-7.

Dar, O., Buckley, E.J., Rokadiya, S., Huda, Q., Abrahams, J., 2014. Integrating health into disaster risk reduction strategies: key considerations for success. Am. J. Public Health 104 (10), 1811–1816, http://doi.org/10.2105/AJPH.2014.302134.

EMDAT, 1999. Marmara Depremi, Country Profile, 1900–2016 Earthquake. Available from: http://www.emdat.be/country_profile/index.html

Gougelet, R.M., 2016. Disaster mitigation. In: Ciottone, G.R., Biddinger, P.D., Darling, R.G., Fares, S., Keim, M.E., Molloy, M.S., Suner, S. (Eds.), Ciottone's Disaster Medicine. second ed. Elsevier, Philadelphia, pp. 160–166, http://doi.org/10.1016/B978-0-323-28665-7.00027-3.

Hall, M.L., Lee, A.C.K., Cartwright, C., Marahatta, S., Karki, J., Simkhada, P., 2017. The 2015 Nepal earthquake disaster: lessons learned one year on. Public Health 145, 39–44, http://doi.org/http://dx.doi.org/10.1016/j.puhe.2016.12.031.

Hogan, D.E., Burstein, J.L., 2010. General concepts: basic perspectives on disasters. In: Hogan, D.E., Burstein, J.L. (Eds.), Disaster Medicine. second ed. Wolters Kluwer & Lippincott Williams & Wilkins, Philadelphia, PA, p. 10.

Hou, S., Lv, Q., Ding, H., Zhang, Y., Yu, B., Liu, Z., Su, B., Liu, J., Yu, M., Sun, Z., Fan, H., 2016. Disaster medicine in China: present and future. Disast. Med. Public Health Prep., 1–9, http://doi.org/10.1017/dmp.2016.71.

Hubloue, I., Debacker, M., 2010. Education and research in disaster medicine and management: inextricably bound up with each other. Eur. J. Emerg. Med. 17 (3), 129–130, http://doi.org/10.1097/MEJ.0b013e32833981c7.

IPKB, 2014. Increasing Emergency Preparation Capacity. Beyaz Gemi Sosyal Proje Ajansı, stanbul.

ISMEP, 2014. Strengthening and Reconstruction Studies 4. Beyaz Gemi Sosyal Proje Ajansı, stanbul.

Jaiswal, R., Donahue, J., Reilly, M.J., 2016. Disaster risk management. In: Ciottone, G.R., Biddinger, P.D., Darling, R.G., Fares, S., Keim, M.E., Molloy, M.S., Suner, S. (Eds.), Disaster Medicine. second ed. Elsevier Inc, pp. 167–177, http://doi.org/10.4135/9781473955516.

James, J.J., Benjamin, G.C., Burkle, F.M., Gebbie, K.M., Kelen, G., Subbarao, I., 2010. Disaster medicine and public health preparedness: a discipline for all health professionals. Disast. Med. Public Health Prep. 4 (2), 102–107, http://doi.org/10.1001/dmp.v4n2.hed10005.

Keim, M.E., 2016. Disaster Preparedness. Ciottone's Disaster Medicine, second ed. Elsevier, Philadelphia, http://doi.org/10.1016/B978-0-323-28665-7.00032-7.

Leaning, J., Guha-Sapir, D., 2013. Natural disasters, armed conflict, and public health. N. Engl. J. Med. 369 (19), 1836–1841, http://doi.org/10.1056/NEJMra1109877.

Manni, C., Magalini, S., 2017. Disaster medicine: a new discipline or a new approach? DOI: https://doi.org/10.1017/S1049023X00029976.

Maya, İ., Çalışkan, C., 2015. Evaluation Disaster Education and Training Programs at the Level of Undergraduate Degree in the World and Turkey Sample. Turk. Studies 11/9 (Spring), 579–604, http://dx.doi.org/10.7827/TurkishStudies.9761.

Merlin, 2009. Health and disaster risk reduction. Available from: http://www.preventionweb.net/publications/view/8746

Miller, K.T., 2010. Emergency medical services scene management. In: Koenig, K.L., Schultz, C.H. (Eds.), Koenig and Schultz's Disaster Medicine Comprehensive Priciples and Practices Disaster Medicine Comprehensive Priciples and Practices. Cambridge University Press, Cambridge, pp. 275–284.

Republic of Turkey Ministry of Health, 2001. Management of Health Services in Disasters, Course notes.

Republic of Turkey Ministry of Health, 2014. CBRN Directive. Ankara.

Ministry of Health of the Republic of Turkey, G.D. of E. H. S., 2017. Hastane Afet ve Acil Durum Planı (HAP) Hazırlama Klavuzu/Hospital Disaster and Emergency Plan (HAP) Preparation Guide. Ankara: Ministry of Health of the Republic of Turkey.

Nagele, P., Hüpfl, M., 2015. Anesthesia and prehospital emergency and trauma care. Miller's Anesthesia, 2460–2478.e4, Chapter 82. http://doi.org/10.1016/B978-0-7020-5283-5.00082-5.

Ragazzoni, L., Ingrassia, P.L., Ripoll, A., Hubloue, I., Debacker, M., Della Corte, F., 2014. European master in disaster medicine: impact analysis on students' professional career. J. Emergency Med. 46 (2), 285–286, http://doi.org/10.1016/j.jemermed.2013.11.030.

Rifino, J.J., Mahon, S.E., 2016. Role of Emergency Medical Services in Disaster Management and Preparedness. Ciottone's Disaster Medicine, second ed. Elsevier, Philadelphia, Chapter 3. http://doi.org/10.1016/B978-0-323-28665-7.00003-0.

Stehrenberger, C.S., Goltermann, S., 2014. Disaster medicine: genealogy of a concept. Soc. Sci. Med. 120, 317–324, http://doi.org/10.1016/j.socscimed.2014.05.017.

Stratton, S.J., 2014. Is there a scientific basis for disaster health and medicine? Prehospital Disast. Med. 29 (3), 221–222, http://doi.org/10.1017/S1049023X14000582.

Suner, S., 2015. History of disaster medicine. Turk. J. Emerg. Med. 15 (Suppl 1), 1–4, http://doi.org/10.5505/1304.7361.2015.69376.

UN, 2009. Hospitals Safe from Disasters—Reduce Risk, Protect Health Facilities, Save Lives.

UNISDR, 2015a. Sendai Framework for Disaster Risk Reduction 2015–2030. Kobe, Hyogo, Japan.

UNISDR, 2015b. Sendai framework for disaster risk reduction 2015–2030. Third World Conference on Disaster Risk Reduction. Sendai, Japan, 14–18 March 2015.

WADEM, 2017. Association overview. Available from: https://wadem.org/about/association-overview/

Wang, Z.G., Zhang, L., Zhao, W.J., 2015. Emergency medicine for disaster rescue. Chin. J. Traumatol.—English Edition 18 (6), 311–313, http://doi.org/10.1016/j.cjtee.2015.12.004.

UN, 2014. Health and Disaster Risk Reduction. Available from: http://www.wcdrr.org/uploads/HEALTH.pdf.

WHO, 2007. Risk reduction and emergency preparedness: WHO six-year strategy for the health sector and community capacity development. Geneva, Switzerland. Available from: http://www.who.int/hac/techguidance/preparedness/emergency_preparedness_eng.pdf.

Chapter 26

Earthquake-Triggered Landslide Modeling and Deformation Analysis Related to 2005 Kashmir Earthquake Using Satellite Imagery

Prashant Kumar Champatiray*, Irshad Parvaiz, Ramakrishna Jayangondaperumal†, Vikram Chandra Thakur†, Vinay Kumar Dadhwal‡**

**Indian Institute of Remote Sensing, Dehradun, India; **Yanbu Industrial College (YIC), Saudi Arabia; †Wadia Institute of Himalayan Geology, Dehradun, India; ‡Indian Institute of Space Science and Technology, Thiruvananthapuram, India*

26.1 INTRODUCTION

The entire stretch of the Himalaya and adjoining hilly ranges starting from the Hindukush region in the west to the Namcha Barua range and the Assam hills in the east is prone to earthquakes and resultant landslides. These landslides, including coseismic and postseismic landslides, constitute one of the most potentially damaging seismic hazards in Himalayan region. In the past, it has caused massive destruction in terms of loss of lives and property as well as infrastructure. It has also emerged as one of the prominent mass-wasting processes active in the Himalaya and plays an important role in landform development and shaping of present-day landforms. Earthquakes as small as magnitude 4.0 may dislodge landslides from susceptible slopes, and larger earthquakes can generate tens of thousands of landslides throughout the affected areas of hundreds of thousands of square kilometers, producing billions of cubic meters of loose, surficial sediments (Keefer, 1994). During the Kashmir earthquake 2005 and Sichuan earthquake 2008, it was observed that earthquake induced landslides have caused tremendous amount of damage and contributed almost 30% of the total loss (Cui et al., 2011; Dunning et al., 2007; Petley et al., 2006). The 2008 Wenchuan earthquake (M_s=8.0; epicenter located at 31.0°N, 103.4°E), with a focal depth of 19.0 km, was triggered by the reactivation of the Longmenshan fault in Wenchuan County, Sichuan Province, China on May 12, 2008. This earthquake directly caused more than 5836 geohazards in the form of landslides, rock falls, and debris flows which resulted in about 20,000 deaths (Cui et al., 2011).

The 7.8 M_w Nepal earthquake of 2015 and associated aftershocks triggered widespread landslides. Even a relatively low-magnitude earthquake of M_w 6.9 in 2011 caused widespread landslides in the Sikkim Himalaya. Similar low-magnitude events such as 1991 Uttarkashi earthquake of M_w 6.6 and 1990 Chamoli earthquake of M_w 6.8 also reported to have widespread landslides as their epicenters and affected areas are located around Main Central Thrust (MCT), which is one of the major fault systems of the Himalaya (Singh and Som, 2016). The mega events such as 1897 Assam earthquake of 8.1 M_w and 1950 Assam earthquake of 8.7 had caused widespread landslides that temporarily blocked the courses of the Subansiri, Dibang, and Dihang rivers (Mathur, 1953). Other mega events of the region such as the 1833 Nepal earthquake of M_w 7.7, 1905 Kangra earthquake of M_w 7.8, and 1934 Bihar Nepal earthquake of M_w 8.1 are also known to have triggered landslides although lacks detailed description of such secondary phenomena (Middlemiss, 1910). In the present context, 2005 Kashmir earthquake has been described in detail.

26.1.1 2008 Kashmir Earthquake 7.6 M_w

A devastating earthquake of magnitude 7.6 M_w occurred on October 8, 2005 (03:50:38 UTC) with epicenter located at 10 km north–northwest of Muzaffarabad within the Hazara syntaxis of the Indus Kohistan Seismic Zone (IKSZ) at distances of 105 km NNE of Islamabad and 125 km WNW of Srinagar (Jayangondaperumal and Thakur, 2008; Mahajan et al., 2006;

Integrating Disaster Science and Management. http://dx.doi.org/10.1016/B978-0-12-812056-9.00026-9

Thakur et al., 2006). It caused death of at least 75,000 persons, injured around 140,000 persons, and made homeless 3.5 million people in Kashmir and adjoining areas of Pakistan, India, and Afghanistan (Dunning et al., 2007; NDMA, 2007). It was one of the most devastating earthquakes in the Himalayan arc in terms of loss of lives and property. The epicenter was located at 34.493N and 73.629E with a focal depth of 26 km. The area had experienced 23 aftershocks of magnitude >5 and 1 aftershock of magnitude 6.2 recorded on the same day of the earthquake (USGS, 2006). The aftershocks define a linear belt which corresponds to the NW–SE trending Balakot–Bagh Fault (BBF, also known as Muzaffarabad and Tanda faults), an active fault (Yeats and Hussain, 2006) which was mapped by earlier workers (Nakata et al., 1991). Teleseismic data indicate reverse faulting on a NW/SE striking fault with 90 × 40 km wide rupture dipping 37° toward NE (Bilham, 2005). In the present context, an attempt was made to map the causative fault and associated deformation mainly landslides using remote sensing with limited ground observation due to inaccessibility of the earthquake affected region (Avouac et al., 2006; Champatiray et al., 2005; Pathier et al., 2006; Sato et al., 2007; Thakur et al., 2006). Multiresolution satellite data products from ASTER (15-m resolution), IRS-LISS-III (23.5-m resolution), IRS-LISS-IV (5.8-m resolution), and Cartosat-1 (2.5-m resolution) were analyzed in conjunction with ground deformation to decipher the nature of the causative fault and its influence on landslides in the region.

26.2 STUDY AREA AND SEISMOTECTONIC SETTING OF THE REGION

The study area of approximately 6300 km^2 centers on the earthquake's epicenter in Kashmir region in the Lesser Himalaya of northeast Pakistan and northwest India bounded by latitude 34–35° and longitude 73–74°. It covers areas that were largely affected by landslides due to 2005 earthquake. The epicenter location is within the Hazara syntaxis, which is interpreted as a major ramp structure of the under thrusting Indian shield rocks below the Himalaya and as a result major Himalayan structures have developed around it on three sides (Fig. 26.1). In the eastern part of syntaxis, the main boundary fault (thrust) (MBF/MBT) and its equivalent in Pakistan, the Muree Thrust and Punjal Thrust (PT-equivalent of the MCT), are oriented in NW–SE direction; gradually, these thrusts take a sharp bend around the ramp structure and follow almost north–south direction. The thrust system to the east of syntaxis has different geometries than those to the west. On the eastern side, the MBF/MBT and the overlying Punjal Thrust (PT) occur parallel to each other with a NW–SE regional strike. On the western side, several northeasterly trending thrusts branch out from the MBT. Although MBT has been considered as inactive in most part, however, near Muzaffarabad, traces of active faults were mapped very close to MBT, oriented in NW–SE/N–S direction parallel to the course of Jhelum. The present causative fault partly coincides with this fault (trending northwest) and thereby suggests its reactivation during the 2005 event (Baig, 2006; Yeats et al., 2006). Similar observation was also made by Nakata and Kumahara (2006) who had reinterpreted the earlier mapped (Nakata et al., 1991) Muzaffarabad and Tanda faults that were now considered to be parts of the causative fault of 2005 earthquake.

26.3 SATELLITE IMAGE PROCESSING

Multiresolution temporal satellite data products such as LANDSAT-TM (30 m), IRS-LISS-III (23.5 m), ASTER (15 m), IRS-LISS-IV (5.8 m), and Cartosat-1 (2.5 m) data products corresponding to pre- and post-earthquake periods were used to map the ground deformation related to the 2005 event (Table 26.1, Fig. 26.2). The 7-band ETM data (2001) and IRS-LISS-III (September 20, 2005) were utilized to assess the seismotectonic setup of the region vis-à-vis the earthquake epicenter. The Cartosat-1 PAN data of 2.5 m resolution (acquired on October 9, 2005) and Resourcesat multispectral data of 5.8 m resolution (acquired on October 4 and 9, 2005) were georeferenced and analyzed especially for mapping landslides and surface ruptures. SRTM DEM (90 m) resolution was used for the generation of ancillary terrain information such as slope, aspect, curvature, and elevation changes due to extensive mass wasting.

ASTER images acquired on November 14, 2000 (pre-event period) and October 27, 2005 (post-event period) with similar viewing geometry were coregistered by subpixel correlation using a method similar to the processing of SPOT (Binet and Bollinger, 2005; Leprince et al., 2007; Michel and Avouac, 2002; Van Puymbroeck et al., 2000) and ASTER (Avouac et al., 2006; Schiek and Hurtado, 2007) images for deformation study. Temporal ASTER data (Band 3N) were ortho-rectified on a common resolution of 15 m with respect to a digital elevation model (DEM) prepared from the stereo pair of ASTER images using Leica Photogrammetry Suite (LPS) of ERDAS Imagine 9.1 (ERDAS IMAGINE, 2006) and reference IRS LISS-III image acquired on September 20, 2005. Spatial off-sets were then measured from local cross-correlation of two ortho-rectified images using Fourier transformation to detect the horizontal pixel offsets between the time-separated images and based on which information on widespread surface deformation due to the earthquake motion was derived. The correlation image was obtained by sliding 32 × 32 pixels window at 8-pixel step resulting in ground resolution of 120 m on the correlation image.

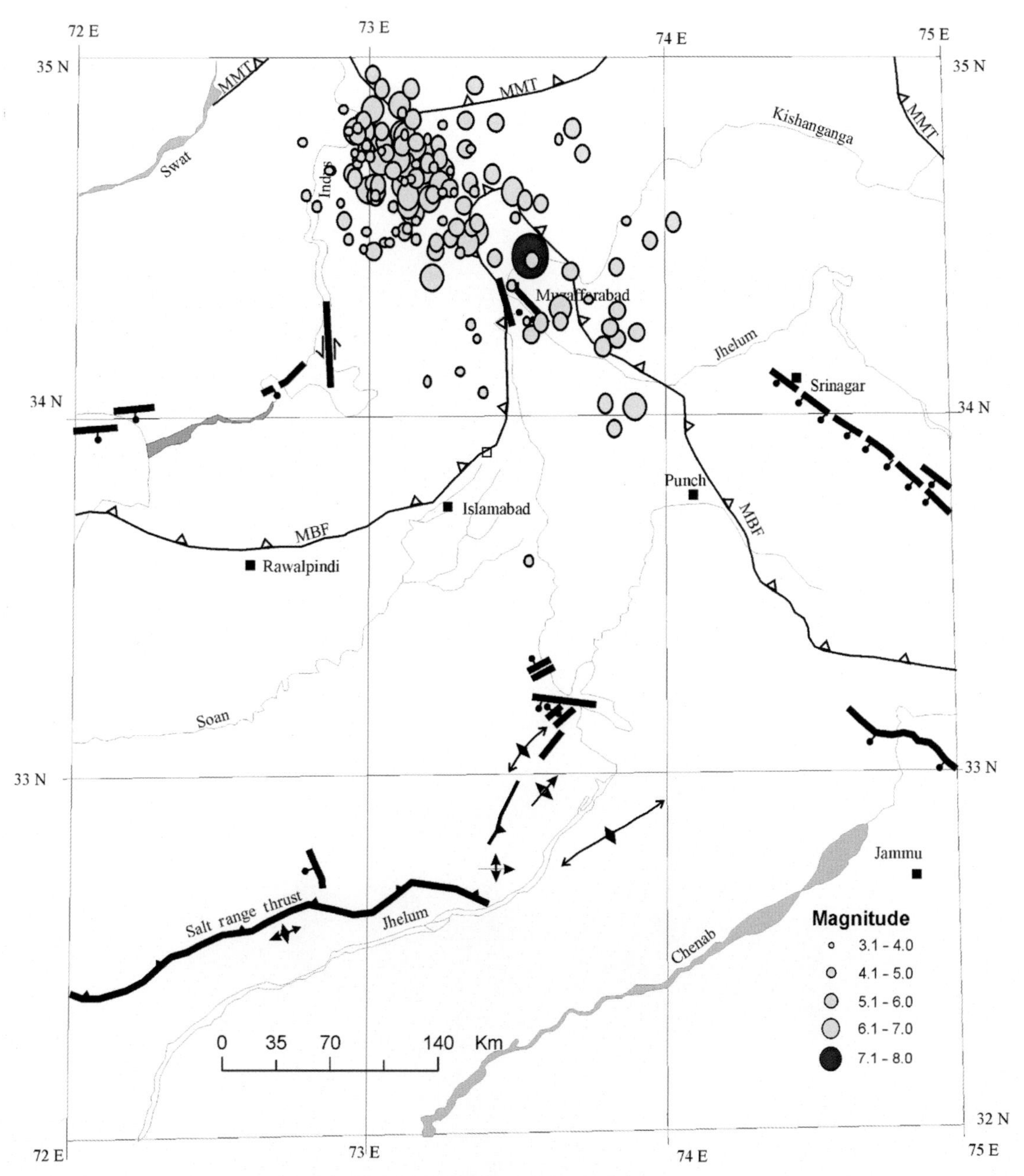

FIGURE 26.1 Simplified regional tectonic map showing active folds and faults of the Pakistan and Kashmir Himalaya adapted from Yeats et al. (1992). The epicenter (*red*) and spatial distribution of aftershocks of the October 8, 2005, earthquake (*yellow*) are also shown. Active faults are shown by *heavy line* with *solid circle* on down thrown side. Abbreviations: *MBF*, main boundary fault (thrust); *MMT*, main mantle thrust.

26.4 CAUSATIVE FAULT MAPPING

The subpixel registration of ASTER images revealed a linear to curvilinear discontinuity that could be traced up to a distance of 86 km from Balakot to SW of Uri on both its north–south and east–west components of the horizontal deformation (Fig. 26.3A and B). Most importantly, it revealed areas that are affected by massive landslides due to loss of correlation between pre- and post-earthquake period images. In general, the deformation zone mostly affected by landslides was limited to few hundred meters on ground for most part of the fault except for the northwest margin where it was more diffused due to the presence of folded scarp rather than ground ruptures (Yeats and Hussain, 2006). The fault extends in a more curvilinear fashion parallel to the Kunhar valley along the previously mapped Muzaffarabad Fault (Nakata and Kumahara, 2006; Nakata et al., 1991). The geomorphic expression of this fault was subtly marked by break in slope and triangular facets on the western bank of a smaller stream after it crossed the Kunhar river valley at Balakot (Fig. 26.4A). Between Balakot

TABLE 26.1 Details of Satellite Data Used During Pre- and Post-earthquake Period

Time period	Satellite/sensor	Ground resolution	Date	ID	Multispectral (MX) / panchromatic (PAN)
Pre-earthquake period	L7/ETM+	30 m	30 Sep. 2001, 07 Oct. 2001, 11 Apr. 2000	149/036 and 037 150/036 and 037	MX/PAN
	IRS P6/LISS-3	23.5 m	20 Sep. 2005	P091-R046	MX
	IRS P6/LISS-IV	5.8 m	7 Nov. 2005	P202- R032	MX
	IRS P6/LISS-IV	5.8 m	4 Oct. 2005	P202-R024, R032	MX
	Terra/ASTER	15 m	14 Nov. 2000	P150-R036	MX/PAN
Post-earthquake period	IRS P6/LISS-IV	5.8 m	9 Oct. 2005	P202-R023, R024	MX
	Terra/ASTER	15 m	27 Oct. 2005, 19 Nov. 2005	P150-R036	MX/PAN
	IRS P5/ Cartosat-1	2.5 m	09 Oct. 2005	P501-R242	PAN

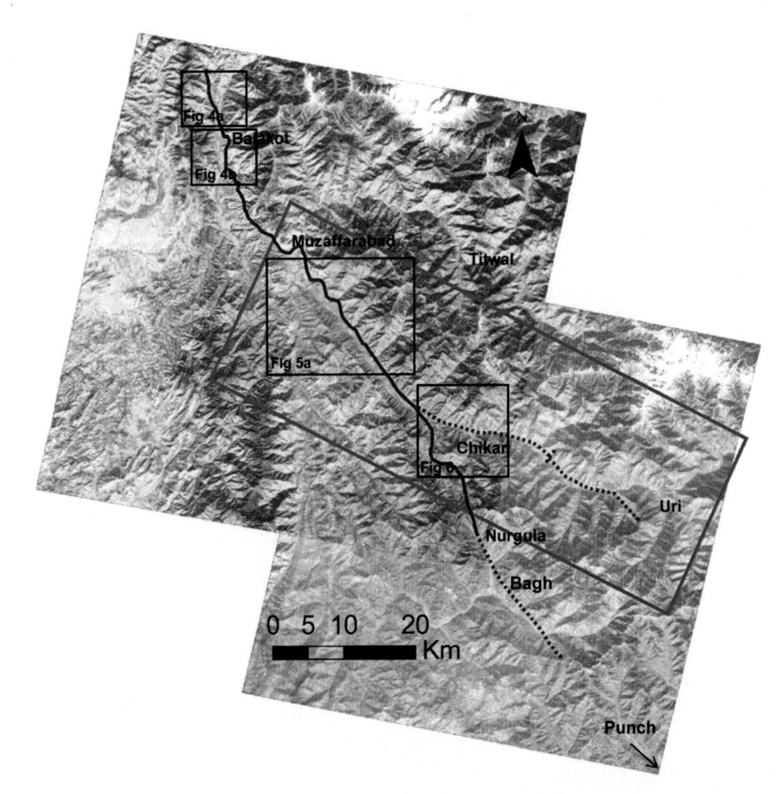

FIGURE 26.2 **Post-earthquake coverage of ASTER images of the study area showing the causative fault (*red*) and its inferred extension toward Uri and Punch (*dotted lines*). *Blue box* indicates extent of geological map referred.**

to Muzaffarabad, the break in slope is sharper and the lower limit of the landslides on the eastern bank of Kunhar valley coincides with the causative fault (Fig. 26.4B).

Toward further southeast in the upper Jhelum valley, the fault extends linearly cutting northeastern bank of the Jhelum river for about 54 km along previously mapped Muzaffarabad-Tanda fault (Nakata et al., 1991), now referred as Balakot–Bagh Fault (BBF) (Yeats and Hussain, 2006). The fault is marked by steep scarplets in the Muree Formation mainly due to juxtaposition of three sets of joints and subtle break in slope along the right bank of Jhelum river as observed on ASTER DEM (Fig. 26.5A and B). After following a linear trend from Muzaffarabad, the fault extends further toward southeast direction where Jhelum valley takes an eastward diversion toward Uri. At Chikar, the fault shows "v"-shaped expression

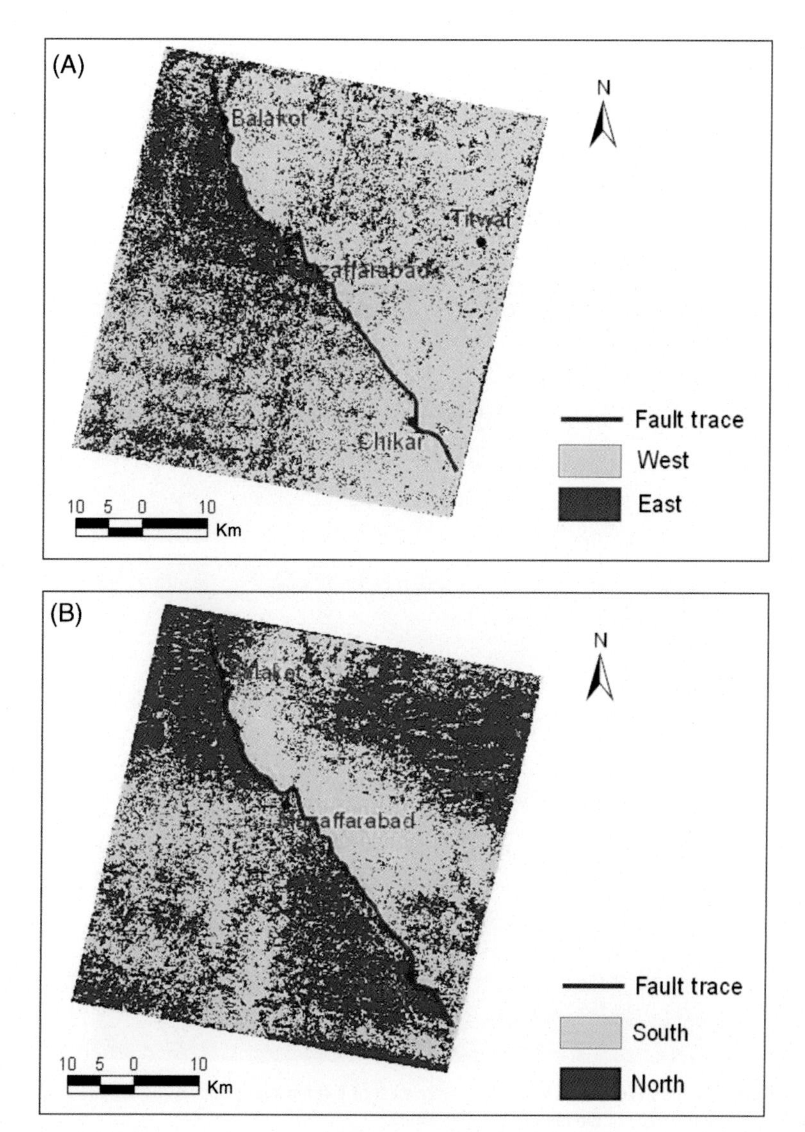

FIGURE 26.3 (A) East–West and (B) North–South components of the ground deformation as detected from the spatial correlation of ASTER images acquired before and after earthquake event. East, West, North, and South indicate movement toward respective directions. Fault trace was mapped based on the overall deformation pattern.

due to a hill and NE dipping fault. At this place, a very large landslide known as Hattian landslide (Owen et al., 2008), the largest in the region, was observed on the southeastern aspect of the hill range (Fig. 26.6). This landslide was interpreted as a rock avalanche by Dunning et al. (2007) and as a sturzstrom by Owen et al. (2008). Toward further southeast, the Tanda Fault continues for some distance and the surface deformation related to the 2005 event was observed up to Nurgala village at a distance of 40 km from Punch and 25 km from Uri. Based on this, the total length of the fault was estimated to be around 86 km which is marginally higher than the reported rupture length (Avouac et al., 2006; Pathier et al., 2006).

26.5 LANDSLIDES MAPPING

Various techniques and inventory of seismicity-induced landslides were demonstrated through number of case studies including the Kashmir earthquake of 2005 (Das et al., 2007; Kamp et al., 2008; Saraf, 2000; Sarkar and Saraf, 1999; Sato et al., 2007; Sudmeier-Rieux et al., 2007a, b; Owen et al., 2008). In the present study, various band combinations of temporal multiresolution data sets were interpreted based on the tonal difference for landslide detection and classification in terms of activity, that is, old, new, and reactivated. The area experienced many landslides and slope failures in rocks, colluvial, and alluvial fan and terrace deposits (Fig. 26.8). The Cartosat-1 offers better resolution and 3D perception but the multispectral LISS-IV data is superior in many cases due to better tonal discrimination.

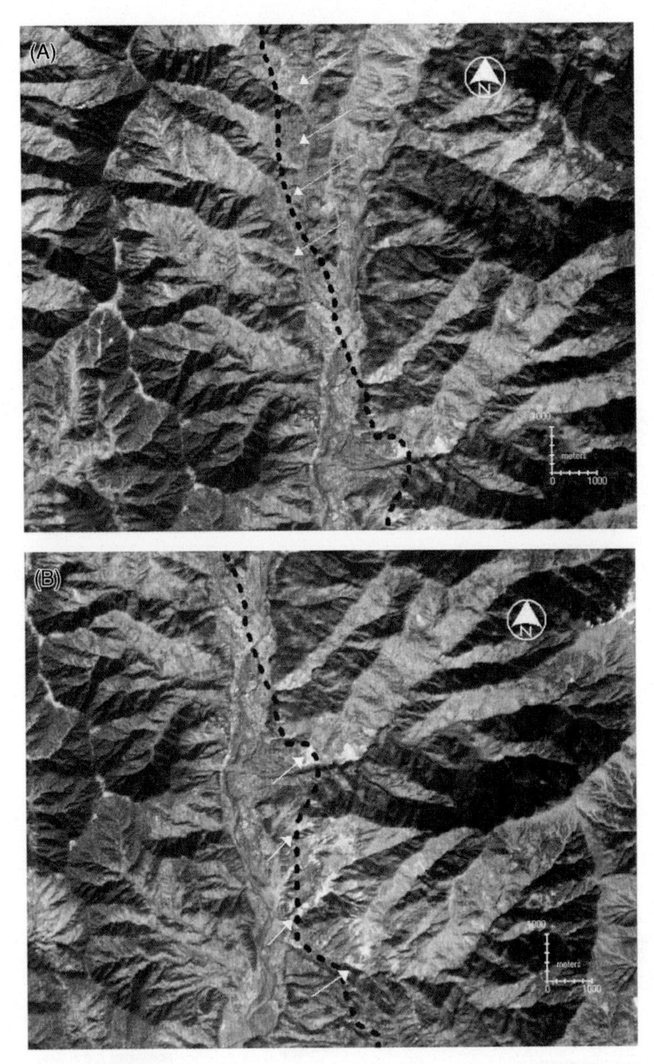

FIGURE 26.4 (A) Triangular facets on the west bank of a smaller stream west of Kunhar valley shows continuation of the Muzaffarabad fault; (B) lower slope limit of landslides coincides with the trace of the fault in Kunhar valley.

In the Jhelum valley section at 8–10 km before Uri, numerous landslides were observed due to failure of the scarp faces of the river terraces and steep slopes along road sections. Toward further west, the occurrences of landslides are found to be more pronounced near the Punjal and Murree thrusts. In this region, the Murree and Punjal thrusts occur very close to each other and are separated by a narrow band of quartzite, siltstone, and slate. North of the Murree Thrust, steeply dipping alternative beds of red shale and sandstone are vulnerable to landslides due to differential weathering. In this area, mostly old slides were reactivated. In the Kashmir earthquake, within the Indian territory, maximum devastation took place in Uri and west of Uri. Landslides were mainly observed on colluviums (Fig. 26.9A and B). The landslides could be classified based on generic material constituent and mechanism of failure, that is, rock and debris falls. The landslides were more pronounced in a linear zone encircling the earlier described fault. In the Jhelum valley, the lower limit of the landslides on valley slopes are demarked by fault extent, thereby suggesting the direct influence of the fault on occurrences of landslides.

26.5.1 Landslide Distribution

We have mapped 776 landslides covering 54.5 km^2 using Cartosat-1, LISS-IV, LISS-III, and ASTER data in approximately 6300 km^2. Cartosat-1 and LISS-IV data have been used in the Indian part of the study area, whereas ASTER and LISS-III data sets were used in the Pakistan part. Landslides larger than the 3 × 3 pixels size were mapped with minimum size varying from 10 × 10 to 45 × 45 m depending on the satellite image used. However, for statistical analysis, we used landslides larger than 45 × 45 m as that was the minimum size we have mapped for most of the study area. In order to assess the

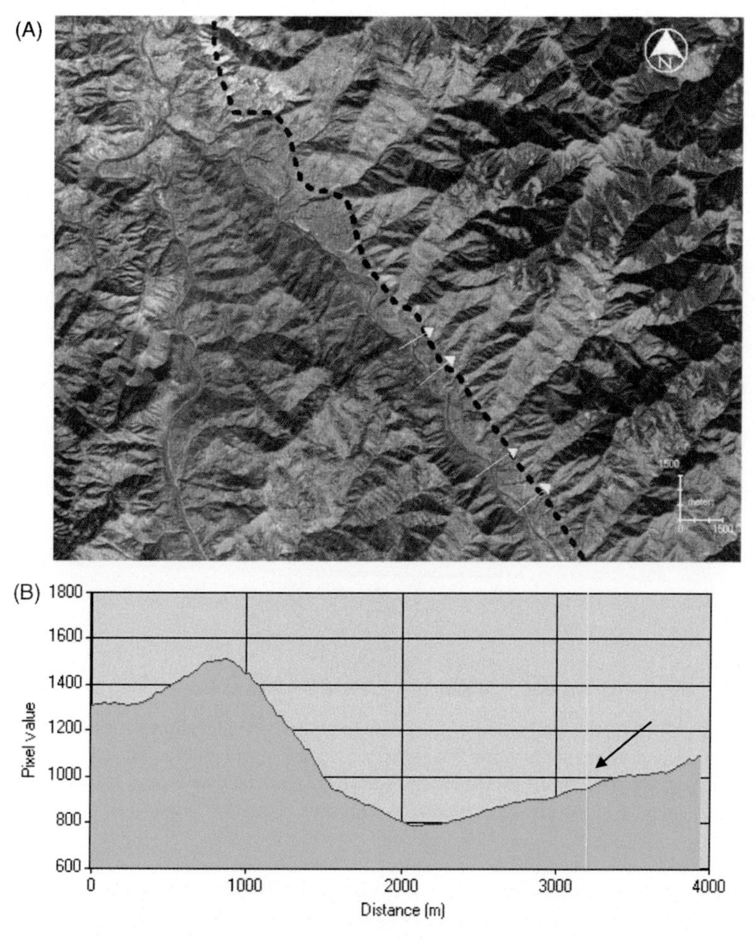

FIGURE 26.5 (A) Asymmetric basin development (larger on east and smaller on west), triangular facets as observed on ASTER data acquired on October 27, 2005; (B) subtle topographic break in slope as observed on the eastern bank of the Jhelum. *Source*: ASTER DEM.

FIGURE 26.6 **After following a linear stretch from Muzaffarabad, the fault moves further in southeast direction and shows "v"-shaped expression due to the northeast dip of the fault.** At the same place, a huge landslide, known as Hattian landslide, had occurred on the southeastern aspect of the hill range.

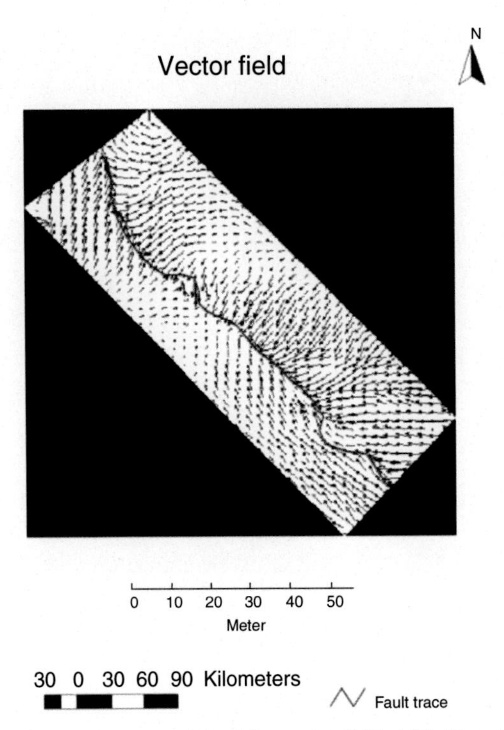

FIGURE 26.7 **The East–West and North–South components of the deformation field yield slip vectors across the fault.** The amplitude of the slip vector varies and reaches maximum around 6–7 m northwest of Muzaffarabad. At the junction of the Muzaffarabad and Tanda segment, minimum slip vector was observed. In the western part, the sense of the movement was thrust and maximum movement was observed on the hanging wall side. In the southeast segment, the slip reduces to around 4 m and gradually orient in the direction of the fault (northwest) showing a relatively right lateral movement.

influence on causative fault on landslides, buffer zones of 5 km on both sides along the fault were generated and superimposed by landslide occurrence map (Fig. 26.10). It was observed that 42% of landslides are located within 5 km from the causative fault on the hanging wall side and 10% on the footwall side (Table 26.2, Fig. 26.11A). Sato et al. (2007) also reported many landslides within 1 km from the active fault. Further from ASTER DEM, it was observed that although relief varies from 450 to 4470 m, 65% of the total landslides were concentrated in elevation range 850–1750 m, out of which around 27% of the total landslides had occurred in the elevation range 1150–1450 m which corresponds to the mountain front and contains the surface expression of the causative fault (Fig. 26.11B).

26.5.2 Influence of Lithology on Landslides

In order to study relation between the geological materials and landslides, lithology (Gansser, 1964) and geomorphic maps were integrated to produce a surface geology map which provided dominant material types present in the area. General rock types include sandstone, shale, siltstone, conglomerate, limestone, slate, phyllite, and gneissic rocks belonging Proterozoic–Tertiary and recent sediments of Quarternary period (Table 26.3). This map has been superposed with the landslide map and it was found that largest percentage (65%) of landslides were observed in the Murree, Subathu formations consisting of red shale, silt stone, and limestone, followed by terrace, colluvium, and scree deposits of recent to subrecent origin. About 9% of landslides occurred in Salkhala Group consisting of phyllite, schist, and limestone (Fig. 26.10C). Similar observations were also made by Sato et al. (2007), who reported around 50% of landslides in Miocene sandstone and siltstone (Murree and Subathu formations) followed by schist and quartzite (Salkhala Group).

26.5.3 Relation of Slope Gradient, Slope Aspect, and Curvature With Landslides

Important topographical parameters such as the slope gradient, aspect and curvature maps were prepared from the ASTER DEM. It was found that maximum landslides were concentrated in the slope class corresponding to 30–40° (35% of landslides). Landslide density gradually increased in slope classes ranging from 30-40° to 60–70° (Fig. 26.11D). The slope aspect map showed maximum number of landslides in south (26.8%), followed by southwestern (21%) and southeastern slopes (17.4 %) which mostly coincided with the free face on the hanging wall side (Fig. 26.11E). Ground movement was also observed mostly in the same direction (Fig. 26.7), thereby suggesting that the fault movement had a direct bearing on

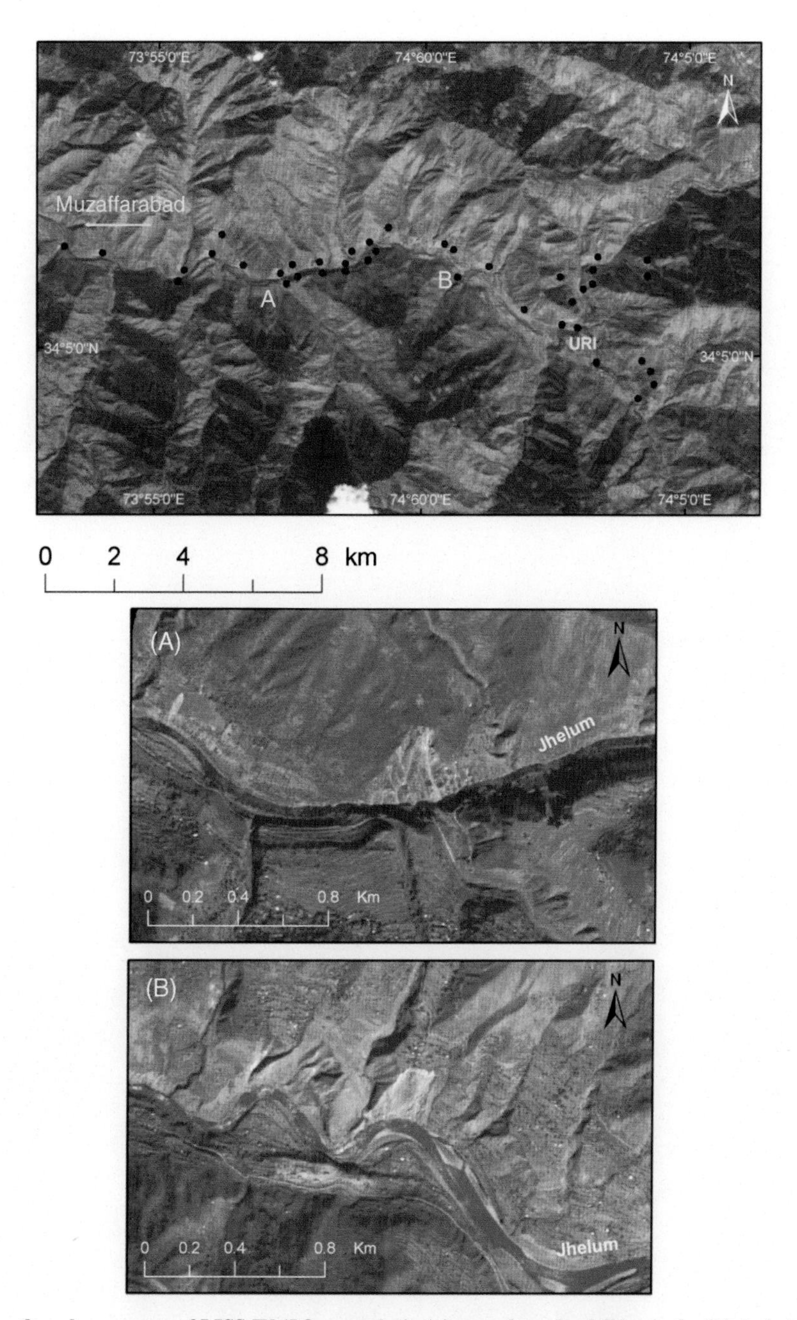

FIGURE 26.8 **The post-earthquake coverage of LISS-IV (5.8 m resolution) image shows landslides (*red solid circles*) along the Jhelum river to the east and west of Uri town.** (A, B) Enlarged view of the recent landslides on post-earthquake coverage of Cartosat-1 image (2.5 m resolution).

the mass movement direction. The slope curvature map showed concentration of landslides on concave slope (61%), suggesting the depletion of mass resulting in concave slope formation (Fig. 26.11E). It was interesting to note that majority of landslides were observed on convex slope on the preearthquake DEM derived from SRTM (90-m resolution), which had developed concave slope due to slope failure (Sato et al., 2007).

26.5.4 Probability Density Function (PDF) for Landslide Distribution

It was shown that the frequency–area statistics of three substantially complete landslide inventories from different parts of world were well approximated by the same probability density function, a three-parameter-based inverse-gamma distribution (Malamud et al., 2004a, b). The common behavior was observed in spite of large differences in landslide types, topography, soil types, and triggering mechanism. A probability density function $p(A_L)$ is defined as

FIGURE 26.9 (A) Seismicity-induced landslide (SIL) in the steeply dipping hard *red color* sandstone of Murree Formation at red bridge (34°06′03.3″, 73°57′29.1″); (B) SIL in the thick colluvial deposit near Urusua village along the national highway leading to Muzaffarabad (34°06′25.9″, 73°56′00.4″).

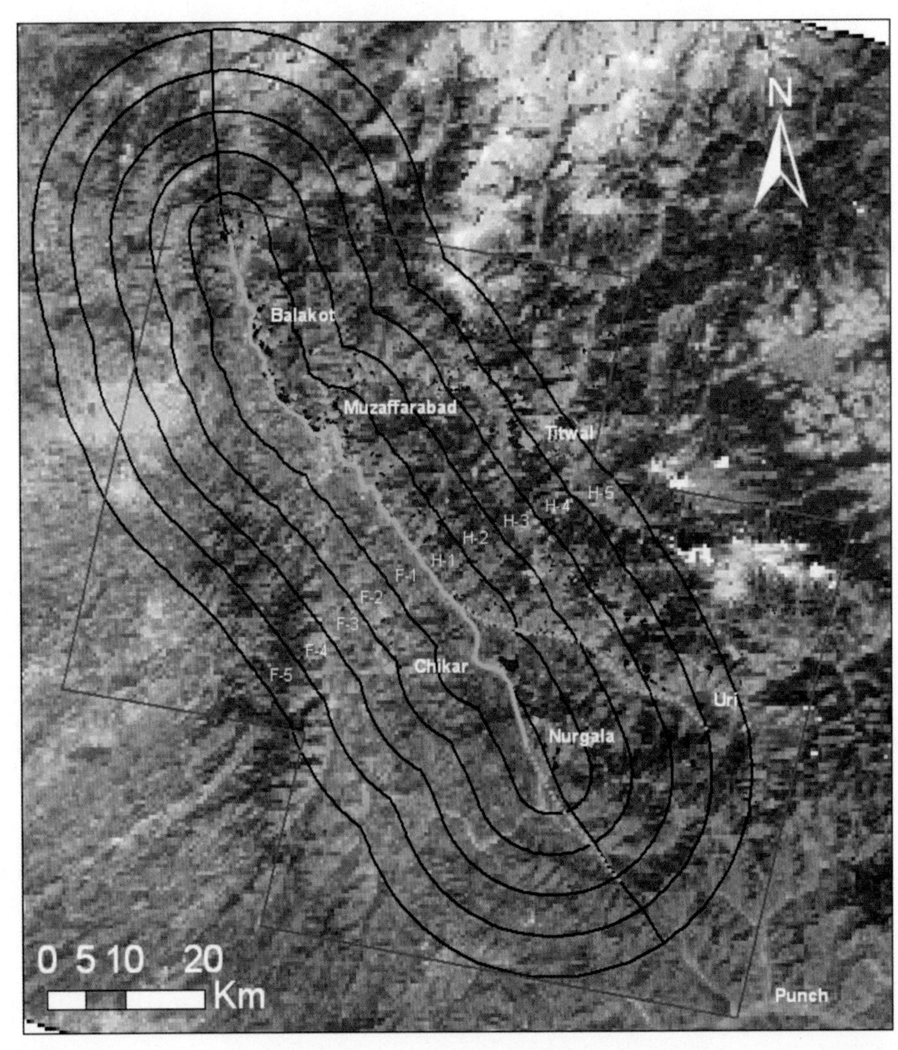

FIGURE 26.10 **Landslides larger than the size of 3 × 3 pixels, that is, 45 × 45 m have been mapped (*red*: new, *orange*: reactivated, *green*: old landslides), out of which a large proportion of landslides have been mapped in the hanging wall side compared to footwall side.** Distance zones (at 5 km interval) in hanging wall (H1, 2, 3, 4, 5) and footwall side (F1, 2, 3, 4, 5) show gradual reduction in landslide occurrences with exception of H4 due to the presence of landslides in Neelam valley due to deep incision and steep slope conditions.

TABLE 26.2 Higher Percentage of Landslides was Observed in Buffer Zones (at 5 km interval) of the Hanging Wall Side Compared to Footwall Side of the Fault

	Zone	Total area (km²)	Landslide (km²)	Landslide (%)
Hanging wall	H1(0–5 km)	454.71	17.93	41.92
	H2 (5–10 km)	507.28	3.37	7.88
	H3 (10–15 km)	568.41	3.28	7.66
	H4 (15–20 km)	629.39	7.68	17.96
	H5 (20–25 km)	691.14	4.89	11.42
Foot wall	F1 (0–5 km)	454.27	4.43	10.13
	F2 (5–10 km)	533.01	1.00	2.35
	F3 (10–15 km)	621.80	0.24	0.57
	F4 (15–20 km)	713.62	0.06	0.13
	F5 (20–25 km)	806.11	0.00	0.00
	TOTAL	5979.73	42.79	100.00

The total landslide area corresponds to landslide occurring within the defined zones; actual inventory consists of larger area and few more landslides outside the zones.

$$p(A_L) = \frac{1}{N_{LT}} \times \frac{\delta N_L}{\delta A_L} \tag{26.1}$$

With normalization condition

$$\int_1^{\infty} p(A_L)\,dA_L = 1 \tag{26.2}$$

where A_L is the landslide area, N_{LT} is the total number of landslides in the inventory, and δN_L is the number of landslides with areas between A_L and $A_L + \delta A_L$. The probability densities of three substantially complete inventories: (i) $N_{LT} = 11{,}111$ landslides triggered by the magnitude $M = 6.7$ Northridge earthquake (California, USA) on January 17, 1994; (ii) $N_{LT} = 4223$ landslides in the Umbria region (Italy), triggered by rapid snowmelt in January 1997; and (iii) $N_{LT} = 9594$ landslides in Guatemala, triggered by heavy rainfall from Hurricane Mitch in the late October and early November 1998 exhibited a characteristic shape (Guzzetti et al., 2002; Malamud et al., 2004a), with densities increasing to a maximum value (most frequently occurring landslide size) and then decreasing with increase in landslide size (Fig. 26.12). Based on these three sets of probability densities, a general probability distribution function is defined as three parameter-based inverse-gamma distribution:

$$p(A_L;\rho,a,s) = \frac{1}{a\Gamma(\rho)}\left[\frac{a}{A_L - s}\right]^{\rho+1} \exp\left[-\frac{a}{A_L - s}\right] \tag{26.3}$$

where $\Gamma(\rho)$ is the gamma function of ρ. The inverse gamma distribution has a power-law decay for medium and large slides, whereas exponential rollover for small landslides (Fig. 26.12). The maximum likelihood fit of the equation yields $\rho = 1.4$, $a = 1.28 \times 10^{-3}$ km², $s = -1.32 \times 10^{-4}$ km², and $\Gamma(\rho) = 0.88726$. Based on the good agreement between three landslide inventories and the inverse-gamma distribution, it was assumed that Eq. (26.3) could be applied to all landslide inventories. It was expected that not all landslide-event inventories would be in very good agreement as the considered database; however, it was considered as a good approximation, and therefore, it was tested on our landslide inventory of Kashmir earthquake. The frequency distribution of three reference landslide inventories (Northridge, Umbria, and Guatemala) are comparable to partial inventory of landslides due to Kashmir earthquake (Fig. 26.12), and therefore, Eq. (26.3) and its derivatives were applied in the present case.

Using the general landslide distribution as mentioned above, Eqs. (26.4)–(26.6) were derived for quantifying the maximum landslide areas and volumes (Malamud et al., 2004a, b). The required integrations were dominated by a power-law tail

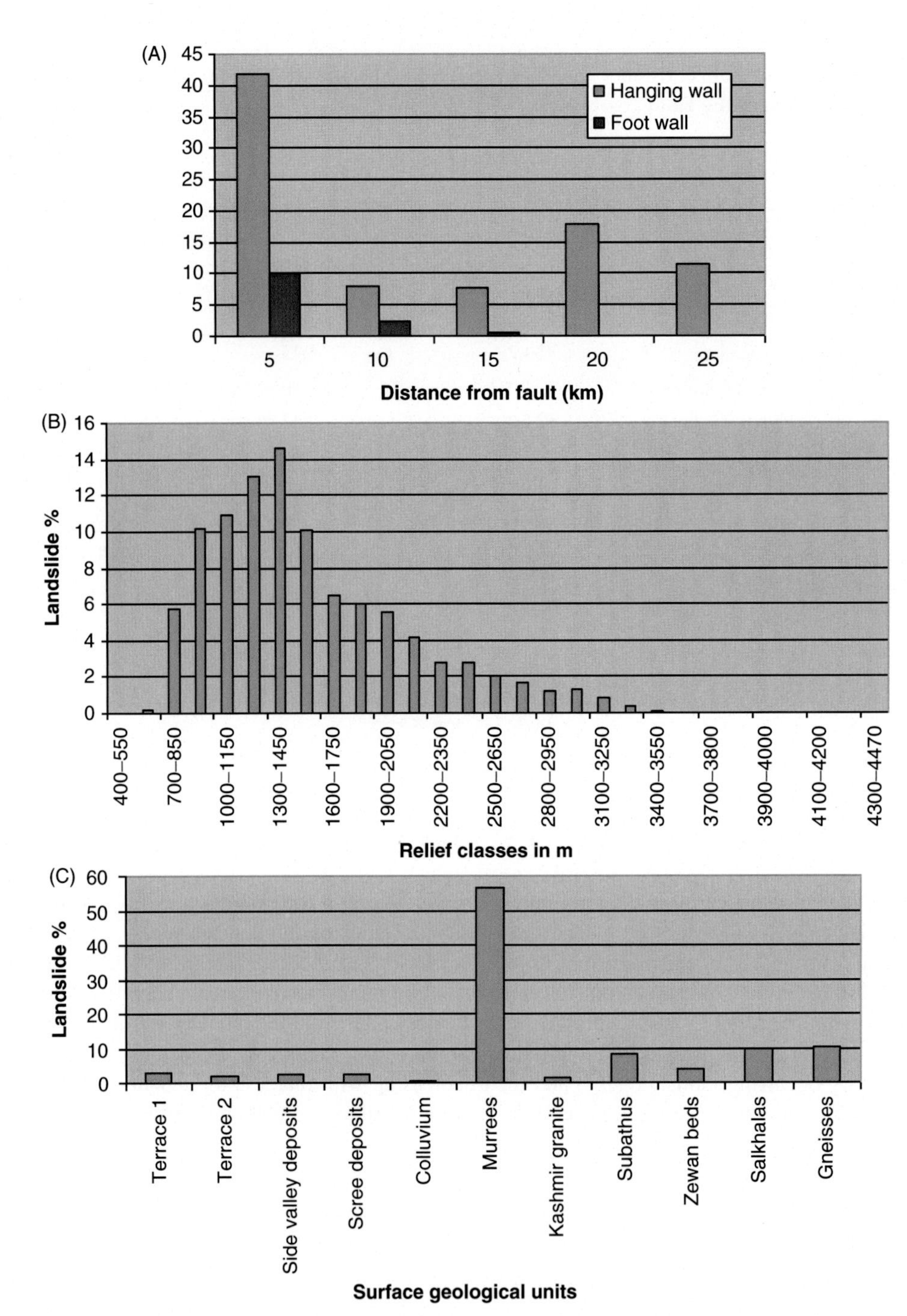

FIGURE 26.11 Landslide distribution in: (A) hanging and footwall of the causative fault; (B) different relief classes (derived from ASTER data); (C) surface geological units; (D) slope gradient classes; (E) slope aspect classes; and (F) vertical curvature classes.

of the medium and large landslides versus the much smaller contribution of the exponential rollover for smaller landslides. This was more significant as our database was complete with respect to medium and large landslides.

Total landslide area:

$$A_{LT}\,(\mathrm{km}^2) = \overline{A}_L . N_{LT} = 3.07 \times 10^{-3} N_{LT} \tag{26.4}$$

Total landslide volume:

$$V_{LT}\,(\mathrm{km}^3) = \overline{V}_L . N_{LT} = 7.30 \times 10^{-6} N_{LT}^{1.122} \tag{26.5}$$

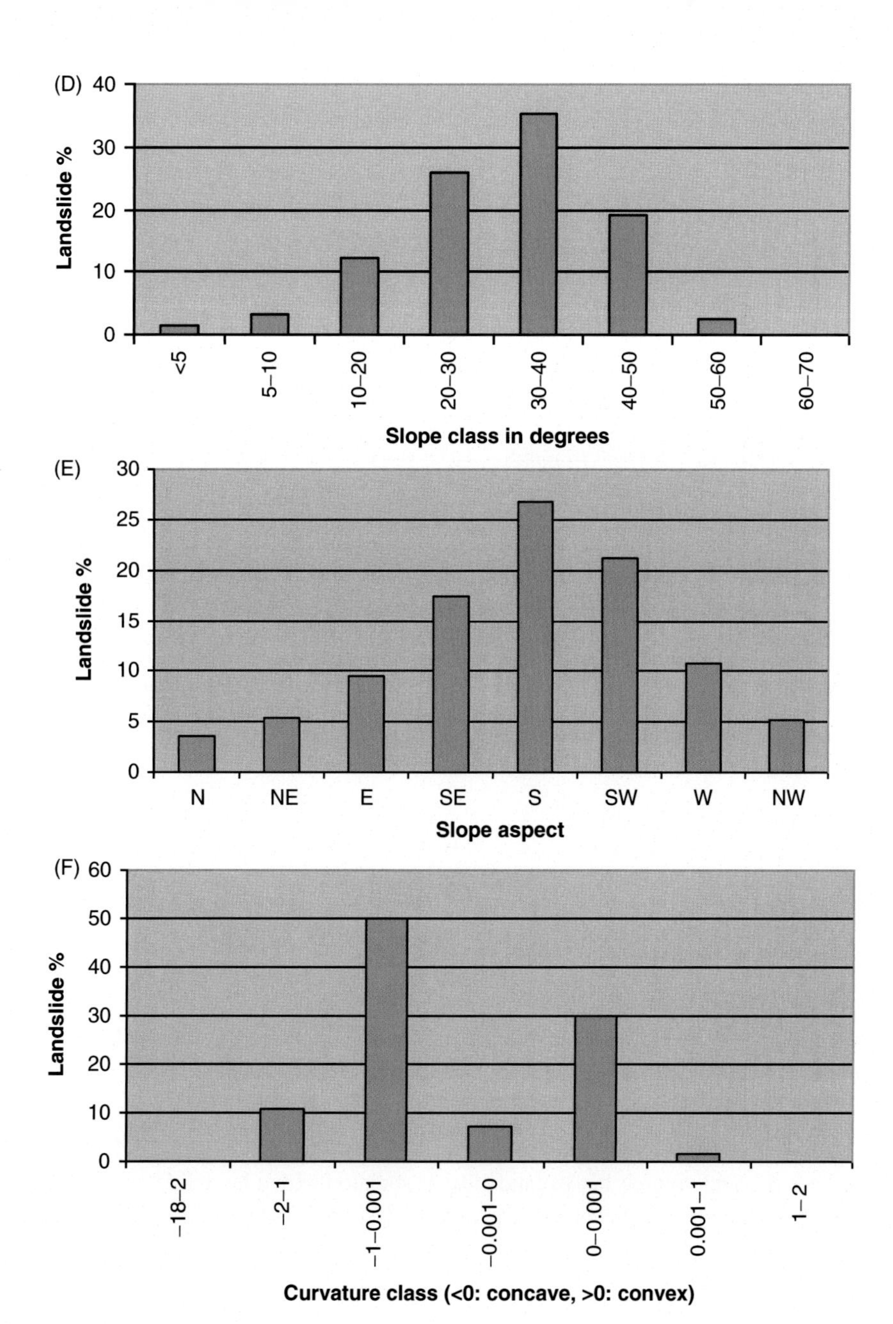

FIGURE 26.11 (*Cont.*)

TABLE 26.3 Characteristics of Lithological Formations/Groups

	Lithology (formation/group)	Characteristics
1.	Surfacial deposits (terrace deposits, talus cones, and deposits on narrow valleys)	Alluvium, river terraces, and unconsolidated deposits
2.	Murree Formation	Highly friable soft to hard red and purple colored sandstone and siltstone with thin intermittent conglomerate and clay layers
3.	Kashmir Granite	White massive granite with joints
4.	Subathu Formation	Friable and brecciated to massive argillaceous, red or brown lithified sandstone, and nummulitic limestone
5.	Zewan Group	Hard massive basaltic rocks
6.	Salkhala Group	Phyllites, schists (chloritic, talcose, serecitic), flaggy quartzites, carbonaceous limestone, and dolomite
7.	Gneiss	Stratified migmatites and gneisses intercalated with quartz–muscovite–biotite schist

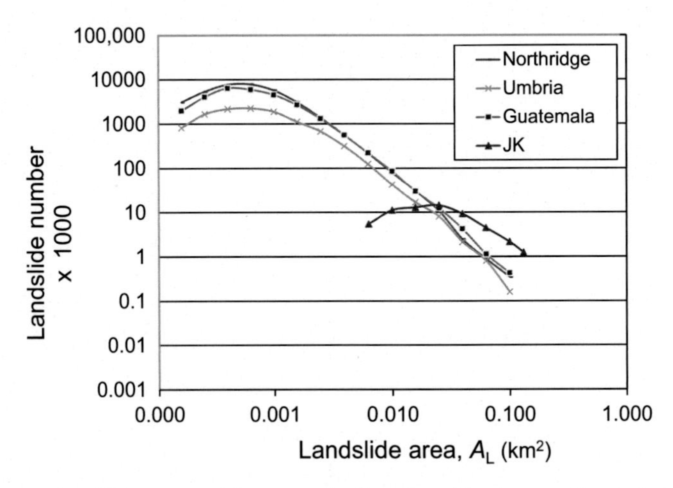

FIGURE 26.12 **Relationship between landslide area and frequency, both on logarithmic axes.** Also included are the frequency distribution of three landslide inventories (Malamud et al., 2004b) and partial inventory of landslides due to 2005 Kashmir earthquake (JK) which shows rollover due to partial inventory.

Maximum landslide area:

$$A_{L\max}\,(\text{km}^2) = 1.10 \times 10^{-3} N_{LT}^{0.714} \tag{26.6}$$

where $\overline{A}_L$ and $\overline{V}_L$ are the average area (km^2) and volume (km^3), respectively.

26.5.4.1 *Landslide Magnitude Scale*

Landslide magnitude, m_L, can be defined for a landslide event based on the logarithm to the base 10 of the total number of landslides associated with the landslide event (Keefer, 1984; Malamud et al., 2004b):

$$m_L = \log(N_{LT}) \tag{26.7}$$

The total measured area A_{LT} and volume V_{LT} of landslides associated with a landslide event can be used to determine a landslide magnitude m_L. Assuming the applicability of general landslide distribution, Eqs. (26.4) and (26.5) are combined with Eq. (26.7) to derive

$$m_L = \log(A_{LT}) + 2.51 = 0.89 \log V_{LT} + 4.58 \tag{26.8}$$

where A_L is in km^2 and V_L is in km^3.

26.5.4.2 *Incomplete Landslide Inventory*

Considering the general applicability of landslide distribution, it was used to extrapolate a partial inventory of landslide areas to give the total landslide area, A_{LT}. Let A_{LC} ($\geq A_L$) be the cumulative area of all landslides in a partial inventory with areas $\geq A_L$. We could find out a relation between A_{LC} ($\geq A_L$) and the total area, A_{LT} based on the probability density function. For each area class (A_L), theoretical A_{LC} ($\geq A_L$)/A_{LT} could be numerically determined (Table 26.4). As mentioned earlier, we mapped landslides larger than 3 × 3 pixel size using ASTER data for most of the inaccessible region in Pakistan. Therefore, landslides of smaller dimension were not included. Additionally, hilly areas further northwest of our study area, beyond Balakot, could not be mapped due to lack of appropriate satellite data and therefore, our database could be considered as partial inventory consisting of landslides larger than 2025 m^2. Considering 3835.5 m^2 as minimum size closer to the bin size of 3980 m^2 for theoretical consideration, we determined cumulative area, A_{LC}, as 45.545 km^2 (total area of landslides in the subset of database) for a total of 691 landslides. Using the theoretical ratio of A_{LC} ($\geq A_L$)/A_{LT} as 0.676 for A_L = 3980 m^2 from Table 26.4, we estimated the total landslide area, A_{LT} as 67.36 km^2 for our partial database. So, 33% of the additional landslide area was attributed to the smaller landslides, not mapped by us in the area.

TABLE 26.4 A_{LC} ($\geq A_L$) **is the Cumulative Area of Inventory Landslides With Areas** $\geq A_L$**, With** A_{LT} **the Total of Landslide Areas in the "Complete" Inventory**

A_L (km^2)	A_L (m^2)	$A_{LC}(>A_L)/A_{LT}$
0.100	100,000	0.207
0.0631	63,100	0.248
0.0398	39,800	0.297
0.0251	25,100	0.354
0.0158	15,800	0.421
0.0100	10,000	0.498
0.00631	6,310	0.584
0.00398	3,980	0.676
0.00251	2,510	0.768
0.00158	1,580	0.852
0.001	1,000	0.919
0.000631	631	0.964
0.000398	398	0.987
0.000251	251	0.996
0.000158	158	0.999
0.0001	100	1

The theoretical ratios A_{LC} ($\geq A_L$)/A_{LT} are obtained from probability distribution (Malamud et al., 2004a).

Based on $A_{LT} = 67.36$ km^2, using Eq. (26.8), landslide magnitude, m_L, is calculated as: 4.328 and total landslide volume, V_{LT}, is calculated as: 0.52 km^3.

26.5.4.3 Correlation Between Earthquake and Landslide Magnitude

Based on 16 historical earthquakes, Keefer (1994) and Malamud et al. (2004b) obtained an empirical correlation between the total volume of landslides triggered by an earthquake, V_{LT}, and the earthquake's moment magnitude, M_w. Despite a wide variety of geological (topography and rock type), geophysical (earthquake type and depth), geotechnical, and climatic parameters associated with earthquake-triggered landslide events, a reasonably good power-law dependence on the total landslide volume V_{LT} to the earthquake's moment magnitude M_w was established. The least-square best-fit line to the log V_{LT} versus M_w data gives

$$\log V_{LT} = 1.42M_w - 11.26 \pm 0.52 \quad (26.9)$$

$$m_L = 1.27M_w - 5.45 \pm 0.46 = \log N_{LT} \quad (26.10)$$

Using $m_L = 4.328$, the earthquake magnitude was calculated as 7.69 (7.7). This is in good agreement with the M_w 7.6 earthquake magnitude estimated by instrumental data. This is a very important finding that supports the hypothesis that landslide occurrences show a direct relationship with ground motion in the case of landslides triggered by large earthquakes on thrust faults such as in the case of Northridge earthquake in California, Chi-Chi earthquake in Taiwan, and Ramu-Markham fault-related earthquake in Papua New Guinea (Meunier et al., 2007). The earthquake magnitude is also related to total landslide area by combining Eqs. (26.8) and (26.10) and the area of largest triggered landslide, $A_{L\max}$ (Eqs. (26.6) and (26.10)):

$$\log A_{LT} = 1.27M_w - 7.96 \pm 0.46 \quad (26.11)$$

$$\log A_{L\max} = 0.91M_w - 6.85 \pm 0.33 \quad (26.12)$$

Based on Eq. (26.11), A_{LT} was estimated as 49.20 km^2, which is lower than the estimate based on the actual inventory, that is, 67.36 km^2. This implies that the A_{LT} based on partial inventory is overestimated. This can be attributed by two factors: (a) landslides and landslide debris were mapped together at many places because it was not possible to differentiate due to lack of tonal difference on satellite image; (b) the area consists of steep slopes underlain by fractured and faulted rocks (vulnerable to slope failures) due to ongoing tectonic activity at the mountain front.

The area of largest landslide, A_{Lmax}, was estimated (Eq. (26.12)) to be 1.164 km^2 within a range of 0.544–2.48 km^2. The largest landslide in the inventory had an area of 1.2 km^2 (without debris) and 1.99 km^2 (with debris) which showed very good agreement with the estimated value.

The total landslide volume, V_{LT}, was estimated from earthquake magnitude (Eq. (26.9)) as: 0.340 km^3 and based on the total landslide area estimate (Eq. (26.11)) of 49.20 km^2, the average landslide thickness was estimated as 6.9 m. Based on landslide inventory, the total landslide volume, V_{LT}, was estimated (Eq. (26.8)) as 0.52 km^3. Based on the estimated total landslide area from inventory, A_{LT} = 67.36 km^2, the average landslide thickness was estimated at 7.7 m. Both the results show good agreement and therefore, the average landslide thickness could be considered to be around 6.9–7.7 m.

26.6 DISCUSSION AND CONCLUSION

Analysis of multitemporal ASTER data of 15 m resolution, LISS-III data of 23.5 m resolution, Cartosat-1 (2.5 m) and 5.8 m multispectral data from IRS-P6 (Resourcesat-1) showed the extent of causative fault. Widespread ground deformation due to relative movement of faulted blocks led to large number of landslides in the mountainous terrain. Based on the tonal differences and temporal satellite data analysis, landslides have been classified in terms of their activity as old, new, and reactivated. In some cases, it was possible to classify the landslides based on generic material constituent and mechanism of failure, that is, rock and debris falls. It was observed that a large proportion of landslides occurred in the hanging wall side of the fault compared to footwall side, particularly within 5 km from the causative fault. As a result, localities lying close to the active fault in the hanging wall side suffered more damage in comparison to those away from the fault, irrespective to the location of epicenter. The occurrences of landslides are also largely influenced by rock formations, for example, largest percentage (65%) of landslides were observed in the Murree and Subathu formations consisting of red shale, silt stone, and limestone, followed by terrace, colluvium, and scree deposits of recent to subrecent origin. Analysis by Sato et al. (2007) and Kamp et al. (2008) showed that most of the landslides were concentrated in the Miocene mudstone, siltstone, and sandstone (50% in Sato et al.'s (2007) study and >60% in Kamp et al.'s (2008) study). Therefore, the present study has brought out two important aspects: mapping of causative fault and mechanism is critical to assessment of ground deformation and associated landslides. Secondly, role of rock type and surface deposits also plays an important role which needs to be considered for future road construction and developmental activities in this region.

Terrain characteristics revealed maximum number of landslides in slope class corresponding to 30–40° (35% of landslides) and slope curvature corresponding to concave slope (61%), suggesting a depletion of mass resulted in concave slope formation as revealed by post-earthquake DEM. In a similar study, Kamp et al. (2008) noted that most of the landslides occurred on slopes ranging from 25 to 35°. It is to be noted that although relief varies from 450 to 4470 m in the earthquake-affected area, 65% of total landslides were concentrated in the elevation ranges from 850 to 1750 m. Of these areas, one-third of landslides occurred in the elevation range 1150–1450 m that corresponds to the mountain front and the surface expression of the causative fault.

Probability density function-based statistical analysis of landslide partial inventory revealed that the 2005 Kashmir earthquake event caused 67 km^2 landslide area around the causative Balakot–Bagh Fault (BBF). The study by Kamp et al. (2008, 2010) showed that the number of landslides increased from 369 in 2001 (covering an area of 8.2 km^2) to 2252 in October 2005 (covering an area of 61.1 km^2). Although the event had caused landslides in only 67 km^2, in many other vulnerable slopes, ground fractures and fissures were observed parallel to hill slope, further enhancing the chances of slope failure in subsequent years as reported by Owen et al. (2008). The study suggests that the causative fault can be mapped using a subpixel registration of optical satellite data, thus it can be used as an alternative or complementary technique to other methods of fault mapping. This study further demonstrated that spatial distribution of landslides was influenced by causative fault, lithology, and terrain attributes and support the hypothesis that in the case of large earthquakes related to thrust faults, landslide densities are related to ground motion and possibly landslides can be used as markers for preparing better shake map in noninstrumented or nonpopulated regions like that exists in many mountainous terrain like Himalaya.

ACKNOWLEDGMENTS

The authors are thankful to the Department of Science and Technology, Government of India, for partly funding the project, Prof. J.P. Avouac for providing a software tool for ASTER data subpixel correlation, LPDAAC for ASTER data covering part of the study area, and Prof. M.I. Bhat, University of Kashmir and Prof. M.A. Mallik, Jammu University, for their support during field investigations, and Dr. S.L. Chattoraj for editing of the manuscript.

REFERENCES

Avouac, J.P., Ayoub, F., Leprince, S., Konca, O., Helmberger, D.V., 2006. The 2005, M_w 7.6 Kashmir earthquake: sub-pixel correlation of ASTER images and seismic waveforms analysis. Earth and Planetary Science Letters 249 (3–4), 514–528. doi: 10.1016/j.epsl.2006.06.025.

Baig, M.S., 2006. Active faulting and earthquake deformation in Hazara-Kashmir syntaxis, Azad Kashmir, northwest Himalaya. In: Kausar et al. (Ed.), International Conference on 8 October 2005 Earthquake in Pakistan: Its Implications and Hazard Mitigationpp. 27–28, January 18–19, 2006, Extended Abstract.

Bilham, R., 2005. http://cires.colorado.edu/~bilham/Kashmir%202005.htm.

Binet, R., Bollinger, L., 2005. Horizontal coseismic deformation of the Bam (Iran) earthquake measured from SPOT-5 THR satellite imagery. Geophys. Res. Lett. 32, L02307. doi: 10.1029/2004GL021897.

Champatiray, P.K., Perumal, R.J.G., Thakur, V.C., Bhat, M.I., Mallik, M.A., Singh, V.K., Lakhera, R.C., 2005. A quick appraisal of ground deformation in Indian region due to the October 8, 2005 earthquake, Muzaffarabad, Pakistan. J. Indian Soc. Remote Sens. 33 (4), 465–473.

Cui, P., Chen, X.-Q., Zhu, Y.-Y., Su, F.-H., Wei, F.-Q., Han, Y.-S., Liu, H.-J., Zhuang, J.-Q., 2011. The Wenchuan earthquake (May 12, 2008), Sichuan Province, China, and resulting geohazards. Nat. Hazards 56, 19–36.

Das, J.D., Saraf, A.K., Panda, S., 2007. Satellite data in a rapid analysis of Kashmir earthquake (October 2005) triggered landslide pattern and river water turbidity in and around the epicentral region. Int. J. Remote Sens. 28 (8), 1835–1842.

Dunning, S.A., Mitchell, W.A., Rosser, N.J., Petley, D.N., 2007. The Hattian Bala rock avalanche and associated landslides triggered by the Kashmir earthquake of 8 October 2005. Eng. Geol. 93, 130–144.

ERDAS IMAGINE, 2006. ERDAS IMAGINE 9.1. LPS project manager, pp. 169–191.

Gansser, A., 1964. Geologic and Tectonic Maps of the Himalayan Region. Interscience Publications/John Wiley, London, Plate 1A.

Guzzetti, F., Malamud, B.D., Turcotte, D.L., Reichenbach, P., 2002. Power-law correlations of landslide areas in central Italy. Earth Planety Sci. Lett. 195, 169–183.

Jayangondaperumal, R., Thakur, V.C., 2008. Co-seismic secondary surface fractures on southeastward extension of the rupture zone of the 2005 Kashmir earthquake, 2008. Tectonophysics 446, 61–76.

Kamp, U., Growley, B.J., Khattak, G.A., Owen, L.A., 2008. GIS-based landslide susceptibility mapping for the 2005 Kashmir earthquake region. Geomorphology 101, 631–642.

Kamp, U., Owen, L.A., Growley, B.J., Khattak, G.A., 2010. Back analysis of landslide susceptibility zonation mapping for the 2005 Kashmir earthquake: an assessment of the reliability of susceptibility zoning maps. Nat. Hazards 54, 1–25.

Keefer, D.K., 1984. Landslides caused by earthquakes. Geol. Soc. Am. Bull. 95, 406–421.

Keefer, D.K., 1994. The importance of earthquake-induced landslides to long-term slope erosion and slope-failure hazards in seismically active regions. Geomorphology 10, 265–284.

Leprince, S., Barbot, S., Ayoub, F., Avouac, J.P., 2007. Automatic precise, ortho-rectification and co-registration for satellite image correlation, application to seismotectonics. IEEE T. Geosci. Remote 45 (6), 1529–1558.

Mahajan, A.K., Kumar, N., Arora, B.R., 2006. Quick look isoseismal map of 8 October 2005 Kashmir earthquake. Curr. Sci. 91 (3), 356–361.

Malamud, B.D., Turcotte, D.L., Guzzetti, F., Reichenbach, P., 2004a. Landslide inventories and their statistical properties. Earth Surf. Proc. Land. 29, 687–711.

Malamud, B.D., Turcotte, D.L., Guzzetti, F., Reichenbach, P., 2004b. Landslides, earthquakes, and erosion. Earth Planet. Sci. Lett. 229, 45–59.

Mathur, L.P., 1953. Assam earthquake of 15th Aug 1950—a short note on factual observations. Cent. Board Geophys. 1, 56–60.

Meunier, P., Hovius, N., Haines, A.J., 2007. Regional patterns of earthquake-triggered landslides and their relation to ground motion. Geophys. Res. Lett. 34 (L20408), 1–5.

Michel, R., Avouac, J.P., 2002. Deformation due to the 17th August 1999 Izmit, Turkey, earthquake measured from SPOT satellite images. J. Geophys. Res. (Solid Earth) 107 (B4), 1–7, 2062 ETG 2.

Middlemiss, C.S., 1910. The Kangra Earthquake of 4th April 1905, Memoirs of the Geological Survey of India, Vol. XXXVIII, Geological Survey of India 1910, reprinted in 1981.

Nakata, T., Kumahara, Y., 2006. Active faults of Pakistan with reference to the active faults in source area of the 2005 North Pakistan earthquake. Extended Abstract. Int. Conference on 8 October 2005 Earthquake in Pakistan: Its Implications and Hazard Mitigation. Geol. Surv. of Pakistan, Islamabad, 18-22.

Nakata, T., Tsutsumi, H., Khan, S.H., Lawrence, R.D., 1991. Special Publication, v. 21. Research Centre for Regional Geography. Hiroshima University, Japan, 141 pp.

NDMA, 2007. Earthquake-8/10: Learning from Pakistan's experience. NDMA Publication, Pakistan.

Owen, L.A., Kamp, U., Khattak, G.A., Harp, E.L., Keefer, D.K., Bauer, M.A., 2008. Landslides triggered by the 8th October 2005 Kashmir earthquake. Geomorphology 94 (1-2), 1–9.

Pathier, E., Fieldings, E.J., Wright, T.J., Walker, R., Parsons, B.E., Hensley, S., 2006. Displacement field and slip distribution of the 2005 Kashmir Earthquake from SAR Imagery. Geophys. Res. Lett. 33, L20310. doi: 10.1029/2006GL027193.

Petley, D.N., Dunning, S.A., Rosser, N.J., Kausar, A.B., 2006. Incipient landslides in the Jhelum Valley, Pakistan following the 8th October 2005 earthquake. Mauri, H. (Ed.), Disaster Mitigation of Debris Flows, Slope Failures and Landslides, 47, Universal Academy Press, Tokyo, Japan, pp. 47–56, Frontiers of Science Series.

Saraf, A.K., 2000. IRS-1C-PAN depicts Chamoli earthquake induced landslides in Garhwal Himalayas, India. Int. J. Remote Sens. 21, 2345–2352.

Sarkar, I., Saraf, A.K., 1999. Some observations of the Chamoli earthquake-induced damage using ground and satellite data. Curr. Sci. 78, 91–97.

Sato, H.P., Hasegawa, H., Fujiwara, S., Tobita, M., Koarai, M., Une, H., Iwahashi, J., 2007. Interpretation of landslide distribution triggered by the 2005 Northern Pakistan earthquake using SPOT 5 imagery. Landslides 4, 113–122.

Schiek, C.G., Hurtado, Jr., J.H., 2007. Slip analysis of the Kokoxili earthquake using terrain change detection and regional earthquake data. Geosphere 2 (2007), 187–194.

Singh, H., Som, S.K., 2016. Earthquake triggered landslide—Indian scenario. J. Geo. Soc. India 87, 150-111.

Sudmeier-Rieux, K., Qureshi, R.A., Peduzzi, P., Jaboyedoff, M.J., Breguet, A., Dubois, J., Jaubert, R., Cheema, M.A., 2007a. An interdisciplinary approach to understanding landslides and risk management: a case study from earthquake-affected Kashmir. Mountain Forum, Mountain GIS e-Conference, January 14–25, 2008. http://www.mtnforum.org/rs/ec/scfiles/Neelum_PAK_landslides_2007.pdf.

Sudmeier-Rieux, K., Qureshi, R.A., Peduzzi, P., Nessi, J., Breguet, A., Dubois, J., Jaboyedoff, M.J., Jaubert, R., Rietbergen, S., Klaus, R., Cheema, M.A., 2007b. Disaster risk, livelihoods and natural barriers, strengthening decision-making tools for disaster risk reduction, a case study from Northern Pakistan. The World Conservation Union (IUCN) Pakistan Programme, Final report, Karachi.

Thakur, V.C., Perumal, R.J.G., Champati Ray, P.K., Bhat, M.I., Mallik, M.A., 2006. Muzaffarabad earthquake and seismic hazard assessment of Kashmir gap in northwest Himalaya. J. Geo. Soc. India 68, 187–200.

USGS (United States Geological Survey), 2006. Magnitude 7.6—Pakistan earthquake. http://earthquake.usgs.gov/eqcenter/eqinthenews/2005/usdyae/#summary.

Van Puymbroeck, N., Michel, R., Binet, R., Avouac, J.P., Taboury, J., 2000. Measuring earthquakes from optical satellite images. App. Opt. Inf. Process. 39, 1–14.

Yeats, R.S., Hussain, A., 2006. Surface features of the Mw 7.6, 8 October 2005 Kashmir earthquake, northern Himalaya, Pakistan: implications for the Himalayan front. Seismol. Res. Lett. 77, 207.

Yeats, R.S., Nakata, T., Farah, A., Fort, M., Mirza, M.A., Pandey, M.R., Stein, R.S., 1992. The Himalayan frontal fault system. Annales Tectonicae 16 (Suppl.), 85–98.

Yeats, R.S., Parson, S.T., Hussain, A., Yuji, Y., 2006. Stress changes with the 8 October 2005 Kashmir earthquake: lessons for future. In: Kausar et al. (Ed.), International Conference on 8 October 2005 Earthquake in Pakistan: Its Implications and Hazard Mitigation. Extended Abstractpp. 16–17, January 18–19, 2006.

Chapter 27

Spatiotemporal Variability of Soil Moisture and Drought Estimation Using a Distributed Hydrological Model

Jayakumar Drisya*, Sathish Kumar D*, Thendiyath Roshni**

**National Institute of Technology, Calicut, India; **National Institute of Technology, Patna, India*

27.1 INTRODUCTION

Drought, one of the major natural disasters, is occurring due to the weather extremities on land surface hydrological cycle. Unlike other natural disasters (flood, hurricanes, earthquakes, and tsunamis) that develop quickly and last for a short time, drought is a creeping phenomenon that accumulates over a period of time across vast area, and the effect lingers for years even after the end of drought. Following publication of the Intergovernmental Panel on Climate Change report on extreme events (Intergovernmental Panel on Climate Change (IPCC), 2012), the issue of quantifying loss and damage from extreme climate events such as droughts has become important for policy implementation, especially with regard to the United Nations Framework Convention on Climate Change agenda. However, drought monitoring and assessment is facing difficulties since different disciplines use water in various ways and thus use different indicators for defining and measuring drought. According to the different viewpoints of defining drought, there exist different drought types.

27.2 TYPES OF DROUGHT AND THEIR ESTIMATION

Wilhite and Glantz (1985) analyzed more than 150 definitions of different droughts and then broadly grouped such drought definitions under four categories: meteorological, agricultural, hydrological, and socioeconomic drought. The four droughts are interrelated in a sequential way. Prolonged meteorological drought results in agricultural drought and hydrological drought. As a consequence, socioeconomic conditions are also affected.

1. *Meteorological drought:* It is defined as a period of prolonged dry weather condition due to continuous precipitation deficits in an area. It is estimated through the analysis of precipitation deficits of the area.
2. *Agricultural drought:* It is defined as the decrease in agricultural/vegetative production caused either due to short-term precipitation shortages or temperature anomaly. It is measured by increased evapotranspiration and reduced soil water conditions of the area.
3. *Hydrological drought:* Overall reduction of water resources like rivers, reservoirs, and groundwater due to precipitation shortfall, temperature increase, and decreased soil moisture results in hydrological drought. Detailed hydrological modeling can be utilized for such estimations.
4. *Socioeconomic drought:* It is associated with the negative socioeconomic balances that occur due to meteorological, agricultural, and hydrological droughts. Its estimation is little complicated and needs a prior estimation of all other types of droughts.

Intensity and duration of different drought types are quantified through various indices. Some of the more common indices are documented and explained in the WMO Handbook published by the National Drought Mitigation Center (NDMC)

Integrating Disaster Science and Management. http://dx.doi.org/10.1016/B978-0-12-812056-9.00027-0

at the University of Nebraska-Lincoln, USA. The handbook is a ready reckoner for various drought indices. Heim, (2002), Keyantash and Dracup (2002), and Zargar et al. (2011) also presented excellent reviews on various drought indices. These reviews help in comparing and selecting the best indices for drought estimation based on the data availability and purpose of study.

27.3 NEED FOR AGRICULTURAL DROUGHT ESTIMATION

India, an agrarian-based economy, is dependent on agriculture by INR 3981.16 billion (17.9%) for its gross domestic product ratio in the year 2016–2017. India is the second largest producer of agriculture product and accounts for 7.68% of total global agricultural output (Central Statistical Office, 2017). However, Indian agriculture is facing serious threat due to urban expansions, transportation infrastructure developments, poor soil and seed health, lesser infrastructure for storage of harvested products, etc. In addition to these, climatic threats like decreased water availability for irrigation, highly erratic rainfall, and reduced soil moisture conditions are adversely affecting the agricultural production. On a long term, this will lead to agricultural drought. Moreover, appropriate information can be disseminated to the farming community if such droughts are estimated in advance.

27.4 METHODS FOR AGRICULTURAL DROUGHT ESTIMATION

Agricultural drought can be estimated by analyzing remote-sensing images or through estimating the parameters that causes the drought like temperature, evapotranspiration, soil moisture deficit, and rainfall deficit.

27.4.1 Agricultural Drought through Remote-Sensing Techniques

In the recent years, the increasing advancements made in remote-sensing technology have proved the great potential of remote sensing for drought monitoring and assessment. The different indices derived from remote-sensing products are described below.

27.4.1.1 Normalized Difference Vegetation Index

The normalized difference vegetation index (NDVI), which is derived from remote-sensing (satellite) data, is closely linked to drought conditions. To determine the density of green on a patch of land, the distinct colors (wavelengths) of visible and near-infrared sunlight reflected by the plants are observed. Range of NDVI is −1 to +1. Higher value of NDVI refers to healthy and dense vegetation. Lower NDVI values show sparse vegetation:

$$NDVI = \frac{NIR - RED}{NIR + RED} \tag{27.1}$$

where RED and NIR stand for the spectral reflectance measurements acquired in the red (visible) and near-infrared regions, respectively.

27.4.1.2 Vegetation Condition Index

The vegetation condition index (VCI) is an NDVI-derived index and separates the short-term weather-related NDVI fluctuations from long-term ecosystem changes (Kogan, 1990):

$$VCI = 100\frac{(NDVI - NDVI_{min})}{(NDVI_{max} - NDVI_{min})} \tag{27.2}$$

The VCI varies from 0 to 100, corresponding to the changes in vegetation conditions from extremely unfavorable to optimal (Kogan, 1995). Given that VCI is based on vegetation condition, it is primarily useful during the growing season and has limited utility during the dormant season. During the growing season, the VCI allows for the detection of drought and measurement of the time of its onset, intensity, duration, and impact of vegetation (Heim, 2002).

27.4.1.3 Temperature Condition Index

Kogan (1995) has developed the temperature condition index (TCI) using the thermal bands of the NOAA-AVHRR to determine the temperature-related vegetation stress as well as stresses caused by the excessive wetness:

$$TCI = 100\frac{(BT_{max} - BT)}{BT_{max} - BT_{min}} \tag{27.3}$$

where BT is brightness temperature and BT_{max} and BT_{min} are the absolute maximum and minimum value, respectively.

27.4.1.4 Vegetation Health Index

The vegetation health index (VHI) has been developed using the VCI and TCI and is found to be more effective compared to other indices in monitoring vegetative drought:

$$VHI = \alpha VCI + (1-\alpha)TCI \tag{27.4}$$

where α determines the contribution of two indices and is generally taken as 0.5.

27.4.2 Agricultural Drought through Hydrological Cycle Modeling

Understanding the drought phenomenon using hydrological cycle has multiple reasons. Hydrological models simulate water flux and storage through various media within the hydrological cycle. Precipitation and temperature are major driving inputs for all hydrological models. Critical droughts occur after prolonged periods of soil moisture deficits. Rainfall, temperature, and soil moisture are the critical parameters affecting the drought situations of a region. There are wide varieties of methods for measuring these parameters in regional scales. However, these are highly variable and depends on many subunits like land use, soil type, etc. On spatial scales, these are difficult to measure or account. Hence, determining drought using these parameters also faces serious issues unless we ought to hydrological models.

The outputs obtained from a process-based or conceptual hydrological model ranges from evaporation, soil moisture, and groundwater recharge to reservoir inflow and river runoff. Many drought indices are based on the outputs that can be obtained from a hydrological model, for example, evaporation deficit index, soil moisture index, and standardized runoff index. Although some observation data may be available for hydrological drought monitoring and analysis of past droughts in some countries, such data are generally limited to few locations and are usually insufficient to cover the spatial variability of droughts. Especially in countries like India, such information is not available. In these situations, the use of a hydrological model, distributed or semidistributed, is extremely useful. Another important aspect is the use of hydrological model as a predicting tool, which is necessary for operational drought forecasting. Therefore, hydrological models can be successfully applied for drought assessment, monitoring, and forecasting. Generally, a detailed process-based, distributed or semidistributed model is preferred with a continuous simulation capability for this purpose. However, some models that are aimed at application for floods may not have a good evaporation component or may only be used for an event simulation (contrary to a continuous simulation). Trambauer et al. (2013) presented a decision tree for selecting the right type of hydrological model for drought forecasting.

27.5 SOIL MOISTURE VARIABILITY AS A MEASURE OF AGRICULTURAL DROUGHT

Soil moisture is an important variable that is controlled by various other subprocesses in hydrological cycle. Although only a small percentage of total precipitation is stored in the soil after accounting for evapotranspiration (ET), surface runoff, and deep percolation, soil moisture reserve is critical for sustaining agriculture and surface water storages. It is an integrated measure of several state variables of climate and physical properties of land use and soil. Due to its high spatial variability, soil moisture acts as a good measure for spatial drought monitoring. The spatial and temporal variability of soil moisture is due to heterogeneity in soil properties, land cover, topography, and nonuniform distribution of precipitation and ET. Huang et al. (1996) found that precipitation anomalies influence soil moisture anomalies and that the soil moisture anomalies have greater persistence during periods of low precipitation. Several drought indices can be derived from simulated soil moisture information. Table 27.1 lists the agricultural drought indices that can be derived from soil moisture information

27.6 SOIL MOISTURE ESTIMATION

27.6.1 Field Techniques

Field scale soil moisture is measured using instruments, such as tensiometers, time domain reflectometry, neutron probes, gypsum blocks, and capacitance sensors. These are often widely spaced, and the averages of these point measurements seldom yield soil moisture information on a watershed scale or regional scale due to the heterogeneity involved.

TABLE 27.1 Soil Moisture-Derived Agricultural Drought Indices

Indices	Input Parameters Needed	Reference
Soil moisture anomaly (SMA)	Precipitation, temperature, and available water content	Bergman et al. (1988)
Soil moisture deficit index (SMDI)	Soil moisture measurement	Narasimhan and Srinivasan (2005)
Soil water storage (SWS)	Rooting depth, available water storage capacity of the soil type, and maximum soil water deficit	British Columbia Ministry of Agriculture (2015)
Aggregate dryness index (ADI)	Precipitation, evapotranspiration, available water content, streamflow, and reservoir flow	Keyantash and Dracup (2004)
Vegetation drought response index (VegDRI)	Precipitation, temperature, available water content, evapotranspiration, land cover, and satellite images	Brown et al. (2008)
Palmer hydrological drought severity index (PHDI)	Precipitation, temperature, and available water content	Palmer (1965)
Standardized soil moisture index (SSI)	Soil moisture	Farahmand and AghaKouchak (2015)

27.6.2 Hydrological Models

Among the many hydrologic models developed in the last decade, the Soil and Water Assessment Tool (SWAT), developed by Arnold et al. (1993), has been used extensively by researchers. QSWAT is the SWAT version working in the QGIS platform. The advantages of the QSWAT model are:

(i) It uses readily available inputs for weather, soil, land, and topography.
(ii) It allows considerable spatial detail for basin scale modeling.
(iii) It is capable of simulating crop growth and land management scenarios.

Due to its popularity, QSWAT has been integrated with GRASS GIS and with ArcView GIS (Srinivasan and Arnold, 1994). The hydrological components of the model have been validated for numerous watersheds under varying hydrologic conditions (Arnold et al., 2012; Harmel et al., 2000; Saleh et al., 2000; Santhi et al., 2001; Spruill et al., 2000). From the literature, it is evident that the model can simulate the hydrological cycle reasonably well encompassing a wide variety of terrains and climatic zones.

Simulation of the hydrology of a watershed can be separated into two major divisions, namely land phase and water phase or routing phase. Land phase of hydrological cycle controls the quantification of water, sediment, nutrient, and pesticide loadings to the main channel in each subbasin. The routing phase can be defined as the movement of water, sediment, etc. through the channel network to the outlet of the watershed. It is essential to have a distributed hydrological model that can simulate both these phases.

QSWAT is an open source, physically based distributed hydrological model that operates on a daily time step. QSWAT has proved to be an effective tool for assessing water resource and nonpoint source pollution problems for a wide range of scales and environmental conditions across the globe. The major model components include weather, hydrology, soil temperature, plant growth, land management practices, and reactive as well as nonreactive contaminant transport modeling.

The distributed hydrological model QSWAT spatially divides the watershed into subwatersheds or subbasins based on the topographical information. The subwatersheds are further divided into smaller spatial modeling units called hydrologic response units (HRUs). HRUs are lumped areas within the subbasin that are comprised of unique land cover, soil, and management practices. The fundamental QSWAT simulations are carried down at the HRU level. The major hydrological processes modeled in QSWAT are surface runoff, soil and root zone infiltration, evapotranspiration, soil and snow evaporation, and base flow (Arnold et al., 1998).

The basic concept behind the land phase of the hydrological cycle is water balance equation:

$$\mathrm{SW_t} = \mathrm{SW_o} + \sum_{i=1}^{t} R_i - Q_i - \mathrm{ET}_i - W_{\mathrm{p}i} - Q_{\mathrm{r}i} \tag{27.5}$$

where $\mathrm{SW_t}$ is the final soil moisture content (mm), $\mathrm{SW_o}$ is the initial moisture content (mm), t is the time (days), R_i is the amount of precipitation on day i (mm), Q_i is the amount of surface runoff on day i (mm), ET_i is the amount of evapotranspiration on day i (mm), $W_{\mathrm{p}i}$ is the amount of percolation and bypass flow exiting soil profile on day i (mm), and $Q_{\mathrm{r}i}$ is the amount of return flow on day i (mm).

SCS curve number is used for accounting rainfall excess. A storage routing technique is used to simulate the flow through each soil layer. QSWAT directly simulates saturated flow only and assumes that water is uniformly distributed within a given layer.

Unsaturated flow between layers is indirectly modeled using depth distribution functions for plant water uptake and soil water evaporation. Downward flow occurs when the soil water in the layer exceeds field capacity and the layer below is not saturated. The rate of downward flow is governed by the saturated hydraulic conductivity. Lateral flow in the soil profile is simulated using a kinematic storage routing technique that is based on slope, slope length, and saturated conductivity. Upward flow from a lower layer to the upper layer is regulated by the soil water to field capacity ratios of the two layers. Percolation from the bottom of the root zone is recharged to the shallow aquifer.

27.7 AGRICULTURAL DROUGHT ESTIMATION—A CASE STUDY OF KALPATHY WATERSHED

27.7.1 Study Area

Kalpathy watershed located in southernmost part of India is one of the water-stressed watershed in Bharathapuzha river basin of Kerala. The watershed is facing acute drought during recent years. The geographical area of watershed is 1695.5 km^2. The watershed has an average slope of 1 in 40, with the highest elevation of 2051 m above mean sea level. The geographical extent of Kalpathy watershed is between 10°37′N and 10°57′N latitudes and 76°25′E and 77°9′E longitudes. The river Kalpathy originates from the upper slopes of the Western Ghats deep inside the Palakkad district, north of Walayar. The watershed has two rainfall-measuring stations and one stream-gauging station. Daily rainfall data of 25 years (1991–2015) is collected from the rainfall measuring stations, and stream-gauging data was obtained from the Central Water Commission. The methodology adopted for data extraction and modeling in shown in Fig. 27.1.

27.7.2 Model Setup

For this study, the QGIS interface for the model was used to extract the model parameters from the GIS layers to delineate subbasins and HRUs. A comparative evaluation of digital elevation models (DEMs) were carried prior to hydrological modeling and CARTOSAT-derived DEM was found best for the study region (Drisya and Sathish Kumar, 2016). Topographic parameters and stream channel parameters were then estimated from the DEM. Land use and soil map details were procured from the state remote-sensing agency. The study watershed constitutes 33 subwatersheds obtained by the QSWAT model. The simulation period for the hydrological model is from 1991 to 2015. The initial 2 years are defined as the warmup period. The next 15 years' data (1993–2007) is used for parameter calibration and rest of the data (2008–2015) is used for model validation. Stream flow measured at Mankara gauge station established by the Central Water Commission is used for calibrating and validating the model.

The Sequential Uncertainty Fitting version-2 (SUFI-2) method (Abbaspour et al., 2007) is used to calibrate the QSWAT model. Nash-Sutcliffe efficiency is used as the objective function for calibration. The details of the parameters used for calibration are listed in Table 27.2.

27.7.3 Spatiotemporal Soil Moisture

Validation of the QSWAT model using the calibrated optimal values for the parameters allows us to ascertain the model performance. This is done by comparing the observed and simulated stream flow at the Mankara gauging station for both the calibration and validation periods (Fig. 27.2). 0.76 and 0.73 efficiency are observed for calibration and validation processes of the model, respectively.

After successful performance evaluations of QSWAT, the hydrological processes are simulated. The results are visualized daily, monthly, or yearly as per the user requirement. It gives estimation of all hydrological components like potential evapotranspiration, surface runoff, soil moisture content, and water yield for each subbasin (Table 27.3). Water balance of the study area is depicted in Fig. 27.3. Potential evapotranspiration is having higher influence on the study region.

27.7.4 Standardized Soil Moisture Index (SSMI)

The daily model output of available soil moisture is averaged over a month to get soil moisture for each subbasin. These were used for deriving the standardized soil moisture index for accounting drought. The standardized soil moisture index

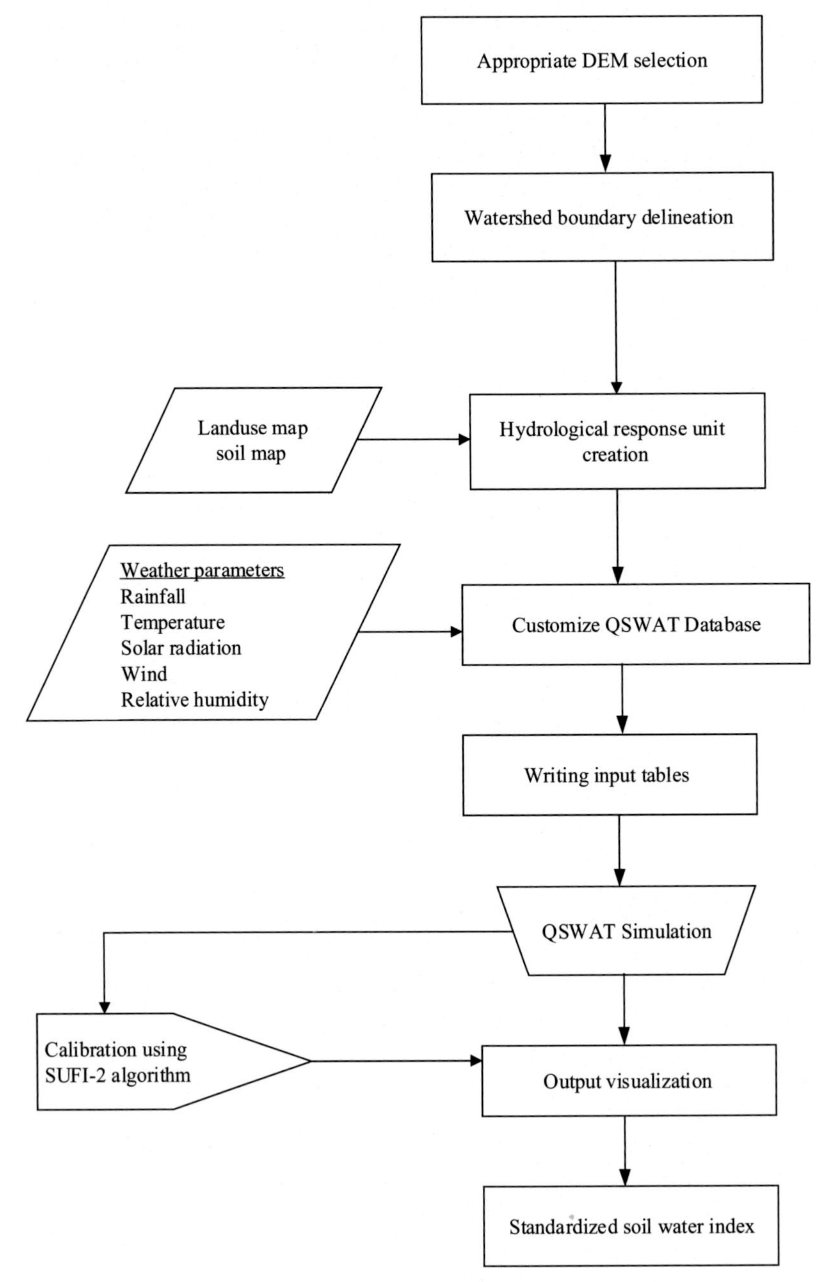

FIG. 27.1 **Flowchart of the methodology.**

TABLE 27.2 Calibration Parameter Details

Parameter	Description	Range	Optimum Value
CN.mgt	SCS runoff curve number	35–98	76.74
ESCO.hru	Soil evaporation compensation factor	0–1	0.25
SURLAG.bsn	Surface lag time	0.05–24	6.037
SOL_AWC.sol	Plant available water	0–1	0.85

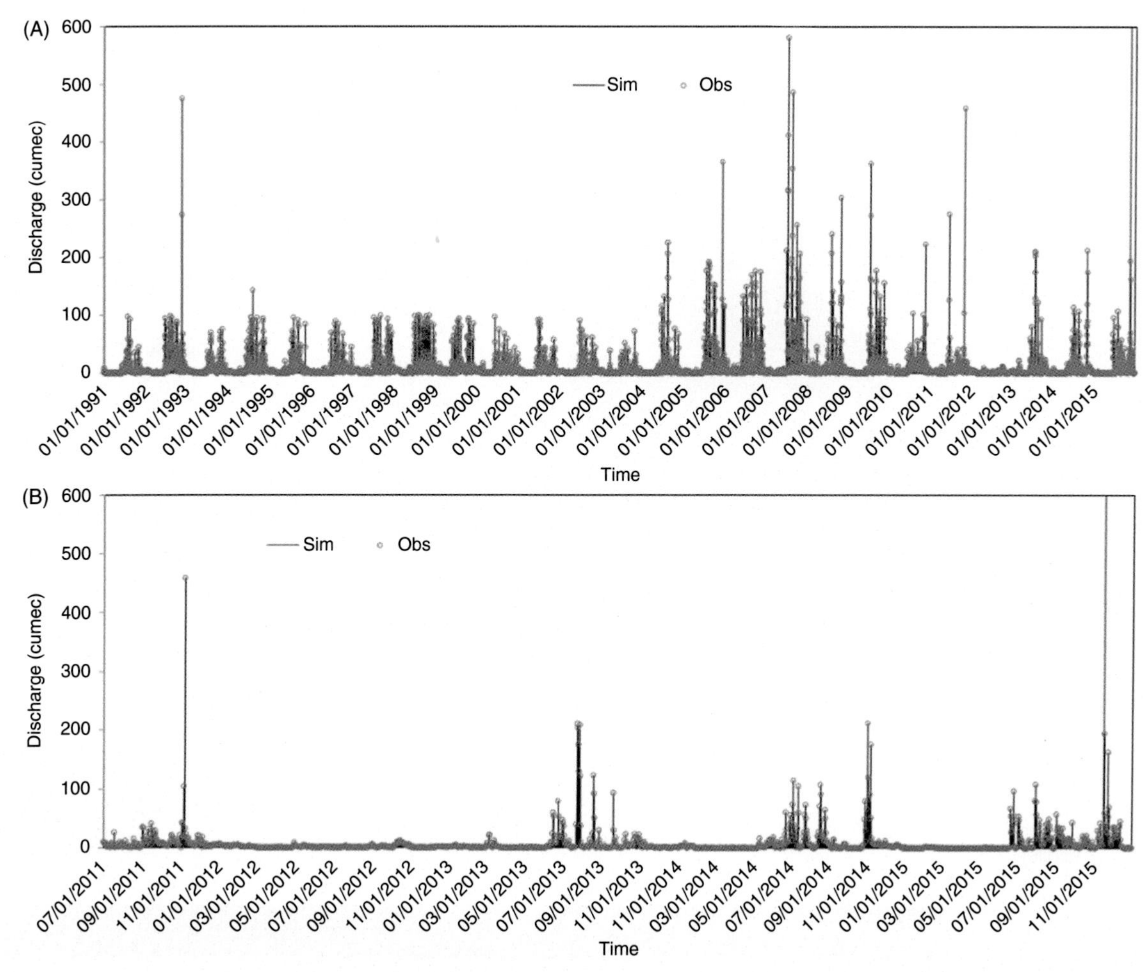

FIG. 27.2 Comparison between the observed and QSWAT simulated runoff (A) for the period 1991–2015 (B) for 7 months (July 2011–December 2011).

TABLE 27.3 Simulated Hydrological Components Details

Month	Precipitation (mm)	Direct Runoff (mm)	Baseflow (mm)	Actual Evapotranspiration (mm)	PET (mm)	Water Yield (mm)
January	4.32	0.59	0.28	11.04	157.68	6.71
February	8.25	2.58	0.25	17.54	168.65	3.80
March	29.16	5.45	0.66	83.06	222	6.83
April	42.3	1.02	0.98	48.26	202.19	2.86
May	58.96	6.65	1.31	36.51	215.58	7.35
June	233.3	63.99	6.12	60.55	190.53	66.5
July	320.7	132.11	10.63	78.33	169.69	156.96
August	169	55.01	6.98	70.91	187.72	102.36
September	117.5	35.21	4.31	55.36	188.64	81.09
October	146.3	46.68	4.62	53.78	153.06	85.57
November	110.6	42.62	4.13	44.35	128.23	79.79
December	9.37	4.17	0.68	17.73	143.06	28.03

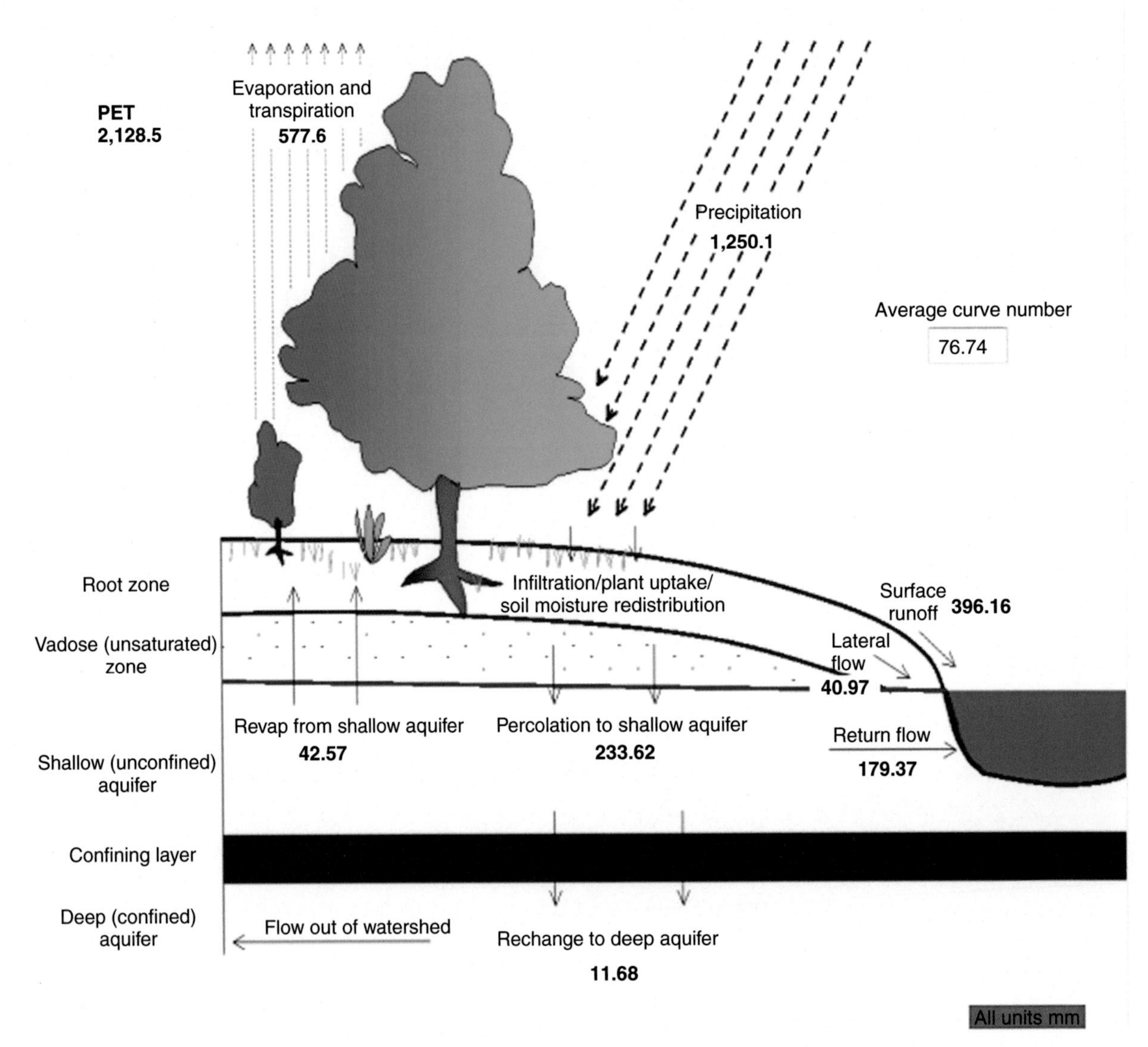

FIG. 27.3 **Water balance of the study area.**

is statistically similar to the other most commonly used standardized precipitation indices. The index for a monthly period is defined as the difference of soil moisture from monthly mean divided to standard deviation calculated on monthly basis.

$$SSMI_{\tau} = \frac{SW_{\upsilon\tau} - \overline{SW}_{\tau}}{\sigma_{\tau}} \tag{27.6}$$

$$\overline{SW}_{\tau} = \frac{1}{n}\sum_{\upsilon=1}^{N} F_{\upsilon,\tau} \tag{27.7}$$

$$\sigma_{\tau} = \sqrt{\frac{1}{n-1}\sum_{\upsilon=1}^{n}(SW_{\upsilon,\tau} - \overline{SW}_{\tau})} \tag{27.8}$$

where τ denotes the interval within year and υ denotes the year. $\bar{F}_{\tau}$ and σ_{τ} the are mean and standard deviation of month τ and it varies between 1 and 12 for monthly calculations.

Using the simulated soil water from QSWAT, the standardized soil moisture index is developed for the time period 1991–2015 (Fig. 27.4). It is found that soil moisture is having a temporal variation throughout the entire study period. Moderate to extreme threat is observed for soil moisture conditions. Since spatial variation in soil moisture is to be examined,

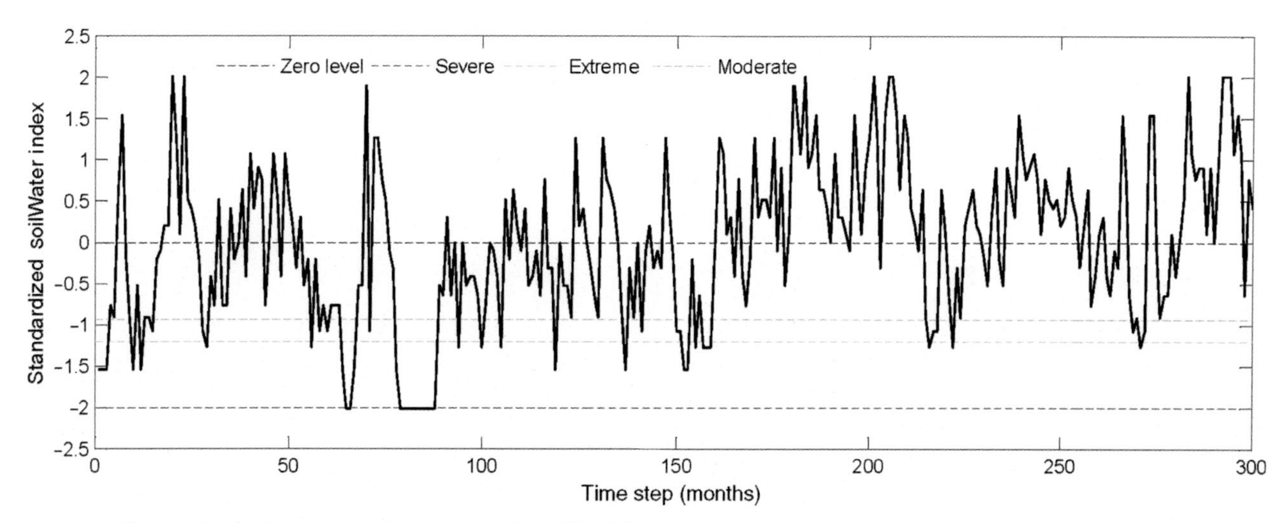

FIG. 27.4 **Standardized soil moisture index for the period 1991–2015.**

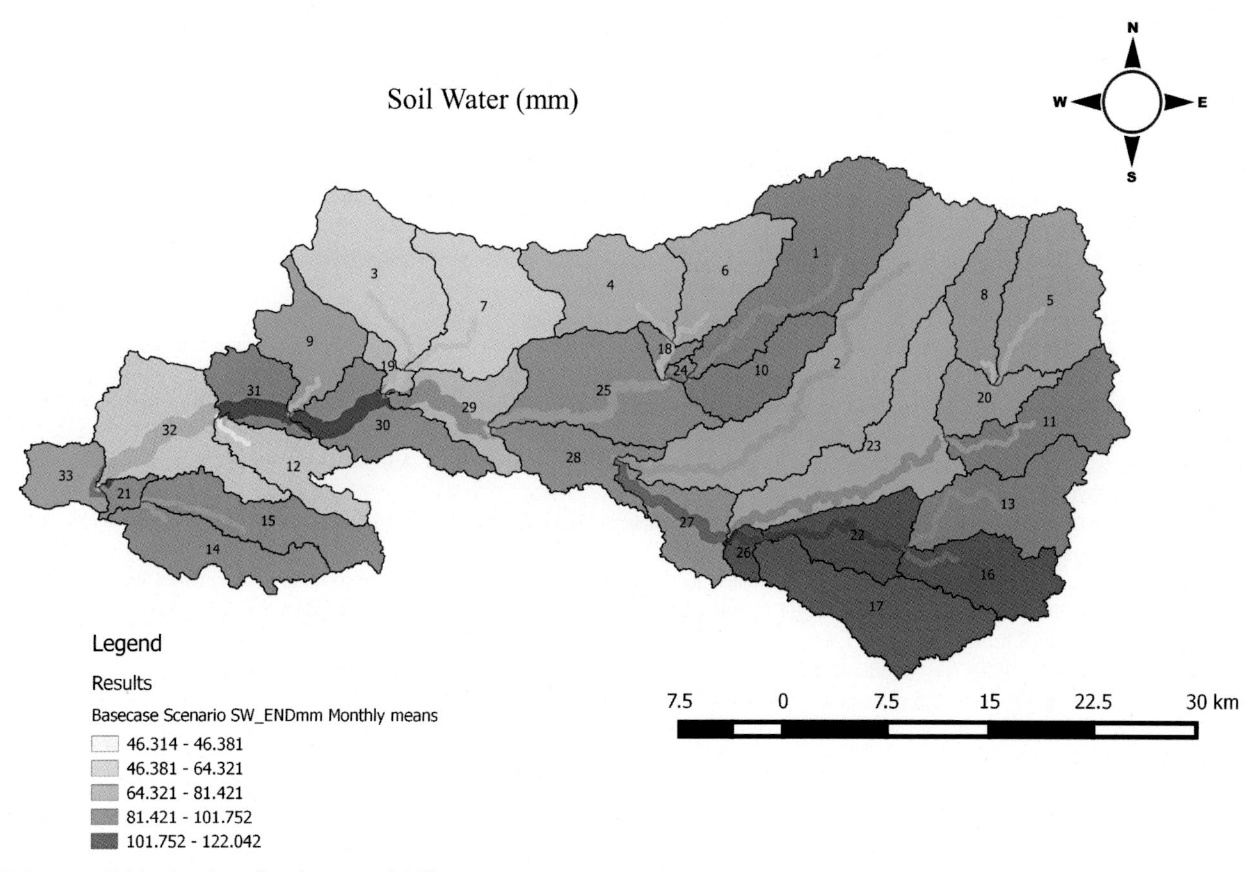

FIG. 27.5 **Subbasin-wise soil moisture availability.**

a subbasin-wise analysis is carried for the soil moisture availability. Fig. 27.5 shows the spatial variation of monthly average soil water for the study area. Subbasins 12 and 32 which constitute the developed urban region are having lesser soil moisture availability compared to other subbasins.

27.8 CONCLUSIONS

This chapter discusses about the challenges in agricultural-related drought assessments and shows the potentials of a process-based hydrological model to simulate the drought conditions. Toward the end of the chapter, a case study on the Kalpathy watershed is presented. The model is calibrated and validated using the Sequential Uncertainty Fitting ver.2

(SUFI-2) with the observed discharge at a gauging station. Calibration and verification results showed good agreement between simulated and measured data discharge. The simulated soil moisture is used to develop the standardized soil moisture index. The validated model is also used to analyze the spatial distribution of the agricultural drought indices. The study showed that QSWAT model, if properly validated, can be used effectively to assess agricultural drought. The proposed methodology seems also suitable for an extended use with other hydrological quantities that display connections with vegetation water stress similar to soil moisture.

ACKNOWLEDGMENTS

The authors thank all the data providers and INSPIRE component of the Department of Science and Technology, Government of India, for funding the research.

REFERENCES

Abbaspour, K.C., Yang, J., Maximov, I., Siber, R., Bogner, K., Mieleitner, J., Zobrist, J., Srinivasan, R., 2007. Modelling hydrology and water quality in the pre-alpine/alpine Thur watershed using SWAT. J. Hydrol. 333, 413–430.

Arnold, J.G., Allen, P.M., Bernhardt, G., 1993. A comprehensive surface groundwater flow model. J. Hydrol. 143 (1–4), 47–69.

Arnold, J.G., Srinivasan, R., Muttiah, R.S., Williams, J.R., 1998. Large area hydrologic modeling and assessment. Part 1: model development. J. Am. Soc. Water Resour. Assoc. 34 (1), 73–89.

Arnold, J.G., Moriasi, D.N., Gassman, P.W., Abbaspour, K.C., White, M.J., Raghavan, S., Chinnasamy, S., Daren, H., van Ann, G., Van Liew, M.W., Narayanan, K., Jha, M.K., 2012. SWAT: model use, calibration, and validation. Trans. ASABE 55, 1491–1508.

Bergman, K.H., Sabol, P., Miskus, D., 1988. Experimental indices for monitoring global drought conditions. In: Proceedings of the 13th Annual Climate Diagnostics Workshop. United States Department of Commerce, Cambridge, MA.

British Columbia Ministry of Agriculture, 2015. Soil Water Storage Capacity and Available Soil Moisture. Water Conservation Fact Sheet. http://www2.gov.bc.ca/assets/gov/farming-natural-resources-and-industry/agriculture-and-seafood/agricultural-land-andenvironment/soil-nutrients/600-series/619000-1_soil_water_storage_capacity.pdf.

Brown, J.F., Wardlow, B.D., Tadesse, T., Hayes, M.J., Reed, B.C., 2008. The Vegetation Drought Response Index (VegDRI): a new integrated approach for monitoring drought stress in vegetation. GIScience Remote Sens. 45, 16–46.

Central Statistical Office, 2017. Ministry of Statistics and Implementation Govt of India.

Drisya, J., Sathish Kumar, D., 2016. Comparison of digitally delineated stream networks from different space borne digital elevation models: a case study based on two watersheds in South India. Arab. J. Geosci. doi: 10.1007/s12517-016-2726-x.

Farahmand, A., AghaKouchak, A., 2015. A generalized framework for deriving nonparametric standardized drought indicators. Adv. Water Resour. 76, 140–145.

Harmel, R.D., Richardson, C.W., King, K.W., 2000. Hydrologic response of a small watershed model to generated precipitation. Trans. ASAE 43 (6), 1483–1488.

Heim, R.R., 2002. A review of twentieth-century drought indices used in the United States. Bull. Am. Meteorol. Soc. 83, 1149–1165, http://dx.doi.org/10.1175/1520-0477(2002)083.

Huang, J., van den Dool, Georgakakos, 1996. Analysis of model-calculated soil moisture over the United States (1931–1993) and applications in long-range temperature forecasts. J. Climatol. 9, 1350–1362.

Intergovernmental Panel on Climate Change (IPCC), 2012. Managing the Risks of Extreme Events and Disasters to Advance Climate Change Adaptation., Special Report of Working Groups I and II of the IPCC (C.B. Field, V. Barros, T.F. Stocker, D. Qin, D.J. Dokken, K.L. Ebi, M.D. Mastrandrea, K.J. Mach, G.-K. Plattner, S.K. Allen).

Keyantash, J., Dracup, J.A., 2002. The quantification of drought: an evaluation of drought indices. Bull. Am. Meteorol. Soc. 83, 1167–1180.

Keyantash, J.A., Dracup, J.A., 2004. An aggregate drought index: assessing drought severity based on fluctuations in the hydrologic cycle and surface water storage. Water Resour. Res. 40 (W09304). doi: 10.1029/2003WR002610.

Kogan, F.N., 1990. Remote sensing of weather impacts on vegetation in non-homogeneous areas. Int J Remote Sens 11, 1405–1419.

Kogan, F.N., 1995. Droughts of the late 1980s in the United States as derived from NOAA polar orbiting satellite data. Bull. Am. Meteorol. Soc. 76, 655e668.

Narasimhan, B., Srinivasan, R., 2005. Development and evaluation of soil moisture deficit index (SMDI) and evapotranspiration deficit index (ETDI) for agriculture drought monitoring. J. Agric. For. Meteorol. 133, 69–88.

Palmer, W.C., 1965. Meteorological Drought. Research Paper No. 45. United States Weather Bureau, Washington, DC.

Saleh, A., Arnold, J.G., Gassman, P.W., Hauck, L.M., Rosenthal, W.D., Williams, J.R., McFarland, A.M.S., 2000. Application of SWAT for the Upper North Bosque River watershed. Trans. ASAE 43 (5), 1077–1087.

Santhi, C., Arnold, J.G., Williams, J.R., Dugas, W.A., Srinivasan, R., Hauck, L.M., 2001. Validation of the SWAT model on a large river basin with point and nonpoint sources. J. Am. Soc. Water Resour. Assoc. 37 (5), 1169–1188.

Spruill, C.A., Workman, S.R., Taraba, J.L., 2000. Simulation of daily and monthly stream discharge from small watersheds using the SWAT model. Trans. ASAE 43 (6), 1431–1439.

Srinivasan, R., Arnold, J.G., 1994. Integration of a basin-scale water quality model with GIS. J. Am. Soc. Water Resour. Assoc. 30 (3), 453–462.

Trambauer, P., Maskey, S., Winsemius, H., Werner, M., Uhlenbrook, S., 2013. A review of continental scale hydrological models and their suitability for drought forecasting in (sub-Saharan) Africa. Phys. Chem. Earth 66, 16e26, http://dx.doi.org/10.1016/j.pce.2013.07.003.

Wilhite, D.A., Glantz, M.H., 1985. Understanding the drought phenomenon: the role of definitions. Water Int. 10 (3), 111–120.

Zargar, A., Sadiq, R., Naser, B., Khan, F.I., 2011. A review of drought indices. Environ. Rev. 19, 333–349.

Index

N

O

P

R

S

T

U

V

W

Printed in the United States
By Bookmasters